FOUNDING FIGURES AND COMMENTATORS IN ARABIC MATHEMATICS

In this unique insight into the history and philosophy of mathematics and science in classical Islamic civilisation, the eminent scholar Roshdi Rashed illuminates the various historical, textual and epistemic threads that underpinned the history of Arabic mathematical and scientific knowledge up to the seventeenth century. The first of five wide-ranging and comprehensive volumes, this book provides a detailed exploration of Arabic mathematics and sciences in the ninth and tenth centuries.

Extensive and detailed analyses and annotations support a number of key Arabic texts, which are translated here into English for the first time. In this volume Rashed focuses on the traditions of celebrated polymaths from the ninth- and tenth century 'School of Baghdad' – such as the Banū Mūsā, Thābit ibn Qurra, Ibrāhīm ibn Sinān, Abū Ja'far al-Khāzin, Abū Sahl Wayjan ibn Rustām al-Qūhī – and eleventh-century Andalusian mathematicians such as Abū al-Qāsim ibn al-Samḥ and al-Mu'taman ibn Hūd. The Archimedean–Apollonian traditions of these polymaths are thematically explored to illustrate the historical and epistemological development of 'infinitesimal mathematics' as it became more clearly articulated in the eleventh-century influential legacy of al-Hasan ibn al-Haytham ('Alhazen').

Contributing to a more informed and balanced understanding of the internal currents of the history of mathematics and the exact sciences in Islam, and of its adaptive interpretation and assimilation in the European context, this fundamental text will appeal to historians of ideas, epistemologists and mathematicians at the most advanced levels of research.

Roshdi Rashed is one of the most eminent authorities on Arabic mathematics and the exact sciences. A historian and philosopher of mathematics and science and a highly celebrated epistemologist, he is currently Emeritus Research Director (distinguished class) at the Centre National de la Recherche Scientifique (CNRS) in Paris, and is the Director of the Centre for History of Medieval Science and Philosophy at the University of Paris (Denis Diderot, Paris VII).

Nader El-Bizri is a Reader at the University of Lincoln, and a Chercheur Associé at the Centre National de la Recherche Scientifique in Paris (CNRS, UMR 7219). He has lectured on 'Arabic Sciences and Philosophy' at the University of Cambridge since 1999. He held a Visiting Professorship at the University of Lincoln (2007–2010), and since 2002 he has been a senior Research Associate affiliated with The Institute of Ismaili Studies, London.

T0199402

CULTURE AND CIVILIZATION
IN THE MIDDLE EAST
General Editor: Ian Richard Netton
Professor of Islamic Studies, University of Exeter

This series studies the Middle East through the twin foci of its diverse cultures and civilizations. Comprising original monographs as well as scholarly surveys, it covers topics in the fields of Middle Eastern literature, archaeology, law, history, philosophy, science, folklore, art, architecture and language. While there is a plurality of views, the series presents serious scholarship in a lucid and stimulating fashion.

PREVIOUSLY PUBLISHED BY CURZON

THE ORIGINS OF ISLAMIC LAW
The Qur'an, the *Muwatta'* and Madinan *Amal*
Yasin Dutton

A JEWISH ARCHIVE FROM OLD CAIRO
The history of Cambridge University's Genizah collection
Stefan Reif

THE FORMATIVE PERIOD OF TWELVER SHI'ISM
Hadith as discourse between Qum and Baghdad
Andrew J. Newman

QUR'AN TRANSLATION
Discourse, texture and exegesis
Hussein Abdul-Raof

CHRISTIANS IN AL-ANDALUS 711–1000
Ann Rosemary Christys

FOLKLORE AND FOLKLIFE IN THE UNITED ARAB EMIRATES
Sayyid Hamid Hurriez

THE FORMATION OF HANBALISM
Piety into power
Nimrod Hurvitz

ARABIC LITERATURE
An overview
Pierre Cachia

STRUCTURE AND MEANING IN MEDIEVAL ARABIC AND PERSIAN LYRIC POETRY
Orient pearls
Julie Scott Meisami

MUSLIMS AND CHRISTIANS IN NORMAN SICILY
Arabic-speakers and the end of Islam
Alexander Metcalfe

MODERN ARAB HISTORIOGRAPHY
Historical discourse and the nation-state
Youssef Choueiri

THE PHILOSOPHICAL POETICS OF ALFARABI, AVICENNA AND AVERROES
The Aristotelian reception
Salim Kemal

PUBLISHED BY ROUTLEDGE

1. THE EPISTEMOLOGY OF IBN KHALDUN
Zaid Ahmad

2. THE HANBALI SCHOOL OF LAW AND IBN TAYMIYYAH
Conflict or concilation
Abdul Hakim I Al-Matroudi

3. ARABIC RHETORIC
A pragmatic analysis
Hussein Abdul-Raof

4. ARAB REPRESENTATIONS OF THE OCCIDENT
East–West encounters in Arabic fiction
Rasheed El-Enany

5. GOD AND HUMANS IN ISLAMIC THOUGHT
Abd al-Jabbār, Ibn Sīnā and al-Ghazālī
Maha Elkaisy-Friemuth

6. ORIGINAL ISLAM
Malik and the Madhhab of Madina
Yasin Dutton

7. AL-GHAZALI AND THE QUR'AN
One book, many meanings
Martin Whittingham

8. BIRTH OF THE PROPHET MUHAMMAD
Devotional piety in Sunni Islam
Marion Holmes Katz

9. SPACE AND MUSLIM URBAN LIFE
At the limits of the labyrinth of Fez
Simon O'Meara

10. ISLAM AND SCIENCE
The intellectual career of Nizam al-Din al-Nizaburi
Robert G. Morrison

11. IBN 'ARABÎ – TIME AND COSMOLOGY
Mohamed Haj Yousef

12. THE STATUS OF WOMEN IN ISLAMIC LAW AND SOCIETY
Annotated translation of al-Ṭāhir al-Ḥaddād's *Imra'tunā fi 'l-sharīʿa wa 'l-mujtamaʿ*, with an introduction
Ronak Husni and Daniel L. Newman

13. ISLAM AND THE BAHA'I FAITH
A comparative study of Muhammad 'Abduh and 'Abdul-Baha 'Abbas
Oliver Scharbrodt

14. COMTE DE GOBINEAU AND ORIENTALISM
Selected Eastern writings
Translated by Daniel O'Donoghue Edited by Geoffrey Nash

15. EARLY ISLAMIC SPAIN
The history of Ibn al-Qūṭīya
David James

16. GERMAN ORIENTALISM
The study of the Middle East and Islam from 1800 to 1945
Ursula Wokoeck

17. MULLĀ ṢADRĀ AND METAPHYSICS
Modulation of being
Sajjad H. Rizvi

18. SCHOOLS OF QUR'ANIC EXEGESIS
Genesis and development
Hussein Abdul-Raof

19. AL-GHAZALI, AVERROES AND THE INTERPRETATION OF THE QUR'AN
Common sense and philosophy in Islam
Avital Wohlman, translated by David Burrell

20. EASTERN CHRISTIANITY IN THE MODERN MIDDLE EAST
Edited by Anthony O'Mahony and Emma Loosley

21. ISLAMIC REFORM AND ARAB NATIONALISM
Expanding the crescent from the Mediterranean to the Indian Ocean (1880s–1930s)
Amal N. Ghazal

22. ISLAMIC ETHICS
Divine command theory in Arabo-Islamic thought
Mariam al-Attar

23. MUSLIM FORTRESSES IN THE LEVANT
Between Crusaders and Mongols
Kate Raphael

24. BEING HUMAN IN ISLAM
The impact of the evolutionary worldview
Damian Howard

25. THE UAE AND FOREIGN POLICY
Foreign aid, identities and interests
Khalid S. Almezaini

26. A HISTORY OF EARLY AL-ANDALUS
The *Akhbār majmū'a*
David James

27. INSPIRED KNOWLEDGE IN ISLAMIC THOUGHT
Al-Ghazali's theory of mystical cognition and its Avicennian foundation
Alexander Treiger

28. SHI'I THEOLOGY IN IRAN
The challenge of religious experience
Ori Goldberg

29. FOUNDING FIGURES AND COMMENTATORS IN ARABIC MATHEMATICS
A history of Arabic sciences and mathematics, Volume 1
Roshdi Rashed (translations edited by Nader El-Bizri)

30. THE MUSLIM CONQUEST OF IBERIA
Medieval Arabic narratives
Nicola Clarke

31. ANGELS IN ISLAM
Jalal al-Din al-Suyuti's *al-Haba'ik fi akhbar al-mala'ik*
Stephen Burge

FOUNDING FIGURES AND COMMENTATORS IN ARABIC MATHEMATICS

A history of Arabic sciences and mathematics

Volume 1

Roshdi Rashed

Edited by

Nader El-Bizri

Translated by Roger Wareham, with Chris Allen and Michael Barany

Routledge
Taylor & Francis Group
LONDON AND NEW YORK

مركز دراسات الوحدة العربية
CENTRE FOR ARAB UNITY STUDIES

First published 2012
by Routledge

2 Park Square, Milton Park, Abingdon, Oxfordshire OX14 4RN
52 Vanderbilt Avenue, New York, NY 10017

Routledge is an imprint of the Taylor & Francis Group, an informa business

Copyright © 2012 Taylor & Francis

The right of Roshdi Rashed to be identified as the author has been asserted in accordance with sections 77 and 78 of the Copyright, Designs and Patents Act 1988.

All rights reserved. No part of this book may be reprinted or reproduced or utilised in any form or by any electronic, mechanical, or other means, now known or hereafter invented, including photocopying and recording, or in any information storage or retrieval system, without permission in writing from the publishers.

Notice:
Product or corporate names may be trademarks or registered trademarks, and are used only for identification and explanation without intent to infringe.

British Library Cataloguing in Publication Data
A catalogue record for this book is available from the British Library

Library of Congress Cataloging in Publication Data
A catalogue record for this book has been requested

ISBN 978-0-415-58217-9 (hbk)
ISBN 978-0-367-86528-3 (pbk)

Typeset in Times New Roman
by Cenveo Publisher Services

This book was prepared from press-ready files supplied by the editor.

CONTENTS

Editor's Foreword xiii
Preface xix
Note xxiv

CHAPTER I : BANŪ MŪSĀ AND THE CALCULATION OF THE VOLUME OF THE SPHERE AND THE CYLINDER

1.1. INTRODUCTION 1
1.1.1. The Banū Mūsā: dignitaries and learned 1
1.1.2. The mathematical works of the Banū Mūsā 7
1.1.3. Treatise on the measurement of plane and spherical figures: a Latin translation and a rewritten version by al-Ṭūsī 10
1.1.4. Title and date of the Banū Mūsā treatise 34

1.2. MATHEMATICAL COMMENTARY 38
1.2.1. Organization and structure of the Banū Mūsā book 38
1.2.2. The area of the circle 40
1.2.3. The area of the triangle and Hero's formula 46
1.2.4. The surface area of a sphere and its volume 47
1.2.5. The two-means problem and its mechanical construction 60
1.2.6a. The trisection of angles and Pascal's *Limaçon* 66
1.2.6b. Approximating cubic roots 69

1.3. TRANSLATED TEXT: *On the Knowledge of the Measurement of Plane and Spherical Figures* 73

CHAPTER II: THĀBIT IBN QURRA AND HIS WORKS IN INFINITESIMAL MATHEMATICS

2.1. INTRODUCTION 113
2.1.1. Thābit ibn Qurra: from Ḥarrān to Baghdad 113
2.1.2. The works of Thābit ibn Qurra in infinitesimal mathematics 122
2.1.3. History of the texts and their translations 124

2.2. MEASURING THE PARABOLA 130
2.2.1. Organization and structure of Ibn Qurra's treatise 130
2.2.2. Mathematical commentary 133
2.2.2.1. Arithmetical propositions 133
2.2.2.2. Sequence of segments and bounding 142
2.2.2.3. Calculation of the area of a portion of a parabola 154
2.2.3. Translated text: *On the Measurement of the Conic Section Called Parabola* 169

2.3. MEASURING THE PARABOLOID 209
2.3.1. Organization and structure of Ibn Qurra's treatise 209
2.3.2. Mathematical commentary 214
2.3.2.1. Arithmetical propositions 214
2.3.2.2. Extension to sequences of segments 218
2.3.2.3. Volumes of cones, rhombuses and other solids 223
2.3.2.4. Property of four segments 230
2.3.2.5. Arithmetical propositions 231

2.3.2.6. Sequence of segments and bounding 233
2.3.2.7. Calculation of the volumes of paraboloids 244
2.3.2.8. Parallel between the treatise on the area of the parabola and the treatise on the volume of the paraboloid 256
2.3.3. Translated text: *On the Measurement of the Paraboloids* 261

2.4. ON THE SECTIONS OF THE CYLINDER AND ITS LATERAL SURFACE 333
2.4.1. Introduction 333
2.4.2. Mathematical commentary 337
2.4.2.1. Plane sections of the cylinder 337
2.4.2.2. Area of an ellipse and elliptical sections 341
2.4.2.3. Concerning the maximal section of the cylinder and concerning its minimal sections 356
2.4.2.4. Concerning the lateral area of the cylinder and the lateral area of portions of the cylinder lying between the plane sections touching all sides 363
2.4.3. Translated text: *On the Sections of the Cylinder and its Lateral Surface* 381

CHAPTER III: IBN SINĀN, CRITIQUE OF AL-MĀHĀNĪ: THE AREA OF THE PARABOLA

3.1. INTRODUCTION 459
3.1.1. Ibrāhīm ibn Sinān: 'heir' and 'critic' 459
3.1.2. The two versions of *The Measurement of the Parabola*: texts and translations 463

3.2. MATHEMATICAL COMMENTARY 466

3.3. TRANSLATED TEXTS
3.3.1. *On the Measurement of the Parabola* 483
3.3.2. *On the Measurement of a Portion of the Parabola* 495

CHAPTER IV: ABŪ JA'FAR AL-KHĀZIN: ISOPERIMETRICS AND ISEPIPHANICS

4.1. INTRODUCTION 503
4.1.1. Al-Khāzin: his name, life and works 503
4.1.2. The treatises of al-Khāzin on isoperimeters and isepiphanics 506

4.2. MATHEMATICAL COMMENTARY 507
4.2.1. Introduction 507
4.2.2. Isoperimetrics 509
4.2.3. Isepiphanics 524
4.2.4. The opuscule of al-Sumaysāṭī 546

4.3. TRANSLATED TEXTS
4.3.1. *Commentary on the First Book of the* Almagest 551
4.3.2. *The Surface of any Circle is Greater than the Surface of any Regular Polygon with the Same Perimeter* (al-Sumaysāṭī) 577

CHAPTER V: AL-QŪHĪ, CRITIQUE OF THĀBIT: VOLUME OF THE PARABOLOID OF REVOLUTION

5.1. INTRODUCTION 579
5.1.1. The mathematician and the artisan 579

5.1.2. The versions of the volume of a paraboloid 583

5.2. MATHEMATICAL COMMENTARY 588
5.3. TRANSLATION TEXTS
5.3.1. *On the Determination of the Volume of a Paraboloid* 599
5.3.2. *On the Volume of a Paraboloid* 609

CHAPTER VI: IBN AL-SAMḤ: THE PLANE SECTIONS OF A CYLINDER AND THE DETERMINATION OF THEIR AREAS

6.1. INTRODUCTION 615
6.1.1. Ibn al-Samḥ and Ibn Qurra, successors to al-Ḥasan ibn Mūsā 615
6.1.2. Serenus of Antinoupolis, al-Ḥasān ibn Mūsā, Thābit ibn Qurra and Ibn al-Samḥ 618
6.1.3. The structure of the study by Ibn al-Samḥ 622

6.2. MATHEMATICAL COMMENTARY 623
6.2.1. Definitions and accepted results 623
6.2.2. The cylinder 626
6.2.3. The plane sections of a cylinder 627
6.2.4. The properties of a circle 628
6.2.5. Elliptical sections of a right cylinder 632
6.2.6. The ellipse as a plane section of a right cylinder 639
6.2.7. The area of an ellipse 645
6.2.8. Chords and sagittas of the ellipse 653
6.3. TRANSLATED TEXT: *On the Cylinder and its Plane Section* 667

CHAPTER VII: IBN HŪD: THE MEASUREMENT OF THE PARABOLA AND THE ISOPERIMETRIC PROBLEM

7.1. INTRODUCTION 721
7.1.1. *Kitāb al-Istikmāl,* a mathematical *compendium* 721
7.1.2. Manuscript transmission of the texts 727

7.2. THE MEASUREMENT OF THE PARABOLA 729
7.2.1. Infinitesimal property or conic property 729
7.2.2. Mathematical commentary on Propositions 18–21 733
7.2.3. Translation: *Kitāb al-Istikmāl* 749

7.3. THE ISOPERIMETRIC PROBLEM 755
7.3.1. An extremal property or a geometric property 755
7.3.2. Mathematical commentary on Propositions 16 and 19 758
7.3.3. Translation: *Kitāb al-Istikmāl* 764

SUPPLEMENTARY NOTES
The Formula of Hero of Alexandria according to Thābit ibn Qurra 767
Commentary of Ibn Abī Jarrāda on *The Sections of the Cylinder* by Thābit ibn Qurra 767

BIBLIOGRAPHY 779

INDEXES
Index of names 793
Subject index 797
Index of works 805

EDITOR'S FOREWORD

The ninth and tenth centuries (third and fourth centuries of the *hijrī* calendar) constituted the foundational classical epoch of the history of scientific ideas in Islamic civilization. While the developments in this period encompassed most fields of rational knowledge that have been handed down to us in the medium of the Arabic language, the writing of a restorative and reasoned history of the science and mathematics, particularly of this era, remains a task requiring great attention and thoughtfulness from historians and epistemologists, given its significance in explicating the development of later scientific and mathematical traditions within Islamic civilization, and the elucidation of their epistemic and conceptual prolongations up to the seventeenth century in Europe, by-passing the Italian Renaissance.

It is with the aim of redressing this state of affairs that the present groundbreaking volume is dedicated. This task of reinstating the writing of the history of mathematics and science in classical Islamic civilization, with appropriate faithfulness, is undertaken within the covers of this volume by the celebrated mathematician, historian and philosopher of mathematics, Professor Roshdi Rashed (Emeritus Research Director [distinguished class], Centre National de la Recherche Scientifique, Paris; Honorary Professor at the University of Tokyo). This voluminous book belongs to a constellation of several of Rashed's texts in which he has endeavoured to rewrite in painstaking detail the history of science and mathematics within the span of the ninth- and tenth-century founding epoch. This research venture aims also to illuminate the various historical, textual and epistemic threads that underpinned the later unfolding of the history of mathematical and scientific knowledge up to the early-modern period in the seventeenth century. Rashed seeks to reliably re-establish the particulars of the historical interpretation of diverse imperative chapters in which the mathematicians of classical Islamic civilization were most innovative and inventive in terms of receiving, transmitting, adapting, developing and renewing the contributions of Hellenic mathematics, at levels that remained unsurpassed until the seventeenth century with the works of figures of the calibre of Descartes, Fermat and Leibniz.

In this regard, Rashed focuses in this present volume on the mathematical and scientific traditions of the polymaths of the ninth- and tenth-century school of Baghdad, such as the famed Banū Mūsā (the illustrious three sons of Mūsā ibn Shākir) and the celebrated Thābit ibn Qurra and his reputed grandson Ibrāhīm ibn Sinān, in addition to remarkable mathematicians such as Abū Jaʿfar al-Khāzin and Abū Sahl Wayjan ibn Rustam al-Qūhī, in addition to the eleventh century mathematicians of Andalusia of the standing of Abū al-Qāsim ibn al-Samḥ and al-Mu'taman ibn Hūd. These prominent scholars belonged to outstanding cross-generational research groups that conducted their work in operational contexts that interlaced the uninterrupted currents of closely interconnected legacies, spanning over more than two centuries and yielding pioneering and progressive investigations in mathematics and the exact sciences, as principally modulated within the Arabic extensions of the Archimedean–Apollonian heritage. The classical traditions in mathematics and science of these polymaths are explored thematically by Rashed in this volume through key treatises in geometry, with special emphasis on detailed and complex demonstrations, constructions and proofs in the domain of conics, and by illustrating the historical and epistemological development of 'infinitesimal mathematics' as it became clearly articulated in the magnificent scientific and mathematical legacy of the polymath, geometer, optician and astronomer, al-Ḥasan ibn al-Haytham (known in the Latinate rendering of his name as 'Alhazen'; d. ca. after 1041 CE, Cairo).

In order to grasp the extent and significance of Ibn al-Haytham's original contributions to the innovative field of infinitesimal mathematics, and his accomplishments within the unfolding of the Apollonian tradition, and 13 centuries after Archimedes, it was obviously necessary to address the history of mathematics and the exact sciences in the ninth and tenth centuries, comprising traditions and practices that underpinned and inspired Ibn al-Haytham's revolutionary research. It is precisely this task that Rashed has undertaken in writing this book. This line of inquiry belongs to Rashed's broader academic and intellectual project of examining the foundations of infinitesimal mathematics in historical and epistemological terms. This series of investigations is not focused on the contributions of isolated individuals and their treatises; rather it accounts for these textual and mathematical legacies within the continual and progressive traditions to which they essentially belonged.

The careful selections of texts in this present volume were ultimately guided by these historical and epistemic dimensions, in terms of revealing their internal chains of interconnections, and the disclosure of the deeper layers of their continuities and channels of transmission. Emphasis is placed particularly on studies in the variegated domains of geometry, with a focus on the investigation of conic sections, their construction, their measurement, and the demonstrations and proofs that pertain to their properties. The selected treatises that are gathered in this present volume comprise investigations on the conic sections, with the associated measurements of volumes and surface areas, and with detailed studies of parabolas and of paraboloid solids, in addition to related analyses of the properties of ellipses, circles, curved surfaces and portions of curved solids, including the multifarious characteristics of the sphere and the cylinder. All these inquiries are considered in the broader context of elucidating the foundational elements that resulted from the emergence of the domain of infinitesimal mathematics, as it was set against the background of significant investigations in arithmetic and algebra, and in the fields of their applications to each other and to geometry, all leading to the development of new chapters and diversified branches in mathematical research.

These studies constitute valuable historical material that is vital to any endeavour that aims at achieving a more informed and balanced understanding of the internal currents of the history of mathematics and the exact sciences within the Islamic history of ideas, and of its adaptive interpretation and assimilation in the European context within mediaeval and Renaissance scientific and mathematical disciplines. These traditions underpinned seventeenth-century positive knowledge by way of the interlinked grand dynamic chain in the unfurling of the classical Greek–Arabic–Latin–Hebrew–Italian lineage. Furthermore, the minutiae of mathematical details are closely explored by Rashed in this regard, in view of authentically restoring numerous traditions in infinitesimal geometry, in the research on conics and spherical geometry, in addition to the investigation of geometric transformations and the associated novel methods of analysis and synthesis.

The noteworthy assemblage of prominent classical treatises in this volume, which are rendered here into the English language, has been painstakingly established through original Arabic critical editions of extant manuscripts of these works, supported by mathematical, epistemic and textual commentaries and analyses, and guided by thoughtful and reflective historiography, while including also some highly informative and supportive studies of biographical–bibliographical sources. This volume offers annotated English translations of the French renderings of these texts, based also on the primary sources in the Arabic critical editions of the manuscripts, which are also presented here in English translation for the first time. This body of work will render great service to historians, philosophers and epistemologists who are interested in the history of mathematics, and to researchers at various advanced stages in these fields of inquiry, in addition to modern mathematicians who will appreciate the elegance of Rashed's interpretations of fundamental historical data through modern formal mathematical notation and by way of intricate geometric figures, models and constructs.

It has been a task of the highest order to oversee and edit the annotated English translation of Rashed's texts, and to achieve a proper balance between the demands of the English language and its stylistic intricacies on the one hand, while maintaining fidelity to the mathematical content, reliably conveying the sense of the original French renditions, and, more essentially, transmitting the intended meanings of the Arabic sources with accuracy. This present volume is the first in a sequence of five lengthy volumes that were originally published in annotated French translations with Arabic critical editions of the primary sources. The original volumes of this colossal series were published by Al-Furqān Islamic Heritage Foundation in Wimbledon, London, and all entitled in their main headings as: *Les mathématiques infinitésimales du IXe au XIe siècle*. This unique undertaking was assumed by Rashed at the time in association with Al-Furqān under the praiseworthy Directorship of the late Professor Y. K. Ibish. It is no exaggeration to affirm that this collection of Arabic critical editions and annotated French translations, with historical and mathematical commentaries, has an epistemic significance that makes them akin in their value as scientific references to the publication by Cambridge University Press of Isaac Newton's multi-volume *Mathematical Papers*. The present book is based on an adapted, revised and updated version of the first volume that was published in 1996, in the Al-Furqān series, under the title *Les mathématiques infinitésimales du IXe au XIe siècle: Fondateurs et commentateurs*. Volume 1: *Banū Mūsā, Ibn Qurra, Ibn Sinān, al-Khāzin, al-Qūhī, Ibn al-Samḥ, Ibn Hūd*. The present volume constitutes the first in a larger project that has been undertaken by Routledge (Taylor & Francis Group), in association with the Centre for Arab Unity Studies in Beirut (*Markaz dirāsāt al-waḥda al-ʿarabiyya*), of publishing the annotated English translations of the remaining four volumes of the *Mathématiques infinitésimales*. This task is now well under way, and this set of publications will constitute a welcome addition to a whole cluster of Rashed's vast *œuvre* that have been translated into the English language. Of these we note the *Encyclopedia of the History of Arabic Science* (editor and co-author) (London/New York, Routledge, 1996), *Omar Khayyam. The Mathematician* (Persian Heritage Series no. 40, New York, Bibliotheca Persica Press, 2000) and the immense *Geometry and Dioptrics in Classical Islam* (Al-Furqān Islamic Heritage Foundation, Wimbledon, London, 2005). Another outstanding example is *Al-Khwārizmī*: *The Beginnings of Algebra* (Saqi Books, London, 2009), a book whose publication I had the privilege of coordinating, as well as revising the annotated English translation of that part which contained the original and first Arabic critical edition of al-Khwārizmī's *Book of Algebra* (*Kitāb al-jabr wa-al-muqābala*). This growing corpus of English translations of Rashed's works will prove to be an excellent source for future endeavours to rewrite manifold chapters of the history of mathematics and the exact sciences in classical Islam, and the subsequent unfurling of several of its rudiments up to the seventeenth century within the European mediaeval, Renaissance, and early-modern *milieus*.

It is my delightful duty in this context to gratefully thank Professor Roshdi Rashed for entrusting me with the momentous responsibility of overseeing and editing the annotated English translations of his precious works, with what such endeavours require in terms of meeting exacting demands, attributes, and criteria in the pursuit of academic excellence. This monumental project could not have been realized without the genuine generosity of the Centre for Arab Unity Studies in Beirut (*Markaz dirāsāt al-waḥda al-ʿarabiyya*) in its munificent sponsoring of the annotated English translations herein and those that are works in progress. Very special thanks are therefore due to the Centre's eminent Director and renowned scholar, Dr Khair El-Din Haseeb, and to the members of the Centre's prestigious Board of Trustees, for their continual endorsement and magnanimous patronage of these long-term initiatives. I express also my deep thankfulness to Mr Joe Whiting, to Ms Emily Davies and to Ms Suzanne Richardson, the editors at Routledge for their enthusiastic adoption of this publication project, and for accompanying us in the long journey of accomplishing it with success. The same also applies to the willingness of Professor Ian R. Netton to include this publication in the distinguished Routledge book series that he edits: *Culture and Civilization in the Middle East*. With profound gratitude, I owe thanks to the translators who contributed to the composition of this volume. I am obliged to Mr Roger Wareham, who undertook the arduous task of rendering the bulk of the draft translations, with contributions made alongside his efforts by Mr Chris Allen and Mr Michael Barany. They may all be assured of my sentiments of gratitude. I am also personally indebted to Mrs Aline Auger, at the Centre National de la Recherche Scientifique in Paris, for preparing this volume for printing. Recognition is also owed to the memory of the late Professor Mohammad Arkoun for his wise counsel in the initial stages of launching this project. I must furthermore acknowledge with gratitude the unremitting commitment of Professor Judith V. Field in continuing to translate Rashed's works into English. Finally, I ought to thank again Professor Roshdi Rashed, with profound appreciation, for encouraging me with thoughtfulness and care in accomplishing the challenging tasks of coordinating and editing the publication of his translated works, and for his patience and diligence in benevolently responding to my queries in view of refining the production of this present volume. It remains to be stated that I enjoyed the dispensing of the expectations behind this stimulating responsibility, despite its trying character, and I affirm that shortcomings in performing the privileges of rendering and presenting this work into the English language, and its niceties, are ultimately mine.

Nader El-Bizri (Editor)

London, 22nd December 2010

PREFACE

Scientific historians are in agreement that the tracing of scientific traditions constitutes a fundamental component of their work. At first glance, this task would appear to be an easy one. These traditions often appear to be obvious and immediately recognizable from the names of the authors and the titles of the manuscripts themselves. However, as soon as the task is begun, these clues are quickly seen to be no more than a deceptive illusion of simplicity. Is it not characteristic of every documentary tradition that it lives, diversifies and reinvents itself with every author and with every question that it raises? There are many other obstacles to be found upon this road, and every historian sooner or later comes up against the thorny question of 'style'. It is certainly this scientific style that, through the variety of forms and transformations that model a tradition, distinguishes it from all others and defines its identity. The difficulty always lies in isolating this identifying note, which can everywhere be sensed, but which inevitably eludes one's grasp. But seize it one must, as it is this elusive note alone that allows us to place an individual work in perspective, and to elucidate its significance. It is therefore this phenomenological process that gives the tradition its true organizational purpose. It reveals the relationships between the works that weave it together, and it protects the historian from falling into the traps set by their own preconceptions, from getting themselves tied up in the search for precursors and from becoming seduced by the illusion of the new.

Essential to the history of science as a whole, this task appears to us to provide a particularly appropriate response to the pressing questions currently facing the history of mathematics and the classical sciences in Islam. The reasons for such urgency have their roots in the fragility of historical research in this field and in the weaknesses of its history. Isolated by language, and often trailing in the wake of other oriental studies, research into the history of classical Islamic mathematics and science is subjected to quality criteria that are still obscure and changing. To this must be added a further obstacle holding back progress in research into these traditions: the question of how to recognize them, buried under a mountain of facts, when so many of the principal players are notable by their absence? How, for example, can we begin to trace the algebraic tradition when we know little more of al-Samaw'al and Sharaf al-Dīn al-Ṭūsī than their names? How can we follow the history of number theory when the works of al-Khāzin, al-Fārisī and others have been lost? The history of optics cannot be fully known without Ibn Sahl, and we have no clear idea of the Marāgha School in relation to astronomy. While it is possible to identify a tradition, even under these conditions, it is quite another question to define its limits, to isolate its unifying elements and to understand the reasons for a succession of changes to a given text. To achieve that, we need the help of a detailed and attentive epistemological study, albeit one that remains, as it has to, discrete. Only through such an analysis can we hope to understand how an epistemic structure mutates and evolves over time.

This approach, which has already guided our work on the history of algebra, number theory, Diophantine analysis and optics, and which remains our preferred approach in this volume, has enabled us to explore a number of promising avenues, and even to follow a few through to their conclusion where their choice appears to be well merited. In our research into the history of classical Islamic mathematics and science, we have never deviated from the one fundamental postulate: that it is impossible to fully understand individual works without anchoring them firmly in the traditions that gave them birth. We have also remained faithful to the necessity of breaking with the historical reduction approach all too common in this field. One cannot succeed by wandering through the garden at random, gathering the odd flower here and there.

In this book, it has been our aim to retrace the tradition of research into 'infinitesimal mathematics', on this occasion with the intention of exploring the main stem, if not all of the lateral branches. This ambition was based partly on the nature of this field of study, but also on the work of those who have preceded us. We have had recourse to a limited number of manuscripts, most of which have survived, dating from the second half of the ninth century – most notably the works of the Banū Mūsā brothers – up to the first half of the eleventh century, at which point the writings of Ibn al-Haytham bring the tradition to an abrupt halt. This topic has also proved attractive to a number of scientific historians whose work, preliminary and provisional though it is, has proved to be extremely useful. Particular cases in point are the translations into German by H. Suter.

So, what exactly does this expression 'infinitesimal mathematics' cover? The question is not merely a rhetorical one. This phrase is not a translation of any Arabic term in use during the classical period and, without further explanation, its meaning may easily be confused with that of 'infinitesimal calculus'. It is but a small step from one to the other, and this step that can all too quickly be taken without noticing the abyss that lies below. In order to address this question, we must first break it down into two parts. The first, and more general one, is the absence of any original name for this branch of mathematics. Can we include such an unnamed body of work within the history of a discipline? Such is the epistemological and historical problem that we have uncovered: that of the status and independence of the knowledge obtained. In inventing a name, we are implying, in this case at least, that a new requirement to distinguish this knowledge from all other knowledge has arisen. However, in our favour it must be accepted that the absence of a name does not necessarily denote the non-existence of the object. No-one today could deny the existence of formal research in combinatorial analysis before that term was invented, or the contributions to elementary algebraic geometry before such an expression become widespread, or the studies in Diophantine analysis carried out before the name of this Alexandrian mathematician was adopted. Our problem is more specific in this case. It is to understand the nature of this 'infinitesimal mathematics', its organization, coherence and unity, together with the links between the various branches making up the whole. In brief, we need to understand the extent of the cleft that separates it from 'infinitesimal calculus'. We believe that we are now in a position to improve our understanding of its origins, and to use this understanding in order to derive its true beginnings.

The first objective of this book is to retrace this tradition of 'infinitesimal mathematics', in order to then be in a position to examine this breach between the history and prehistory of infinitesimal calculus. We begin by tracing, translating and commenting on all the surviving manuscripts relating to the measurement of the areas and volumes of curved solids – lunes, circles, parabolas, ellipses, spheres, cylinders and paraboloids – together with the determination of their limiting values, isepiphanics or isoperimeters. Our reason for confining ourselves to these manuscripts alone is that they all share a common consistency and a progressive unity, due to rather than despite their successive corrections and additions. Each of the mathematicians who laid their own individual brick in this wall, without exception, built on the work of their predecessors in order to derive their improved proofs and new extensions. Is this not the mark of a living tradition? The reason for excluding other works from this *corpus mathematicorum* despite their relationships and similarities to these treatises on infinitesimal mathematics is not a circumstantial one. While these works on astronomy, statics and numerical analysis include certain infinitesimal considerations, they do not form part of the organic structure of this tradition. Wherever I have made reference to one of these works, it is either to provide clarification to the reader or to illustrate the basis for a history of editions. Their appearance in the footnotes and appendices is not to imply that they constitute no more than addenda to the history of infinitesimal mathematics. Rather, it is to indicate that here they complement it, without in any way detracting from the fact that they merit a similar volume to this in their own right.

The first volume of this book is dedicated to the beginnings of research into infinitesimal mathematics, up to the point at which it may be considered to be almost complete; in other words, to the founders of this branch of mathematics. It therefore traces, translates and comments on texts written between the second half of the ninth century and the end of the tenth century, including those of the Banū Mūsā brothers, Thābit ibn Qurra, al-Khāzin, Ibrāhīm ibn Sinān, al-Qūhī and Ibn al-Samḥ. It is regretted that the works of al-Māhānī, Ibn Sahl and doubtless many others have been lost, either temporarily or for good. We have, however, thought it appropriate to include a chapter on Ibn Hūd, a successor to Ibn al-Haytham and a commentator on both his work and that of Ibn Sinān.

In the second volume, which appeared in its original French version in 1993, we traced, translated and commented on the works of Ibn al-Haytham, the mathematician who finalized this tradition and who marks its end-point.

Ibn al-Haytham was the last to carry out innovative research in this field. Eleven centuries after Archimedes, and in a totally different mathematical and cultural context, the history of analytical mathematics was to repeat itself. Two attempts to break through in this field were each brought to an abrupt halt after having enjoyed periods of great success. This phenomenon, of intense interest to mathematical historians and a rich source of material for the epistemologist, will be the subject of a final volume once we have covered all the necessary preliminary stages and reverses essential to a restoration of the Archimedean tradition.

In order to understand the research of Ibn al-Haytham in the field of infinitesimal mathematics and to identify his innovations within the Archimedean tradition, we have found it necessary in preparing this final volume to trace, translate and analyse his separate contribution to the Apollonian tradition. The research carried out by Ibn al-Haytham in the field of infinitesimal mathematics can then take its place within the larger body of his work. The third volume will therefore be largely dedicated to his studies on the conics and their applications. Taken together with our previous publications (*Les Connus, L'analyse et la synthèse*) these two volumes will, for the first time, bring together the entire mathematical works of Ibn al-Haytham with the exception of his commentaries on Euclid.

Some of the texts edited, translated and commented on in this volume were thought to have been lost, but have been rediscovered during our work. Others have been the subject of confusion and misunderstanding, and this we have sought to clarify. Most of them have never before been edited or translated. Those few texts that have been traced have never, with one exception, appeared in a critical edition. A strict translation of all these works is given in this volume.

We have discussed the method used to edit the texts many times in many of our works. The original French translations adhered strictly to these principles. These translations are literal, as faithful to the letter of the original as to its meaning, without ever violating the sensibilities of the reader. We have deliberately set out to be selective in our choice of citations, and these are in no way exhaustive. We trust that our readers will be generous enough to attribute any omissions to our selectivity rather than to our ignorance. Finally, we hope that the experts in this field will find something of use in our work, and that they will pardon any errors that we have made. For our part, we are satisfied that we have done our best.

I would like to thank Christian Houzel for taking on the onerous task of proof-reading the original French version of this work in line with the conventions of the series in which it first appeared, a task that he completed with all the scientific knowledge and erudition for which he is well known. I would also like to thank Philippe Abgrall, Maroun Aouad, Hélène Bellosta, Pascal Crozet and Régis Morelon for proof-reading the final drafts. My thanks are also due to Aline Auger, Ingénieur d'Études at the Centre National de la Recherche Scientifique for her dedicated and effective collaboration during the difficult preparation of the manuscript, and for preparing the index. I am grateful to the al-Furqān Islamic Heritage Foundation and its President, Sheikh Ahmed Zaki Yamani, for the support enabling the publication of this work. I thank the members of the Council of the al-Furqān Islamic Heritage Foundation for having selected this book, and I take this opportunity of expressing my gratitude to the Secretary General, Dr H. Sharifi, and the Coordinator, Dr A. Hamilton for their work on its publication.

Roshdi Rashed
Paris, 1996

Director of Research,
Centre National de la Recherche Scientifique, Paris
Professor, Department of the History and Philosophy of Science, University of Tokyo

NOTE

< > These brackets are introduced in order to isolate an addition to the Arabic text that is necessary for understanding the English text.

In the mathematical commentaries, we have used the following abbreviations:

per.: perimeter
p.: polygon
port.: portion
sg.: segment
tp.: trapezium
tr.: triangle.

CHAPTER I

BANŪ MŪSĀ AND THE CALCULATION OF THE VOLUME OF THE SPHERE AND THE CYLINDER

1.1. INTRODUCTION

1.1.1. *The Banū Mūsā: dignitaries and learned*

The three brothers Muḥammad, Aḥmad and al-Ḥasan, sons of Mūsā ibn Shākir, are usually referred to collectively by their patronymic alone. Early biobibliographers simply entitled their articles Banū Mūsā (the sons of Moses),[1] and their modern counterparts have invariably followed suit – where they have not just copied their work wholesale.[2] Aspects of this tradition continued in the Latin texts, with Gerard of Cremona referring to them in this way: 'Filii Sekir, *i.e.* Maumeti, Hameti, Hasen'.[3] It should be understood that referring to the lives of the Banū Mūsā in this manner has not prevented biographers from acknowledging their existence as independent individuals from one another, nor the occasional mention of one of them without mentioning the other two. They have also managed to highlight a number of individual differences between the brothers, which are of great importance to us, including Muḥammad's interest in astronomy and

[1] Al-Nadīm, *Kitāb al-Fihrist*, ed. R. Tajaddud, Tehran, 1971, pp. 330–1; al-Qifṭī, *Ta'rīkh al-ḥukamā'*, ed. J. Lippert, Leipzig, 1903, pp. 315–16 and 441–3; Ibn Abī Uṣaybiʿa, *'Uyūn al-anbā' fī ṭabaqāt al-aṭibbā'*, ed. A. Müller, 3 vols, Cairo/Königsberg, 1882–84, vol. I, pp. 187, 9–12; 207, 22–208, 17; ed. N. Riḍā, Beirut, 1965, pp. 260, 11–13; 286, 19–287, 15. Ibn Abī Uṣaybiʿa speaks, however, of the sons of Shākir.

[2] C. Brockelmann, *Geschichte der arabischen Literatur*, 2nd ed., I, Leiden, 1943, p. 216; H. Suter, *Die Mathematiker und Astronomen der Araber und ihre Werke*, Leipzig, 1900, pp. 20–1; F. Sezgin, *Geschichte des arabischen Schrifttums*, V, Leiden, 1974, pp. 246–52; M. Steinschneider, 'Die Söhne des Musa ben Schakir', *Bibliotheca Mathematica* 1, 1887, pp. 44–8, 71–6; J. al-Dabbagh, 'Banū Mūsā', *Dictionary of Scientific Biography*, vol. I, New York, 1970, pp. 443–6. The Arabic introduction by Ahmad Y. al-Hasan to the *Kitāb al-ḥiyal* edition of the Banū Mūsā, Aleppo, 1981, pp. 18–30.

[3] M. Clagett, *Archimedes in the Middle Ages*, vol. I, Madison, 1964, p. 238.

mathematics, Aḥmad's talents in the field of mechanics, and al-Ḥasan's genius in geometry.[4] They have even attributed at times writings composed under the forenames of all three Banū Mūsā to a single brother.[5]

Biobibliographers and historians are unanimous in affirming the importance of the scientific works of the Banū Mūsā and acknowledging their contribution to the scientific movement of the period.[6] In the political arena, they also seem to agree that the eldest, Muḥammad, was the most important, with the other two playing significantly paler roles.

While it is important for us to recognize these aspects, they are not included herein for their anecdotal value, but rather as an indication that the three brothers clearly worked together as a team. Yet, within this collaborative effort, their collective works did not exclude individual compositions. A closer look at their works shows that the three brothers did not simply constitute what would today be described as a research team; rather their working relationship was much closer to being constitutive of the solid core of a school. Moreover, their collective efforts were not restricted purely to scientific research; rather, they were active in the politics of scientific activity, and in politics *per se*. Their team also appears to have been in competition with other scientific groups, apparently including the more loosely constituted team of al-Kindī. All of these aspects, which became obvious to us when studying various accounts and testimonies regarding the Banū Mūsā, al-Kindī, and their period in general, lead us to raise a new question: What does such teamwork structure represent in the ninth century?

[4] In the case of al-Ḥasan for example, his own brothers tell of his erudition in geometry – *vide infra*, p. 8. The biobibliographers tell a story that, while of doubtful authenticity, does serve to illustrate the contemporary reputation of al-Ḥasan in the field of geometry. Having read no further than the first six books of Euclid's *Elements*, he was able by himself to work out the contents of the remaining seven books. The caliph al-Ma'mūn would have personally criticized him should he have failed in reading such a fundamentally important book in full, regardless of his need to do so (al-Qifṭī, *Ta'rīkh al-ḥukamā'*, p. 443). For further particulars, regarding the importance of the contributions of Muḥammad to astronomy, see G. Saliba, 'Early Arabic critique of Ptolemaic cosmology', *Journal for the History of Astronomy* 25, 1994, pp. 115–41.

[5] Al-Nadīm, for example, attributes the composition of *Kitāb al-ḥiyal* (*Book of Ingenious Devices*) to Aḥmad alone. He attributes the book on *The Elongated Circular Figure* to al-Ḥasan, an attribution confirmed by Thābit ibn Qurra at the start of his treatise *On the Sections of the Cylinder and its Lateral Surface.* He also attributes several treatises to Muḥammad alone.

[6] For example, al-Nadīm, *al-Fihrist*, pp. 304 and 331; Ibn Abī Uṣaybiʿa, *'Uyūn al-anbā'*, ed. Müller, I, pp. 187, 9–12; 205, 29–31; 215, 29–31; ed. Riḍā, pp. 260, 11–13; 283, 9–11; 295, 9–11.

The instinctive mutual understanding, or complicity, between brothers cannot be the only answer. The story of Johann and Jacob Bernoulli, centuries later, affords a prime counter-example. This team could not have worked together without the school that they led, and of which they formed the heart. The three brothers also worked closely with some of the greatest translators of the time, including Ḥunayn ibn Isḥāq and Hilāl ibn Hilāl al-Ḥimṣī.[7] They were also able to recruit collaborators of the class of Thābit ibn Qurra.[8] Their school divided its efforts between innovative research and the translation of older work passed down from the Greeks: two activities that were complementary and interdependent, as we have shown on more than one occasion.[9] Finally, the Banū Mūsā were also interested in the institutionalization of science. Hence, we find them associated with the famous House of Wisdom in Baghdad, making astronomical calculations, and working on the problems of hydraulics. The engagement of the Banū Mūsā with the scientific and cultural activities of their time was only equalled by their participation in politics (at least in the case of Muḥammad) and in administrative roles (as was the case with both Muḥammad and Aḥmad). There are many indications of their links to the circles of power and learning in Baghdad, the centre of an immense empire at the peak of its glory in the first half of the ninth century. An entire book could be written on the intrigues and dealings that took place there; and such a line of investigation would be well merited, since the story of the Banū Mūsā did not constitute an isolated exclusive case.

This portrayal, although painted in broad strokes, is still sufficient to give some understanding of the background to their work. It clarifies the writings of the early biobibliographers, and forms an initial attempt at a critical

[7] Al-Nadīm wrote about Hilāl ibn Hilāl al-Ḥimṣī that 'he translated the first four books of the Conics of Apollonius in the presence of Aḥmad ibn Mūsā' (*al-Fihrist*, p. 326):

وترجم الأربع المقالات الأولة بين يدي أحمد بن موسى، هلال بن أبي هلال الحمصي.

This fact is confirmed by the translation manuscripts. The sentence by al-Nadīm is effectively copied from the introduction to the translation of the *Conics*, in which can be found: 'the one who was in charge of the translation of the first four books of the Conics of Apollonius in the presence of Aḥmad ibn Mūsā, Hilāl ibn Hilāl al-Ḥimṣī'. See R. Rashed, *Apollonius*: *Les Coniques,* tome 1.1: *Livre I*, Berlin, New York, 2008, p. 507, 12–14.

[8] See the next chapter.

[9] R. Rashed, 'Problems of the transmission of Greek scientific thought in Arabic: examples from mathematics and optics', *History of Science* 27, 1989, pp. 199–209; reprinted in *Optique et mathématiques: Recherches sur l'histoire de la pensée scientifique en arabe*, Variorum CS388, Aldershot, 1992, I.

examination of the descriptions that have come down to us. We can now begin to understand how a single book (in particular the one discussed here) can include both geometrical problems and new mechanical constructions. We can also see how research begun by one of the three brothers, al-Ḥasan for example, could be continued by another, namely Aḥmad. We can also begin to understand the fictionalized nature of their biography, which we do not take to be certain, but which is widely accepted today, effectively without proper examination.

What would be more favourable for a novelist than the story of these three wise men developing their ideas against a background of headlong scientific advance and political tumult? Victims of biobibliographers with unrestrained imaginations, the Banū Mūsā became the heroes of fantasy fiction. We have noted this tendency on more than one occasion in the works of the ancient biobibliographer al-Qifṭī,[10] our main source of information relating to the Banū Mūsā. He was fond of embellishing his stories in order to draw in his readers and entertain them. Al-Qifṭī tells that the father of the Banū Mūsā,[11] that is to say Mūsā ibn Shākir, had nothing to do with 'the sciences and the letters' during his youth, but that he lived as a bandit, robbing travellers on the roads of Khurāsān. As we shall see, the choice of this region was by no means a random one, given the conclusion to his story. Al-Qifṭī is sparing neither in the details of the deviousness of this character nor in the tricks that he got up to in order to cheat those around him. He is even able to describe the face of Mūsā ibn Shākir, his horse and other attributes in detail; three and a half centuries after the events took place![12] All this wealth of detail throws considerable doubt on the account of al-Qifṭī, or at the very least on his sources.

The reason for the choice of Khurāsān as the location becomes evident later in the story, when the bandit teams up with another robber who eventually becomes the caliph al-Ma'mūn. The region of Khurāsān was bequeathed to al-Ma'mūn by Hārūn al-Rashīd, and al-Ma'mūn lived there for a while before deposing his brother al-Amīn and becoming the seventh of the Abbasid caliphs. The story told by al-Qifṭī continues and ends in storybook fashion: the bandit repents, and becomes the companion of the future ruler, and then dies at just the right time (the exact date being

[10] See *Les mathématiques infinitésimales du IXe au XIe siècle*, vol. II: *Ibn al-Haytham*, London, 1993, pp. 5–8; R. Rashed and B. Vahabzadeh, *Al-Khayyām mathématicien*, Paris, 1999; English version (without the Arabic texts): *Omar Khayyam. The Mathematician*, Persian Heritage Series no. 40, New York, 2000.

[11] Al-Qifṭī, *Ta'rīkh al-ḥukamā'*, pp. 441–3.

[12] *Ibid.* This story is often retold by both early and modern historians. One example is Ibn al-ʿIbrī, *Tārīkh mukhtaṣar al-duwal,* ed. O.P. A. Ṣāliḥānī, 1st ed., Beirut, 1890; reprinted 1958, pp. 152–3.

conveniently left vague) to confide his three children to the care of the caliph. This opportune demise sets the three brothers firmly on the path to royalty. At first, they were protected by their new guardian, the caliph in person, then they were left in the care of Isḥāq ibn Ibrāhīm al-Muṣʿabī, who was for a time the governor of Baghdad. He becomes their tutor, and arranges for them to enter the House of Wisdom, under the aegis of the famous astronomer Yaḥyā ibn Abī Manṣūr (died in 217/832).

This then is the narrative as told by al-Qifṭī. This account will be relegated by Ibn al-ʿIbrī (also known as Bar Hebraeus), and it has since been relentlessly repeated over and over by everyone else, right up to the present day. At the present time, we know of no source, independent of al-Qifṭī himself, that can confirm his story as a whole or, indeed, in any part. On the contrary, al-Qifṭī often contradicts himself. For example, elsewhere in his book he paints a portrait of Mūsā ibn Shākir that is barely compatible with the preceding one; this time describing him as a member of the most advanced group of mathematicians and astronomers![13]

In the absence of other sources, we can only dismiss the story told by al-Qifṭī, especially as it appears to have been a late addition, tacked to the end of his book.[14] However, without it the history of the Banū Mūsā fades and diminishes. Very little remains as the basis for a biography, and what little is left lies dispersed among the annals and other bibliographies. In the *Annals* of al-Ṭabarī,[15] Muḥammad and Aḥmad appear in the course of events as members of the entourage of a number of successive caliphs. The two brothers appear in turn as wealthy individuals, as counsellors to the caliphs, and as managers of major civil engineering projects. In the year 245/859, Muḥammad and Aḥmad appear on the list of rich citizens required to provide the caliph al-Mutawakkil[16] with the funds needed to build his new city of al-Jaʿfariyya.[17] This list consists of around twenty names, including a number of famous ministers such as Ibn Farrūkhānshāh and Ibn Mukhlad.

[13] This is what al-Qifṭī wrote without drawing attention to the flagrant contradiction with his earlier assertion: 'Advanced in geometry, he [Mūsā ibn Shākir] and his sons – Muḥammad ibn Mūsā, his brother Aḥmad and their brother al-Ḥasan – were all advanced in the field of mathematics, the configuration of orbs and the movements of the stars. This Mūsā ibn Shākir was famous among the astronomers of al-Ma'mūn and his sons were among those with the greatest insight into geometry and the science of ingenious procedures', *Ta'rīkh al-ḥukamā'*, p. 315. This portrait and the dates given contradict the other story in every respect.

[14] Actually, the penultimate article.

[15] *Tārīkh al-rusul wa-al-mulūk*, ed. Muḥammad Abū al-Faḍl Ibrāhīm, Cairo, 1967, vol. 9, p. 413.

[16] *Ibid.*, p. 215.

[17] *Ibid.*, p. 216.

Three years later, in 248/862, Muḥammad ibn Mūsā was also present among those who listened to the caliph al-Muntaṣir[18] telling of his dream. In 251/865-6, this same Muḥammad is ordered by the commander of the army of caliph al-Mustaʿīn to undertake an intelligence mission to assess the strength of the enemy forces.[19] In the same year, Muḥammad ibn Mūsā was part of the delegation sent to negotiate the abdication of the caliph.[20]

The context and form of these reports by al-Ṭabarī indicate that they are authentic, and they have also been confirmed by other historians. Both al-Masʿūdī[21] and Ibn Khurdādhbih[22] describe the relationships between the Banū Mūsā and the caliph al-Wāthiq (842–847), and Ibn Abī Uṣaybiʿa repeats the often-told story in which the Banū Mūsā take advantage of their position at the court of the caliph al-Mutawakkil in order to plot and intrigue against their colleague al-Kindī.[23] All agreed on one point: the two brothers Muḥammad and Aḥmad were present at the court of the Abbasid caliphs during at least the period from the epoch of al-Mutawakkil (847) to at least that of al-Mustaʿīn (866); namely, before the death of Muḥammad, which, according to al-Nadīm, took place in 873. Aḥmad ibn Mūsā himself confirms their privileged status, telling how he was posted to Damascus as administrator of the diwān responsible for the postal service.[24]

The eminence of this rank supports the assertion by al-Nadīm that the Banū Mūsā themselves financed missions to search for Greek manuscripts in what remained of the Byzantine empire,[25] and that they recruited a group of translators who were each very well paid. Ibn Abī Uṣaybiʿa supports the work of al-Nadīm, citing a number of translators, including Isḥāq ibn Ḥunayn, Ḥubaysh, and Thābit ibn Qurra, who received a regular salary from the Banū Mūsā.

[18] *Ibid.*, p. 253.

[19] *Ibid.*, p. 292.

[20] *Ibid.*, p. 344.

[21] *Al-Tanbīh wa-al-ishrāf*, ed. M. J. de Goeje, Bibliotheca Geographorum Arabicorum VIII, Leiden, 1894, p. 116.

[22] *Al-Masālik wa-al-mawālik*, ed. M. J. de Goeje, Bibliotheca Geographorum Arabicorum VI, Leiden, 1889; reproduced by al-Muthanna publishers in Baghdad, undated, p. 106.

[23] Ibn Abī Uṣaybiʿa, *'Uyūn al-anbā'*, ed. Müller, pp. 207, 22–208, 17; ed. Riḍā, pp. 286, 9–287, 15.

[24] In the treatise by the Banū Mūsā entitled *Muqaddamāt Kitāb al-Makhrūṭāt* (*Lemmas of the Book of Conics*), ed. R. Rashed in *Les Coniques,* tome 1.1: *Livre I*, p. 505, 16-17:

ثم تهيأ لأحمد بن موسى الشخوص إلى الشام واليًا لبريدها.

[25] See al-Nadīm, *al-Fihrist*, pp. 330–1.

Other reliable sources depict the Banū Mūsā making astronomical observations and working on civil engineering projects. Ibn Khallikān[26] gives a precise report of work they carried out at the personal request of al-Ma'mūn to verify the length of the circumference of the Earth.[27] The astronomical historian C. Nallino[28] has concluded from the statements of Ibn Khallikān, and based on the age of the three brothers and existing knowledge regarding this important scientific event, that the Banū Mūsā could only have acted as assistants to the astronomers of the time, and not as the principal investigators in charge of this experiment. In relation to their civil engineering works, al-Ṭabarī describes a canal dug under their direction and Ibn Abī Uṣaybiʿa echoes what is noted regarding this project of hydraulics.[29]

These then were the Banū Mūsā: three wealthy brothers, moving in the corridors of power and surrounded by a team of advanced researchers, working not only in the mathematical sciences, but also in the applied mathematics of their time, particularly hydraulics and mechanics; major contributors to the scientific community of which they were a part; and founders of a school that also counted Thābit ibn Qurra among its members. We shall now consider their mathematical legacy.

1.1.2. *The mathematical works of the Banū Mūsā*

The early biobibliographers, and al-Nadīm and al-Qifṭī in particular, provide two lists of the works of the Banū Mūsā in the fields of mechanics, astronomy, music, meteorology and mathematics. These lists are not exhaustive, and do not provide a definitive record of their written works. In geometry, the branch of mathematics that interests us here, Aḥmad himself mentions works missing from the lists of the two biobibliographers, and other later mathematicians proceeded likewise. We know of five mathematical texts attributed to the Banū Mūsā, of which only two are known to exist at the present time.

[26] *Wafayāt al-aʿyān*, ed. Iḥsān ʿAbbās, vol. 5, Beirut, 1977, pp. 161–2.

[27] Al-Bīrūnī, *al-Athār al-bāqiya ʿan al-qurūn al-khāliya*, ed. C.E. Sachau under the title: *Chronologie Orientalischer Völker*, Leipzig, 1923, p. 151; also al-Bīrūnī, 'Kitāb taḥdīd nihāyat al-amākin', edited by P. Bulgakov and revised by Imām Ibrāhīm Aḥmad in *Majallat Maʿhad al-Makhṭūṭāt*, May-November 1962, p. 85.

[28] C. Nallino, *Arabian Astronomy, its History during the Medieval Times*, [Conferences at the Egyptian University], Rome, 1911, pp. 284–6.

[29] Ibn Abī Uṣaybiʿa, *'Uyūn al-anbā'*, ed. Müller, pp. 207, 27–208, 17; ed. Riḍā, pp. 286, 23–287, 15.

1° The first work is entitled *The Elongated Circular Figure* (*al-Shakl al-mudawwar al-mustaṭīl*). It is attributed by al-Nadīm and al-Qifṭī to al-Ḥasan ibn Mūsā, and this is confirmed by the late tenth century mathematician al-Sijzī. Not only does he quote the title, when he writes that the Banū Mūsā composed a book 'on the properties of the ellipse, which they called the elongated circle (*al-dā'ira al-mustaṭīla*)', he also summarizes the procedure used by them to trace a continuous ellipse making use of the bifocal property.[30]

In their short treatise on the *Lemmas of the Book of Conics*, Muḥammad and Aḥmad ibn Mūsā mentioned that their brother al-Ḥasan had written a treatise on the generation of elliptical sections and the determination of their areas:

> Drawing on his powerful understanding of geometry, and on his superiority over all others in this field, al-Ḥasan ibn Mūsā was able to study cylindrical sections; namely, those plane figures formed when a cylinder is intersected by a plane that is not parallel to its base, in such a way that the outline of the section forms a continuous enclosing curve. He found out its science and the science of the fundamental proprieties relative to the diameters, the axes, and the chords, and he has found out the science of its area.[31]

According to al-Sijzī, the treatise on *The Elongated Circular Figure* also dealt with the generation of elliptical sections. All indicates that these constitute one and the same treatise, but that is our only certainty; everything else remains conjecture: The treatise seems to have been written before the author had gained an in-depth understanding of the *Conics* of Apollonius; perhaps he had read the book by Serenus of Antinoupolis *On the Section of a Cylinder.*[32] The treatise must have been a substantial work,

[30] This is what al-Sijzī wrote in *The Description of Conic Sections* (ed. R. Rashed in *Œuvre mathématique d'al-Sijzī*. Volume I: *Géométrie des coniques et théorie des nombres au Xe siècle*, Les Cahiers du Mideo, 3, Louvain-Paris, 2004, p. 247, 5–7):

وطريق آخر غريب مستخرج من خواصه. وعمل على هذه الخاصة وبنى عليها بنو موسى بن شاكر كتاباً في خواص القطع الناقص وسموه الدائرة المستطيلة.

which may be rendered as 'Another strange pathway resulting from its [*i.e.* the ellipse] properties. The Banū Mūsā ibn Shākir constructed, from this property, and composed a book on the properties of the ellipse that they called: *the elongated circle*'; this was the bifocal property.

[31] Banū Mūsā, *Muqaddamāt Kitāb al-Makhrūṭāt*, ed. R. Rashed in *Les Coniques,* tome 1.1: *Livre I*, p. 505.

[32] We take up this question later in the analysis of the treatise by Ibn al-Samḥ, *vide infra*, Chapter VI.

forming the basis, along with a deep understanding of the *Conics* this time, of the magisterial development of this study by Thābit ibn Qurra.[33]

The treatise has not survived, but we believe that a part of the text may have influenced the writings of Ibn al-Samḥ, part of whose work does still exist in a Hebrew version.[34] The importance of this treatise in the history in the theory of conics and of infinitesimal mathematics in Arabic, along with the allusions made by Aḥmad ibn Mūsā, in addition to the information supplied by al-Sijzī, and our own conjectures, all can only encourage us to address this question for its own sake.

2° The second text is that mentioned earlier by al-Nadīm and al-Qifṭī, namely *The Lemmas of the Book of Conics*; and several manuscript copies of this text have survived. Nine of the lemmas are established: 'which are required in order to facilitate the comprehension of the *Conics* of Apollonius'.[35]

3° In the introduction to their preceding *opusculum*, Muḥammad and Aḥmad ibn Mūsā retrace the history of their studies of the *Conics*, mentioning a commentary written by Aḥmad on seven books of the oeuvre of Apollonius. This sibylline allusion is the only information we have on this commentary.[36]

4° A book entitled *On a Geometric Proposition Proved by Galen*, of which no copies are currently known to exist.

5° The treatise that we establish in the next section.

Finally, another short text on the trisection of angles carries the names of the brothers, but there appear to be a number of serious problems in this attribution.[37]

All these titles share a common factor that, like a faint watermark, seems to run through all the research begun through the Arabic language by the Banū Mūsā; namely, their simultaneous interest in the geometry of conics

[33] The treatise by Ibn Qurra *On the Sections of the Cylinder and its Lateral Surface* is discussed later.

[34] See the analysis of the text by Ibn al-Samḥ later.

[35] R. Rashed, *Les Coniques,* tome 1.1: *Livre I*, p. 509, 1:

يحتاج إليها في تسهيل فهم الكتاب.

[36] *Ibid.*, p. 507, 1-2, contains the following: 'He prepared himself to depart from Syria and go to Iraq; once in Iraq, he saw again the commentary (*tafsīr*) of the seven books which reached us at present'.

وتهيأ انصرافه من الشام إلى العراق، فلما صار إلى العراق عاد إلى تفسير بقية السبع المقالات التي وقعت إلينا.

[37] Cf. mss Oxford, Bodleian Library, Marsh 207, fol. 131^{v} and Marsh 720, fol. 260^{v}.

and the measurement of areas and volumes delimited by curves; as a combination of the two traditions of Apollonius and Archimedes.

1.1.3. ***Treatise on the measurement of plane and spherical figures: a Latin translation and a rewritten version by al-Ṭūsī***

The fate of this treatise has been a strange one. Two fragments of the original Arabic text have been found (see Table II). But it survives through an edited rewritten version composed by Naṣīr al-Dīn al-Ṭūsī in the thirteenth century. Luckily, Gerard of Cremona's Latin translation of the original Arabic text has survived, and this has been transcribed and translated into several languages.[38]

These are the bare facts. It appears to all intents and purposes that the Ṭūsī version simply replaced the original. One can even imagine a scenario that describes how this would have happened – taking into account that this shifts away from the truth: while this important original text remained available to students, as can be seen by comparison with the works of later writers on these topics, the treatise was chosen by al-Ṭūsī to be included with a number of other mathematical works in the edited collection known as the *mutawassiṭāt* (namely, 'the abridged astronomies', to which were added some books in mathematics). Originally intended to be used as textbooks for teaching purposes, these collections were very successful; judging by the number of manuscript copies that have survived. This work by al-Ṭūsī ensured that the thoughts of the Banū Mūsā reached a wide audience. However, this success was to the detriment of the original work. The popularity of the Ṭūsī version was such that no-one bothered to copy the original Banū Mūsā treatise; and, despite our best efforts, no trace of it has ever been found!

The examination of the Latin translation reveals the omission of a long passage from it, which was quoted by al-Ṭūsī from the original, in which the Banū Mūsā describe the mechanical device that they designed to determine two segments lying between two given segments such that the four segments form a proportional sequence. This missing passage also discusses the trisection of angles.[39] There is no doubt as to the authenticity of this

[38] M. Curtze, 'Verba Filiorum Moysi, Filli Sekir, id est Maumeti, Hameti et Hasen. Der Liber trium fratrum de Geometria, nach der Lesart des Codex Basileenis F. II. 33 mit Einleitung und Commentar', *Nova Acta der Ksl. Leop.-Carol. Deutschen Akademie der Naturförscher*, vol. 49, Halle, 1885, pp. 109–67; H. Suter, 'Über die Geometrie der Söhne des Mûsâ ben Shâkir', *Bibliotheca Mathematica* 3, 1902, pp. 259–72; Clagett, *Archimedes in the Middle Ages*, I, pp. 223–367. See also W. Knorr, *Textual Studies in Ancient and Medieval Geometry*, Boston, Basel, Berlin, 1989, pp. 267–75.

[39] Cf. further in Banū Mūsā's text, Proposition 18.

passage. Perhaps Gerard was simply defeated by its real linguistic and technical difficulty. In attempting to understand the contribution of the Banū Mūsā, it is necessary to refer to the translation by Gerard of Cremona. However, in order to fully achieve our objective, it is also essential to read the Ṭūsī version. One further merit of the Latin translation is that it clarifies the meaning that al-Ṭūsī gave to the word *taḥrīr* (in the sense of re-editing, re-writing or re-composing), thereby providing us with a measure of the distance separating his text from that of the Banū Mūsā. In return, the Ṭūsī version also throws light on the Latin translation, or at least its lexical characteristics. The historian attempting to track down the original thoughts of the Banū Mūsā in this field is therefore faced with the double problem of having nothing to go on but an indirect translation and a text re-written three centuries after the original. Having clearly indicated the dangerous rocks waiting to wreck the efforts of the historian under these circumstances, we can now begin to try to understand, provisionally at least, what al-Ṭūsī has to tell us about what he meant by this *taḥrīr* (re-editing, re-writing).

The first clue we have relates to the writing style employed by the Banū Mūsā and their contemporaries. They were writing for the mathematicians of their time, and for students of mathematics and astronomy. All these readers would have been familiar with a range of other books, including Euclid's *Elements* and *Data*. The Banū Mūsā were therefore able to use propositions from these books without repeating them explicitly; since these were assumed to be common knowledge. No criticism of the authors is implied in stating this fact. This has been common practice from the earliest period right through to the present day. Even al-Ṭūsī, who understood Euclid better than anyone, and easily recognized the tacit references made by the Banū Mūsā, never considered it necessary to expand them. He saw no omission that needed rectifying. To interpret this practice as an attempt to hide their sources is to misunderstand the mathematical traditions of the time. It is hardly necessary to point out that they often used propositions that they themselves had proved without explicitly citing them.

The second drawback relates to the edited version composed by al-Ṭūsī, who was also writing for advanced students of mathematics. These apprentices would have been familiar with the works of Euclid, and would have been quite capable of filling in the gaps in any of the basic proofs. When al-Ṭūsī omitted these steps, he was not carelessly failing to include them. He was taking a deliberate decision that they were not necessary.

We already know that one characteristic of the version edited by al-Ṭūsī is its economy. Throughout the Banū Mūsā text, he ruthlessly cuts out everything that is not strictly necessary to mathematical exposition. Whether or not we agree with his editorial decisions, this economy with words remains for al-Ṭūsī one of the principles of an elegant text, giving it an implicit didactic value.

Let us now consider what al-Ṭūsī meant by his *taḥrīr* (re-editing, re-writing). Although of considerable importance, this question has never, to our knowledge, been addressed. We only consider it here in the case of the Banū Mūsā treatise, and we shall begin by identifying a few general traits, before moving on to a precise analysis of a full example. By 're-editing', al-Ṭūsī intends to provide us with a condensed text, from which all the arguments that he considered unnecessary have been excised. His main technique was to eliminate all repetition and redundancy, and to reformulate the sentences, introducing pronouns to reinforce the longer expressions. A number of specific instances may be noted in this regard:

1. Al-Ṭūsī has cut large portions from the sections in which the Banū Mūsā explain their reasons for writing the text. This is especially true in the introduction, where they also describe the methods they adopted in preparing the work. He also makes significant cuts in the conclusion in which the various results obtained are summarized. There is no need to point out how interesting these sections would have been for the historian.

2. Al-Ṭūsī also cut out any sections that he considered to be repetitious. At the beginning of Proposition 16, the Banū Mūsā explain how the determination of two magnitudes lying between two others in such a way that all four progress in proportion can be used also to solve the problem of extracting cubic roots. This technique is later summarized at the end of the book,[40] and this summary does not appear in the version edited by al-Ṭūsī.

3. In the mathematical sections, al-Ṭūsī removed all except the essential text. The expressions used to describe the proposition and the proof, such as *mithāl* (example), *aqūlu inna* (I say), *burhān* (proof), *in amkana dhālika* (if this were possible), *hādhihi ṣūratuhu* (this is its figure) and *wa-dhālika mā aradnā an nubayyin* (this is what we wanted to show), all have been eliminated, either in whole or in part.

However, throughout his entire edited version of the text, al-Ṭūsī makes no alterations to the sense of the strictly mathematical sections of the work, usually quoting these in full. He takes great care to distinguish his own notes and comments from those of the Banū Mūsā by preceding each of them with the remark: 'I say'. Comparing his edition with the Latin translation shows that he made no changes to the structure of the lines of reasoning, or to those of explanation. It can therefore rightly be said that he captured the quintessence of the Banū Mūsā text.

The situation is therefore less serious than we had feared, and we have to admit that, to all intents and purposes, we have the original Banū Mūsā

[40] Cf. Proposition XIX in the Latin translation, p. 348.

text. If some still need convincing, let us consider the example of Proposition 14, and let us attempt to 'reconstitute' its occurrence in the Arabic text that was translated into Latin. This conjectural reconstitution will, no doubt, differ from the original in the choice of some terms and syntactical expressions. However, it is our contention that these differences will not be significant. Such an exercise will, in any event, enable an assessment to be made of the differences between the edited al-Ṭūsī version and the original text. For the purposes of the comparison, it should be noted that the geometrical letters *ṭa', zayn, wāw* and *jīm* have been rendered as *G*, *U*, *Z* and *T* by Gerard of Cremona and as *C*, *F*, *G* and *I* respectively by us (see Table I).

One possible objection to the above argument is that no-one, least of all us, can claim a precise knowledge of the accuracy of the Latin translation. It is true that this can never be known unless and until the actual text written by the Banū Mūsā themselves, or one or more fragments of it, comes to light. Our search for such fragments has led to our discovery of two propositions that, on analysis, seem to confirm the results already obtained.[41] Apart from one or two transcription errors, it can be seen that Gerard of Cremona produced a literal translation of the Arabic text, and that al-Ṭūsī edited the text in the way that we have described. Before giving a further table of comparisons to demonstrate these assertions, it should be remembered that the rediscovered citations appear in an anonymous commentary on Euclid's *Elements*,[42] in which the author cites, among others, Thābit ibn Qurra, al-Nayrīzī, al-Anṭākī, Ibn al-Haytham, Ibn Hūd, al-Dimashqī and al-Fārābī. This same author cites the Banū Mūsā when discussing the trisection of angles. He writes: 'the angle may be divided into three parts following what the Banū Mūsā have indicated. A *lemma* ought to precede this'.[43] In this way, he cites Proposition 12 of the Banū Mūsā treatise, before going on to quote Proposition 18 (see Table II).

This comparison reassures us that the Latin translation is faithful to the original text written by the Banū Mūsā. Two different propositions, and sufficiently widely separated from each other, demonstrate that Gerard of Cremona made a literal translation of the Arabic text. They also confirm our description of the nature of the editing process employed by al-Ṭūsī, by way of an analysis that we carried out before discovering these citations.

[41] It should be noted that the citation of these fragments demonstrates at least that the Banū Mūsā treatise was still in circulation at the end of the thirteenth century.

[42] Ms. Hyderabad, Osmania University 992.

[43] *Ibid.*, fol. 50^r:

وقد تقسم الزاوية بثلاثة أقسام على ما ذكره بنو شاكر. ويقدم لذلك مقدمة.

A first-rate mathematician himself, al-Ṭūsī also found time to rewrite a number of fundamental treatises. The manner in which he approached this task is now clear: he did not hesitate to excise portions of the original text and he makes no attempt to convey the author's style. However, he does not alter the mathematical ideas or the structure of the treatise. He preserves the original reasoning, and adds nothing to the book that was not there in the original. Editing a text in this way is not easy, and only a mathematician of the stature of al-Ṭūsī could have accomplished the task so well, albeit, not uniformly performed. The most mathematically and technically complex of the propositions are the hardest to deal with, and it is for this reason that the propositions of this kind in al-Ṭūsī's text are those that remain truest to the Banū Mūsā original. This can easily be seen by comparing al-Ṭūsī's text with the Latin translation. This is particularly the case with Propositions 17 and 18, in which the mathematics is accompanied by a description of the technical instruments used. It is precisely at this point that three pages are missing from the surviving copies of the Latin text, but, luckily, this section appears in full in the edited al-Ṭūsī version.

This section also contains the most disconcerting assertions. We can begin by revealing that, contrary to the statements of some commentators, the method proposed by the Banū Mūsā is different to that cited by Eutocius and attributed to Plato. It should also be noted that nothing in this section throws any doubt on its authenticity, or on its attribution to the Banū Mūsā. Al-Ṭūsī himself, who always took great pains to distinguish between his own work and that of the Banū Mūsā, leaves us in no doubt on this point. Moreover, the history of the Arabic text also confirms the attribution of this section to the Banū Mūsā. Finally, both the Arabic text and the Latin translation provide a clear response to this question. This is easy to understand if we consider that the text in question appears at the end of Proposition 17 in the Banū Mūsā text. In Proposition 18 of the same text, the Banū Mūsā refer explicitly to the mechanical procedure described in this fragment. The enunciation of Proposition 18 may be translated from the Arabic as: 'Using this ingenious procedure, we may divide any angle into three equal parts', while a corresponding translation from the Latin version reads: 'Et nobis quidem possibile est cum hoc ingenium sit inventum ut dividamus quemcunque angulum volumus in tres divisiones equales' (p. 344, 1–3). However, a reader of the Latin text alone would not have access to the 'ingenious procedure' that the Banū Mūsā referred to, hence demonstrating that it could not have been inserted by al-Ṭūsī. The Banū Mūsā wrote later in the text, based on al-Ṭūsī's version: 'Now, move *GH* using the ingenious procedure described ...' (*infra*, p. 109), which Gerard translated as: 'Et quoniam possibile est nobis per ingenium quod narravimus in eis que premissa sunt et per ea que sunt ei similia ut moveamus lineam

ZH …' (pp. 346–8, 33–35), 'And since by means of the device which we have described in connection with the propositions previously proved and by means of things which are similar to it it is possible for us to move line *GH* ...' (Clagett's translation, pp. 347–9).

It is therefore patently obvious that this ingenious procedure was described earlier by the Banū Mūsā, in a passage not translated by Gerard of Cremona.

There is no escaping it – anyone wishing to study this contribution from the Banū Mūsā is obliged to tackle the problem on two fronts: the edited al-Ṭūsī version and the Latin translation. Each serves to illuminate the other. The Latin translation clarifies the Ṭūsī edition, while the Ṭūsī edition defines the boundaries for the Latin translation. In some respects, the edited al-Ṭūsī version provides a more faithful rendering of the quintessence of the text, despite his interventions, but one cannot deny that the Latin translation provides more of the detail, declarations and repetitions – all integral parts of the original text, and cut out by al-Ṭūsī. Both versions have contributed to the preservation of the Banū Mūsā text, and have assured its historical position as the principal reference to Archimedean teaching for many centuries.

TABLE I

<table>
<tr><th>I
TRANSLATION BY GERARD
(Clagett, pp. 328–31)</th><th>II
RECONSTITUTION OF THE ARABIC TEXT OF I (PROPOSITION 14)</th></tr>
<tr><td>(1) Embadum superficiei omnis medietatis spere est duplum embadi superficiei maioris circuli qui cadit in ea.</td><td>كل نصف كرة فإن مساحة سطحه (أو بسيطه) ضعف مساحة سطح الدائرة العظيمة التي تقع فيها .</td></tr>
<tr><td colspan="2">The surface area of every hemisphere is double the area of the greatest circle which falls in it.</td></tr>
<tr><td>(2) Verbi gratia, sit medietas spere BCAD, et maior circulus qui cadit in ea sit circulus ABC, et punctum D sit polus huis circuli.</td><td>مثـال ذلك : فليكن ا ب جـ د نصف كـرة ، ودائرة ا ب جـ عظيمة تقع فيها ونقطة د قطب هذه الدائرة .</td></tr>
<tr><td colspan="2">For example, let there be the hemisphere BCAD and circle ABC the greatest circle falling in it, and let point D be the pole of this circle.</td></tr>
<tr><td>(3) Dico ergo quod embadum superficiei medietatis spere ABCD est duplum embadi superficiei circuli ABC, quod sic probatur.</td><td>فأقول إن : مساحة سطح (أو بسيط) نصف كرة ا ب جـ د ضعف مساحة سطح دائرة ا ب جـ . وبرهانه أن ...</td></tr>
<tr><td colspan="2">I say, therefore, that the surface area of hemisphere ABCD is equal to double the area of circle ABC.</td></tr>
<tr><td>(4) Si non fuerit duplum embadi circuli ABC equale superficiei medietatis spere ABCD, tunc sit duplum eius aut minus superficie medietatis spere ABCD aut maius ea.</td><td>فإن لم يكن ضعف مساحة سطح دائرة ا ب جـ مساوياً لمساحة سطح نصف كرة ا ب جـ د فهو إما أن يكون أقل منها وإما أن يكون أكثر منها .</td></tr>
<tr><td colspan="2">Proof: If double the area of circle ABC is not equal to the area of hemisphere ABCD, then it is less than the area of hemisphere ABCD or greater than it.</td></tr>
<tr><td>(5) Sit ergo in primis duplum embadi circuli ABC minus embado superficiei medietatis spere ABCD, si fuerit illud possibile. Et sit duplum embadi circuli ABC equale superficiei medietatis spere minoris medietate spere ABCD, que sit medietas spere EHIK.</td><td>فليكن أولاً ضعف مساحة سطح دائرة ا ب جـ أقل من مساحة سطح نصف كرة ا ب جـ د ، إن أمكن ذلك ؛ وليكن ضعف مساحة سطح دائرة ا ب جـ مساوياً لمساحة سطح نصف كرة أقل من نصف كرة ا ب جـ د ، وليكن نصف كرة هـ ح ط ك .</td></tr>
<tr><td colspan="2">First, let double the area of circle ABC be less than the area of hemisphere ABCD, if that is possible. And let double the area of circle ABC be equal to the area of a hemisphere smaller than hemisphere ABCD, namely, hemisphere EHIK.</td></tr>
</table>

III AL-ṬŪSĪ'S EDITION	COMMENTS
سطح نصف الكرة المستدير ضعف سطح الدائرة العظيمة التي هي قاعدتها . The surface area of a hemisphere is double the area of the greatest circle which is its base.	It can be seen that the meaning is conserved and that the expression used by al-Ṭūsī is a little shorter.
فليكن ا ب جـ د نصف كرة، ودائرة ا ب جـ عظيمة تقع فيها وهي قاعدتها ود قطبها . Therefore, let the hemisphere be *ABCD*, the greatest circle *ABC* falling in it and which is its base, and let *D* be its pole.	The only difference is that in text (III) the greatest circle is the base of the hemisphere, which is only implied in (I and II).
	This phrase has been removed by al-Ṭūsī.
فإن لم يكن ضعف سطح دائرة ا ب جـ مساويًا لسطح نصف الكرة، If double the area of circle *ABC* is not equal to the area of the hemisphere,	The Latin translator then only retains one of the two terms *embadum* and *superficies*. Al-Ṭūsī takes off the second part, continuing directly with the alternative.
فليكن أولاً أصغر منه، وليكن مساويًا لسطح نصف كرة أصغر من نصف كرة ا ب جـ د، وهو نصف كرة هـ ح ط ك. first, let it be less than it, and let it be equal to the area of a hemisphere that is smaller than the hemisphere *ABCD*, namely the hemisphere *EHIK*.	The two texts are identical except for the fact that al-Ṭūsī has used pronouns in place of the subjects, and has removed the phrase 'if that is possible', which is implied in the exposition. It is these stylistic differences that distinguish the 'edition' from this section of the text.

<table>
<tr><th>I
TRANSLATION BY GERARD</th><th>II
RECONSTITUTION OF THE ARABIC TEXT</th></tr>
<tr><td>(6) Cum ergo fiet in medietate spere ABCD corpus compositum ex portionibus piramidum columnarum, cuius basis sit superficies circuli ABC et cuius caput sit punctum D, et ponetur ut corpus non tangat medietatem spere EHIK,</td><td>فإذا عمل في نصف كرة ا ب جـ د مجسم من قطع من مخروطات الأساطين مركب بعضها على بعض، قاعدته دائرة ا ب جـ ورأسه نقطة د بحيث لا يماس نصف كرة هـ ح ط ك،</td></tr>
<tr><td colspan="2">When, therefore, there is described in hemisphere ABCD a body composed of segments of cones, the base of which body is the surface of circle ABC and its vertex is point D, and it is posited that the body does not touch hemisphere EHIK,</td></tr>
<tr><td>(7) tunc oportebit ex eis que premisimus ut embadum superficiei corporis ABCD sit minus duplo embadi superficiei circuli ABC. Sed embadum superficiei corporis ABCD est maius embado superficiei medietatis spere EHIK, quoniam continet ipsam. Ergo embadum superficiei medietatis spere EHIK est multo minus duplo embadi superficiei circuli ABC. Et iam fuit ei equalis. Hoc vero contrarium est et impossibile.</td><td>فمما بينا آنفًا تكون مساحة سطح مجسم ا ب جـ د أقل من ضعف مساحة سطح دائرة ا ب جـ. ولكن مساحة سطح مجسم ا ب جـ د أكثر من مساحة سطح نصف كرة هـ ح ط ك لأن الأول يحيط بالآخر. فمساحة سطح نصف كرة هـ ح ط ك أقل كثيراً من ضعف مساحة سطح دائرة ا ب جـ. وقد كان مثله، هذا خلف لا يمكن.</td></tr>
<tr><td colspan="2">then from what we have proved before it will follow that the surface area of body ABCD is less than double the area of circle ABC. But the surface area of body ABCD is greater than the surface area of hemisphere EHIK, since the one contains the other. Therefore, the surface area of hemisphere EHIK is much less than double the area of circle ABC. But it was posited as equal to it. This indeed is a contradiction and is impossible.</td></tr>
<tr><td>(8) Et iterum sit duplum embadi superficiei circuli ABC maius embado superficiei medietatis spere ABCD, si fuerit possibile illud. Et sit equale superficiei medietatis spere maioris medietate spere ABCD, que sit medietas spere FGLM.</td><td>ثم ليكن ضعف مساحة سطح دائرة ا ب جـ أكثر من مساحة سطح نصف كرة ا ب جـ د، إن أمكن ذلك؛ وليكن مساويًا لمساحة سطح نصف كرة أعظم من نصف كرة ا ب جـ د، وليكن نصف كرة و ز ل م.</td></tr>
<tr><td colspan="2">Now again let double the area of circle ABC be greater than the surface area of hemisphere ABCD, if that is possible. Let it be equal to the area of a hemisphere greater than hemisphere ABCD, namely, hemisphere FGLM.</td></tr>
<tr><td>(9) Cum ergo fiet in medietate spere FGLM corpus compositum ex portionibus piramidum columpnarum, cuius basis sit superficies circuli FGLM et cuius caput sit punctum D, et non sit corpus tangens medietatem spere ABCD,</td><td>فإذا عمل في نصف كرة و ز ل م مجسم من قطع من مخروطات الأساطين مركب بعضها على بعض، قاعدته دائرة ا ب جـ ورأسه نقطة د بحيث لا يماس نصف كرة ا ب جـ د.</td></tr>
</table>

III AL-ṬŪSĪ'S EDITION	COMMENTS
فإذا عمل في نصف كرة ا ب جـ د مجسم – كما وصفنا – قاعدته دائرة ا ب جـ ورأسه نقطة د بحيث لا يماس نصف كرة هـ ح ط ک، If, as we have described, we inscribe within the hemisphere *ABCD* a solid whose base is the circle *ABC* and whose vertex is the point *D*, such that it does not touch hemisphere *EHIK*	Identical. Al-Ṭūsī has simply replaced 'composed of segments of cones' with 'as we have described' in order to avoid a repetition. This appears to be one of the motives for him writing his 'edition'.
كان سطحه أصغر من ضعف سطح دائرة ا ب جـ وأعظم من سطح نصف كرة هـ ح ط ک. فضعف سطح دائرة ا ب جـ المساوي لسطح نصف كرة هـ ح ط ک أعظم كثيراً منه؛ هذا خلف. Then its surface area will be less than double the area of the circle *ABC* and greater than the surface area of the hemisphere *EHIK*. Twice the area of the circle *ABC,* which is equal to the surface area of the hemisphere *EHIK* is much greater than it. This is contradictory.	
ثم ليكن ضعف سطح دائرة ا ب جـ أعظم من سطح نصف كرة ا ب جـ د، وليكن مساوياً لسطح نصف كرة و ز ل م. Now, let double the area of the circle *ABC* be greater than the surface area of the hemisphere *ABCD*, and let it be equal to the surface area of the hemisphere *FGLM*.	In this way, al-Ṭūsī has combined two steps in the proof into a single step.
ونعمل فيه مجسمًا – كما وصفنا – غير مماس لنصف كرة ا ب جـ د،	As before, al-Ṭūsī has omitted the description of the solid, simply reminding us that it has already been described, thereby removing an unnecessary repetition from this section of the text.

<table>
<tr><th>I
TRANSLATION BY GERARD</th><th>II
RECONSTITUTION OF THE ARABIC TEXT</th></tr>
<tr><td colspan="2">When, therefore, there is inscribed in hemisphere FGLM a body composed of segments of cones, the base of which body is circle FGLM and its vertex is point D and the body does not touch hemisphere ABCD,</td></tr>
<tr><td>(10) tunc oportebit ex eo quod premisimus ut sit embadum superficiei corporis FGLM maius duplo embadi circuli ABC.</td><td>فيكون مساحة سطح مجسم و ز ل م أكثر من ضعف مساحة سطح دائرة ا ب ج، لما مرّ.</td></tr>
<tr><td colspan="2">then it will follow from what we have proved before that the surface area of body FGLM is greater than double the area of circle ABC.</td></tr>
<tr><td>(11) Verum embadum superficiei medietatis spere FGLM est maius embado superficiei corporis FGLM.</td><td>ومساحة سطح نصف كرة و ز ل م أعظم من مساحة سطح مجسم و ز ل م لكونه محيطًا به.</td></tr>
<tr><td colspan="2">But the surface area of hemisphere FGLM is greater than the surface area of body FGLM..</td></tr>
<tr><td>(12) Ergo embadum medietatis spere FGLM est maius duplo embadi superficiei circuli ABC. Sed iam fuit ei equale. Hoc vero est contrarium et impossibile.</td><td>فمساحة سطح نصف كرة و ز ل م أكثر كثيرًا من ضعف مساحة سطح دائرة ا ب ج، وقد كان مثله؛ هذا خلف لا يمكن.</td></tr>
<tr><td colspan="2">Therefore, the surface area of hemisphere FGLM is greater than double the area of circle ABC. But it was posited as equal to it. This indeed is a contradiction and is impossible.</td></tr>
<tr><td>(13)</td><td>فليس مساحة سطح نصف كرة ا ب ج د بأقل من ضعف مساحة سطح دائرة ا ب ج، وقد كنا بينا أنها ليست بأكثر منها، فهي إذن مثلها؛ وذلك ما أردنا أن نبين.
Therefore, the surface area of the hemisphere ABCD is not smaller than double the area of circle ABC; but we have proved before that is not greater than it; it is then equal to it. This is what we wanted to prove.</td></tr>
<tr><td>(14) Iam ergo ostensum est quod embadum superficiei omnis spere est quadruplum embadi superficiei maioris circuli cadentis in ea. Et illud est quod declarare voluimus. Et hec est forma eius.</td><td>وهنالك تبين أن كل كرة فإن مساحة سطحها أربعة أمثال مساحة سطح أعظم دائرة تقع فيها، وهذا ما أردنا بيناه. وهذه صورته.</td></tr>
<tr><td colspan="2">Therefore, it has now been demonstrated that the surface area of any sphere is quadruple the area of the greatest circle falling in it. And this is what we wished to show. And this is its form.</td></tr>
</table>

III AL-ṬŪSĪ'S EDITION	COMMENTS
Let us inscribe within this a solid – as we have already described – such that it does not touch the hemisphere *ABCD*.	
فيكون سطح المجسم أعظم من ضعف دائرة ا ب جـ لما مرّ. The area of the solid is greater than double the area of ABC, according to what precedes.	The only difference between this and (III) is the presence of the word 'area' and the naming of the solid.
وسطح نصف كرة و ز ل م أعظم من سطح المجسم لكونه محيطاً به and the area of the hemisphere $FGLM$ is greater than the surface area of the solid as it surrounds it.	The final phrase, 'as it surrounds it', is missing in the Latin version, and al-Ṭūsī has not named the solid.
فسطح نصف كرة و ز ل م أعظم كثيراً من <ضعف> سطح دائرة ا ب جـ، وكان مثله؛ هذا خلف. The area of the hemisphere $FGLM$ is therefore much greater than double the area of ABC. Now, it is equal to it; this is contradictory.	
فإذن الحكم ثابت، وذلك ما أردناه The assertion is therefore proved. This is what we required.	The phrase by al-Ṭūsī: 'The assertion is therefore proved. This is what we required', has no counterpart in the Latin translation. However, given their known style of writing, it is extremely unlikely that the Banū Mūsā would have forgotten to include such a conclusion. It is more likely – and the remainder of their treatise supports this – that the conclusion is missing, either omitted by Gerard or not present in the manuscript that he was translating.
وقد بان منه أن سطح الكرة أربعة أمثال سطح أعظم دائرة تقع فيها. It has been shown from this assertion that the surface area of a sphere is four times the area of the greatest circle that can be found within it.	

TABLE II

<table>
<tr><th>I
GERARD’S TRANSLATION
(Clagett, pp. 310–15)</th><th>II
ARABIC TEXT OF PROPOSITION 12</th></tr>
<tr><td>Cum fuerit circulus cuius diameter sit protracta, et protrahitur ex centro ipsius linea stans super diametrum orthogonaliter et perveniens ad lineam continentem et secatur una duarum medietatum circuli in duo media, tunc cum dividitur una harum duarum quartarum in divisiones equales quotcunque sint, deinde protrahitur corda sectionis cuius una extremitas est punctum super quod secant se linea erecta super diametrum et linea continens et producitur linea diametri in partem in quam concurrunt donec concurrunt et protrahuntur in circulo corde equidistantes linee diametri ex omnibus punctis divisionum per quas divisa est quarta circuli, tunc linea recta que est inter punctum super quod est concursus duarum linearum protractarum et inter centrum circuli est equalis medietati diametri et cordis que protracte sunt in circulo equidistantibus diametro coniunctis.</td><td>إذا كانت دائرة وأخرج من مركزها خطٌ يقوم على القطر على زاوية قائمة وينتهي إلى خط المحيط ويفصل نصف الدائرة بنصفين، فإنه إذا قسم أحد هذين الربعين بأقسام متساوية كم كانت، ثم أخرج وتر القسم الذي أحد طرفيه نقطة تفاصل نصف قطر الدائرة القائم مع الخط المحيط، وأخرج القطر في الجهة التي يلتقيان فيها، وأخرج في الدائرة أوتار موازية لخط القطر من جميع نقط الأقسام التي قسم بها ربع الدائرة، فإن الخط المستقيم الذي بين النقطة التي التقي عليها الخطان المخرجان وبين مركز الدائرة مثلُ نصف قطر الدائرة والأوتار التي أخرجت في الدائرة الموازية للقطر مجموعة.</td></tr>
<tr><td colspan="2">When there is a circle whose diameter is drawn and there is drawn from its center a line perpendicular to the diameter and terminating at the circumference so that one of the two halves of the circle is bisected, and then when one of the two quadrants is divided into any number of equal parts and the chord of the segment, one of whose extremitites is the point of intersection of the line erected on the diameter and the circumference, is produced while the diameter is produced in the direction of their intersection until the two lines intersect, and there are drawn in the circle from the points at which the quadrant arc of the circle is divided chords parallel to the diameter, then the straight line between the point where the two extended lines meet and the center of the circle is equal to the sum of the radius plus the chords drawn in the circle parallel to the diameter.</td></tr>
</table>

III **AL-ṬŪSĪ'S EDITION**	**COMMENTS**
	It can be seen that al-Ṭūsī, in line with his common practice, has omitted the *protasis* from his edition, beginning directly at the *ekthesis*. Gerard of Cremona translated the Arabic text literally, and the sole difference is probably due to a copying error occurring at some point in the manuscript tradition. This concerns the phrase 'Cuius diameter sit protracta', which is a translation of the Arabic *wa-ukhrija quṭruhā*. This is most probably a *saut du même au même*, and the original was more likely to have been إذا كانت دائرة وأخرج قطرها وأخرج من مركزها ... Critical apparatus for Text II: 3 أحد : احدى – 5 طرفيه : طرفين / تفاصل : تفاضل – 7 وأخرج : واخرجت.

<table>
<tr><th>I
GERARD'S TRANSLATION</th><th>II
ARABIC TEXT OF PROPOSITION 12</th></tr>
<tr><td>Verbi gratia, sit circulus ABC, cuius diameter sit linea AC et cuius centrum sit punctum D. Et protrahatur ex eo linea DB erecta super lineam AC orthogonaliter et dividat arcum ABC in duo media. Et dividam quartam circuli super quam sunt A, B in divisiones equales quot voluero et ponam eas divisiones AG, GL, LB. Et protraham cordam BL et faciam ipsam penetrare. Et elongabo iterum lineam AC, que est diameter, secundum rectitudinem donec concurrant super punctum E. Et protraham ex duobus punctis G, L duas cordas GI, LH equidistantes diametro AC. Dico ergo quod linea DE est equalis medietati diametri et duabus cordis GI, LH coniunctis, cuius hec est demonstratio.</td><td dir="rtl">مثاله: دائرة ا ب جـ، قطرها ا جـ ومركزها نقطة د، وقد أخرج منه خط د ب يقوم على خط ا جـ على زاويتين قائمتين، ويقسم قوس ا ب جـ بنصفين. ثم نقسم ربع الدائرة الذي عليه ا ب بأقسام متساوية كم شئنا، وهي ا ز ز ل ل ب. ونصل ل ب، ونخرج خطي ا جـ ل ب حتى يلتقيا على نقطة هـ، ونخرج من نقطتي ز ل وتري ز ط ل ح يوازيان قطر ا جـ. فأقول: إن خط د هـ مثل نصف القطر ووتري ز ط ل ح مجموعة.</td></tr>
<tr><td colspan="2">For example, let there be a circle ABC whose diameter is line AC and whose center is point D. And from the center let line DB be drawn perpendicular to AC, thus bisecting arc ABC. And I shall divide the quadrant AB into as many equal parts as I wish, and I shall assume these parts to be AG, GL, LB. And I shall draw chord BL and makc it continue. And I shall also extend line AC, the diameter, rectilinearly until they meet at point E. And I shall draw from the two points G and L the two chords GI et LH parallel to diameter AC. I say, therefore, that line DE is equal to the sum of the radius plus the two chords GI and LH.</td></tr>
<tr><td>Protraham lineam IA et protraham lineam HG et faciam ipsam penetrare secundum rectitudinem donec occurrat linee EC super F. Et similiter faciam, si quarta circuli super quam sunt A, B fuerit divisa in divisiones plures istis divisionibus. Linee ergo IG, HL sunt equidistantes, quoniam taliter sunt protracte. Et linee IA, HF, BE sunt equidistantes propterea quod due divisiones IH, HB sunt equales duabus divisionibus AG, GL. Ergo quadratum IAFG est equidistantium laterum. Ergo linea IG est equalis AF. Et iterum quadratum HFEL est equidistantium laterum. Ergo linea HL est equalis FE. Ergo tota linea ED est equalis duabus lineis IG, HL et linee erecte que est medietas diametri coniunctis.</td><td dir="rtl">برهانه: أنا نخرج خط ط ا، ونخرج خط ح ز وننفذه على استقامة حتى يلقى / خط هـ جـ على نقطة و. وكذلك ندبر إن كانت الأقسام أكثر. فخطوط <جـ هـ> ط ز ح ل متوازية لأنها كذلك أخرجت في الوضع، وخطوط ا ط و ح هـ ب متوازية من أجل أن قسمي ط ح ب ح مساويتان لقسمي ا ز ز ل، فمربع ط ا و ز متوازي الأضلاع وخط ط ز مثل ا و. وأيضًا مربع ح و هـ ل متوازي الأضلاع، فخط ح ل مثل خط و هـ، فجميع خط هـ د مساوٍ لخطي ط ز ح ل ولنصف القطر مجموعة.</td></tr>
<tr><td colspan="2">Proof: I shall draw line IA and I shall draw line HG, continuing the latter rectilinearly until it meets line EC at F. I shall proceed in a similar way if the quadrant AB is divided into more parts than these. Hence lines IG and HL are parallel, since they are so drawn. And lines IA, HF and BE are parallel, since IH is equal to AG and HB is equal to GL. Therefore, the quadrilateral IAFG is a parallelogram. Therefore, line IG is equal to AF. And also quadrilateral HFEL is a parallelogram. Therefore, line HL is equal to FE. Therefore, the whole line ED is equal to the sum of IG and HL plus the radius.</td></tr>
</table>

III AL-ṬŪSĪ'S EDITION	COMMENTS
ليكن ا ب جـ دائرة قطرها ا جـ ومركزها د، وقد قام عمود د ب منه على القطر، ولنقسم ربع ا ب بأقسام متساوية كم كانت، وهي ا ز ز ل ل ب، ولنخرج وتر ب ل وننفذه، وننفذ قطر جـ ا إلى أن يلتقيا على هـ، ونخرج من نقطتي ز ل وتري ز ط ل ح موازيين لقطر جـ ا. فأقول: إن خط د هـ يساوي نصف قطر جـ ا ووتري ز ط ل ح جميعًا. Let *ABC* be a circle of diameter *AC* with its centre at *D*. The perpendicular *DB* is drawn to the diameter at the point *D*. Let us divide the quarter *AB* into any number of equal parts, which are *AG, GL* and *LB*. Let us draw the chord *BL* and extend it. We also extend the diameter *CA* until they meet at *E* and we draw the two chords *GI* and *LH* from the points *G* and *L* such that these are parallel to the diameter *CA*. I say that the straight line *DE* is equal to the sum of one half of the diameter *CA* and the two chords *GI* and *LH*.	After having ignored the word *mithāluhu* (example), al-Ṭūsī here recasts the Banū Mūsā text in a language that is more compact and often more elegant. The Latin translation does not depart from the original. The minor variation in the phrase 'Et protraham … punctum *E*' is doubtless due to the translation.
فنخرج ط ا ح ز وننفذ ح ز إلى أن يلقى جـ هـ على و، وبمثل ذلك ندبر إن كانت الأقسام أكثر. فخطوط جـ هـ ط ز ح ل متوازية، وخطوط ط ا ح و ب هـ متوازية، لأن قوسي ط ح ح ب مساويتان لقوسي ا ز ز ل، فسطح ط ا و ز متوازي الأضلاع وط ز مثل ا و. وبمثل ذلك ح ل مثل و هـ، فـ د هـ مثل د ا ط ز ح ل جميعًا؛ وذلك ما أردناه. We draw *IA* and *HG* and extend *HG* until it meets *CE* at *F*. We proceed in a similar way if more parts have been used. The straight lines *CE, IG* and *HL* are parallel, and the straight lines *IA*, *HF* and *BE* are parallel as the two arcs *IH* and *HB* are equal to the two arcs *AG* and *GL*. The surface *IAFG* is therefore a parallelogram and *IG* is equal to *AF*. In the same way, *HL* is equal to *FE*, and therefore *DE* is equal to the sum of *DA, IG* and *HL*. That is what we required.	Al-Ṭūsī ignores the word *burhān* (demonstration), and condenses the Banū Mūsā text, changing some of the terms, for example: *wa-mithlu dhālika, li-anna, saṭḥ* in place of *wa-kadhālika, min ajli an, murabba'*. These are in fact synonyms. More importantly, when the line of reasoning (for *ED*), is followed exactly, it also follows the reasoning employed by the Banū Mūsā. The Latin translation is completely faithful, with the exception of a minor variation 'si quarta circuli …', which could be an explanation inserted by Gerard or by the copyist of the manuscript that he translated, as it is implied by the text. Critical apparatus for Text II: 3 ندبر: نريد – 4 كذلك: لذلك – 5 قسمي: قد تقرأ قسي، وفي هذه الحالة يكون الصواب "قوسي" – 6 ب ح: با / لقسمي: لقسي: قد تقرأ لقسي، والصواب يكون "لقوسي" / ط ا و ز: طا دال الف واو زاي – 8 ح و هـ ل: ها واو ها لام.

<table>
<tr><th>I
GERARD'S TRANSLATION</th><th>II
ARABIC TEXT OF PROPOSITION 12</th></tr>
<tr><td>Si ergo nos protraxerimus in hac figura lineam ex centro et secuerit unam cordarum divisionum quarte circuli in duo media, sicut lineam DM, tunc secatur linea LB super duo media super punctum M in duo media. Tunc iam scietur ex eo quod narravimus in hac figura quod multiplicatio medietatis corde BL in duas cordas equidistantes diametro et in medietatem diametri coniunctas est minor multiplicatione medietatis diametri in se et maior multiplicatione linee DM in se, propterea quod triangulus DMB est similis triangulo EDB et est similis triangulo EMD. Ergo proportio linee MB ad BD est sicut proportio DB ad BE.
Et propter illud erit multiplicatio linee DB, que est medietas diametri, in se equalis multiplicationi linee MB in lineam BE. Verum linea BE est longior duabus cordis GI, LH et medietate diametri coniunctis, propterea quod iste coniuncte sunt DE, et linea BE est longior DE. Ergo multiplicatio linee MB in duas cordas GI, LH et in medietatem diametri coniunctas est minor multiplicatione medietatis diametri in se. Et quoniam triangulus DMB est similis triangulo EMD, erit proportio BM ad MD sicut proportio MD ad ME. Et similiter erit multiplicatio linee BM in lineam ME equalis multiplicationi linee MD in se. Sed linea ME est minor duabus cordis GI, LH et medietate diametri coniunctis, propterea quod iste omnes sunt equales linee DE, et linea DE est longior EM. Ergo multiplicatio MB in duas cordas GI, LH et in medietatem diametri coniunctas est maior multiplicatione DM in se.</td><td>وإن نحن أخرجنا في هذا الشكل خطًا من المركز وقطع وتراً من أوتار ربع الدائرة بنصفين، مثل خط د م يقطع ب ل على نقطة م بنصفين، فقد نعلم مما وصفنا أن تضعيف نصف وتر ب ل بالأوتار الموازية للقطر <ونصف قطر الدائرة> مجموعةً أقلّ من تضعيف نصف القطر بمثله وأعظم من تضعيف د م بمثله، من أجل أن مثلث د م ب يشبه مثلث هـ د ب ويشبه مثلث هـ م د ونسبة خط م ب إلى ب د كنسبة د ب إلى ب هـ.
فلذلك يكون تضعيف خط د ب الذي هو نصف القطر بمثله مثل تضعيف خط م ب بخط ب هـ. ولكن خط هـ ب أطول من وتري ز ط ل ح ونصف القطر مجموعة. فتضعيف خط م ب بخطوط ل ح وز ط ود ب <مجموعة> أقلّ من تضعيف نصف القطر بمثله. ولأن مثلث د م ب يشبه مثلث د م هـ، يكون نسبة ب م إلى م د كنسبة م د إلى م هـ. ولذلك يكون تضعيف خط ب م بخط م هـ مثل تضعيف خط م د بمثله. ولأن خط م هـ أصغر من وتري ز ط ل ح ونصف القطر مجموعة، من أجل أن هذه جميعًا مثل خط د هـ وخط د هـ أطول / من خط م هـ، فتضعيف م ب بوتري ز ط ل ح ونصف القطر مجموعةً أعظم من تضعيف د م بمثله.</td></tr>
<tr><td colspan="2">Hence in this figure we draw a line, e.g., line DM, from the center thus bisecting one of the chords of the quadrant, LB being the line bisected at point M. Then it will already be known, from what we have recounted concerning this figure, that the multiplication of one half of the chord BL by the sum of the two chords parallel to the diameter plus the radius is less than the square of the radius and is greater than the product of DM with itself, because of the fact that the three triangles DMB, EDB, and EMD are similar. Therefore the ratio of MB to BD is equal to the ratio of DB to BE.</td></tr>
</table>

III **AL-ṬŪSĪ'S EDITION**	**COMMENTS**
وإن أخرجنا د م عموداً على وتر ب ل، كان سطح نصف ب ل في د هـ أصغر من مربع نصف القطر وأكثر من مربع د م، وذلك لأن مثلثي د ب م ب هـ د متشابهان لكون زاويتي د م ب هـ د ب قائمتين وزاوية ب مشتركة، فنسبة ب م إلى م د كنسبة ب د إلى د هـ. فـ ب م – أعني نصف ب ل – في د هـ مساوٍ لـ ب د في م د. وب د في م د أصغر من مربع ب د وأعظم من مربع م د. فإذن نصف ب ل في نصف القطر وفي وتري ط ز ح ل جميعًا أصغر من مربع نصف القطر وأعظم من مربع د م. If we draw *DM* perpendicular to the chord *BL*, then half of the product of *BL* and *DE* is less than the square of the half-diameter and greater than the square of *DM*, as the two triangles *DBM* and *BED* are similar, given that the angles *DMB* and *EDB* are right angles and the angle *B* is common to both. The ratio of *BM* to *MD* is therefore equal to the ratio of *BD* to *DE*.	This time, al-Ṭūsī summarizes the Banū Mūsā text without changing any of the reasoning. In place of 'wa-in … al-shakl', he writes simply 'wa-in akhrajnā …'; and similarly, instead of 'khaṭṭan … *EBD*', he writes "amūdan … *BED*'. In the latter phrase, he uses a ratio that is different from that used by the Banū Mūsā. The Latin version is a literal translation of the Arabic, with the exception of a few small variations. 'in hac figura' (*fī hādhā al-shakl*) does not appear in the Arabic, and the dual 'in duas cordas' is a plural in the Arabic. Al-Ṭūsī also adds a justification for the triangles being similar that appears in neither the Arabic nor the Latin. He then summarizes the Banū Mūsā text, which he must doubtless have found excessively long. Gerard follows the Arabic text word for word except for a few small variations. The phrase 'propterea … longior *DE*', which must be a translation of 'min ajli anna hādhihi jamī'an mithlu *DE* wa-khaṭṭ *BE* aṭwal min *DE*', does not appear in the Arabic. It is difficult to judge whether this is an omission or whether it is a superfluous addition inserted either by the translator or by one of the copyists. In the second variation, in place of 'the straight lines *LH*, *GI* and *DB*' in Arabic, the Latin version is more explicit, repeating 'duas cordas … diametri'. Finally, Gerard writes 'Et similiter …', which must be a translation of *kadhālika*, and which is a copyist error. The text should read *wa-lidhālika*.

I GERARD'S TRANSLATION	II ARABIC TEXT OF PROPOSITION 12

Hence, the product of *DB* with itself is equal to the product of *MB* and *BE, DB* being the radius. Now line *BE* is greater than the sum of *GI* and *LH* plus *BD*, since the sum of *GI* and *LH* plus *BD* is equal to *DE*, and *BE* is greater than *DE*. Hence, the product of line *MB* and the sum of *GI* and *LH* and *BD* is less than the product of *DB* with itself. And since triangle *DMB* is similar to triangle *EMD*, the ratio of *BM* to *MD* is equal to the ratio of *MD* to *ME*. And similarly the product of *BM* and *ME* is equal to the product of *MD* with itself. But line *ME* is less than the sum of *GI* and *LH* plus *BD*, since the sum of *GI* and *LH* plus *BD* is equal to *DE*, and *DE* is greater than *EM*. Therefore, the product of *MB* and the sum of *GI*, *LH* and *BD* is greater than the product of *DM* with itself.

Iam ergo ostensum est quod in omni circulo in quo protrahitur ipsius diametrus deinde dividitur una duarum medietatum ipsius in duo media, postea dividitur una duarum quartarum in divisiones equales quotcunque fuerint et protrahuntur ex punctis divisionum omnium corde in circulo equidistantes diametro, tunc multiplicatio medietatis corde unius sectionum quarte circuli in medietatem diametri et in omnes cordas que protracte sunt in circulo equidistantes diametro coniunctim est minor multiplicatione medietatis diametri in se et maior multiplicatione linee que egreditur ex centro et pervenit ad unam cordarum divisionum quarte circuli et dividit eam in duo media in se. Et illud est quod declarare voluimus.	فقد استبان أن ... تضعيف نصف وتر قسم من أقسام ربع الدائرة بنصف القطر وبجميع الأوتار الموازية للقطر أقلُّ من تضعيف نصف القطر بمثله وأعظمُ من تضعيف الخط الذي خرج من المركز وينتهي إلى وتر من أوتار أقسام ربع الدائرة ويقسمه بنصفين بمثله؛ وذلك ما أردنا بيانه.

Therefore it has now been demonstrated that in every circle where the diameter is drawn and one of the two halves of the circle is bisected and one of the two quadrants <thus formed> is then divided into any number of equal parts and from the <dividing> points of the parts are drawn chords in the circle parallel to the diameter, then the multiplication of one half of the chord of one of the segments of the quadrant by the sum of the radius plus all the chords drawn in the circle parallel to the diameter is less than the square of the radius and greater than the square of the line going out from the center which meets and bisects the chord of one of the parts of the quadrant. And this is what we wished to show.

III AL-ṬŪSĪ'S EDITION	COMMENTS
Therefore <the product of> *BM*, *i.e.* half of *BL*, and *DE* is equal to <the product of> *BD* and *MD*. But <the product of> *BD* and *MD* is less than the square of *BD* and greater than the square of *MD*. Consequently, <the product of> half of *BL* and the sum of the half-diameter and the two chords *IG* and *HL* is less than the square of the half-diameter and greater than the square of *DM*.	
فكل دائرة يخرج قطر فيها وينصف نصفها ويقسم أحد الربعين بأقسام متساوية كم كانت، ويخرج من نقط الأقسام أوتار في الدائرة موازية للقطر، كان سطح نصف وتر أحد تلك الأقسام في نصف القطر وفي جميع الأوتار أصغر من مربع نصف القطر وأعظم من مربع العمود الخارج من المركز الواقع على أحد أوتار تلك الأقسام، وذلك هو المطلوب. Therefore, for any circle in which is drawn a diameter, if one half of it is divided into two halves, and one of the two quarters is divide into any number of equal parts, and if chords are drawn from the points of this division that are parallel to the diameter, then the product of half the chord from one of these parts and the half-diameter, plus its product with the sum of the chords is less than the square of the half-diameter and greater than the square of the perpendicular drawn from the centre of one of the chords from these parts. That is what was required.	Al-Ṭūsī here rephrases the conclusion of the Banū Mūsā in his own words, but without omitting any portion of it. It should be noted that throughout, he replaces the term *taḍ'īf* with the term *saṭḥ*, which has a slightly more geometric connotation. The citation is missing a phrase here that is included by al-Ṭūsī and translated by Gerard, and which may have been "كل دائرة إذا أخرج قطرها وقسم أحد نصفيها بنصفين، وقسم أحد الربعين بأقسام كم كانت، وأخرج من نقط الأقسام أوتار موازية للقطر..." This phrase has most probably been omitted by the anonymous author who supplied the citation.

<table>
<tr><th>I
GERARD'S TRANSLATION
(Clagett, pp. 344–9)</th><th>II
ARABIC TEXT OF PROPOSITION 18</th></tr>
<tr><td>Sit itaque angulus ABC in primis minor recto. Et accipiam ex duabus lineis BA, BC duas quantitates equales, que sint quantitates BD, BE. Et revolvam super centrum B et cum mensura longitudinis BD circulum DEL. Et extendam lineam DB usque ad L. Et protraham lineam BG erectam super lineam LD orthogonaliter. Et lineabo lineam EG et extendam ipsam usque ad H. Et non ponam linee GH finem determinatum.</td><td>فلتكن الزاوية المفروضة زاوية ا ب جـ؛ ونأخذ من خطيها مقدارين متساويين وهما ب هـ ب د، وذلك بأن نتخذ نقطة ب مركزاً وندير ببعدهما دائرة د ل هـ. ونخرج خط د ب إلى ل. ولتكن أوّلاً أقلّ من قائمة. ونخرح ب ز يقوم على خط د ل على زاويتين قائمتين، ونخطّ خطّ هـ ز وننفذه إلى ح، ولا نجعل له غاية محدودة.</td></tr>
<tr><td colspan="2">And so let angle angle ABC at first be less than a right angle. And I shall take from its two lines BA and BG two equal quantities BD and BE. And I shall describe circle DEL on center B with a radius BD. And I shall extend line DB up to L and erect line BG perpendicularly on line LD. Further, I shall draw line EG, extending it to H, but without assuming GH to have any fixed length.</td></tr>
<tr><td>Et accipiam de linea GH equale medietati diametri circuli, quod sit linea GO. Quando ergo ymaginamus quod linea GEH movetur ad partem puncti L et punctum G adherens est margini circuli in motu suo et linea GH non cessat transire super punctum E circuli DEL et ymaginamus quod punctum G non cessat moveri donec fiat punctum O super lineam BG, oportet tunc ut sit arcus qui est inter locum ad quem pervenit punctum G et inter punctum L tertia arcus DE; cuius demonstratio est:</td><td>ونأخذ من خط ز ح مثل نصف قطر الدائرة، وهو ز ع. فإذا توهمنا أن خطاً ز ع يتحرك على محيط الدائرة إلى ناحية ل <و>نقطةَ ز لازمة لمحيط الدائرة في حركتها وخط ز هـ ح لا يزال يتحرك على نقطة هـ <من دائرة د هـ ل>، وتوهمنا نقطة ز لا تزال تتحرك حتى تصير نقطة ع على خط ب ز، حينئذٍ وجب أن يكون القوس الذي بين الموضع الذي انتهت إليه نقطة ز وبين نقطة ل هو ثلث قوس د هـ.</td></tr>
<tr><td colspan="2">And I shall cut from line GH a line equal to the radius of the circle, namely, line GO. Therefore, when we imagine that line GEH is moved in the direction of point L and that point G <continually> adheres to the circumference in the course of its motion, and that line GH continues to pass through point E of circle DEL, and we imagine that point G continues to be moved until point O falls on line BG, then it is necessary for the arc between the point at which G arrives and point L to be one third of arc DE.</td></tr>
</table>

III AL-ṬŪSĪ'S EDITION	COMMENTS
فلتكن الزاوية ا ب جـ، ولتكن أولاً أقل من قائمة. ونأخذ من خطي ب ا ب جـ مقداري ب د ب هـ متساويين. ونرسم على مركز ب وببعدهما دائرة د هـ ل، ونخرج د ب إلى ل، ونقيم ب ز عموداً على ل د، ونصل هـ ز ونخرجه إلى ح لا إلى غاية. Let the angle be *ABC* and let it be first less than a right angle. On the straight lines *BA* and *BC*, we take two equal magnitudes *BD* and *BE*. With its centre at *B* and at their distance, we draw the circle *DEL* and we extend *DB* as far as *L*. We draw *BG* perpendicular to *LD*, we join *EG*, and we extend it to *H* without an end.	The texts by al-Ṭūsī and Gerard are so similar that one has the impression that the beginning of the citation has simply been copied verbatim. The very first words in both the Latin version and the Ṭūsī edition state that we begin by considering an acute angle. This expression appears later in the citation. In addition, the first two refer to the sides of the angle 'Et accipiam … equales', while the cited text simply reads 'its two lines'. However, these differences make it no less certain that both are taken from the same text.
ونفصل من ز ح ز ع مثل نصف قطر الدائرة. فإذا توهمنا أن ز ح يتحرك إلى ناحية نقطة ل ونقطة ز لازمة للمحيط في حركتها وخطّ ز هـ ح في حركته لا يزال يمرّ على نقطة هـ من دائرة د هـ ل، وتوهمنا نقطة ز لا تزال تتحرك حتى تصير نقطة ع على خط ب ز، وجب حينئذ أن تكون القوس التي بين الموضع الذي انتهت إليه نقطة ز وبين نقطة ل هي ثلث قوس د هـ. والزاوية التي توترها هذه القوس ثلث زاوية د ب هـ. And from *GH*, we separate out *GO* equal to the half-diameter of the circle. If we imagine that *GH* moves in the direction of the point *L* while the point *G* remains on the circumference in the course of its motion, and that the straight line *GEH* continues to pass through the point *E* on the circle *DEL* in the course of its motion, and if we imagine that the point *G* continues to move until the point *O* arrives at the straight line *BG*, then the arc between the final position of the point *G* and the point *L* must be one third of the arc *DE*. The angle intercepted by this arc is one third of the angle *DBE*.	Apart from a few negligible variations, al-Ṭūsī here follows the Banū Mūsā text, which Gerard translates literally. He only omits one phrase after 'ad partem puncti *L*' in order to say 'on the circumference of the circle' (*'alā muḥīṭ al-dā'ira*). It should be noted that al-Ṭūsī wrote: 'wa-al-zāwiya … thulth zāwiya *DBE'*, which is missing in both the citation and the Latin version. Critical apparatus for Text II: 2 يتحرك : يحرك – 3 لمحيط الدائرة : لخط با زاي – 4 ز هـ ح : زاي / تزال تتحرك : نزال يتحرك – 6 خط ب ز : محيط الدائرة – 7 الذي : الذين / ز : عين.

<table>
<tr><th>I
GERARD'S TRANSLATION</th><th>II
ARABIC TEXT OF PROPOSITION 18</th></tr>
<tr><td>Quod ego ponam locum ad quem pervenit punctum G apud cursum puncti O super lineam BG apud punctum I. Et protraham lineam IE secantem lineam BG super punctum S. Ergo linea IS est equalis medietati diametri circuli, propterea quod est equalis linee GO. Et protraham ex B lineam equidistantem linee IS, que sit linea MBK. Et protraham lineam ex I ad M. Ergo linea MI et linea IS sunt equidistantes duabus lineis MB, BS et equales eis. Ergo linea MI est equidistans linee BS et equalis ei. Sed linea BS est perpendicularis super diametrum LD. Ergo corda arcus IM erigitur ex diametro LD super duos angulos rectos. Ergo dividit diametrus LD cordam MI in duo media et dividit propter illud arcum MI in duo media super punctum L. Verum arcus ML est equalis arcui DK. Ergo arcus DK est equalis medietati arcus MI. Sed arcus MI est equalis arcui EK, propterea quod linea IE equidistat linee MK. Ergo arcus DK est tertia arcus DE. Et similiter angulus DBK est tertia anguli ABC.</td><td dir="rtl">برهان أنّا نجعل الموضع الذي انتهت إليه نقطة زَ عند نقطة طَ، ونخرج ط هـ يقطع خط ب ز / على نقطة سَ، فخط ط س مساوٍ لنصف قطر الدائرة من أجل أنه مساوٍ لخط ز ع. ونخرج من بَ خطًا موازيًا لخط ط س وهو م ک، ونخرج خطًا من طَ إلى مَ؛ فخطا م ط ط س موازيان لخطي م ب ب س ومساويان لهما. وخط ب س عمود على قطر ل د، فوتر قوس م ط يقوم على قطر ل د على زاويتين قائمتين. فقد قسم قطرُ ل د وترَ م ط بنصفين، وقسم لذلك قوسَ م ط بنصفين على نقطة لَ. ولكن قوس م ط مساوية لقوس هـ ک من أجل أن ط هـ موازٍ لخط م ک، إذاً <قوس د ک> ثلث قوس د هـ. وكذلك زاوية ک ب د ثلث زاوية هـ ب د.</td></tr>
<tr><td colspan="2">Proof: For I posit point I as the place at which point G arrives as point O meets line BG. And I shall draw line IE cutting line BG at point S. Therefore, line IS is equal to the radius of the circle since it is equal to line GO. And I shall draw through B a line parallel to line IS, namely, line MBK. And I shall draw a line from I to M. Therefore, lines MI and IS are <respectively> parallel and equal to the two lines BS and MB. Therefore, line MI is parallel and equal to line BS. But line BS is perpendicular to the diameter LD. Therefore, the chord of arc IM forms two right angles with diameter LD. Therefore, diameter LD bisects chord MI and <therefore> it also bisects arc MI at point L. But arc ML is equal to arc DK. Therefore, arc DK is equal of half the arc MI. But arc MI is equal to arc EK, since line IE is parallel to line MK. Therefore, arc DK is one third of arc DE. Therefore angle DBK is one third of angle ABC.</td></tr>
</table>

III
AL-ṬŪSĪ'S EDITION

برهانه: ليكن الموضع الذي انتهت إليه ز نقطة ط، ونخرج ط هـ يقطع ب ز على س، فخط ط س مساوٍ لنصف قطر الدائرة لكونه مساويًا ز ع. ونخرج من المركز قطراً يوازي ط هـ وهو م ب ك. ونخرج م ط، فـ ط س مساوٍ وموازٍ لـ م ب، وم ط موازٍ ومساوٍ لـ ب س، وب س عمود على ل د، فـ م ط عمود على ل د، ولذلك يكون منصفًا بالقطر، ويكون م ل مثل ل ط ود ك مثل م ل وم ط مساوٍ لـ ك هـ فـ د ك مثل نصف ك هـ و<مثل> ثلث د هـ، وزاوية ك ب د ثلث زاوية ا ب جـ؛ وذلك ما أردناه.

Proof: Let the position at which <the point> *G* arrives be the point *I*. We draw *IE* cutting *BG* at *S*. The straight line *IS* is then equal to the half diameter of the circle as it is equal to *GO*. From the centre, we draw a diameter parallel to *IE*, and let this be *MBK*. We draw *MI*. Then, *IS* is equal and parallel to *MB*, *MI* is parallel and equal to *BS*, and *BS* is perpendicular to *LD*. Therefore, *MI* is perpendicular to *LD*. It is for this reason that it is divided into two equal parts by the diameter. <The arc> *ML* is equal to <the arc> *LI*, <the arc> *DK* is equal to <the arc> *ML*, and <the arc> *MI* is equal to <the arc> *KE*. Therefore, <the arc> *DK* is equal to half of <the arc> *KE*, and is equal to one third of <the arc> *DE*. The angle *KBD* is therefore one third of the angle *ABC*. That is what we required.

COMMENTS

In the first part of this section, it can be seen that al-Ṭūsī follows the Banū Mūsā text very closely. The first sentence is identical to that of the Banū Mūsā with the exception of two insignificant alterations: *liyakun* in place of *annā naj'al* and *li-kawnihi* in place of *min ajli*. Al-Ṭūsī then completes the section keeping very close to the Banū Mūsā text. Gerard's translation remains literal throughout. However, it does contain one phrase which does not appear in the Arabic text: 'apud cursam puncti *O* super lineam *BG*' – 'as point *O* meets the line *BG*' (*'indamā taṣīru nuqṭa* O *'alā khaṭṭ* BG).

The second phrase that is missing in the Arabic text is 'Ergo linea *MI* est equidistans linee *BS* et equalis ei' (*fa-khaṭṭ* MI *muwāzin wa-musāwin li-khaṭṭ* BS), which is clearly an addition inserted by Gerard himself or appearing in the manuscript he used. Finally, the Latin version also includes the phrase 'Verum … *MI*' (*wa-lākin qaw* ML *musāwiya li-qaws* DK, *fa-qaw* DK *musāwiya li-qaws* MI), which is evidently a *saut du même au même* in the manuscript cited by the anonymous author, committed either by the author himself or by the copyist of his manuscript.

Critical apparatus for Text II:

1 ز: عين.

1.1.4. *Title and date of the Banū Mūsā treatise*

Let us now consider the title of the treatise. This time, the Latin version provides no help, as it is entitled simply *Verba filiorum Moysi filii Sekir ....* The edition of al-Ṭūsī indicates that the title may have been *Kitāb fī ma'rifat misāḥat al-ashkāl al-basīṭa wa-al-kuriyya, i.e.* 'Book on the knowledge of the measurement of plane and spherical figures'. However, the early biobibliographers gave a slightly different title. In the tenth century, al-Nadīm used the title 'Book on the measurement of spheres, the trisection of the angle and the way in which two magnitudes can be set between two other magnitudes such that the four progress in the same ratio'. Later, al-Qifṭī, after having quoted the list of writings by the Banū Mūsā established by al-Nadīm, gives the title carelessly as 'Book on the measurement of the sphere and the trisection of the angle'. In fact, the title given by al-Nadīm is a true reflection of the content of the Banū Mūsā book in the correct order, as they themselves describe it in the conclusion removed by al-Ṭūsī but retained in the Latin version, while the title given by al-Ṭūsī seems to derive from the first two lines of the book, also retained in the Latin version. At the very beginning of their book, the Banū Mūsā speak of '[...] scientie mensure figurarum superficialium et magnitudinis corporum', *i.e.* '[...] knowledge of the measure of plane figures and the volume of bodies'. In this case, the bodies referred to are essentially spherical. We need more information before we can explain the differences between these two titles, each of which seems equally valid.

We are hardly in a better position when it comes to the date of the treatise. Muḥammad ibn Mūsā, the eldest of the brothers, died in 873. His younger brother al-Ḥasan died before him. All we know for sure is that the treatise was written after the translation of the *Spherics* of Menelaus, and *The Measurement of the Circle* and *The Sphere and the Cylinder* of Archimedes. But we know that the *Spherics* was translated before 862, as the translator Qusṭā ibn Lūqā dedicated his translation to Prince Aḥmad, who became the Caliph Aḥmad in that year. We have already shown that an initial translation of *The Measurement of the Circle* was in existence prior to 856.[44] No other data is available to reduce this interval further with any certainty.

With regard to the text under discussion here, al-Ṭūsī's edition of the Banū Mūsā treatise, we know from the colophons on an entire family of manuscripts that it was written either in 653/1255 or in 658/1260, depending on whether one reads خنجه or خنج , an expression in *jummal*

[44] R. Rashed, 'Al-Kindī's Commentary on Archimedes' *The Measurement of the Circle*', *Arabic Sciences and Philosophy* 3.1, 1993, pp. 7–53.

used to designate the years.[45] This indicates that al-Ṭūsī wrote the text either 14 or 19 years before his death. This edition exists in a number of surviving manuscripts. This is not surprising, as the work took part of the 'Intermediate Books' (*al-mutawassiṭāt*), intended, as we have shown, for a much wider public than just the most eminent mathematicians. The popularity of these 'Intermediate Books' ensured their survival, a fate not always shared by works of advanced research. A large number of these manuscripts have survived, and every major library, and even some minor ones, possesses one of more copies of these 'Intermediate Books'. They also exist in many private collections. Under present conditions, it is a vain hope even to identify the location of all these manuscripts, and only the unreasonably ambitious would attempt to bring them all together in one place. Of the several dozen manuscripts of these texts that have passed through my hands, I have only been able to obtain copies of 25, for a number of reasons that it is inappropriate to list here. While not inconsiderable, this number represents a small fraction of these manuscripts in existence throughout the world. However, with these 25 manuscripts, dispersed across three continents, it should be possible to establish a faithful version of the text. I do not, therefore, run much of a risk in asserting that access to additional manuscripts would not reveal anything new that would bring any real improvement to the edition, unless of course someone were to discover the original written in the hand of al-Ṭūsī, or even better, the original Banū Mūsā text. The reason that I have reproduced all the variations in these manuscripts in the Critical Apparatus is so that others may go further and increase the number of copies. While to some it may seem that all this effort is a total waste of time, it may one day make it possible, given sufficient resources and perseverance, to identify the locations of all the existing manuscripts and to make a collated set of copies that will reveal the history of the manuscript tradition. However, this is not a project for now, or even for the immediate future.

While we can be sure that the text given here is accurate, its history remains a subject for conjecture. We have attempted simply to list the 25 manuscripts but, given the nature of this book, we shall not include the numerous tables that were necessary to identify them.

Here is the list of these manuscripts:

1 - [A] Istanbul, 'Atif 1712/14, fols 97^v–104^v.

2 - [B] Berlin, Staatsbibliothek, or. quart. 1867/13, fols 156^v–164^v.

[45] We have a set of five letters that could mean one of two possible dates: Monday 27th July 1260 or Monday 20th September 1255. The latter date seems to us to be the more likely of the two, taking into account the complete set of manuscripts.

3 - [C] Istanbul, Carullah 1502, fols 42^{v}–47^{v}.[46]
4 - [D] Istanbul, Topkapi Sarayi, Ahmet III 3453/13, fols 148^{r}–152^{v}.[47]
5 - [E] Istanbul, Topkapi Sarayi, Ahmet III 3456/15, fols 61^{v}–64^{v}.[48]
6 - [F] Vienna, Nationalbibliothek, Mixt 1209/13, fols 163^{v}–173^{r}.
7 - [G] London, India Office 824/3 (No. 1043), fols 36^{r} –39^{r}, 50^{r}–52^{v}.[49]
8 - [H] Tehran, Sepahsalar 2913, fols 86^{v}–89^{v}.
9 - [I] Tehran, Milli Malik 3179, fols 256^{v}–261^{v}, 264^{r}–267^{v}.
10 - [J] Paris, Bibliothèque Nationale 2467, fols 58^{v}–68^{r}.[50]
11 - [K] Istanbul, Köprülü 930/14, fols 214^{v}–227^{r} (or 215^{v}–228^{r} according to a second numbering scheme).[51]
12 - [L] Istanbul, Carullah 1475/3, fols 1^{v}–14^{v} (folios not numbered).
13 - [M] Meshhed, Astān Quds 5598, fols 18–33.[52]
14 - [N] New York, Columbia University, Plimpton Or 306/13, fols 116^{r}–122^{v}.[53]
15 - [O] Oxford, Bodleian Library, Marsh 709/8, fols 78^{r}–89^{v}.[54]

[46] This is a collection transcribed from the copy belonging to the famous astronomer Quṭb al-Dīn al-Shīrāzī, according to the copyist Ibn Maḥmūd ibn Muḥammad Muḥammad al-Kunyānī. The text is written in *naskhī*. The page size is 25.5 × 17.9 cm. Each page contains 25 lines of text occupying an area of 17.2 × 11.2 cm.

[47] Manuscript copied by ʿAbd al-Kāfī ʿAbd al-Majīd ʿAbd Allāh al-Tabrīzī in 677 in Baghdad. Fatḥ Allāh al-Tabrīzī had possession of this manuscript in 848. It is written in *naskhī* (page: 17.1 × 13.2 cm, text: 13.9 × 9.6 cm). The numeration of the folios is recent.

[48] One of the texts in this collection was copied on the 12 Rabīʿ al-awwal 651 (see fol. 81^{v}). It is written in *nastaʿlīq* (page: 25.5 × 11.3 cm, text: 19.4 × 8.9 cm). The numeration is early.

[49] This manuscript only contains the proof of Proposition 7 by al-Khāzin (fols 36^{r}–37^{r}), followed by Proposition 7 of the Banū Mūsā (fols 37^{r}–39^{r}), and Proposition 16 (fols 50^{r}–52^{v}). There are a large number of interlinear comments by Aḥmad ibn Sulaymān, who is none other than the grandson of the copyist Muḥammad Riḍā ibn Ghulān Muḥammad ibn Aḥmad ibn Sulaymān. This collection is dated to Dhū al-Ḥijja 1134 H. See Otto Loth, *A Catalogue of the Arabic Manuscripts in the Library of the India Office*, London, 1877, pp. 297–9.

[50] See M. Le Baron de Slane, *Catalogue des manuscrits arabes de la Bibliothèque Nationale*, Paris, 1883–1895.

[51] See *Catalogue of Manuscripts in the Köprülü Library*, prepared by Dr Ramazan Şeşen, Cevat Izgi, Cemil Akpinar and presented by Dr Ekmeleddin Ihsanoğlu, Research Centre for Islamic History, Art and Culture, 3 vols, Istanbul, 1986, vol. I, pp. 463–7. Note that this manuscript belonged to the mathematician and astronomer Taqī al-Dīn al-Maʿrūf.

[52] See Aḥmad G. Maʿānī, *Fihrist Kutub Khaṭṭī Kitābkhāna Astān Quds*, Meshhed, 1350/1972, vol. VIII, no 403, pp. 366–7.

[53] The writing is in *naskhī* (page: 20 × 15 cm, 27 lines per page).

[54] See Joanne Uri, *Bibliothecae Bodleianae Codicum Manuscriptorum Orientalium*, Oxonii, 1787, p. 208.

16 - [P] Istanbul, Köprülü 931/14, fols 129^{r}–136^{v}.[55]
17 - [Q] Cairo, Dār al-Kutub, Riyāḍa 41, fols 26^{v}–33^{v}.[56]
18 - [R] Tehran, Majlis Shūrā 209/3, fols 33–54.[57]
19 - [S] Istanbul, Süleymaniye, Esad Effendi 2034, fols 4^{v}–15^{v}.[58]
20 - [T] Tehran, Majlis Shūrā 3919, fols 272–298.
21 - [U] Tehran, Danishka 2432/13, fols 123–137 (144^{v}–151^{v} according to a second numbering scheme).[59]
22 - [V] Istanbul, Süleymaniye, Aya Sofya 2760, fols 177^{r}–183^{v}.
23 - [W] Istanbul, Haci Selimaga 743, fols 71^{v}–81^{v}.
24 - [X] Istanbul, Beşiraga 440/14, fols 162^{v}–171^{v}.[60]
25 - [Y] Krakow, Biblioteka Jagiellonska, fols 183^{v}–194^{v}.[61]
26 - [Z] Manchester, John Rylands University Library 350.

A study of the variations in these manuscripts, taken two by two, and their copying accidents – omissions, additions, errors, etc. – has enabled the following *stemma* to be determined:

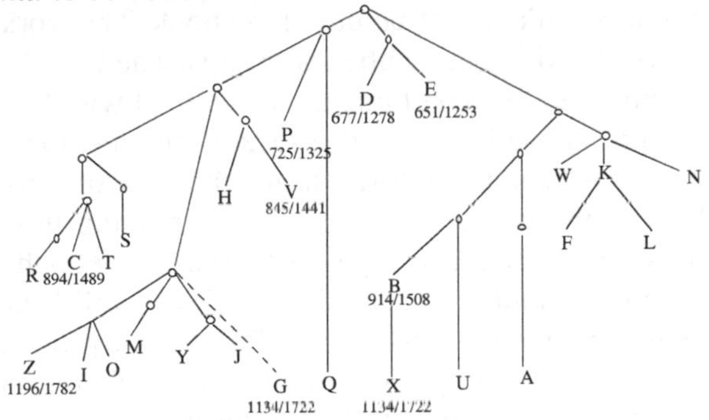

[55] See *Catalogue of Manuscripts in the Köprülü Library*, vol. I, pp. 467–72.

[56] For a description of this manuscript, see *Géométrie et dioptrique*, p. CXXXVI. The manuscript is incomplete and ends at the start of Proposition 16.

[57] See *Catalogue of the Arabic and Persian Manuscripts in the Madjless Library,* by Y. E. Tessami, Publications of the Library, Tehran, 1933, vol. II, pp. 117–18. Some of this treatise is missing between Propositions 6 and 7.

[58] The text is in a different hand to the rest of the collection, and the paper is also different. It is therefore an addition. The name of the mathematician Ibn Ibrāhīm al-Ḥalabī appears on the first page. It is written in *naskhī* (page: 22.2 × 12.7 cm, text: 14.3 × 6.2 cm).

[59] See *Catalogue of the Manuscripts*, University of Tehran, IX, pp. 1100–1.

[60] The copy dates from early Dhū al-Qaʿda 1134 H. The writing is in *naskhī*, and very carefully done (page: 28.2 ×15.7 cm).

[61] This manuscript corresponds to ms. Berlin, Staatsbibliothek, no 5938 (= Or. fol. 258), which disappeared from the library when the contents were being evacuated during the Second World War. We owe this information to Dr Hars Kurio, to whom we extend our grateful thanks. For a description of this manuscript, see W. Ahlwardt, *Handschriften der Königlichen Bibliothek zu Berlin XVII*, Berlin, 1893, p. 313.

1.2. MATHEMATICAL COMMENTARY

1.2.1. *Organization and structure of the Banū Mūsā book*

While the Banū Mūsā book on the measurement of plane and spherical figures is firmly rooted in the Archimedean tradition, it does not follow the model of *The Sphere and the Cylinder* or any other treatise by Archimedes. While the fundamental ideas are essentially the same as those of Archimedes, the Banū Mūsā followed a simpler and more direct path in arriving at them. It is only in this sense that their book can be described as Archimedean. Both the structure of the book and the methods followed by the Banū Mūsā are different from any that can be found in works by Archimedes on the same topic. This constellation of ideas, combined with these differences in structure, in composition and the methods of proof, highlight the unique position of this early research in Archimedean mathematics in Arabic.

We will first consider the structure of the book. The work consists of 18 propositions organized into a number of groups. The first three are lemmas in plane geometry and the next three are concerned with the measurements of circles and the calculation of π. The seventh proposition provides a new proof of Hero of Alexandria's formula for the area of a triangle, and the eighth deals with the uniqueness of a sphere passing through four non-coplanar points. The three following propositions relate to the lateral area of a cone of revolution and a truncated cone. The twelfth proposition is a lemma in plane geometry, and this is followed by three propositions concerning the surface area and volume of a sphere. The final three propositions are devoted to the determination of two means and the trisection of an angle. The logical connections between these propositions can be represented by the diagram on the facing page.

This clearly shows how the Banū Mūsā addressed four main themes in their book: namely, the measurement of the circle; Hero of Alexandria's formula for the area of a triangle; the surface area and volume of a sphere; and the two means and the trisection of an angle. At first glance, the inclusion of the seventh and the final three propositions may appear surprising, especially as they mark a radical departure from the main theme of the book, defined in its title as a compendium dedicated to interesting or difficult measurements of plane and spherical figures. In spite of this, there can be no doubt as to the authenticity of these propositions, or in relation to their inclusion as an integral part of this book. Firstly, their presence is attested not only in the Arabic manuscript tradition, but also in that of the Latin translation produced by Gerard of Cremona in the twelfth century. As further confirmation, this Latin translation contains a historically important final section in which the Banū Mūsā summarize the main results obtained.

Needless to say, this final summary includes references to all these propositions. Moreover, at the very end of this final section of the Latin translation, the Banū Mūsā make the following particularly important declaration:

> Everything that we describe in this book is our own work, with the exception of knowing the circumference from the diameter, which is the work of Archimedes, and the position of two magnitudes in between two others such that all <four> are in continued proportion, which is the work of Menelaus as stated earlier (*infra*, p. 109).

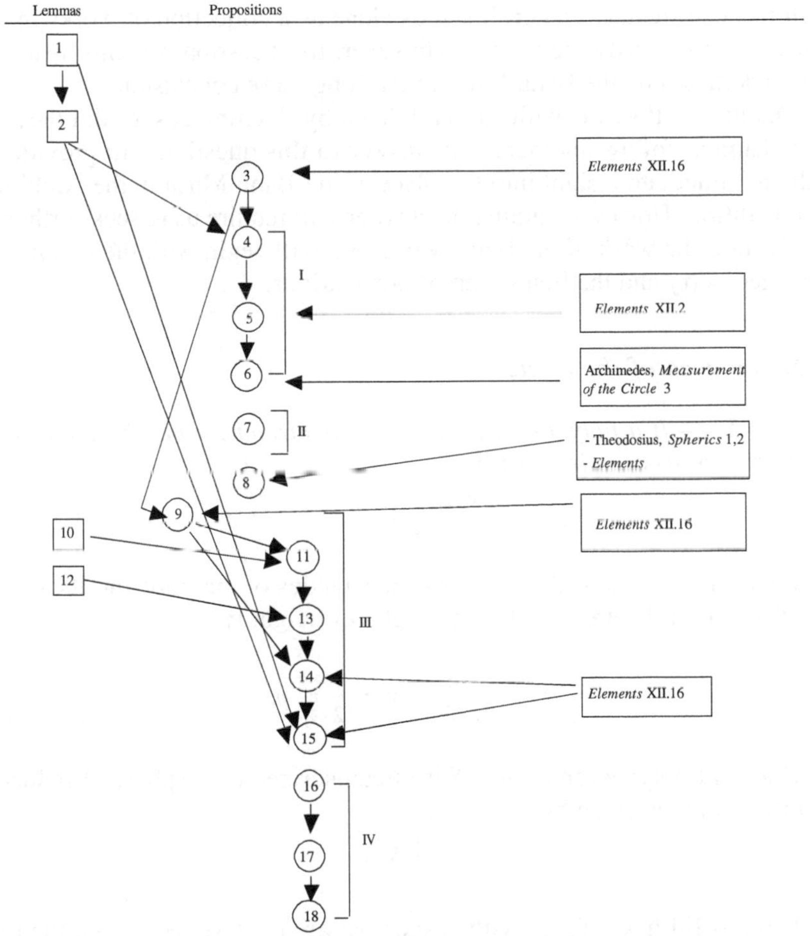

We shall examine after the full meaning of the appreciation of the Banū Mūsā of their own contribution in this regard, but for the moment it is sufficient to note that it confirms the presence of Proposition 6 and the final group of propositions. The presence of Hero of Alexandria's formula is confirmed by the manuscript traditions and by the Banū Mūsā themselves through the Latin translation, in addition also to an appendix commonly associated with the book in the Arabic tradition. This contains another proof of the same formula attributed to al-Khāzin in the middle of the tenth century.

It can therefore be seen that the Banū Mūsā book is not based on any Archimedean treatise; rather, it stands alone as a collection of works in the four areas previously mentioned. However, the question remains as to the path undertaken by the Banū Mūsā in reaching these conclusions.

Did they follow the pathway laid down by Archimedes, or did they, as they claimed, follow another? The answer to this question will provide us with an immediate insight into the place of the Banū Mūsā in the Archimedean tradition. However, finding the answer will require us to seek further, if briefly, into the work of the Banū Mūsā. We will begin with the lemmas of plane geometry and the first group of propositions.

1.2.2. *The area of the circle*

Lemma 1. — *If a polygon of perimeter* p *is circumscribed by a circle of radius* r, *then its area is given by*

$$S = \frac{1}{2} p \cdot r.$$

Let $a_1, a_2, \ldots, a_n$ be the lengths of the n sides of the polygon. Its area is then the sum of the areas of the n triangles of height r:

$$S = \sum_{i=1}^{n} \frac{1}{2} a_i \,.\, r = \frac{1}{2} r \,.\, p.$$

If a solid polyhedron of area S is circumscribed by a sphere of radius r, then its volume is given by

$$V = \frac{1}{3} S \cdot r.$$

If the solid has n faces with respective areas of $s_1, s_2, \ldots, s_n$, then its volume is the sum of the volumes of the n pyramids of height equal to r:

$$V = \sum_{i=1}^{n} \frac{1}{3} s_i \,.\, r = \frac{1}{3} r \,.\, S.$$

Comment – The formula for obtaining the volume of a pyramid, regardless of the shape of the base, is assumed to be known. This formula can be found in the *Elements* XII.

Lemma 2. — *If a polygon of perimeter* p *is inscribed by a circle of radius* r, *then its area is given by*

$$S < \frac{1}{2}\, p \cdot r < \text{the area of the circle.}$$

Let $a_1, \ldots, a_n$ be the lengths of the n sides of the polygon and let h_i be the length of the perpendicular dropped from the centre of the circle onto the side of length a_i, and let s_i be the area of the corresponding sector. We then have

$$\frac{1}{2}\, a_i\, h_i < \frac{1}{2}\, a_i\, r < s_i$$

from which

$$\frac{1}{2} \sum_{i=1}^{n} a_i\, h_i \; < \; \frac{1}{2}\, r \sum_{i=1}^{n} a_i \; < \; \sum_{i=1}^{n} s_i$$

and thence the result.

Similarly, if a solid polyhedron of n faces, with a total surface area of S, is inscribed within a sphere of radius r, then

$$\text{volume of the solid} < \frac{1}{3}\, S \cdot r < \text{volume of the sphere.}$$

The Banū Mūsā then go on to prove the following proposition.

Proposition 3. — *Consider a circle of circumference* p *and a line segment of length* l. *Then*

1. *If* $l <$ p, *then it is possible to inscribe a polygon of perimeter* p_n *within the circle such that*

$$l < p_n < p.$$

2. If $l >$ p, *then it is possible to circumscribe a polygon of perimeter* q_n *outside the circle such that*

$$p < q_n < l.$$

The proofs of statements 1 and 2 are based on the existence of a circle of circumference l and a regular polygon. The Banū Mūsā admit the existence of this circle. For the polygon, they make use of Proposition XII.16 of Euclid's *Elements*:

> Given two circles about the same centre, to inscribe in the greater circle an equilateral polygon with an even number of sides which does not touch the lesser circle.[1]

It may be noted that, for a regular n-sided polygon to comply with the criteria for the problem, it is a necessary and sufficient condition that its apothem a_n satisfies the following:

$$r_1 < a_n < r_2 \Leftrightarrow r_1 < r_2 \cos\frac{\pi}{n} < r_2 \Leftrightarrow \frac{p_1}{p_2} < \cos\frac{\pi}{n} < 1$$

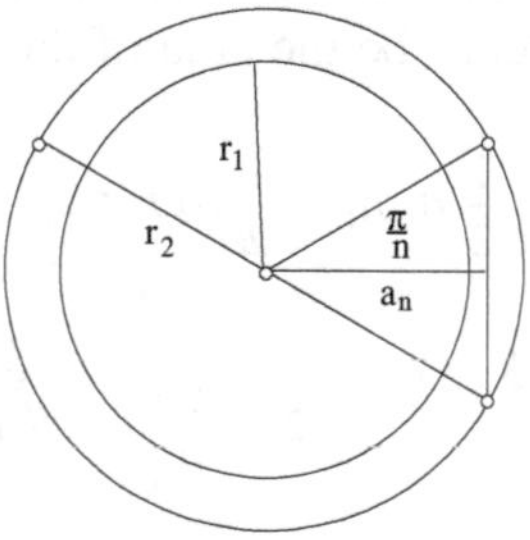

Fig. 1.2.1

where r_1 and r_2 are the radii of the two concentric circles, and p_1 and p_2 are their circumferences (the existence of the integer n depends on the continuity of the cosine function).

The proof given by the Banū Mūsā is as follows: Consider two concentric circles *ABC* and *DEG*.

1) $l < p$: Let p be the circumference of *ABC* and l the circumference of *DEG*.
2) $l > p$: Let l be the circumference of *ABC* and p the circumference of *DEG*.

In both cases, *ABC* is therefore the larger of the two circles, and any regular or irregular polygon inscribed within circle *ABC* with the sides not touching circle *DEG* will have a perimeter lying between l and p.

[1] *The Thirteen Books of Euclid's Elements*, translated with introduction and commentary by Th. L. Heath, New York, Dover, vol. 3, p. 423.

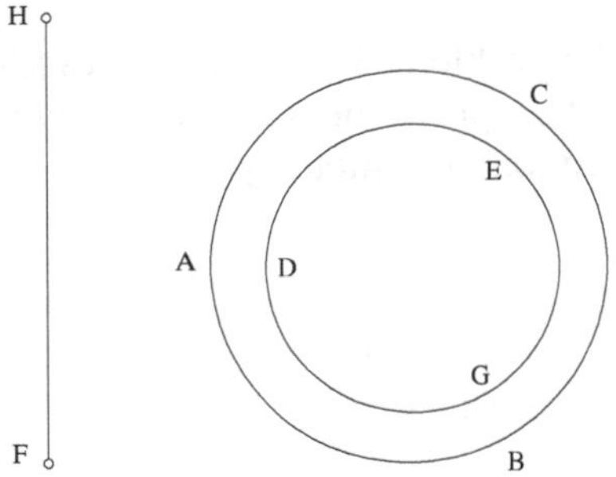

Fig. 1.2.2

However, in order to prove exactly the second statement, when $l > p$, it is necessary to consider a polygon circumscribed around the circle *EDG* of perimeter p whose sides do not cut circle *ABC*. This can be proved from Proposition XII.16 of the *Elements* together with a homothety.

In Proposition 3, the Banū Mūsā begin with a circle $\mathbf{C}_1$ of circumference p and effectively *admit* the existence of a circle $\mathbf{C}_2$ with the given circumference l. They then go on to consider two cases:

a) $l < p$: $\mathbf{C}_2$ and $\mathbf{C}_1$ are concentric, with $\mathbf{C}_2$ lying inside $\mathbf{C}_1$. They then wish to *inscribe* a polygon P_n of perimeter p_n inside $\mathbf{C}_1$ such that

$$l < p_n < p,$$

the polygon P_n defined in the *Elements* XII.16 – P_n inscribed within $\mathbf{C}_1$ and not touching $\mathbf{C}_2$ – is a solution to the problem.

b) $l > p$: $\mathbf{C}_1$ lying within $\mathbf{C}_2$. Using *Elements* XII.16, it is possible to inscribe a polygon P_n within $\mathbf{C}_2$ and not touching $\mathbf{C}_1$ such that

$$p < p_n < l.$$

If one wishes to *circumscribe* a polygon P_n' of perimeter p_n' around $\mathbf{C}_1$ such that $p < p_n' < l$, this can be done by deducing P_n' from P_n by means of a homothety. Thus, if $OH = a_1$, the apothem of P_n, then

$$r_1 < OH < r_2.$$

In the homothety $\left(O, \frac{r_1}{a_1}\right)$, the image of P_n is P_n' such that

$$p < p_n' < p_n < l;$$

P'_n is the solution to the problem. It *circumscribes* $\mathbf{C}_1$ and does not touch $\mathbf{C}_2$ (see Fig. 1.2.3.). The proof begins with *Elements* XII.16, and is completed by the application of homothety.

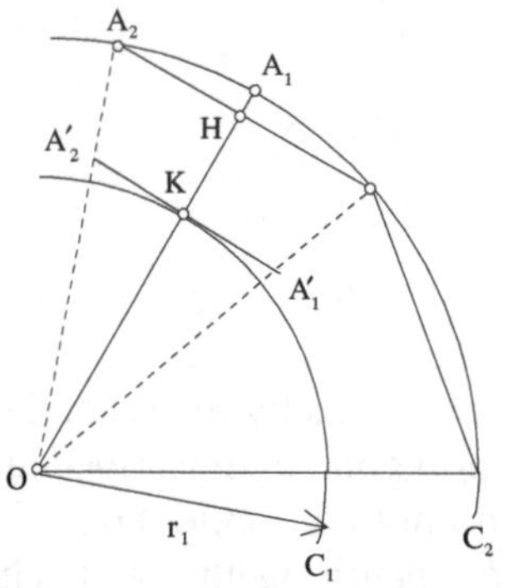

Fig. 1.2.3

In the next proposition, the Banū Mūsā prove that the area of a circle is the product of half its diameter multiplied by half its circumference using an apagogic method.

Proposition 4. — *For any circle of radius* r *and circumference* p, *the area is given by*

$$\mathrm{S} = \frac{1}{2}\,\mathrm{p} \cdot \mathrm{r}.$$

If $S < \frac{1}{2} p \cdot r$, then $S = \frac{1}{2} l \cdot r$, where $l < p$, and it is possible to inscribe a polygon of perimeter p' within the circle such that $l < p' < p$ (using Proposition 3). From Lemma 2, the area S_1 of this polygon is such that

$$S_1 < \frac{1}{2} p' \cdot r < S.$$

However, $l < p'$ implies that $\frac{1}{2} l \cdot r < \frac{1}{2} p' \cdot r$, *i.e.* $\frac{1}{2} p' \cdot r > S$, which is clearly absurd.

If $S > \frac{1}{2} p \cdot r$, then $S = \frac{1}{2} l \cdot r$ where $l > p$. It is possible to circumscribe the circle with a polygon of perimeter p'' such that $p < p'' < l$. We then have

$$\frac{1}{2} r \cdot l > \frac{1}{2} r \cdot p'';$$

this is also absurd, as $\frac{1}{2}\, r \cdot p''$ is the area of the polygon and this area is greater than that of the circle given by

$$S = \frac{1}{2}\, l \cdot r.$$

Note that, unlike Archimedes, who gives the area of a circle by comparing it with the area of another figure, a right triangle with the two sides enclosing the right angle equal in length to half the diameter and the circumference respectively, the Banū Mūsā give the area as the product of two quantities. In their proof of the proposition, they compare p with two lengths, $p' < p$ and $p'' > p$, in order to show that each results in a contradiction; and this contrasts with Archimedes' use of two areas. Finally, their approach differs also from that of Archimedes in the way in which the exhaustion method is applied. The Banū Mūsā avoid the greatest problem with this method,[2] what we would call the 'taking to the limit', by making use of Proposition XII.16 in the *Elements*, which is proved using the limit ($\lim_{n\to\infty} \cos \frac{\pi}{n} = 1$).

At the end of this proposition, the Banū Mūsā give an expression for the area of a sector of a circle, without giving a corresponding proof. This could be done using a method similar to that used to prove Proposition 4 itself, by inscribing a polygonal sector within the sector of the circle, or simply by noting that the length p' of an arc of a circle is proportional to the angle α subtended at the centre and that the area S' of a sector of a circle is also proportional to this angle. Therefore, if S and p are the area and circumference respectively of the circle, and S' and p' are the area of the sector and the length of the corresponding arc, then

$$\frac{S}{S'} = \frac{P}{P'} = \frac{360}{\alpha} \quad \text{(if } \alpha \text{ is measured in degrees);}$$

as $S = \frac{1}{2}\, p \cdot r$, then $S' = \frac{1}{2}\, p' \cdot r$.

Proposition 5. — *The ratio of the diameter to the circumference is the same for all circles.*

The Banū Mūsā based their proof on Proposition XII.2 of the *Elements*: The ratio of the areas of two circles is equal to the ratio of the squares of

[2] See the article by J. al-Dabbagh, 'Banū Mūsā', *D.S.B*, vol. 1, pp. 443–6.

their radii. A proof by *reductio ad absurdum* is not necessary as the brothers have already shown in their previous proposition that $S = \frac{1}{2} p \cdot r$. However, in this proposition the Banū Mūsā have used just such a proof.

In Proposition 6, they continue by calculating this ratio using the method developed by Archimedes, which they acknowledge. Ultimately, Archimedes' method enables the upper and lower bounds of this ratio to be obtained to any desired degree of approximation.

This group of six propositions is followed by two unrelated propositions, and before the book moves on to another important group of propositions relating to the sphere. The first of these two propositions concerns the formula proposed by Hero of Alexandria.

1.2.3. *The area of the triangle and Hero's formula*

Proposition 7. — *If* p *is the perimeter of a triangle with sides* a, b, *and* c, *then the area of the triangle satisfies the following:*

$$S^2 = \frac{p}{2}\left(\frac{p}{2} - a\right)\left(\frac{p}{2} - b\right)\left(\frac{p}{2} - c\right).$$

The Banū Mūsā did not attribute this to Hero, or to any other mathematician. Later mathematicians, including al-Bīrūnī, attributed the formula to Archimedes.[3] The Banū Mūsā derived the formula using a different proof from that of Hero. This proof was copied by a large number of later authors, including Fibonacci and Luca Pacioli.[4] However, this proof did not find favour with others, including al-Khāzin (who gave an alternative proof that is often included as an appendix to the Banū Mūsā book), or later al-Shannī.[5]

Proposition 8. — *If a point* G *is equidistant from four non-coplanar points on a given sphere, then* G *is the centre of that sphere.*

This proposition effectively demonstrates the uniqueness of a sphere passing through four non-coplanar points. In order to prove this proposition, the Banū Mūsā turned once again to the *Elements* and the first two

[3] Al-Bīrūnī, *Istikhrāj al-awtār fī al-dā'ira*, ed. Aḥmad Sa'īd al-Dimerdash, Cairo, n.d., p. 104.

[4] M. Clagett, *Archimedes in the Middle Ages*, vol. 1, Appendix IV, pp. 635–40.

[5] This demonstration has been reported by al-Bīrūnī, *Istikhrāj al-awtār fī al-dā'ira*.

propositions in the *Spherics* of Theodosius in the translation made by Qusṭā ibn Lūqā.[6] It should be noted that their proof does not make the assumption that G lies within the sphere. The proof may be summarized as follows:

Let B, C, D and E be the four non-coplanar points. The plane (B, C, E) cuts the sphere, forming a circle whose axis passes through the centre of the sphere and point G, as $GB = GC = GE$.

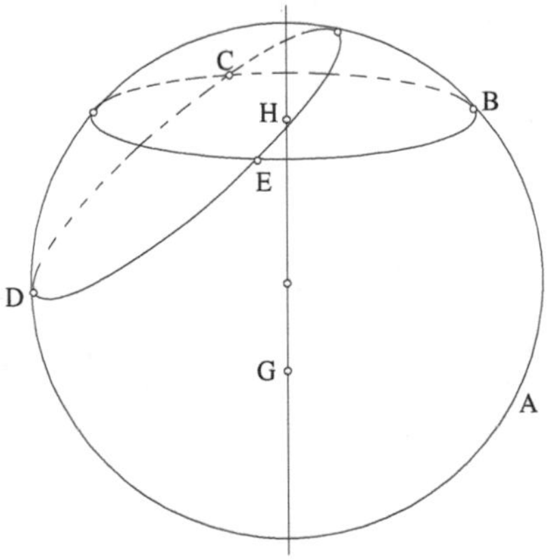

Fig. 1.2.4

Similarly, the axis of circle ECD also passes through the centre of the sphere and point G. These two axes are different, having only a single point in common at the centre of the sphere. Point G must therefore lie at the centre of the sphere.

1.2.4. *The surface area of a sphere and its volume*

The following group of seven propositions form the core of the Banū Mūsā book. Their aim is to enable the determination of the surface area and the volume of a sphere. We have already highlighted a number of differences between the methods adopted by Archimedes and the Banū Mūsā when discussing the measurement of circles. The question now is to determine whether the path taken by the Banū Mūsā was chosen deliberately, or simply by chance. In other words, are we going to discover the same deviations from the Archimedean method in the case of the sphere? To

[6] See the edition by al-Ṭūsī of the translation by Qusṭā ibn Lūqā of the *Kitāb al-ukar* of Theodosius, printed by Osmania Oriental Publications Bureau, Hyderabad, 1358/1939.

answer that question, we need to look at this group of seven propositions in more detail.

Proposition 9. — *The lateral area* S *of a cone of revolution is given by* $S = \frac{1}{2} p \cdot l$, *where* p *is the circumference of the base circle and* l *is the length of the generator.*

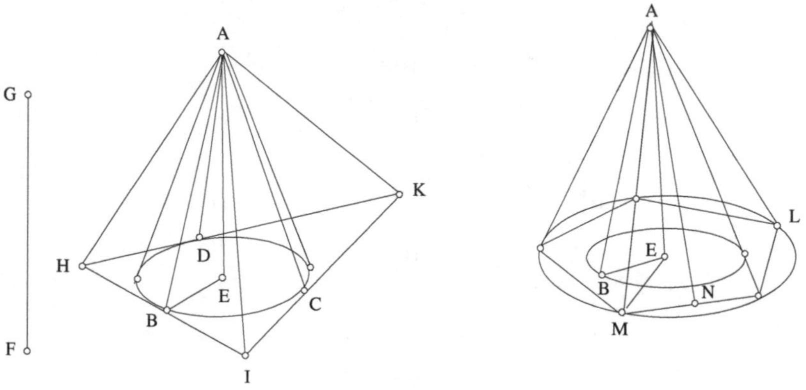

Fig. 1.2.5

Let the cone be (A, BCD), with A being the vertex, BCD the base, AE the axis, and $AB = l$ the generator.

1° If $S > \frac{1}{2} p \cdot l$, then $S = \frac{1}{2} p' \cdot l$, with $p' > p$.

The circle BCD may be circumscribed with a polygon of perimeter p_1, where $p' > p_1 > p$. This has been proved to be possible in Proposition 3. This may then be considered the base of a pyramid with its vertex at A that circumscribes the cone. However, $EB \perp HI$ and $AE \perp (HIK)$; therefore $AB \perp HI$. Similarly, $AC \perp IK$ and $AD \perp HK$. The lateral area of the pyramid is therefore given by

$$\frac{1}{2} p_1 \cdot l < \frac{1}{2} p' \cdot l.$$

However, $S = \frac{1}{2} p' \cdot l$, which is impossible.

2° $S < \frac{1}{2} p \cdot l$. The Banū Mūsā therefore admit the existence of a cone of revolution with a vertex A, axis AE and lateral area $S' = \frac{1}{2} p \cdot l > S$. Let ML be its base circle; then $AM > AB$ and $EM > EB$.

A regular polygon may be inscribed within the circle ML, which does not touch the circle ABC. If the circumference of this circle is p_1, then $p_1 > p$. A regular pyramid on this polygon base will have a lateral area of

$$S_1 = \frac{1}{2} p_1 \cdot AN,$$

where N is the midpoint of one side of the polygon. However,

$$AN > AB,$$

from which

$$S_1 > \frac{1}{2} p \cdot l.$$

Therefore

$$S_1 > S'.$$

This is also impossible as the cone with lateral area S' encloses the pyramid with lateral area S_1.

The result can be derived from 1° and 2°.

In both cases, the Banū Mūsā extend Postulate 2 of Archimedes' *The Sphere and the Cylinder* relating to convex curves to include, by analogy, convex surfaces.

The Banū Mūsā then introduce a technical lemma:

Lemma 10. — *The intersection of the lateral surface of a cone of revolution and a plane parallel to the base is a circle centred on the axis of the cone.*

It should be noted that the two parallel planes correspond in the homothety $\left(A, \frac{AH}{AE}\right)$. The figure IGH is therefore homothetic to the circle centred on E. It is therefore a circle centred on H. However, the reasoning proposed by the Banū Mūsā does not include such a transformation.

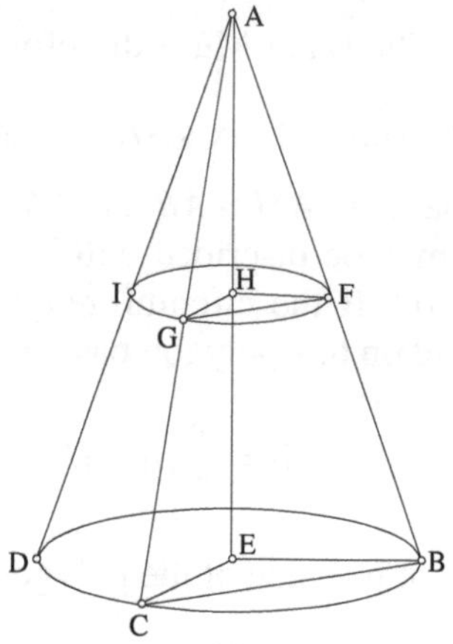

Fig. 1.2.6

Proposition 11. — *The lateral area of a truncated cone of revolution with parallel bases is given by*

$$S = \frac{1}{2}(p_1 + p_2)\, l,$$

where p_1 *and* p_2 *are the circumferences of the two bases respectively, and* l *is the length of the generator.*

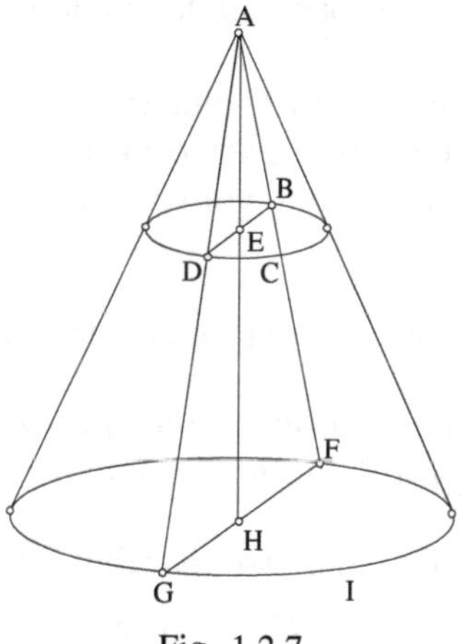

Fig. 1.2.7

We have

$$\text{area } (A, GIF) = S_1 = \frac{1}{2} AF \cdot p_1 \text{ and area } (A, BCD) = S_2 = \frac{1}{2} AB \cdot p_2.$$

The area of the truncated cone is therefore

$$S = \frac{1}{2}(AF \cdot p_1 - AB \cdot p_2) = \frac{1}{2}(p_1 - p_2)\, AB + \frac{1}{2}\, BF \cdot p_1.$$

However,

$$\frac{AB}{p_2} = \frac{AF}{p_1} = \frac{BF}{p_1 - p_2};$$

therefore

$$AB\,(p_1 - p_2) = BF \cdot p_2.$$

From this, we can deduce that

$$S = \frac{1}{2}\, BF\,(p_1 + p_2).$$

But we know that BF is the generator of the truncated cone, $BF = l$, hence the result.

The Banū Mūsā go on to determine the lateral area of a solid of revolution formed by a truncated cone and a full cone sharing the same base, and with generators of the same length l:

$$S = \frac{1}{2}\, l\,(p_1 + p_2) + \frac{1}{2}\, l\, p_2 = \frac{1}{2}\, l\, p_1 + l\, p_2$$

where p_1 and p_2 are the circumferences of the bases.

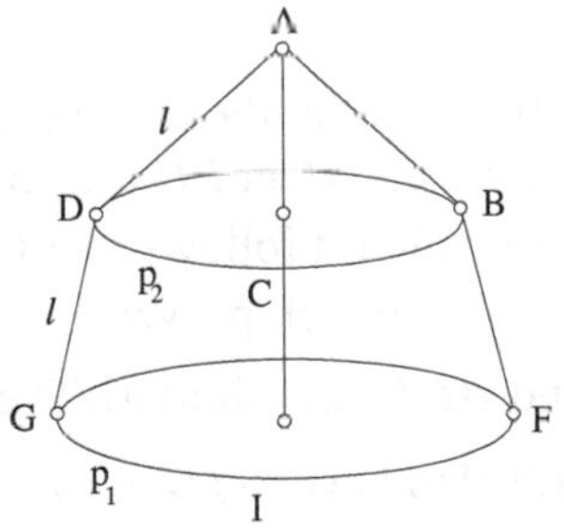

Fig. 1.2.8

The Banū Mūsā then generalize the previous result for a solid of revolution formed by any number of truncated cones and a complete cone, all with generators of the same length:

$$S = \frac{1}{2} l \sum_{k=2}^{n} (p_{k-1} + p_k) + \frac{1}{2} l p_n = \frac{1}{2} l \left(p_1 + 2 \sum_{k=2}^{n} p_k \right) = \pi\, l \left(r_1 + 2 \sum_{k=2}^{n} r_k \right).$$

Next, the Banū Mūsā introduce another plane geometry lemma.

Lemma 12. — *Let a circle of centre* D *have a diameter that is perpendicular to* AC, *and let* DB *be such that* DB ⊥ AC. *If we then assume that*

$$\widehat{BL} = \widehat{LG} = \widehat{GA}$$

and

$$HL \parallel AC,\ GI \parallel AC,\ DM \perp BL,\ BL \cap AC = \{E\},$$

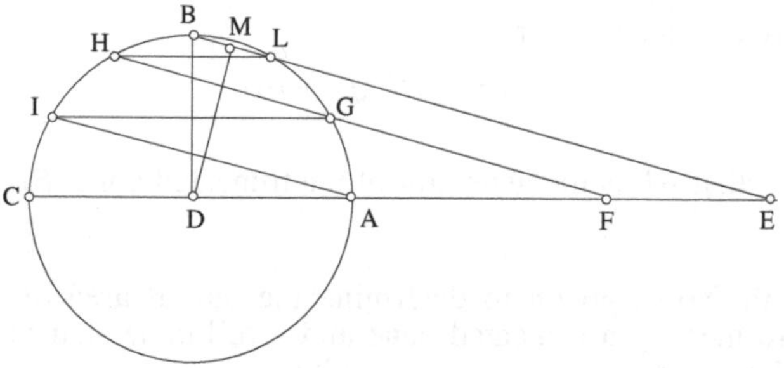

Fig. 1.2.9

then

1° $DE = DA + IG + HL$

2° $DA^2 > \frac{1}{2} BL\,(DA + IG + HL) > DM^2.$

From symmetry, it can be seen that the arcs *LG* and *HI* are equal to the *BL* and *BH*. Therefore the two arcs *GL* et *BH* are equal and $BL \parallel GH$.

Similarly, $\widehat{AG} = \widehat{IH}$, from which it follows that $GH \parallel AI$. If *HG* cuts *DE* at *F*, then $HL = FE$ and $IG = AF$, which proves 1°.

The triangles *BMD* and *BDE* are similar. Therefore $\frac{BM}{MD} = \frac{BD}{DE}$, and hence $BM \cdot DE = MD \cdot BD$. However $MD < BD$, and therefore $MD^2 < MD \cdot BD < BD^2$, from which $MD^2 < \frac{1}{2} BL \cdot DE < DA^2$, which proves 2°.

The result for the three equal arcs *AG*, *GL,* and *LB* can be extended to any number of equal arcs. We can then rewrite this lemma for the general case, bringing out the underlying trigonometric concepts.

If a quarter circle A_1B *is divided into* n *equal arcs by the points* A_2, A_3, …, A_n, *then*

1° $$A_1B_1 + 2\sum_{k=2}^{n} A_k B_k = B_1E$$

2° $$B_1M^2 < \frac{1}{2} BA_n \left[B_1A_1 + 2\sum_{k=2}^{n} B_kA_k \right] < B_1B^2.$$

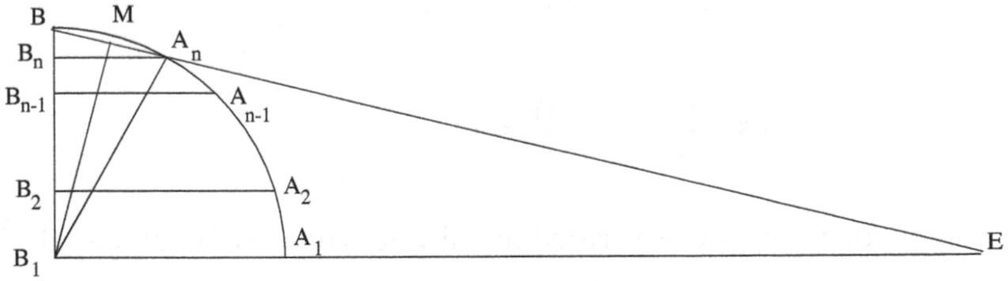

Fig. 1.2.10

We then have

$$\widehat{BA_n} = \frac{\pi}{2n},\ \widehat{BA_{n-1}} = 2\cdot\frac{\pi}{2n},\ \ldots,\ \widehat{BA_2} = (n-1)\frac{\pi}{2n},$$

from which

$$A_nB_n = R\sin\frac{\pi}{2n},\ A_{n-1}B_{n-1} = R\sin 2\cdot\frac{\pi}{2n},\ \ldots,\ A_2B_2 = R\sin(n-1)\frac{\pi}{2n}.$$

If $B_1M \perp BA_n$, then

$$B\hat{B}_1M = \frac{\pi}{4n} = B_1\hat{E}B,$$

from which

$$B_1E = R\cot\frac{\pi}{4n}.$$

If we allow $R = 1$, then 1° may be rewritten as

(1) $$2\sum_{k=1}^{n-1} \sin k\cdot\frac{\pi}{2n} = \cot\frac{\pi}{4n} - 1,$$

which can also be written as follows (by adding 2 to each side):

(2) $$2\sum_{k=1}^{n} \sin k\cdot\frac{\pi}{2n} = \cot\frac{\pi}{4n} + 1,$$

which may be verified by multiplying both sides by $\sin\frac{\pi}{4n}$.

In 2°, we have

$$B_1M = R \cos \frac{\pi}{4n} \quad \text{and} \quad \frac{1}{2} BA_n = BM = R \sin \frac{\pi}{4n}.$$

If we allow $R = 1$, then 2° may be rewritten as

$$\text{(3)} \qquad \cos^2 \frac{\pi}{4n} < \sin \frac{\pi}{4n} \cdot \cot \frac{\pi}{4n} < 1,$$

i.e.

$$\cos^2 \frac{\pi}{4n} < \cos \frac{\pi}{4n} < 1.$$

This relationship may be verified for all values of n, as for all $\alpha \in \left]0, \frac{\pi}{2}\right[$, we have $\cos^2 \alpha < \cos \alpha < 1$. we may therefore make n arbitrarily large, enabling the use of an apagogic method. In modern terms, this is equivalent to evaluating the integral $\int_0^{\frac{\pi}{2}} \sin x dx$. It should be noted, however, that the Banū Mūsā proceeded in an altogether different way.

These are exactly the same as the sums and inequalities used to determine the surface area and volume of a sphere.

Proposition 13. — In Proposition 13, the Banū Mūsā consider a semicircle ABD with centre M and radius R_2, within which is inscribed a regular polygonal line with an even number of sides. A second semicircle is inscribed within this line. Rotating this figure produces a hemisphere, a solid of revolution consisting of a cone and several truncated cones, and a second hemisphere inscribed within this solid of revolution and concentric with the first hemisphere. They go on to prove that

$$2\,\pi\,R_1^2 < S < 2\,\pi\,R_2^2\,;$$

where R_1 and R_2 are the radii of the inscribed and circumscribed circles respectively, and S is the lateral area of the solid.

It should be noted that this solid satisfies the conditions laid down in Proposition 11, and that the assumptions relating to the plane figure in the plane ABD are the same as those in Proposition 12. Therefore

$$\text{(1)} \qquad \frac{1}{2} BE\,(MB + HE + GF) < MB^2;$$

and, from Proposition 11,

(2) $$S = \pi\, EB\, (MB + HE + GF).$$

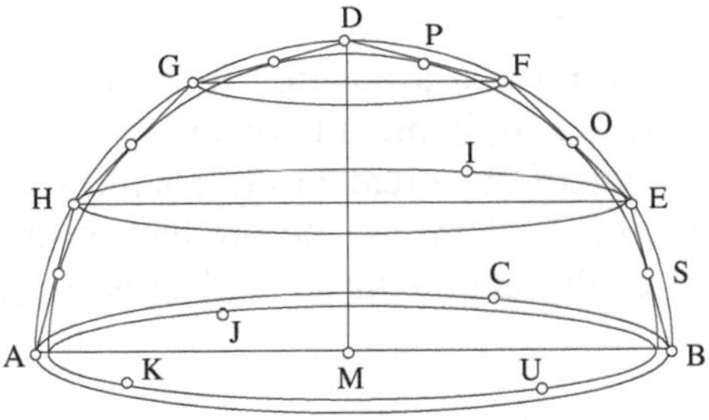

Fig. 1.2.11

Combining (1) and (2), we have

$$S < 2\,\pi\, MB^2 = 2\,\pi\, R_2^2.$$

If S, O and P are the midpoints of the chords BE, EF and FD, then

$$MS = MO = MP = MU = R_1,$$

the radius of the inscribed sphere.
From Lemma 12, we have

(3) $$MS^2 < \frac{1}{2}\, BE\, (MB + HE + GF).$$

Combining (2) and (3), we have

$$S > 2\,\pi\, MS^2 = 2\,\pi\, R_1^2$$

and hence we obtain the result.

In other terms: consider a semicircle $\mathbf{C}(M, R_2)$, a regular polygonal line with $2n$ sides inscribed within $\mathbf{C}$, and a semicircle $\mathbf{C}'(M, R_1)$ inscribed within the polygonal line. From these, the Banū Mūsā construct:

- a hemisphere $\Sigma(M, R_2)$;
- a solid Γ formed from cones and truncated cones *inscribed* within Σ and satisfying the conditions of Proposition 11;
- a hemisphere $\Sigma'\,(M, R_1)$ inscribed within this solid.

They then show that, if S is the lateral area of solid Γ, then

$$2\,\pi\, R_1^2 < S < 2\,\pi\, R_2^2$$

using Propositions 11 and 12 and *making no reference to* Proposition XII.16 in the *Elements*.

The Banū Mūsā are now in a position to apply the apagogic method twice: firstly in Proposition 14, in order to obtain the lateral surface area of a hemisphere, 'twice that of its great circle'; and, secondly in order to determine the volume of a sphere as 'the product of its half-diameter and one third of the area of the lateral surface'. The proof given by the Banū Mūsā is as follows:

Proposition 14. — *The surface area* S *of a hemisphere is twice that of its great circle.*

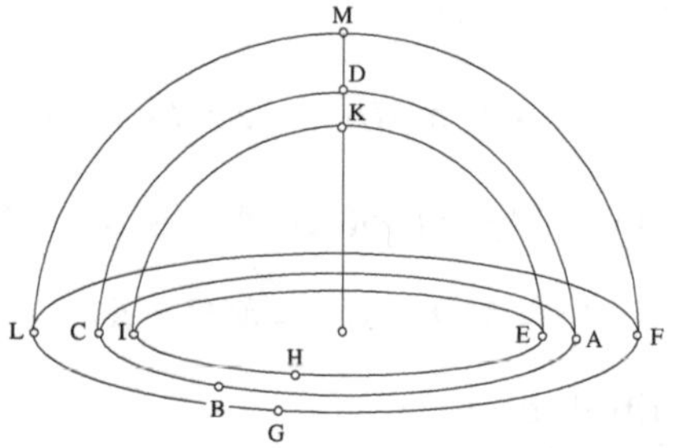

Fig. 1.2.12

Let s be the area of circle ABC, and let S be the area of the hemisphere $ABCD = \Sigma$.

a) $S > 2s$. If $2s = S_1$, then $S_1 < S$. The Banū Mūsā admit the existence of a hemisphere $EHIK = \Sigma_1$ inside and concentric with Σ. The area of this hemisphere is S_1.

The Banū Mūsā then proceed in a similar way to Proposition 13 by considering a solid Γ *inscribed* within Σ. This solid comprises cones and truncated cones, and its surface does not touch Σ_1. Such a solid is derived from a regular polygonal line that is *inscribed* within the great semicircle of the hemisphere Σ and does not touch the great semicircle $\mathbf{C}_1$ of the hemisphere Σ_1, basing their argument on Proposition XII.16 of the *Elements* and not on XII.17 as some have claimed.

b) $S < 2s$. If $2s = S_2$, then $S_2 > S$. The Banū Mūsā consider a sphere Σ_2 with area S_2 outside Σ, together with a solid Γ' that is inscribed within Σ_2 and does not touch the sphere Σ. This solid is derived from Proposition XII.16 of the *Elements* in the same way as in the first case.

Using the inequalities established in Proposition 13 leads to an impossible result in both cases, a) and b). Therefore, for a hemisphere, $S = 2s = 2\,\pi\,R^2$.

Proposition 15. — *The volume of a sphere* Σ *of radius* R *and surface area* S *is given by*

$$V = \frac{1}{3}R \cdot S = \frac{4}{3}\pi R^3.$$

Let *ABCD* be the given sphere Σ. There are two possibilities:

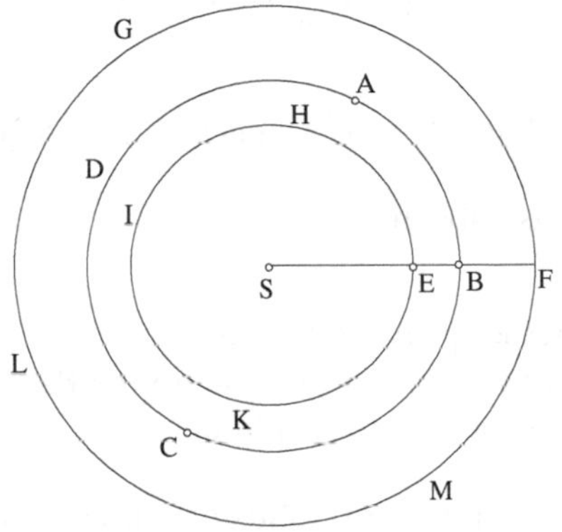

Fig. 1.2.13

• If $\frac{1}{3} R \cdot S < V$, then the Banū Mūsā admit the existence of a concentric sphere $FGLM = \Sigma_1$ with surface area S_1 such that

$$\frac{1}{3} R \cdot S_1 = V, \qquad \text{where } S_1 > S.$$

The Banū Mūsā are therefore considering a sphere Σ_1 concentric with Σ and with a surface area of $S_1 > S$. Therefore, Σ lies *inside* Σ_1. They then consider a polyhedron that is *circumscribed* around Σ and does not touch Σ_1 and apply Lemma 1. If S_2 and V_2 are the area and volume of this solid respectively, then, from Lemma 1,

$$V_2 = \frac{1}{3} R \cdot S_2.$$

We know that $S_2 < S_1$, therefore $V_2 < V$. This is absurd, as the solid with volume V_2 surrounds the sphere with volume V.

• If $\frac{1}{3} R \cdot S > V$, the Banū Mūsā consider a concentric sphere $EHIK = \Sigma'_1$, smaller than $ABCD$ and with surface area S'_1 such that

$$V = \frac{1}{3} R \cdot S'_1.$$

They then consider a polyhedron that is inscribed within Σ and does not touch Σ'_1, and apply Lemma 2. If its area is S'_2 and its volume is V'_2, then $V'_2 < V$ and, from Lemma 2,

$$V'_2 < \frac{1}{3} R \cdot S'_2 < V.$$

However, we know that $S'_2 > S'_1$; hence $\frac{1}{3} R \cdot S'_2 > \frac{1}{3} R \cdot S'_1$, *i.e.* $\frac{1}{3} \mathrm{R} \cdot S'_2 > V$. This is also absurd.

The two cases together prove the result.

In neither the first nor the second case do the Banū Mūsā question the existence of the polyhedron that they introduce.

Comment — The only volumes of solids discussed in this text are those in Lemmas 1 and 2; namely, the volume of a polyhedron of surface area S *circumscribed* around a sphere of radius R_1,

$$V = \frac{1}{3} S \cdot R_1,$$

and the volume of a polyhedron of surface area S *inscribed* within a sphere of radius R_2,

$$V < \frac{1}{3} S \cdot R_2 < \text{the volume of the sphere.}$$

It is these results that the Banū Mūsā used in proving Proposition 15, which leads one to suppose that the solids they are considering are *polyhedra*. There remains the question of which polyhedra may be chosen while still complying with the conditions of the two cases defined in Proposition 15.

Other commentators have shown that the problem may be resolved by using the solid P_n defined in Proposition XII.17 of the *Elements*, a solid *inscribed* within a sphere. However, a sphere cannot be inscribed within such a solid and Lemma 1 of the Banū Mūsā therefore cannot be used.

In the second case in Proposition 15, the polyhedron P_n *inscribed* within the sphere Σ of radius R must lie outside the sphere Σ'_1 of radius R'_1. The value of n must therefore be chosen such that the shortest distance h from the centre of the two spheres to each face of P_n is such that $h > R_1$. The volume V_n of P_n is then such that $V_n > \frac{1}{3} S_n \cdot R_1$, from Lemma 2.

It should be noted at this point that al-Khāzin discusses these distances in Proposition 19 (as noted later).

In the first case, the two spheres under consideration are Σ of radius R and Σ_1 of radius $R_1 > R$. Instead of considering a polyhedron *circumscribed* around Σ and lying within Σ_1, we should rather consider a polyhedron Γ_n inscribed within Σ_1 such that the shortest distance h from the centre to each of its faces satisfies $h > R$. Its volume is therefore such that $V_n > \frac{1}{3} S_n \cdot R$ (from Lemma 2). It is then possible to reach the desired conclusion.

Finally, we should note that the original text includes the phrases 'let us circumscribe, as we have described, a solid around the sphere *ABCD* ...' and 'let us inscribe, as we have described, a solid within the sphere *ABCD* ...' In neither case do they specify the exact nature of the solid. One possibility is to consider solids formed from cones and truncated cones as was done for the area of a sphere in Proposition 14. However, the Banū Mūsā do not discuss the volumes of such solids anywhere in this book. They were doubtless aware that, in Propositions 26 and 31 of *The Sphere and the Cylinder*, Archimedes had shown that, if a solid of this type with a surface area of S *circumscribes* a sphere of radius R_1, then $V = \frac{1}{3} S \cdot R_1$, and that in Proposition 27, if the solid is *inscribed* within a sphere of radius R_2, then $V < \frac{1}{3} S \cdot R_2$.

The reasoning described by the Banū Mūsā could then be applied to this type of solid. It is possibly for this reason that they felt it unnecessary to discuss the nature of the solid in detail.

In this group of propositions relating to the lateral area and volume of a sphere, we can find the same differences between the approaches of Archimedes and the Banū Mūsā that we saw in relation to the measurement of a circle. The first relates to the exhaustion method used. They begin by establishing the double inequality

$$\cos^2 \frac{\pi}{4n} < \sin \frac{\pi}{4n} \left(1 + 2 \sum_{k=1}^{n-1} \sin \frac{k\pi}{4n} \right) < 1.$$

Then, as we have explained, they go on to apply propositions from Book XII of the *Elements,* which enable them to avoid the requirement to evaluate the sine series already referred to 'to the limit'. Here, once again, they apply an apagogic method to the lateral areas rather than the volumes when determining the volume of a sphere. Finally, the volume of the sphere is not given in terms of another volume as in Archimedes – 'a cone with a base equivalent to the great circle of one of the spheres and a height equal to the radius of the sphere' – but as the product of two variables. These differences show that the Banū Mūsā were intent on exploring a different path to that of Archimedes in their search for the area of a circle, the surface area of a sphere and the volume of a sphere, while they were content to adopt Archimedes' method for approximating π.

We have seen that the Banū Mūsā found space in their book to address some of the classic problems of Hellenistic mathematics, especially the two famous problems found in Eutocius' commentary on *The Sphere and the Cylinder*: the two means and the trisection of angles.

1.2.5. *The two-means problem and its mechanical construction*

Proposition 16. — In seeking to determine two magnitudes X and Y lying between two given magnitudes M and N, the Banū Mūsā begin by describing the solution derived by 'one of the ancients whose name was Menelaus; he set forth it in one of his books on geometry'. They also highlight the usefulness of this method in the calculation of cubic roots. The book by Menelaus that best fits this description is *On the Elements of Geometry*, translated by Thābit ibn Qurra (*fī uṣūl al-handasa*) and quoted by al-Nadīm.[7] There are no known copies of this book currently in existence. It is also the case that the solution attributed by the Banū Mūsā to Menelaus is actually that which, according to Eutocius,[8] was attributed by Eudemus to Archytas. The task then, given two lengths M and N, is to find two lengths X and Y such that

$$\frac{M}{X}=\frac{X}{Y}=\frac{Y}{N}.$$

[7] Al-Nadīm, *al-Fihrist,* p. 327. Under the name of Menelaus is found a 'work on the elements of geometry, made by Thābit ibn Qurra, in three books' (*Kitāb uṣūl al-handasa, 'amalahu Thābit ibn Qurra, thalāth maqālāt*).

[8] *Archimidis Opera Omnia,* iterum edidit I.L. Heiberg, vol. 3 corrigenda adiecit E.S. Stamatis, Teubner, 1972, pp. 84–8.

If $M = 1$, and N is the volume of a cube, then X is the length of one of the edges of the cube.

Suppose that $M > N$ and construct a circle of diameter $AB = M$, a chord $AC = N$ and a tangent at B that cuts the straight line AC at G.

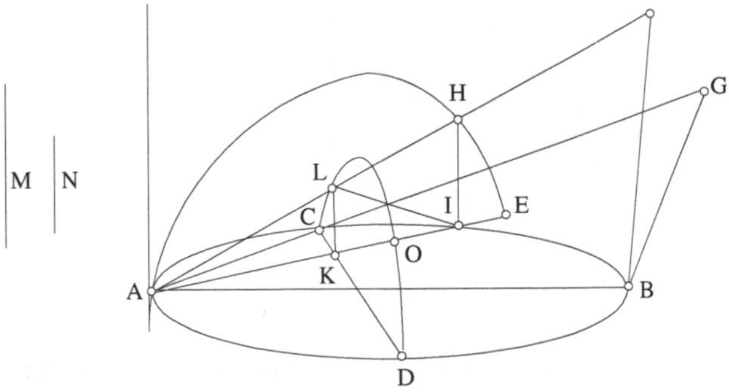

Fig. 1.2.14

Then consider a half cylinder of revolution standing on the semicircle ACB with its generators perpendicular to the plane ABC. A semicircle of diameter AB is drawn on the plane perpendicular to ABC passing through AB, and this is rotated about the axis Az ($Az \perp ABC$) to a position defined by the semicircle AHE at which position the straight line AE cuts the arc ACB at I and the semicircle AHE cuts the cylinder at H. IH is a generator of the cylinder. During the rotation, I describes the arc ACB and H describes a curve $\mathbf{C}$ on the surface of the cylinder.

The triangle ABG is now rotated about the line AB. The point C describes a semicircle COD and, at each position, the straight line AG cuts COD at a point L and cuts the cylinder at a point H'. During the rotation, H' describes a curve $\mathbf{C'}$ on the surface of the cylinder.

The semicircle AHE and the triangle ABG are fixed in such a position that $H = H'$. In this case $H \in \mathbf{C} \cap \mathbf{C'}$.

The intersection of planes COD and AHI is LK. We know that $LK \perp CD$ and $LK^2 = KC \cdot KD$ as CLD is a right-angled triangle. However, $KC \cdot KD = KA \cdot KI$ (power of the point K), and hence $LK^2 = KA \cdot KI$. The triangle ALI therefore has a right angle at L. The triangles AHE, AIH and ALI are all right-angled and similar; hence

$$\frac{AE}{AH} = \frac{AH}{AI} = \frac{AI}{AL}.$$

However, $AE = AB = M$ and $AL = AC = N$. We then have

$$X = AH \text{ and } Y = AI.$$

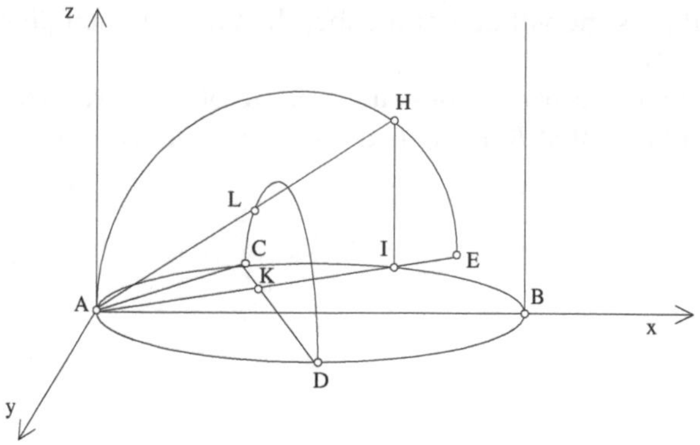

Fig. 1.2.15

In other words, the solution attributed to Menelaus is obtained from the intersection of a right cylinder: $x^2 + y^2 = ax$, a right cone: $b^2 (x^2 + y^2 + z^2) = a^2x^2$, and a torus: $x^2 + y^2 + z^2 = a\sqrt{x^2 + y^2}$ (where $a = M$ and $b = N$).

If $H (x_0, y_0, z_0)$ is the point of intersection, then

$$X = \sqrt{x_0^2 + y_0^2 + z_0^2} \text{ and } Y = \sqrt{x_0^2 + y_0^2}.$$

The Banū Mūsā point out, with good reason, the difficulty of constructing this solution and propose a mechanical method for this purpose. It has been claimed that this mechanical system was similar to one described by Eutocius under the name of Plato. It was nothing of the kind. We have already noted that this subtle and difficult-to-describe mechanism was omitted by Gerard of Cremona, and that it does not appear in the Latin translation.

The procedure is as follows:

Proposition 17. — Let A and B be the two given lengths and X and Y the two lengths to be found such that

$$\frac{A}{X} = \frac{X}{Y} = \frac{Y}{B}.$$

Let DC and DE be two straight perpendicular lines such that $DC = A$ and $DE = B$. The line perpendicular to CE and passing through E cuts DC at F, and the line parallel to EF extended through C cuts ED at M. Let U be a point on the extension to MC such that $MU = FE$.

We now define a movement of the line segment FE and a further movement of the line segment MU with the length of each line segment remaining constant:

F slides along the straight line *DC* towards *D*.
FE rotates about the point *E*.
Simultaneously, *MU* remains parallel to *FE*,
M slides along the straight line *ED*, moving away from *D*,
and *MU* rotates about *C*.

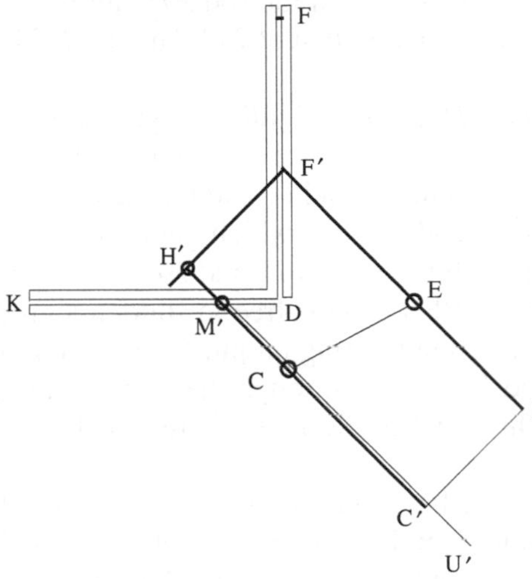

Fig. 1.2.16

The movement is halted as soon as the straight line perpendicular to *FE* at *E* cuts the straight line *MU* at point *U*. Let F_1E_1 and M_1U_1 be the positions of the two line segments at this stage. The figure $F_1E_1U_1M_1$ is a rectangle. The triangles CM_1F_1 and M_1F_1E are both right-angled triangles; therefore

$$M_1D^2 = DC \cdot DF_1 \text{ and } DF_1^2 = DM_1 \cdot DE,$$

from which

$$\frac{DC}{DM_1} = \frac{DM_1}{DF_1} = \frac{DF_1}{DE}.$$

However, $DC = A$ and $DE = B$; therefore DM_1 and DF_1 are the two line segments *X* and *Y* that were sought.

There still remains the question of an easy method of finding the two line segments DM_1 and DF_1. The Banū Mūsā introduced the point *H* defined by $CH = EF$ (*H* lies on the extension of *CM*). *FECH* is then a rectangle, and *H* moves along the straight line *DE* to point M_1 as *F* moves to F_1. It is therefore possible to imagine a mechanism that moves an arrangement of metal rods forming the figure *EFHC*.

The three rods *EF*, *CH* and *MU* have equal lengths l defined from the data as

$$l = \frac{A}{B}\sqrt{A^2 + B^2}.$$

The length of rod *EC* is $\sqrt{A^2 + B^2}$, and the rod *FH* can have any arbitrary length provided it is at least equal to that of *EC*. The rod *EC* is the only one to be fixed.

The two rods *EF* and *FH* are held rigidly at right angles, and the point *F* is fitted with a pin, the tip of which moves along the straight line *FD*. Pins are placed at the two fixed points *E* and *C*, and the head of each pin carries a ring that is free to rotate, and through which passes one of the moveable rods. Rod *EF* passes through the ring at *E*, and *HC* passes through *C*. The rod *MU* is thinner than the others and is free to slide in a groove along the top of rod *HC*, passing through the ring on pin *C*. A ringed pin is attached to rod *MU* at *M*. The rod *HC* passes through the ring and the tip of the pin moves along the straight line *DK*. Rod *FH* is free to slide through a ring attached to rod *HC* at *H*.

A flat base could than be placed under the plane of the moveable rectangle *HFEC* with pins *E* and *C* securely attached to this base and two slides provided for the moveable pins *F* and *M*. The slide *FD*, for example, could consist of two parallel guides placed either side of the straight line *FD*, with a similar arrangement provided for *MK*.

The system of articulated rods could then be fitted to the base, with the rods *FE* and *HC* passing through the rings on pins *E* and *C* respectively, and pins *F* and *M* placed in the appropriate slides.

Comments.

1) The diagram in Fig. 1.2.16 shows an intermediate position of the moveable rectangle, *i.e.* $E'F'H'C'$. The thin rod sliding along the top of *HC* does not appear to be necessary.

2) The movement is halted when the point H' arrives at point M', $H' = M' = M_1$. At this stage, $C' = U_1$.

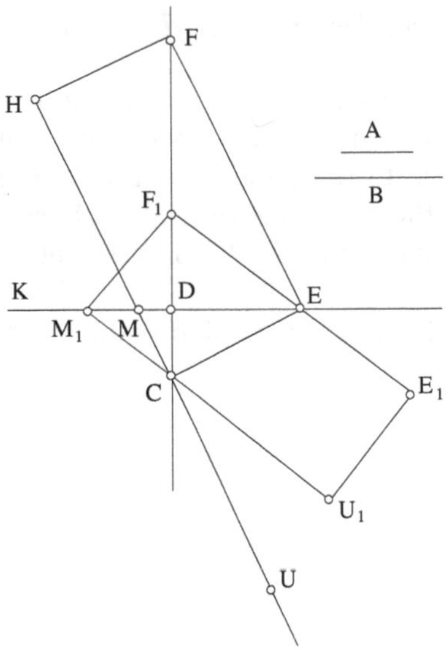

Fig. 1.2.17

3) As the two sides of the right angle *CDE* are known (*CD* = *A*, and *DE* = *B*), Problem 17 becomes the determination of *F* on the extension to *CD* and *M* on the extension to *ED* such that triangle *ECM* has a right angle at *C* and triangle *MFE* has a right angle at *F*.

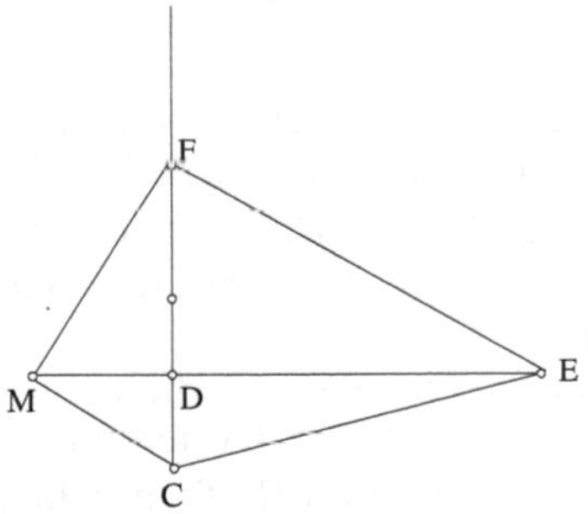

Fig. 1.2.18

The problem was discussed by Plato,[9] but the mechanical apparatus attributed to Plato and that described here by the Banū Mūsā are *different*.

[9] *Archimidis Opera Omnia*, pp. 56–9.

1.2.6a. *The trisection of angles and Pascal's* Limaçon

Proposition 18. — In this proposition, the Banū Mūsā return to the problem of the trisection of angles, but only to present their own solution and a mechanical device to trace the trisecting curve. This curve is the conchoid of a circle, the same curve that Roberval[10] called Pascal's *Limaçon*. The solution is obtained by finding the intersection of this spiral curve with a half-line.

In the original text, the Banū Mūsā address the problem as follows:

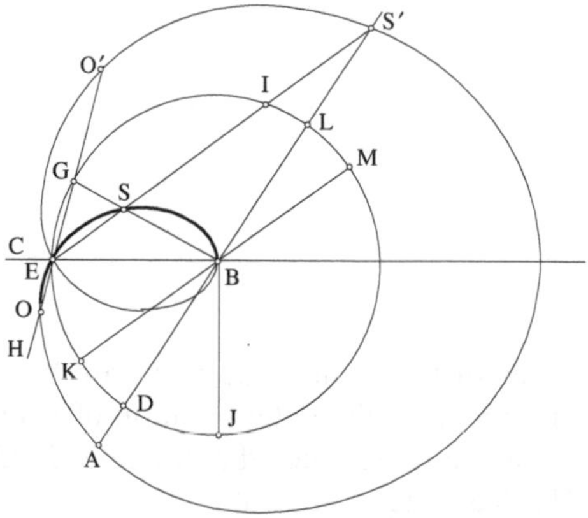

Fig. 1.2.19

Let ABC be an acute angle, and let a circle of centre B cut BA and BC at points D and E respectively. Let $BG \perp BD$. Let GH be a half-line joining G, E and O on GH such that $GO = BD$. Imagine that the straight line GH moves as follows: the line continues to pass through E while G describes a circle in the direction of L.

If I is the position reached by point G when point O reaches the straight line BG, then $IS = IB = BE$. If the diameter is now moved such that $KM \parallel EI$, we have $SI \parallel MB$ and $SI = MB$, from which $IM \parallel SB$ or $IM \perp BL$

[10] Roberval, 'Observations sur la composition du mouvement et sur les moyens de trouver les touchantes des lignes courbes', in *Mémoires de l'Académie Royale des Sciences*, ed. 1730, vol. 6, pp. 1–79, a course by Roberval edited by his pupil François du Verdus. See also P. Dedron and J. Itard, *Mathématiques et mathématiciens*, Paris, 1959, pp. 400–1, in which the text by Roberval is cited. According to P. Tannery, E. Pascal designed this curve as a conchoid of a circle in 1636–1637; see *Mémoires scientifiques*, vol. 13, pp. 337–8. See also M. Clagett, *Archimedes in the Middle Ages*, vol. 1, Appendix VI entitled 'Jordanus and Campanus on the trisection of an angle', pp. 666–70.

and BL is the bisector of angle IBM. Therefore the arcs $\widehat{IL} = \widehat{LM} = \widehat{DK}$, and $\widehat{IM} = \widehat{KE}$; therefore $\widehat{KE} = 2\widehat{KD}$. The straight line BK is therefore the line being sought.

$$D\hat{B}K = \frac{1}{3} D\hat{B}E.$$

If $A\hat{B}C$ is obtuse, then we draw the bisector and take the third of its half. Two thirds of this half gives a third of the obtuse angle.

Comment 1. — As the point G describes the arc GL, the associated point O ($GO = R$) describes an arc of a conchoid and S is the intersection of this with the straight line GB.

In other words, the point S lies on both the spiral and on the straight line BG. The equation of BG in polar coordinates relative to a pole at point E may be written as follows:

$$\rho = \frac{a \cos \alpha}{\cos (\theta - \alpha)}, \qquad \text{where } a = BE \text{ and } \alpha = D\hat{B}C.$$

The equation of thc spiral may be written as

$$\rho = a\,(2 \cos \theta - 1).$$

The coordinates of point S are (ρ, θ); hence

$$\frac{\cos \alpha}{\cos (\theta - \alpha)} = 2\cos\theta - 1.$$

Now, $\theta = \frac{2\alpha}{3}$ is a solution of this equation. Therefore angle BES equals $\frac{2}{3}$ α, angle $BS'E$ equals $\frac{1}{3}\alpha$, and angle DBK equals $\frac{1}{3}\alpha$.

The angle ABC has therefore been trisected using the intersection of an arc of the conchoid of circle ($GO = GO' = R$) with the half-line BG.

The Banū Mūsā then go on to describe a mechanical device to trace Pascal's *Limaçon*. Their device consists of a circular groove in which is placed a ringed pin at point E. A rod is passed through the ring and a pin is attached to one end at G. This pin is free to move in the circular groove. A second pin is fixed at a point O on this rod, where $GO = R$. This pin is used to trace the arc of the conchoid. The intersection of this arc with the perpendicular BG gives the point S that was sought.

If the user wishes to trace the entire conchoid, an additional rod is required that extends beyond G by a length equal to $GO = R$.

This part of the conchoid may also be used to trisect the angle DBE. If IS is extended to point S' on the conchoids, then $IS = IS' = IB = R$, and hence SBS' is a right angle at point B. Point S' is the intersection of the conchoids with the straight line BD.

Comment 2. – In Proposition 8 of the book of lemmas,[11] attributed to Archimedes, the author draws a chord AB from a point A on a circle of centre D, which is then extended to a point C such that $BC = AD = R$. The straight line CD cuts the circle at E and F, and we have $\widehat{AE} = 3\widehat{BF}$, thus completing the discussion of the trisection of angles.

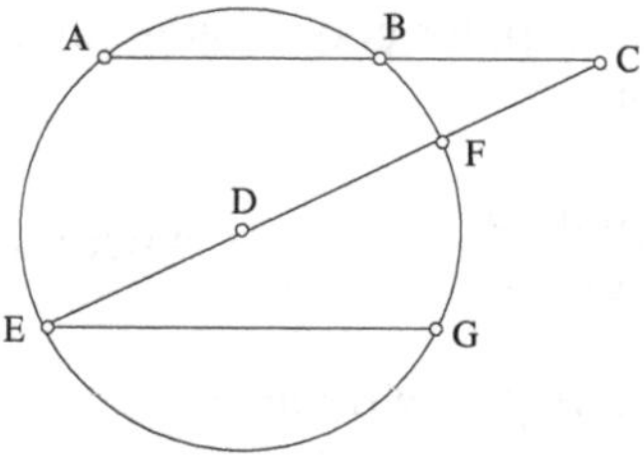

Fig. 1.2.20

This proposition, which could have been originated by Archimedes, is also associated with a conchoid: as point B describes an arc on the given circle, point C describes an arc of the external conchoid of this circle.

Could the Banū Mūsā have been inspired by an Arabic translation of this text? Given our present state of knowledge, we cannot answer this question with any degree of certainty. There is a difference between the two discussions in that the Banū Mūsā use an arc of the internal conchoid of the circle, while the text attributed to Archimedes implies the use of an arc of the external conchoid of the circle.

Comment 3. — While the origins of the solution to this problem proposed by the Banū Mūsā remain obscure, its path can clearly be traced in later works. Their solution was copied in the *Liber de triangulis*.[12] It should also be noted that, according to Roberval, Etienne Pascal designed his *Limaçon*

[11] *Archimidis Opera Omnia*, Liber assumptorum, vol. 3, p. 518; Archimède, transl. Mugler, vol. 3, pp. 148–149.

[12] M. Clagett, *Archimedes in the Middle Ages*, vol. 5, Philadelphia, 1984, pp. 146–7; 297 *sqq.* and especially 324–5.

in the same way – as a conchoid of a circle, and that he also applied it to the trisection of angles.

1.2.6b. *Approximating cubic roots*

The Banū Mūsā end their treatise with a discussion of an approximation of the cubic root of an integer N in which they give an expression equivalent to

$$\sqrt[3]{N} = \frac{1}{60^k} \sqrt[3]{N \,.\, 60^{3k}},$$

from which an approximation of the cubic root of N can be determined within an accuracy of the order of k.

1.3. *Translated text*

Banū Mūsā

On the Knowledge of the Measurement of Plane and Spherical Figures

In the Name of God, the Merciful, the Compassionate

THE BOOK OF THE BANŪ MŪSĀ, MUḤAMMAD, AL-ḤASAN AND AḤMAD

On the knowledge of the measurement of plane and spherical figures

In eighteen propositions

Introduction to the book

The length is the first of all the magnitudes[1] that define[2] figures,[3] and it is what extends along a straight line in both directions simultaneously,[4] and from that which is extended only length can be obtained. If a surface[5] is extended in a direction other than the length, then this extension is the breadth.[6] The breadth is not, as many believe, the line which surrounds the surface in a direction other than the length. If this were the case, then the

[1] 'The first of all the magnitudes' is a translation of the Arabic *awwal al-aqdār*, translated by Gerard as *prima quantitatum* ('quantities' in Clagett's translation, p. 238, 34). We prefer the term 'magnitude' in order to distinguish between *ʿiẓam* (dimension) and *kammiya* (quantity).

[2] We have translated *yaḥuddu* as 'define'. Gerard translated it as *terminare* ('delimit' in Clagett's translation), which also means 'to define'.

[3] *Al-ashkāl*; Gerard translated it as *corporis* ('body' in Clagett's translation), giving it a more concrete sense than that expressed here.

[4] Reading the corresponding Arabic text, one could conjecture that a *saut du même au même* has occurred. This should then read:

... وهو ما امتد على استقامة في الجهتين جميعًا <وما امتد على استقامة في الجهتين جميعًا> فإنه لا يكون ...

and we then have: '... extends along a straight line in both directions simultaneously, and from that which is extended along a straight line in both directions simultaneously, only length can be obtained ...'. This could also easily be simply an abridgement, albeit an ambiguous one, in which the pronoun in *fa-innahu* refers to *mā*. Gerard has translated this text as *Et longitudo est prima quantitatum que terminant illud. Et est illud quod extenditur secundum rectitudinem in duas partes simul. Nam non fit ex eo nisi longitudo tantum* (pp. 240–2, 34–36).

[5] *Al-saṭḥ*. Here, Gerard translates *ṭūl* as *longitudo* (p. 240, 36).

[6] Gerard also includes the phrase *Et tunc provenit superficies* (pp. 242, 38–39) which is missing in al-Ṭūsī's text. It should be noted that this phrase does not appear in the remainder of either the Latin or the Arabic text.

surface would not have a length and breadth alone[7] and the breadth would also be a length, as for them the breadth is a line and a line is a length.[8]

Euclid was correct in this regard when he stated: A line is only length, and a surface is that which has only length and breadth. As for depth, this is an extension[9] in a direction other than those of the length and the breadth. However, those who believe that the breadth is a line also believe that the depth is a line. They are as wrong in one as in the other.[10] *These three

[7] See the French edition, Supplementary note p. 1029.

[8] It is clear that al-Ṭūsī has severely edited the introduction, removing anything that appeared to him to be non-mathematical, including all the historic and theoretical sections in which the Banū Mūsā explained the reasons that led them to write this treatise. The removed text runs to some thirty lines in the Clagett edition of the translation by Gerard of Cremona (pp. 238–40, 4–34). This is Clagett's translation of the Latin text: 'Because we have seen that there is fitting need for the knowledge of the measure of surface figures and of the volume of bodies, and we have seen that there are some things, a knowledge of which is necessary for this field of learning but which – as it appears to us – no one up to our time understands, and that there are some things we have pursued because certain of the ancients who lived in the past had sought understanding of them and yet knowledge has not come down to us, nor does any one of those we have examined understand, and that there are some things which some of the early savants understood and wrote about in their books but knowledge of which, although coming down to us, is not common in our time – for all these reasons it has seemed to us that we ought to compose a book in which we demonstrate the necessary part of this knowledge that has become evident to us. And if we consider some of those things which the ancients posed and the knowledge of which has become public among men of our time but which we need for the proof of something we pose in our book, we shall merely call it to mind and it will not be necessary for us in our book to describe it [in detail], since knowledge of it is common; for this reason we seek only a brief statement. On the other hand, if we consider something which the ancients posed and which is not well remembered nor excellently known but the explanation of which we need in our book, then we shall put it in our book; relating it to its author. It will be evident from what we shall recount concerning the composition of our book that one who wishes to read and understand it must be well instructed in the books of geometry in common usage among men of our time. The common property of every surface is the possession of length and breadth alone, while the property of a corporeal figure is the possession of length, breadth, and height. Length, breadth, and height are quantities which delimit the magnitude of every body.'

[9] *Imtidād*. In the Latin text, we have *extensio superficiei*, which indicates that he was translating the Arabic *imtidād al-saṭḥ*. The final part of the phrase is *scilicet extensio eius in altum* (p. 242, 47–48), which could be translation of *a'anī imtidādan fī al-irtifā'*.

[10] In the Latin translation, this is followed by *Iam ergo ostensum est quid sit longitudo et quid latitudo et quid altitudo* (p. 242, 51–53), which is likely to be a translation of the Arabic: *fa-qad tabayyana idhān mā al-ṭūl wa-mā al-'arḍ wa-mā al-samk*, which indicates that this section has been summarized by al-Ṭūsī.

magnitudes define the dimension of every body and the extension of every surface. The procedure for estimating their quantities is based on the unit plane and the unit solid*.[11]

The unit plane used to measure a surface is a surface whose length is one, whose breadth is one and whose angles are right angles. The unit solid, used to measure a solid, is a solid whose length is one, whose breadth is one, and whose depth is one, and wherein each of the surfaces is at right angles to the others. The magnitude used to measure surfaces and solid bodies requires that its parts be brought together, one against the other, in such a way as to leave no void, without filling the surface or body. It also requires there to be an obvious distinction between that which has been measured completely and that which has not been measured. There is no more effective method of obtaining this distinction than <to ensure that> the unit rule used to make the measurements is the same, regardless of whether it is marked with units taken singularly or repeated,[12] so that the effort required to distinguish that which has been measured from that which has not should be the same in all cases. This exists in no other figure than the quadrilateral as, if a quadrilateral is doubled, only its quantity is changed, and its squareness remains.[13] However, of all the quadrilateral figures <with equal perimeters>, that with angles that are right angles is the largest.[14] *It is for this reason that we propose this as a measure, and no other*.[15]

[11] *...*: As this section of text is longer in the Latin translation, this seems to be the translation by al-Ṭūsī. This section therefore contains an idea that is missing in the Arabic, that a fourth magnitude is not required in order to define a body. *Et declaratur iterum quod non est aliquid corporum indigens quantitate alia quarta qua eius magnitudo terminetur* (pp. 242–3, 54–56).

[12] 'Taken singularly or repeated' (*fī afrādihi wa-fī taḍā'īfihi*), given in Latin as *in singularitate sua in sua duplatione* (pp. 244–6, 76–77).

[13] There is an entire paragraph in Latin (p. 246, 79–88) that does not appear in Arabic.

[14] Lit.: has the greatest perimeter.

[15] *...*: The Latin text contains a number of lines (p. 246, 91–95) to describe the same idea: *Iam ergo manifestum est propter quam causam ponitur quadratum orthogonium ex superficiebus et corporibus esse quantitas qua comparantur superficies et corpora. Et ita verificatur sermo in eo cuius narrationem voluimus in hoc nostro libro. Incipiamus ergo nunc narrare illud quod volumus.*

The propositions[16]

– **1** – For any polygon[17] circumscribed around a circle, the product of the half-diameter of this circle and half the sum of the sides of this polygon is its area.

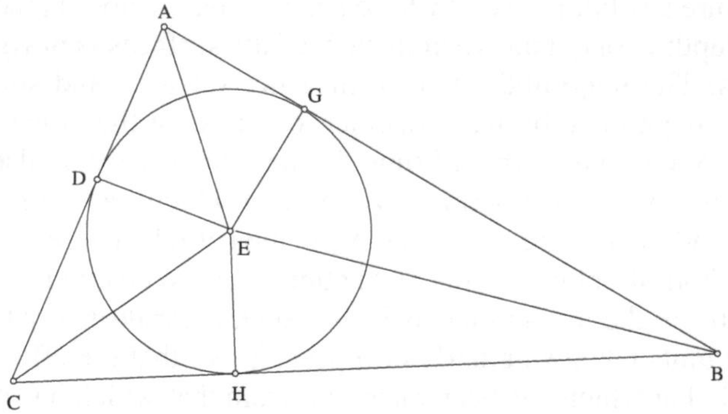

Fig. I.1

Let[18] the figure *ABC* be circumscribed around the circle *DHG* whose centre is at *E* and whose half-diameter is *EH*.[19] Let us join *EA*, *EB* and *EC*.[20] It is clear that *EH* is the height of the triangle *EBC* and that the product of *EH* and half *BC* is the area of the triangle *EBC*. The same rule may be applied to the two triangles *AEB* and *AEC*.[21] The <product of the>

[16] *Al-ashkāl*: missing in the Latin text.

[17] A regular polygon is implied here, as throughout the text. We shall not indicate any further occurrences.

[18] As is usual in a mathematical exposition, the line in the Latin text begins: *Verbi gratia*, a translation of the Arabic *mithāl dhālika*. Al-Ṭūsī omitted this expression throughout the text. We shall not indicate any further occurrences.

[19] The Latin version continues with this expression: *Dico ergo quod multiplicatio linee* EH *in medietatem omnium laterum figure* ABG *est embadum superficiei* ABG. *Cuius hec est demonstratio* (p. 248, 7–9), which is a translation of the Arabic:

فأقول: إن سطح خط هـ ح في نصف جميع أضلاع مضلع ا ب جـ هو مساحة شكل ا ب جـ. برهان ذلك: أن...

This type of expression has been omitted by al-Ṭūsī in his edition. We shall not indicate any further occurrences.

[20] 'Let us join *EA*, *EB* and *EC*'. The corresponding Latin text is *Protraham duas lineas* BED, GEZ (p. 248, 10).

[21] The preceding phrase appears to be a summary of a longer one, translated by Gerard as: *Et per huiusmodi proprietatem sciemus quod multiplicatio medietatis diametri circuli* ZDH *in medietatem linee* AB *aut in medietatem linee* AG *est embadum duorum triangulorum* GEA, AEB. *Et illud est quod declarare voluimus* (p. 248, 12–15).

half-diameter of the circle by half the sum of the sides is therefore the area of the triangle *ABC*.

We know in a similar manner that for any solid circumscribed around a sphere, the product of the half-diameter of the sphere and one third of the area of the surface of the circumscribed solid is equal to the volume of the solid, and that this volume is always greater than the volume of the sphere.[22]

I say that this can be shown by imagining the solid divided into pyramids whose vertices are the centre of the sphere, and whose bases are the bases of the solid, and which are arranged such that a half-diameter of the sphere is perpendicular to the base*[23] *of each of them. The volume of the solid is therefore equal to the volume of these pyramids.[24]

– **2** – For any polygon inscribed within a circle, the product of the half-diameter of this circle and half the sum of the sides of this polygon is less than the area of the circle.

Let a triangle be inscribed within the circle *ABC*, and let *E* be the centre. We join *EB* and *EC*; let *ED* be perpendicular to *BC*. We produce it to *G* and we join *BG* and *CG*. The product of *EG* and half of *BC* is equal to the area of the two triangles *EBC* and *GBC*; *this area is less than the area of the sector *EBGC* and greater than the area of the triangle *EBC**.[25] *We shall show that the same applies to the remainder of the figure, and we shall show that the area of the circle is much greater than the area of the triangle *ABC*.*[26]

ونعلم من مثل ذلك أن سطح هـ ح في نصف اب هو مساحة مثلث اهـ ب وأن سطح هـ ح في نصف از هو مساحة مثلث ب هـ ز.

[22] The Latin translation gives *corporis* (p. 248, 21). The section ends with the expression used in this case: *Et illud est quod declarare voluimus* (*ibid.*), which is a translation of the Arabic *wa-dhālika mā aradnā an nubayyin*, omitted by al-Ṭūsī. We shall not indicate any further occurrences.

[23] Lit.: their bases.

[24] Understood to mean: to the sum of the volumes of the pyramids.

…: This is a long commentary by al-Ṭūsī, clearly indicated by the introductory 'I say …'.

[25] *…*: This expression is missing in the Latin text. It should also be noted that the Latin text uses letters for the geometric figures that are different from those used in the Arabic text.

[26] *…*: In the Latin text: *Et per modum similem huic scitur quod multiplicatio medietatis diametri circuli* ABG *in medietatem laterum* AG, BG, AB *est minor embado circuli* ABG. *Iam ergo declaratum est quod multiplicatio medietatis diametri circuli* ABG <*in medietatem omnium laterum figure*> *est minor embado circuli* ABG. *Iam ergo ostensum est quod multiplicatio medietatis diametri circuli in medietatem*

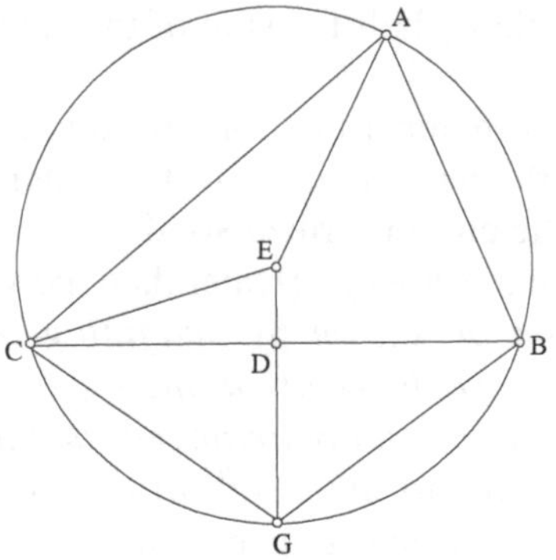

Fig. I.2

We know in a similar manner that the solid inscribed within a sphere is such that the product of the half-diameter of the sphere and one third of the surface area of the solid is less than the volume of the sphere.

– **3** – Consider a segment of a straight line and a circle. If the segment is less than the circumference of the circle, then it is possible to inscribe within the circle a polygon, the sum of whose sides is greater than this segment. If the segment is longer than the circumference of the circle, then it is possible to circumscribe a polygon around the circle, the sum of whose sides is less than the segment.

Let the circle be *ABC* and the segment *HF*, which is firstly shorter than the circumference of <the circle> *ABC*. Let the circumference of circle *DGE* be equal to the segment *HF*. If a polygon is inscribed within the circle *ABC* without touching the circumference of *EDG*,[27] then the sum of its sides is greater than the circumference of *EDG*, *i.e.* the segment *HF*.[28]

---

omnium laterum figure est minor embado circuli (p. 250, 12–19), which is a fairly close translation of the Arabic:

وبمثله نبين أن سطح نصف قطر دائرة ا ب جـ في نصف جميع أضلاع ا جـ ب جـ ا ب أقل من مساحة دائرة ا ب جـ. فقد تبين أن سطح نصف قطر دائرة ا ب جـ في نصف جميع أضلاع المضلع الذي تحيط به الدائرة أقل من مساحة الدائرة.

It is very likely that al-Ṭūsī considered this to be too long and summarized it in a single phrase, leaving the explanation to the reader.

[27] See Euclid, *Elements*, XII.16.

[28] The Latin version reads: *Sed linea* EDZ *est equalis linee* HU. *Iam ergo ostensum est quod possibile est ut faciamus in circulo* ABG *figuram lateratam et angulosam et latera eius agregata sint longius linea* HU. *Et illud est quod declarare voluimus* (p. 254, 19–23), which is doubtless a translation of the Arabic:

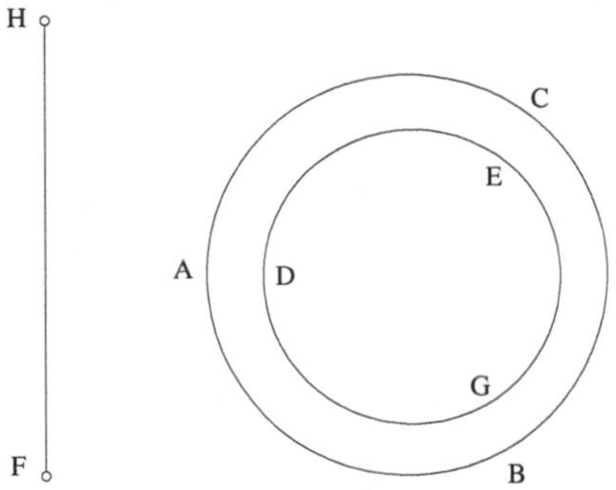

Fig. I.3

Now, let the circle be *EDG* and a segment *HF* be longer than its circumference, and let the circumference of *ABC* be equal to the segment *HF*. If a polygon is inscribed within the circle *ABC* without touching the circumference of *EDG*, then the sum of its sides will be less than the circumference of *ABC*, *i.e.* the segment *HF*. If a similar polygon to that mentioned is then circumscribed around and touching[29] the circle *EDG*, then the sum of its sides will be much less than the segment *HF*. This is what we required.[30]

I say that this is founded upon the existence of a circle whose circumference is equal to any given segment. This has not been proved anywhere else.[31]

– **4** – For any circle, the product of the half-diameter and half the circumference is its area.

Let the circle be *ABC*, its centre *E* and *EC* the half-diameter. If the product of *EC* and half the circumference of *ABC* is not equal to the area of the circle, then the product of *EC* and a straight line either longer than half the circumference of *ABC* or shorter than it is equal to its area.

لكن خط هـ د ز مساوٍ لخط ح و، فقد تبين أنه يمكن أن يعمل في دائرة ا ب جـ مضلع ويكون جميع أضلاعه أطول من خط ح و؛ وذلك ما أردنا أن نبين.

[29] See commentary.

[30] The Latin text continues: *Et hec est forma figure* (p. 254, 36), which is a translation of the Arabic: *wa-hādhihi ṣūra al-shakl*, omitted by al-Ṭūsī.

[31] *…*: Commentary by al-Ṭūsī.

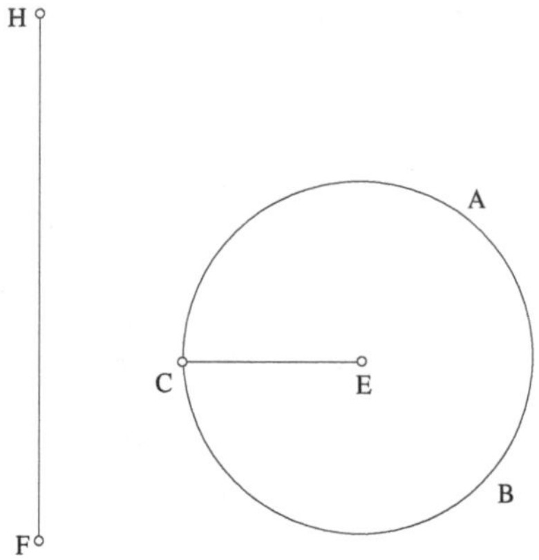

Fig. I.4

Firstly, let the product of *EC* and a straight line shorter than half the circumference of *ABC* be equal to the area of the circle. Let this straight line be the straight line *HF*. Twice *HF* is therefore less than the circumference of *ABC*. It is possible to inscribe a polygon within the circle *ABC* such that the sum of its sides is greater than twice *HF* and half of that is longer than *HF*.[32] <The product> of the half-diameter *EC* and half the sum of the sides of this polygon is less than the area of the circle.[33] The product of *EC* and *HF* is therefore much less than the area of the circle. But it is equal, therefore this is contradictory.

Now, let the product of *EC* and a straight line longer than half the circumference of *ABC* be equal to the area of the circle.[34] Let this straight line be *HF*. Twice *HF* is therefore longer than the circumference of the circle. It is possible to circumscribe a polygon around the circle *ABC* such that the sum of its sides is less than twice *HF* and half of that is less than *HF*. The product of the half-diameter *EC* and half the sum of the sides of this polygon is greater than the area of the circle. The product of *EC* and *HF* is therefore much greater than the area. But it is equal; therefore this is contradictory. The product of *EC* and half the circumference of *ABC* must therefore be equal to the area of the circle *ABC*. This is what we required.

[32] From Proposition 3.

[33] From Proposition 2.

[34] *i.e.* the circumference. The Latin text continues: *Ergo multiplicatio linee* EG *in lineam* HU *est embadum circuli* ABG (p. 258, 12–13), which is a translation of the Arabic:

فيكون سطح خط هـ ج في خط ح و مساويًا لدائرة ا ب ج

It becomes clear from this that the product of the half-diameter and half of any given arc is equal to the area of the sector enclosed by this arc and the two half-diameters passing through its two extremities.

– **5** – For any circle, the ratio of the diameter to the circumference is the same.

Let the two circles *ABC* and *DEG* be different, and let *BC* be the diameter of *ABC* and *DE* the diameter of *DEG*.

If this is not as we have stated,[35] then let the ratio of *BC* to the circumference of *ABC* be equal to the ratio of *DE* to *HF*, with *HF* being either longer than the circumference of *DEG* or shorter than it.

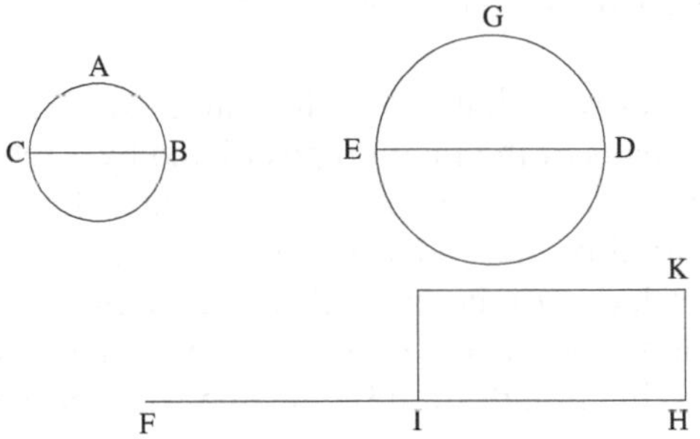

Fig. I.5

We firstly assume that it is shorter. We divide *HF* into two halves at *I*. Let the perpendicular *HK* to *HF* be equal to half of *DE*. We now complete the surface *KI*.[36] The surface *KI* is therefore less than the area of the circle *EDG*. However, the ratio of *KH* to *HI* is equal to the ratio of half of *BC* to half the circumference of *ABC*, and the product of *KH* and *HI* is the area of *KI*, and the product of half of *BC* and half of the circumference of *ABC* is

[35] In Gerard's translation, after the expression 'I say that …', omitted as usual by al-Ṭūsī together with the word 'Proof', he continues: *Si non fuerit proportio amborum una, tunc …* (p. 260, 9), which is probably a translation of the Arabic: *fa-in lam takun al-nisbatayn wāḥida, fa-…*

[36] The Latin text then continues: *Et quoniam linea* HK *est equalis medietati linee* EZ, *et linea* HT *est brevior medietate linee* DZE, *erit quadratum* KT *minus superficie circuli* DEZ (p. 262, 15–17). It can be seen that al-Ṭūsī has almost certainly omitted the intermediate step: 'As the straight line *HK* is equal to half the line *DE*, and the straight line *HI* is less than half the line *DGE*, then the area *KI* is less than the area of the circle *DEG*', which is a translation of the Arabic:

ولأن خط ح ک مساوٍ لنصف خط د هـ وخطاً ح ط أصغر من نصف خط د ز هـ يكون سطح ک ط أصغر من مساحة دائرة د هـ ز.

the area of the circle ABC. The ratio of the surface KI to the circle ABC is therefore equal to the square of the ratio of KH, that is half of DE, to half of BC, which is the square of the ratio of DE to BC. Now, Euclid has shown that the square of the ratio of DE to BC is equal to the ratio of the circle DEG to the circle ABC.[37] Therefore, the ratio of the surface KI to the circle ABC is equal to the ratio of the circle DEG to this circle, and the surface KI is therefore equal to the circle DEG. Now, it was less, so this is contradictory. The straight line HF is therefore not less than the circumference of DEG.

Using a similar procedure, we can show that it is also not longer. It therefore follows that the ratio of DE to the circumference of DEG is equal to the ratio of BC to the circumference of ABC, and that this holds for any other pair of circles. This is what we required.

– **6** – Let us now calculate[38] the ratio of the diameter to the circumference by means of the method postulated by Archimedes. No other method discovered by any other person has come down to us, up to the present day. While this method does not lead to a knowledge of the magnitude of one relative to the other that is exactly the true magnitude, it does allow the magnitude of one relative to the other to be determined to the degree of approximation desired by the one who seeks it.[39]

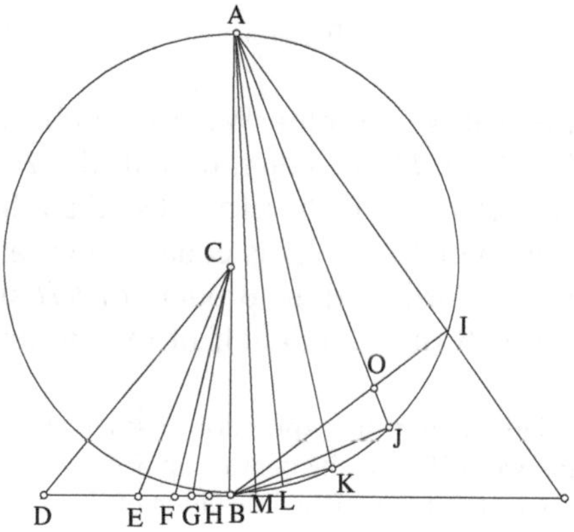

Fig. I.6

[37] Euclid, *Elements*, XII.2.

[38] Lit.: Let us show. The translation by Gerard of Cremona uses the verb *ostendere*.

[39] The Latin translation includes a section in which the Banū Mūsā return to the same idea, but expressed differently (see p. 262, 15–25).

In order to show this,[40] let a circle be AIB, its diameter AB and the centre C. We draw a straight line CD from C which includes with CB a third

[40] This, in similar terms, is the procedure followed in the first part: Let BD be a tangent to the circle at B. Beginning with the angle at the centre $B\hat{C}D = \frac{1}{3} \times \frac{\pi}{2}$, and taking successively the half, quarter, eighth and sixteenth, we have $B\hat{C}H = \frac{1}{48} \times \frac{\pi}{2} = \frac{4}{192} \times \frac{\pi}{2}$. Therefore, BH is half the side of the regular polygon with 96 sides circumscribed around the circle. $BH = \frac{1}{2} C_{96}$. In the triangle CBD, we have $DB = \frac{1}{2} CD$ and $CB^2 = \frac{3}{4} CD^2$. Now, as E is the foot of the bisector, we can write

$$(1) \qquad \frac{ED}{EB} = \frac{CD}{CB} \Leftrightarrow \frac{EB+ED}{EB} = \frac{CB+CD}{CB} \Leftrightarrow \frac{CB+CD}{DB} = \frac{CB}{EB}.$$

If we let $CD = 306$ and $BD = 153$, then $CB = 265.0037736 > 265$, which is a good approximation, as can be seen. This leads to

$$CB + CD > 571 \text{ and, from (1), } \frac{CB}{EB} > \frac{571}{153}.$$

The Banū Mūsā go on to link the segments and the numbers: If $EB = 153\,u$ (where u is the unit, of which $EB = 153$), then $CB > 571$ and $CB^2 + EB^2 = CE^2$, hence $CE > 591 + \frac{1}{8}$. Similarly, in triangle CEB, we have

$$(2) \qquad \frac{CE+CB}{EB} = \frac{CB}{FB} \Rightarrow \frac{CB}{FB} > \frac{1162+\frac{1}{8}}{153};$$

if $FB = 153\,u$ (where u is the unit, of which $FB = 153$), then $CB > 1\,162 + \frac{1}{8}$ and $CF > 1\,172 + \frac{1}{8}$. Similarly, in triangle CFB, we have

$$(3) \qquad \frac{CF+CB}{FB} = \frac{CB}{GB} \Rightarrow \frac{CB}{GB} > \frac{2334+\frac{1}{4}}{153},$$

and, if $GB = 153$, we have $CB > 2\,334 + \frac{1}{4}$ and $CG > 2\,339 + \frac{1}{4}$. Similarly, in triangle CGB, we have

$$\frac{CG+CB}{GB} = \frac{CB}{HB} \Rightarrow \frac{CB}{HB} > \frac{4673+\frac{1}{2}}{153}.$$

Now, CB is the half-diameter and HB is half of one of the sides of C_{96}. Therefore, if P_{96} is the perimeter of the polygon with 96 sides, then

$$\frac{P_{96}}{d} < \frac{96 \times 153}{4673+\frac{1}{2}} < 3 + \frac{1}{7}.$$

of a right angle, and we draw a perpendicular *BD* from *B* on to *CB*. The arc intercepted by the angle *BCD* is a half-sixth of the circle *AIB* and the straight line *BD* is half the side of a hexagon circumscribed around the circle *AIB*. We now divide the angle *BCD* into two halves by means of a straight line *CE*, we divide the angle *BCE* into two halves by a straight line *CF*, divide the angle *BCF* into two halves by the straight line *CG* and we divide the angle *BCG* into two halves by the straight line *CH*. It is clear that the arc intercepted by the angle *BCH* is one part of 192 <parts> around the circumference of *AIB* and that the straight line *BH* is half the side <of a polygon> with ninety-six sides circumscribed around the circle *AIB*. Let us assume in order to facilitate the procedure, as we have shown, that *CD* is 306, and its square will be 93,636. *BD* is 153 as the angle *BCD* is one third of the angle *CBD*, which is a right angle, and the square of *BD* is 23,409, and the square of *CB* is 70,227. The straight line *CB* is therefore longer than 265. However, the ratio of *BC* and *CD*, which together make up *BD*, is equal to the ratio of *CB* to *BE* as *CE* bisects the angle *BCD*. But *BC* and *CD* together are greater than 571 and *BD* is equal to 153. Therefore, the ratio of *CB* to *BE* is greater than the ratio of 571 to 153. Therefore, as the magnitude of *BE* is 153, *CB* is greater than 571, its square is greater than 326,041, the square of *BE* is 23,409, and the square of *CE* is greater than 349,450. The straight line *CE* is therefore longer than 591 and one eighth.

Similarly, we can show that the ratio of *CB* to *BF* is greater than the ratio of 1,162 and one eighth to 153. If *BF* is 153, then *CB* is greater than 1162 and one eighth, its square is greater than 1,350,534,[41] the square of *BF* is 23,409 and the square of *CF* is greater than 1,373,943.[42] Therefore, the straight line *CF* is longer than 1,172 and one eighth.

Similarly, we can show that the ratio of *CB* to *BG* is greater than the ratio of 2,334 and a quarter to 153. If *BG* is 153, then *CB* is greater than 2,334 and a quarter, its square is greater than 5,448,723, the square of *BG* is 23,409 and the square of *CG* is greater than 5,472,132. The straight line *CG* is therefore longer than 2,339 plus one quarter.

Similarly, we can show that the ratio of *CB* to *BH* is greater than the ratio of 4,673 and a half to 153. If the straight line *BH* is 153, then *CB* will be greater than 4,673 plus a half. This is the magnitude of <the ratio of> the side <of the polygon> of ninety-six sides to the diameter. The magnitude of <the ratio of> the diameter to the sum of the sides <of the polygon> with ninety-six sides circumscribed around the circle is greater than the magnitude of <the ratio of> 4,673 plus a half to 14,688. <The ratio of the magnitude of the sum of the sides of the polygon with ninety-six sides

[41] Gerard's translation includes '*et quarta*'.
[42] *Idem*.

circumscribed round the circle to the diameter is therefore less than the magnitude of the ratio of 14,688 to 4,673 plus a half>, which is less than three plus one seventh part of unity.[43]

We then draw[44] the one-sixth chord in the circle AIB, that is IB, and extend AI. Let us divide the angle IAB into two halves by a straight line AJ

[43] The Arabic text translated here appears to be incomplete. However, as both the manuscripts agree, we have left it unaltered. We believe that the missing phrase that we have inserted in English must have been:

... عند ١٤٦٨٨ <فقدر جميع أضلاع ذي ستة وتسعين ضلعًا يحيط بالدائرة عند القطر أقلّ من قدر ١٤٦٨٨ عند ٤٦٧٣ ونصف>، وهو ...

Gerard's translation includes: *Et hec quidem est proportio lateris figure habentis nonaginta sex latera continentis circulum ad diametrum. Ergo proportio diametri ad omnia latera figure habentis nonaginta sex latera continentis circulum est maior proportione quattuor millium et sexcentorum et septuaginta trium et medietatis ad quattuordecim millia et sexcenta et octoginta octo. Iam ergo ostensum est quod proportio omnium laterum figure habentis nonaginta sex latera ad diametrum est minor tribus et septima unius* (pp. 270–2, 87–94).

We arrive at

$$\frac{4673+\frac{1}{2}}{14688} = 0,318525326 < 0.$$

[44] Let us take up the proof of the second part: Let I be a point on the circle such that $B\hat{A}I = \frac{1}{3}\times\frac{\pi}{2}$. By successively dividing this angle in half, then a quarter, eighth and sixteenth, we also have the points J, K, L, M on the circle, and the chord BM is the side of the inscribed polygon with 96 sides. The bisector AJ cuts IB at O, and we have

$$\frac{OB}{OI} = \frac{AB}{AI} \Leftrightarrow \frac{OB+OI}{OI} = \frac{AB+AI}{AI} \Leftrightarrow \frac{AI+AB}{IB} = \frac{AI}{OI} = \frac{AJ}{JB},$$

as AIO and AJB are similar. We set $AB = 1560$, $BI = 780$; from this we deduce $AI < 1351$, which is also a good approximation. We then have:

$$(1) \qquad \frac{IA+AB}{IB} = \frac{AJ}{JB} \Rightarrow \frac{AJ}{JB} < \frac{2911}{780}.$$

If we now set $JB = 780\ u$ (where u is the unit of which JB is 780), then $AJ < 2911$, $AB^2 = AJ^2 + JB^2 < 9\,082\,321$ and $AB < 3\,013 + \frac{3}{4}$.

Similarly, in the triangle AJB, the bisector is AK, and we have

$$(2) \qquad \frac{JA+AB}{JB} = \frac{AK}{KB} \Rightarrow \frac{AK}{KB} < \frac{5924+\frac{3}{4}}{780} = \frac{1823}{240}$$

(simplified by multiplying both sides by $\frac{4}{13}$). This gives $KB = 240$; hence

$$AK < 1823,\ AB^2 = AK^2 + KB^2 < 3380929 \text{ and } AB < 1838 + \frac{9}{11}.$$

and we join *JB*. We divide the angle *JAB* into two halves by a straight line *AK* and we join *KB*. We divide the angle *KAB* into two halves by a straight line *AL* and we join *LB*. We divide the angle *LAB* into two halves by a straight line *AM* and we join *MB*. *MB* will then be the side <of the polygon> with ninety-six sides inscribed within the circle. We now assume in order to facilitate the procedure that *AB* is equal to 1,560. Then the chord *BI* will be 780, the square of *AB* will be 2,433,600, the square of *BI* will be 608,400 and the square of *IA* will be 1,825,200. Therefore, straight line *IA* is less than 1,351. However, the ratio of *IA* and *AB* together to *IB* is equal to the ratio of *AI* to *IO*, which is also the ratio of *AJ* to *JB*. But the two straight lines *IA* and *AB* together are less than 2,911 and *IB* is 780. If therefore *JB* is 780, then *AJ* is less than 2,911, the square of *AJ* is less than 8,473,921 and the square of *JB* is 608,400 and the square of *AB* is less than 9,082,321. Therefore, the straight line *AB* is less than 3,013 and three quarters of unity.

Similarly, in the triangle *AKB*, the bisector is *AL*, and we have

$$(3) \qquad \frac{KA+AB}{KB}=\frac{AL}{LB}\Rightarrow\frac{AL}{LB}<\frac{3661+\frac{9}{11}}{240}=\frac{1007}{66}$$

(simplified by multiplying both sides by $\frac{11}{40}$). This gives $LB = 66$, and therefore $AL < 1007$, $AL^2+LB^2=AB^2<1018405$ and $AB<1009+\frac{1}{6}$.

Similarly, in the triangle *LAB*, the bisector is *AM*, and we have

$$(4) \qquad \frac{LA+AB}{LB}=\frac{AM}{MB}\Rightarrow\frac{AM}{MB}<\frac{2016+\frac{1}{6}}{66}.$$

This gives $MB = 66$, and therefore $AM < 2016 + \frac{1}{6}$, $AM^2+MB^2=AB^2<4069284$, and hence $AB<2017+\frac{1}{4}$. But *MB* is a side of C'_{96}, and therefore $P'_{96} = 66 \times 96 = 6336$ and *AB* is the diameter of the circle; hence

$$\frac{P'_{96}}{d}>\frac{6336}{2017+\frac{1}{4}}>3+\frac{10}{71}.$$

If *P* is the circumference of the circle, we therefore have

$$P'_{96}<P<P_{96};$$

hence

$$\frac{P'_{96}}{d}<\frac{P}{d}<\frac{P_{96}}{d},$$

and hence

$$3+\frac{10}{71}<\frac{P}{d}<3+\frac{1}{7}.$$

Similarly, we can show that the ratio of *AK* to *KB* is less than the ratio of 5,924 and three quarters of unity to 780. If, therefore, the straight line *KB* is 780, then *AK* will be less than 5,924 and three quarters of unity. However, the magnitude of the ratio of 5,924 and three quarters of unity to 780 is equal to the magnitude of the ratio of 1,823 to 240. If, therefore, *KB* is 240, then *AK* will be less than 1,823 and the square of *AK* is less than 3,323,329 and the square of *KB* is 57,600. Therefore, the square of *AB* is less than 3,380,929, and the straight line *AB* is less than 1,838 and nine of eleven <parts> of unity.

Similarly, we can show that the ratio of *AL* to *LB* is less than the ratio of 3,661 and nine elevenths to 240, and the magnitude of the ratio of 3,661 and nine elevenths to 240 is equal to the magnitude of the ratio of 1,007 to 66. If *LB* is 66, then *AL* is less than 1,007, the square of *AL* is less than 1,014,049, the square of *LB* is 4,356 and the square of *AB* is less than 1,018 405. Therefore, the straight line *AB* is less than 1,009 and one sixth of unity.

Similarly, we can show that the ratio of *AM* to *MB* is less than the ratio of 2,016 and one sixth of unity to 66. If, therefore, *MB* is 66, then *AM* is less than 2,016 and a sixth, the square of *AM* is less than 4,064,928, the square of *MB* is 4,356 and the square of *AB* is less than 4,069,284. The straight line *AB* is therefore less than 2,017 plus one quarter of unity. However, the straight line *MB* has a magnitude of 66 and the straight line *MB* is the side <of the polygon> with ninety-six sides inscribed within the circle. The ratio of the diameter to <the sum of> the sides <of the polygon> with ninety-six sides inscribed within the circle is less than the ratio of 2,017 plus one quarter of unity to 6,336.

It has therefore been shown that the ratio of the sum of the sides <of the polygon> with ninety-six sides inscribed within the circle to the diameter is greater than the ratio of three plus ten parts of seventy-one parts to unity. The circumference of the circle is greater than the sum of the sides of the polygon with ninety-six sides inscribed within the circle and less than the sum of the sides <of the polygon> with ninety-six sides circumscribed around the circle. From that which we have described, it has therefore been proved that the ratio of the circumference of a circle to its diameter is greater than the ratio of three, plus ten parts of seventy-one <parts>, to unity, and less than the ratio of three, plus one seventh, to unity. This is what we required.

It is possible, using this method, to achieve any required degree of accuracy in this procedure.

– **7** – For any triangle, multiplying half the sum of the sides by the amount by which this exceeds each of the sides, multiplying by the excess over one of the sides, then by the second, and then by the third, the result will be equal to the product of the area by itself.[45]

Let the triangle be *ABC*. We draw the largest circle that can be inscribed within it, and let that circle be *DGF*, and let its centre be at *E*. We draw *ED*, *EF* and *EG* to the points of contact, and we extend *AE*. It is clear that *AD* and *AF* are equal. This also applies to *BD* and *BG*, and *CF* and *CG*.[46] It is

[45] See Supplementary note (The formula of Hero).

[46] These inequalities are proved in the Latin translation. Al-Ṭūsī seems to consider it too simple to be left there (p. 280, 22–29). Although his version contains the same ideas and proof as the Latin, al-Ṭūsī expresses them in a far more condensed form. Here is a brief résumé:

Let *P* be the perimeter of the triangle *ABC* with sides *a*, *b*, and *c*. It is required to prove that the area *S* of this triangle satisfies

$$S^2 = \frac{p}{2}\left(\frac{p}{2}-a\right)\left(\frac{p}{2}-b\right)\left(\frac{p}{2}-c\right).$$

Let *E* be the centre of a circle of radius *r* inscribed within the triangle, and let *D*, *F* and *G* be the points of contact between the circle and the sides of the triangle, *AB*, *AC* and *BC* respectively. Let *H* be a point on *AB*, and let *K* be a point on *AC*, such that $BH = CG$ and $CK = BG$. Then, $AH = AK = \frac{a+b+c}{2} = \frac{P}{2}$. The bisector *AE* is an axis of symmetry of the triangle *HAK*. The perpendicular to *AH* at *H* and that to *AK* at *K* therefore meet *AE* at a single point *I*, and $IH = IK$.

If $BL = BH = CG$, then $CL = GB = CK$, and $BI^2 - CI^2 = BH^2 - CK^2 = BL^2 - CL^2$. Therefore, $IL \perp BC$ and $IL = IH = IK$, $H\hat{B}I = I\hat{B}L$, as the right-angled triangles *HBI* and *IBL* are congruent.

In addition, $H\hat{I}L = D\hat{B}G$ and therefore $E\hat{B}D = B\hat{I}H$, and the right-angled triangles *BDE* and *BHI* are similar. From this, we deduce

$$\frac{DE}{DB} = \frac{HB}{HI} \Rightarrow \frac{DE}{GB} = \frac{GC}{HI} \Rightarrow DE \times HI = GB \times GC.$$

Also, $\frac{DE}{HI} = \frac{DE^2}{DE \times HI} = \frac{DE^2}{GB \times GC}$, but $\frac{DE}{HI} = \frac{AD}{AH}$, hence $DE^2 \times AH^2 = GB \times GC \times AD$, $DE^2 \times AH^2 = GB \times GC \times AD \times AH$. However, from Proposition 1, $DE \times AH = \frac{1}{2}pr = S$ (as $AH = \frac{1}{2}p$). Also,

$$GB = CK = \frac{1}{2}p - b,\ GC = BH = \frac{1}{2}p - c$$

and

$$AD = \frac{1}{2}p - (BH + BD) = \frac{1}{2}p - a;$$

hence

manifest that one of the two straight lines *AD* and *AF* is the difference between the half-sum of the sides and *BC*, and one of the two straight lines *BD* and *BG* is the difference between the half-sum and *AC*, and that one of the two straight lines *CF* and *CG* is the difference between the half-sum and *AB*. We then extend *AE* as far as *I*, and *AB* as far as will make *BH* equal to *CG*, and *AC* as far as will make *CK* equal to *BG*. Each <of the segments> *AH* and *AK* will be equal to the half-sum of the sides.

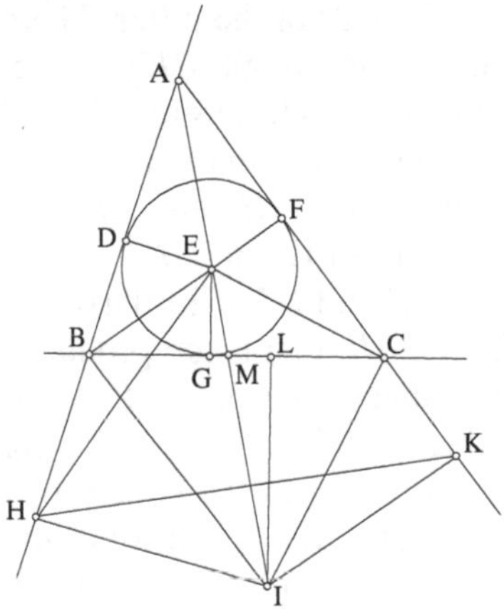

Fig. I.7

We then raise two perpendiculars *HI* and *KI* from the points *H* and *K*. They will necessarily meet at a single point on *AI*, such as the point *I* for example, such that *IH* and *IK* are equal. If we so wish, we could draw the perpendicular *HI*, join *IK* and show that it is also perpendicular due to the equality of the two sides *AK* and *AH* and given that *AI* is common and that the angles *HAI* and *KAI* are equal. We join *BI* and *IC*, we separate *BL* equal to *BH* from *BC*, and join *IL*. This is perpendicular to *BC* as the difference between the squares of the two straight lines *BI* and *IC* is equal to the difference between the squares of the two straight lines *BH* and *CK*, and *BH* is equal to *BL* and *CK* is equal to *CL*. Therefore, the difference between the

---

$$S^2 = \frac{1}{2}p\left(\frac{1}{2}p-a\right)\left(\frac{1}{2}p-b\right)\left(\frac{1}{2}p-c\right).$$

The other method proposed by the Banū Mūsā is based on

$$\frac{ED}{DB} = \frac{BH}{HI} \Rightarrow \frac{ED}{HI} = \frac{ED}{DB} \times \frac{DB}{HI} = \frac{ED^2}{DB \times BH} = \frac{ED^2}{BG \times CG};$$

but $\frac{ED}{HI} = \frac{AD}{AH}$, and therefore $\frac{AD}{AH} = \frac{ED^2}{BC \times CG}$.

The remainder of the proof is the same. See also the Supplementary note.

squares of the two straight lines *BI* and *IC* is equal to the difference between the squares of the two straight lines *BL* and *LC*. It is for this reason that *IL* is perpendicular to *BC*. But it is equal to *IH*, given that *BH* is equal to *BL*, *BI* is common, and the two angles *H* and *L* are right angles. Therefore, the two angles *LBI* and *HBI* are equal. We join *EB*. The two angles *GBE* and *DBE* are then equal. However, given that the angle *LBH* plus the angle *LIH* are equal to two right angles, the angle *GBD* is equal to the angle *LIH* and the half of one is equal to the half of the other. Therefore, the angle *EBD* in the triangle *BDE* is equal to the angle *BIH* in the triangle *BHI*. But the angles *BDE* and *BHI* are both right angles. Therefore, the two triangles *BDE* and *BHI* are similar, and the ratio of *ED* to *DB* is equal to the ratio of *BH* to *HI*. However, *DB* is equal to *GB*, and *BH* is equal to *GC*, the ratio of *ED* to *GB* is equal to the ratio of *GC* to *HI*, and the product of *ED* and *HI* is equal to the product of *BG* and *GC*. Similarly, the ratio of the square of *ED* to the product of *ED* and *HI*, that is the product of *BG* and *GC*, is equal to the ratio of *ED* to *HI*, that is the ratio of *AD* to *AH*. The ratio of the square of *ED* to the product of *BG* and *GC* is therefore equal to the ratio of *AD* to *AH*. Therefore, the product of the square of *ED* and *AH* is equal to the product of *BG* and *GC* and *AD*. If we multiply them by *AH*, then the square of *ED* multiplied by the square of *AH* will be equal to the product of *BG* and *GC* and *AD* and *AH*. However, given that *ED* multiplied by *AH* is equal to the area of the triangle,[47] the square of *ED* multiplied by the square of *AH* will be equal to the square of the area of the triangle. It follows that the square of the area of the triangle is equal to the product of *BG* and *GC* and *AD* and *AH*, that is to the <product> of the three differences and the half-sum of the sides. This is what we required.

Similarly, by another method, if after having established that the ratio of *ED* to *DB* is equal to the ratio of *BH* to *HI*, we place the second, at a mean between the first and the fourth, then the ratio of the first to the fourth will be compounded of the ratio of the first to the second and of the ratio of the second to the fourth, that is the ratio of the first to the third. The ratio of *ED* to *IH* is therefore compounded of the ratio of *ED* to *DB* and of the ratio of *ED* to *BH*. However, *DB* is equal to *BG* and *BH* is equal to *GC*. Therefore, the ratio of *ED* to *HI*, that is the ratio of *AD* to *AH*, is compounded of the ratio of *ED* to *BG* and of the ratio of *ED* to *GC*. Therefore, the product of *AD* and *BG* and *GC* is equal to the product of the square of *ED* and *AH*, and the proof is completed as before.

[47] This is the triangle *ABC*.

– **8** – If four equal straight lines are produced from any point within a sphere to the surface of the sphere at points which do not lie on the same plane, then that point is the centre of the sphere.

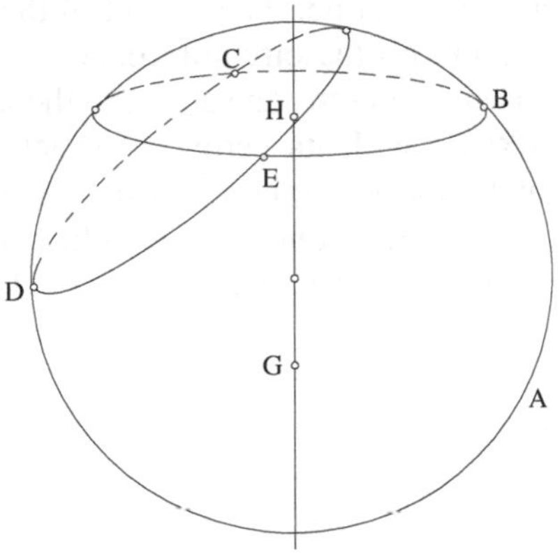

Fig. I.8

Let the sphere be *ABCDE*, the internal point *G* and the straight lines joining the point to the surface of the sphere *GB*, *GC*, *GD* and *GE*, which are equal and which are not on the same plane.[48] Actually, three of these points will be on the same plane, as proved in the book by Euclid. We describe a circle *BCE* through the points *B*, *C* and *E*, and a circle *ECD* through the points *E*, *C* and *D*. We draw the perpendicular *GH* from *G* to the plane of the circle *BCE*, passing through the centre of the circle *BCE* as, if we join the straight lines *BH*, *CH* and *EH*, they are equal as the straight lines *GB*, *GC* and *GE* are equal, given that *GH* is common and that all the angles at *H* are right angles. As the circle *BCE* is on the surface of the sphere *ABCDE* and the perpendicular *HG* has been drawn through its centre, this must pass through the centre of the sphere as has been shown in the second proposition of the book of *Spherics* by Theodosius.[49] We can similarly show that the perpendicular from the centre of the circle *ECD* also passes through the centre of the sphere. However, these two perpendiculars

[48] As they are coplanar, the four points *B*, *C*, *D*, *E* will also be coplanar, which is contrary to the hypothesis. Three of the straight lines may be coplanar, e.g. *GB*, *GC* and *GD*, if *G* is in the plane *BCD*. However, in this case, the fourth *GE* will intersect the plane.

[49] See Theodosius, Propositions 1 and 2 in the Arabic version of the *Spherics* translated by Qusṭā ibn Lūqā, *Kitāb al-ukar*, edited by Naṣīr al-Dīn al-Ṭūsī, published by Osmania Oriental Publications Bureau, Hyderabad, 1358 H., pp. 3 and 4. This reference to Theodosius is omitted from Gerard's Latin translation. Could it have been added by al-Ṭūsī?

only meet at G; therefore G is the centre of the sphere. This is what we required.

– **9** – For any circular right cone, the product of the straight line joining its vertex to any given point on the circumference of the base and half the circumference of the base is equal to <the area of> the lateral surface.[50]

Let the cone be $ABCD$, with its vertex at A; let the circular base be BCD with its centre at E, and the axis AE, which is perpendicular to the plane of the base in order for the cone to be a right cone. We join AB. The product of AB and half the circumference of BCD is the area of the lateral surface of the cone.[51]

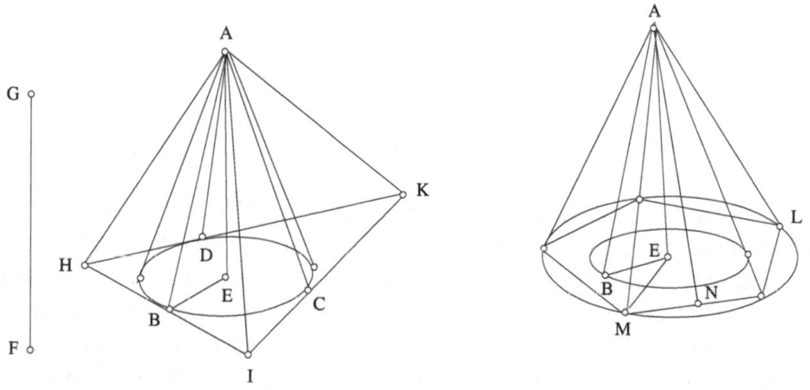

Fig. I.9

If it is not so, first let it be <the product of> AB and a straight line longer than half the circumference, and let this straight line be FG. We circumscribe a polygon around the circumference of BCD such that the sum of its sides is less than twice FG. Let this polygon be HIK and let it touch the circle at points B, C and D.[52] We draw the straight lines AH, AI and AK

[50] Lit.: The circular surface (*saṭḥahu al-mustadīr*). The Latin version is: *Cum linea que protrahitur ex puncto capitis omnis piramidis columpne ad centrum basis eius est perpendicularis super basim ipsius, tunc linee que protrahuntur ex puncto capitis eius ad circulum continentem superficiem basis eius secundum rectitudinem sunt equales*… (p. 292, 1–6).

Even accounting for the effects of the translation, it is clear that al-Ṭūsī has not thought it necessary to retain such a long phrase simply to remind us that the straight lines are equal. This would be assumed to be known to anyone with an interest in mathematics at the time.

[51] Lit.: The circular surface surrounding the cone.

[52] 'let it touch the circle at points *B*, *C*, and *D*': omitted from the Latin text.

and join *AC* and *AD*. Then the straight lines *AB*, *AC* and *AD*, which are equal, will be perpendicular to the sides *HI*, *IK* and *KH* as *AE* is perpendicular to the plane of the circle *BCD* and the straight lines joining its centre and the points of contact are perpendicular to the sides.[53] It is for this reason that the product of *AB* and half the sum of the sides is equal to the area of the polygon circumscribed around the circular cone and it is greater than the area of the circular cone.[54] However, half the sum of the sides is less than the straight line *FG*, and the product of *AB* and *FG* is the area of the circular cone. Therefore, the area of the circular cone is greater than the area of that which circumscribes it; this is contradictory.[55]

Now, let *FG* be shorter than half the circumference, and let <the product of> *AB* and *FG* be the area of the circular cone, and let <the product of> *AB* and half the circumference of *BCD*, which is greater, be equal to the area of the circular cone whose base is the circle *ML* and whose vertex is at *A*. We inscribe a regular polygon within the circle *ML* such that the sides do not touch the circle *BCD*. We produce straight lines from its angles to *A*. The lateral surface of the solid thus formed is less than the surface area of the circular cone whose base is *ML*, given that the cone contains it. However, the product of a straight line drawn from *A* to the mid-point of one of the sides of the figure that does not touch the circle *BCD* and half the sum of the sides is equal to the lateral surface of this solid.[56] Hence, the straight line drawn from *A* to the mid-point of this side is

[53] This rather general phrase appears in the Latin version as: *Tunc linee que protrahuntur ex punctis* B, G, D *ad centrum eriguntur super lineas* HT, TK, KH *orthogonaliter, quoniam sunt contingentes circulum* (p. 294, 36–38), which is a translation of the Arabic:

فالخطوط الواصلة بين نقط ب ز د والمركز أعمدة على خطوط ح ط ط ك ك ح لأنها تماس الدائرة.

i.e.: 'The straight lines joining the points *B*, *G* and *D* to the centre are perpendicular to the straight lines *HI*, *IK* and *KH* as these are tangents to the circle'.

It can be seen that al-Ṭūsī has read the same text, but from a more general point of view.

[54] The following phrase appears in the Latin version: *quoniam ipsum continet illud* (p. 294, 43), which must have been a translation of *li-anna aḥadahumā yuḥīṭu bi-al-ākhar*, omitted by al-Ṭūsī, as it is obvious from the figure.

[55] The Latin version repeats the conclusion: *Hoc est contrarium; ergo non est possibile ut multiplicatio linee* AB *in lineam que sit longior medietate circuli* BGD *sit embadum piramidis* ABGD (p. 296, 46–48). It should be noted that, as usual, the Arabic letter *jim* is transcribed by Gerard as *G* and here as *C*.

[56] A further illustration of the editing technique employed by al-Ṭūsī can be seen from the Latin translation: *corporis cuius basis est figura habens latera facta in circulo* ML *et cuius caput est punctum* A… (p. 296, 58–60), which must be a translation of the Arabic:

longer than the straight line AB, and half the sum of the sides of the figure is greater than half the circumference of the circle BCD. Therefore, the lateral area of the circular cone with the base ML is less than the lateral area of the solid which is inscribed in it. This is contradictory.

Consequently, the product of AB and half the circumference of the circle BCD is equal to the lateral area of the cone $ABCD$. This is what we required.

– **10** – If any circular right cone whose base is a circle is cut by a plane parallel to the base, then the intersection is a circle with the axis passing through its centre.

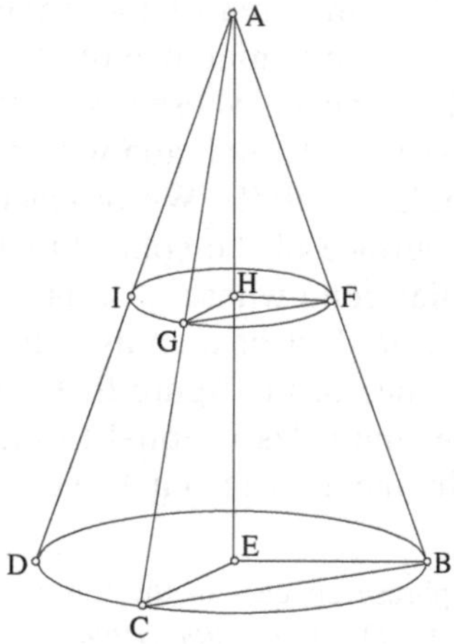

Fig. I.10

Let there be a cone, whose vertex is A and whose base is BCD with centre E. Let the intersecting plane be FIG and let the axis AE pass through point H on the intersecting plane. We mark two points B and C on BCD such that the arc BC is shorter than a semicircle. We draw EB, EC, BA, CA and BC. Then the triangle ABE passes through the intersection FH on the intersecting plane, triangle AEC through the intersection GH, and triangle ABC through the intersection FG. These form the triangle FHG whose sides are parallel to the sides of triangle BEC, each side parallel to the corresponding side in the other. The triangles are therefore similar. The ratio

---

المجسم الذي قاعدته الشكل ذو الأضلاع والزوايا المتساوية الذي تحيط به دائرة م‌ل وراسه نقطة آ.

i.e.: 'the solid whose base is the regular polygon inscribed within the circle ML and whose vertex is at the point A'. This agrees perfectly with the style employed by the mathematicians of the time.

of *BE* to *EC* will be equal to the ratio of *FH* to *HG*; yet *BE* and *EC* are equal, it is for this reason that *FH* and *HG* are equal, as is any straight line drawn from *H* to the circumference *FGI*. Therefore *FGI* is a circle with center *H*. This is what we required.[57]

– **11** – If, for any segment of a right circular cone between two parallel circles, two parallel diameters are drawn across the circles and their extremities joined by two opposite straight lines, then the product of one of these straight lines and half the sum of the circumferences of the two circles is equal to the lateral surface of the segment of the circular <cone>.

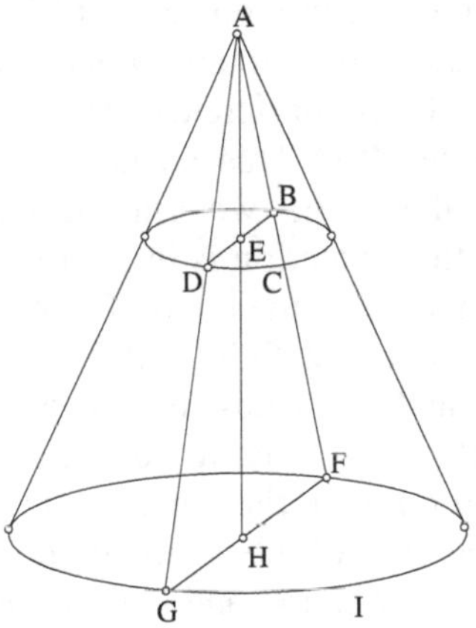

Fig. I.11

Let the segment of the cone be *BCDFIG* with the base *FIG* and the other segment closer to the vertex of the cone being *BCD*. Let *EH* be the segment of the axis between the two segments, and perpendicular to the two circles. Draw the two parallel diameters *BD* and *FG* and join them by *BF* and *DG*.[58]

We say that:[59] The product of *BF* and half the sum of the circumferences of the two circles *BCD* and *FIG* is the area of the surface enclosing the segment of the cone.

[57] Al-Ṭūsī arrives at the conclusion more rapidly than the Banū Mūsā. Compare with the Latin text, p. 300, 22–27.

[58] The Latin text continues: *que sunt equales, propterea quod linea* EH *iam secuit unamquamque duarum linearum* BD, UZ *in duo media, et est orthogonaliter erecta super unamquamque earum* (p. 304, 22–24), which is a translation of the Arabic:

فهما متساويان لأن خط هـ ح يلقى خطي ب د و ز وهو عمود عليهما

[59] Note that al-Ṭūsī retains this expression on this occasion.

Let us complete the cone up to its vertex at *A*, and let us produce *HE* as far as *A*.[60] Similarly for *FB* and *GD*. We know that the product of *AF* and half the circumference of *FIG* is the lateral area of the entire cone, and that the product of *AB* and half the circumference of *BCD* is the lateral area of the cone *ABCD*. The amount by which the first exceeds the second is the area of the surface enclosing the segment, which is the product of *BF* and half the circumference of *FIG*, plus the product of *AB* and the difference between half the circumference of *FIG* and half the circumference of *BCD*.[61] However, the product of *AB* and the difference between half the circumference of *FIG* and half the circumference of *BCD* is equal to the product of *BF* and half the circumference of *BCD* as the ratio of *AB* to *BF* is equal to the ratio of half <the circumference of> the circle *BCD* to the difference between half <the circumference of> the circle *FIG* and half <the circumference of> the circle *BCD*. This is what we required.

From this, we know that if the two straight lines *FB* and *BA* are equal – regardless of whether their junction is in a straight line or not – then the product of one of them and half <the circumference of> the circle *FIG* and the <circumference of> the circle *BCD* is the area of the lateral surface of the solid whose vertex is at *A* and whose base is the circle *FIG*.

And from this, we also know that if we have a number of segments of cylindrical cones stacked on upon the other such that the upper base of the lower segment is also the base of the segment above it, and that the vertex of the uppermost segment is a point, and that all the bases are parallel, and that all the straight lines drawn in all the segments from their bases to the base above them are equal straight lines, then the product of any one of these straight lines and half the circumference of the base of the lower segment and the sum of the circumferences of all the bases above it is the lateral area of the solid composed of all the segments, regardless of whether or not the surfaces of these segments arc joincd in a straight line.[62]

[60] In the Latin text, the Banū Mūsā justify this operation in the following terms: *Propter illud quod ostendimus quod linea que egreditur ex puncto* A *ad punctum* H *transit per punctum* E, *ergo linea* AH *egreditur ex capite piramidis ad centrum basis eius et cadit perpendicularis super basim* (p. 304, 30–33), which is a translation of the Arabic:

لأنا نبين أن الخط الخارج من نقطة آ إلى نقطة ح يمر بنقطة هـ فخط آح يخرج من رأس المخروط إلى مركز قاعدته عموداً على القاعدة.

[61] Al-Ṭūsī seems to have skipped a number of steps in the calculation, which have been retained in the Latin version.

[62] *i.e.* The generators may or may not be in a straight line.

– **12** – Let[63] *ABC* be a circle of diameter *AC* with its centre at *D*; *DB* is drawn from *D* perpendicular to the diameter.[64] Let us divide the quarter-circle *AB* into any number of equal parts, say *AG*, *GL* and *LB*. Let us draw the chord *BL* and extend it, also extending the diameter *CA* until they meet at *E*, and let us draw chords *GI* and *LH* from points *G* and *L* parallel to the diameter *CA*.

I say that the straight line DE *is equal to the sum of the half-diameter* CA *and the two chords* GI *and* LH.

We draw *IA* and *HG* and extend *HG* until it meets *CE* at *F*. We proceed in a similar manner if there are more parts. The straight lines *CE*, *IG* and *HL* are parallel, and the straight lines *IA*, *HF* and *BE* are parallel, as the two arcs *IH* and *HB* are equal to the two arcs *AG* and *GL*. Therefore, the surface *IAFG* is a parallelogram and *IG* is equal to *AF*. Similarly, *HL* is equal to *FE*, and therefore *DE* is equal to the sum of *DA*, *IG* and *HL*. This is what we required.

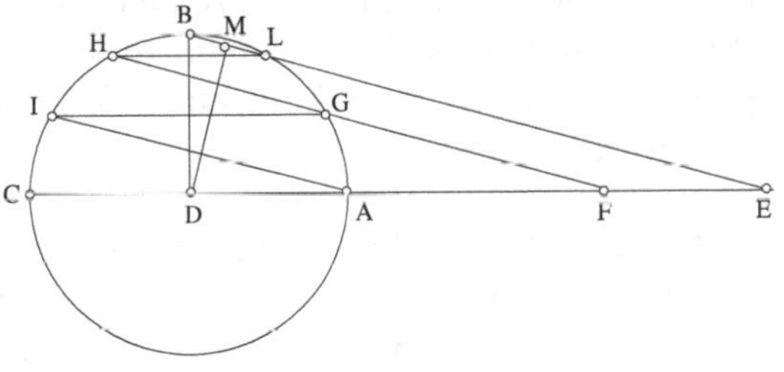

Fig. I.12

[63] Here, al-Ṭūsī has omitted the statement, which is retained in the Latin version: *Cum fuerit circulus cuius diameter sit protracta, et protrahitur ex centro ipsius linea stans super diametrum orthogonaliter et perveniens ad lineam continentem et secatur una duarum medietatum circuli in duo media, tunc cum dividitur una harum duarum quartarum in divisiones equales quotcunque sint, deinde protrahitur corda sectionis cuius una extremitas est punctum super quod secant se linea erecta super diametrum et linea continens et producitur linea diametri in partem in quam concurrunt donec concurrunt et protrahuntur in circulo corde equidistantes linee diametri ex omnibus punctis divisionum per quas divisa est quarta circuli, tunc linea recta que est inter punctum super quod est concursus duarum linearum protractarum et inter centrum circuli est equalis medietati diametri et cordis que protracte sunt in circulo equidistantibus diametro coniunctis* (p. 310, 1–20). As usual, he also omits *mithāl dhālika* (*Verbi gratia*).

[64] The Latin version continues: *et dividat arcum* ABG *in duo media* (p. 310, 23–24), which is a translation of the Arabic *fa-huwa yunaṣṣifu qaws ABC* ('thus dividing the arc *ABC* into two halves'), which is obvious. This is yet another example of al-Ṭūsī's editing style.

If we draw *DM* perpendicular to the chord *BL*, the product of half of *BL* and *DE* is less than the square of the half-diameter and larger than the square of *DM*. This is because the two triangles *DBM* and *BED* are similar,[65] given that the angles *DMB* and *EDB* are right angles and angle *B* is common. The ratio of *BM* to *MD* is therefore equal to the ratio of *BD* to *DE*. Therefore, <the product of> *BM*, that is half of *BL*, and *DE* is equal to <the product of> *BD* and *MD*. However, <the product of> *BD* and *MD* is less than the square of *BD* and greater than the square of *MD*. Consequently, <the product of> half of *BL* and the sum of the half-diameter plus the two chords *IG* and *HL* is less than the square of the half-diameter and greater than the square of *DM*.

Therefore, for any circle in which a diameter is drawn, if the semicircle is divided into two halves and each of the two quarters is divided into any number of equal parts, and if chords parallel to the diameter are drawn from each of the dividing points, then the product of half of the chord from one of these parts and the half-diameter plus its product with the sum of the chords is smaller than the square of the half-diameter and greater than the square of the perpendicular drawn from the centre to one of the chords of these parts. This is what was sought.

– **13** – If a solid is inscribed within a hemisphere, and if this solid is composed of any number of segments of circular cones, such that the upper base of each segment forms the base of the segment above it, and if the base of the lowest segment is the base of the hemisphere, and if the vertex of the uppermost segment of the cone is at the point formed by the pole[66] of the hemisphere, and if the bases are parallel, and if the straight lines drawn from the bases of the segments to their upper parts are equal, and if a hemisphere is then inscribed within this solid, of which the base is a circle in the plane of the base of the first hemisphere, then the lateral surface of the solid is less than twice <the area> of the base of the first hemisphere and greater than twice <the area> of the base of the second hemisphere.

Let the hemisphere be *ABCD* whose base is the great circle *ABC* and whose pole is *D*.[67] Let a solid be inscribed within this hemisphere consisting

[65] In the remainder of this section al-Ṭūsī version differs slightly from the Latin text (see pp. 312–14, 50–65).

[66] ‘and if the vertex … pole’. In Latin, this is given as: *et fuerit portio superior piramidis piramis capitis, et punctum capitis eius est polus…* (p. 316, 9–10). ‘The upper portion of the vertex of the cone will be a cone, and the point at the vertex will be a pole…’.

[67] The Latin text is: *Et signabo in medietate spere in primis corpus compositum ex portionibus quot voluero piramidum columpnarum secundum modum quem*

of three segments as we have described it. The first of these segments extends from the circle *ABC* as far as the circle *EIH*, the second extends from this circle as far as the circle *FLG*, and the third extends from this circle as far as the point *D*.

We say that the sum of the areas of these circular surfaces surrounding this solid is less than twice the area of the circle ABC.

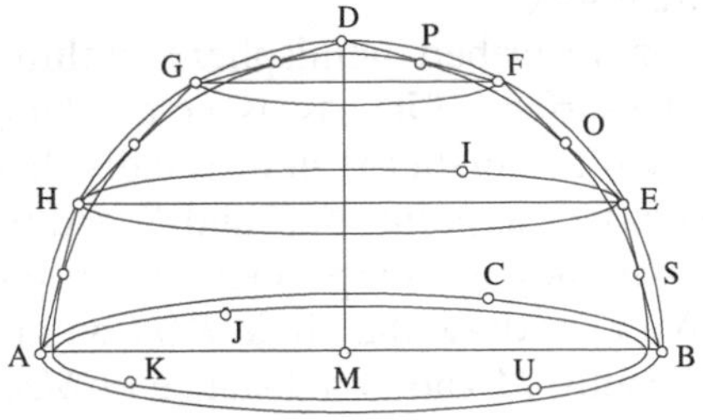

Fig. I.13

Let us draw half a great circle on the hemisphere *ABCD* passing through the pole, and let this be *ADB*. Let us draw the diameter of the sphere, *AB*, and divide this into two halves at *M*. Let us draw *HE* and *GF*. These will be parallel to *AB* as they are the intersections of the great circle *ADB* and the three circles, and they are the two diameters of circles *EHI* and *FGL*. We raised the straight lines *BE*, *EF* and *FD* from the bases. These are equal as stated in the hypothesis, and the product of half of any one of them and half *AB* and the sum of *EH* and *FG* is less than the square of half of *AB*, as was proved above.[68] Similarly, the product of any one of them and half the circumference of the circle *ABC* and the sum of the circumferences of the circles *HEI* and *GFL* is equal to the area of the surface surrounding the solid, as was proved above.[69] However, the product of any one of them and half of *AB* and the sum of *EH* and *FG*, and that which, when multiplied by the diameter, gives the circumference, is equal to the product of any one of these and half the circumference of the circle *ABC* and the sum of the circumferences of the circles *HEI* and *GFL*, that is, equal to the area of the surface surrounding the solid, which is less than twice the result obtained from the product of the square of half of *AB* and that which, when multiplied by the diameter, gives the circumference. But

---

narravimus (p. 318, 24–26), which must have been a translation of an original text along the lines of:

فليقع أولاً في نصف الكرة مجسم مركب من قطع مخروطات مستديرة كم كانت على الوجه الذي وصفنا .

[68] See Proposition 12.

[69] See Proposition 11.

the product of the square of half of *AB* and that which, when multiplied by the diameter, gives the circumference, is equal to the surface area of the circle, as the product of half of *AB* and that which, when multiplied by the diameter, gives the circumference, is half the circumference, and hence its further multiplication by half of *AB* gives the surface area of the circle. The area of the surface surrounding the solid is therefore less than twice the surface area of the circle *ABC*.[70]

Now, we draw an inscribed hemisphere within the solid *ABCD*. However, given that the surface of its base is a circle lying within the surface of the circle *ABC*, it will be smaller than it. We divide each of the straight lines *BE*, *EF* and *FD* in half at points *S*, *O* and *P*, and we join *MS*, *MO* and *MP*, which are equal as they are perpendiculars dropped from the centre onto equal chords. We then draw the circle *KUJ* within the surface of the circle *ABC* from the centre *M* and at a distance of *MS*, and a straight line *MU* in the plane of this circle, which is not in the plane of the circle *ADB*. As the four equal straight lines *MS*, *MO*, *MP* and *MU*, which are not in the same plane, all connect point *M* to the surface of the inner sphere, then *M* will be its centre, *MS* its half-diameter, and the circle *KUJ* will be its base. However, the square of *MS* is less than the product of half of *BE* and half of *AB* and the sum of *EH* and *FG*. Therefore, the <product of the> square of *MS* and the magnitude which, when multiplied by the diameter, gives the circumference, that is the area of the surface of the circle *KUJ*, is less than the product of half of *BE* and half of *AB* and the sum of *EH* and *FG* and the magnitude which, when multiplied by the diameter, gives the circumference, that is half of the area of the surface surrounding half of the inner sphere. Therefore, the area of the entire surface[71] surrounding the solid is greater than twice the area of the surface of the circle *KUJ*. This is what we required.[72]

[70] This text has been considerably abridged by al-Ṭūsī, as can be seen from the Latin version.

[71] This refers to the lateral surface.

[72] Here al-Ṭūsī follows his usual practice of leaving the reader to conclude the argument. The Latin text reads: *Iam ergo ostensum est quod embadum superficiei corporis* ABGD *est minus duplo embadi basis medietatis spere que continet corpus et maius duplo embadi basis medietatis spere quam continet corpus* ABGD. *Et illud est quod declarare voluimus Et hec est forma eius* (p. 328, 150–154), which is a translation of an Arabic expression of the type:

فقد تبين أن سطح مجسم ا ب جـ د أقلّ من ضعف سطح قاعدة نصف الكرة الذي يحيط بالمجسم وأعظم من ضعف سطح قاعدة ‹نصف› الكرة الذي يحيط به مجسم ا ب جـ د ؛ وذلك ما أردنا أن نبين. وهذه صورته.

– **14** – The lateral surface of a hemisphere is twice the surface area of the great circle forming its base.

Therefore let *ABCD* be a hemisphere, with its great circle *ABC* within it forming its base, and *D* its pole. If twice the surface area of the circle *ABC* is not equal to the surface of the hemisphere, let it first be smaller, and let it be equal to the surface of a hemisphere that is smaller than the hemisphere *ABCD*. Let this hemisphere be *EHIK*. If, as we have described, a solid is inscribed within the hemisphere *ABCD* with the base of this solid being the circle *ABC* and its vertex being at point *D*, without it touching the hemisphere *EHIK*, then its surface area will be less than twice the surface area of the circle *ABC* and greater than the surface area of the hemisphere *EHIK*. Twice the surface area of the circle *ABC*, which is equal to the surface area of the hemisphere *EHIK*, is much greater than this, which is contradictory.

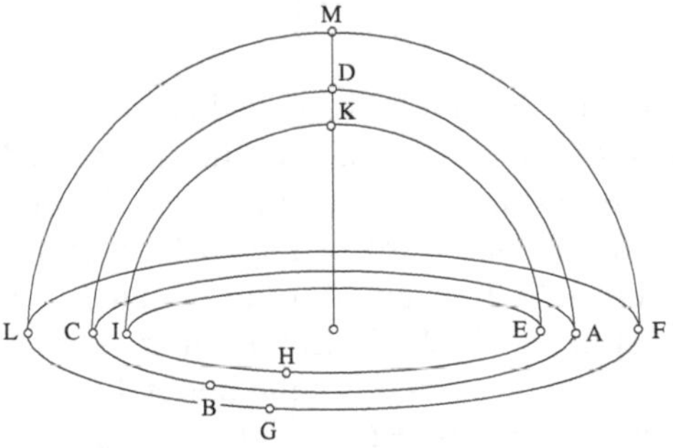

Fig. I.14

Now, let the surface area of the circle *ABC* be greater than the surface area of the hemisphere *ABCD*, and let it be equal to the surface area of the hemisphere *FGLM*. We inscribe a solid within this – as we have described – without it touching the hemisphere *ABCD*. The surface of the solid will be greater than twice the surface area of the circle *ABC*, as shown above, and the surface area of the hemisphere *FGLM* is greater than the surface area of the solid as it surrounds it. *The surface area of the hemisphere *FGLM* is therefore much greater than twice[73] the surface area of the circle *ABC*, or it is equal to it, which is contradictory*.[74] The assertion is therefore proved. This is what we required.

It has been shown using this assertion that the surface area of a sphere is four times that of the largest circle that can be found within it.

[73] Omitted in the Arabic text, but present in the Latin version as *duplo*.
[74] *...*: We believe this section to be a citation by al-Ṭūsī.

– **15** – For any sphere, the product of its half-diameter and one third of its lateral surface area is equal to its volume.[75]

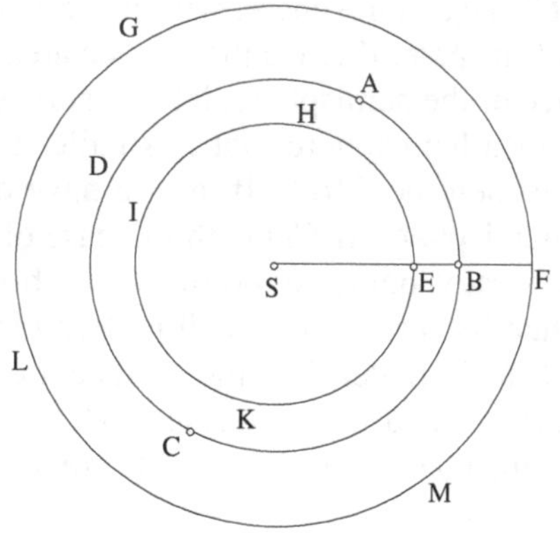

Fig. I.15

Let the sphere be *ABCD* and let its half-diameter be *SB*. If <the product of> *SB* and one third of the surface area of the sphere *ABCD* is not equal to its volume, then let us assume firstly that it is less than the volume and that <the product of> *SB* and one third of the surface area of a sphere that is larger than the sphere *ABCD* is equal to the volume of the sphere *ABCD*, for example the sphere *FGLM*. Let their centre be the same. We circumscribe, as we have described, a solid about the sphere *ABCD* without allowing it to touch the sphere *FGLM*. It necessarily follows from that proved earlier,[76] that <the product of> *SB* and one third of the surface area of the solid is equal to the volume of the solid and that it is greater than the sphere *ABCD*. From this, it necessarily follows that one third of the surface area of the solid is greater than one third of the surface area of the sphere *FGLM* surrounding it. This is contradictory.

Now, let <the product of> *SB* and one third of the surface area of the sphere *ABCD* be greater than its volume, and let <the product of> *SB* and one third of the surface area of a sphere that is smaller than the sphere *ABCD*, such as the sphere *EHIK*, be equal to the volume of the sphere *ABCD*. We inscribe, as we have described, a solid within the sphere *ABCD* without allowing it to touch the sphere *EHIK*. It is necessary, from that proved earlier,[77] that <the product of> *SB* and one third of the surface area of the solid is less than the volume of the sphere *ABCD*. Therefore, one

[75] Lit.: its greatness (*'iẓam*). We shall translate it as such in the remainder of the text.
[76] From Proposition 1.
[77] From Proposition 2.

third of the surface area of *EHIK* is greater than one third of the surface area of the solid which surrounds it, which is impossible.[78]

The assertion is therefore proved. This is what we required.

– **16** – Find two magnitudes lying between two given magnitudes such that all four are in continued proportion.

A knowledge of how to do this is useful to a student of geometry as it is needed in order to calculate the side of a cube. In fact, if we know two magnitudes between unity and the cube related by the same ratio, the second magnitude after unity will be the side of the cube.[79] This procedure is due to one of the ancients whose name was Menelaus; he set forth it in one of his books on geometry and we shall now describe it.

Let the two magnitudes be two straight lines *M* and *N*, and let *M* be greater than *N*. We now draw a circle *ABC*, and set its diameter, which is *AB*, equal to *M*. We then draw a chord *AC* equal to the magnitude *N* within this circle, and draw from *B* a perpendicular to *AB*. We produce *AC* until it meets it at *G* and raise from the arc *ACB* a right circular half-cylinder; I mean that its sides are perpendicular to the plane of the circle *ACB*. We describe a semicircle on the straight line *AB*[80] such that its plane is perpendicular to the plane of *ABC*; this semicircle is the arc *AHE*. We fix the point *A* of the arc *AHE* in its position as a centre of rotation, and rotate the arc *AHE* around the centre *A* such that, during its rotation, its plane remains erected on the plane of *ABC* at right angles, so that the arc *AHE* cuts the surface of the right half-cylinder according to the arc *ACB*.[81] We then fix the straight line *AB* as an axis of rotation, and rotate the triangle *AGB* around the axis *AB* until the straight line *AG* meets the intersection[82] of the surface of the half-cylinder, and so that the point *C* on the straight line *AG* describes during the rotation the semicircle *COD* erected on the plane of *ABC* at right angles. We mark the point *H* where the straight line *AG* meets the intersection[83] of the surface of the half-cylinder.

[78] See commentary.

[79] The Latin translation is a little obscure, and it appears that Gerard has not translated the Arabic text: *Et hac eadem operatione extrahatur latus cubi, quod est quoniam quando illud quod est in cubo de unitatibus et partibus est notum et ponuntur inter numerum cubi et inter unum duo numeri continui secundum proportionem <unam>, tunc ille qui sequitur unum ex duobus numeris mediis est latus cubi* (p. 334, 6–10).

[80] The segment *AB* is taken to be the diameter.

[81] The straight line *AE* cuts the arc *ACB* at *I*, and this point describes the arc *ACB*.

[82] This refers to the curve described by the point of intersection of the circle *AHE* and the cylinder.

[83] See previous note.

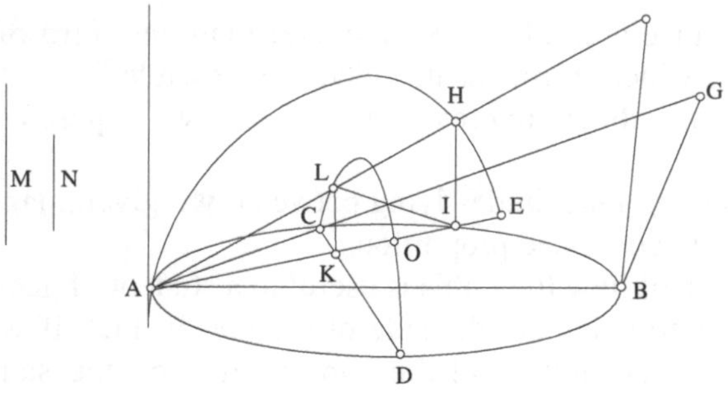

Fig. I.16

We now fix the arc *AHE* on its trajectory at the point *H* and draw the two straight lines *AH* and *AE*.[84] We mark the point *L* where the straight line *AH* meets the arc *COD*. *Now, we draw a perpendicular from the point *H* to the plane of the circle *ABC*, that is the straight line *HI**.[85] We draw *LK* perpendicular to the plane of the circle *ABC* as this is the intersection of the plane of the triangle *AHE* and the semicircle *COD*, both of which are perpendicular to the plane *ABC*. We draw the straight line *LI* and show[86] that it is perpendicular to *AL*, as the product of *CK* and *KD* is equal to the square of *LK*.[87] But the product of *CK* and *KD* is equal to the product of *IK* and *KA*.[88] Therefore, the product of *IK* and *KA* is equal to the square of *LK*, and the angle *ILA* is therefore a right angle. Now, we have shown that the angle *AHE* is a right angle as it is inscribed[89] within the semicircle *AHE*, that the angle *AIH* is a right angle – as *HI* is perpendicular to the plane of the circle *ABC* and the straight line *IA* is in the plane of the circle *ABC* – and that the angle *ALI* is a right angle from that was proved earlier. Therefore, the triangles *AHE*, *AIH* and *ALI*[90] each have a right angle and one common acute angle; so they are similar. The ratio of *EA* to *AH* is thus equal to the ratio of *AH* to *AI* and is equal to the ratio of *AI* to *AL*. But the straight line *AE* is equal to the magnitude *M* and the straight line *AL* is equal to the magnitude *N*. The two magnitudes *AH* and *AI* therefore lie between them and they are in continued proportion. This is what we required.

[84] *HE* in the Arabic text, *AE* in the Latin version.

[85] *…*: Omitted from the Latin text.

[86] In Arabic: *wa-nubayyin*. However, the Latin translation is *manifestum est* ('it is clear'). It is possible that the Banū Mūsā text included *wa-tabayyan* or *istibāna*, a word in very common use at the time.

[87] Euclid, *Elements*, VI.8 (right-angled triangle *CLD*).

[88] Euclid, *Elements*, III.35 (power of the point *K*).

[89] This word is used to translated *murakkaba* (fixed on).

[90] The triangle *AKL* is added in the Latin version.

– **17** – As the methods[91] used by Menelaus, even if they are true,[92] are either not constructible, or too difficult, we have sought for an easier method.

Let the two magnitudes be *A* and *B*. We draw *CD* equal to *A* and raise a perpendicular *DE* equal to *B*. We join *EC* and extend *CD* and *ED* with no limit. From *E*, we draw a line perpendicular to *EC* and extend it until it meets *CD* at *F*. We draw a straight line from *C* that is parallel to that just drawn until it meets *ED* at *M*. This line is then *MC*. We then produce it until *MU* is equal to *EF*. We now imagine the straight line *FE* moving away from point *F* towards point *D*, such that, during this movement, the extremity of the line at *F* remains attached on the line *FD*, and that the straight line, during this movement, continues to pass through point *E* on the straight line *CE* so that, if the straight line *FE* moves as we have described, with the extremity on the straight line *FD*, then the straight line *FE*, under these conditions, extends between the point at its extremity and the point *E* on the straight line *EC*. We now draw the straight line *EDK* on the extended <section> and we imagine that the straight line *MU* moves away from the point *M* towards the point *K*, such that the extremity of the line at <point> *M* remains attached, during its movement, to the straight line *MK*, and that the straight line *MU*, during its movement, continues to pass through the point *C* on the straight line *EC*, as we have described in relation to the movement of the straight line *FE*. We imagine that the two straight lines *FE* and *MU* remain parallel during their movement. We imagine, at the extremity of the straight line *FE*, at the point *E*, a straight line perpendicular to the straight line *FE*, and fixed relative to that line during its movement, and we do not assign any defined end to this straight line so that this straight line always cuts the straight line *MU* when the two straight lines *FE* and *MU* are moving. Therefore, if the two straight lines *FE* and *MU* move, if they remain parallel during their movement, and if their extremities remain on the two straight lines *FD* and *MK*, as we have described, then it is necessary that the straight line perpendicular to the straight line *FE* that moves with it and that cuts the straight line *MU*, should end at the point *U*. Therefore, if the straight line perpendicular to *FE* ends at <the point> *U*, we fix the two straight lines *FE* and *MU* in this position and we draw the two straight lines *EU* and *FM*. We then know that the straight line *EU* is held perpendicular to each of the straight lines *FE* and *MU*, as it is the straight line that we placed perpendicular to the straight line *FE* and which moves with it until it ends at the point *U*.

[91] Lit.: things (*al-ashyā'*).

[92] *sit demonstratio certa erecta in mente …* (p. 340, 2–3).

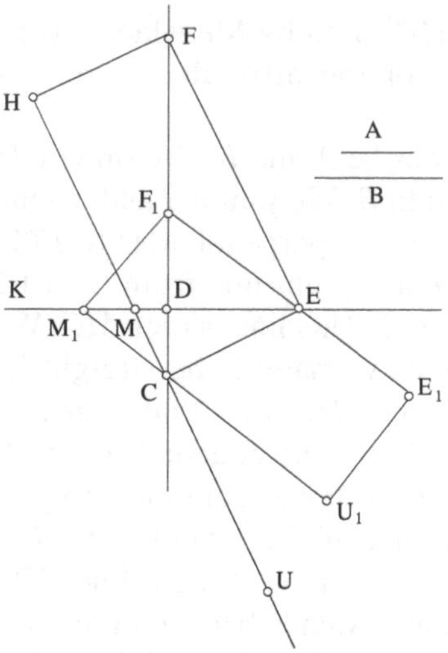

Fig. I.17

I say that the two straight lines DM *and* DF *lie between the two magnitudes* CD *and* DE*: the ratio of* CD *to* DM *is equal to the ratio of* DM *to* DF *and is equal to the ratio of* DF *to* DE.

Proof: The two straight lines *FE* and *MU* are parallel and equal, and the two angles *FEU* and *MUE* are right angles. Therefore, the straight line *FM* is equal to the straight line *EU* and each of the angles *EFM* and *UMF* is a right angle. But *MD* is perpendicular to the straight line *FC*, and the straight line *FD* is perpendicular to the straight line *EM*. Therefore, the ratio of the straight line *CD* to *DM* is equal to the ratio of *DM* to *DF* and is equal to the ratio of *DF* to *DE*. But the straight line *CD* is equal to *A* and the straight line *DE* is equal to *B*. Therefore, the two straight lines *DM* and *DF* lie between *A* and *B* and are in continued proportion. This is what we required.[93]

In order to make easier in practice the existence of this, let us replace the straight line *EF* perpendicular to *EC* with a rule, and let us also replace *EC* by another rule which is linked to the rule *EF* at the point *E* by a pin fixed in its position with the rule *EF* free to pivot around it. We produce the straight line *CM* perpendicular to *EC* as far as the point *H* and make *CH* equal to *EF*. We replace the straight line *CH* by a rule linked to the rule *EC*

[93] The passage beginning hereafter ('In order to …') until the end of this proposition is omitted from the Latin version. There can be no doubt as to authenticity of this text, or to its attribution to al-Ṭūsī. The Banū Mūsā also refer to this mechanical procedure later in the text. The Latin translation includes: *Et quoniam possibile est nobis per ingenium quod narravimus in eis que permissa …* (pp. 346–8, 33–34).

at the point *C* by a pin fixed in its position with the rule *HC* free to pivot around it. Rule *EC* is fixed and does not move. In this configuration, the two rules *EF* and *CH* pivot around the two pins *E* and *C*. Let us now place a rule between the two points *F* and *H* and link this to the rule *FE* by a pin at the point *F*, and to the rule *CH* by a pin at the point *H*, such that these two pins are free to move and not fixed, and such that the three rules – that is, the rules *EF*, *FH* and *HC* – pivot about the rule *EC* fixed to the pins *E* and *C*. Let us place a thin rod on the back of the rule *EF* and let this rod slide along the back of the rule in a groove. Let us arrange for the middle of the rod to lie on the straight line *FE* and let us make its length equal to the length of the rule *EF*. At the extremity of this rod at the point *F*, we place a pin whose centre is at the point *F*, and we construct two planes on each side of *FD* such that their intersections with the plane *EH* are parallel to the straight line *FD*[94] and we position the two planes such that they touch the pin which is on the rod so that, if the three sides of the rectangle[95] *EH* are moved around the side of *EC* that is fixed, this pin remains between the two planes, the centre of this pin remains in contact with the straight line *FD*, and the extremity of the rod is extended from point *E* following the extension of the straight line joining the centre of the pin to the point *E*. We place another rod on the back of the rule *CH* which slides on the back of the rule. We position the start of this rod at the point *M* and its far extremity at the point *U* such that the length of the rod is equal to the length of the rod mounted on the rule *EF*. We place a pin at the extremity of this rod at <the point> *M*, and we arrange it by the procedure that we have already described so that, if we rotate the three sides of the rectangle[96] *EH* around the fixed side *EC*, then the centre of this pin moves along the straight line *MK* and the extremity of this rod approaches the point *K*. Let us then attach another rod to the rod mounted on the back of the rule *EF* at the extremity which is at the point *E*, and let this additional rod make a right angle with the first and move with it, and let us arrange for this rod to end at the rod mounted on the ruler *CH*, cutting it such that, if we rotate the three sides of the rectangle[97] *EH* around the side *EC* which remains fixed, then the extremity of this intermediate rod between the other two rods must cut the rod mounted on the rule *CH*.

By virtue of the proof shown earlier concerning the lines in this proposition, we know that, if the rules and the rods that slide upon them are fixed in the position at which the intermediate rod lies at the extremity of

[94] See commentary.
[95] Lit.: square.
[96] Lit.: square.
[97] Lit.: square.

the rod mounted on the rule *CH*, then we have achieved that which we wished to construct.

– **18** – Using this ingenious procedure, we may divide any angle into three equal parts.

Let the angle be *ABC* and let us initially assume that it is less than a right angle. We take two equal magnitudes *BD* and *BE* on the straight lines *BA* and *BC*. We describe the circle *DEL* at their distance with its centre at *B* and we produce *DB* as far as *L*. Then we raise a perpendicular *BG* on *LD*, join *EG* and produce it as far as *H* with no extremity. From *GH*, we separate *GO* equal to the half-diameter of the circle. If we now imagine that *GH* moves in the direction of the point *L*, and that the point *G* remains on the circumference during its movement, such that the straight line *GEH* continues as it moves to pass through the point *E* on the circle *DEL*, and if we imagine that the point *G* continues to move until the point *O* arrives on the straight line *BG*, then the arc between the position at which the point *G* arrives and the point *L* must be one third of the arc *DE*. The angle which intercepts this arc is one third of the angle *DBE*.

Proof: Let the final position of <the point> *G* be the point *I*. Let us draw *IE*, which cuts *BG* at *S*. The straight line *IS* is therefore equal to the half-diameter of the circle, given that this is equal to *GO*. Let us draw a diameter through the centre parallel to *IE*, that is *MBK*. Draw *MI*. Then, *IS* is equal and parallel to *MB*, *MI* is parallel and equal to *BS*, and *BS* is perpendicular to *LD*. Therefore, *MI* is perpendicular to *LD*. It is for this reason that it is divided into two equal parts by the diameter, and hence <the arc> *ML* is equal to <the arc> *LI*, <the arc> *DK* is equal to <the arc> *ML* and <the arc> *MI* is equal to < the arc > *KE*. Therefore, <the arc> *DK* is equal to half of <the arc> *KE* and equal to one third of <the arc> *DE*. The angle *KBD* is therefore one third of the angle *ABC*. This is what we required.

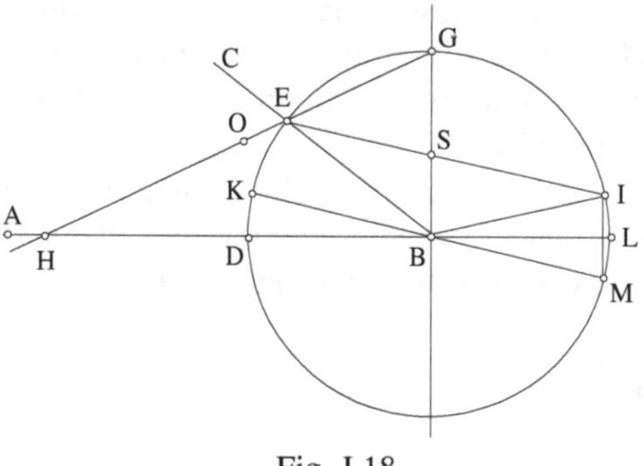

Fig. I.18

Now, move *GH* using the ingenious procedure described, with the condition that *G* moves on the circumference without leaving it, and that the straight line *GH*, during its movement, continues to pass through the point *E*, until the point *O* falls on the straight line *BG*, and we have achieved that which we sought.

If the angle is obtuse, we divide it into two equal parts, find the third of each half, and then two of these thirds are one third of the obtuse <angle>.

We must describe after that the approximation to the side of a cube so that it becomes rational[98] in case of need. We will now do this using an <approximation> method which is superior to all other approximation methods, that is to say that if we wish to make the error between the approximation and the truth less that one minute or one second, then we would be able to do so. The procedure is to break the cube down into parts: thirds, sixths, ninths, and so on.[99] We then look for a cube equal to this number if there is one. If not, we look for the cube closest to it and, when found, we note its side. If the parts are thirds, then it is minutes, and if they are sixths, it is seconds. The problems are treated in a similar manner.

Everything that we describe in this book is our own work, with the exception of knowing the circumference from the diameter, which is the work of Archimedes, and the position of two magnitudes in between two others such that all <four> are in continued proportion, which is the work of Menelaus as stated earlier.

The book is finished.

[98] Lit.: so that one can say it.

[99] Lit.: and other than that.

There is another proof of the seventh proposition in the book by the Banū Mūsā that is a general method for the area of triangles. I believe that it is the work of al-Khāzin. It is the following:

For any triangle, if one multiplies half the sum of the sides by the amount by which this exceeds the first, then by the amount by which this exceeds the second, and then by the amount by which this exceeds the third, and then takes the square root, then one will have the area of the triangle.

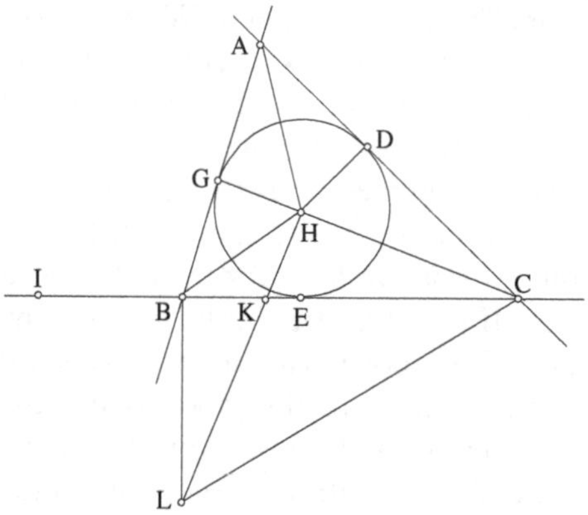

Fig. I.19

Proof: Let the triangle be *ABC*, within which we inscribe a circle *DEG* whose centre is at *H*. Let us join the centre to the points of contact by the straight lines *HD*, *HE* and *HG*. These will be perpendicular to the sides and equal, and *CE* and *CD* will be equal, as will *BG* and *BE*, and *AD* and *AG*. Let us join *CB* and draw *BI* equal to *AD*. The straight line *CI* is therefore equal to half <the sum of> the sides, and *IB* is therefore the excess over the side *BC*, *BE* the excess over the side *AC*, and *EC* the excess over the side *AB*. The assertion is <to show> that the product of *IC* and *IB*, and *BE* and *EC*, is equal to the square of the area of the triangle, which is the product of *EH* and *IC*. Let us draw *BL* from *B* perpendicular to *CB*, and *HK* from *H* perpendicular to *CH*, and let us extend them until they meet at *L*. Now, let us join *CL*. But, given that the two angles *CHL* and *CBL* are right angles, the quadrilateral *CHBL* is inscribed within a circle, whose diameter is *CL*. It is for this reason that <the sum of> the two opposite angles *CHB* and *CLB* is equal to two right angles. But the sum of the angle *CHB* and the angle *AHD* is equal to two right angles as they are half of the six angles surrounding the point *H* and which are four right angles. It is for this reason that the angle *AHD* is equal to the angle *CLB* and the two angles *CBL* and *HDA* are two right angles. The triangle *CBL* is therefore similar to the triangle *HDA*, and therefore the ratio of *CB* to *AD*, that is *BI*, is equal to the ratio of *BL* to *DH*, that is *EH*, and is also equal to the ratio of *BK* to *KE*. If we compound, the ratio of *CI* to *IB* is equal to the ratio of *BE* to *EK*. If we make *CI* the common height for the first two and *EC* the common height for the last two, then the ratio of the square of *CI* to *CI* multiplied by *IB* is equal to the ratio of *BE* multiplied by *EC* to *EK* multiplied by *EC*, that is the square of *EH*. But the product of the square of *CI*, the first, and the square of *EH*, the fourth, is equal to the product of *CI* and *IB* and *BE* and *EC*. But the ratio of the square of *CI* to the product of *CI* and *EH* is equal to the ratio of the product of *CI* and *EH* to the square of *EH*, and hence the product of *CI* and *EH* is the proportional mean between the two squares of *CI* and *EH*. It is for this reason that the product of the square of *CI* and the square of *EH*, which is equal to the product of *CI* and *IB*, and *BE* and *EC* will be equal to the square of the product of *CI* and *EH*, which is the area. This is what we required.

CHAPTER II

THĀBIT IBN QURRA AND HIS WORKS IN INFINITESIMAL MATHEMATICS

2.1. INTRODUCTION

2.1.1. *Thābit ibn Qurra: from Ḥarrān to Baghdad*

The little that we know about Thābit ibn Qurra derives mainly from the biobibliographical details provided on him by al-Nadīm, al-Qifṭī and Ibn Abī Uṣaybiʿa.[1] These accounts are by no means all of equal importance. The

[1] Al-Nadīm, *Kitāb al-Fihrist*, ed. R. Tajaddud, Teheran, 1971, p. 331. Al-Nadīm cites only four titles that relate to Thābit's mathematical writings: the *Treatise on Numbers* (*Risāla fī al-aʿdād*; probably his treatise on amicable numbers), the *Treatise on the Defining of Geometrical Problems* (*Risāla fī istikhrāj al-masā'il al-handasiyya*), the *Treatise on the Sector-Figure* (*Kitāb fī al-shakl al-qaṭṭāʿ*) and, finally, the *Treatise on the Proof Attributed to Socrates* (*Risāla fī al-ḥujja al-mansūba ilā Suqrāṭ*). Al-Nadīm writes that Thābit 'was born in the year two hundred and twenty-one and died in the year two hundred and eighty-eight; his age was seventy-seven solar [*sic*] years'. He also refers to the privileged relationship Thābit enjoyed with the Caliph al-Muʿtaḍid.

Al-Qifṭī, *Ta'rīkh al-ḥukamā'*, ed. J. Lippert, Leipzig, 1903, pp. 115–22. This is what he says about the life of Thābit ibn Qurra: 'A Sabian from the people of Ḥarrān, he moved to the city of Baghdad and made it his own. With him, it was philosophy that came first. He lived in the reign of al-Muʿtaḍid. We are indebted to him for numerous books on different branches of knowledge such as logic, arithmetic, geometry, astrology and astronomy. We owe to him an amazing book: the *Introduction to the Book of Euclid* (*Kitāb mudkhil ilā K. Uqlīdis*), and a book: the *Introduction to Logic* (*Kitāb al-Mudkhal ilā al-manṭiq*). He translated the book on *al-Arithmāṭīqī* and summarized the book on *The Art of Healing* (*Kitāb Ḥīlat al-bur'*). In his knowledge he ranks among the most outstanding. He was born in the year two hundred and twenty-one at Ḥarrān, where he worked as a money-changer. Muḥammad ibn Mūsā ibn Shākir brought him back when he returned from the country of the Byzantines, for he had found him eloquent. He is said to have gone to live with Muḥammad ibn Mūsā and to have pursued his studies in his house. He thus had some influence over his career. Muḥammad ibn Mūsā put him in touch with al-Muʿtaḍid, and introduced him to the astronomers' circle. He it was [Thābit] who introduced Sabian management to Iraq. In this way their social position was determined, their status raised, and they attained distinction. Thābit ibn Qurra achieved so prestigious a rank and so eminent a position at the court of al-Muʿtaḍid that he would even

one that we owe to al-Nadīm, invaluable by reason of its date – the end of the tenth century – is, however, very thin. But that of al-Qifṭī, thanks to a happy accident, provides everything that posterity knows about Thābit. Good luck placed in al-Qifṭī's path papers deriving from Thābit's family that related more to his work than to his life. Al-Qifṭī's book was the source drawn on by subsequent biobibliographers, for example Ibn Abī Uṣaybiʿa. Even Ibn al-ʿIbrī (alias Bar Hebraeus),[2] who apparently had at his disposal

---

sit down in his presence at any time he wished, speak with him at length and joke with him, and come to see him even when his ministers or his intimates were not there.' (pp. 115–16).

Ibn Abī Uṣaybiʿa, *'Uyūn al-anbā' fī ṭabaqāt al-aṭibbā'* ed. A. Müller, 3 vols, Cairo/Königsberg, 1882–84, vol. I, pp. 215, 26–220, 29; ed. N. Riḍā, Beirut, 1965, pp. 295, 6–300, 23.

[2] Ibn al-ʿIbrī, *Tārīkh mukhtaṣar al-duwal,* ed. O. P. A. Ṣāliḥānī, 1st ed., Beirut, 1890; repr. 1958, p. 153.

Thābit ibn Qurra's biography is reproduced in various publications, without anything new being added. The following are a representative selection:

Ibn Kathīr, *al-Bidāya wa-al-nihāya*, ed. Būlāq, 14 vols, Beirut, 1966, vol. XI, p. 85; the account is borrowed from Ibn Khallikān.

Ibn Khallikān, *Wafayāt al-a'yān*, ed. Iḥsān ʿAbbās, 8 vols, Beirut, 1978, vol. I, pp. 313–15.

Ibn al-Athīr, *al-Kāmil fī al-tārīkh*, ed. C. J. Tornberg, 12 vols, Leiden, 1851–71, vol. VII (1865), p. 510; repr. 13 vols, Beirut, 1965–67.

Al-Masʿūdī, *Murūj al-dhahab (Les Prairies d'or)*, ed. C. Barbier de Meynard and M. Pavet de Courteille, revised and corrected by Charles Pellat, Publications de l'Université Libanaise, Section des études historiques XI, Beirut, 1966, vol. II, § 835, 1328, 1382.

Ibn al-Jawzī, *al-Muntaẓam fī tārīkh al-mulūk wa-al-umam,* 10 vols, Hyderabad, 1357–58/1938–40, vol. VI, p. 29.

Ibn Juljul, *Ṭabaqāt al-aṭibbā' wa-al-ḥukamā'*, ed. F. Sayyid, Publications de l'Institut Français d'Archéologie Orientale du Caire. Textes et traductions d'auteurs Orientaux, 10, Cairo, 1955, p. 75.

Al-Nuwayrī, *Nihāyat al-arab fī funūn al-adab,* 31 vols, Cairo, 1923–93, vol. II, p. 359.

Ibn al-ʿImād, *Shadharāt al-dhahab fī akhbār man dhahab*, ed. Būlāq, 8 vols, Cairo, 1350–51 H., (in the year 288), vol. II, p. 196–8. Repeats Ibn Khallikān.

Al-Ṣafadī, *al-Wāfī bi-al-Wafayāt*, 24 vols published (1931–1993); vol. X, Wiesbaden, 1980, ed. Ali Amara and Jacqueline Sublet, pp. 466-467.

Al-Dhahabī, *Tārīkh al-Islām* (years 281–290), ed. ʿUmar ʿAbd al-Salam Tadmūrī, Beirut, 1989–1993, pp. 137–8. Borrowed from Ibn Abī Uṣaybiʿa.

Al-Sijistānī, *The Muntakhab Ṣiwān al-ḥikmah*, Arabic Text, Introduction and Indices ed. D. M. Dunlop, The Hague, Paris, New York, 1979, pp. 122–5.

M. Steinschneider, 'Thabit ("Thebit") ben Korra. Bibliographische Notiz', *Zeitschrift für Mathematik u. Physik*, XVIII, 4, 1873, pp. 331–8.

Syriac sources that are of no great importance as far as Thābit goes, adds nothing substantial to al-Qifṭī's account. Should we then content ourselves with that? The paucity of the documentary evidence seems to me to impose the obligation to consult all of it, if only to compare the various versions.

Meagre though they are, the bibliographers' accounts in broad outline place the man in the circle in which he moved at one of the most important moments in the history of mathematics and science: the second half of the ninth century in Baghdad. The town had not only become the political centre of the world as it was then, it was also its cultural heart, and by that token a magnet for every talent. For the young people of the day who wanted to secure themselves a first-class education, the watchword was, 'Go up to Baghdad!' The city was an established scientific centre that housed a settled community of scholars whose links with the seat of power had long since been forged. For the more mature, 'going up to Baghdad' meant meeting their intellectual equals, making a name for themselves and guaranteeing themselves a career.[3] Somewhere in this barely sketched landscape we shall have to try to locate one of the crucial events in the life of Thābit ibn Qurra, his departure from Ḥarrān in Upper Mesopotamia, the town of his birth and one of the remaining centres in which elements of Hellenism were still to be found,[4] for Baghdad where he was to spend the rest of his life.

What were the particular circumstances that led to the decision that fixed the future course of Thābit's life? This is where a second event comes in, one whose effects on his destiny and his career as a scholar were by no means negligible: his meeting with the eldest of the three brothers Banu Mūsā, Muḥammad ibn Mūsā. From al-Nadīm onwards, all the biographers agree in linking these two facts: the departure from Ḥarrān and this meeting. Muḥammad ibn Mūsā had just completed a mission to Byzantine territory in

---

See also D. Chwolsohn, *Die Ssabier und der Ssabismus*, vol. I, St. Petersburg, 1856; repr. Armsterdam, 1965, pp. 546–67; E. Wiedemann, 'Über Ṯābit ben Qurra, sein Leben und Wirken', *Aufsätze zur arabischen Wissenschafts-Geschichte*, Hildesheim, 1970, vol. II, p. 548–78. *Thābit ibn Qurra. Œuvres d'astronomie*, text ed. and transl. Régis Morelon, Collection Sciences et philosophies arabes. Textes et études, Paris, 1987, p. XI–XIX.

[3] To gain some idea of the number of scholars engaged in disciplines such as literature, history, theology, etc. see al-Khaṭīb al-Baghdādī, *Tārīkh Baghdād*, ed. Muḥammad Amīn al-Khānjī, 14 vols, Cairo, 1931; repr. Beirut with an additional index volume: *Fahāris Tārīkh Baghdād li-al-Khaṭīb al-Baghdādī*, Beirut, 1986. See also A. A. Duri's article, 'Baghdād', *E.I.*^{2}, t. I, pp. 921–36.

[4] The description given by al-Masʿūdī dans *Murūj al-dhahab* shows that the traces of Hellenism in Ḥarrān towards the end of the third century of the Hegira were essentially religious. Cf. the revised edition by Ch. Pellat, vol. II, § 1389–1398, pp. 391–6.

search of manuscripts when he came across Thābit, then simply a money-changer with linguistic skills impressive enough for Muḥammad to decide to take him back with him to Baghdad. This story is quite plausible for several reasons. For one thing, there is the unanimity of the sources, certainly not in itself a compelling argument; for another, there is the privileged connection that Thābit maintained throughout his life with the Banū Mūsā, and particularly the eldest brother; and finally there is the undoubted fact that he was a gifted linguist. We have only to read his translations and his scholarly work to be convinced that this man, whose mother tongue was Syriac, had also mastered Arabic and Greek. There was perhaps an additional reason that influenced his departure – he may have had to quit his native town because of differences with his co-religionists. The only report in Arabic of this event comes from a late biobibliographer, Ibn Khallikān,[5] who mentions these quarrels and also that Thābit was forced to leave Ḥarrān for the neighbouring locality of Kafr Tūtha, in which his meeting with Muḥammad ibn Mūsā took place. Whether these reported differences actually occurred or were dreamed up by the biobibliographers is of little significance here, for even if they were a factor in his decision to leave, they were scarcely the main reason for his departure for Baghdad.

As to the date of Thābit's meeting with Muḥammad ibn Mūsā, we know nothing, just as we know nothing either of the individuals or of the circumstances that were instrumental in bringing the meeting about. But we do know that Muḥammad died in 873, and that Thābit was engaged before that date in the education of his children.[6] It is then a reasonable hypothesis that Thābit came to Baghdad relatively early and that he very likely lived there for at least 30 years, given that we know he died in 901.

The early biobibliographers have passed on some invaluable details on the relationship between Thābit and Muḥammad and his brothers. We learn that Muḥammad had accommodated Thābit in his own house on his arrival in Baghdad, where he took charge not only of his career but of his scientific education too. He was also responsible for introducing him to the circle of the Caliph's astronomers. All the early biobibliographers agree on this point. The celebrated astronomer al-Bīrūnī, a century and a half after Thābit's

[5] Ibn Khallikān, *Wafayāt al-a'yān*, vol. I, p. 313.

[6] The list of Thābit's writings cited by al-Qifṭī from Abū 'Alī al-Muḥassin al-Ṣābi' (Thābit's great-grandson; cf. Yāqūt, *Mu'jam al-Udabā',* Beirut, s.d., vol. 8, p. 152), refers to 'several summaries on astronomy and geometry that I have seen in his handwriting: these he identifies in his own hand as "what Thābit composed for the young people"; he means the children of Muḥammad ibn Mūsā ibn Shākir' (*Ta'rīkh al-ḥukamā'*, p. 120). See R. Rashed and Ch. Houzel, *Recherche et enseignement des mathématiques au IXe siècle. Le Recueil de propositions géométriques de Na'īm ibn Mūsā*, Les Cahiers du Mideo, 2, Louvain/Paris, 2004.

death,[7] alone casts a shadow of doubt on the roles played by the various parties and indeed their places in the hierarchical structure. According to him, Thābit was the corner-stone of the school of the Banū Mūsā. But we know in another connection that al-Bīrūnī, with his own acute sense of justice, had no love for the Banū Mūsā, who sometimes showed scant respect for it. In any case, there is no real contradiction here, since there is nothing to stop us from imagining that Thābit took over the leadership of the school after the death of Muḥammad ibn Mūsā, the more so in that al-Ḥasan ibn Mūsā, the brilliant geometer, was already deceased, and their brother Aḥmad ibn Mūsā was more concerned with mechanics. On the other hand, there is nothing in what has come down to us from Thābit ibn Qurra himself to suggest that he had any such role. Whenever he has occasion to mention Muḥammad, al-Ḥasan, he does so with the consideration owed to an elder. This is further exemplified in his writings by the respectful attitude he adopts towards al-Ḥasan ibn Mūsā in his research on the measure of the lateral surface of the cylinder and on the elliptical sections, and by the terms in which he refers to Muḥammad ibn Mūsā on the subject of calculating the position of the stars for the astronomical tables.

If therefore Thābit ibn Qurra had overtaken the Banū Mūsā in his research in mathematics and astronomy, that in no way contradicts the fact that it was to them that he owed his education. There is not the slightest indication to suggest that he came by any scientific education whatever in his native Ḥarrān, before he entered the school presided over by the Banū Mūsā.[8] We know of no mathematical work of his written in his native

[7] Al-Bīrūnī writes that Thābit ibn Qurra was 'the protégé of these people (the Banū Mūsā), lived among them, and was the man who steered their scientific work back to the right course', in *al-Āthār al-bāqiya 'an al-qurūn al-khāliya*, *Chronologie orientalischer Völker*, ed. C. E. Sachau, Leipzig, 1923, p. 52. It is worth noting that al-Bīrūnī, fair-minded as he was, further on had no hesitation in paying tribute to the Banū Mūsā for their observation on the mean moon, declaring that, of all his predecessors, they were the ones whose statement on the subject one should opt for (p. 151):

لبذلهم المجهود في إدراك الحق وتفردهم في عصرهم بالمهارة في عمل الرصد، والحذق به، ومشاهدة العلماء منهم ذلك ...

On the other hand, in *al-Istī'āb*, he finds fault with their attitude towards al-Kindī. The story is well known and often reported.

[8] It would be a matter of great interest for the history of philosophy, mathematics and the sciences if we knew exactly how much activity there was in these fields at Ḥarrān in the eighth and especially the ninth century. Such knowledge is obviously indispensable to a better understanding of how Arabic became a vehicle for the transmission of the legacy of Greece, and how particular disciplines came to be established in that language. In the

language, Syriac. The two mathematical books in Syriac cited by Ibn al-ʿIbrī are mentioned by al-Qifṭī[9] and repeated along with the list as a whole

---

absence of such information, it can often happen that people offer conclusions before embarking on the relevant research. They extrapolate from the most remote periods and it is none other than Thābit ibn Qurra that they call on for their evidence and at the same time treat as the main proof of their case. This sort of reasoning is quite clearly marred by circularity: what they would have needed to do first was to set out what Thābit owed to the philosophy and science that was going on at Ḥarrān during his formative years. We shall look in vain in his biography or in his writings for any shred of evidence, any scintilla of support for the notion that he had received any such education before his meeting with the Banū Mūsā and his arrival in Baghdad. Thus two questions remain open. Was there any such activity at that time in Ḥarrān as scientific and philosophic teaching in any shape or form other than that hallowed by tradition in religion and the occult sciences? Could Ḥarrān claim to possess genuine libraries, and not mere repositories of old books that were now beyond the comprehension of the Sabians of the time? This is a perfectly reasonable question in view of the situation described by Ibn Waḥshiyya affecting a comparable community, whose members could no longer understand their ancestors' books, but, even so, piously and jealously guarded them (*al-Filāḥa al-nabaṭiyya*, ms. Istanbul, Topkapi Saray, Ahmet III 1989, fol. 1^r–2^r; critical edition by Toufic Fahd, vol. I, Damascus, 1993). Let us now turn to Thābit himself, whose praise of Ḥarrān and the Sabians is recorded for us by Ibn al-ʿIbrī: 'Many <Sabians> were constrained to forsake the true path for fear of persecution. Our fathers, by contrast, were able to withstand what they withstood with the help of the Most High, and achieved their salvation through their own courage. The blessed town of Ḥarrān was never sullied by its Christians' straying from the true path. *We* are the Sabians' heirs, and they are *our* heirs, dispersed throughout the world. Anyone who bears the burdens borne by the Sabians with confident hope will be held to enjoy a happy destiny. Oh, please Heaven! Who but the best of the Sabians and their kings brought civilisation to the land, built the towns? Who constructed the harbours and canals? Who explained the occult sciences? Who were the people to whom the divine power that made known the art of divination and taught the future was revealed? Were they not the renowned Sabians? It was they who elucidated all that and who wrote on the art of medicine for souls and on their deliverance, and who published also on medicine for the body, and filled the world with good and wise deeds that are the bulwark of virtue. Without the Sabians and their knowledge, the world would be deserted, empty, and sunk in destitution.' From Thābit's own words – at any rate according to Ibn al-ʿIbrī – it emerges clearly that the Sabians of his time excelled in practical skills, occult sciences and medicine. But there is no mention either of mathematics or of mathematical sciences. All that might count as philosophy is 'medicine for souls'. These, however, are the very areas that were given prominence by the early historians and biobibliographers. Al-Nadīm for example tells us that astrolabes were first manufactured at Ḥarrān before the craft was taken up elsewhere and became widespread under the Abassids (*al-Fihrist*, p. 342). See also note 9 below. On Ḥarrān, see Tamara M. Green, *The City of the Moon God*, Leiden, 1992, which contains a bibliography.

[9] The list of Thābit's writings drawn up by his great-grandson and reproduced by al-Qifṭī includes nine titles in Syriac. On the other hand, Ibn al-ʿIbrī, in a book written in

by Ibn Abī Uṣaybiʿa in Arabic. Both deal with Euclid's fifth postulate, and there is nothing in either to support the contention that the Syriac version was the first to be written. On the contrary, the reverse may well be true, especially since, at the time, it was common practice for Arabic texts to be translated into Syriac.

All things considered, the following conclusion may, then, be put forward: this man of outstanding intellect came to Baghdad with Muḥammad ibn Mūsā, joined the school of the Banū Mūsā and lost no time in becoming one of its active members. He followed the way opened by al-Ḥasan ibn Mūsā, particularly in his work on the measure of curved planes and solids, and on the properties of conic sections. He collaborated with Aḥmad ibn Mūsā, translated the last three books of Apollonius' *Conics*, and, in astronomy and also in philosophy, carried on certain aspects of the work of Muḥammad, with whom he maintained a close and enduring relationship. From the prestigious town of his birth, Ḥarrān, he seems to have taken with

---

Syriac, *Tārīkh al-zamān* (Arabic translation by Father Isḥāq Armala, Beirut, 1980, pp. 48–9), refers to Thābit and attributes to him 'about one hundred and fifty books in Arabic' and 'sixteen books in Syriac, the majority of which we have read'. The two Syriac lists have seven titles in common and thus provide us with an indirect way of assessing Thābit ibn Qurra's output in that language, and how much his scientific and philosophic education might possible owe to his native town of Ḥarrān.

Eleven of the sixteen titles are devoted to religion and Sabian rites; one to a history of the ancient Syriac kings, that is to say the Chaldaeans, one to a 'history of the famous members of his family, and the lineage of his forefathers', 'the book on music', and finally, 'a book on: if two straight lines are drawn following <two angles> that are equal to less than two right angles, they meet, and another book on the same subject'. Now this last title (comprising two books), listed in *Tārīkh al-zamān* as being in Syriac, turns up almost word for word in the list of Thābit's works in Arabic drawn up by al-Qifṭī (*Ta'rīkh al-ḥukamā'*, p. 116), and later in that of Ibn Abī Uṣaybiʿa (*'Uyūn al-anbā'*, ed. Müller, vol. I, p. 219, 4; ed. Riḍā, p. 299, 3–4); furthermore, the Arabic manuscript of this work has, happily, come down to us, confirming its title (mss Istanbul, Aya Sofya 4832, fol. 51^{r}–52^{r}; Carullah 1502, fol. 13^{r}–14^{v}; Paris, BN 2457, fol. 156^{v}–159^{v}). So it turns out that the only mathematical title cited as being in Syriac also exists in Arabic. Given that there was a corresponding Syriac version, it might then be thought that Thābit himself translated his Syriac text into Arabic. But nothing is less certain, nor is there any lexical, stylistic, let alone mathematical indication that might lend weight to a conjecture on these lines. Exactly the opposite assumption, on the other hand, that the work was translated from Arabic into Syriac, is not only possible, but would reflect a practice that was current at the time. Finally, the hypothesis that both versions were produced at the same time should not be excluded. The author was, after all, completely bilingual, as he had demonstrated when he was engaged in revising the translation of the *Elements*. Sooner than allowing ourselves to be swamped by conjectures, let us stick to this one, negative certainty: there is nothing to show that he received any scientific instruction at all in Ḥarrān.

him only his religion, his knowledge of languages and perhaps some philosophy, while it was in Baghdad that he learned mathematics and astronomy.

Like his fellow-townsmen, Thābit ibn Qurra was of the Sabian persuasion, a follower of a Hellenistic faith that needed all the hypocrisy its exegetes could muster in order to qualify for the status of a recognized 'religion of the Book', which alone could guarantee its free practice in Islamic territory. This meant not just that he was tolerated as a member of a subordinate community, but also that he enjoyed full citizen rights, including the right to seek and to attain the highest positions in the land. In this he was far from unique, and many other scholars emanating from religious minorities secured themselves most exalted posts. All the biobibliographers recount episodes from his life at court, where the caliph heaped favours upon him. This 'promotion', which is often referred to anecdotally, seems to me to deserve very much closer attention from historians. It was neither exceptional nor ephemeral and throws light for us on the social status to which a scholar could lay claim in the second half of the ninth century in an Islamic city, and at the same time exemplifies the esteem in which the light of knowledge was held by the ruling authorities.

The career of Thābit ibn Qurra provides a good illustration of the power to attract talent that marked Baghdad's prestige at this period, and an example too of how membership of a religious minority was no bar to achieving the highest offices of state; and there is a third reason for choosing him as a representative case: he is a source of information on the development of the schools and their scientific traditions. As an active member of the school of the Banū Mūsā, and tutor to the sons of Muḥammad ibn Mūsā, he was able to ensure the school's continued existence after the deaths of Muḥammad, Aḥmad and al-Ḥasan. In due course, his own descendants and pupils took over from him. Thābit's children and grandchildren, including the mathematician Ibrāhīm ibn Sinān, and pupils of his such as Naʿīm ibn Mūsā, traces of whom we have only recently brought to light,[10] were to carry on the work for three generations at least. We are not yet in a position to give anything like a complete account of the ramifications in the structure and tradition of the school, but, even as it is, through Ibn Qurra, we can glimpse their outline.

One further aspect of Thābit ibn Qurra's career, when added to the three already mentioned, will allow us to complete the picture of our mathematician: he was also a translator. We are aware of the considerable number of Greek treatises he translated into Arabic, including Archimedes'

[10] See R. Rashed and Ch. Houzel, *Recherche et enseignement des mathématiques au IXe siècle*.

Sphere and Cylinder as well as the last three books of Apollonius' *Conics* and the *Arithmetical Introduction* of Nicomachus of Gerasa. He also revised many other translations, including, among others, Euclid's *Elements* and Ptolemy's *Almagest*. These few titles are enough in themselves to illustrate the wide range of topics Thābit covered and to bring out the closeness of the ties that link innovative research and translation, and indeed their mutual dependency – something that I have been at pains to emphasize.[11] The example of Thābit himself proves my point in full measure, since in his case the two activities are combined in one and the same person.

As a gifted translator and one of the most eminent mathematicians that ever lived, Thābit ibn Qurra's status as a luminary remains unchallenged; indeed, down the centuries, no one has cast the slightest doubt on his importance. His renown in the East as well as the Muslim West, the translation of some of his works into Latin and others into Hebrew, are eloquent enough testimony.[12] From the point of view of the history of mathematics, to overlook Thābit's contribution is quite simply to forego the possibility of understanding the development of the subject over the following two centuries, especially in the field that concerns us here.

Let us now return to the early biobibliographers to take up two particular points: Ibn Qurra's name and his dates. All of them report his name in the same way, Thābit ibn Qurra, and give his lineage from the sixth generation of his ancestors. Al-Qifṭī confirms the accuracy of this information, on the basis of the family papers to which he had been able to gain access. He had got his hands on the evidence written down by Abū ʿAlī al-Muḥassin ibn Ibrāhīm ibn Hilāl al-Ṣābiʾ, who was none other than Thābit's great-grandson. Abū ʿAlī's father, as we know,[13] had in 981 copied a manuscript in Thābit's own hand, and it appears to have been this branch

[11] Cf. my 'Problems of the Transmission of the Greek Scientific Thought into Arabic: Examples from Mathematics and Optics', *History of Science*, 27, 1989, pp. 199–209; repr. in *Optique et mathématiques: Recherches sur l'histoire de la pensée scientifique en arabe*, Variorum Reprints, London, 1992, p. 199–209.

[12] On his impact in Latin, see for example F. J. Carmody, *The Astronomical Works of Thābit b. Qurra*, Berkeley/Los Angeles, 1960. See also A. Björnbo, 'Thābit's Werk über den Transversalensatz', *Abhandlungen zur Geschichte der Naturwissenschaften und der Medezin,* 7, 1924; and also F. Buchner, 'Die Schrift über den Qarastûn von Thabit b. Qurra', *Sitzungsberichte der physikalisch-medizinischen Sozietät in Erlangen*, Bd 52-53, 1920/21, pp. 141–88.

[13] Attention was drawn to this Istanbul manuscript, Köprülü 948, by H. Ritter, according to K. Garbers, *Ein Werk Ṯābit b. Qurra's über ebene Sonnenuhren*, Dissertation, Hamburg/Göttingen, 1936, p. 1. See also E. Bessel-Hagen, O. Spies 'Ṯābit b. Qurra's Abhandlung über einen halbregelmässigen Vierzehnflächner', in *Quellen und Studien zur Geschichte der Math. und Phys.*, B. 2.2, Berlin, 1932, pp. 186–98; and *Thābit ibn Qurra. Œuvres d'astronomie*, ed. Régis Morelon, p. 301.

of the family that preserved the family papers. In his book dating from 647/1249 al-Qifṭī writes:

> As to the titles of his [Thābit's] written works, I have found pages in the hand of Abū ʿAlī al-Muḥassin ibn Ibrāhīm ibn Hilāl al-Ṣābi' that included mention of the lineage of this Abū al-Ḥasan Thābit ibn Qurra ibn Marwān, and likewise mention of the books he had written, in an exhaustive and complete fashion [...] which I append below, since it is a proof of that matter.[14]

This invaluable piece of evidence leaves no room for doubt either about Thābit's name or as to his writings. But when it comes to his date of birth, we are a long way from the same degree of certainty. Al-Nadīm in fact notes the year as 221/836, and then goes on to tell us that he died at the age of 77 solar years. Now, if this date were accepted for his birth, he would have lived for only 65 solar years, or 67 lunar years, since he died on Thursday 26 Ṣafar, in the year 288 of the Hegira, *i.e.* Thursday 19th February, 901. Al-Qifṭī repeats the birth date given by al-Nadīm without noticing the discrepancy. Still today, there are historians who follow al-Qifṭī and fail to observe the contradiction this dating entails. Ibn Abī Uṣaybiʿa, on the other hand, states that he was born on Thursday 21 Ṣafar 211 of the Hegira, *i.e.* 1st June 826, which is indeed a Thursday.[15] This date seems reasonable, in that it fixes his lifespan at 77 *lunar* instead of solar years, as al-Nadīm would have it. This is also the date given by the late biobibliographer al-Ṣafadī.[16]

2.1.2. *The works of Thābit ibn Qurra in infinitesimal mathematics*

A considerable body of purely mathematical work can be attributed to Thābit ibn Qurra, even if we omit his research in astronomical mathematics and statics. This work covers geometry, geometric algebra and number theory.[17] Thābit left his mark on every field of mathematics. Our discussion herein is limited to his works in infinitesimal mathematics. However, we should be clear on one point. Infinitesimal mathematics can be found throughout the works of Ibn Qurra. In astronomy, he uses infinitesimal processes

[14] Al-Qifṭī, *Ta'rīkh al-ḥukamā'*, p. 116.

[15] This date is confirmed by the Paris Observatory, thus providing irrefutable proof that renders superfluous all attempts to base conclusions solely on the often contradictory evidence contained in the bibliographical and historical sources.

[16] Al-Ṣafadī, *al-Wāfī bi-al-Wafayāt*, vol. 10, p. 466–7.

[17] For an overview, see the article 'Thābit ibn Qurra', *Dictionary of Scientific Biography*, vol. XIII, 1976, pp. 288–95, by B. A. Rosenfeld and A. T. Grigorian.

to examine the problem of the 'visibility of crescents',[18] and they also appear when he discusses 'how the speed of movement on the ecliptic can appear slower, average, or faster, depending on the point of the eccentric at which it occurs'.[19] Thābit also applies infinitesimal processes to statics in his book *al-Qarasṭūn*.[20] However, we know of only three works on infinitesimal geometry, all of which have fortunately survived.

The early biobibliographers refer only to the following three titles: *The Measurement of the Conic Section known as the Parabola*; *The Measurement of the Paraboloids*; and *On the Sections of the Cylinder and its Lateral Surface*. These three titles are given in the list reproduced by al-Qifṭī, and in the article on Thābit written by Ibn Abī Uṣaybiʿa.

The information given by the biobibliographers agrees with what was presented by Thābit himself, who confirms that he only determined the areas and volumes of these curved figures:

> With regard to the plane figures, it is like that which resembles a circle without being a circle, given that its length is greater than its width, and which is called an ellipse, together with other conic sections and the cylinder. I have shown this in the books that I have composed, describing my findings and determinations on this subject. With regard to the solid figures, these are those formed by rotating the plane figures.[21]

This refers exactly to the figures discussed in the three books mentioned earlier.

Finally, the internal references in the works of Thābit provide further confirmation. In his treatise on *The Measurement of the Paraboloids*, he cites the text on *The Measurement of the Parabola*, and in another, obviously later work, he makes reference to the third treatise as follows: 'With regard to the lateral surface area of a cylinder, I determined this and proved in my book on the sections of a cylinder and its lateral surface that ...'.[22] No other publication in the field of infinitesimal mathematics has been attributed by anyone to Thābit, or has been cited by Thābit himself.

[18] R. Morelon, *Œuvres d'astronomie*, pp. 93–112.

[19] *Ibid*., pp. 68–9.

[20] See E. Wiedemann, 'Die Schrift über den Qarasṭūn', *Bibliotheca Mathematica*, 12, 3, 1911–12, pp. 21–39, a translation of the Arabic text of the Thābit book. A defective edition of this text, with a French translation, has been published by Kh. Jaouiche (*Le livre du Qarasṭūn de Ṭābit ibn Qurra*, Leiden, 1976). See also W. R. Knorr, 'Ancient sciences of the mediaeval tradition of mechanics', in *Supplemento agli Annali dell'Istituto e Museo di Storia della Scienza*, Fasc. 2, Firenze, 1982.

[21] See my edition and French translation of this treatise, *On the Measurement of Plane and Solid Figures*, in *Thābit ibn Qurra. Science and Philosophy in Ninth-Century Baghdad*, p. 208, Arabic text p. 209, 13–17.

[22] *Ibid*., p. 199, 7–8.

The early biobibliographers also cited two other titles by Thābit, one of which has given rise to the incorrect supposition that he contributed to a book by the Banū Mūsā. This was a treatise 'On the Measurement of Plane and Solid Figures', *Fī misāḥat al-ashkāl al-musaṭṭaḥa wa-al-mujassama*. It is true that the similarity of this title to a work of the Banū Mūsā could lead to this conjectural conclusion. If one considers the strong ties between Thābit and the Banū Mūsā, the assumption that he contributed to their work does not require any great deductive leap. However, closer examination of the text reveals that it does not contain proofs. The author simply gives the formulae for determining the areas of plane, rectilinear and curvilinear figures, together with the volumes of certain solids, including the cube and the sphere. It therefore has nothing in common with the work of the Banū Mūsā and, moreover, it does not touch on the problems of infinitesimal mathematics.

There remains the one enigmatic work by Thābit, of which we know nothing. Given that it is entitled 'The Measurement of Line Segments', (*Kitābihi fī misāḥat qaṭ' al-khuṭūṭ*), there is very little chance that it had anything to do with infinitesimal mathematics. We should finally note the famous philosophical correspondence with Abū Mūsā 'Īsā ibn Usayyid in which Thābit defends the concept of a true infinity. [23]

2.1.3. *History of the texts and their translations*

The history of the manuscript tradition of the works of Thābit ibn Qurra that we are about to describe is somewhat paradoxical. In one sense it is sparse, as only single copies survive of two of the treatises. In another sense it is very rich, as the earliest copies are all found in valuable collections. *The Measurement of the Parabola* alone survives in five copies, including the two mentioned earlier. We will now consider each of these texts in detail.

THE MEASUREMENT OF THE PARABOLA

Five manuscript copies of this book by Thābit ibn Qurra have survived.

1) The first manuscript, referred to here as copy A, occupies folios 26^v–36^v in the 4832 Aya Sofya collection in the Süleymaniye Library in Istanbul. This collection includes a large number of works by Thābit. It was part of the estate left by the Sultan al-Ghāzī Maḥmūd Khān. The history of

[23] See Marwan Rashed, 'Thābit ibn Qurra sur l'existence de l'infini: les *Réponses aux questions posées par Ibn Usayyid*', in *Thābit ibn Qurra. Science and Philosophy in Ninth-Century Baghdad*, pp. 619–73.

the collection is described in folio 1^{r}: 'It has been said that this book belonged to Abū ʿAlī al-Ḥusayn ibn ʿAbd Allāh ibn Sīnā'.

Impossible to verify, this claim may well be legendary but it bears witness to the prestige once enjoyed by this collection – and which it continues to enjoy. For us, the important point to note is that it mentions the name of one previous owner, a certain Ibn al-Ḥamāmī, who bought it on 'the nineteenth day of Rajab in the year five hundred and sixty eight' (of the Hegira), *i.e.* 6th March 1173. The copy cannot therefore be later than the sixth century of the Hegira, and is very possibly up to a century older. The following line also appears in folio 1^{r}: 'It has been mentioned that this book is the work of al-Shaykh al-Ra'īs ... Abī ʿAlī al-Ḥusayn ʿAbd Allāh ibn Sīnā. May God have Mercy upon him'. This claim is no more verifiable than the first, but it does indicate a strong belief that the collection is very old. The text is written with care, using *naskhī* calligraphy, and the paper is smooth with a slight red tint. The paper size is (21.8/11.6), and then text (17.9/9.1). All the sheets of paper come from the same manufacturer. The copyist left some pages blank, and some of these were used later by other copyists; such as folio 57, copied in the year 700 of the Hegira. The pages have been numbered in a later hand. The text is copied in black ink, while the figures are carefully drawn in red ink. It is bound in reinforced board, and the spine is in brown leather that has been recently restored.

Ibn Qurra's text is in the hand of the copyist, without any addition or commentary. There are a few marginal notes in the same hand, all of words or phrases omitted during the copying process.

2) The second manuscript, referred to here as copy B, forms part of Collection 2457 in the Bibliothèque Nationale de Paris. *The Measurement of the Parabola* occupies folios 122^{v}–134^{v}. This manuscript consists of 219 paper sheets (18/13.5).[24] The section of this collection that interests us was copied by the geometer Aḥmad ibn ʿAbd al-Jalīl al-Sijzī in Shīrāz in 359/969. This volume was taken to Cairo at the beginning of the nineteenth century by a pupil of Caussin de Perceval named Reiche. Al-Sijzī corrected the original when making his copy. On page 132^{v} appears the word بلغ with a corresponding ∴ sign in the margin. This indicates one of the steps in al-Sijzī's revision. There are no additions or commentaries in any other hand. The only marginal notes are words or phrases omitted during the

[24] It has been described by G. Vajda, 'Quelques notes sur les fonds de manuscrits arabes de la Bibliothèque Nationale de Paris', *Rivista degli Studi Orientali*, 25, 1950, 1 to 10; *Index général des manuscrits arabes musulmans de la Bibliothèque Nationale de Paris*, Publications of the Institut de recherche et d'histoire des textes IV, Paris, 1953, p. 481.

copying process. The text is written in *naskhī* calligraphy, with drawn geometric figures.

3) The third manuscript, referred to here as copy Q, forms part of Collection 40 in the Dār al-Kutub in Cairo. *The Measurement of the Parabola* occupies folios 165^v–181^r. This 226-sheet manuscript is relatively recent, having been made in the eighteenth century by the copyist Muṣṭafā Ṣidqī, a name that we have come across on a number of occasions.[25] He completed the copy on the 12th day of Dhū al-Qaʿda in year 1159 of the Hegira, *i.e.* 26th November 1746. The text is written using *naskhī* calligraphy, and the copy contains no additions or commentaries. There is also nothing to indicate that the copyist made any corrections to the original. This collection includes other works by Thābit, together with a number of texts by Ibn Sinān and al-Qūhī.

4) The fourth manuscript, referred to here as copy M, forms part of Collection 5593 in the Astān Quds library in Meshhed. This collection consists of 156 sheets (16.5/8), and it was copied in the year 867 of the Hegira, *i.e.* 1462/3. The text of *The Measurement of the Parabola* occupies folios 26–42. The text is written in *nastaʿlīq* calligraphy and the geometric figures have not been drawn in, although blank spaces have been left in the text for them. There are no additions or commentaries, and no indication of any corrections to the original.

5) The fifth manuscript, referred to here as copy D, forms part of Collection 5648 in Damascus.

Comparing these manuscripts shows that the Damascus copy, D, was taken from the Cairo copy, Q, and no other manuscript. In discussing the origins of the text, we can therefore discount copy D. Conversely, the Paris manuscript, B, belongs to a separate manuscript tradition, independent of all the others. Manuscript B omits 19 sentences and 90 words, including 14 occurrences of the word *ʿadad,* 7 occurrences of the word *khaṭṭ,* and 2 occurrences of the word *ḍarb*. The omissions common to manuscripts B and Q are a single occurrence of the word *ʿadad* and a single occurrence of the word *khaṭṭ*, which is insignificant. This comparison of omissions is confirmed by a comparison of mechanical errors. It can only be concluded that copy B belongs to a different manuscript tradition from those of A and Q.

[25] R. Rashed, *Geometry and Dioptrics in Classical Islam*, London, 2005, Chap. I.

Manuscript M has been copied from a precursor of A. All the terms and expressions omitted from A are also omitted from M, with the exception of the two words: 35/22 and 44/10, which the copyist of M could easily have inferred from the context. On the other hand, there are a few grammatical errors in A that do not appear in M, and some repetitions in A that are avoided in M. Given that the copyist of M is known to have been careless, it is not likely that M is a descendent of A. In any event, we have only cited the variants of M where it is different from A.

The manuscripts Q and A share a common ancestor. Copy Q has the following unique omissions: 1 sentence and 15 words, including 1 occurrence of the word *khaṭṭ*, 4 occurrences of the word *ḍarb,* and 4 occurrences of the word *'adad*. It shares the following omissions with copy A: 10 sentences and 55 words, including 7 occurrences of the word *ḍarb*, 13 occurrences of the word *'adad* and 44 occurrences of the word *khaṭṭ*. As we have seen, there are hardly any omissions shared with copy B. Finally, there is one sentence (2/6) copied incorrectly in both A and Q. However, the errors are different, suggesting that the sentence in question was unclear in their common ancestor. The sentence in A reads as follows: *fa-'adad K akthar min 'adad aṣghar min AB* ('The number *K* is therefore larger than a number smaller than *AB*'). The same sentence in copy Q reads: *fa-'adad K akthar min C wa-aṣghar min 'adad AB, fa-'adad K akthar min 'adad aṣghar min AB*. The copyist has then crossed out the final phrase, leaving: 'The number *K* is larger than *C* and smaller than the number *AB*'. The sentence in copy A has therefore been crossed out by the copyist Muṣṭafā Ṣidqī. However, we know that the latter had the mathematical knowledge to understand what he was copying.

The results of all these comparisons enable the construction of the following *stemma*:

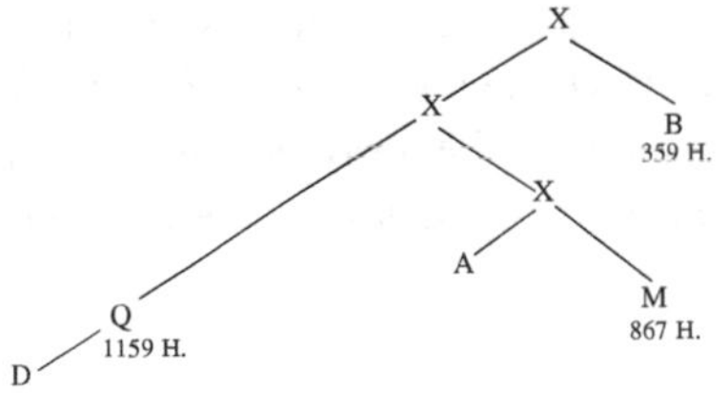

The true text of *The Measurement of the Parabola* can therefore be derived from copies A, B and Q, and from M, when M differs from A. This gives us the *princeps* edition of this treatise, together with its translation. It should be noted that H. Suter has made a partial and free translation of manuscript B of Thābit's text. This translation only includes certain passages and does not attempt a literal translation of the original. While this provisional work is little more than a freely translated abstract, it is however useful as a means of providing mathematical historians with access to the contents of this treatise by Thābit.[26]

THE MEASUREMENT OF THE PARABOLOIDS

The only surviving manuscript of this text, copied in 358/969 is held in Collection B in Paris, folios 95^{v}–122^{r}. A serious mistake was made during the copying process, which has not been noticed until now. Al-Sijzī repeated three folios, 110^{v}–113^{r}. We have called this fragment M, the first letter of the Arabic word meaning 'repeated'. Strange as it may seem, this mistake effectively gives us a second copy, providing us with an insight into some of the details of al-Sijzī's technique as a copyist. Compared with M, B includes one omission, one repetition and five errors. Al-Sijzī has also made one correction to B that he failed to make to M. Compared with B, M omits two sentences and one word. Three sentences are repeated in M. This comparison of the two fragments is reassuring, as it shows that the mathematician is acting as a true copyist, capable of making a few small mistakes. All the comments we have made regarding al-Sijzī in our discussion of *The Measurement of the Parabola* remain word for word applicable to *The Measurement of the Paraboloids*.

For this text also, we have provided the *princeps* edition. H. Suter[27] produced an abstract of this treatise similar to that for the first.

ON THE SECTIONS OF THE CYLINDER AND ITS LATERAL SURFACE

As with the previous treatise, only one manuscript copy of *On the Sections of the Cylinder and its Lateral Surface* survives. It occupies folios 4^{r}–26^{r} in the 4832 Aya Sofya collection A. Everything that we have said with reference to *The Measurement of the Parabola* applies equally to this text. However, in this case, the copyist has left space for almost all the references Thābit makes to the *Conics* of Apollonius. Did he intend to add

[26] H. Suter, 'Über die Ausmessung der Parabel von Thâbit b. Ḳurra al-Ḥarrânî', *Sitzungsberichte der phys.– med. Soz. in Erlangen*, 48, 1916, pp. 65–86. This text has also been translated into Russian from manuscript B by J. al-Dabbagh and B. Rosenfeld. See Thābit B. Qurra, *Matematitcheskie traktaty* (in Russian), Coll. nautchnoie nasledstro, vol. 8, Moscow, 1984.

[27] H. Suter, 'Die Abhandlungen Thâbit b. Ḳurras und Abû Sahl al-Kûhîs über die Ausmessung der Paraboloide', *Sitzungsberichte der phys.– med. Soz. in Erlangen*, 49, 1917, pp. 186–227. This text has also been translated into Russian by J. al-Dabbagh and B. Rosenfeld; see *Matematitcheskie traktaty,* pp. 157–96.

them later in a different ink? Or did they not appear in the original that he was copying?

There also exists an edition of Thābit's treatise made by the mathematician Ibn Abī Jarrāda in the thirteenth century, which has been erroneously assumed to be another copy of the original treatise. The Ibn Abī Jarrāda text is a complete re-write of Thābit's treatise, making it useless in tracing its history. Ibn Abī Jarrāda also added a number of lemmas to the original Thābit text, together with one completely new proof. In his favour, it has to be said that he took care to distinguish his own additions from the re-written version of Thābit's text. However, while the re-written version doubtless preserves the spirit of the original, it is not a faithful copy. Ibn Abī Jarrāda also gives all the references to the *Conics* of Apollonius. Did he find them in his copy of the Thābit treatise, or did he copy them from his own copy of the *Conics*?

Only one copy of the Ibn Abī Jarrāda edition now exists, in folios 36ᵛ–64ᵛ of Collection 41 in the Dār al-Kutub in Cairo. It is stated in this copy that Ibn Abī Jarrāda composed the text in 691/1292. We can recognize the hand of Muṣṭafā Ṣidqī, even though he is not named explicitly. Muṣṭafā Ṣidqī completed his copy on the 25th day of Rabīʿ al-awwal 1153, *i.e.* 20th June 1740. This edition by Abī Jarrāda demonstrates the continuing interest in this field at the end of the thirteenth century. We have included the lemmas and proofs that he added in the Supplementary notes, and we have used his text to restore the references that Thābit made to the *Conics*. This is acknowledged on each occasion.

We provide here the annoted English translation of the *princeps* edition of this text.[28]

[28] The commentary by Ibn Abī Jarrāda has been translated into Russian by J. al-Dabbagh and B. Rosenfeld, *Matematitcheskie traktaty,* pp. 196–236, as though it were the work of Thābit. See previous note.

2.2. MEASURING THE PARABOLA

2.2.1. *Organization and structure of Ibn Qurra's treatise*

The *Treatise on the Measurement of the Parabola* occupies a particularly important place in Thābit ibn Qurra's own work, in the history of infinitesimal mathematics and in the historiography of the *Archimedes Arabus.* Indeed, it is the mathematician's first book dedicated to the areas and volumes of curved surfaces and solids. In this book, Thābit ibn Qurra introduces essential ideas that he doesn't hesitate to recall in his second treatise *On the Measurement of the Paraboloids.* Moreover, the book had elicited in itself a wave of research on the parabola's measurement lasting nearly a century after his death, involving several leading mathematicians: al-Māhānī, Ibrāhīm ibn Sinān and Ibn Sahl. The first tried to pare down Ibn Qurra's 20 preliminary propositions. The second, who didn't want to let anyone overtake his grandfather without being passed in turn by another family member, proceeded to reduce the number of preliminary propositions to two. The last, in all likelihood, wished to improve the method itself; his book unfortunately has not reached us, but it was cited by his contemporary al-Qūhī, and one could, like him, find traces of it in Ibn al-Haytham's work on the measurement of the paraboloid and the measurement of the sphere. Ultimately, Ibn Qurra's *Treatise on the Measurement of the Parabola* allows us to appreciate the state of knowledge of the Archimedean corpus in Arabic, and specifically to know whether the contributions of the mathematician from Syracuse were known at all. Briefly discussed here, these questions will be revisited in detail at several points, particularly in the third volume. We note for now that Ibn Qurra, who manifestly ignored the work of Archimedes as much on the parabola as on conoids and spheroids, is seen to be constrained to cut a new path, and to forge the necessary conceptual tools for determining the area of a portion of a parabola. Throughout, we portray and analyse this path and the means brought into play: globally, a tendency toward arithmetization surpassing that of which we can observe in Archimedes, but treated somewhat less nimbly; an explicit use of the properties of the upper bound of a convex set; and a recourse to the famous Proposition X.1 from Euclid's *Elements*, both to guarantee the necessary approximation for the method of exhaustion and to settle the question of existence. We shall see how Ibn Qurra, in the *Treatise on the Measurement of the Paraboloids*, had thought to extend the usage of this Euclidean proposition.

These features, clarified by a phenomenological description of the work, in fact emerge from the very structure of the book. It is in making manifest this latent structure that we will show these to be Ibn Qurra's aims. We will have recourse to a method that has proven fruitful in other circumstances:[1] we establish the graph of the logical relations of implication between the different propositions on the basis of the proofs given by Ibn Qurra. We then try to grasp which semantic structure superimposes itself upon this syntactic structure in order to understand how these two structures co-determine each other and, together, determine the book's organization. In addition, our method offers the advantage of being the precious auxiliary of the philologico-historical method, localizing in the text the interpolations and eventual omissions of propositions. It keenly draws our attention to the isolated propositions, and incites us to a supplementary and narrow examination of the text in its context. But a mere glimpse suffices, it seems, to convince ourselves that, in the case of Ibn Qurra's book, this risk is non-existent. (See the graph of implications between propositions from Ibn Qurra's *Measurement of the Parabola*.)

Ibn Qurra's book, as he presents it, is composed of two lemmas, twenty preliminary propositions, and one theorem, which divide up, as one can see from the graph, into three groups. The first consists of two lemmas and twelve propositions, which all pertain to integers and sequences of integers. The second group is of four propositions dedicated to segments and sequences of segments. The third group is composed of four propositions, and the theorem, pertaining to the parabola. We already clearly see the importance of arithmetical propositions in Ibn Qurra's book. We further observe that the graph consists of three levels: the first, on the arithmetical propositions, is the foundation for the second, devoted to segments; now the latter also depends on the introduction of the Axiom of Archimedes in order to proceed with the necessary bounds. The third level, on the parabola, rests on its two predecessors, but also on the propositions from Apollonius's *Conics*, and on Proposition X.1 from Euclid's *Elements*, or rather on the properties therein recounted: those of the parabola as a conic section and of the method of approximation.

One already glimpses at least the shadow of the semantic structure that superimposes itself on this syntactic structure. It will appear clearly if one reads the graph another way. Here, it is split into two levels, where one concerns equalities and the other inequalities. In the first, Ibn Qurra establishes the propositions that bear upon equalities between sequences of integers in order to pass to equalities between ratios of sequences of integers, as well as ratios of sequences of segments, and to take us directly back to Proposition 18. Thanks to the Axiom of Archimedes, he turns the preceding

[1] R. Rashed, 'La mathématisation des doctrines informes dans la science sociale', in G. Canguilhem (ed.), *La mathématisation des doctrines informes*, Paris, 1972, pp. 73–105.

equalities between ratios into inequalities, as one notes in Proposition 15, in order to return directly to Proposition 20. Now these are precisely the Propositions – 18 and 20 – that, with Proposition 19 introduced *ad hoc*, allow the final proof of the theorem. The structure of the significations brings out the syntactic structure; the one always assures the realization, but also the scope, of the other: the arithmetical propositions are there in order to prepare the partitions of the diameter of the portion of the parabola, just as the inequalities on the sequences of segments prepare for the introduction of the properties of the upper bound; which is to say that the polygons created in the wake of these partitions have for an upper bound the area of the portion of the parabola.

This description might seem somewhat less succinct; the given analysis of the propositions from this book will finish the clarification. We must, however, begin by recalling the explicit definitions, denoted D, along with the propositions used in the course of the proof, and which have been considered as axioms – here denoted A – and lemmas – here denoted L – proven by contradiction.

D_1 consecutive integers
D_2 consecutive odd numbers
D_3 consecutive even numbers
D_4 consecutive squares

A_0 The difference between two consecutive integers is one.
A_1 The difference between two consecutive even numbers is two.
A_2 The difference between two consecutive odd numbers is two.
A_3 Between two consecutive even numbers is an odd number.
A_4 The product of an integer and two is an even number.
A_5 Every odd number increased by one gives an even number.

L_1 Two consecutive squares are the squares of two consecutive integers (lemma proven in the first proposition).

L_6 Two consecutive odd squares are the squares of two consecutive odd numbers (lemma proven in Proposition 6).

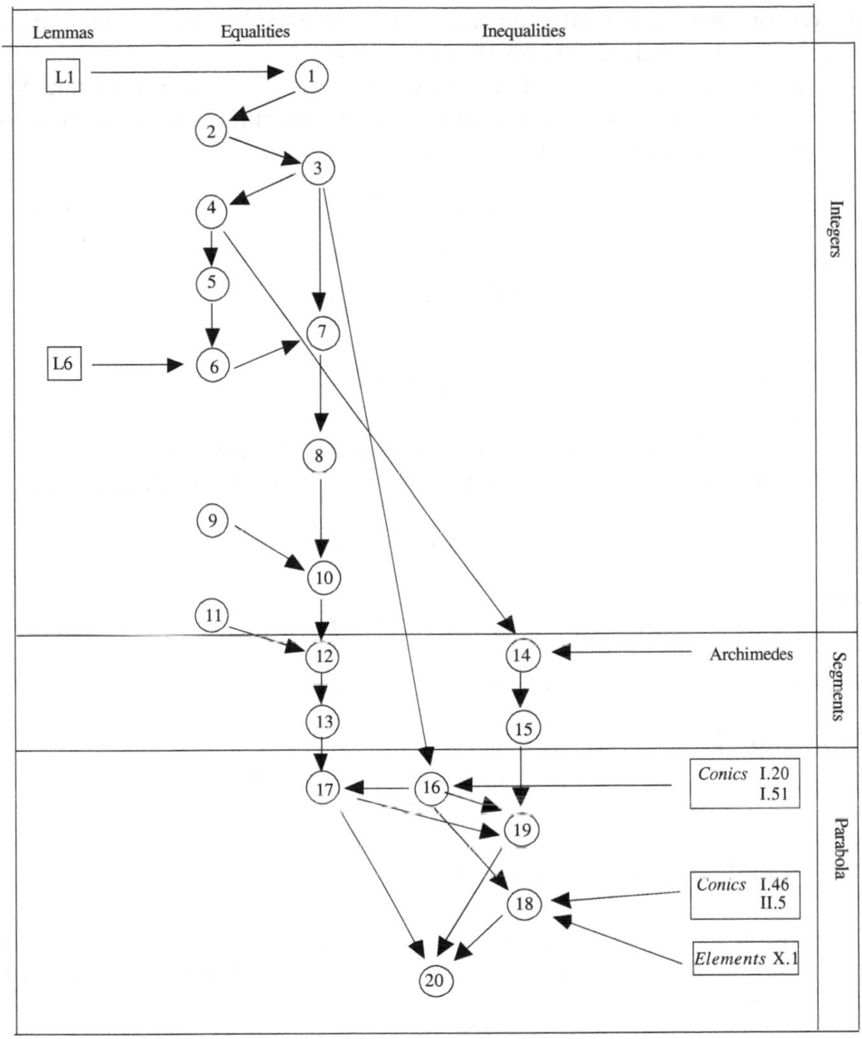

Note: Proposition 17 is relative to a property of the parabola.

2.2.2. *Mathematical commentary*

2.2.2.1. *Arithmetical propositions*

Proposition 1.

$$\forall\ n \in \mathbf{N}^*, n^2 - (n-1)^2 = 2n - 1.$$

Thābit proved this proposition with the help of Lemma 1: Two integer squares a and b, with $a > b$, are consecutive if and only if they are the squares of two consecutive integers. To first establish the lemma, it is necessary and sufficient to show that $\sqrt{a} - \sqrt{b} = 1$.

Suppose that $\sqrt{a} - \sqrt{b} \neq 1$. It follows that $\sqrt{a} - \sqrt{b} > 1$, for $\sqrt{a}$ and $\sqrt{b}$ are two integers (the definition used by Thābit: the difference between two consecutive integers is 1). Let

$$\sqrt{a} - \sqrt{b} = 1 + c, \qquad \text{with } c \text{ an integer.}$$

Hence

$$\sqrt{b} < \sqrt{b} + 1 < \sqrt{a}$$

and

$$b < (\sqrt{b} + 1)^2 < a,$$

which is absurd, since b and a are two consecutive integer squares.

The proof of the proposition is then immediate. By the lemma, we get

$$a = 1 + b + 2\sqrt{b}.$$

Hence

$$a - b = 2\sqrt{b} + 1;$$

but since $\sqrt{b}$ is an integer, $2\sqrt{b}$ is even, from which the result follows.

Proposition 2.

$$\forall \; n \in \mathbf{N}^*, (n + 1)^2 - n^2 > n^2 - (n - 1)^2$$
$$(n + 1)^2 - n^2 = n^2 - (n - 1)^2 + 2.$$

The proof follows from Proposition 1.

Proposition 3. — *Let* $(\mathrm{u}_n)_{n\geq 1}$ *be a sequence of consecutive squares such that* $\mathrm{u}_1 = 1$ *and* $(\mathrm{v}_n)_{n\geq 1}$ *a sequence of consecutive odd numbers such that* $\mathrm{v}_1 = 3$. *Then*

$$\forall \; n \in \mathbf{N}^*, \; (\mathrm{u}_{n+1} - \mathrm{u}_n) = \mathrm{v}_n.$$

Unlike with the first proposition, Thābit wanted to show not only that the difference between two consecutive integer squares is an odd number, but further that the odd numbers so obtained for the pairs of consecutive squares were consecutive. The proof was done by an archaic induction:

The proposition is true for $n = 1$, that is,

$$u_2 - u_1 = v_1 = 3.$$

Suppose that the proposition is true up to p:

$$u_p - u_{p-1} = v_{p-1}.$$

Then by Proposition 2, we have

$$u_{p+1} - u_p = u_p - u_{p-1} + 2,$$

and thus

$$u_{p+1} - u_p = v_{p-1} + 2 = v_p$$

by the 'implicit' definition of consecutive odd numbers.

Proposition 3′. — *Let* $(u_n)_{n\geq1}$ *be a sequence of integers such that* $u_1 = 1$ *and* $(v_n)_{n\geq1}$ *the sequence of consecutive odd integers such that* $v_1 = 3$. *If* $u_{n+1} - u_n = v_n$, *then* $(u_n)_{n\geq1}$ *is the sequence of consecutive squares such that* $u_1 = 1$.

This is the converse of the preceding proposition and the proof can be done with the help of the same ideas as in the previous proof:

$$u_2 - u_1 = u_2 - 1 = v_1 = 3.$$

Hence

$$u_2 = v_1 + 1 = 2^2.$$

Suppose that the proposition holds for *n*; in other words,

$$u_n = n^2;$$

so

$$u_{n+1} - u_n = v_n = 2n + 1.$$

Thus

$$u_{n+1} = u_n + (2n + 1) = n^2 + 2n + 1 = (n + 1)^2.$$

Proposition 4. — *Let* $(u_k)_{1\leq k\leq n}$ *be a sequence of consecutive odd numbers such that* $u_1 = 1$. *Then*

$$\sum_{k=1}^{n} u_k = \left(\frac{u_n + 1}{2}\right)^2.$$

$$\left(\sum_{k=0}^{n}(2k + 1) = \left(\frac{(2n + 1) + 1}{2}\right)^2 = (n + 1)^2.\right)$$

The proof is by *finite descent*.

Let the sequence $(v_k)_{1\le k\le n}$ be such that $v_k = \frac{u_k+1}{2}$ for $(1 \le k \le n)$. We have

$$u_{k+1} - u_k = 2 \qquad \text{for } 1 \le k \le n-1,$$

and hence

$$\tfrac{1}{2}(u_{k+1}+1) - \tfrac{1}{2}(u_k+1) = v_{k+1} - v_k = 1 \qquad \text{for } 1 \le k \le n-1.$$

Thus $(v_k)_{1\le k\le n}$ is a sequence of consecutive integers starting at 1. By Lemma 1, $\left(v_k^2\right)_{1\le k\le n}$ is a sequence of consecutive squares starting at 1; and we obtain by Proposition 3

$$w_k = v_{k+1}^2 - v_k^2 \qquad \text{for } 1 \le k \le n-1,$$

a sequence of consecutive odd numbers starting at 3, *i.e.* the sequence (u_k) for $2 \le k \le n$. Hence

$$\sum_{k=2}^{n} u_k = \sum_{k=1}^{n-1} w_k = v_n^2 - v_1^2,$$

and hence

$$\sum_{k=1}^{n} u_k = v_n^2$$

as $v_1 = u_1$.

The schema for the *finite descent* used by Thābit is the following:

1. $w_1 = v_2^2 - v_1^2$ (Proposition 3);

2. suppose that $\sum_{k=1}^{n-1} w_k = v_n^2 - v_1^2$,

3. so $\sum_{k=1}^{n} w_k = \left(v_n^2 - v_1^2\right) + \left(v_{n+1}^2 - v_n^2\right)$,

by Proposition 3. Hence

$$\sum_{k=1}^{n} w_k = v_{n+1}^2 - v_1^2.$$

Proposition 5. — *Let* $(v_k)_{1\le k\le n}$ *be the sequence of consecutive even numbers starting at* 2 *and* $(u_k)_{1\le k\le n}$, *the sequence of consecutive odd numbers starting at* 1. *Then*

$$\sum_{k=1}^{n} v_k^2 = \sum_{k=1}^{n} u_k^2 + \frac{v_n^2}{2} + n.$$

$$\left(\sum_{k=1}^{n} (2k)^2 = \sum_{k=1}^{n} (2k - 1)^2 + \frac{(2n)^2}{2} + n.\right)$$

The result follows immediately from Proposition 4.

Proposition 6. — *Let* $\left(v_k^2\right)_{1\le k\le n}$ *be the sequence of consecutive squares starting at* 1 *and* $\left(u_k^2\right)_{1\le k\le n}$ *the sequence of consecutive odd squares starting at* 1. *Then*

$$2\sum_{k=1}^{n} v_k^2 = \frac{1}{2}\sum_{k=1}^{n} u_k^2 + v_n^2 + \frac{n}{2}.$$

$$\left(2\sum_{k=1}^{n} k^2 = \frac{1}{2}\sum_{k=1}^{n} (2k - 1)^2 + n^2 + \frac{n}{2}.\right)$$

Thābit first proves by *reductio ad absurdum* Lemma 6, stated as:

The consecutive odd squares starting at 1 *are the squares of consecutive odd numbers starting at* 1.

Let us come to the proof of the proposition. By Lemma 1, $(v_k)_{1\le k\le n}$ is the sequence of consecutive integers such that $v_1 = 1$. Let

$$w_k = 2\, v_k, \qquad 1 \le k \le n.$$

Then $(w_k)_{1\le k\le n}$ is the sequence of consecutive even numbers with $w_1 = 2$; hence

$$\sum_{k=1}^{n} w_k^2 = 4 \sum_{k=1}^{n} v_k^2.$$

By Lemma 6, $(u_k)_{1\le k\le n}$ is the sequence of consecutive odd numbers starting at 1. By Proposition 5, we have

$$\sum_{k=1}^{n} w_k^2 = \sum_{k=1}^{n} u_k^2 + \frac{w_n^2}{2} + n;$$

hence

$$2\sum_{k=1}^{n} v_k^2 = \frac{1}{2}\sum_{k=1}^{n} w_k^2 = \frac{1}{2}\left(\sum_{k=1}^{n} u_k^2\right) + v_n^2 + \frac{n}{2}.$$

Proposition 7. — *Let* $(u_k)_{1\le k\le n}$ *be a sequence of consecutive odd numbers starting at* 1; *we thus have*

$$\sum_{k=1}^{n} u_k + \sum_{k=1}^{n-1} 2s_k = \frac{1}{2}\sum_{k=1}^{n} u_k^2 + \frac{n}{2}, \textit{ with } s_k = \sum_{p=1}^{k} u_p.$$

$$\left(\sum_{k=1}^{n}(2k-1)+2\sum_{k=2}^{n}(2k-3)+2\sum_{k=3}^{n}(2k-5)+\ldots+2(1+3)+2.1=\frac{1}{2}\sum_{k=1}^{n}(2k-1)^2+\frac{n}{2}.\right).$$

We get

$$s_k - s_{k-1} = u_k \qquad \text{for } 2 \le k \le n;$$

therefore $(s_k - s_{k-1})$ is the sequence of consecutive odd numbers starting at 3.

By Proposition 3′, the sequence $(s_k)_{1\le k\le n}$ is the sequence of consecutive squares starting at 1, and by Proposition 6, we have

$$2\sum_{k=1}^{n} s_k = \frac{1}{2}\sum_{k=1}^{n} u_k^2 + s_n + \frac{n}{2};$$

hence

$$s_n + 2\sum_{k=1}^{n-1} s_k = \frac{1}{2}\sum_{k=1}^{n} u_k^2 + \frac{n}{2}.$$

Proposition 8. — *Let* $(u_k)_{1\le k\le n}$ *be the sequence of consecutive odd numbers starting at* 1; *we thus have*

$$\sum_{k=1}^{n} u_k \cdot u_{n-k+1} = \frac{1}{2}\sum_{k=1}^{n} u_k^2 + \frac{n}{2}.$$

$$\left(\sum_{k=1}^{n} (2k-1)\,[2(n-k)+1] = \frac{1}{2}\sum_{k=1}^{n} (2k-1)^2 + \frac{n}{2}\right).$$

Thābit ibn Qurra proves this proposition by incomplete induction. We follow this proof step by step.

Let $(u_k)_{1\le k\le n}$ be the sequence of the n prime consecutive odd numbers starting at 1; the sequence $(u_k - 1)$ for $2 \le k \le n$ is the sequence of the $(n-1)$ prime consecutive even numbers starting at 2 and the sequence $(u_k - 1 - 2)$ for $3 \le k \le n$ is the sequence of the $(n-2)$ prime consecutive even numbers starting at 2. Continuing in this manner, we have

$$u_k - 1 - 2(k-2), \ldots, u_n - 1 - 2(k-2),$$

the sequence of the $(n-k+1)$ prime consecutive even numbers starting at 2; and ultimately

$$u_n - 1 - 2(n-2) = 2.$$

Hence

$$(1) \qquad u_n = 2n - 1.$$

Moreover,

$$\begin{aligned}
1 \cdot u_1 + \ldots + 1 \cdot u_p + \ldots + 1 \cdot u_{n-1} + 1 \cdot u_n &= 1 . \sum_{k=1}^{n} u_k , \\
2 \cdot u_1 + \ldots + 2 \cdot u_p + \ldots + 2 \cdot u_{n-1} &= 2 . \sum_{k=1}^{n-1} u_k , \\
\ldots & \\
2 \cdot u_1 + \ldots + 2 \cdot u_p &= 2 . \sum_{k=1}^{p} u_k , \\
\ldots & \\
2 \cdot u_1 &= 2 \cdot u_1 .
\end{aligned}$$

By summing each column, we obtain

$$[1 + 2(n-1)]\, u_1 + \ldots + [1 + 2(n-p)]\, u_p + \ldots + 1 \cdot u_n$$

$$= 1 . \sum_{k=1}^{n} u_k + 2 . \sum_{p=1}^{n-1} \sum_{k=1}^{p} u_k .$$

Hence by (1) and Proposition 7

$$\sum_{k=1}^{n} u_k \cdot u_{n-k+1} = \frac{1}{2} \sum_{k=1}^{n} u_k^2 + \frac{n}{2}.$$

Proposition 9. — *Let* $(u_k)_{1 \le k \le n}$ *be the sequence of the* n *prime consecutive odd numbers starting at* 1 *and* $(v_k)_{1 \le k \le n}$ *the sequence of the* n *prime consecutive even numbers starting at 2. Then*

$$w_k = v_n - u_k \qquad \textit{for } 1 \le k \le n$$

is the decreasing sequence of the n *prime consecutive odd numbers starting at* $w_1 = v_n - 1 = u_n$ *and terminating at* 1.

Thābit ibn Qurra proves this proposition by *finite descent*. We have

$$v_k - u_k = 1 \qquad \text{for } 1 \le k \le n.$$

Moreover,

$$u_n + w_n = v_n$$

and

$$u_{n-1} + w_{n-1} = v_n\,;$$

hence

$$u_n - u_{n-1} = w_{n-1} - w_n = 2.$$

Likewise, we may show, for each p, $2 \le p \le n-1$, that

$$w_{n-p-1} - w_{n-p} = 2.$$

The sequence $(w_k)_{1 \le k \le n}$ is thus the decreasing sequence of the n prime consecutive odd numbers starting at $w_1 = v_n - 1 = u_n$ and terminating at 1.

Proposition 10. — *Let* $(u_k)_{1 \le k \le n}$ *be the sequence of the* n *prime consecutive odd numbers starting at* 1 *and* $(v_k)_{1 \le k \le n}$ *the sequence of the* n *prime consecutive even numbers starting at* 2; *then*

$$\sum_{k=1}^{n} u_k^2 + \frac{n}{3} = \frac{2}{3}\left(\sum_{k=1}^{n} u_k\right) . \, v_n .$$

$$\left(\sum_{k=1}^{n} (2k-1)^2 + \frac{n}{3} = \frac{2}{3}\left(\sum_{k=1}^{n} (2k-1)\right) . \, 2n.\right)$$

Let

(1) $$w_k = v_n - u_k \qquad \text{for } 1 \le k \le n.$$

By Proposition 9, we have

(2) $$w_k = u_{n-k+1} \qquad \text{for } 1 \le k \le n.$$

So by Proposition 8 and (2), we have

$$\sum_{k=1}^{n} u_k w_k = \frac{1}{2} \sum_{k=1}^{n} u_k^2 + \frac{n}{2};$$

hence

$$\sum_{k=1}^{n} u_k w_k + \sum_{k=1}^{n} w_k^2 = \frac{3}{2} \sum_{k=1}^{n} u_k^2 + \frac{n}{2},$$

and hence

$$\sum_{k=1}^{n} (u_k + w_k) w_k = \frac{3}{2} \sum_{k=1}^{n} u_k^2 + \frac{n}{2}.$$

Hence by (1)

$$v_n \cdot \sum_{k=1}^{n} w_k = \frac{3}{2} \sum_{k=1}^{n} u_k^2 + \frac{n}{2},$$

and by (2)

$$v_n \cdot \sum_{k=1}^{n} u_k = \frac{3}{2} \sum_{k=1}^{n} u_k^2 + \frac{n}{2}.$$

The result is obtained on multiplying by $\frac{2}{3}$.

Note that the result can be rewritten as

$$\sum_{k=1}^{n} (2k - 1)^2 = \frac{4n^3}{3} - \frac{n}{3}.$$

Proposition 11. — *Let* $(v_k)_{0 \le k \le n}$ *be the sequence of the* n *prime consecutive even numbers with* $v_0 = 0$ *and* $v_1 = 2$, *and let* $(w_k)_{1 \le k \le n}$ *be the sequence defined by*

$$w_k = \frac{v_{k-1} + v_k}{2} \qquad \textit{for } 1 \le k \le n.$$

Then $(w_k)_{1 \le k \le n}$ *is the sequence of the* n *prime consecutive odd numbers starting at* 1.

We have

$$w_1 = 1 \text{ and } v_k - v_{k-1} = 2 \qquad \text{for } 1 \le k \le n;$$

thus

$$v_k - \frac{v_{k-1} + v_k}{2} = 1 \qquad \text{for } 1 \le k \le n,$$

and hence

$$v_k - w_k = 1 \qquad \text{for } 1 \le k \le n.$$

As $(v_k)_{1 \le k \le n}$ is the sequence of consecutive even numbers starting at 2, $(w_k)_{1 \le k \le n}$ is the sequence of consecutive odd numbers starting at 1.

2.2.2.2. *Sequence of segments and bounding*

Proposition 12. — *Let* $(u_k)_{1 \le k \le n}$ *be the sequence of* n *consecutive odd numbers starting at* 1 *and* $(a_k)_{1 \le k \le n}$ *an increasing sequence of* n *segments satisfying*

$$\frac{a_{k-1}}{a_k} = \frac{u_{k-1}}{u_k} \qquad \textit{for } 2 \le k \le n.$$

Let $(v_k)_{1 \le k \le n}$ *be the sequence of* n *consecutive even numbers starting at* 2 *and* $(b_k)_{1 \le k \le n}$ *an increasing sequence of* n *segments satisfying*

$$\frac{b_{k-1}}{b_k} = \frac{v_{k-1}}{v_k} \qquad \textit{for } 2 \le k \le n.$$

If $a_1 = \frac{b_1}{2}$, *then* $\sum_{k=1}^{n} a_k \cdot \frac{b_{k-1} + b_k}{2} + \frac{n}{3} a_1 \cdot \frac{b_1}{2} = \frac{2}{3} b_n \cdot \sum_{k=1}^{n} a_k.$

Let

$$c_k = \frac{b_{k-1} + b_k}{2}, \qquad w_k = \frac{v_{k-1} + v_k}{2} \quad \text{for } 1 \le k \le n$$

and

$$v_0 = b_0 = 0.$$

We have

$$\frac{a_1}{b_1} = \frac{u_1}{v_1}$$

and for $2 \le k \le n$

$$\text{(1)} \qquad \frac{a_{k-1}}{a_k} = \frac{u_{k-1}}{u_k}, \qquad \frac{b_{k-1}}{b_k} = \frac{v_{k-1}}{v_k}; \qquad \text{hence } \frac{a_k}{b_k} = \frac{u_k}{v_k}.$$

Hence

$$\text{(2)} \qquad \frac{a_k}{c_k} = \frac{a_k}{\frac{1}{2}(b_{k-1} + b_k)} = \frac{u_k}{\frac{1}{2}(v_{k-1} + v_k)} = \frac{u_k}{w_k} \quad \text{for } 1 \le k \le n.$$

But (w_k) is the sequence of the n prime consecutive odd numbers starting at 1, by Proposition 11; thus

$$u_k = w_k \qquad \text{for } 1 \le k \le n.$$

Hence

$$a_k = c_k \qquad \text{for } 1 \le k \le n,$$

(2′)
$$a_k^2 = a_k c_k \qquad \text{for } 1 \le k \le n.$$

Likewise,

$$\frac{a_{k-1}^2}{a_k^2} = \frac{u_{k-1}^2}{u_k^2} \qquad \text{for } 2 \le k \le n;$$

thus

(3)
$$\frac{a_k^2}{a_n^2} = \frac{u_k^2}{u_n^2} \qquad \text{for } 1 \le k \le n.$$

Hence

(4)
$$\frac{\sum_{k=1}^{n} a_k^2}{a_n^2} = \frac{\sum_{k=1}^{n} u_k^2}{u_n^2}.$$

However,

$$\frac{u_n^2}{u_n \cdot v_n} = \frac{a_n^2}{a_n \cdot b_n};$$

hence

(5)
$$\frac{\sum_{k=1}^{n} a_k^2}{a_n \cdot b_n} = \frac{\sum_{k=1}^{n} u_k^2}{u_n \cdot v_n}.$$

But

(6)
$$\frac{u_n \cdot v_n}{\left(\sum_{k=1}^{n} u_k\right) v_n} = \frac{a_n \, . \, b_n}{\left(\sum_{k=1}^{n} a_k\right) b_n}.$$

Thus, by (5) and (6), we get

(7)
$$\frac{\sum_{k=1}^{n} u_k^2}{\left(\sum_{k=1}^{n} u_k\right) v_n} = \frac{\sum_{k=1}^{n} a_k^2}{\left(\sum_{k=1}^{n} a_k\right) b_n}.$$

But by (3)

$$\frac{1}{\sum_{k=1}^{n} u_k^2} = \frac{a_1^2}{\sum_{k=1}^{n} a_k^2};$$

hence by (7) and the property of equal ratios

$$\frac{\left[\sum_{k=1}^{n} u_k^2 + \frac{n}{3}\right]}{\left(\sum_{k=1}^{n} u_k\right) v_n} = \frac{\left(\sum_{k=1}^{n} a_k^2 + \frac{n}{3}\, a_1^2\right)}{\left(\sum_{k=1}^{n} a_k\right) b_n}.$$

Thus by (2′)

$$\frac{\left[\sum_{k=1}^{n} u_k^2 + \frac{n}{3}\right]}{\left(\sum_{k=1}^{n} u_k\right) v_n} = \frac{\left(\sum_{k=1}^{n} a_k \cdot \frac{b_{k-1} + b_k}{2} + \frac{n}{3} \cdot a_1 \cdot \frac{b_1}{2}\right)}{\left(\sum_{k=1}^{n} a_k\right) b_n}.$$

But by Proposition 10, we have

$$\left(\sum_{k=1}^{n} u_k^2 + \frac{n}{3}\right) = \frac{2}{3}\left(\sum_{k=1}^{n} u_k\right) v_n;$$

hence the result.

Comment. — Proposition 12 reduces to Proposition 10 by the choice of a unit segment a_1. In fact, if we let

$$a_k = u_k \cdot a_1$$

and with the hypothesis $a_1 = \frac{b_1}{2}$, which otherwise is not fundamental, as we will see in the next proposition, we have

$$\begin{aligned}\sum_{k=1}^{n} a_k \frac{b_{k-1} + b_k}{2} + \frac{n}{3}\, a_1 \cdot \frac{b_1}{2} &= \sum_{k=1}^{n} u_k a_1 \cdot \frac{v_k a_1 + v_{k-1} a_1}{2} + \frac{n}{3}\, a_1^2 \\ &= a_1^2 \left(\sum_{k=1}^{n} u_k^2 + \frac{n}{3}\right) \\ &= a_1^2 \cdot \frac{2}{3}\left(\sum_{k=1}^{n} u_k\right) v_n \text{ (by Proposition 10)} \\ &= \frac{2}{3}\left(\sum_{k=1}^{n} a_k\right) b_n.\end{aligned}$$

Proposition 13. — *Let* $(u_k)_{1 \le k \le n}$ *be the sequence of* n *consecutive odd numbers starting at* 1, $(a_k)_{1 \le k \le n}$ *an increasing sequence of* n *segments satisfying*

$$\frac{a_{k-1}}{a_k} = \frac{u_{k-1}}{u_k} \qquad \textit{for } 2 \le k \le n,$$

$(v_k)_{1 \le k \le n}$ *the sequence of* n *consecutive even numbers starting at* 2 and $(b_k)_{1 \le k \le n}$ *an increasing sequence of* n *segments satisfying*

$$\frac{b_{k-1}}{b_k} = \frac{v_{k-1}}{v_k} \qquad \textit{for } 2 \le k \le n.$$

If $a_1 \neq \frac{b_1}{2}$, *then*

$$\sum_{k=1}^{n} a_k \cdot \frac{b_{k-1} + b_k}{2} + \frac{n}{3} a_1 \cdot \frac{b_1}{2} = \frac{2}{3} b_n \cdot \sum_{k=1}^{n} a_k .$$

Let the sequence $(c_k)_{1 \le k \le n}$ be defined as follows:

$$c_1 = 2a_1 \text{ and } \frac{c_{k-1}}{c_k} = \frac{b_{k-1}}{b_k} \qquad \text{for } 2 \le k \le n;$$

we obtain by permutation

$$(1) \qquad \frac{b_{k-1}}{c_{k-1}} = \frac{b_k}{c_k} \qquad \text{for } 2 \le k \le n.$$

Hence

$$(2) \qquad \frac{\frac{b_{k-1}}{2}}{\frac{c_{k-1}}{2}} = \frac{\frac{b_k}{2}}{\frac{c_k}{2}} \qquad \text{for } 2 \le k \le n.$$

Moreover,

$$(3) \qquad \frac{a_k\left(\frac{b_{k-1} + b_k}{2}\right)}{a_k\left(\frac{c_{k-1} + c_k}{2}\right)} = \frac{\frac{b_{k-1} + b_k}{2}}{\frac{c_{k-1} + c_k}{2}} \qquad \text{for } 1 \le k \le n,$$

with $b_0 = c_0 = 0$.

But by (2) and (3), we have

$$\frac{\sum_{k=1}^{n} a_k \left(\frac{b_{k-1}+b_k}{2}\right)}{\sum_{k=1}^{n} a_k \left(\frac{c_{k-1}+c_k}{2}\right)} = \frac{\sum_{k=1}^{n} \frac{b_{k-1}+b_k}{2}}{\sum_{k=1}^{n} \frac{c_{k-1}+c_k}{2}} = \frac{\frac{b_k}{2}}{\frac{c_k}{2}} = \frac{b_n}{c_n} \quad \text{for } 1 \le k \le n,$$

$$= \frac{b_n \left(\sum_{k=1}^{n} a_k\right)}{c_n \left(\sum_{k=1}^{n} a_k\right)}.$$

Hence

$$\text{(4)} \qquad \frac{\sum_{k=1}^{n} a_k \frac{(b_{k-1}+b_k)}{2}}{b_n \left(\sum_{k=1}^{n} a_k\right)} = \frac{\sum_{k=1}^{n} a_k \frac{(c_{k-1}+c_k)}{2}}{c_n \left(\sum_{k=1}^{n} a_k\right)},$$

with $1 \le k \le n$ and $b_0 = c_0 = 0$.

Moreover, we have

$$\frac{a_1 \cdot \frac{b_1}{2}}{b_n \cdot \sum_{k=1}^{n} a_k} = \frac{a_1}{\sum_{k=1}^{n} a_k} \cdot \frac{\frac{b_1}{2}}{b_n}.$$

But by (1)

$$\frac{b_1}{b_n} = \frac{c_1}{c_n}.$$

Hence

$$\frac{a_1 \cdot \frac{b_1}{2}}{b_n \cdot \sum_{k=1}^{n} a_k} = \frac{a_1}{\sum_{k=1}^{n} a_k} \cdot \frac{\frac{c_1}{2}}{c_n},$$

and hence

$$\text{(5)} \qquad \frac{\frac{n}{3} \cdot a_1 \cdot \frac{b_1}{2}}{b_n \cdot \sum_{k=1}^{n} a_k} = \frac{\frac{n}{3} \cdot a_1 \cdot \frac{c_1}{2}}{c_n \cdot \sum_{k=1}^{n} a_k}.$$

From (4) and (5), we deduce

$$(6)\ \frac{\frac{n}{3}\cdot a_1\cdot\frac{b_1}{2}}{b_n\cdot\sum_{k=1}^{n}a_k}+\frac{\sum_{k=1}^{n}a_k\frac{(b_{k-1}+b_k)}{2}}{b_n\cdot\sum_{k=1}^{n}a_k}=\frac{\frac{n}{3}\cdot a_1\cdot\frac{c_1}{2}}{c_n\cdot\sum_{k=1}^{n}a_k}+\frac{\sum_{k=1}^{n}a_k\frac{(c_{k-1}+c_k)}{2}}{c_n\cdot\sum_{k=1}^{n}a_k},$$

with $1 \leq k \leq n$ and $b_0 = c_0 = 0$.

Now, $a_1 = \frac{c_1}{2}$; so, by Proposition 12, the second member of (6) is equal to $\frac{2}{3}$; therefore the first member of (6) is equal to $\frac{2}{3}$; hence the result.

Comments.

1) In Proposition 12, the ratio between a_1 and b_1 is equal to $\frac{1}{2}$, whereas in Proposition 13 the ratio is unspecified.

$a_1 \neq \frac{b_1}{2}$ means that the two sequences (a_k) and (b_k) are not given with respect to the same unit of length, but each has a different unit length. Thābit ibn Qurra's idea is to introduce a sequence (c_k) that on the one hand is given as a function of the same unit length as the sequence (a_k) and on the other hand has the ratios between its terms the same as the ratios between terms of the sequence (b_k). It is in this manner that he avoids the difficulty arising with the hypothesis $a_1 \neq \frac{b_1}{2}$.

Moreover, had one reduced the given sequences to their respective units of measure, this would have allowed one to avoid Proposition 12 and to reduce Propositions 10 and 13 to a single proposition, for in this case one would only have invoked the lone numerical sequences. Otherwise, in expressing these sequences with respect to their respective units of measure, one relies only on the relations between the even and odd sequences underlying the proof of Proposition 10.

2) Had Ibn Qurra explicitly expressed the choice of unit length, he would have been able to directly deduce Proposition 13 from Propositions 10 and 11. In fact, since $a_k = u_k a_1$, $b_k = \frac{v_k}{2}\, b_1$,

$$\sum_{k=1}^{n} a_k\,\frac{b_{k-1}+b_k}{2}+\frac{n}{3}\,a_1\cdot\frac{b_1}{2}=\sum_{k=1}^{n} u_k a_1\cdot\frac{1}{2}\left[\frac{v_{k-1}}{2}\cdot b_1+\frac{v_k}{2}\cdot b_1\right]+\frac{n}{3}\,a_1\cdot\frac{b_1}{2}$$

$$=a_1\cdot\frac{b_1}{2}\left[\sum_{k=1}^{n} u_k\cdot\frac{v_{k-1}+v_k}{2}+\frac{n}{3}\right]$$

$$= a_1 \cdot \frac{b_1}{2}\left[\sum_{k=1}^{n} u_k^2 + \frac{n}{3}\right] \qquad \text{by Proposition 11}$$

$$= a_1 \cdot \frac{b_1}{2}\left[\frac{2}{3} v_n \cdot \sum_{k=1}^{n} u_k\right] \qquad \text{by Proposition 10}$$

$$= \frac{2}{3} b_n \cdot \sum_{k=1}^{n} a_k .$$

Finally, Proposition 12 appears as a technical lemma in order to obtain the general result of Proposition 13.

Proposition 14. — Let a *and* b *be two segments such that* $\frac{a}{b}$ *is known; then there exists* $n \in \mathbf{N}^*$ *such that the sequence* $(u_k)_{1 \leq k \leq n}$ *of* n *consecutive odd numbers starting at* 1 *and the sequence* $(v_k)_{1 \leq k \leq n}$ *of* n *consecutive even numbers starting at* 2 *satisfy*

$$\frac{n}{v_n \cdot \sum_{k=1}^{n} u_k} < \frac{a}{b}.$$

By the axiom of Archimedes, there exists $n \in \mathbf{N}$ such that

$$n\, a > b \qquad \text{with } n \geq 1.$$

So let $(v_k)_{1 \leq k \leq n}$ be the sequence of consecutive even numbers starting with 2; we thus have

$$v_n = 2\, n.$$

Let

$$u_k = v_k - 1 \qquad \text{for } 1 \leq k \leq n.$$

The sequence $(u_k)_{1 \leq k \leq n}$ is the sequence of n consecutive odd numbers starting at 1.

By Proposition 4, we get

$$(1) \qquad \left(\frac{v_n}{2}\right)^2 = \sum_{k=1}^{n} u_k ;$$

thus

$$\frac{\frac{v_n}{2}}{\left(\frac{v_n}{2}\right)^2} = \frac{\frac{v_n}{2}}{\sum_{k=1}^{n} u_k}.$$

But as

$$\frac{v_n}{2} \cdot \sum_{k=1}^{n} u_k \geq \sum_{k=1}^{n} u_k,$$

since $v_n = 2n$ by hypothesis and $n \geq 1$, we have

$$\frac{\frac{v_n}{2}}{\frac{v_n}{2} \cdot \sum_{k=1}^{n} u_k} \leq \frac{\frac{v_n}{2}}{\sum_{k=1}^{n} u_k}.$$

But by (1), we have

$$\frac{\frac{v_n}{2}}{\sum_{k=1}^{n} u_k} = \frac{1}{\frac{v_n}{2}};$$

and moreover

$$\frac{1}{\frac{v_n}{2}} = \frac{1}{n};$$

hence

$$\frac{\frac{v_n}{2}}{\frac{v_n}{2} \cdot \sum_{k=1}^{n} u_k} \leq \frac{1}{n}.$$

But

$$\frac{\frac{v_n}{2}}{v_n \cdot \sum_{k=1}^{n} u_k} < \frac{\frac{v_n}{2}}{\frac{v_n}{2} \cdot \sum_{k=1}^{n} u_k} \quad \text{and} \quad \frac{1}{n} < \frac{a}{b};$$

hence

$$\frac{n}{v_n \cdot \sum_{k=1}^{n} u_k} < \frac{a}{b}.$$

Proposition 15. — *Let* AB, H *be two given segments,*[2] *and* a *and* b *two segments such that* $\frac{a}{b}$ *is given. For any given* n,

1) *there exists a partition* $(A_k)_{0\leq k\leq n}$ *with* $A_0 = A$, $A_n = B$ *and such that*

$$\frac{A_kA_{k+1}}{A_{k+1}A_{k+2}} = \frac{u_{k+1}}{u_{k+2}} \qquad \textit{for } 0 \leq k \leq n-2,$$

with $(u_k)_{1\leq k\leq n}$ *the sequence of consecutive odd numbers starting at* 1;

2) *there exists a sequence of segments* $(H_j)_{1\leq j\leq n}$ *with* $H_n = H$ *and such that*

$$\frac{H_j}{H_{j+1}} = \frac{v_j}{v_{j+1}} \qquad \textit{for } 1 \leq j \leq n-1,$$

with $(v_j)_{1\leq j\leq n}$ *the sequence of consecutive even numbers starting at* 2.

If n *satisfies the condition*

$$\frac{n}{v_n \cdot \sum_{k=1}^{n} u_k} < \frac{a}{b},$$

then

$$\frac{n\, A_0 A_1 \cdot \frac{H_1}{2}}{AB \cdot H} < \frac{a}{b}.$$

By Proposition 14, we know that there exists $n \in \mathbf{N}^*$ satisfying the condition

$$(1) \qquad \frac{n}{v_n \left[\sum_{p=1}^{n} u_p\right]} < \frac{a}{b}.$$

Let $(A_k)_{0\leq k\leq n}$ be a sequence of points from the segment AB (with $A_0 = A$, $A_n = B$) such that

$$(2) \qquad \frac{A_k A_{k+1}}{A_k A_n} = \frac{u_{k+1}}{\sum_{p=k+1}^{n} u_p} \qquad \text{for } 0 \leq k \leq n-2.$$

Modifying Thābit ibn Qurra's language, we may write

[2] In the MSS, the segments are denoted *CD* and *E*.

$$\frac{A_0A_1}{u_1}=\frac{A_1A_2}{u_2}=\ldots=\frac{A_kA_{k+1}}{u_{k+1}}=\ldots=\frac{A_{n-1}A_n}{u_n}. \tag{3}$$

We have thus constructed a partition of AB following the ratios of consecutive odd numbers.

Let $(H_j)_{1\leq j\leq n}$ be a sequence of segments (with $H_n = H$) such that

$$\frac{H_1}{v_1}=\frac{H_2}{v_2}=\frac{H_k}{v_k}=\frac{H_n}{v_n}; \tag{4}$$

this is possible if we take $H_1 = \frac{H_n}{n}$. By (3), we deduce

$$\frac{A_0\,A_1}{u_1}=\frac{A_{n\text{-}1}B}{u_n}=\frac{\sum\limits_{k=0}^{n\text{-}1} A_k\,A_{k+1}}{\sum\limits_{p=1}^{n} u_p}=\frac{AB}{\sum\limits_{p=1}^{n} u_p}; \tag{5}$$

hence

$$\frac{u_1}{\sum\limits_{p=1}^{n} u_p}=\frac{A\,A_1}{AB}. \tag{6}$$

But by (5), we have

$$\frac{\left[\sum\limits_{p=1}^{n} u_p\right]^2}{u_n\sum\limits_{p=1}^{n} u_p}=\frac{AB^2}{AB\,.\,A_{n\text{-}1}\,B}. \tag{7}$$

Hence [squaring both sides of (6) and multiplying the respective sides of (6) and (7)]

$$\frac{u_1^2\,.\,n}{u_n\sum\limits_{p=1}^{n} u_p}=\frac{(A\,A_1)^2\,.\,n}{AB\,.\,A_{n\text{-}1}\,B}. \tag{8}$$

1st case. — Suppose

$$\frac{AA_1}{H_1}=\frac{u_1}{v_1}. \tag{9}$$

Then

(10) $$\frac{u_1 \cdot \frac{v_1}{2}}{u_1^2} = \frac{A\,A_1 \cdot \frac{H_1}{2}}{A\,A_1^2}$$

and

(11) $$\frac{n}{u_n\left[\sum_{p=1}^{n} u_p\right]} = \frac{n\,A\,A_1 \cdot \frac{H_1}{2}}{AB \cdot A_{n-1}\,B}.$$

But

$$\frac{u_n}{u_1} = \frac{A_{n-1}B}{AA_1} \qquad \text{by (3)},$$

$$\frac{u_1}{v_1} = \frac{AA_1}{H_1} \qquad \text{by hypothesis},$$

$$\frac{v_1}{v_n} = \frac{H_1}{H} \qquad \text{by (4)}.$$

Whence, multiplying the respective sides of the last three equalities, we have

(12) $$\frac{u_n}{v_n} = \frac{A_{n-1}B}{H}.$$

And multiplying the respective sides of (11) and (12), we obtain

$$\frac{n}{v_n\left[\sum_{p=1}^{n} u_p\right]} = \frac{n \cdot A\,A_1 \cdot \frac{H_1}{2}}{AB \cdot H}.$$

Thus by (1), we have

$$\frac{n \cdot A\,A_1 \cdot \frac{H_1}{2}}{AB \cdot H} < \frac{a}{b},$$

which completes the proof for this case.

2nd case. — Suppose $\frac{AA_1}{H_1} \neq \frac{u_1}{v_1}$.

Let $G_1, G_2, \ldots, G_n$ be n segments satisfying

(13) $$\frac{AA_1}{G_1} = \frac{u_1}{v_1}$$

and

(14) $$\frac{G_1}{v_1} = \ldots = \frac{G_k}{v_k} = \ldots = \frac{G_n}{v_n}.$$

By the 1st case, we have

(15) $$\frac{n \,.\, A\,A_1 \,.\, \dfrac{G_1}{2}}{AB \,.\, G_n} < \frac{a}{b}.$$

Meanwhile,

$$\frac{A\,A_1 \cdot \dfrac{G_1}{2}}{A\,A_1 \cdot \dfrac{H_1}{2}} = \frac{\dfrac{G_1}{2}}{\dfrac{H_1}{2}}.$$

But by (4) and (14), we have

$$\frac{G_1}{H_1} = \frac{G_n}{H};$$

hence

$$\frac{A\,A_1 \,.\, \dfrac{G_1}{2}}{A\,A_1 \,.\, \dfrac{H_1}{2}} = \frac{G_n}{H}.$$

But

$$\frac{G_n}{H} = \frac{AB \cdot G_n}{AB \cdot H}.$$

Hence

$$\frac{A\,A_1 \cdot \dfrac{G_1}{2}}{A\,A_1 \cdot \dfrac{H_1}{2}} = \frac{AB \cdot G_n}{AB \cdot H};$$

hence

$$\frac{A\,A_1 \cdot \dfrac{G_1}{2}}{AB \cdot G_n} = \frac{A\,A_1 \cdot \dfrac{H_1}{2}}{AB \cdot H};$$

hence

$$\frac{n \cdot AA_1 \cdot \frac{G_1}{2}}{AB \cdot G_n} = \frac{n \cdot AA_1 \cdot \frac{H_1}{2}}{AB \cdot H}.$$

But by (15), we have

$$\frac{n \cdot AA_1 \cdot \frac{G_1}{2}}{AB \cdot G_n} < \frac{a}{b}.$$

Hence finally

$$\frac{n . AA_1 . \frac{H_1}{2}}{AB . H} < \frac{a}{b}.$$

Comment. — The proof relies on the partition of a given segment into a sequence of segments proportional to the numbers of a given sequence, as well as on the generalization of Proposition 14 (namely, the one that introduces the approximation) with regard to a sequence of segments, and consequently the generalization of the bounding of a sequence of ratios of segments.

In order to partition the segment AB into a sequence of n segments proportional to the numbers u_k of a sequence of n numbers, Thābit ibn Qurra proceeds once again by *finite descent*: we construct A_1 such that

$$\frac{AA_1}{A_1B} = \frac{u_1}{\sum_{k=2}^{n} u_k}$$

and we are thus lead to partition A_1B into a sequence of $n - 1$ segments proportional to the $(u_k)_{2 \leq k \leq n}$.

2.2.2.3. *Calculation of the area of a portion of a parabola*

Proposition 16. — *Let* ABC *be a portion of a parabola of diameter* BD. *Let* $E_1 G_1 F_1, \ldots, E_{n-1} G_{n-1} F_{n-1}$ *be the ordinates of the diameter* BD *intersecting it at* $G_1, G_2, \ldots, G_{n-1}$.

If $BG_1, G_1G_2, \ldots, G_{n-1}D$ *are such that*

(1) $$\frac{G_kG_{k+1}}{G_{k+1}G_{k+2}} = \frac{2k+1}{2k+3} \qquad \textit{for } 0 \leq k \leq n-2,$$

with $B = G_0$, $D = G_n$, *then the ordinates* $E_1 F_1, \ldots, E_{n-1} F_{n-1}$, AC *are such that*

(2) $$\frac{E_kF_k}{E_{k+1}F_{k+1}} = \frac{2k}{2k+2} \qquad \textit{for } 1 \leq k \leq n-1,$$

with $E_n = A$, $F_n = C$.

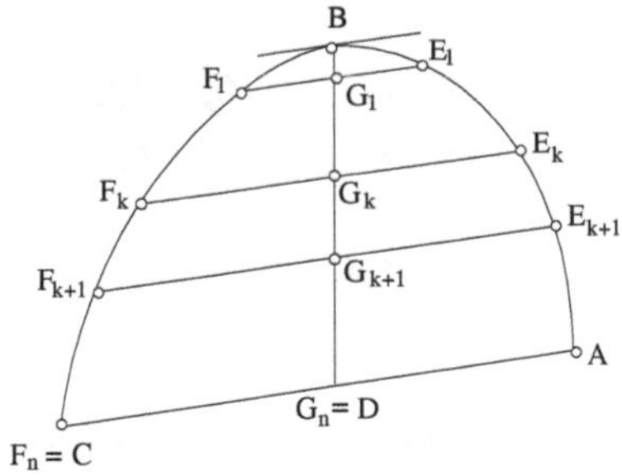

Fig. 2.2.1

Let

$$s_1 = 1, \ldots, s_k = \sum_{p=1}^{k} (2p - 1),$$

so

$$2k - 1 = s_k - s_{k-1} \qquad \text{for } 2 \leq k \leq n.$$

The sequence $(s_k - s_{k-1})_{2\leq k\leq n}$ is thus a sequence of consecutive odd numbers starting at 3 and, by proposition 3′, $(s_k)_{1\leq k\leq n}$ is a sequence of consecutive squares starting at 1.

Moreover, we have by hypothesis (1)

$$\frac{1}{3} = \frac{BG_1}{G_1G_2}.$$

Hence

$$\frac{1}{1+3} = \frac{BG_1}{BG_2};$$

so

$$\frac{s_1}{s_2} = \frac{BG_1}{BG_2}.$$

But by Proposition 20 of book I of Apollonius's *Conics*

$$\frac{B\,G_1}{B\,G_2} = \frac{G_1\,F_1^2}{G_2\,F_2^2},$$

hence

$$\frac{s_1}{s_2} = \frac{G_1 F_1^2}{G_2 F_2^2}.$$

We likewise show that

$$\frac{s_{k-1}}{s_k} = \frac{G_{k-1} F_{k-1}^2}{G_k F_k^2} \qquad \text{for } 3 \leq k \leq n;$$

hence

$$\frac{G_1 F_1^2}{s_1} = \frac{G_2 F_2^2}{s_2} = \ldots = \frac{G_{k-1} F_{k-1}^2}{s_{k-1}} = \frac{G_k F_k^2}{s_k} = \ldots = \frac{G_{n-1} F_{n-1}^2}{s_{n-1}} = \frac{G_n F_n^2}{s_n}.$$

But since $s_1, \ldots s_n$ are consecutive squares starting at 1, then $s_1^{\frac{1}{2}}, \ldots s_n^{\frac{1}{2}}$ are consecutive integers starting at 1. Thus

$$G_1 F_1, \ldots, G_n F_n$$

are proportional to successive integers starting at 1. Since

$$E_k F_k = 2\, G_k F_k \qquad \text{for } 1 \leq k \leq n,$$

then $E_1 F_1, \ldots, E_n F_n$ are proportional to the consecutive even numbers starting at 2.

Comment. — Let us note that Thābit takes as the ordinate the entire chord, that is to say double of the classical ordinate. Ultimately, if the abscissae studied are proportional to consecutive squares, then the ordinates to which they are associated are proportional to consecutive integers, and for Thābit their doubles are proportional to the consecutive even numbers. Thus to a subdivision of the diameter *BD* into *n* segments proportional to consecutive odd numbers there corresponds a subdivision of *DA* into *n* equal parts, and vice-versa. Thābit will use the converse in proposition 18.

Proposition 17. — *Let* P *be a portion of a parabola of diameter* BD. *If* $BG_1, G_1 G_2, \ldots, G_{n-1} D$ *is a subdivision of* BD *such that*

$$(1) \qquad \frac{G_k G_{k+1}}{G_{k+1} G_{k+2}} = \frac{2k+1}{2k+3} \qquad \textit{for } 0 \leq k \leq n-2$$

(*with* $B = G_0$, $D = G_n$), *and if* $E_1 G_1 F_1, \ldots, E_{n-1} G_{n-1} F_{n-1}$, ADC *are the corresponding ordinates and* BR *the perpendicular dropped from* B *onto* AC *and* F *its point of intersection with* $E_1 F_1$, *then if we designate by* S_n *the area of the polygon* $AE_{n-1} \ldots E_1 BF_1 \ldots F_{n-1} C$, *we have*

$$\frac{2}{3}\,AC \cdot BR - S_n \;=\; \frac{n}{3}\,BF \cdot G_1\,F_1.$$

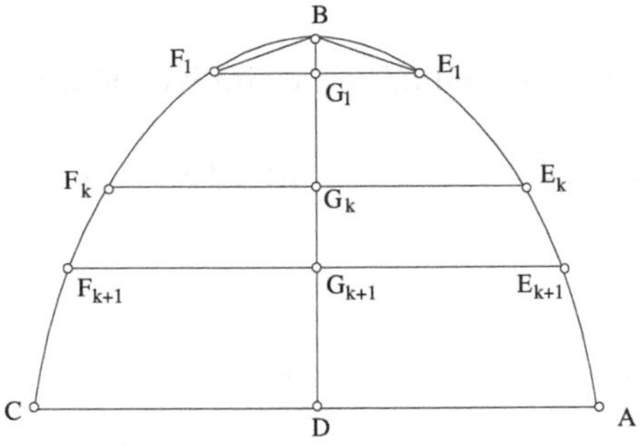

Fig. 2.2.2

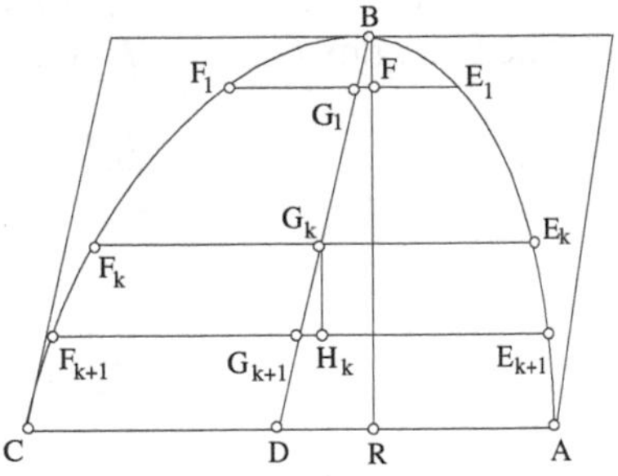

Fig. 2.2.3

1st case. — The diameter BD is the axis of symmetry of the parabola ($BD = BR$; $G_1 = F$).

By (1) and Proposition 16, we have

$$\frac{E_k F_k}{2k} = \frac{E_{k+1} F_{k+1}}{2k+2} \qquad \text{for } 1 \le k \le n-1$$

(with $E_n = A$, $F_n = C$).

By Proposition 13, we have

$$(2) \qquad \sum_{k=0}^{n-1} G_k\,G_{k+1} \cdot \frac{E_k\,F_k + E_{k+1}\,F_{k+1}}{2} + \frac{n}{3}\,G_0\,G_1 \cdot \frac{E_1\,F_1}{2} = \frac{2}{3}\,AC \cdot BD$$

(with $E_0 = F_0 = B$). But

$$G_k\,G_{k+1} \cdot \frac{E_k\,F_k + E_{k+1}\,F_{k+1}}{2}$$

is the area of the trapezoid with vertices $E_{k+1}\,E_k\,F_k\,F_{k+1}$, for which the height is $G_k\,G_{k+1}$; hence

$$S_n + \frac{n}{3}\,G_0\,G_1 \cdot \frac{E_1 F_1}{2} = \frac{2}{3}\,AC \cdot BD;$$

hence the result, since $G_0\,G_1 = BF$; $G_1\,F_1 = \dfrac{E_1 F_1}{2}$.

2nd case. — The diameter BD is not the axis of symmetry of the parabola; we have $BD \neq BR$.

From the point G_k we drop the perpendicular $G_k H_k$ onto the ordinate $E_{k+1} F_{k+1}$ $(0 \leq k \leq n-1;\ H_0 = F)$.

The triangles $G_k G_{k+1} H_k$ $(0 \leq k \leq n-1)$, and BDR are similar. Hence

$$(3) \qquad \frac{G_0H_0}{G_0G_1} = \frac{G_kH_k}{G_kG_{k+1}} = \frac{BR}{BD} \qquad (0 \leq k \leq n-1),$$

and hence

$$(4) \qquad \frac{\sum_{k=0}^{n-1} G_k H_k \cdot \frac{1}{2}\left(E_k F_k + E_{k+1} F_{k+1}\right)}{\sum_{k=0}^{n-1} G_k G_{k+1} \cdot \frac{1}{2}\left(E_k F_k + E_{k+1} F_{k+1}\right)} = \frac{BR \cdot AC}{BD \cdot AC}.$$

Yet

$$(5) \qquad \frac{BR \cdot AC}{BD \cdot AC} = \frac{\frac{n}{3} G_0H_0 \cdot \frac{E_1F_1}{2}}{\frac{n}{3} G_0G_1 \cdot \frac{E_1F_1}{2}}.$$

We may observe that the numerator of the left side of (4) is the area S_n of the polygon $AE_{n-1} \ldots E_1BF_1 \ldots F_{n-1}C$.

From (4) and (5), we obtain

$$\frac{BR \cdot AC}{BD \cdot AC} = \frac{S_n + \frac{n}{3} G_0H_0 \cdot \frac{E_1F_1}{2}}{\sum_{k=0}^{n-1} G_kG_{k+1} \cdot \frac{1}{2}\left(E_kF_k + E_{k+1}F_{k+1}\right) + \frac{n}{3} G_0G_1 \cdot \frac{E_1F_1}{2}}.$$

But by Proposition 13, the denominator of the right side is equal to $\frac{2}{3} BD \cdot AC$; hence

$$\frac{S_n + \frac{n}{3} G_0H_0 \cdot \frac{E_1F_1}{2}}{\frac{2}{3} BD \cdot AC} = \frac{BR \cdot AC}{BD \cdot AC}.$$

Hence the result follows.

Comments.

1) To explain the construction of a polygon of $2n + 1$ vertices inscribed in a portion of a parabola, for whatever value of n, Thābit ibn Qurra uses Proposition 16 in order to apply Proposition 13 in the proof.

2) Thābit ibn Qurra gives the expression for the difference between two

thirds of the area of the parallelogram associated with the parabola and the area S_n of the inscribed polygon.

3) The second case is treated directly, without using the first, which is nothing but a particular case with $BR = BD$, whence $H_k = G_{k+1}$.

4) Let us note that the product $BR \cdot AC$ is the area S of the parallelogram of base AC associated with the portion of the parabola. It is defined by the tangent at B and the parallels of the diameter described by A and C. The product $BF \cdot F_1G_1$ is the area of the triangle BE_1F_1.

Proposition 18. — *Let* ABC *be a portion of a parabola,* BD *its diameter and* **S** *its area. For all* $\varepsilon > 0$, *there exists a subdivision* (G_k) *of diameter* BD, $0 \le k \le 2^{n-1}$, *with* $G_0 = B$, $G_{2^{n-1}} = D$, *satisfying*

$$\frac{G_kG_{k+1}}{G_{k+1}G_{k+2}} = \frac{2k+1}{2k+3}$$

such that the area S_n *of the polygon* $\mathbf{P}_n$ *associated with that subdivision satisfies*

$$\mathbf{S} - S_n < \varepsilon.$$

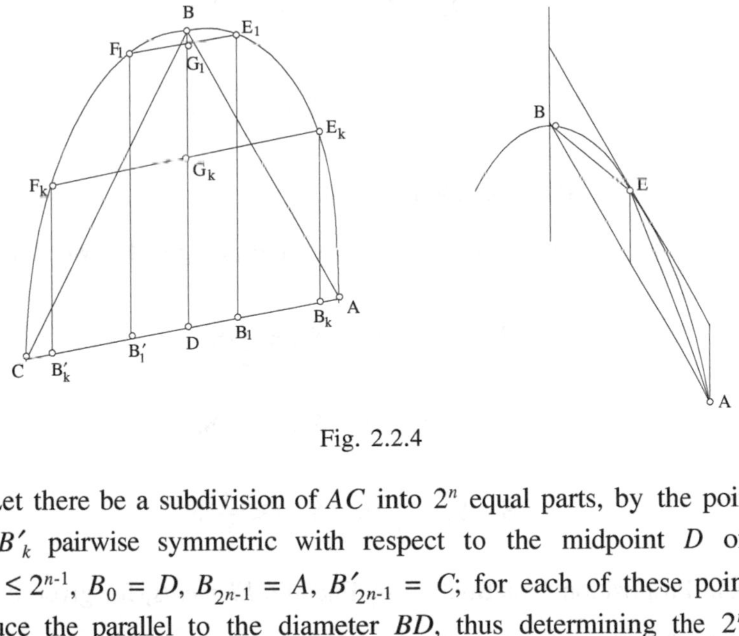

Fig. 2.2.4

Let there be a subdivision of AC into 2^n equal parts, by the points B_k and B'_k pairwise symmetric with respect to the midpoint D of AC, $0 \le k \le 2^{n-1}$, $B_0 = D$, $B_{2^{n-1}} = A$, $B'_{2^{n-1}} = C$; for each of these points we produce the parallel to the diameter BD, thus determining the $2^n + 1$ vertices of the polygon $\mathbf{P}_n$, $A \ldots E_k \ldots E_1BF_1 \ldots F_k \ldots C$; let S_n be its area. We want to find n so that $\mathbf{S} - S_n < \varepsilon$.

Construct $\mathbf{P}_1$, which is the triangle ABC; let S_1 be its area. We have

$$S_1 > \frac{1}{2}\mathbf{S};$$

hence

$$\mathbf{S} - S_1 < \frac{1}{2}\mathbf{S}.$$

a) If $\frac{1}{2}\mathbf{S} < \varepsilon$, then $\mathbf{S} - S_1 < \varepsilon$, $\mathbf{P}_1$ solves the problem.

b) If $\frac{1}{2}\mathbf{S} > \varepsilon$, we double the subdivision, and we construct $\mathbf{P}_2$ of area S_2. Using the following lemma, if E is the vertex associated with a chord AB of a parabola, then tr. $(AEB) > \frac{1}{2}$ portion (AEB), we show that

$$S_2 - S_1 > \frac{1}{2}(\mathbf{S} - S_1).$$

But

$$\mathbf{S} - S_2 = (\mathbf{S} - S_1) - (S_2 - S_1);$$

hence

$$\mathbf{S} - S_2 < \frac{1}{2}(\mathbf{S} - S_1) < \frac{1}{2^2}\mathbf{S}.$$

a) If $\frac{1}{2^2}\mathbf{S} < \varepsilon$, $\mathbf{P}_2$ solves the problem.

b) If $\frac{1}{2^2}\mathbf{S} > \varepsilon$, we iteratively construct the polygon $\mathbf{P}_3$; we thus have successively

$$\mathbf{S} - S_3 < \frac{1}{2}(\mathbf{S} - S_2) < \frac{1}{2^3}\mathbf{S},$$

$$\mathbf{S} - S_4 < \frac{1}{2}(\mathbf{S} - S_3) < \frac{1}{2^4}\mathbf{S},$$

$$\ldots$$

$$\mathbf{S} - S_n < \frac{1}{2}(\mathbf{S} - S_{n-1}) < \frac{1}{2^n}\mathbf{S},$$

and from Proposition 1 of Book X of Euclid's *Elements*, for given ε, there exists n such that $\frac{1}{2^n}\mathbf{S} < \varepsilon$; hence $\mathbf{S} - S_n < \varepsilon$.

The corresponding polygon $\mathbf{P}_n$ is the desired polygon.

It remains to show that the polygon $\mathbf{P}_n$ thus determined for given ε corresponds to a partition of the diameter BD into segments proportional to consecutive odd numbers starting at one.

On the segment DA the points $B_1 \dots B_k \dots B_{2n-1}$ give a partition of DA into segments $DB_1 \dots DB_k \dots DA$ proportional to the consecutive integers from 1 to 2^{n-1}, the segments $B_1B'_1, \dots B_kB'_k \dots AC$ are then proportional to consecutive even numbers. The points E_k and F_k having equal ordinates DB_k and DB'_k, E_kF_k is parallel to AC, by II.5 of Apollonius's *Conics*, and intersecting BD at G_k, for $0 \le k \le 2^{n-1}$; we thus obtain on BD the points B, $G_1, \dots G_k \dots G_{2n-1}$. And, by the converse of 16, the segments BG_1, $G_1G_2 \dots G_{2^{n-1}-1}G_{2^{n-1}}$ are proportional to consecutive odd numbers starting at 1; we thus have

$$\frac{G_kG_{k+1}}{G_{k+1}G_{k+2}} = \frac{2k+1}{2k+3} \qquad (0 \le k \le 2^{n-1} - 2).$$

Proposition 19. — *Let* ABC *be a portion of the parabola and* S *the area of the parallelogram with base* AC *associated with the parabola. Then for all* ε > 0, *there exists a polygon* $\mathbf{P}_n$ *of area* S_n, *inscribed in the portion of the parabola and such that*

$$\frac{2}{3}S - S_n < \varepsilon.$$

Let ABC be the portion of the parabola of diameter BD and base AC. Let any $\varepsilon > 0$ be given.

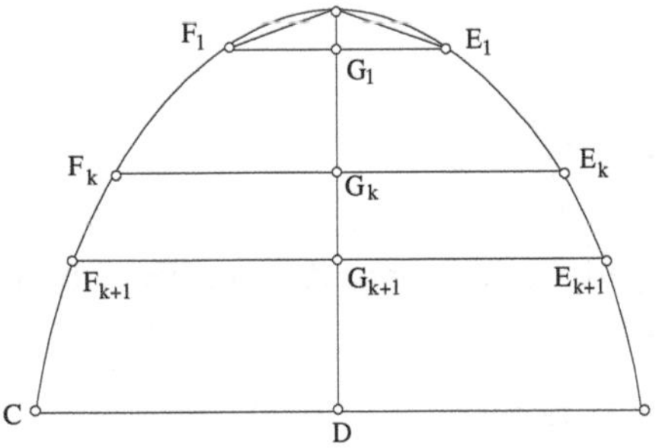

Fig. 2.2.5

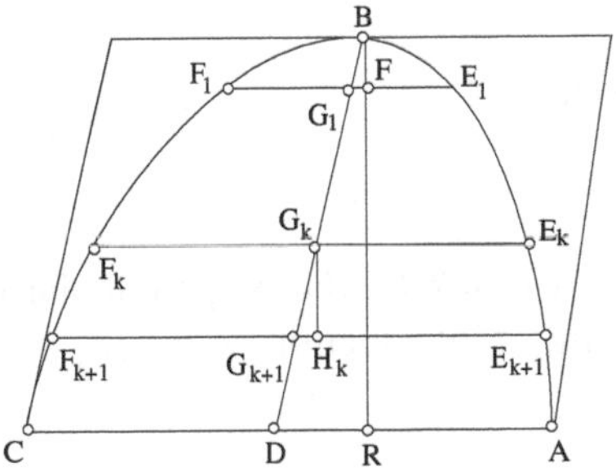

Fig. 2.2.6

By Proposition 15, there exists a partition $(G_k)_{0 \le k \le n}$ of BD ($G_0 = B$, $G_n = D$) such that

$$\frac{G_kG_{k+1}}{G_{k+1}G_{k+2}} = \frac{2k+1}{2k+3} \qquad (0 \le k \le n - 2)$$

and a sequence of segments (H_j) $1 \le j \le n$ $(H_n = AC)$ such that

$$\frac{H_j}{H_{j+1}} = \frac{2j}{2j+2} \qquad (1 \le j \le n-1)$$

satisfying

$$(1) \qquad \frac{n \cdot G_0G_1 \cdot \frac{H_1}{2}}{BD \cdot AC} < \frac{\varepsilon}{BD \cdot AC}.$$

But by Proposition 16, we can make to correspond to the partition $(G_k)_{0\le k\le n}$, the sequence of ordinates $(E_kF_k)_{1\le k\le n}$ $(E_nF_n = AC)$ such that

$$\frac{E_kF_k}{E_{k+1}F_{k+1}} = \frac{2k}{2k+2} \qquad (1 \le k \le n-1),$$

as

$$H_n = E_n\,F_n = AC,$$

and as, on the other hand, the H_j are unique, then $H_1 = E_1F_1$ and (1) is rewritten

$$(2) \qquad \frac{n \cdot G_0G_1 \cdot \frac{E_1F_1}{2}}{BD \cdot AC} < \frac{\varepsilon}{BD \cdot AC}.$$

We thus have

$$n \cdot G_0G_1 \cdot \frac{E_1F_1}{2} < \varepsilon;$$

hence

$$\frac{n}{3} \cdot G_0G_1 \cdot \frac{E_1F_1}{2} < \varepsilon.$$

Let BR be the perpendicular dropped from B onto AC and let F be its point of intersection with E_1F_1; we have

$$BF < G_0G_1;$$

hence

$$\frac{n}{3}\, BF \cdot \frac{E_1F_1}{2} < \varepsilon.$$

But by Proposition 17, we have

$$\frac{2}{3}S - S_n = \frac{n}{3}BF \cdot \frac{E_1F_1}{2};$$

hence

$$\frac{2}{3}S - S_n < \varepsilon.$$

Comment I.

1) Proposition 15 guarantees:

a) the existence of the partition $(G_i)_{0 \leq i \leq n}$ and the proportionality of the obtained segments to the consecutive odd numbers starting at 1;

b) the existence and uniqueness of a sequence of segments proportional to the consecutive even numbers $(H_j)_{1 \leq j \leq n}$ with $H_n = BC$ satisfying

$$(1) \qquad n \cdot G_0G_1 \cdot \frac{H_1}{2} < \varepsilon.$$

2) Proposition 16 shows that if a) is satisfied, then the terms of the sequence (E_jF_j) of the ordinates associated with the partition (G_i) are proportional to the consecutive even numbers starting at 2; as

$$E_n\,F_n = BC = H_n,$$

the uniqueness of H_j allows one to rewrite (1) thus:

$$n \cdot G_0G_1 \cdot \frac{E_1F_1}{2} < \varepsilon.$$

3) By a supplementary bounding and by Proposition 17, we obtain the result.

Comment II. — Proposition 17 shows that $\frac{2}{3}S$ is an upper bound of S_n for all n.

Proposition 19 shows that $\frac{2}{3}S$ is the least upper bound.

In fact, by Proposition 17, for all n,

$$\frac{2}{3}S - S_n = \alpha_n \qquad (\alpha_n > 0),$$

and by Proposition 19, for all $\varepsilon > 0$, there exists N such that for $n > N$,

$$0 < \alpha_n < \varepsilon.$$

Comment III. — We may observe that Thābit uses the ε effortlessly; that is, starting from an arbitrary fixed ε, he introduces an $\varepsilon' = \frac{\varepsilon}{\alpha}$ with $\alpha = BD \cdot AC$, so α allows the effective use of Proposition 15.

Proposition 20. — *The area of the parabola is infinite, but the area of any of its portions is equal to two thirds the area of the parallelogram associated with the parabola.*

Let $\mathbf{S}$ be the area of the portion of the parabola $\mathbf{P}$ and S the area of the parallelogram associated with this portion.

If $\frac{2}{3}\, S \neq \mathbf{S}$, there are two cases:

1) $\mathbf{S} > \frac{2}{3}\, S.$

Let $\varepsilon > 0$ be such that

$$\mathbf{S} - \frac{2}{3}\, S = \varepsilon. \tag{1}$$

By Proposition 18, for this ε, there exists N such that for $n > N$, the polygon $\mathbf{P}_n$ of area S_n satisfies

$$\mathbf{S} - S_n < \varepsilon\,. \tag{2}$$

By (1) and (2), we have

$$(\frac{2}{3}\, S + \varepsilon) - S_n < \varepsilon\,;$$

hence

$$\frac{2}{3}\, S < S_n.$$

But by Proposition 17, we have

$$\frac{2}{3}\, S > S_n,$$

giving a contradiction.

2) $\mathbf{S} < \frac{2}{3}\, S.$

Let $\varepsilon > 0$ be such that

$$(3) \qquad \frac{2}{3} S - \mathbf{S} = \varepsilon.$$

By Proposition 19, for this ε, there exists N such that for $n > N$, the polygon $\mathbf{P}_{\mathrm{n}}$ of area S_n satisfies

$$(4) \qquad \frac{2}{3} S - S_n < \varepsilon.$$

By (3) and (4), we have

$$(\mathbf{S} + \varepsilon) - S_n < \varepsilon\,;$$

hence

$$\mathbf{S} < S_n.$$

But $\mathbf{P}_{\mathrm{n}}$ is inscribed in $\mathbf{P}$, thus $S_n < \mathbf{S}$. This gives a contradiction. Therefore

$$\frac{2}{3} S = \mathbf{S}.$$

Comment. — This theorem returns to establish the uniqueness of the upper bound and essentially uses the properties of the upper bound in the proof.

In fact, we want to show that $\frac{2}{3} S = \mathbf{S}$, knowing that

1) $\mathbf{S}$ = upper bound $(S_n)_{n\geq 1}$;
2) $\frac{2}{3} S$ = upper bound $(S_n)_{n\geq 1}$.

Suppose $\mathbf{S} \neq \frac{2}{3} S$. We have two cases:

a) $\mathbf{S} > \frac{2}{3} S$: there thus exists $\varepsilon > 0$ such that $\mathbf{S} - \frac{2}{3} S = \varepsilon$. But by 1), $\mathbf{S}$ is the least upper bound of S_n ; thus for this ε, there exists S_n such that

$$S_n > \mathbf{S} - \varepsilon\,;$$

thus

$$\frac{2}{3} S < S_n,$$

which is absurd because by (2), $\frac{2}{3}S$ is an upper bound of the S_n.

b) $\mathbf{S} < \frac{2}{3}S$: there thus exists $\varepsilon > 0$ such that $\frac{2}{3}S - \mathbf{S} = \varepsilon$. But by 2), $\frac{2}{3}S$ is the least upper bound of the S_n; thus for this ε, there exists S_n such that

$$S_n > \frac{2}{3}S - \varepsilon;$$

thus

$$\mathbf{S} < S_n,$$

which is absurd because by 1), $\mathbf{S}$ is an upper bound of the S_n.

Naturally, we do not pretend that Thābit ibn Qurra, any more than his predecessors or successors up to the eighteenth century, had defined the concept of the upper bound. But it seems to us that he uses the properties of the upper bound as a guiding idea in the measurement of convex sets.

2.2.3. *Translated text*

Thābit ibn Qurra

On the Measurement of the Conic Section Called Parabola

In the Name of God, the Merciful, the Compassionate

THE BOOK OF THĀBIT IBN QURRA AL-ḤARRĀNĪ

On the Measurement of the Conic Section Called Parabola

Introduction

Successive numbers are such that there is no other number between them. Successive odd numbers are such that there is no other odd number between them. Similarly, successive even numbers are such that there is no other even number between them. Successive square numbers are also such that there is no other square number between them.

I say, in general, that: Successive <elements> of any species are such that there is no other element of the same species between them.

Propositions

– **1** – The difference between any two successive square numbers is an odd number.

Let two successive square numbers be *AB* and *C*, and let *AD* be their difference.

I say that AD *is an odd number.*

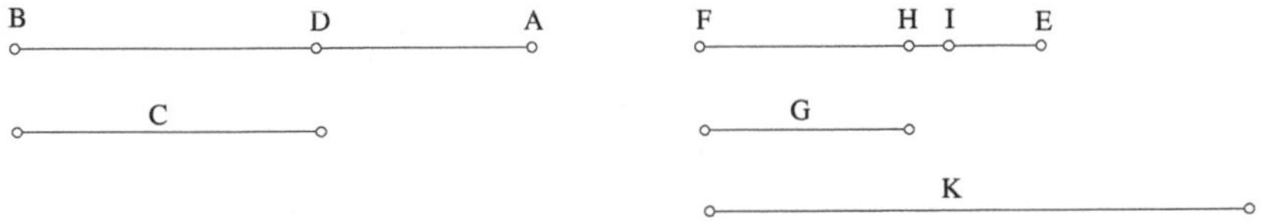

Fig. II.1.1

Proof: Let the side of *AB* be the number *EF* and the side of *C* be the number *G*. From *EF*, we take away the equal of *G*, that is *FH*. I say that *EH* is one.

If this is not the case, it is therefore greater than one, as it is the difference between two <integer> numbers. Let its unity be *HI* and let the number *K* be the square of the number *FI*. The number *FI* is greater than *G* and less than the number *EF*. The number *K* is therefore greater than the square of *G* and less than the square of the number *EF*. The number *K* is therefore greater than *C* and less than the number *AB* which is a square number. There is therefore a square number between the successive square numbers *AB* and *C*. This is contradictory.

Consequently, *HE* is one and the square of the number *EF* is equal to the squares obtained from *EH* and *HF*, plus twice the product of *EH* and *HF*. The square obtained from *FH* is *C* as *FH* is equal to *G*. The square of the number *EF* is *AB* and their difference is *AD*. The result of twice the product of *EH* and *HF*, plus the square obtained from *EH*, is therefore equal to the number *AD*. The result of the product of the number *EH* and *HF* is any number, and the result of twice the product of *EH* and *HF* is an even number. The square obtained from *EH* is one. If one is added to an even <number>, then the sum is an odd <number>. The result of twice the product of *EH* and *HF*, plus the square obtained from *EH*, is an odd number equal to the number *AD*. This is what we wanted to prove.

From what we have said, it has also been proved that, if *C* is one, then *AD* is three.

– **2** – In any three successive square numbers, the difference between the largest and the middle number exceeds the difference between the middle number and the smallest number by two.

Let the three successive square numbers be *AB*, *CD* and *E*, of which the largest is *AB*. Let the amount by which *CD* exceeds *E* be the number *CG*, and let the amount by which *AB* exceeds *CD* be the number *AF*.

I say that AF *exceeds* CG *by two.*

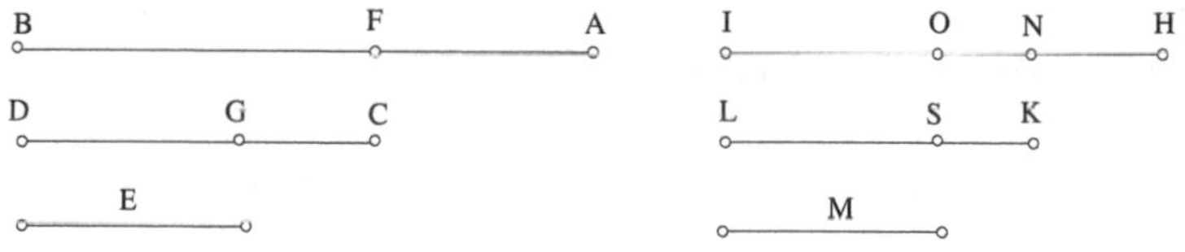

Fig. II.1.2

Proof: We let *HI* be the side of *AB*, we let *KL* be the side of *CD* and we let *M* be the side of *E*. From *HI*, we take away the equal of *KL*, that is *IN*, and from *KL* we take away the equal of *M*, that is *LS*. As in the previous proposition, we can show that each of <the numbers> *KS* and *HN* is one, and that twice the product of *HN* and *NI*, plus the square obtained from *HN*, is equal to the number *AF*, and that twice the product of *KS* and *SL*, plus the square obtained from *KS*, is equal to the number *CG*.

We set *IO* equal to *M*; there remains *NO* equal to *KS*. *NO* will be one. Twice the product of *NO* and *OI*, plus the square obtained from *NO*, is equal to the number *CG*. But twice the product of *HN* and *NI*, plus the square obtained from *HN*, was equal to the number *AF*. The difference between the number *AF* and the number *CG* is therefore equal to the difference between twice the product of *HN* and *NI*, plus the square obtained from *HN*, and twice the product of *NO* and *OI*, plus the square obtained from *NO*. Removing the two equal squares which are the square obtained from *HN* and the square obtained from *NO*, there remains twice the product of *HN* and *NI* and twice the product of *NO* and *OI*, equal to the difference between the two numbers *AF* and *CG*. But *HN* is equal to *NO*. Therefore there remains the difference between twice the product of *NO* and *NI* and twice the product of *NO* and *OI*, equal to the difference between the two numbers *AF* and *CG*. Yet, twice the product of *NO* and *NI* is greater than twice the product of *NO* and *OI* by twice the square obtained from *NO*. The number *AF* is therefore greater than the number *CG* by twice the square obtained from *NO*. But twice the square obtained from *NO* is two, as *NO* is one. The number *AF* thus exceeds the number *CG* by two. This is what we wanted to prove.

– **3** – The differences[1] between successive square numbers beginning with one are successive odd numbers beginning with three.

Let the successive square numbers be *A*, *B*, *C*, *D* and *E*, of which *A* is one, and let the successive odd numbers be *F*, *G*, *H* and *I*, of which the number *F* is three.

I say that the difference between B *and* A *is* F, *that the difference between* C *and* B *is* G, *that the difference between* D *and* C *is* H, *and that the difference between* D *and* E *is* I, *and so on in the same way.*

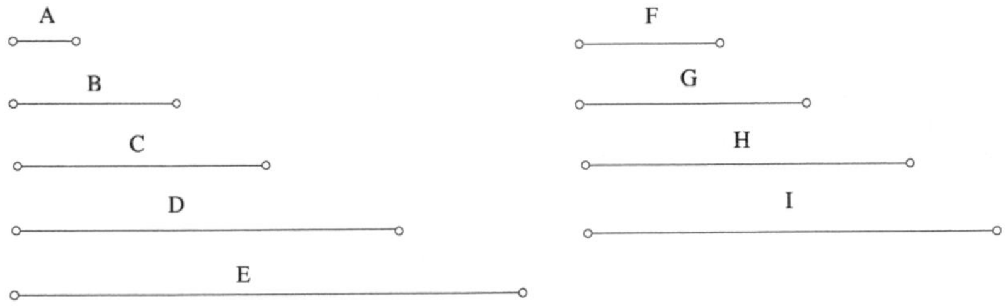

Fig. II.1.3

Proof: *A* is one, therefore *B* exceeds it by three, which is equal to the number *F*. The difference between *C* and *B* is greater than the difference

[1] Lit.: the difference.

between *A* and *B* by two, as the numbers *A*, *B* and *C* are successive square numbers. The difference between *C* and *B* is therefore equal to the number *F* plus two, which is the number *G*. The difference between *D* and *C* exceeds the difference between *C* and *B* by two, and the difference between *C* and *B* is the number *G*. Therefore, the difference between *D* and *C* exceeds the number *G* by two. But the number *H* also exceeds the number *G* by two, as these are two successive odd <numbers>. The difference between *D* and *C* is therefore the number *H*. Similarly, we can show that the difference between *E* and *D* is the number *I*, and so on in the same way. This is what we wanted to prove.

From this it is clear that, if the numbers *A*, *B*, *C*, *D* and *E* are numbers beginning with one, such that the successive differences between them are successive odd numbers beginning with three, then these are successive square numbers beginning with one.

– **4** – Given the successive odd numbers beginning with one. If one is added to the largest of these numbers and half of this sum taken which is then multiplied by itself, the result is equal to the sum of these odd numbers.

Let the successive odd numbers be *A*, *B*, *C*, *D* and *E*, of which *A* is one, and let the number *E*, added to one, be equal to the number *F*. Then the number *F* is even, as the number *E* is odd. Let the number *N* be half the number *F*, and let the number *H* be the square obtained from *N*.

I say that the number H *is equal to the sum of the odd numbers* A, B, C, D and E.

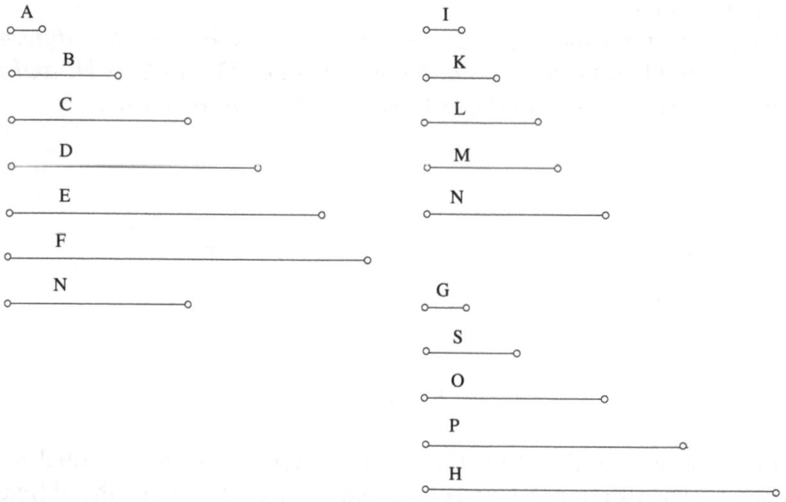

Fig. II.1.4

Proof: Adding one to *A* gives an even <number>. Take half of the sum, which is one; let it be *I*. Adding one to *B* also gives an even <number>. Take half of that sum; let it be *K*. Similarly we can derive the number *L* from the number *C*, and the number *M* from the number *D*. The difference between each of the numbers *A*, *B*, *C*, *D* and *E* and the number that follows it is two, as these are successive odd <numbers>. The differences are the same if one is added to each number. If they are halved, the differences between the halves is half of two, which is one.

The numbers *I*, *K*, *L*, *M* and *N* are successive numbers beginning with one. We set the numbers *G*, *S*, *O*, *P* and *H* their squares, which are successive square numbers beginning with one. The differences between these are equal to the successive odd numbers beginning with three, which are *B*, *C*, *D* and *E*. The number *H* therefore exceeds the number *G* by a number equal to the sum of the numbers *B*, *C*, *D* and *E*. But *G* is one and equal to *A*. The difference between *H* and *G*, to which *G* is added, is equal to the sum of the numbers *A*, *B*, *C*, *D* and *E*. But the difference between *H* and *G*, with *A* added, is equal to the number *H*. Therefore, the number *H* is equal to <the sum of> the odd numbers *A*, *B*, *C*, *D* and *E* beginning with one. This is what we wanted to prove.

It is also clear, from that which we have said, that the halves of successive even numbers are successive numbers.

– **5** – Given successive even numbers beginning with two and an equal number of successive odd numbers beginning with one. Then, the sum of the squares of the successive even numbers is equal to the sum of the squares of the successive odd numbers plus half the square of the greatest even number plus units equal to the number of odd numbers.

Let the successive even numbers beginning with two be *A*, *B*, *C* and *D*, of which the greatest is *D*, and let the same number of successive odd numbers beginning with one be *E*, *F*, *G* and *H*, of which the greatest is *H*.

I say that the sum of the squares of the numbers A, B, C *and* D *is equal to the sum of the squares of the numbers* E, F, G *and* H, *plus half the square of the number* D, *plus units equal to the number of odd numbers* E, F, G *and* H.

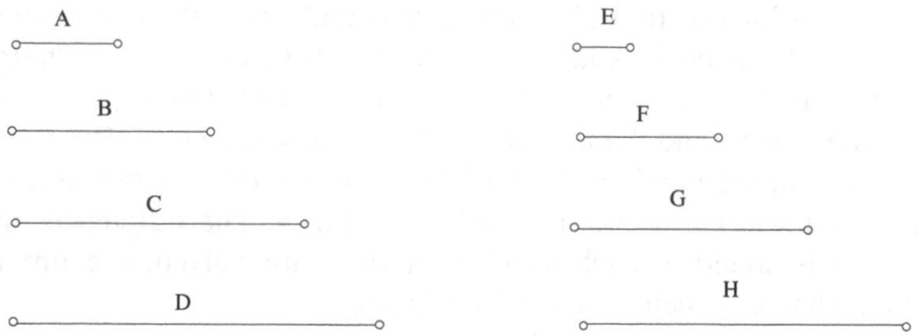

Fig. II.1.5

Proof: Each of the numbers *A*, *B*, *C* and *D* exceeds the associate number in the numbers *E*, *F*, *G* and *H* by one. Therefore, the square of each of them exceeds the square of the associate odd number by twice the product of one and this odd number plus the square of one. The sum of the squares of the even numbers *A*, *B*, *C* and *D* exceeds the sum of the squares of the odd numbers *E*, *F*, *G* and *H* by twice the product of one and the sum of the numbers *E*, *F*, *G* and *H*, plus the squares of the equal number of units. But twice the product of one and the numbers *E*, *F*, *G* and *H* is equal to twice the numbers *E*, *F*, *G* and *H*, and the squares of the units are units. The sum of the squares of the numbers *A*, *B*, *C* and *D* therefore exceeds the sum of the squares of the numbers *E*, *F*, *G* and *H* by twice the sum of the numbers *E*, *F*, *G* and *H*, plus the equal number of units. But, if one is added to the number *H* and half of the sum is then multiplied by itself, then the result is equal to the sum of the numbers *E*, *F*, *G* and *H*, as these are successive odd <numbers> beginning with one. The sum of the squares of the numbers *A*, *B*, *C* and *D* therefore exceeds the sum of the squares <of the numbers> *E*, *F*, *G* and *H* by twice the product of half of one thing by itself, which is the number *H* and one, plus the number of units equal to the number of *E*, *F*, *G* and *H*. But, if one is added to the number *H*, this gives <a number> equal to the number *D*. The sum of the squares of the numbers *A*, *B*, *C* and *D* therefore exceeds the sum of the squares of the numbers *E*, *F*, *G* and *H* by twice the product of half of the number *D* by itself, which is equal to half the square of the number *D*, plus the number of units equal to the number of *E*, *F*, *G* and *H*. The sum of the squares of the numbers *A*, *B*, *C* and *D* is therefore equal to the sum of the squares of the numbers *E*, *F*, *G* and *H*, plus half the square of the number *D*, plus the number of units equal to the number of *E*, *F*, *G* and *H*. This is what we wanted to prove.

– **6** – Given successive square numbers beginning with one, and an equal number of successive square odd numbers beginning with one, then twice the sum of the successive square <numbers> beginning with one is equal to half the squares of the successive odd <numbers> plus the greatest of the successive squares plus half the units equal in number to the successive odd squares.

Let the successive squares beginning with one be *A*, *B*, *C* and *D* of which *D* is the greatest, and let the equal number of successive odd squares beginning with one be *E*, *F*, *G* and *H*.

I say that twice the sum of the squares A, B, C *and* D *is equal to half the sum of the squares* E, F, G *and* H *to which is added the square* D *plus half the units equal in number to the numbers* E, F, G *and* H.

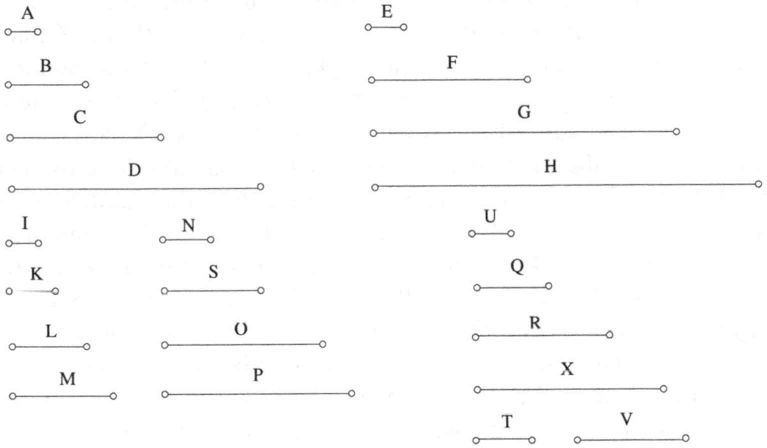

Fig. II.1.6

Proof: We let the numbers *I*, *K*, *L* and *M* be the sides of the squares *A*, *B*, *C* and *D*. The numbers *I*, *K*, *L* and *M* are successive beginning with one. Let the numbers *N*, *S*, *O* and *P* be twice these numbers. Twice <a series of> successive numbers is <a series of> successive even numbers. The numbers *N*, *S*, *O* and *P* are therefore successive even numbers beginning with two, and each of the numbers *N*, *S*, *O* and *P* is twice the associated number in the numbers *I*, *K*, *L* and *M*. The sum of their squares is therefore four times the sum of the squares of the numbers *I*, *K*, *L* and *M*. The squares of the numbers *N*, *S*, *O* and *P* are therefore four times the numbers *A*, *B*, *C* and *D*. Similarly, the numbers *E*, *F*, *G* and *H* are successive odd squares beginning with one. Let the numbers *U*, *Q*, *R* and *X* be their sides, and *U* will be one.

I say that the numbers U, Q, R *and* X *are successive odd numbers beginning with one.*

It is clear that they must be odd as, if any of them were even, its square would also be even; and they are successive. If it is possible that they were not successive, there would have to be another odd number between them.

Let *T* be an odd <number> between *U* and *Q*, and let the number *V* be its square. But the number *T* is less than the number *Q*, and greater than the number *U*. Therefore, the square *V* is less than the square *F* and greater than the square *E*; it is odd as it is the product of an odd number and itself. The odd squares *E* and *F* are therefore not successive. If they were, it would be contradictory. The numbers *U*, *Q*, *R* and *X* are therefore successive odd numbers beginning with one. But we have shown that the numbers *N*, *S*, *O* and *P* are successive even numbers beginning with two. But the number of <the numbers> *U*, *Q*, *R* and *X* is equal to the number of <the numbers> *N*, *S*, *O* and *P*. Therefore, the sum of the squares of the numbers *N*, *S*, *O* and *P* is equal to the sum of the squares of the numbers *U*, *Q*, *R* and *X* plus half the square of the number *P*, plus units equal in number to the numbers *U*, *Q*, *R* and *X*. But the squares of the numbers *U*, *Q*, *R* and *X* are the numbers *E*, *F*, *G* and *H*. The sum of the squares of the numbers *N*, *S*, *O* and *P* is therefore equal to the sum of the numbers *E*, *F*, *G* and *H*, plus half the square of the number *P*, plus units equal in number to the numbers *U*, *Q*, *R* and *X*. But we have shown that the sum of the squares of the numbers *N*, *S*, *O* and *P* is equal to four times the sum of the squares of the numbers *I*, *K*, *L* and *M*, which are the numbers *A*, *B*, *C* and *D*. Four times the sum of the square numbers *A*, *B*, *C* and *D* is therefore equal to the sum of the numbers *E*, *F*, *G* and *H*, plus half the square of the number *P*, plus units equal in number to the numbers *U*, *Q*, *R* and *X*. But half the square of the number *P* is equal to twice the square of the number *M*, as the number *P* is twice the number *M*. Four times the sum of the numbers *A*, *B*, *C* and *D* is equal to the sum of the numbers *E*, *F*, *G* and *H*, to which is added twice the square of the number *M* and units equal in number to the numbers *U*, *Q*, *R* and *X*. But the square of the number *M* is the number *D*. Four times the sum of the numbers *A*, *B*, *C* and *D* is therefore equal to the sum of the numbers *E*, *F*, *G* and *H*, to which is added twice the number *D* and units equal in number to the numbers *U*, *Q*, *R* and *X*. Halving everything mentioned above, it is clear that twice the sum of the numbers *A*, *B*, *C* and *D* is equal to half the sum of the numbers *E*, *F*, *G* and *H*, to which is added the number *D* and half the units equal in number to the numbers *E*, *F*, *G* and *H*, as their number is equal to that of <the numbers> *U*, *Q*, *R* and *X*. This is what we wanted to prove.

– **7** – Given successive odd numbers beginning with one, if we add them, then multiply <the sum> by one, then subtract from their sum the greatest of them, and if we multiply the remainder by two, and then subtract from this remainder the number that follows the greatest number, and if we multiply this remainder also by two, and if we then subtract from the remainder the number that follows the last to be subtracted, and then multiply the remainder again by two, and continue to proceed in the same way until we arrive at one, and if we add all of this, then this sum is equal to half the sum of the squares of the odd numbers to which is added half of the units equal in number to their number.

Let the successive odd numbers beginning with one be the numbers *A*, *B*, *C* and *D*, of which the number *D* is the greatest. Let the sum of the numbers *A*, *B*, *C* and *D* equal the number *E*, and let the sum of the numbers *A*, *B* and *C* equal the number *F*, and let the sum of the two numbers *A* and *B* equal the number *G*, and let the number *A*, which is one, equal the number *H*.

I say that the sum of the product of one and E, *and the product of two and* F, *and* G *and* H, *is equal to half the sum of the squares of the numbers* A, B, C *and* D, *to which is added half the units equal in number to the numbers* A, B, C *and* D.

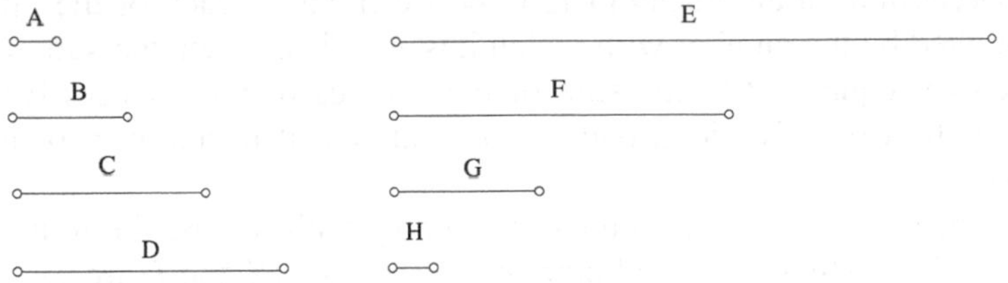

Fig. II.1.7

Proof: The sum of the numbers *A*, *B*, *C* and *D* is equal to the number *E*, the sum of the numbers *A*, *B* and *C* is equal to the number *F*, and the sum of the two numbers *A* and *B* is equal to the number *G*. The difference between the number *E* and the number *F* is therefore the number *D*. Similarly, we show that the difference between the number *F* and the number *G* is equal to the number *C*, and that the difference between the number *G* and the number *H* is equal to the number *B*. The numbers *H*, *G*, *F* and *E* are numbers that begin with one, which is *H*, and the successive differences between them are the numbers *B*, *C* and *D*, which are successive odd numbers beginning with the number *B*, which is three. The numbers *H*, *G*, *F* and *E* are successive square numbers beginning with one, and the squares of the numbers *A*, *B*, *C* and *D* are successive odd square numbers beginning with one, the number of which is equal to that of the numbers *H*, *G*, *F* and *E*. Twice the sum of the numbers *H*, *G*, *F* and *E* is therefore equal to half of the sum of the squares of the numbers *A*, *B*, *C* and *D*, to which is added the number *E* and half the units equal in number to that of the numbers H, G, F and E. Removing from both parts[2] the number E, there remains the sum of the number E and twice the numbers H, G and F equal to half the sum of the squares of the numbers A, B, C and D, to which is added half the units equal in number to that of the numbers A, B, C and D. But twice the numbers H, G and F is the product of two and H, G and F, and the number E is the product of E by one. The sum of the product of E and one, and the products of the numbers H, G and F and two, is therefore equal to half the sum of the squares of the numbers A, B, C and D, to which is added half of the units equal in number to that of the numbers A, B, C and D. This is what we wanted to prove.

– **8** – Given successive odd numbers beginning with one that is associated with an equal number of numbers, the greatest of which is associated with <the number> one of the first numbers, and the smallest of which, that is one, is associated with the greatest of the first numbers, and so on in succession for all the numbers in between, and if each of the numbers is multiplied by the number with which it is associated, then the sum <of the products> is equal to half the sum of the squares of the <successive> odd numbers, to which is added half of the units equal in number to the odd numbers.

Let the successive odd numbers beginning with one be the numbers A, B, C and D, to which the equal number of EF, GH, IK and L are associated. Let A be equal to L, the number B equal to IK, the number C equal to GH and the number D equal to EF. Let the product of A and EF equal the number M, the product of B and GH equal the number N, the product of C and IK equal the number S, and the product of D and L equal the number O.

I say that the sum of the numbers M, N, S *and* O *is equal to half the sum of the squares of the numbers* A, B, C *and* D, *to which is added half of the equal number of units.*

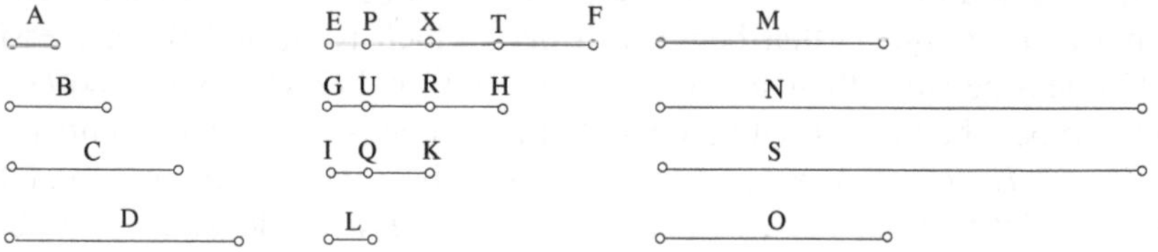

Fig. II.1.8

Proof: We let EP, GU and IQ each be equal to L, which is one. There remain the numbers QK, UH and PF of the successive even <numbers>

[2] Lit.: from the two.

beginning with two, and the difference between each of them and the number that succeeds it is therefore two. We subtract from each of these numbers <a number equal to> QK which is two. The subtracted numbers are the numbers UR and PX. The numbers RH and XF also become two successive even <numbers> beginning with two. We subtract from the remaining number XF a number equal to RH, which is two, that is XT. There remains TF, which is two. The sum of the products of A and EP, B and GU, C and IQ, and D and L, is equal to the product of the sum of the numbers A, B, C and D and one. The sum of the products of A and PX, B and UR, and C and QK, is equal to the product of the sum of the numbers A, B and C and two. The sum of the products of A and XT, and of B and RH, is equal to the product of the sum of the two numbers A and B and two. The product of A and TF is equal to the product of A and two. The product of A and EP, and PX, and XT and TF is the number M. The product of B and GU, and UR and RH is the number N. The product of C and IQ and QK is the number S. The product of D and L is the number O. If the sum of the numbers A, B, C and D is multiplied by one, and the sum of the numbers A, B and C by two, and also the two numbers A and B, and the number A <by two>, and if all these are added, then the sum will be equal to the sum of the numbers M, N, S and O. But the sum of the products of the numbers A, B, C and D and one, and the numbers A, B and C, and the numbers A and B, and the number A and two, is equal to half the sum of the squares of the numbers A, B, C and D, plus half the equal number of units, as the numbers A, B, C and D are successive odd numbers beginning with one. The sum of the numbers M, N, S and O is therefore equal to half the sum of the squares of the numbers A, B, C and D, to which is added half of the equal number of units. This is what we wanted to prove.

– **9** – Given successive odd numbers beginning with one, and an equal number of successive even numbers beginning with two, and if we take other numbers equal to the difference between the greatest even number and each of the odd <numbers>, then these numbers will be equal to the odd numbers, each of them to its associate.

Let the successive odd numbers beginning with one be A, B, C and D, and let the equal number of associated even numbers beginning with two be E, F, G and H. Let the difference between the number H and the number A be the number I, and let the difference between it and the number B be the number K, and let the difference between it and the number C be the number L, and let the difference between it and the number D be the number M.

I say that I *is equal to* D, *that* K *is equal to* C, *that* L *is equal to* B *and that* M *is equal to* A.

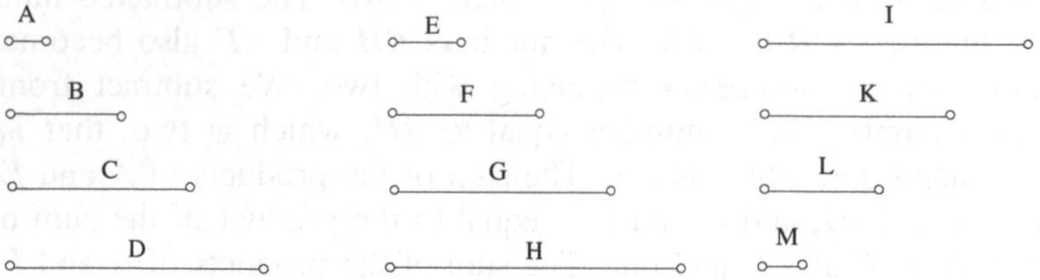

Fig. II.1.9

Proof: The numbers *A*, *B*, *C* and *D* are successive odd numbers beginning with one, and the numbers *E*, *F*, *G* and *H* are successive even numbers beginning with two. The difference between each of the numbers *A*, *B*, *C* and *D* and its associate in the numbers *E*, *F*, *G* and *H* is one. The difference between *H* and *D* is *M*. The number *M* is therefore one. The sum of the two numbers *D* and *M* is then equal to *H*, and the sum of the two numbers *C* and *L* is also equal to *H*. The amount by which *D* exceeds *C* is equal to the difference between *M* and *L*. The amount by which *D* exceeds *C* is two, as they are two successive odd numbers. Therefore, the amount by which *L* exceeds *M* is two. Similarly, we can also show that the amount by which the number *K* exceeds *L*, and the number *I* exceeds *K* is also two. The numbers *M*, *L*, *K* and *I* are therefore successive odd numbers beginning with one, as are the numbers *A*, *B*, *C* and *D*. Consequently, they are equal. *A* is equal to *M*, *B* is equal to *L*, *C* is equal to *K*, and *D* is equal to *I*. This is what we wanted to prove.

– **10** – Given successive odd numbers beginning with one and an equal number of successive even numbers beginning with two, then the sum of the squares of the odd numbers, to which is added one third of an equal number of units, is equal to two thirds of the product of the sum of these odd numbers and the greatest of the even numbers.

Let the successive odd numbers beginning with one be the numbers *A*, *B*, *C* and *D*, the greatest of which is *D*, and let the equal number of successive even numbers beginning with two be *E*, *F*, *G* and *H*, the greatest of which is the number *H*.

I say that the sum of the squares of the numbers A, B, C *and* D, *to which is added one third of an equal number of units, is equal to two thirds of the product of the sum of the numbers* A, B, C *and* D *and the number* H.

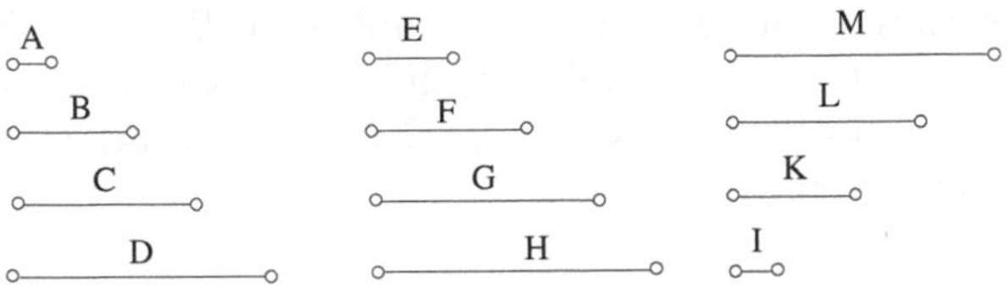

Fig. II.1.10

Proof: We let the amount by which the number *H* exceeds the number *D* be equal to *I*, the amount by which it exceeds *C* equal to *K*, the amount by which it exceeds *B* equal to *L*, and the amount by which it exceeds *A* equal to *M*. The numbers *I*, *K*, *L* and *M* are therefore equal to the numbers *A*, *B*, *C* and *D* which are the successive odd <numbers> beginning with one, of which the greatest is *M* and the smallest is *I*. The sum of the products of *A* and *M*, *B* and *L*, *C* and *K*, and *D* and *I* is equal to half the sum of the squares of the numbers *A*, *B*, *C* and *D*, to which is added half of an equal number of units. We add the squares of the numbers *A*, *B*, *C* and *D* on both sides. The sum of the products of *A* and *M*, *B* and *L*, *C* and *K*, and *D* and *I*, plus the squares of the numbers *A*, *B*, *C* and *D* is equal to one and one half times the sum of the squares of the numbers *A*, *B*, *C* and *D*, to which is added half of the units equal in number to that of the numbers *A*, *B*, *C* and *D*. The product of *A* and *M*, plus the square obtained from *M* is equal to the product of the sum of *A* and *M*, and *M*. The product of *B* and *L*, plus the square obtained from *L* is equal to the product of the sum of *B* and *L*, and *L*. The product of *C* and *K*, plus the square obtained from *K*, is equal to the product of the sum of *C* and *K*, and *K*. The product of *D* and *I*, plus the square obtained from *I*, is equal to the product of the sum of *D* and *I*, and *I*. The sum of the product of the sum of the numbers *A* and *M*, and *M*, the product of the sum of *B* and *L*, and *L*, the product of the sum of *C* and *K*, and *K*, and the product of the sum of *D* and *I*, and *I*, is equal to one and one half times the sum of the squares of the numbers *A*, *B*, *C* and *D*, to which is added half of the units, equal in number to that of the numbers *A*, *B*, *C* and *D*. The sum of the two numbers *A* and *M* is equal to the number *H*. The same applies to the two numbers *B* and *L*, the two numbers *C* and *K*, and the two numbers *D* and *I*. The product of the number *H* and the sum of the numbers *M*, *L*, *K* and *I* is equal to one and one half times the sum of the squares of the *A*, *B*, *C* and *D*, to which is added half of the units equal in number to the numbers *A*, *B*, *C* and *D*. If this is the case, then the sum of the squares of the numbers *A*, *B*, *C* and *D*, to which is added one third of the equal number of units, is equal to two thirds of the product of the number *H* and the sum of the numbers *M*, *L*, *K* and *I*. But the sum of the numbers *M*, *L*, *K* and *I* is equal to the sum of the numbers *A*, *B*, *C* and *D*.

Therefore, the sum of the squares of the numbers *A*, *B*, *C* and *D*, to which is added one third of the equal number of units, is equal to two thirds of the product of the sum of the numbers *A*, *B*, *C* and *D*, and the number *H*. This is what we wanted to prove.

– **11** – Given successive even numbers of which the first is two, if we take other numbers such that the first is half of the first of these numbers, the second is half the sum of the first number and the second number, the third is half the sum of the second and the third, and so on in the same way, then the new considered numbers will be successive odd numbers beginning with one.

Let the successive even numbers be *A*, *B*, *C* and *D*, of which the number *A* is two. Let *E* be half of this number, let the number *F* be half of the sum of the two numbers *A* and *B*, let the number *G* be half the sum of the two numbers *B* and *C*, and let the number *H* be half of the sum of the two numbers *C* and *D*.

I say that the numbers E, F, G *and* H *are successive odd numbers beginning with one.*

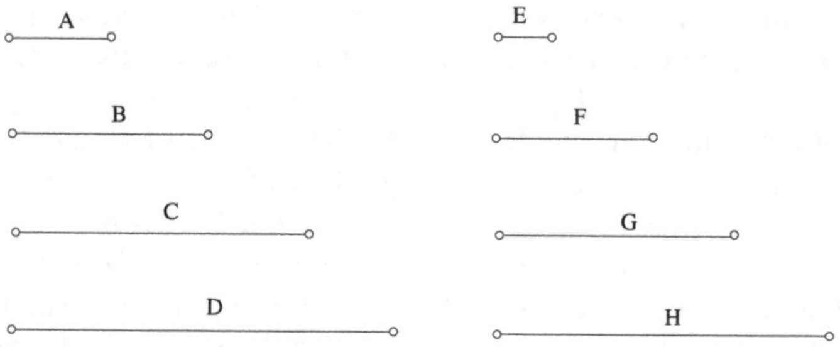

Fig. II.1.11

Proof: The number *A* is two and the number *E* is half of this, therefore one. The difference between each of the numbers *A*, *B*, *C* and *D*, taken in succession, is two. If two successive <numbers> from these are added together, and the sum halved, the difference between this half and each of them is one. The amount by which *B* exceeds *F* is therefore one, as is the amount by which *C* exceeds *G* and that by which *D* exceeds *H*. But the numbers *A*, *B*, *C* and *D* are successive even numbers beginning with two. Therefore, the numbers *E*, *F*, *G* and *H* are successive odd numbers beginning with *E*, which is one. This is what we wanted to prove.

– **12** – Given a set of straight lines[3] such that their ratios between them, taken in succession, are equal to the ratios of the successive odd numbers beginning with one, each to the others, and such that the first straight line is the smallest, and given another set of associated straight lines, equal in number, such that the ratios of these, each to the others and taken in succession, are equal to the ratios between the successive even numbers beginning with two, and such that the first of the first set of straight lines is half of the first of the second set of straight lines, then if the first of the first set of straight lines, that with ratios equal to the odd numbers, is multiplied by half of its associate in the second set of straight lines, and if the second of the first set <of straight lines> is multiplied by half of the sum of the first and second of the second set of straight lines, and if the third of the first is multiplied by half of the sum of the second and third of the second set of straight lines, and so on in the same way, and if these products are added together and added to the sum of the planes, each of which is equal to the product of the first of the first set of straight lines and half of the first of the second set of straight lines, as many times as one third of the number of straight lines in the first set, then the sum obtained is equal to two thirds of the product of the sum of the straight lines whose ratios are those of the odd numbers and the greatest of the straight lines whose ratios are those of the even numbers.

Let the straight lines whose ratios are those of the successive odd numbers beginning with one be the straight lines *A*, *B*, *C* and *D*, of which *A* is the smallest. Let the equal number of associated straight lines whose ratios are those of the successive even numbers beginning with two be *E*, *F*, *G* and *H*. Let *A* be half of *E*.

I say that the sum of the products of the straight line A *and half of the straight line* E, *of the straight line* B *and half of the sum of the two straight lines* E *and* F, *of the straight line* C *and half of the sum of the two straight lines* F *and* G, *and of the straight line* D *and half of the sum of the two straight lines* G *and* H, *to which is added the product of the straight line* A *and half of the straight line* E, *as many times as one third of the number of straight lines* A, B, C *and* D, *then the sum is equal to two thirds of the product of the sum of the straight lines* A, B, C *and* D *and the straight line* H *which is the greatest of the straight lines* E, F, G *and* H.

[3] Lit.: straight lines. We add 'a set of' for the needs of the translation.

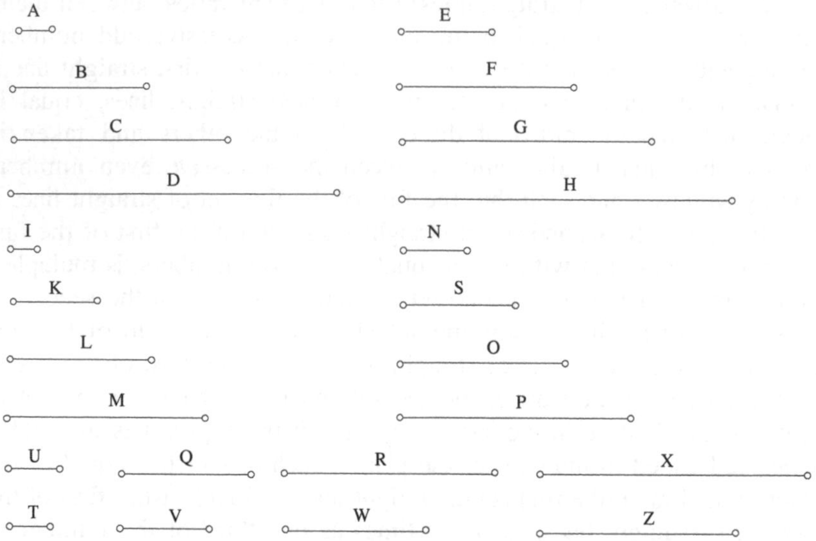

Fig. II.1.12

Proof: Let the successive odd numbers beginning with one be *I*, *K*, *L* and *M*, and the successive even numbers beginning with two be *N*, *S*, *O* and *P*. Let the straight line *U* be half the straight line *E*, let the straight line *Q* be half the two straight lines *E* and *F*, let the straight line *R* be half the two straight lines *F* and *G*, and let the straight line *X* be half the two straight lines *G* and *H*. Let the number *T* be half the number *N*, let the number *V* be half the two numbers *N* and *S*, let the number *W* be half the two numbers *S* and *O*, let the number *Z* be half the two numbers *O* and *P*, and let the straight line *A* be half the straight line *E*. The ratio of the straight line *A* to the straight line *E* is equal to the ratio of *I* to the number *N*. It is for this reason that the ratio of *I* to *T*, which is half the number *N*, is equal to the ratio of the straight line *A* to the straight line *U* which is half the straight line *E*. Similarly, the ratio of the straight line *B* to *A* is equal to the ratio of the number *K* to *I*, the ratio of *A* to *E* is equal to the ratio of *I* to *N*, and the ratio of *E* to *F* is equal to the ratio of *N* to *S*. The ratios of the straight lines *B*, *A*, *E* and *F* are therefore equal to the ratios of the numbers *K*, *I*, *N* and *S*. It is for this reason that the ratio of the straight line *B* to *E* and to *F*, and to the sum of *E* and *F* is equal to the ratio of the number *K* to *N* and to *S* and to the sum of *N* and *S*, and the ratio of the straight line *B* to half the sum of *E* and *F*, which is *Q*, is equal to the ratio of *K* to half the sum of *N* and *S* which is *V*. Similarly, the ratio of the straight line *C* to *B* is equal to the ratio of the number *L* to *K*, the ratio of *B* to *F* is equal to the ratio of *K* to *S*, and the ratio of *F* to *G* is equal to the ratio of *S* to *O*. The ratio of *C* to half of *F* and *G*, which is *R*, is then equal to the ratio of *L* to half of *S* and *O*, which is *W*. Similarly, we can show that the ratio of *D* to *X* is equal to the ratio of *M* to *Z*. The ratios of the straight lines *A*, *B*, *C* and *D* to the straight lines *U*, *Q*, *R* and *X*, each to its homologue, are therefore equal to the ratios of the number *I*, *K*, *L* and *M* to the numbers *T*, *V*, *W* and *Z*, each to its homologue. But the number *T* is half of the number *N*, the number *V* is half of the two numbers *N* and *S*, the number *W* is half of the two numbers *S* and *O*, and the number *Z* is half of the two numbers *O* and *P*. The numbers *N*, *S*, *O* and *P* are successive even numbers beginning with two, and the numbers *T*, *V*, *W* and *Z* are successive odd numbers beginning with one. The same applies to the numbers *I*, *K*, *L* and *M*. The numbers *T*, *V*, *W* and *Z* are equal to the numbers *I*, *K*, *L* and *M*, each to its homologue. Similarly, the straight lines *U*, *Q*, *R* and *X* are equal to the straight lines *A*, *B*, *C* and *D*, each to its homologue. Hence, the product of *A* and *U*, which is half of *E*, is equal to the square of the straight line *A*, the product of *B* and *Q*, which is half of *E* and *F*, is equal to the square of the straight line *B*, the product of *C* and *R*, which is half of *F* and *G*, is equal to the straight line *C*, and the product of *D* and *X*, which is half of *G* and *H*, is equal to the square of *D*.

Similarly, the ratios of the squares of the numbers *I*, *K*, *L* and *M*, each to the others, are equal to the ratios of the squares of the straight lines *A*, *B*, *C* and *D*, each to the others. The ratio of the sum of the squares of the numbers *I*, *K*, *L* and *M* to the square of the number *M* is therefore equal to the sum of the squares of the straight lines *A*, *B*, *C* and *D* to the square of the straight line *D*. We know that the ratio of the square of the number *M* to the product of *M* and *P*, which is equal to the ratio of *M* to *P*, is equal to the ratio of the square of the straight line *D* to the product of *D* and *H*, which is equal to the ratio of *D* to *H*. Using the equality ratio (*ex aequali*), the ratio of the sum of the squares of the numbers *I*, *K*, *L* and *M* to the product of *M* and *P* is equal to the ratio of the sum of the squares of the straight lines *A*, *B*, *C* and *D* to the product of *D* and *H*. But the ratio of the product of *M* and *P* to the product of the sum of the numbers *I*, *K*, *L* and *M*, and *P*, which is equal to the ratio of the number *M* to the sum of the numbers *I*, *K*, *L* and *M*, is equal to the ratio of the product of *D* and *H* to the product of the sum of the straight lines *A*, *B*, *C* and *D* and the straight line *H*, which is equal to the ratio of the straight line *D* to the sum of the straight lines *A*, *B*, *C* and *D*, as the ratio of the number *M* to the sum of the numbers *I*, *K*, *L* and *M* is equal to the ratio of the straight line *D* to the sum of the straight lines *A*, *B*, *C* and *D*. Using the equality ratio, the ratio of the sum of the squares of the numbers *I*, *K*, *L* and *M* to the product of the sum of the numbers *I*, *K*, *L* and *M* and the number *P* is equal to the ratio of the sum of the squares of straight lines *A*, *B*, *C* and *D* to the product of the sum of the straight lines *A*, *B*, *C* and *D* and the straight line *H*. If we take the square obtained from *I*, which is one, as many times as one third of the number of the numbers *I*, *K*, *L* and *M*, and the square of the straight line *A* as many times as one third of the number of the straight lines *A*, *B*, *C* and *D*, then the ratio of this considered multiple of the square obtained from *I* to the sum of the squares of the numbers *I*, *K*, *L* and *M* is equal to the ratio of the considered multiple of the square of the straight line *A* to the sum of the squares of the straight lines *A*, *B*, *C* and *D*. Therefore, the ratio of the sum of the squares of the numbers *I*, *K*, *L* and *M*, plus the squares obtained from the product of *I* and itself, of which the number is equal to one third of the number of the numbers *I*, *K*, *L* and *M*, to the product of the sum of these numbers and the number *P*, is equal to the ratio of the sum of the squares of the straight lines *A*, *B*, *C* and *D*, plus the squares obtained from the straight line *A*, of which the number is equal to one third of the number of straight lines *A*, *B*, *C* and *D*, to the product of the sum of the straight lines *A*, *B*, *C* and *D*, and the straight line *H*. We have shown that the sum of the squares of the straight lines *A*, *B*, *C* and *D* is equal to the product of *A* and half of *E*, plus the product of *B* and half of *E* and *F*, plus the product of *C* and half of *F* and *G*, plus the product of *D* and half of *G* and *H*. Therefore, the ratio of the sum of the squares of the numbers *I*, *K*, *L* and *M*, plus the squares obtained from the product of *I* and itself, of which the number is equal to one third of the number of the numbers *I*, *K*, *L* and *M*, to the product of the sum of the numbers *I*, *K*, *L* and *M*, and the number *P*, is equal to the product of *A* and half of *E*, plus the product of *B* and half of *E* and *F*, plus the product of *C* and half of *F* and *G*, plus the product of *D* and half of *G* and *H*, plus the squares obtained from the straight line *A*, of which the number is equal to the number of the numbers *A*, *B*, *C* and *D*, to the product of the sum of the straight lines *A*, *B*, *C* and *D*, and the straight line *H*. But the numbers *I*, *K*, *L* and *M* are successive odd numbers beginning with one, and the numbers *N*, *S*, *O* and *P* are successive even numbers beginning with two. Therefore, if the squares of the numbers *I*, *K*, *L* and *M* are added and then added to one third of the units equal in number, then this sum will be equal to two thirds of the product of the sum of the numbers *I*, *K*, *L* and *M*, and the number *P*.[4] But one third of the units equal in number to the number of the numbers *I*, *K*, *L* and *M* is equal to the squares obtained from the product of *I* by itself, equal in number to one third of the number of the numbers *I*, *K*, *L* and *M*, as the square obtained from *I* is one. Therefore, the sum of the product of *A* and half of *E*, the product of *B* and half of *E* and *F*, the product of *C* and half of *F* and *G*, and the product of *D* and half of *G* and *H*, to which is added the squares equal

[4] From Proposition 10.

to the square of A, of which the number is equal to one third of the number of the straight lines A, B, C and D, is equal to two thirds of the product of the sum of the straight lines A, B, C and D and the straight line H. But the square of A is equal to the product of A and half of E. The sum of the product of A and half of E, the product of B and half of E and F, the product of C and half of F and G, and the product of D and half of G and H, to which is added the product of A and half of E as many times as one third of the number of straight lines A, B, C and D, is therefore equal to two thirds of the product of the sum of the straight lines A, B, C and D and the straight line H. This is what we wanted to prove.

– **13** – Given a set of straight lines such that their ratios between them, taken in succession, are equal to the ratios of the successive odd numbers beginning with one, each to the others, and such that the first straight line is the smallest of them, and given another set of associated straight lines, equal in number, such that the ratios of these, each to the others and taken in succession, are equal to the ratios between the successive even numbers beginning with two, and such that the first of the first set of straight lines is not equal to half of the first of the second set of straight lines, then the sum of the product of the first of the first set of straight lines, those whose ratios are those of the odd numbers, and half of its associate in the second set of straight lines, the product of the second of the first <set of straight lines> and half of the first and second straight lines in the other set, the product of the third of the first <set of straight lines> and half of the second and third straight lines in the other set, and so on in the same way, plus the planes each of which is equal to the product of the first of the first set of straight lines and half of the first of the second set of straight lines, the number of which is equal to one third of the number of the first straight lines, is equal to two thirds of the product of the sum of the straight lines whose ratios are those of the odd numbers, and the greatest of the straight lines whose ratios are those of the even numbers.

Let the successive straight lines whose ratios are those of the odd numbers beginning with one be the straight lines A, B, C and D, of which A is the smallest. Let the equal number of associated straight lines whose ratios are those of the successive even numbers beginning with two be E, F, G and H, such that A is not half of E.

I say that the sum of the product of A *and half of the straight line* E, *the product of* B *and half of the sum of the two straight lines* E *and* F, *the product of* C *and half of the sum of the two straight lines* F *and* G, *and the product of* D *and half of the sum of the two straight lines* G *and* H, *to which is added the product of* A *and half of* E, *as many times as one third of the number of straight lines* A, B, C *and* D, *then the sum is equal to two thirds of the product of the sum of the straight lines* A, B, C *and* D *and the straight line* H *which is the greatest of the straight lines* E, F, G *and* H.

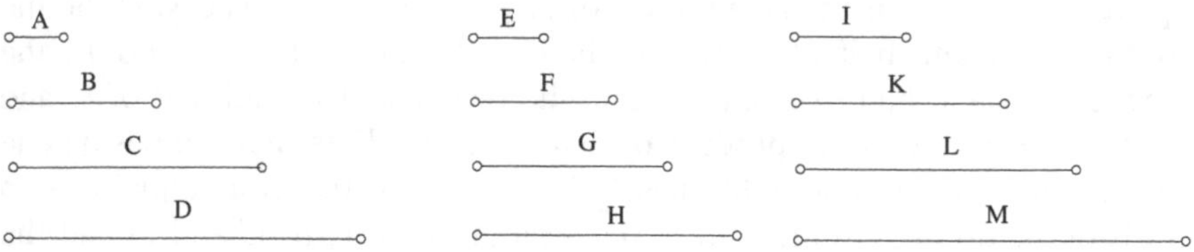

Fig. II.1.13

Proof: Let the straight line I be equal to twice the straight line A, and let the ratios of the straight lines I, K, L and M, each to the others and taken in succession, be equal to the ratios of the straight lines E, F, G and H taken in succession. Therefore, the ratio of E to I is equal to the ratio of F to K, equal to the ratio of G to L, equal to the ratio of H to M, and equal to the ratio of their halves to their halves. But the ratio of the product of A and half of E to the product of A and half of I is equal to the ratio of half of E to half of I. The ratio of the product of B and half of E and F to its product with half of I and K is equal to the ratio of half of E and F to half of I and K. The ratio of the product of C and half of F and G to its product with half of K and L is equal to the ratio of half of F and G to half of K and L. The ratio of the product of D and half of G and H to its product with half of L and M is equal to the ratio of half of G and H to half of L and M.

The ratio of all, that is the sum of the product of A and half of E, the product of B and half of E and F, the product of C and half of F and G, and the product of D and half of G and H, to the sum of the product of A and half of I, the product of B and half of I and K, the product of C and half of K and L, and the product of D and half of L and M, is equal to the ratio of half of the straight line E to half of the straight line I. But the ratio of E to I is equal to the ratio of H to M and the ratio of H to M is equal to the ratio of the product of the sum of the straight lines A, B, C and D and the straight line H to its product with the straight line M. Therefore, the ratio of the sum of the product of A and half of E, the product of B and half of E and F, the product of C and half of F and G, and the product of D and half of G and H, to the sum of the product of A and half of I, the product of B and half of I and K, the product of C and half of K and L, and the product of D and half of L and M, is equal to the ratio of the product of the sum of the straight lines A, B, C and D and the straight line H to its product with the straight line M. Applying a permutation (*permutendo*), the ratio of the sum of the product of A and half of E, the product of B and half of E and F, the product of C and half of F and G, and the product of D and half of G and H, to the product of the sum of the straight lines A, B, C and D and the straight line H, is equal to the ratio of the sum of the product of A and half of I, the product of B and half of I and K, the product of C and half of K and L, and the product of D and half of L and M, to the product of the sum of the straight lines A, B, C and D and the straight line M.

Similarly, the ratio of the product of A and half of E to the product of the sum of the straight lines A, B, C and D and the straight line H is compounded of the ratio of the straight line A to the sum of the straight lines A, B, C and D and the ratio of half of the straight line E to the straight line H. But the ratio of half of E to H is equal to the ratio of half of I to M. Therefore, the ratio of the product of A and half of E to the product of the sum of the straight lines A, B, C and D and the straight line H is compounded of the ratio of the straight line A to the sum of the straight lines A, B, C and D and the ratio of half of I to M. The ratio compounded of these two ratios is equal to the ratio of the product of A and half of the straight line I to the product of the sum of the straight lines A, B, C and D and the straight line M. The ratio of the product of the straight line A and half of the straight line E to the product of the sum of the straight lines A, B, C and D and the straight line H is equal to the product of A and half of the straight line I to the product of the sum of the straight lines A, B, C and D and the straight line M. It is for this reason that the ratio of the product of the straight line A and half the straight line E, taken as many times as the number of straight lines A, B, C and D, to the product of the sum of the straight lines A, B, C and D and the straight line H, is equal to the ratio of the product of the straight line A and half of I, taken as many times as the number of straight lines A, B, C and D, to the product of the sum of the straight lines A, B, C and D and the straight line M.

But we have shown that the ratio of the sum of the product of A and half of E, the product of B and half of E and F, the product of C and half of F and G, and the product of D and half of G and H, to the product of the sum of the straight lines A, B, C and D and the straight line H, is equal to the ratio of the sum of the product of A and half of I, the product of B and half of I and K, the product of C and half of K and L, and the product of D and half of L and M, to the product of the sum of the straight lines A, B, C and D and the straight line M. The ratio of the sum of the product of A and half of E, the product of B and half of E and F, the product of C and half of F and G, the product of D and half of G and H, and the product of A and half of E, taken as many times as the number of the straight lines A, B, C and D, to the product of the sum of the straight lines A, B, C and D and the straight line H, is equal to the ratio of the sum of the product of A and half of *I*, the product of *B* and half of *I* and *K*, the product of *C* and half of *K* and *L*, the product of *D* and half of *L* and *M*, and the product of *A* and half of *I*, taken as many times as the number of the straight lines *A*, *B*, *C* and *D*, to the product of the sum of the straight lines *A*, *B*, *C* and *D* and the straight line *M*.

But the ratios of the straight lines *A*, *B*, *C* and *D*, each to the others and taken in succession, are the ratios of the successive odd numbers beginning with one, each to the others, and the ratios of the straight lines *I*, *K*, *L* and *M*, each to the others, are the ratios of the successive even numbers beginning with two, each to the others, as they are equal to the ratios of the straight lines *E*, *F*, *G* and *H*, each to the others. The sum of the product of *A* and half of *I*, the product of *B* and half of *I* and *K*, the product of *C* and half of *K* and *L*, the product of *D* and half of *L* and *M*, and the product of *A* and half of *I*, taken as many times as one third of the number of straight lines *A*, *B*, *C*, *D*, is therefore equal to two thirds of the product of the sum of the straight lines *A*, *B*, *C* and *D* and the straight line *M*. It is for this reason that the sum of the product of *A* and half of *E*, the product of *B* and half of *E* and *F*, the product of *C* and half of *F* and *G*, the product of *D* and half of *G* and *H*, and the product of *A* and half of *E*, taken as many times as one third of the number of straight lines *A*, *B*, *C* and *D*, is equal to two thirds of the product of the sum of the straight lines *A*, *B*, *C* and *D* and the straight line *H*. This is what we wanted to prove.

– **14** – Given two magnitudes, of which the ratio of one to the other is known, it is possible to find a set of successive odd numbers beginning with one, and an equal number of successive even numbers beginning with two, such that the ratio of the number of units equal in number to the odd numbers to the product of the sum of these odd numbers and the greatest of the even numbers considered is less than the known ratio.

Let the known ratio be that of *A* to *B*. If the magnitude *A* has a ratio to the magnitude *B*, then it is possible to multiply it until its multiples become greater than the magnitude *B*. Let its multiples, which are greater than the magnitude *B*, be the magnitude *C*.[5] Let the number of units contained in the number *D* be equal to the number of times *A* is contained in *C*. Let twice the number *D* be the number *E*. Therefore the number *E* is even. Let the successive even numbers beginning with two and ending with the number *E* be the numbers *F*, *G* and *E*. We subtract one from each of them. The remaining numbers will then be the numbers *H*, *I* and *K*. The numbers *H*, *I* and *K* are therefore successive odd numbers beginning with one and their

[5] Implying: one of its multiples.

number is the same as the number of the even numbers *F*, *G* and *E*. Let the number *L* contain as many units as the number of the numbers *H*, *I* and *K*[6].

I say that the ratio of the number L *to the product of the sum of the numbers* H, I *and* K *and the number* E *is less than the ratio of* A *to* B.

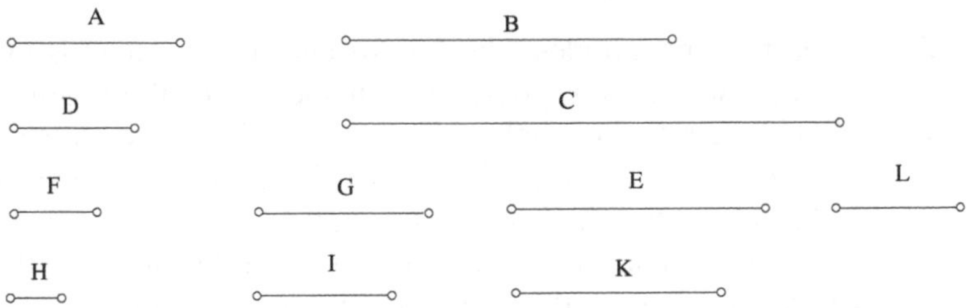

Fig. II.1.14

Proof: The numbers *H*, *I* and *K* are successive odd numbers beginning with one, of which the greatest is the number *K*. The number *E* is greater than the number *K* by one. The square of half of the number *E* is therefore equal to the sum of the numbers *H*, *I* and *K*. The ratio of half of the number *E* to its square is therefore equal to its ratio to the sum of the numbers *H*, *I* and *K*, and its ratio to its product with the sum of the numbers *H*, *I* and *K* is less than its ratio to the sum of the numbers *H*, *I* and *K*, as the product of half of the number *E* and the sum of the numbers *H*, *I* and *K* is greater than the sum of the numbers *H*, *I* and *K*. The ratio of half of the number *E* to the product of half of the number *E* and the sum of the numbers *H*, *I* and *K* is therefore less than the ratio of this number to its square. The ratio of half of the number *E* to the square of half of the number *E* is equal to the ratio of one to half of the number *E*. The ratio of half of the number *E* to the product of half of the number *E* and the sum of the numbers *H*, *I* and *K* is less than the ratio of one to half of the number *E*, which is equal to the ratio of *A* to *C*. The ratio of half of the number *E* to the product of half of the number *E* and the sum of the numbers *H*, *I* and *K* is therefore less than the ratio of *A* to *C*. The product of half of the number *E* and the sum of the numbers *H*, *I* and *K* is less than the product of the number *E* and the sum of the numbers *H*, *I* and *K*. The magnitude *C* is greater than the magnitude *B*. The ratio of half of the number *E* to the product of the number *E* and the sum of the numbers *H*, *I* and *K* is therefore very much less than the ratio of *A* to *B*. But the numbers *F*, *G* and *E* are successive even numbers beginning with two. Therefore, the difference between each of them and that which succeeds it is two. The number *E* contains as many twos as the

[6] Therefore $L = D = E/2$.

number of the numbers *F*, *G* and *E*, and half of the number *E* contains as many units as the number of the numbers *F*, *G* and *E*. The same applies to the units contained in the number *L*. The ratio of the number *L* to the product of the number *E* and the sum of the numbers *H*, *I* and *K* is less than the ratio of *A* to *B*. This is what we wanted to prove.

– **15** – Given two magnitudes whose ratio one to the other is known, and two known straight lines, it is possible to divide one of the straight lines into parts, such that the ratios of each to the others, taken in succession, are equal to the ratios of the successive odd numbers beginning with one, and to consider with the other straight line further straight lines, such that their number plus this one is equal to the number of parts of the first straight line, and that the greatest of these is this other straight line, and that their ratios, each to the others, taken in succession, are equal to the ratios of the successive even numbers beginning with two, such that the ratio of the product of the smallest of the parts of the first straight line and half of the smallest straight line considered in addition to the second straight line as many times as the number of parts in the first straight line, to the product of the first straight line and the second straight line, is less than the ratio of one of the <known> magnitudes having a known ratio to the other magnitude.

Let the two magnitudes be *A* and *B* and let the ratio of *A* to *B* be known. Let the two known straight lines be *CD* and *E*. If we wish to divide *CD* into parts such that the ratios of these, each to the others and taken in succession, are equal to the ratios of the successive odd numbers beginning with one, and <to take a number of> straight lines such that their number, including the straight line *E* is equal to the number of parts of the straight line *CD*, and such that the ratios of these, each to the others and taken in succession, are equal to the ratios of the successive even numbers beginning with two, and such that the greatest of these is the straight line *E*, and such that the ratio of the product of the smallest of the parts of the straight line *CD* and half the smallest of the straight lines which, including *E*, are equal in number to the number of parts of the straight line *CD*, to the product of the straight line *E* and the straight line *CD*, is less than the ratio of *A* to *B*, then we take the successive odd numbers beginning with one to be *F*, *G* and *H*, of which *F* is one, and an equal number of successive even numbers beginning with two to be *I*, *K* and *L*, of which *I* is two. Let the ratio of the units equal in number to the number of the numbers *F*, *G* and *H* to the product of the sum of the numbers *F*, *G* and *H* and the number *L*, be less than the ratio of the magnitude *A* to the magnitude *B*.[7] Let the ratio of the magnitude *CM* to the magnitude *CD* be equal to the ratio of *F* to the sum

[7] This assumption was not mentioned in the statement, but this is possible.

of the numbers F, G and H, and let the ratio of MN to MD be equal to the ratio of G to the sum of the two numbers G and H. In this way, we have divided the straight line CD in the same ratios as the numbers F, G and H, taken in succession. The smallest of these parts is CM. Let the ratio of E to S be equal to the ratio of L to K, and let the ratio of S to O be equal to the ratio of K to I.

I say that the ratio of the product of CM *and half of the straight line* O, *as many times as the number of parts in the straight line* CD, *to the product of* CD *and* E *is less than the ratio of* A *to* B.

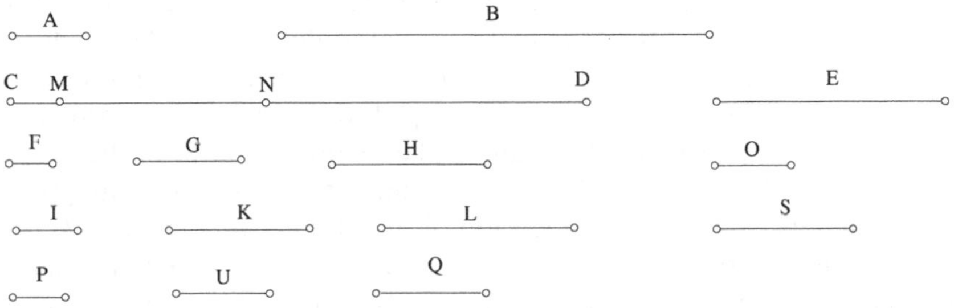

Fig. II.1.15

Proof: The ratios of the numbers F, G and H, each to the others and taken in succession, are equal to the ratios of the straight lines CM, MN, and ND, each to the others and taken in succession. The ratio of F to the sum of the numbers F, G and H is equal to the ratio of the straight line CM to the straight line CD. It is for this reason that the ratio of the square obtained from F to the square <obtained from > the sum of the numbers F, G and H is equal to the ratio of the square of the straight line CM to the square of the straight line CD. But the ratio of the square of the sum of the numbers F, G and H to the product of the sum of the numbers F, G and H and the number H is equal to the ratio of the square of the straight line CD to the product of CD and ND. Using the equality ratio, the ratio of the square obtained from F to the product of the sum of the numbers F, G and H and the number H, then equals the ratio of the square of the straight line CM to the product of CD and ND. But the number of the numbers F, G and H is equal to the number of the parts CM, MN and ND. Therefore, if the square obtained from F, which is one, is multiplied by the number of the numbers F, G and H, then its ratio to the product of the sum of the numbers F, G and H and the number H is equal to the ratio of the square of the straight line CM, multiplied as many times as the number of parts in the straight line CD, to the product of the straight lines CD and ND.

But the ratio of CM to O will either be equal to the ratio of F to I, or it will not be so. If we first let it be equal, then the ratio of the product of F, which is one, and half of I, which is also one, to the square obtained from F is equal to the ratio of the product of CM and half of O to the square of CM. The ratio of the units equal in number to the number of the numbers F, G and H to the product of the sum of the numbers F, G and H and the number H is equal to the ratio of the product of CM and half of O, as many times as the number of parts in the straight line CD, to the product of CD and ND. Similarly, the ratio of H to F is equal to the ratio of ND to CM, and the ratio of F to I is equal to the ratio of CM to O, and the ratio of I to L is equal to the ratio of O to E. The ratio of H to L is therefore equal to the ratio of ND to E. The ratio of H to L is equal to the ratio of the product of the sum of the numbers F, G and H and the number H to its product with the number L. The ratio of ND to E is equal to the ratio of the product of CD and ND to the product of CD and E. Therefore, the ratio of the product of the sum of the numbers F, G and H and the number H to its product with the number L is equal to the ratio of the product of CD and ND to its product with E. However, we have shown that the ratio of the units equal in number to the number of the numbers F, G and H to the product of the sum of the numbers F, G and H and the number H is equal to the ratio of the product of CM and half of O, as many times as the number of parts in the straight line CD, to the product of CD and ND. Using the equality ratio, the ratio of the units equal in number to the numbers F, G and H to the product of the sum of the numbers F, G and H and the number L is equal to the ratio of the product of CM and half of O, as many times as the number of parts in the straight line CD, to the product of CD and E. But the ratio of the units equal in number to the number of the numbers F, G and H to the product of the sum of the numbers F, G and H and the number L is less than the ratio of A to B. Therefore, the ratio of the product of CM and half of O, as many times as the number of parts in the straight line CD, to the product of CD and E is less than the ratio of A to B.

Similarly, let us now assume that the ratio of CM to O is not equal to the ratio of F to I, but the ratio of CM to P is equal to the ratio of F to I. Let the ratios of the straight lines P, U, and Q, each to the others and taken in succession, be equal to the ratios of the numbers I, K, and L, each to the others and taken in succession. The ratio of the product of CM and half of P, as many times as the number of parts of the straight line CD, to the product of CD and Q is less than the ratio of A to B.

Similarly, the ratio of the product of CM and half of P to its product with half of O is equal to the ratio of half of P to half of O, which is equal to the ratio of P to O. But the ratio of P to O is equal to the ratio of Q to E, as the ratios of the straight lines *O*, *S*, and *E*, each to the others, are equal to the ratios of the straight lines *P*, *U*, and *Q*, each to the others. Therefore, the ratio of the product of *CM* and half of *P* to the product of *CM* and half of *O* is equal to the ratio of *Q* to *E*. But the ratio of *Q* to *E* is equal to the ratio of the product of *CD* and *Q* to the product of *CD* and *E*. The ratio of the product of *CM* and half of *P* to its product with half of *O* is therefore equal to the ratio of the product of *CD* and *Q* to its product with *E*. Applying a permutation, the ratio of the product of *CM* and half of *P* to the product of *CD* and *Q* is equal to the ratio of the product of *CM* and half of *O* to the product of *CD* and *E*. It is for this reason that the ratio of the product of *CM* and half of *P*, as many times as the number of parts of the straight line *CD*, to the product of *CD* and *Q* is equal to the ratio of the product of *CM* and half of *O*, as many times as the number of parts of the straight line *CD*, to the product of *CD* and *E*. But it has already been shown that the ratio of the product of *CM* and half of *P*, as many times as the number of parts of the straight line *CD*, to the product of *CD* and *Q*, is less than the ratio of *A* to *B*. Therefore, the ratio of the product of *CM* and half of *O*, as many times as the number of parts of the straight line *CD*, to the product of *CD* and *E* is less than the ratio of *A* to *B*. This is what we wanted to prove.

– **16** – If we produce in a parabola one of its diameters and ordinates to this diameter such that the ratios of the parts of the diameter into which it is divided by the ordinates, taken in succession, are equal to the ratios of the successive odd numbers beginning with one, taken in succession, then the ratios of the ordinates within the parabola, each to the others and taken in succession, are equal to the ratios of the successive even numbers beginning with two, taken in succession.

Let *ABC* be a parabola, let *BD* be one of its diameters, and let the ordinates to this diameter within the parabola be *EFG*, *HIK*, *LMN*, and *ADC*. Let the numbers *S*, *O*, *P* and *U* be successive odd numbers beginning with one, and let the ratios of *BF*, *FI*, *IM*, and *MD*, each to the others and taken in succession, be equal to the ratios of the numbers *S*, *O*, *P*, and *U*, taken in succession. Let the equal number of successive even numbers beginning with two be *Q*, *R*, *X*, and *T*.

I say that their ratios, each to the others and taken in succession, are equal to the ratios of the straight lines EFG, HIK, LMN, ADC, *each to the others and taken in succession.*

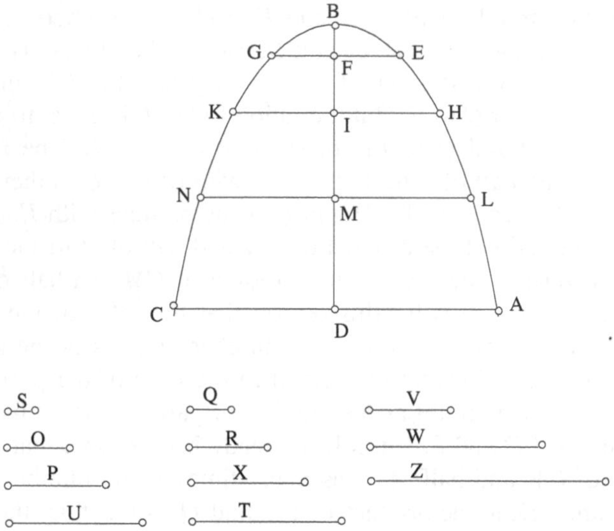

Fig. II.1.16

Proof: Let the number *V* be equal to the sum of the two numbers *S* and *O*, let the number *W* be equal to the sum of the numbers *S*, *O* and *P*, and let the number *Z* be equal to the sum of the numbers *S*, *O*, *P* and *U*. The numbers *S*, *V*, *W* and *Z* begin with one, and the differences between one and the other, considered in succession, are the numbers *O*, *P* and *U* which are successive odd numbers beginning with three. The numbers *S*, *V*, *W* and *Z* are therefore the successive squares beginning with one. The ratio of *S* to *O* is equal to the ratio of *BF* to *FI*. The ratio of *S* to the sum of *S* and *O* is therefore equal to the ratio of *BF* to *BI*. But the sum of *S* and *O* is equal to the number *V*. Therefore the ratio of *S* to *V* is equal to the ratio of *BF* to *BI*. But it has been shown in Proposition 20 of the first book of the work of Apollonius on the *Conics*, as generalized at the end of Proposition 51 of the first book, that the ratio of *BF* to *BI* is equal to the ratio of the square of the straight line *EF* to the square of the straight line *HI*. The ratio of *S* to *V* is therefore equal to the ratio of the square of the straight line *EF* to the square of the straight line *HI*.

Similarly, we can also show that the ratio of *V* to *W* is equal to the ratio of the square of the straight line *HI* to the square of the straight line *LM*, and that the ratio of *W* to *Z* is equal to the ratio of the square of the straight line *LM* to the square of the straight line *AD*. The ratios of the squares of the straight lines *EF*, *HI*, *LM* and *AD*, each to the others, are equal to the ratios of the numbers *S*, *V*, *W* and *Z*, each to the others. But we have shown that the numbers *S*, *V*, *W* and *Z* are successive square numbers beginning with one. Therefore, the ratios of the squares of the straight lines *EF*, *HI*, *LM* and *AD*, each to the others and taken in succession, are equal to the ratios of the successive square numbers beginning with one. It is for this reason that the ratios of these same straight lines, each to the others, are equal to the ratios of the successive numbers beginning with one. Consequently, the doubles of these numbers are successive even numbers beginning with two, which are the numbers *Q*, *R*, *X* and *T*, and that the doubles of these straight lines mentioned are the straight lines *EG*, *HK*, *LN* and *AC*. The ratios of the successive even numbers which are *Q*, *R*, *X* and *T*, each to the others and taken in succession, are equal to the ratios of the straight lines *EG*, *HK*, *LN* and *AC*, each to the others and taken in succession. This is what we wanted to prove.

From this, it can clearly be seen that, if the ratios of the straight lines *EG*, *HK*, *LN* and *AC*, each to the others and taken in succession, are equal to the ratios of the successive even numbers beginning with two, then the ratios of the straight lines *BF*, *FI*, *IM* and *MD*, each to the others and taken in succession, are equal to the ratios of the successive odd numbers beginning with one, each to the others.

– **17** – If we produce in a parabola its diameters and ordinates to this diameter such that the ratios of the parts of the diameter divided by the ordinates, each to the others and taken in succession, are equal to the ratios of the successive odd numbers beginning with one, each to the others and taken in succession, and if the smallest of these parts is the part adjacent to the vertex of the parabola, and if the extremities of the ordinates on any one side and the vertex of the portion and the two ends of the smallest of the ordinates are joined by straight lines, then the polygon thus formed within this portion of a parabola is less than two thirds of the area of the parallelogram whose base is the base of this portion and whose height is equal to its height by the product of the perpendicular dropped from the vertex of the portion onto the smallest of the ordinates drawn in this portion and half of this smallest straight line, as many times as one third of the number of parts of the diameter.

Let the portion of the parabola be *ABC*, let its diameter be *BD* and let its base be *AC*. Let the ordinates to the diameter *BD* within this portion be *EFG*, *HIK* and *ADC*. Let the ratios of the straight lines *BF*, *FI* and *ID*, each to the others and taken in succession, be equal to the ratios of the successive odd numbers beginning with one, which are *L*, *M* and *N*, and let *L* be the smallest of these. We join the straight lines *AH*, *HE*, *EB*, *BG*, *GK* and *KC*, draw the two straight lines *AS* and *CO* parallel to the straight line *BD*, and we make a straight line *SO* passing through the point *B* parallel to the straight line *AC*.

I say that the polygon AHEBGKC *is less than two thirds of the area of the parallelogram* ASOC *by the product of the perpendicular dropped from the point* B *onto the straight line* EG *and half of* EG, *as many times as the number of* BF, FI *and* ID.

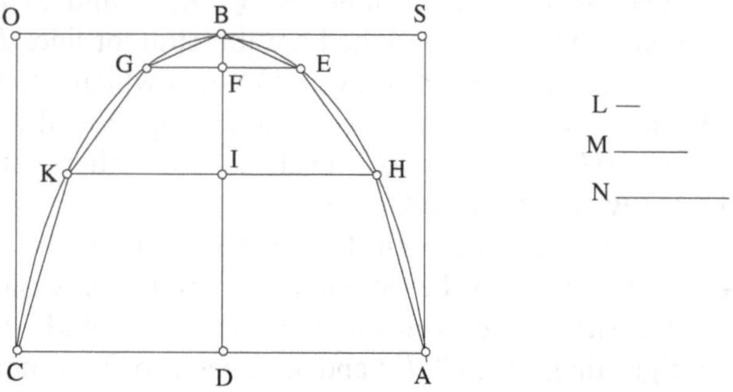

Fig. II.1.17a

Proof: Let the ratios of the straight lines *BF*, *FI* and *ID*, each to the others and taken in succession, are equal to the ratios of the successive odd numbers beginning with one, *L*, *M* and *N*. The ratios of the straight lines *EG*, *HK* and *AC*, each to the others and taken in succession, are therefore equal to the ratios of the successive even numbers beginning with two. If this is the case, then the product of *BF* and half of *EG*, plus the product of *FI* and half of *EG* and *HK*, plus the product of *ID* and half of *HK* and *AC*, plus the product of *BF* and half of *EG* as many times as one third of the number of parts of the diameter *BD* is equal to two thirds of the product of *BD* and *AC*.[8]

In addition, the ordinates must either be perpendicular to the diameter *BD*, or not so. First, let them be perpendicular. The product of *BF* and half of *EG* is therefore equal to the triangle *BEG*, the product of *FI* and half of *EG* and *HK* is equal to the trapezium *EGKH*, and the product of *ID* and half of *HK* and *AC* is equal to the trapezium *HKCA*. The product of *BD* and *AC* is therefore equal to the area *ASOC*. But we have already shown that the product of *BF* and half of *EG*, plus the product of *FI* and half of *EG* and *HK*, plus the product of *ID* and half of *HK* and *AC*, plus the product of *BF* and half of *EG*, as many times as one third of the number of parts of the diameter *BD*, is equal to two thirds of the product of *BD* and *AC*. The polygon *AHEBGKC* is therefore less than two thirds of the area of the

[8] From Proposition 13.

parallelogram *ASOC* by the product of *BF* and half of *EG*, as many times as one third of the number of parts of the diameter *BD*.

Similarly, let us now assume that the ordinates are not perpendicular to the diameter *BD*. Draw the perpendicular *BP* from the point *B* onto *EG*, the perpendicular *FU* from the point *F* onto *HK*, the perpendicular *IQ* from the point *I* onto *AC*, and the perpendicular *BR* from the point *B* onto *AC*. The triangles *BFP*, *FIU*, *IDQ* and *BDR* are all right-angled triangles, and the angles *BFP*, *FIU*, *IDQ* and *BDR* are equal as the ordinates are parallel. The triangles are therefore similar. It is for this reason that the ratio of *BP* to *BF* is equal to the ratio of *FU* to *FI*, and equal to the ratio of *IQ* to *ID*, and equal to the ratio of *BR* to *BD*, and equal to the ratio of the product of *BP* and half of *EG* to the product of *BF* and half of *EG*, and equal to the ratio of the product of *FU* and half of *EG* and *HK* to the product of *FI* and half of *EG* and *HK*, and equal to the ratio of the product of *IQ* and half of *HK* and *AC* to the product of *ID* and half of *HK* and *AC*, and equal to the ratio of the product of *BR* and *AC* to the product of *BD* and *AC*. Adding, we have the ratio of the sum of the product of *BP* and half of *EG*, the product of *FU* and half of *EG* and *HK*, and the product of *IQ* and half of *HK* and *AC*, to the sum of the product of *BF* and half of *EG*, the product of *FI* and half of *EG* and *HK*, and the product of *ID* and half of *HK* and *AC*, equal to the ratio of the product of *BR* and *AC* to the product of *BD* and *AC*. The sum of the product of *BF* and half of *EG*, the product of *FU* and half of *EG* and *HK*, and the product of *IQ* and half of *HK* and *AC* is equal to the polygon *AHEBGKC*. The ratio of the polygon *AHEBGKC* to the sum of the product of *BF* and half of *EG*, the product of *FI* and half of *EG* and *HK*, and the product of *ID* and half of *HK* and *AC* is equal to the ratio of the product of *BR* and *AC* to the product of *BD* and *AC*. But the ratio of the product of *BR* and *AC* to the product of *BD* and *AC* is equal to the ratio of the product of *BP* and half of *EG*, as many times as one third of the number of parts of the diameter *BD*, to the product of *BF* and half of *EG*, as many times as one third of the number of parts of the diameter *BD*. Adding, the ratio of the polygon *AHEBGKC* plus the product of *BP* and half of *EG*, as many times as one third of the number of parts of the diameter *BD*, to the sum of the product of *BF* and half of *EG*, the product of *FI* and half of *EG* and *HK*, the product of *ID* and half of *HK* and *AC*, and the product of *BF* and half of *EG*, as many times as one third of the number of parts of the diameter *BD*, is equal to the ratio of the product of *BR* and *AC*, which is the area *ASOC*, to the product of *BD* and *AC*. Applying a permutation, the ratio of the polygon *AHEBGKC* plus the product of *BP* and half of *EG* as many times as one third of the number of parts of the diameter *BD* to the area *ASOC* is equal to the ratio of the sum of the product of *BF* and half of *EG*, the product of *FI* and half of *EG* and *HK*, the product of *ID* and half of *HK* and *AC*, and the product of *BF* and half of *EG*, as many times as one third of the number of parts of the diameter *BD* to the product of *BD* and *AC*. But we have already shown that the product of *BF* and half of *EG* plus the product of *FI* and half of *EG* and *HK* plus the product of *ID* and half of *HK* and *AC* plus the product of *BF* and half of *EG*, as many times as one third of the number of parts of the diameter *BD* is equal to two thirds of the product of *BD* and *AC*. The polygon *AHEBGKC* plus the product of *BP* and half of *EG*, as many times as one third of the number of parts of the diameter *BD*, is therefore equal to two thirds of the area *ASOC*. The polygon *AHEBGKC* is therefore less than two thirds of the area *ASOC* by the product of *BP*, which is perpendicular to *EG*, and half of *EG*, as many times as one third of the number of parts of the diameter *BD*, which are *BF*, *FI* and *ID*. This is what we wanted to prove.

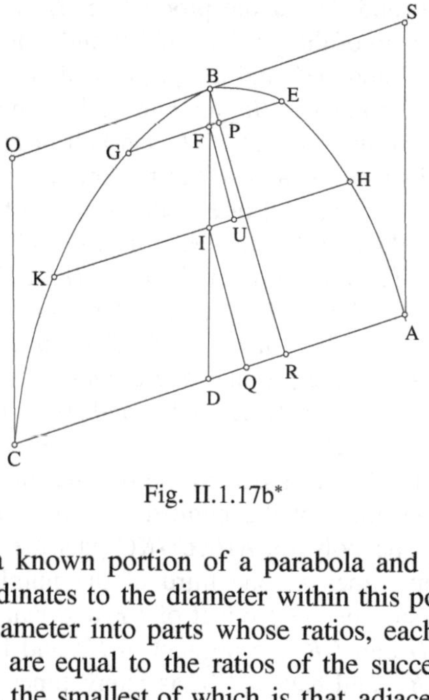

Fig. II.1.17b*

– **18** – Given a known portion of a parabola and a known area, it is possible to draw ordinates to the diameter within this portion of a parabola which divide the diameter into parts whose ratios, each to the others and taken in succession, are equal to the ratios of the successive odd numbers beginning with one, the smallest of which is that adjacent to the vertex of the parabola. If the ends of the ordinates, the vertex of the parabola and the extremities of the smallest of the ordinates are joined with straight lines in

* The manuscript only gives a single figure showing both cases. It is not therefore accurate.

such a way as to generate an inscribed polygon within the portion, then the amount by which this portion of a parabola exceeds the inscribed figure is less than the known area.

Let the given portion of the parabola be *ABC*, let its diameter be *BD*, let its base be *AC* and let the known area be *E*.

I say that it is possible for us to draw ordinates within the portion of a parabola ABC *which divide the diameter* BD *in the ratios of the successive odd numbers beginning with one such that the amount by which the portion of the parabola* ABC *exceeds the figure generated within it by joining the extremities of the ordinates and the vertex of the parabola to the extremities of the smallest ordinate that has been drawn with straight lines, is less than the area* E.

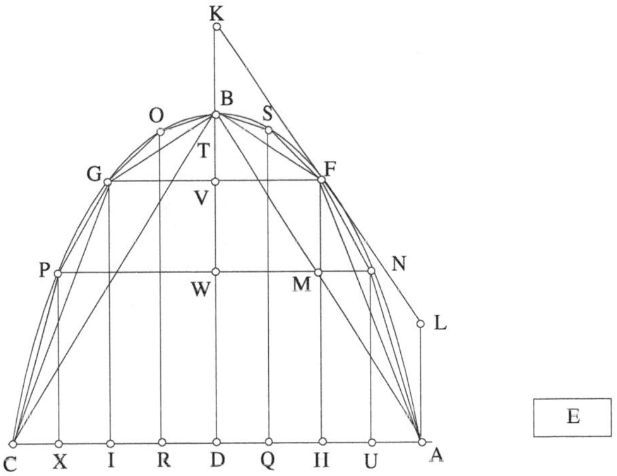

Fig. II.1.18a

Proof: We join the two straight lines *AB* and *BC*. If the two portions *AFB* and *BGC* of the parabola are less than the area *E*, <then we have found that which we sought>; if not, we divide each of the two straight lines *AD* and *DC* in half at the two points *H* and *I* respectively. We draw two straight lines *HF* and *IG* from these points parallel to the diameter *BD*. We join the straight lines *AF*, *FB*, *BG* and *GC*. Through the point *F*, we draw a straight line *KFL* tangent to the parabola, and through the point *A*, we draw a straight line *AL* parallel to the diameter *BD*. The straight line *HF* is parallel to the diameter *BD*. Apollonius has shown in Proposition forty-six of the first book of his work on the *Conics*[9] that, if this is the case, then *HF* is one of the diameters of the parabola. The ratio of *AH* to *HD* is equal to the ratio

[9] It actually follows from Proposition I.46 of the *Conics*.

of *AM* to *MB*. But the straight line *AH* is equal to *HD*. Therefore, the straight line *AM* is equal to *MB* and the straight line *MF* is one of the diameters of the parabola and it divides *AB* into two halves. Apollonius has shown in Proposition 5 of Book 2 of his work on the *Conics* that, if this is the case, then the straight line *KFL* is parallel to the straight line *AB*, as the straight line *KFL* is a tangent to the portion *AFB* of the parabola at the point *F* which is the vertex of its diameter, and the straight line *AL* is parallel to the straight line *BK*. The surface *ABKL* is therefore a parallelogram which surrounds the portion *AFB* of the parabola. It is therefore greater than the portion. The triangle *AFB* is half of the area *ABKL*. Therefore, the triangle *AFB* is greater than half of the portion *AFB* of the parabola.

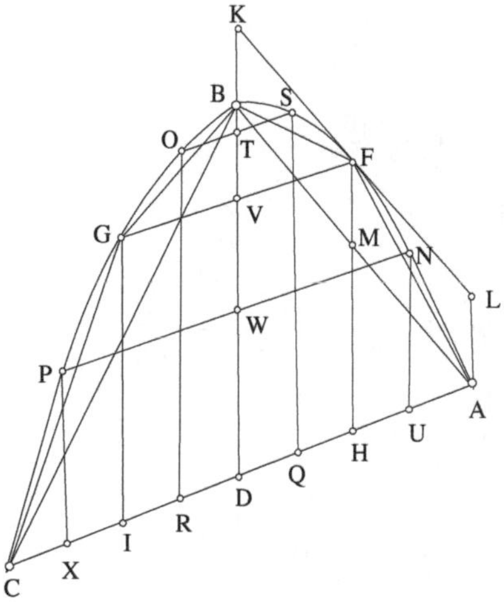

Fig. II.1.18b*

Similarly, we can show that the triangle *BGC* is greater than half of the portion *BGC* of the parabola. If the portions *ANF*, *FSB*, *BOG* and *GPC* of the parabola are less than the area *E*, <then we have found that which we sought>, otherwise also divide <each of> the parts *AH*, *HD*, *DI* and *IC* into two halves at the points *U*, *Q*, *R* and *X* and draw the straight lines *UN*, *QS*, *RO* and *XP* from <each of> these points parallel to the diameter *BD*. Join the straight lines *AN*, *NF*, *FS*, *SB*, *BO*, *OG*, *GP* and *PC*. As before, we can show that the triangles *ANF*, *FSB*, *BOG* and *GPC* are greater than the halves of the portions *ABF*, *FSB*, *BOG* and *GPC* <respectively>. If the remaining portions *ANF*, *FSB*, *BOG* and *GPC* of the parabola are less than

* This figure is not in the manuscript.

the area E, <then we have found that which we sought>, otherwise it is necessary to apply the same procedure several times until we arrive at a remainder of the portion of the parabola that is less than the area *E*. For any two magnitudes, one of which is greater than the other, if we subtract from the greatest magnitude more than its half, from the remainder more than its half, and from the remainder more than its half, and so on in the same way, we must at some point arrive at a remainder of the greatest magnitude that is less than the smallest magnitude. <Let us assume> then that the remainder of the portion which is less than the area *E* are the portions *AN*, *NF*, *FS*, *SB*, *BO*, *OG*, *GP* and *PC*. Join the straight lines *SO*, *FG* and *NP*. The two straight lines *QS* and *RO* are parallel to the diameter *BD*, and the straight line *QD* is equal to the straight line *RD*. Therefore, the straight line *ST* is equal to the straight line *TO*. It has been shown according to Apollonius in Proposition 5 of Book 2 of his work on the *Conics*, that, if this is the case, then the straight line *SO* is an ordinate to the diameter *BD*. Similarly, it can be shown that the two straight lines *FG* and *NP* are ordinates to the diameter *BD*, and that the straight line ordinates *SO*, *FG* and *NP* are equal to the straight lines *QR*, *HI* and *UX*, each to its homologue. Similarly, the parts *AU*, *UH*, *HQ* and *QD* are equal. Therefore, the ratios of the straight lines *DQ*, *DH*, *DU* and *DA*, each to the others and taken in succession, are equal to the ratios of the successive numbers beginning with one. If each of these is doubled, then the ratios of the doubles, each to the others and taken in succession, are equal to the ratios of the successive even numbers beginning with two, as each of these numbers is twice its homologue in the successive numbers. Twice *DQ* is *RQ*, twice *DH* is *IH*, twice *DU* is *XU*, and twice *DA* is *CA*. The ratios of *RQ*, *IH*, *XU* and *CA*, taken in succession, are equal to the ratios of the successive even numbers beginning with two. We have already shown that the straight lines *RQ*, *IH* and *XU* are equal to the straight lines *SO*, *FG* and *NP*. The ratios of the straight lines *SO*, *FG* and *NP*, taken in succession, are therefore equal to the ratios of the successive even numbers beginning with two. It is for this reason that the ratios of the straight lines *BT*, *TV*, *VW* and *WD*, taken in succession, are equal to the ratios of the successive odd numbers beginning with one. We have therefore constructed the polygon *ANFSBOGPC* within the portion *ABC* such that the portion *ABC* of the parabola exceeds an area less than *E*. This is what we wanted to prove.

– **19** – Given a known portion of a parabola and a known area, it is possible to construct a polygon within the portion of a parabola such that the difference between it and two thirds of the surface whose base is that of the portion and whose height is also the height of the portion is a magnitude less than the given area.

Let the given portion of the parabola be *ABC*, let its diameter be *BD*, let its base be *AC*, and let the given area be *E*. Let the parallelogram[10] *AFGC* have a base *AC* and a height equal to that of the portion *ABC*.

I say that it is possible to construct an inscribed polygon within the portion ABC *of the parabola that is less than two thirds of the area* AFGC *by a magnitude that is less than the area* E.

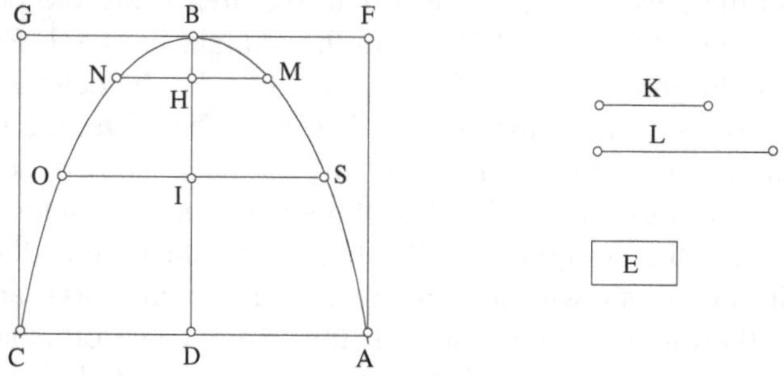

Fig. II.1.19a

Proof: The ratio of the area *E* to the product of *BD* and *AC* is known. The two straight lines *BD* and *AC* are known. We divide *BD* into parts such that their ratios, taken in succession, are equal to the ratios of the successive odd numbers beginning with one, and such that the smallest of them is that adjacent to the point *B*. We find straight lines which, taken with the straight line *AC*, are straight lines in the ratios of the successive even numbers beginning with two, and the greatest of which is the straight line *AC*, such that the ratio of the product of smallest part of the straight line *BD* and half of the smallest of the other straight lines taken with the straight linc *AC*, as many times as the number of parts of the straight line *BD*, to the product of *BD* and *AC*, is less that the ratio of the area *E* to the product of *BD* and *AC*.

Let the parts of the straight line *BD* be the straight lines *BH*, *HI* and *ID*, and let the straight lines taken with *AC* be the two straight lines *K* and *L*, the smallest of which is *K*. We make the two ordinates *MHN* and *SIO* passing through the points *H* and *I* on the diameter *BD*. Join the straight lines *AS*, *SM*, *MB*, *BN*, *NO* and *OC*. The ratios of the straight lines *BH*, *HI* and *ID*, each to the others and taken in succession, are equal to the ratios of the successive odd numbers beginning with one, each to the others, and the ratios of the straight lines *MN*, *SO* and *AC*, each to the others and taken in

[10] Lit.: surface.

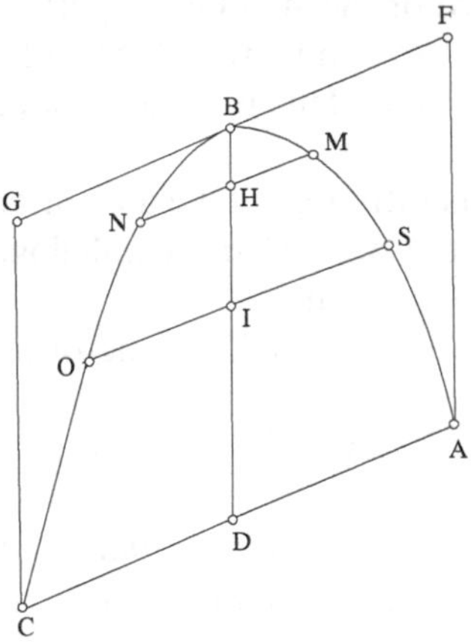

Fig. II.1.19b*

succession, are equal to the successive even numbers beginning with two, the greatest of which is the straight line *AC*. The same applies to the ratios of the straight lines *K*, *L* and *AC*. The two straight lines *MN* and *SO* are equal to the two straight lines *K* and *L*, each to its homologue. The ratio of the product of *BH* and half of the straight line *K*, as many times as the number of parts of the diameter *BD*, to the product of *BD* and *AC* is therefore less than the ratio of the area *E* to the product of *BD* and *AC*. The ratio of the product of the straight line *BH* and half of the straight line *MN*, as many times as the number of parts of the diameter *BD*, to the product of *BD* and *AC* is less than the ratio of the area *E* to the product of *BD* and *AC*. It is for this reason that the product of the straight line *BH* and half of the straight line *MN* as many times as one third of the number of parts of the straight line *BD* is less than the area *E*. If *BH* is perpendicular to *MN*, <we have found that which we sought>*, if not, the perpendicular is less than *BH*. The product of the perpendicular dropped from the point *B* onto *MN* and half of <the straight line> *MN*, as many times as one third of the number of parts of the diameter *BD*, is less than the area *E*, if *BH* is perpendicular to *MN*. If this is not the case, the product of the perpendicular dropped from the point *B* onto the straight line *MN* and half of the straight line *MN*, as many times as one third of the number of parts of the diameter *BD*, is very much less than the area *E*. The polygon *ASMBNOC* is less than two thirds of the area *AFGC* by the product of the perpendicular dropped from the point *B* onto the straight line *MN* and half of the straight line *MN*,

* This figure is not in the manuscript.

* With the help of Proposition 17.

as many times as one third of the number of parts of the diameter *BD*. The polygon *ASMBNOC* is less than two thirds of the area *AFGC* by a magnitude less than the area *E*. This is what we wanted to prove.

– **20** – The parabola is infinite, but the area of any portion of a parabola is equal to two thirds of the area of the parallelogram with the same base and the same height as the portion.

Let *ABC* be the parabola, let *DBE* be one of its portions, let *BF* be the diameter of this portion, and let *DFE* be its base. Let the parallelogram be *DGHE*, whose base is *DFE*, and whose height is the height of the portion *DBE* of the parabola.

I say that the entire parabola is infinite, and that the area of the portion DBE *of the parabola is equal to two thirds of the area of the parallelogram* DGHE.

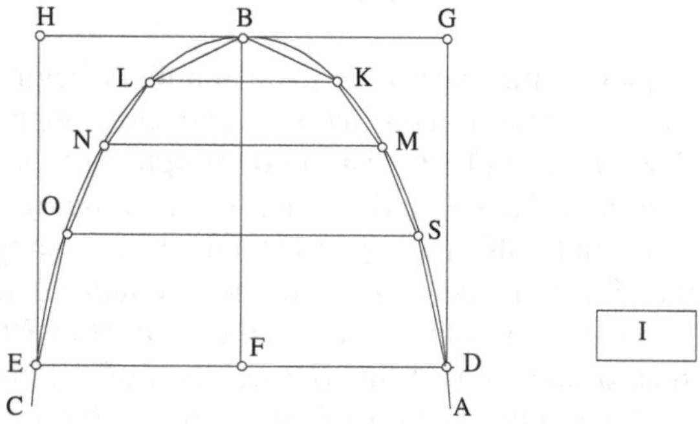

Fig. II.1.20a

Proof: The parabola *ABC* may be extended to infinity and the two lines *BA* and *BC* will never meet at the side of *AC* so as to form a surface. The parabola therefore has no limits.

I say that the portion *DBE* of the parabola is equal to two thirds of the parallelogram *DGHE*.

If this were not the case, then it would be either greater than two thirds or less than two thirds. Let us assume first that it is greater than two thirds and that the amount by which it exceeds two thirds is equal to the area *I*. It is possible to draw ordinates within the portion *DBE* dividing the diameter in the ratios of the successive odd numbers beginning with one. If their extremities are joined by straight lines and the vertex of the parabola joined to the extremities of the smallest of them, a polygon is generated within the portion which the portion exceeds by a magnitude less than the area *I*. Let *KL*, *MN*, *SO* and *DE* be the ordinates mentioned above, and let the straight lines joining <the extremities> be the straight lines *DS*, *SM*, *MK*, *KB*, *BL*,

LN, *NO* and *OE*. The polygon *DSMKBLNOE*, to which is added the area *I*, is greater than the portion *DBE* of the parabola. But the portion *DBE* of the parabola is equal to two thirds of the parallelogram *DGHE* to which is added the area *I*. Therefore, the polygon *DSMKBLNOE*, to which is added the area *I*, is greater than two thirds of the parallelogram *DGHE* to which is added the area *I*. Eliminating the area *I*, which is common to both, the remaining polygon *DSMKBLNOE* is greater than two thirds of the parallelogram *DGHE*. We have shown in the earlier propositions that it is less than two thirds of the parallelogram, so this is contradictory. The portion *DBE* is therefore not greater than two thirds of the parallelogram *DGHE*.

I say that the portion *DBE* is not less than two thirds of the parallelogram *DGHE*.

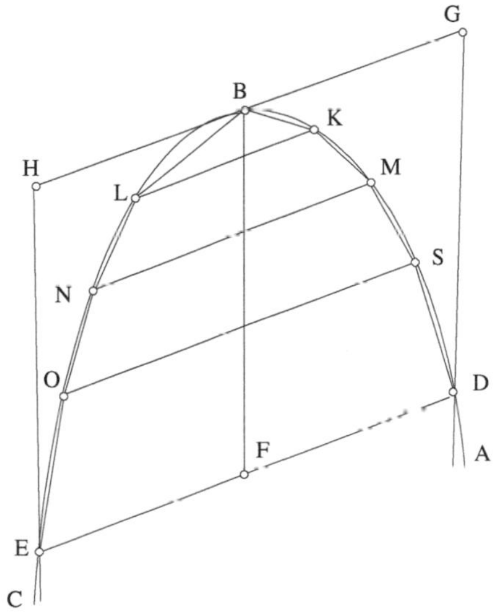

Fig. II.1.20b*

If this is the case, then let it be less than two thirds by a magnitude equal to the area *I*. It is possible to construct an inscribed polygon within this portion of the parabola that is less than two thirds of the parallelogram *DGHE* by a magnitude that is less than the area *E*. Let this be the polygon *DSMKBLNOE*. The polygon *DSMKBLNOE*, plus the area *I*, is greater than two thirds of the parallelogram *DGHE*. But the portion *DBE*, plus the area *I*, is equal to two thirds of the parallelogram *DGHE*. The polygon

* This figure is not in the manuscript.

DSMKBLNOE, plus the area *I*, is therefore greater than the portion *DBE*, plus the area *I*. Eliminating the area *I*, which is common to both, the remaining polygon *DSMKBLNOE* is greater than the portion *DBE* of the parabola. It is therefore greater than anything inscribed within it, which is contradictory. Therefore, the portion *DBE* is not less than two thirds of the parallelogram *DGHE*. We have already shown that it is not greater than two thirds. Consequently, it must be equal to two thirds of the parallelogram *DGHE*. This is what we wanted to prove.

The book of Thābit ibn Qurra al-Ḥarrānī
on the measurement of the conic section called parabola is completed.

2.3. MEASURING THE PARABOLOID

2.3.1. *Organization and structure of Ibn Qurra's treatise*

At the time when he was writing the treatise on *The Measurement of the Parabola*, had Thābit ibn Qurra figured out, at least in his thoughts, the treatise on *The Measurement of the Paraboloids*? The question quite naturally imposes itself in the lesson of the former: the same thought, the same language, with the exception that this time it treats space rather than the plane. Better yet, in the former treatise Thābit explicitly evokes three propositions of the latter. We are somewhat struck by such similarities not only in reasoning, but also, we shall see, in structure; and for this, let us follow Thābit as he determines the volume of the paraboloid.

This treatise is composed of 36 propositions, which are divided into several groups. The first consists of the first 11 propositions, which all pertain to numerical equalities concerning integers. They are established with the help of two lemmas, and two propositions of the same nature borrowed from the treatise on *The Measurement of the Parabola*. This group of arithmetical propositions forms the foundation for Propositions 12 and 13, which extend the result of Proposition 11 to magnitudes, that is to say they generalize the result to real numbers. It is this generalized result that will play a role in Proposition 32.

A little later, Thābit introduces a group of arithmetical propositions that pertain this time to numerical inequalities, thereby preparing for the introduction of the Axiom of Archimedes and the necessary bounds. This group's 11 propositions divide into three subgroups. From 22 to 27, the propositions pertain to numerical inequalities; from 28 to 31, they interpret these inequalities as magnitudes or real numbers – the two last propositions study sequences of real numbers (increasing sequences in 30 and decreasing in 31), and invoke the Axiom of Archimedes. Proposition 21 constitutes a subgroup in itself, and has bearings on a relation of equality between four magnitudes.

These two groups – Propositions 1–11 and Propositions 12, 13, and 21–31 – represent in themselves two levels of the graph of this treatise: the first, on the arithmetical propositions, concerns equalities or inequalities; the second, built on the first, is dedicated to magnitudes, and also depends on the introduction of the Axiom of Archimedes.

Next come a group of lemmas necessary for the last level of the graph, which consists of Propositions 14–20. Proposition 14 is there for the purposes of calculation, in the three ensuing propositions, of the volumes of the frustum of the cone, of the frustum of the 'hollow' cone and of the frustum of the solid rhombus. The results obtained in Propositions 15–17 are used in the proof of Proposition 32. Proposition 18 is required for the study of a property of the tangent to the parabola. In Proposition 19, Thābit shows that the solids made by rotating two parallelograms of equal height around their common base are equivalent. Thus a 'hollow' cylinder – which Ibn al-Haytham will later call 'conic' – is equivalent to a right cylinder. Ultimately, in Proposition 20, Thābit studies the volume of the solid made by a parallelogram rotated around a parallel to one of its bases, a solid called a torus. The results of Propositions 18, 19 and 20 will serve in the establishment of Propositions 33 and 34.

All is now in place in order to establish the principal propositions of the third level of the graph, and to determine the volume of paraboloids.

One sees in this cursory description – and one will verify it – that here the syntactic structure also superimposes a semantic structure, both analogous to that which we have been able to see in the case of the parabola. But one equally observes a similar tendency toward arithmetization, the exploitation of properties of the upper bound of a convex set as well as its uniqueness, a recourse to Euclid's X.1, but generalized to apply to the case of the paraboloid. In short, we will show the analogy of the method in the case of the parabola and in that of the paraboloid, which underlies the structural similarity.

This book of Thābit ibn Qurra has had a historic destiny, to the point where it has founded a tradition of research in which one will find al-Qūhī[1] and then Ibn al-Haytham.[2]

If one now indulges in a detailed analysis of the treatise, one first finds definitions of different parabolic solids. Thus, Thābit begins by distinguishing the different types of paraboloids of revolution. He starts by considering a first group, for which the axis of rotation is a diameter. He then defines three types, according to whether the angle between the diameter and the demi-chord considered is right, obtuse or acute. In the three cases, the engendered solid is called a 'parabolic dome', where the vertex is the point shared by the axis of rotation and the arc of the used parabola. One then has, respectively, a dome with a regular, a pointed and a sunken vertex.

In the second group, the axis of rotation is the base of the section, that is to say the chord of the parabola. The engendered solid is called a 'parabolic sphere', and the extremities of the fixed chord are its poles. There are then two types: the first for which the chord is perpendicular to the axis of the parabola – the parabolic sphere is called 'like a melon'; the second for which the chord is arbitrary – the parabolic sphere is called 'like an egg'.

Thābit finally introduces the definition of the 'hollow cone' and 'solid

[1] See Chapter V: al-Qūhī.
[2] See Vol. II, Chapter II.

rhombus'. From the rotation of a triangle having an obtuse angle around the side incident to that angle, one obtains a hollow cone, which is to say the difference between two cones having the same base, whereas the rotation of a triangle around a side incident to one of its accute angles gives the solid called a 'solid rhombus', which is to say the sum of two cones joined at the base. Thābit then passes to the arithmetical propositions. The treatise comprises in total 17 propositions that pertain to the integers.

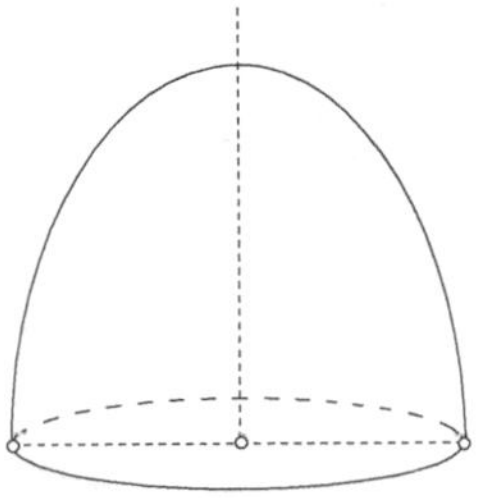

Dome with a regular vertex

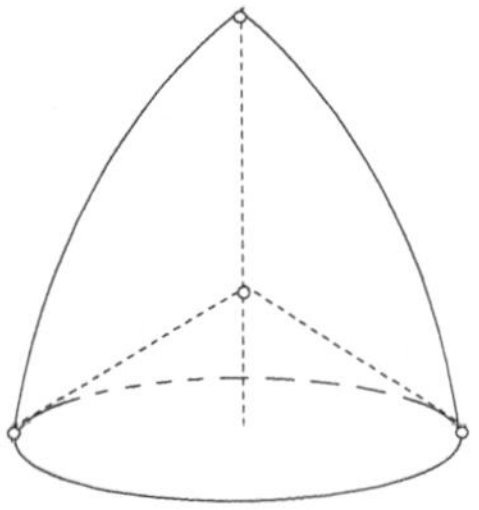

Dome with a pointed vertex

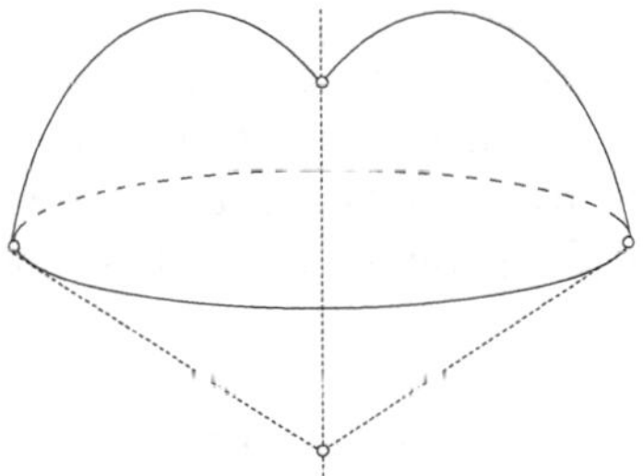

Dome with a sunken vertex

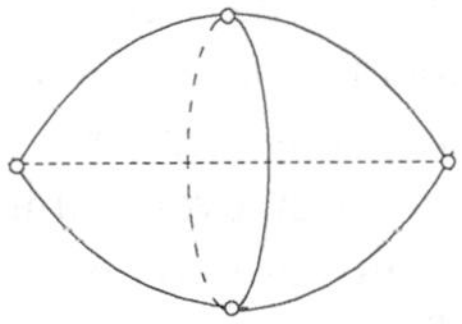

Parabolic sphere 'like a melon'

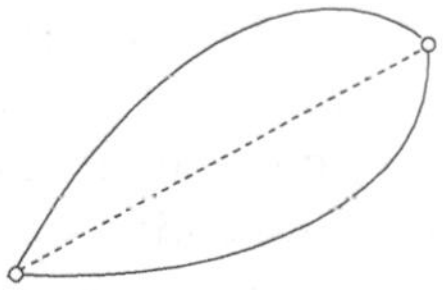

Parabolic sphere 'like an egg'

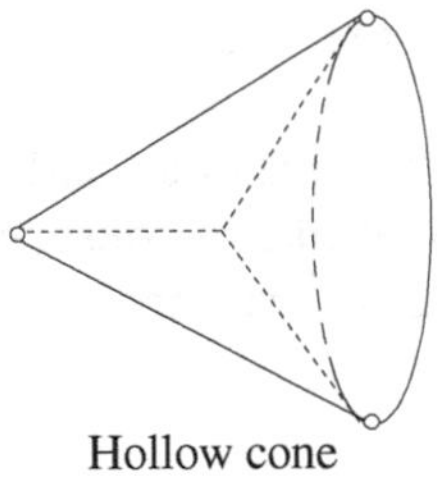

Hollow cone

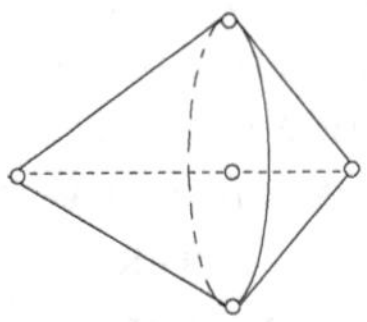

Solid rhombus

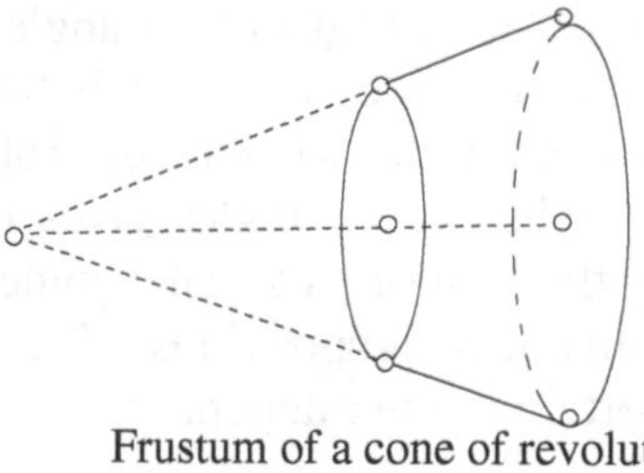

Frustum of a cone of revolution

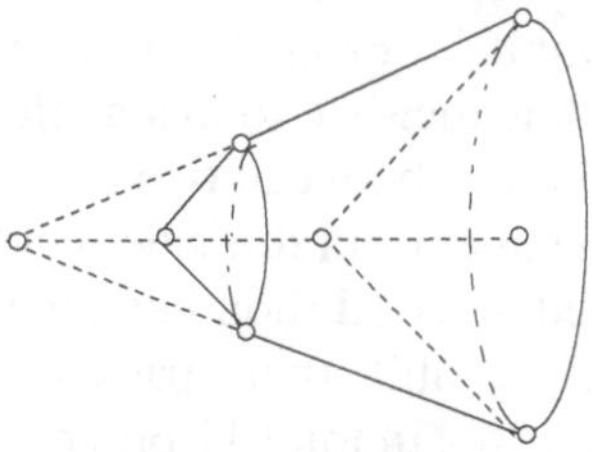

Frustum of a hollow cone

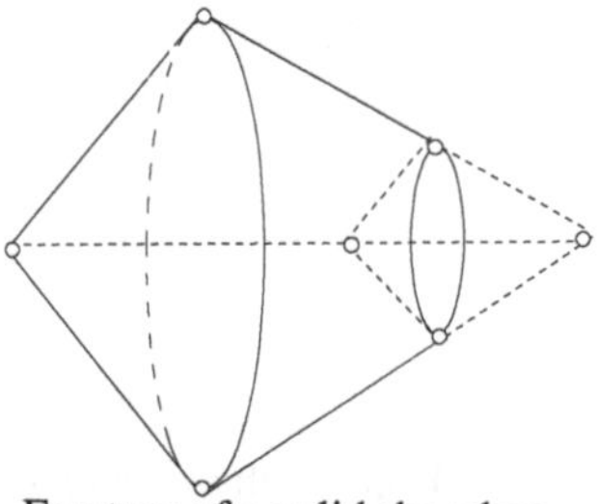

Frustum of a solid rhombus

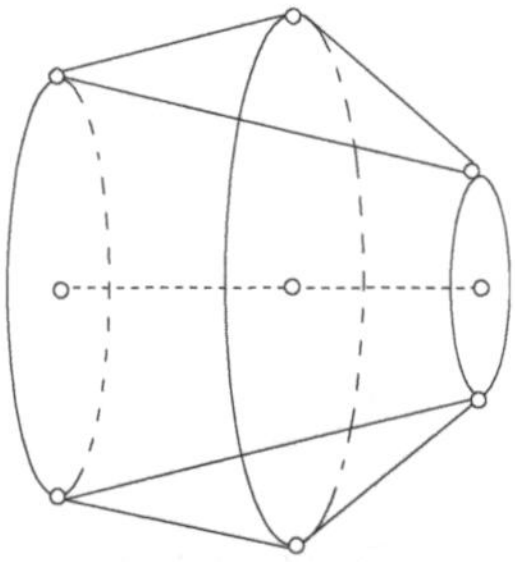

Triangular torus

Recall the properties that the author uses, considering them, for the most part, as axioms, which we denote by A; two of them are lemmas, denoted by L, proven by *reductio ad absurdum*. The propositions of the treatise on the area of the parabola that Thābit ibn Qurra uses here will be denoted by p.

A_0: The difference between two consecutive integers is 1.
A_1: The difference between two consecutive even numbers is 2.
A_2: The difference between two consecutive odd numbers is 2.
A_3: Between two consecutive even numbers, there is an odd number.
A_4: The product of an integer and 2 is an even number.
A_5: Every odd number increased by 1 gives an even number.
L_6: Two consecutive squares are the squares of two consecutive integers – Lemma proven in p_1 and here in the first proposition.
A_7: A square is odd if, and only if, it is the square of an odd number.
L_8: Two consecutive odd squares are the squares of two consecutive odd numbers, proven in p_6.
A_9: A cube is odd if, and only if, it is the cube of an odd number.
A_{10}: Two consecutive cubes are the cubes of two consecutive integers.
A_{11}: Two consecutive squared-squares are the squares of two consecutive squares.

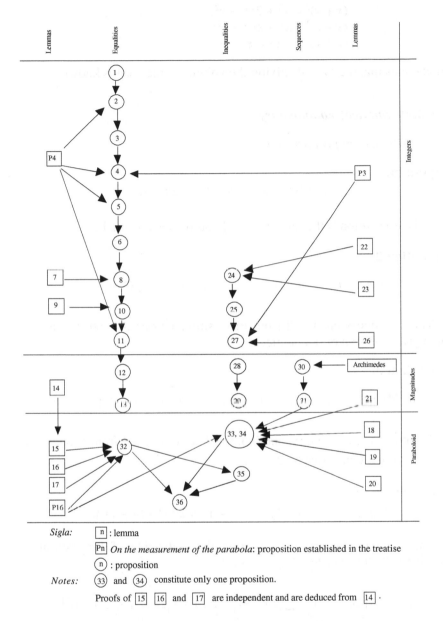

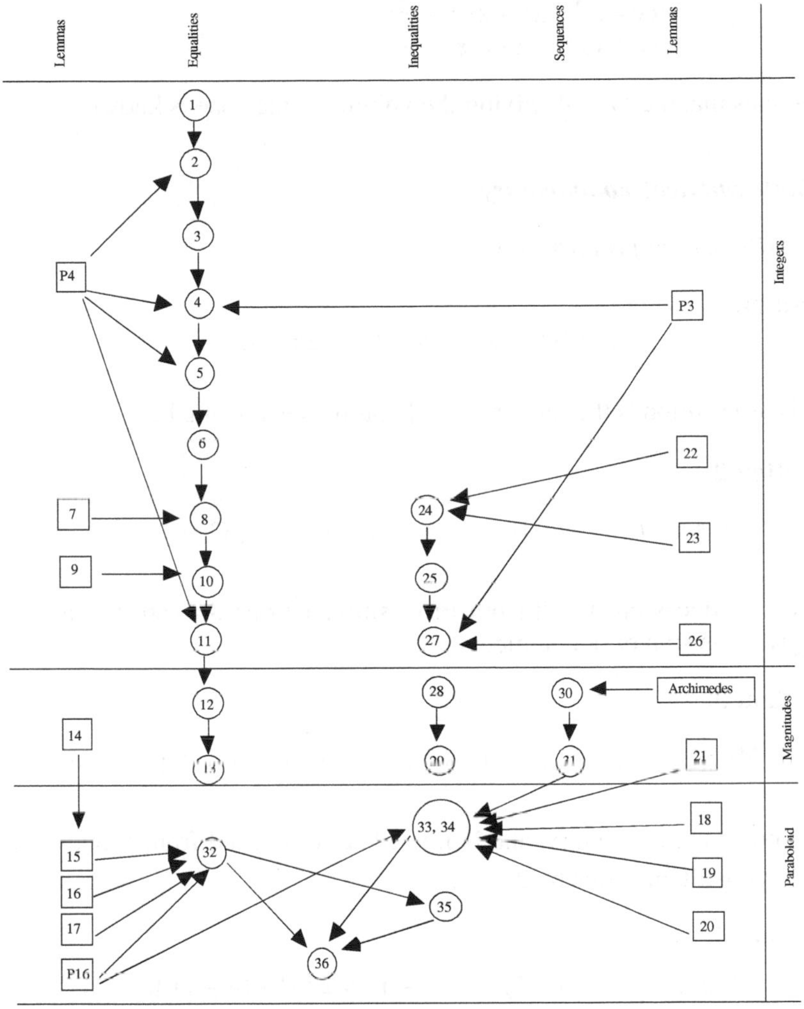

Sigla: [n] : lemma

[Pn] *On the measurement of the parabola*: proposition established in the treatise

(n) : proposition

Notes: (33) and (34) constitute only one proposition.

Proofs of [15] [16] and [17] are independent and are deduced from [14].

The usage of these properties appears clearly, as much in the text as in the segments depicting numbers (Fig. II.2.7, p. 269 for example), but we cite only those of which the author himself makes mention.

Note as well that Thābit ibn Qurra uses the identities

$$(x+y)^2 = x^2 + 2xy + y^2;$$
$$(x-y)^2 = x^2 - 2xy + y^2;$$
$$(x+y)(x-y) = x^2 - y^2$$

and assumes that the formula giving the volume of the cone is known.

2.3.2. *Mathematical commentary*

2.3.2.1. *Arithmetical propositions*

Proposition 1.

$$\forall n \in \mathbf{N}^*, \quad n^2 - (n-1)^2 = 2n - 1.$$

This proposition is the same as p_1. Thābit proves it using L_6.

Proposition 2.

$$\forall n \in \mathbf{N}^*, \quad (2n-1)^2 + 1 = 2(2n-1) + 4\sum_{p=1}^{n-1}(2p-1).$$

Thābit deduces this result from Proposition 1 using A_7 and A_5 and p_4, which gives the sum of odd numbers.

Proposition 3.

$$\forall n \in \mathbf{N}^*, (2n-1)^3 + (2n-1) = 2(2n-1)\left[2n - 1 + 2\sum_{p=1}^{n-1}(2p-1)\right].$$

The proof uses A_7 and A_9 and is immediately deduced from Proposition 2, multiplying all the terms by $2n - 1$.

Proposition 4.

$$\forall n \in \mathbf{N}^*, \quad (2n-1)^3 + (2n-1) = 2\,[n^4 - (n-1)^4].$$

The result is obtained by rewriting the right side of Proposition 3. Effectively, taking account of p_4, we have

$$2(2n-1)\left[2n - 1 + 2\sum_{1}^{n-1}(2p-1)\right] = 2(2n-1)\left[2n - 1 + 2(n-1)^2\right].$$

But, by Proposition 1,

$$2n - 1 = n^2 - (n-1)^2,$$

hence

$$2n - 1 + 2\,(n - 1)^2 = n^2 + (n - 1)^2;$$

yet

$$[n^2 - (n - 1)^2]\,[n^2 + (n - 1)^2] = n^4 - (n - 1)^4.$$

The proof thus uses Proposition 3, A_9, A_{11}, and p_3 and p_4.

Comment. — Up to now, Thābit ibn Qurra expresses the sum of the n prime odd numbers by the square of half the even number that follows the largest of these numbers; he expresses it here by the square of n, that is to say the square of the integer of the same position.

Proposition 5.

$$\forall\, n \in \mathbf{N}^*, \qquad \sum_{1}^{n}(2p-1)^3 + \sum_{p=1}^{n}(2p-1) = 2\left[\sum_{p=1}^{n}(2p-1)\right]^2.$$

Thābit ibn Qurra applies Proposition 4 for p from 1 to n:

$$s_1 = 1^3 + 1 = 2 \cdot 1^4 \Rightarrow \sigma_1 = 2 \cdot 1^4,$$
$$s_2 = (2 \cdot 2 - 1)^3 + (2 \cdot 2 - 1) = 2(2^4 - 1^4) \Rightarrow \sigma_2 = s_1 + s_2 = \sigma_1 + s_2 = 2 \cdot 2^4,$$
$$s_3 = (2 \cdot 3 - 1)^3 + (2 \cdot 3 - 1) = 2(3^4 - 2^4) \Rightarrow \sigma_3 = s_1 + s_2 + s_3 = \sigma_2 + s_3 = 2 \cdot 3^4.$$

Suppose that up to order $p - 1$, we have $\sigma_{p-1} = 2(p - 1)^4$; as we have

$$s_p = (2p - 1)^3 + (2p - 1) = 2\,[p^4 - (p - 1)^4],$$

it follows that

$$\sigma_p = \sigma_{p-1} + s_p = 2p^4.$$

The result is thus true for all order p, so we have $\sigma_n = 2n^4$; but by p_4

$$n^4 = \left[\sum_{p=1}^{n}(2p-1)\right]^2;$$

hence the result, which is thus obtained from Proposition 4 and p_4.

Thābit proceeds by an archaic form of finite induction;[3] he shows in deducing σ_p from σ_{p-1} that the result established for σ_{p-1} is true for σ_p.

[3] Cf. R. Rashed, 'L'induction mathématique: al-Karajī - as-Samaw'al', *Archive for History of Exact Sciences*, 9.1, 1972, pp. 1–21; reprinted in *The Development of Arabic Mathematics: Between Arithmetic and Algebra*, Boston Studies in Philosophy of Science 156, Dordrecht/Boston/London, 1994, pp. 62–84.

Proposition 6.

$$\forall n \in \mathbf{N}^*, \quad \sum_{p=1}^{n}(2p-1)\left[3\,(2p-1)^2+3\right]=6\left[\sum_{p=0}^{n}(2p-1)\right]^2.$$

This result follows immediately from Proposition 5 by multiplying both sides of the equation by 3 and pulling out the common factor of $(2p-1)$ from $(2p-1)^3$ and $(2p-1)$, recalling A_9.

Proposition 7.

$$\forall n \in \mathbf{N}^*, \quad (2n-2).2n+1=(2n-1)^2.$$

The proof is immediate, invoking A_1 and A_3. Proposition 7 is a lemma for passing from Proposition 6 to Proposition 8.

Proposition 8.

$$\forall n \in \mathbf{N}^*, \quad 6+\sum_{p=2}^{n}(2p-1)\,[3\,(2p-2)\cdot 2p+6]=6\left[\sum_{p=1}^{n}(2p-1)\right]^2.$$

By Proposition 7, for $1 \le p \le n$,

$$(2p-2)\cdot 2p+1=(2p-1)^2;$$

hence

$$3\,(2p-2)\cdot 2p+6=3\,(2p-1)^2+3=3\,[(2p-1)^2+1].$$

We then deduce

$$\sum_{p=1}^{n}(2p-1)\,[3\,(2p-2).2p+6]=3\left\{\sum_{p=1}^{n}(2p-1)\left[(2p-1)^2+1\right]\right\}.$$

Thus by Proposition 6

$$(1) \qquad \sum_{p=1}^{n}(2p-1)\,[3\,(2p-2).2p+6]=6\left[\sum_{p=1}^{n}(2p-1)\right]^2.$$

But, since $p = 1$,

$$(2p-1)\,[3\,(2p-2)\cdot 2p+6]=6,$$

(1) can thus be rewritten in the form given by Thābit:

$$6+\sum_{p=2}^{n}(2p-1)\,[3\,(2p-2).2p+6]=6\left[\sum_{p=1}^{n}(2p-1)\right]^2.$$

Proposition 9.

$$\forall\, n \in \mathbf{N}^*, \qquad (2n-2)^2 + (2n)^2 + 2n\,(2n-2) = 3\cdot 2n\,(2n-2) + 4.$$

The result follows from the identity $a^2 + b^2 = 2ab + (b - a)^2$ by the addition of ab to both sides, with $a = 2n - 2$, $b = 2n$; hence $b - a = 2$ (by A_1).

Proposition 10.

$$\forall\, n \in \mathbf{N}^*, \quad 6+\sum_{p=2}^{n}(2p-1)\left[(2p-2)^2+(2p)^2+2p(2p-2)+2\right]=6\left[\sum_{p=1}^{n}(2p-1)\right]^2.$$

By Proposition 9, we have, for $1 \le p \le n$,

$$(2p-2)^2 + (2p)^2 + 2p\,(2p-2) + 2 = 3\cdot 2p\,(2p-2) + 6$$

and Proposition 8 can thus be written in the form of Proposition 10.

Proposition 11.

$$\forall\, n \in \mathbf{N}^*, \qquad \frac{1}{3}\sum_{p=1}^{n}(2p-1)\left[(2p-2)^2+(2p)^2+2p(2p-2)\right]+\frac{2}{3}\sum_{p=1}^{n}(2p-1)$$
$$=\frac{1}{2}(2n)^2\sum_{p=1}^{n}(2p-1).$$

Proposition 10 can be written

$$(1) \qquad \sum_{p=1}^{n}(2p-1)\left[(2p-2)^2+(2p)^2+2p(2p-2)+2\right]=6\left[\sum_{p=1}^{n}(2p-1)\right]^2,$$

since for $p = 1$ we have

$$(2p-1)\,[(2p-2)^2 + (2p)^2 + 2p\,(2p-2) + 2] = 6.$$

But

$$\sum_{p=1}^{n}(2p-1) = n^2 = \frac{1}{4}.(2n)^2 \qquad \text{(by } p_4\text{)};$$

hence

$$6\left[\sum_{p=1}^{n}(2p-1)\right]^2 = \frac{3}{2}(2n)^2 \cdot \sum_{p=1}^{n}(2p-1).$$

Dividing both sides of (1) by 3, we obtain the result.

2.3.2.2. *Extension to sequences of segments*

Proposition 12. — *For* $1 \leq p \leq n$, *let* $(b_p)_{p\geq 1}$ *be a sequence of segments proportional to the terms in the same position in the sequence* $(2p - 1)_{p\geq 1}$ *of consecutive odd numbers, and let* $(a_p)_{p\geq 1}$ *be a sequence of segments proportional to the terms in the same position in the sequence* $(2p)_{p\geq 1}$ *of consecutive even numbers, if we write* $a_0 = 0$ *by convention and if we suppose* $b_1 = \frac{a_1}{2}$, *we get*

$$\frac{1}{3}\sum_{p=1}^{n} b_p\left(a_{p-1}^2 + a_{p-1}\cdot a_p + a_p^2\right) + \frac{2}{3}\left(\frac{a_1}{2}\right)^2 \sum_{p=1}^{n} b_p = \frac{1}{2}a_n^2 \sum_{p=1}^{n} b_p.$$

For $1 \leq p \leq n$, we have

$$\frac{a_p}{a_1} = \frac{2p}{2}, \qquad \frac{b_p}{b_1} = \frac{2p-1}{1}, \qquad \text{with } b_1 = \frac{a_1}{2};$$

hence

$$\frac{b_p}{a_p} = \frac{2p-1}{2p}.$$

On the other hand,

$$\frac{a_{p-1}}{a_p} = \frac{2p-2}{2p}$$

and

$$\frac{a_{p+1}}{a_p} = \frac{2p+2}{2p}.$$

Therefore

$$\frac{b_p\cdot\left(a_{p-1}^2 + a_{p-1}\cdot a_p + a_p^2\right)}{a_p^3} = \frac{(2p-1)\left[(2p-2)^2 + 2p(2p-2) + (2p)^2\right]}{(2p)^3}$$

(for $p = 1$, $2p - 2 = 0$ and $a_0 = 0$). But we have

$$\frac{a_p^3}{a_n^3} = \frac{(2p)^3}{(2n)^3}$$

and

$$\frac{a_n^3}{a_n^2 \cdot \sum_{p=1}^{n} b_p} = \frac{(2n)^3}{(2n)^2 \sum_{p=1}^{n} (2p-1)}.$$

Hence

$$(1) \quad \frac{b_p\left(a_{p-1}^2 + a_{p-1} \cdot a_p + a_p^2\right)}{a_n^2 \sum_{p=1}^{n} b_p} = \frac{(2p-1)\left[(2p-2)^2 + 2p \cdot (2p-2) + (2p)^2\right]}{(2n)^2 \sum_{p=1}^{n} (2p-1)}.$$

On the other hand,

$$(2) \quad \frac{\frac{2}{3}\left(\frac{a_1}{2}\right) \cdot \sum_{p=1}^{n} b_p}{a_n^2 \sum_{p=1}^{n} b_p} = \frac{\frac{2}{3} \cdot \sum_{p=1}^{n} (2p-1)}{(2n)^2 \sum_{p=1}^{n} (2p-1)}.$$

If we denote by A and A' respectively the left sides of Propositions 11 and 12, (1) and (2) give

$$\frac{A'}{a_n^2 \sum_{p=1}^{n} b_p} = \frac{A}{(2n)^2 \sum_{1}^{n} (2p-1)}.$$

But, by Proposition 11, we have

$$A = \frac{1}{2}(2n)^2 \sum_{1}^{n} (2p-1).$$

Hence

$$A' = \frac{1}{2} a_n^2 \sum_{p=1}^{n} b_p.$$

Comment. — The hypothesis $b_1 = \frac{a_1}{2}$ is equivalent to the choice of a unit segment, which would be b_1. The segments of the two sequences can then be expressed with respect to b_1 as $b_p = (2p-1)\, b_1$ and $a_p = p \cdot a_1 = 2p \cdot b_1$; hence

$$A' = b_1^3 \cdot A = b_1^3 \cdot \frac{1}{2}(2n)^2 \sum_1^n (2p-1)$$
$$= \frac{1}{2}(2n \cdot b_1)^2 \cdot \sum_1^n b_1(2p-1)$$
$$= \frac{1}{2} a_n^2 \sum_1^n b_p.$$

But this approach is not that of Thābit, who based his proof on equalities between ratios. The hypotheses give

$$\frac{a_{p-1}}{a_p} = \frac{2p-2}{2p} \text{ and } \frac{a_{p+1}}{a_p} = \frac{2p+2}{2p},$$

the denominators a_p and $2p$ being the same in the two proportions.

In order to obtain the ratio $\frac{b_p}{a_p}$, we begin with the proportions

$$\frac{a_1}{a_p} = \frac{2}{2p} \text{ and } \frac{b_p}{b_1} = \frac{2p-1}{1};$$

hence

$$\frac{b_p}{a_p} \cdot \frac{a_1}{2b_1} = \frac{(2p-1)}{2p}.$$

Letting $\frac{a_1}{2b_1} = 1$, that is $b_1 = \frac{a_1}{2}$, we get the proportion $\frac{b_p}{a_p} = \frac{2p-1}{2p}$, with denominators a_p and $2p$.

This explains the choice of the condition, $b_1 = \frac{a_1}{2}$, for an immediate application of Proposition 11.

Proposition 13. — The statement follows that of Proposition 12, with the same conclusion, but assuming that $b_1 \neq \frac{a_1}{2}$.

So let the sequence (c_p), $1 \leq p \leq n$, be defined by

$$b_1 = \frac{c_1}{2} \text{ and } \frac{c_1}{a_1} = \frac{c_p}{a_p}.$$

We have

$$\frac{a_p^2}{c_p^2} = \frac{a_{p-1}\, a_p}{c_{p-1}\, c_p} = \frac{a_1^2}{c_1^2} \qquad \text{for } 1 \leq p \leq n,$$

from which we deduce

$$\frac{\frac{1}{3}\sum_{p=1}^{n} b_p\left(a_{p-1}^2 + a_{p-1}\,.\,a_p + a_p\right)^2 + \frac{2}{3}\left(\frac{a_1}{2}\right)^2 \sum_{p=1}^{n} b_p}{\frac{1}{3}\sum_{p=1}^{n} b_p\left(c_{p-1}^2 + c_{p-1}\,.\,c_p + c_p\right)^2 + \frac{2}{3}\left(\frac{c_1}{2}\right)^2 \sum_{p=1}^{n} b_p} = \frac{a_1^2}{c_1^2} = \frac{a_n^2}{c_n^2}.$$

But, by Proposition 12, the denominator is equal to

$$\frac{1}{2}\, c_n^2 \sum_{p=1}^{n} b_p,$$

from which we deduce that the numerator is equal to

$$\frac{1}{2}\, a_n^2 \sum_{p=1}^{n} b_p.$$

We thus find the same result as in Proposition 12.

Comment. — The method used is based again on equalities between ratios.

The terms b_p and a_p are here related to different unit segments, b_1 and a_1. Thābit thus introduces a sequence (c_p) such that

$$b_1 = \frac{c_1}{2}, \qquad \frac{c_1}{a_1} = \frac{c_p}{a_p} = \frac{c_n}{a_n};$$

so the sequences (b_p and c_p) satisfy Proposition 12:

$$A\,(b, c) = \frac{1}{2}\, c_n^2 \sum_{p=1}^{n} b_p.$$

Writing $A\,(b, a)$ for the left side of the sought Proposition, we have

$$\frac{A\,(b, c)}{A\,(b, a)} = \frac{c_n^2}{a_n^2};$$

hence

$$A\,(b, a) = \frac{1}{2}\, a_n^2 \sum_{p=1}^{n} b_p.$$

The distinction $b_1 = \frac{a_1}{2}$ and $b_1 \neq \frac{a_1}{2}$ is not necessary; we can give a single proof.

Proposition 11 can be written for all $n \in \mathbf{N}^*$:

$$\frac{4}{3}\sum_{p=1}^{n}(2p-1)\left[(p-1)^2 + p^2 + (p+1)^2\right] + \frac{2}{3}\sum_{p=1}^{n}(2p-1) = \frac{4n^2}{2}\sum_{p=1}^{n}(2p-1).$$

The hypotheses from Propositions 12 and 13 can be written for $1 \leq p \leq n$ as

$$b_p = (2p - 1)\, b_1,$$

$$2a_p = 2p \cdot a_1 \Leftrightarrow a_p = p \cdot a_1,$$

from which we deduce

$$\frac{1}{3}\sum_{p=1}^{n} b_p\left[a_{p-1}^2 + a_p \,.\, a_{p-1} + a_p^2\right] = \frac{1}{3}\, b_1 \,.\, a_1^2 \sum_{p=0}^{n}(2p-1)\left[(p-1)^2 + p\,.(p-1) + p^2\right],$$

$$\frac{2}{3}\left(\frac{a_1}{2}\right)^2 \sum_{p=1}^{n} b_p = \frac{2}{3}\, b_1 \,.\, \frac{a_1^2}{4}\sum_{p=1}^{n}(2p-1),$$

$$\frac{1}{2}\, a_n^2 \sum_{p=1}^{n} b_p = \frac{1}{2}\, n^2 \,.\, b_1 \,.a_1^2 \sum_{p=1}^{n}(2p-1).$$

Hence, multiplying both sides of Proposition 11 by $b_1 \,.\, \frac{a_1^2}{4}$, we have

$$\frac{1}{3}\sum_{p=1}^{n} b_p\left[a_{p-1}^2 + a_p \cdot a_{p-1} + a_p^2\right] + \frac{2}{3}\left(\frac{a_1}{2}\right)^2 \sum_{p=1}^{n} b_p = \frac{1}{2}\, a_n^2 \sum_{p=1}^{n} b_p\,.$$

Proposition 14. — *If five magnitudes* a_1, a_2, a_3, a_4, a_5 *are such that* $\frac{a_1}{a_2} = \frac{a_3}{a_4} = \frac{a_4}{a_5}$ *and* $a_1 < a_2$, *then* $a_1\,(a_5 - a_3) = (a_2 - a_1)\,(a_3 + a_4)$.

The hypothesis $a_1 < a_2$ implies $a_3 < a_4 < a_5$. We have

$$\frac{a_1}{a_2 - a_1} = \frac{a_3}{a_4 - a_3} = \frac{a_4}{a_5 - a_4};$$

hence

$$\frac{a_1}{a_2 - a_1} = \frac{a_3 + a_4}{a_5 - a_3},$$

and therefore

$$a_1\,(a_5 - a_3) = (a_2 - a_1)\,(a_3 + a_4).$$

Comments.

1) The author does not specify the nature of the magnitudes. It is necessary to take a_1 and a_2 of the same nature, since the hypothesis brings their ratio into play and the conclusion invokes their difference, and likewise to take a_3, a_4 and a_5, of the same nature, for the same reasons. If, for example, a_1 and a_2 are lengths and a_3, a_4 and a_5 areas, the conclusion bears upon volumes (this will be the case in Propositions 15–17).

2) The hypothesis $a_1 < a_2$ is needed in the expression of the differences $a_2 - a_1$ and $a_5 - a_3$.

If $a_1 > a_2$, the conclusion is $a_1(a_3 - a_5) = (a_1 - a_2)(a_3 + a_4)$.

3) If we designate by $\frac{1}{k}$, $k \in \mathbf{R}^+ - \{0\}$, the common value of the ratios, we have $a_2 = k\, a_1$, $a_4 = k\, a_3$, $a_5 = k^2\, a_3$; the proposition follows from the identity

$$k^2 - 1 = (k-1)\,(k+1).$$

2.3.2.3. *Volumes of cones, rhombuses and other solids*

In Propositions 15–17, the solid of revolution under consideration is the difference between two homothetic solids, with the ratio of homothety being that between given radii that are presupposed as being known. The volume of the cone of revolution being known, the sought volume is expressed in the three cases as sums or differences of volumes of cones of revolution and the formula is the same for the three solids studied.

Proposition 15. — *Volume of the frustum of a cone of revolution.*

The figure is drawn in a meridian plane. The centres L and M of the base circles and the point K of intersection between AD and BE are aligned.

The volume of the frustum of a cone of revolution of height h *and with base circles of radius* r *and* R *is* $\mathrm{V} = \frac{1}{3}\pi\,\mathrm{h}\,(\mathrm{R}^2 + \mathrm{r\,R} + \mathrm{r}^2)$.

The sought volume is $V = V\,(KDE) - V\,(KAB)$.

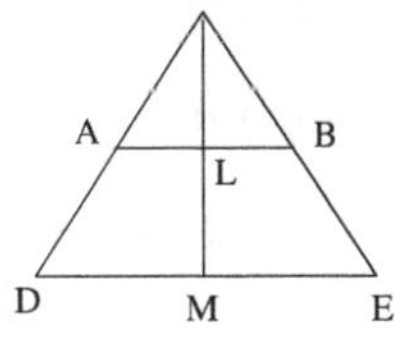

Fig. 2.3.1

Let $H = KM$; we have $H - h = KL$. We have $\frac{r}{R} = \frac{H-h}{H}$, as $LA \parallel MD$. Thābit introduces an auxiliary circle of radius $r' = \sqrt{rR}$; we thus have

$$\frac{r^2}{rR} = \frac{rR}{R^2} = \frac{r}{R} = \frac{H-h}{H},$$

$$\frac{H-h}{H} = \frac{\pi r^2}{\pi rR} = \frac{\pi rR}{\pi R^2};$$

and, by applying Proposition 14,

$$(H - h)\,(\pi R^2 - \pi r^2) = h\,(\pi r^2 + \pi r R).$$

If we add $\pi h \cdot R^2$ to both sides, we have

$$\pi H R^2 - \pi(H - h) r^2 = h(\pi r^2 + \pi r R + \pi R^2).$$

Hence

$$V = \frac{1}{3}\, h\,(\pi r^2 + \pi r R + \pi R^2).$$

Comment. — In fact, for the homothety $\left(K, \frac{r}{R}\right)$, we have

$$\frac{r}{R} = \frac{H-h}{H};$$

hence

$$\frac{r}{R-r} = \frac{H-h}{h},$$

and thus

$$H - h = \frac{h \cdot r}{R-r} \text{ and } H = \frac{R \cdot h}{R-r}.$$

Hence

$$V = \frac{1}{3}\ \pi \cdot h \cdot \frac{R^3 - r^3}{R-r} = \frac{1}{3}\ \pi \cdot h.\ (R^2 + r\,R + r^2).$$

Proposition 14 replaces the identity

$$\frac{R^3 - r^3}{R-r} = R^2 + r\,R + r^2.$$

Proposition 16. — *Volume of the hollow cone and of its frustum.*

The figure is drawn in a meridian plane. The two vertices *H* and *G*, the centres of the circles, the intersection *M* of the straight lines *DA* and *EB* are aligned, and, moreover, one has *AB* ∥ *DE* and *AG* ∥ *DH* (see the definitions in the introduction).

The volume of the frustum of a hollow cone with base circles of radius R *and* r *and axial height* h *is*

$$\mathrm{V} = \frac{1}{3}\ \pi \cdot \mathrm{h}\,(\mathrm{R}^2 + \mathrm{r\,R} + \mathrm{r}^2).$$

Thābit first calculates the volume of the two hollow cones:

$$V(MEHD) = \frac{1}{3}\pi MS \cdot R^2 - \frac{1}{3}\pi HS \cdot R^2 = \frac{1}{3}\pi H_1 \cdot R^2, \text{ with } H_1 = MH,$$
$$V(MAGB) = \frac{1}{3}\pi MN \cdot r^2 - \frac{1}{3}\pi GN \cdot r^2 = \frac{1}{3}\pi(H_1 - h) \cdot r^2.$$

The volume of the frustum of the hollow cone is then

$$V = \frac{1}{3}\pi h \cdot R^2 + \frac{1}{3}\pi(H_1 - h)(R^2 - r^2).$$

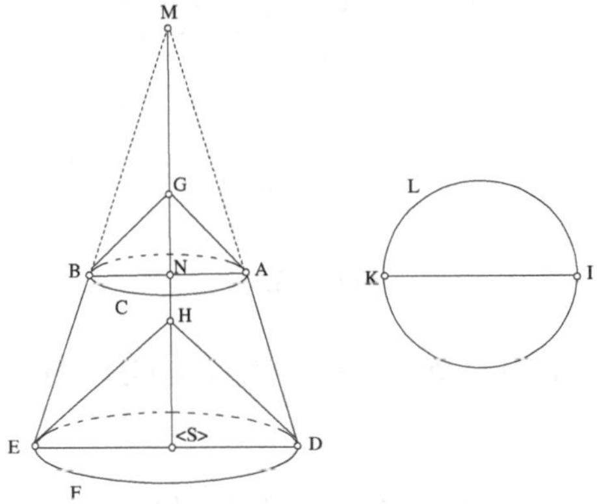

Fig. 2.3.2

If we set, as in Proposition 15, $r'^2 = r R$, then

$$\frac{r}{r'} = \frac{r'}{R} \text{ and } \frac{r^2}{r'^2} = \frac{r'^2}{R^2} = \frac{r}{R} = \frac{H_1 - h}{H_1}.$$

Hence

$$\frac{H_1 - h}{H_1} = \frac{\pi r^2}{\pi r R} = \frac{\pi r R}{\pi R^2}.$$

As in Proposition 15, one finishes by applying Proposition 14.

Proposition 17. — *Volume of a solid rhombus and of its frustum.*

The solid rhombus consisting, by definition, of the sum of two cones of the same base, the method is the same as in Proposition 16 and the given formula for the volume V is the same:

$$V = \frac{1}{3}\pi \cdot h(R^2 + rR + r^2).$$

Proposition 18. — *Let* AB *be an arc of a parabola of diameter* CD, and let E *and* F be two *points on the arc* AB. *From* A, E *and* F *we produce, on* *the one hand, three parallel straight lines between them that intersect the diameter at the points* G, H *and* I *respectively and, on the other hand, three straight lines parallel to the diameter which intersect the tangent at* E *at the points* K, E *and* L *respectively.*

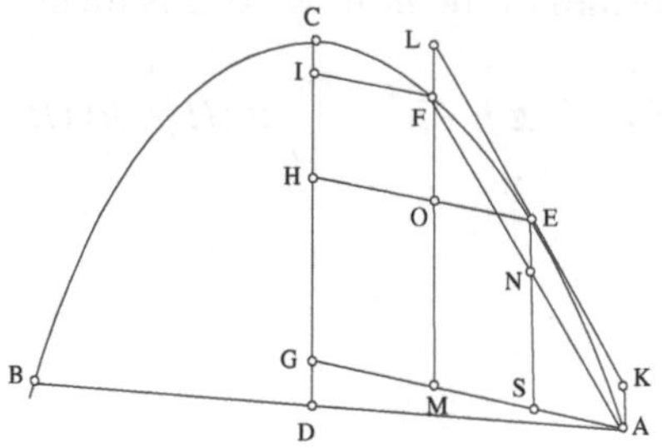

Fig. 2.3.3

If AG – EH = EH – FI, *then* AK = FL = $\frac{1}{2}$(GH – HI).

We have

$$IF \parallel AG \text{ and } IG \parallel FM;$$

hence

$$IF = GM.$$

Likewise,

$$HG \parallel ES \text{ and } HE \parallel AG;$$

hence

$$HE = GS.$$

We then have

$$AG - EH = EH - FI \Leftrightarrow EH = \frac{1}{2}(AG + FI) \Leftrightarrow GS = \frac{1}{2}(AG + GM);$$

thus S is the midpoint of MA, and ES intersects AF at its midpoint N. But $ES \parallel CD$, so ES is the diameter associated with the chord AF, and AF is parallel to the tangent at E. We thus have

$$AK = LF = NE = SE - SN = GH - \frac{1}{2}MF = GH - \frac{1}{2}(GH + HI)$$
$$= \frac{1}{2}(GH - HI).$$

In the particular case where F coincides with C, $H = O$ and $M = G$, so if $MS = \frac{1}{2} MA$, we have

$$LF = AK = \frac{1}{2} (MO - OF) = \frac{1}{2} (GH - HI).$$

Comments.

1) The established result does not depend on the common direction of the straight lines AG, EH and FI.

2) The equality $GS = \frac{1}{2} (AG + GM)$ says that the three diameters from the points A, E and F are equidistant parallels.

Reconsider the parabola with the reference defined by the diameter DC and the tangent at C, let $y^2 = ax$ be its equation, with a the *latus rectum* relative to DC. We have

$$y_F = y_E - \alpha, \quad y_A = y_E + \alpha, \quad x_N = \frac{x_A + x_F}{2}.$$

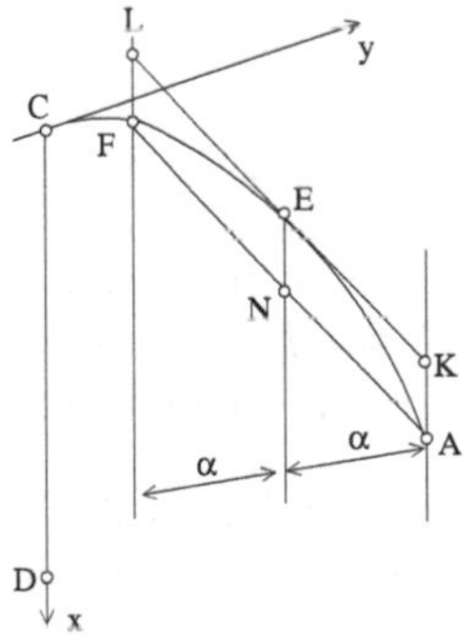

Fig. 2.3.4

But

$$x_A + x_F = \frac{1}{a}\left(y_A^2 + y_F^2\right) = \frac{1}{a}\left(2y_E^2 + 2\alpha^2\right),$$

hence

$$x_N = \frac{x_A + x_F}{2} = \frac{1}{a}\left(y_E^2 + \alpha^2\right).$$

But

$$x_E = \frac{1}{a}\, y_E^2;$$

hence

$$EN = x_N - x_E = AK = \frac{1}{a} \cdot \alpha^2.$$

Thus for every arc AF of a parabola, the tangent parallel to the chord AF determines on the diameters passing through A and F two equal segments:

$$AK = FL = \frac{(y_F - y_A)^2}{4a}.$$

Proposition 19. — *If two parallelograms* ABCD *and* AEFD *with common base* AD *have their bases* BC *and* EF *on the same straight line* Δ *parallel to* AD, *then by rotation about* Δ, *the two parallelograms produce solids of equal volume.*

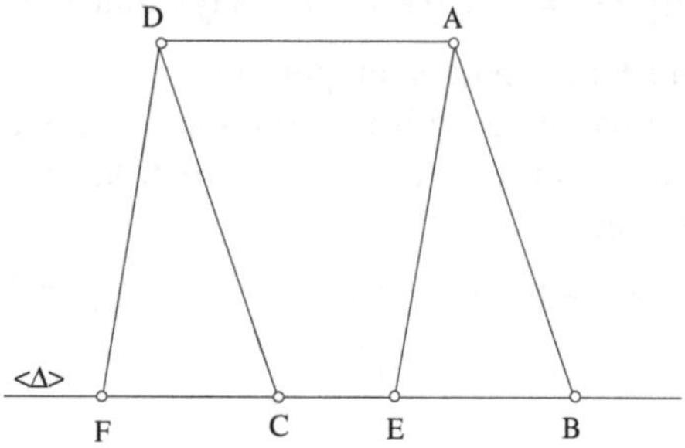

Fig. 2.3.5

The figure from the text gives the points along Δ in the order $BECF$. We have $BC = EF$; hence $BE = CF$. The triangles EAB and FDC are equal and give equal volumes upon rotation about Δ; hence

$$\text{vol.}\,(ABCD) = \text{vol.}\,(ABFD) - \text{vol.}\,(DCF),$$
$$\text{vol.}\,(AEFD) = \text{vol.}\,(ABFD) - \text{vol.}\,(ABE).$$

Hence

$$\text{vol.}\,(ABCD) = \text{vol.}\,(AEFD).$$

Comments.

1) The order of the points E, B, F, C along Δ may be different from that of the text. In every case, the triangles DFC and AEB correspond to each other by translation by the vector AD, and whether they produce solid rhombuses or hollow cones, their volumes are equal.

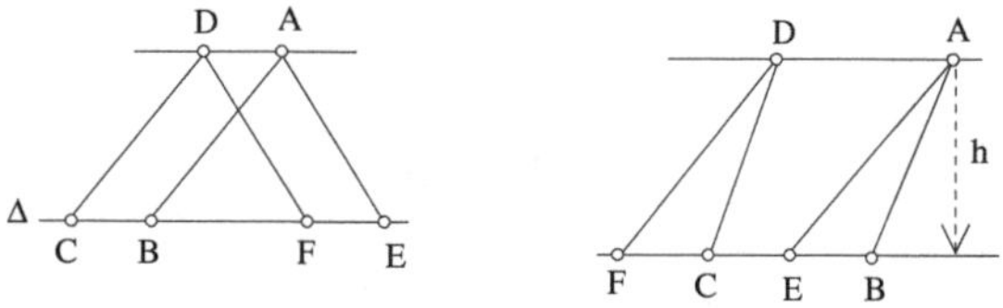

Fig. 2.3.6

2) The volume described by each of the parallelograms, whether in the form of a right cylinder or a hollow one,[4] is equal to a right cylinder. In every case, the volume is

$$V = \pi AD \cdot h^2,$$

if h is the distance between the segments AD and Δ, that is, the height of each of the parallelograms.

Proposition 20. — *Let* ABCD *and* EFGH *be two parallelograms placed in the same plane on the same side of the straight line* Δ *that contains the two bases* BC *and* FG; *if* BC = FG, *then* ADHE *is a parallelogram and the solids produced by rotating the three parallelograms about* Δ *satisfy*

vol. (ADHE) = vol. (ABCD) – vol. (EFGH).

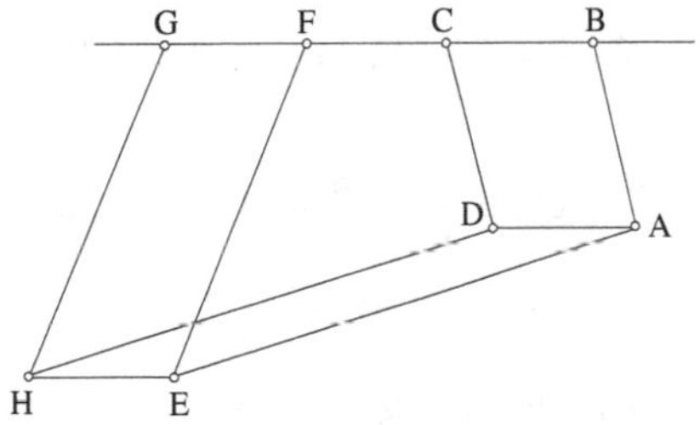

Fig. 2.3.7

The method used is that same as in Proposition 19, proceeding by sums or differences of volumes. It is clear that the quadrilaterals $BAEF$ and $CDHG$ are equal, since $CDHG$ is deduced from $BAEF$ by the translation of vector BC. The solids produced by rotating these two quadrilaterals about Δ correspond according to the same translation, and their volumes are thus equal:

$$\text{vol. }(ABFE) = \text{vol. }(CDHG).$$

But

$$\text{vol. }(BAEHG) - \text{vol. }(CDHG) = \text{vol. }(ABCD) + \text{vol. }(ADHE)$$

and

$$\text{vol. }(BAEHG) - \text{vol. }(ABFE) = \text{vol. }(EFGH);$$

hence

$$\text{vol. }(ADHE) = \text{vol. }(EFGH) - \text{vol. }(ABCD).$$

[4] Note that Ibn al-Haytham calls the 'conic cylinder' الأسطوانة المنخرطة .

Comment. — In the figure from the text, the height h with respect to FG is greater than the height h' with respect to CB; hence vol. ($EFGH$) > vol. ($ABCD$) and the equalities are written according to this hypothesis. But we can have $h \leq h'$; hence vol. ($EFGH$) ≤ vol. ($ABCD$). The general result is thus

$$\text{vol. }(ADHE) = \left|\, \text{vol. }(ABCD) - \text{vol. }(EFGH) \,\right|.$$

2.3.2.4. *Property of four segments*

Proposition 21. — *If* a, b, c, d *are four segments such that* $\mathrm{a} = \frac{\mathrm{b}}{3}$ *and* $\mathrm{c} = \frac{\mathrm{d}}{2}$, *then* $\mathrm{ac^2 + b\,(c^2 + d^2 + cd) - (a + b)\,d^2 > ad^2}$.

$$c = \frac{d}{2} \Rightarrow c^2 = \frac{d^2}{4} \qquad a = \frac{b}{3} \Rightarrow a = \frac{a+b}{4};$$

hence

$$\frac{c^2}{d^2} = \frac{a}{a+b} \text{ or } c^2\,(a + b) = ad^2.$$

But by hypothesis $\frac{c}{d} > \frac{a}{b}$; hence $\frac{c.d}{d^2} > \frac{a}{b}$ and consequently $bcd > ad^2$; hence

$$bcd + (a + b) \cdot c^2 > 2\,a \cdot d^2,$$

and, subtracting $a \cdot d^2$ from both sides, we have

$$ac^2 + b\,(c^2 + cd) - ad^2 > ad^2.$$

Hence we have the conclusion

$$ac^2 + b\,(c^2 + cd + d^2) - (a + b)\,d^2 > ad^2.$$

Comment.

1) Thābit placed himself at the particular case $\frac{b}{a} = 3$, $\frac{d}{c} = 2$; these are the values that come up in Proposition 33. But these numerical values only arise in expressing the conditions

$$(1) \quad \frac{b}{a} > \frac{d}{c} \qquad \text{and} \qquad (2) \quad 1 + \frac{b}{a} = \frac{d^2}{c^2}.$$

The proposition is thus true under hypotheses (1) and (2), which are stronger than those of its statement above. Note that for n an integer, $n \geq 2$, $\frac{d}{c} = n$ and $\frac{b}{a} = n^2 - 1$ satisfy conditions (1) and (2).

2.3.2.5. *Arithmetical propositions*

Proposition 22.

$$\forall p \in \mathbf{N}^*, \quad p\,(p+1) = (p-1)^2 + (p-1) + 2p.$$

The result is immediate. Thābit cites A_0.

Proposition 23.

$$\forall p \in \mathbf{N}^*, \quad (p+1)^2 + (p-1)^2 = p\,[(p-1)+(p+1)] + 2.$$

If we set $a = p - 1, b = p, c = p + 1$, then

$$\begin{aligned} &c = b + 1, \text{ hence } c^2 = cb + c \\ &b = a + 1, \text{ hence } ba = a^2 + a \\ &c^2 + a^2 = cb + ba + (c - a) \\ &\qquad\quad = b\,(c + a) + 2 \end{aligned}$$

(use of A_0).

In Propositions 24, 25 and 27, the hypotheses are the same. They pertain to three consecutive integers $B = p - 1$, $C = p$, $D = p + 1$ to which are associated the odd numbers of the same position:

$$F = 2(p-1) - 1, \qquad G = 2p - 1, \qquad H = 2(p+1) - 1.$$

Observe that the numbers $A = 1$ and $E = 1$ mentioned by Thābit do not appear in the proofs. They arise only to specify that he here means the sequence of natural integers and the sequence of odd numbers, both starting with 1, and that B, C, D on the one hand, and F, G, H on the other, have the same position in their respective sequences.

Proposition 24.

$$\begin{aligned} \forall p \in \mathbf{N}^*, \quad & p(2p-1)\,[(p-1)+(p+1)] + 2p\,(p+1) > \\ & (2p-1)\,[(p-1)^2 + (p+1)^2] + 2(p-1)^2. \end{aligned}$$

Call the two sides of the inequality I and II; taking account of Propositions 22 and 23, we have

$$\mathrm{I} = (2p-1)\,[(p+1)^2 + (p-1)^2] - 2(2p-1) + 2(p-1)^2 + 2(p-1) + 4p.$$

But

$$2(p-1) + 4p - 2(2p-1) = 2p,$$

$$\mathrm{I} = \mathrm{II} + 2p;$$

hence

$$\mathrm{I} > \mathrm{II}$$

(use of A_2 and A_4).

Proposition 25.

$$\forall\, p \in \mathbf{N}^*,\ (2p-1)\,[(p-1)^2 + p^2 + p.(p-1)] + (2p+1)\,[p^2 + (p+1)^2 + p.(p+1)] > [(2p-1) + (2p+1)]\,[(p-1)^2 + p^2 + (p+1)^2].$$

Proposition 24 can be written

$$p(p-1)\,(2p-1) + p(p+1)\,(2p+1) > (2p+1)\,(p-1)^2 + (2p-1)\,(p+1)^2,$$

and adding to both sides the expression

$$(2p-1)\,[(p-1)^2 + p^2] + (2p+1)\,[p^2 + (p+1)^2],$$

we obtain Proposition 25.

Proposition 26.

$$\forall\, p \in \mathbf{N}^*, \quad (p-1)\,(p+1) + 1 = p^2.$$

This proposition is a lemma for which the proof is immediate (use of A_0).

Proposition 27.

$$\forall\, p \in \mathbf{N}^*,\ (2p-1)\,[(p-1)^2 + p^2 + p.(p-1)] + (2p+1)\,[p^2 + (p+1)^2 + p.(p+1)] - [(2p-1) + (2p+1)]\,[(p-1)^2 + (p-1)\,(p+1) + (p+1)^2] > (p+1)^2 - (p-1)^2.$$

We start from Proposition 25 wherein we transform the right side, taking account of Proposition 26 and the equality

$$(2p-1) + (2p+1) = 4p = (p+1)^2 - (p-1)^2,$$

which is deduced from p_3.

2.3.2.6. *Sequence of segments and bounding*

Proposition 28. — *For* $1 \leq p \leq n$, *let* (a_p) *be a sequence of segments proportional to the consecutive integers* p *and let* (b_p) *be a sequence of segments proportional to the consecutive odd numbers* $2p - 1$. *If one supposes that* $a_1 = b_1$, *then*

$$b_p\left(a_{p-1}^2 + a_p \cdot a_{p-1} + a_p^2\right) + b_{p+1}\left(a_p^2 + a_p \cdot a_{p+1} + a_{p+1}^2\right)$$
$$-\left(b_p + b_{p+1}\right)\left(a_{p-1}^2 + a_{p-1} \cdot a_{p+1} + a_{p+1}^2\right) > b_1\left(a_{p+1}^2 - a_{p-1}^2\right).$$

We have for $1 \leq p \leq n$

$$\frac{a_1}{a_p} = \frac{1}{p} \text{ and } \frac{b_1}{b_p} = \frac{1}{2p-1}, \text{ with } a_1 = b_1;$$

hence

$$\frac{b_p}{a_p} = \frac{2p-1}{p}.$$

We also have

$$\frac{a_{p-1}}{a_p} = \frac{p-1}{p}, \qquad \text{setting } a_0 = 0,$$

and

$$\frac{a_{p+1}}{a_p} = \frac{p+1}{p}.$$

We deduce

$$\frac{b_p\left(a_{p-1}^2 + a_p \cdot a_{p-1} + a_p^2\right)}{a_p^3} = \frac{(2p-1)\left[(p-1)^2 + p\cdot(p-1) + p^2\right]}{p^3}$$

$$\frac{b_{p+1}\left(a_p^2 + a_p \cdot a_{p+1} + a_{p+1}^2\right)}{a_p^3} = \frac{(2p+1)\left[p^2 + p.(p+1) + (p+1)^2\right]}{p^3}$$

$$\frac{\left(b_p + b_{p+1}\right)\left(a_{p-1}^2 + a_{p-1} \cdot a_{p+1} + a_{p+1}^2\right)}{a_p^3} =$$

$$\frac{[(2p-1) + (2p+1)]\left[(p-1)^2 + (p-1)(p+1) + (p+1)^2\right]}{p^3};$$

likewise

$$\frac{a_p^3}{b_1\left(a_{p+1}^2 - a_{p-1}^2\right)} = \frac{p^3}{(p+1)^2 - (p-1)^2}.$$

Thus, designating the left side of the sought inequality by A and that of the inequality from Proposition 27 by A', we have

$$\frac{A}{b_1\left(a_{p+1}^2 - a_{p+1}^2\right)} = \frac{A'}{(p+1)^2 - (p-1)^2}.$$

But, by Proposition 27,

$$A' > (p+1)^2 - (p-1)^2,$$

and we thus have

$$A > b_1\left(a_{p+1}^2 - a_{p-1}^2\right).$$

Proposition 29. — The statement is the same as that of Proposition 28, but we suppose that $a_1 \neq b_1$.

Thābit then introduces a sequence (c_p), $1 \leq p \leq n$, such that

$$\frac{a_1}{c_1} = \frac{a_p}{c_p} \text{ and } c_1 = b_1;$$

the two sequences (c_p) and (b_p) then satisfy the inequality from Proposition 28:

$$b_p\left(c_{p-1}^2 + c_p \cdot c_{p-1} + c_p^2\right) + b_{p+1}\left(c_p^2 + c_p \cdot c_{p+1} + c_{p+1}^2\right)$$
$$-\left(b_p + b_{p+1}\right)\left(c_{p-1}^2 + c_{p-1} \cdot c_{p+1} + c_{p+1}^2\right) > b_1\left(c_{p+1}^2 - c_p^2\right).$$

Designate by C and D the two sides of this inequality, and by A and B the two sides of the sought inequality.

From $\frac{a_1}{c_1} = \frac{a_p}{c_p}$ for $1 \leq p \leq n$, we deduce that

$$\frac{a_p^2}{c_p^2} = \frac{a_{p-1}^2}{c_{p-1}^2} = \frac{a_{p+1}^2}{c_{p+1}^2} = \frac{a_{p-1} \cdot a_p}{c_{p-1} \cdot c_p} = \frac{a_p \cdot a_{p+1}}{c_p \cdot c_{p+1}} = \frac{a_{p-1} \cdot a_{p+1}}{c_{p-1} \cdot c_{p+1}}.$$

Hence

$$\frac{A}{C} = \frac{B}{D}.$$

But we have $C > D$ and hence $A > B$.

Comment. — Propositions 28 and 29 can be deduced from Proposition 27 without distinguishing $a_1 = b_1$ and $a_1 \neq b_1$. In fact, we have for $1 \leq p \leq n$, setting $a_0 = 0$,

$$p = \frac{a_p}{a_1},\ p-1 = \frac{a_{p-1}}{a_1},\ p+1 = \frac{a_{p+1}}{a_1},\ 2p-1 = \frac{b_p}{b_1},\ 2p+1 = \frac{b_{p+1}}{b_1}.$$

If we transfer these expressions into the inequality in Proposition 27, we let a denominator of $b_1 a_1^2$ appear on the left side and a_1^2 on the right. Multiplying both sides by $a_1^2\, b_1$, we have the sought inequality (by analogy with Propositions 12 and 13).

Reasoning geometrically, however, we understand why Thābit separated the two cases: the first corresponding to a homothety, whereas the second requires an affine transformation.

Proposition 30. — Let a, b, c *be three magnitudes such that* a < b < c. *We take as given the pairwise ratios of these magnitudes. We consider the increasing sequence* $(\mathrm{a}_p)_{p \geq 1}$ *defined by* $\frac{\mathrm{a}_p}{\mathrm{a}_{p+1}} = \frac{\mathrm{a}}{\mathrm{b}}$ *with* $\mathrm{a}_1 = \mathrm{a}$ *and* $\mathrm{a}_2 = \mathrm{b}$; *then there exists* n *such that* $\mathrm{a}_{n+1} > \mathrm{c}$.

If $\frac{c}{a}$ and $\frac{b}{a}$ are known, then $\frac{b-a}{a}$ is also. We have $a < b < c$, hence $c - a > b - a$; thus there exists $n > 1$ such that $n\,(b - a) > c - a$.

Moreover,

$$\frac{a}{b} = \frac{a_p}{a_{p+1}} \Rightarrow \frac{b-a}{a} = \frac{a_{p+1} - a_p}{a_p};$$

but $a < a_p$ and hence $a_{p+1} - a_p > b - a$, for $p \geq 2$.

We then deduce

$$b - a + \sum_{p=2}^{n} \left(a_{p+1} - a_p\right) > n(b-a) > c - a;$$

hence

$$a_{n+1} - a > c - a \text{ and } a_{n+1} > c.$$

Conclusion: if $n(b - a) > c - a$, then $a_{n+1} > c$.

Comments.

1) The existence of the number n follows from the Axiom of Archimedes. If n is the smallest integer solution to the problem, we have $a_n < c < a_{n+1}$.

2) The author proceeds by iteration (supposing $n = 3$).

3) The terms of the increasing sequence $(a_p)_{p\geq 1}$ are in continuous proportion:

$$\frac{a_1}{a_2} = \frac{a_2}{a_3} = \ldots = \frac{a_n}{a_{n+1}} = \ldots,$$

$\frac{a}{b}$ being the common value of the ratios. Hence

$$\forall\, n \geq 2, \text{ we have } \quad \frac{a}{a_{n+1}} = \left(\frac{a}{b}\right)^n, \quad \text{with } \frac{a}{c} < \frac{a}{b} < 1;$$

and we have the equivalence of the two results:

$$\exists\, n > 1,\ a_{n+1} > c \iff \exists\, n > 1,\ \left(\frac{a}{b}\right)^n < \frac{a}{c}.$$

The sequence $u_p = \left(\frac{a}{b}\right)^p$ is decreasing and $\lim\limits_{n\to\infty} u_n = 0$.

Let us recall that the Axiom of Archimedes takes two forms: additive and multiplicative. The first states that if α and β are two arbitrary magnitudes of the same kind, then there exists an integer n such that $n\,\alpha > \beta$; the second states that if a, b, c are magnitudes of the same kind such that $b > a$, then there exists an integer n such that $\left(\frac{b}{a}\right)^n > \frac{c}{a}$. We derive the second form from the first by setting $\frac{b}{a} = 1 + \theta$, where $\theta = \frac{b-a}{a}$, and by showing that $(1 + \theta)^n > 1 + n\theta$; in fact, we then have

$$\left(\frac{b}{a}\right)^n = (1+\theta)^n > 1 + n\theta > \frac{c}{a}$$

when $n\theta a > c - a$.

Thābit proceeds as follows to prove Proposition 30: he establishes that

$$a_{n+1} - a > n(b - a)$$

for sufficiently large n, the sequence (a_p) being defined according to

$$\frac{a_p}{a_{p+1}} = \frac{a}{b}.$$

The construction of this geometric sequence points to the multiplicative aspect, while the additive aspect is found in the consideration of the differences $a_{p+1} - a_p$ and the inequality $a_{p+1} - a_p > b - a$, which comes from

$$\frac{a_{p+1} - a_p}{a_p} = \frac{b-a}{a} \text{ and } a_p > a.$$

This approach applied to $(1 + \theta)^n > 1 + n\theta$ is expressed thus:

$$(1+\theta)^n - 1 = \sum_{p=0}^{n-1} \left((1+\theta)^{p+1} - (1+\theta)^p\right),$$
$$(1 + \theta)^{p+1} - (1 + \theta)^p = (1 + \theta)^p\, \theta > \theta.$$

This commentary shows that Thābit was able to move away from Euclid's Proposition X.1, which is less general, as it assumes $\frac{a}{b} < \frac{1}{2}$; Thābit uses only the hypothesis that $\frac{a}{b} < 1$.

Proposition 31. — *Let* AB, CD, E *and* FG *be magnitudes such that* AB > CD *and* E < FG < AB. *If we subtract from* AB *a magnitude* X_1 *such that*

$$\frac{X_1}{AB} \geq \frac{E}{FG},$$

from the remainder $(AB - X_1)$ *a magnitude* X_2 *such that*

$$\frac{X_2}{AB - X_1} \geq \frac{E}{FG},$$

and if we continue likewise, we necessarily reach a remainder smaller than CD.

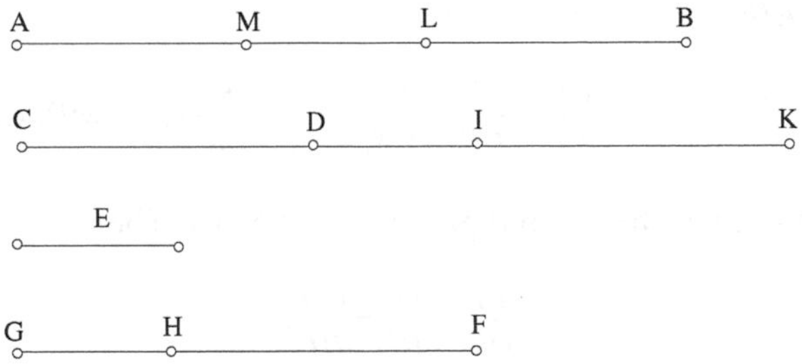

Fig. 2.3.8

Let H be upon GF such that $GH = E$ and let I be upon CD beyond D such that

$$\frac{DI}{DC} = \frac{HG}{HF}.$$

We have

$$CI > CD \quad \text{and} \quad \frac{CI}{CD} = \frac{CD + DI}{DC} = \frac{HG + HF}{HF} = \frac{GF}{HF}.$$

We may have either 1) $CD < AB < CI$
or 2) $CD < CI < AB$.

• If $CI > AB$,

$$\frac{AB}{CD} < \frac{CI}{CD};$$

thus

$$\frac{AB}{CD} < \frac{FG}{FH}.$$

We take L along AB such that

$$\frac{BL}{AB} = \frac{GH}{GF}.$$

Then

$$\frac{AB}{AL} = \frac{FG}{FH}.$$

Thus

$$\frac{AB}{AL} > \frac{AB}{CD},$$

and hence $AL < CD$; AL is the sought segment.

• If $CD < CI < AB$, we can by Proposition 30 find magnitudes in continuous proportion, starting with CD and CI and reaching a magnitude greater than AB:

$$\frac{CD}{CI} = \frac{CI}{CK} = \frac{CK}{CK_1} = \ldots = \frac{CK_{n-1}}{CK_n}, \qquad \text{with } CK_n > AB.$$

Suppose that CK has the desired property, $CK > AB$. Then

$$\frac{DI}{DC} = \frac{IK}{IC} = \frac{GH}{HF}.$$

Let L and M be on AB such that

(1) $$\frac{BL}{BA} \geq \frac{E}{FG}$$

and

(2) $$\frac{LM}{LA} \geq \frac{E}{FG}.$$

Then

$$\frac{BL}{AL} \geq \frac{GH}{HF}.$$

Hence

(3) $$\frac{BL}{AL} \geq \frac{IK}{IC} = \frac{DI}{DC}.$$

Likewise, we have

(4) $$\frac{ML}{MA} \geq \frac{DI}{DC}.$$

From (4) we deduce

$$\frac{AL}{MA} \geq \frac{CI}{CD}.$$

But

$$\frac{BM}{MA} = \frac{BL}{MA} + \frac{LM}{MA} = \frac{BL}{AL} \cdot \frac{AL}{MA} + \frac{LM}{MA};$$

hence

$$\frac{BM}{MA} \geq \frac{DI}{DC}\left(\frac{CI}{CD} + 1\right).$$

But

$$\frac{CI}{CD} = \frac{CK}{CI} = \frac{IK}{ID};$$

hence

$$\frac{CI}{CD} + 1 = \frac{IK}{ID} + 1 = \frac{DK}{ID}.$$

We thus have

(5) $$\frac{BM}{MA} \geq \frac{KD}{DC}.$$

Hence

$$\frac{BA}{AM} \geq \frac{KC}{CD};$$

that is to say,

(6)
$$\frac{BA}{KC} \geq \frac{AM}{CD}.$$

• If $\frac{BA}{KC} = \frac{AM}{CD}$, since we have assumed $KC > BA$, then $AM < CD$, AM is the desired remainder.

• If $\frac{BA}{KC} > \frac{AM}{CD}$, there exists on AB a point N such that $\frac{BA}{KC} = \frac{AN}{CD}$; thus $AN > AM$. But $KC > BA$ implies $AN < CD$ and *a fortiori* $AM < CD$; AM is the desired remainder.

Comments. — After having established (3) and (4), Thābit says: 'From that, one shows that the ratio of BM to MA is not smaller than the ratio of KD to CD'; he thus goes without proof from (3) and (4) to (5). We have preferred to go from (3) and (4) as Thābit indicates.

But we may go directly from (1) and (2) to (6) without using (3) and (4). In fact

$$\frac{BL}{BA} \geq \frac{E}{FG} \Rightarrow \frac{AL}{BA} \leq \frac{FH}{FG},$$

$$\frac{LM}{LA} \geq \frac{E}{FG} \Rightarrow \frac{AM}{LA} \leq \frac{FH}{FG}.$$

Hence

$$\frac{AM}{BA} \leq \left(\frac{FH}{FG}\right)^2.$$

But

$$\frac{FH}{FG} = \frac{CD}{CI} = \frac{CI}{CK};$$

hence

$$\left(\frac{FH}{FG}\right)^2 = \frac{CD}{CK}.$$

We thus have

$$\frac{AM}{BA} \leq \frac{CD}{CK},$$

and hence

$$\text{(6)} \qquad \frac{BA}{KC} \geq \frac{AM}{CD}.$$

We then finish as Thābit indicated.

The interesting aspect of the second method is that it can be generalized to the case where it is necessary to consider the continuous proportion up to order n to obtain $CK_n > AB$. In fact, we then have $\frac{CD}{CK_n} = \left(\frac{FH}{FG}\right)^{n+2}$.

To the point K_n we associate the point M_n such that

$$\frac{AM_n}{AB} \leq \left(\frac{FH}{FG}\right)^{n+2};$$

hence

$$\frac{AM_n}{BA} \leq \frac{CD}{CK_n}.$$

We then have

$$\frac{AB}{CK_n} \geq \frac{AM_n}{CD}$$

and we conclude as in the preceding; AM_n is the desired remainder.

Thābit had not shown the role of successive powers of $\frac{FH}{FG}$, associated with the successive terms CI, CK, … , CK_n of the continuous proportion, whereas he had used such successive powers in the proof of Proposition 30. Perhaps he simply made a concession to a more archaic style of writing for Proposition 31.

In his statement, Thābit considers two magnitudes AB and CD, $AB > CD$, and two other magnitudes E and FG such that $E < FG < AB$. The hypothesis $E < FG$ serves to define a ratio $\frac{E}{FG} = k < 1$. But the condition $FG < AB$ does not show up in the reasoning.

So we may put the problem in the following form:

Let a *and* b *be two magnitudes, with* a < b, *and let* k < 1 *be a ratio; if we consider the sequence* (b_p) *defined by*

$$1 > \frac{\mathrm{b}_1}{\mathrm{b}} \geq \mathrm{k},\ 1 > \frac{\mathrm{b}_2}{\mathrm{b}-\mathrm{b}_1} \geq \mathrm{k}\ \ldots\ 1 > \frac{\mathrm{b}_p}{\mathrm{b}-\sum_{i=1}^{p-1}\mathrm{b}_i} \geq \mathrm{k}\ \ldots .$$

Then there exists n ∈ **N*** *such that*

$$\mathrm{b}-\sum_{i=1}^{n}\mathrm{b}_i < \mathrm{a}.$$

Let a_0 be defined by

$$\frac{a_0}{a} = \frac{k}{1-k}.$$

Hence

$$\frac{a+a_0}{a} = \frac{1}{1-k}.$$

• If $a + a_0 > b$, then $a > b - kb$, but $b_1 \geq kb$; hence $a > b - b_1$. The desired result is then attained for $n = 1$.

• If $a + a_0 \leq b$, we consider the sequence (a_p) defined by

$$a_1 = a,\ \ a_2 = a + a_0 \text{ and } \frac{a_1}{a_2} = \frac{a_2}{a_3} = \ldots = \frac{a_p}{a_{p+1}}.$$

By Proposition 30, there exists $n \in \mathbf{N}^*$ such that $a_n < b < a_{n+1}$. But

$$\frac{a_1}{a_2} = \frac{a}{a+a_0} = 1-k.$$

Hence

$$\frac{a_p}{a_{p+1}} = 1-k \qquad \text{for } p \text{ from 1 to } n,$$

from which we deduce

$$\frac{a_1}{a_{n+1}} = (1-k)^n, \qquad \text{with } a_1 = a.$$

Moreover, by the definition of the sequence b_p, we have

$$\frac{b_1}{b} \geq k \Rightarrow \frac{b-b_1}{b} \leq 1-k,$$

$$\frac{b_2}{b-b_1} \geq k \Rightarrow \frac{b-(b_1+b_2)}{b-b_1} \leq 1-k \Rightarrow \frac{b-(b_1+b_2)}{b} \leq (1-k)^2.$$

Suppose the result holds up to order $(p-1)$, *i.e.* that

$$(1) \qquad \frac{b-\sum_{i=1}^{p-1} b_i}{b} \leq (1-k)^{p-1}.$$

For order p, we have

$$(2) \qquad \frac{b_p}{b-\sum_{i=1}^{p-1} b_i} \geq k \;\Rightarrow\; \frac{b-\sum_{i=1}^{p} b_i}{b-\sum_{i=1}^{p-1} b_i} \leq (1-k),$$

so

$$(1) \text{ and } (2) \Rightarrow \frac{b-\sum_{i=1}^{p} b_i}{b} \leq (1-k)^p.$$

The result is thus true for $1 \leq p \leq n$, and hence for $p = n$; we have

$$\frac{b-\sum_{i=1}^{n} b_i}{b} \leq \frac{a}{a_{n+1}}, \qquad \text{as } (1-k)^n = \frac{a}{a_{n+1}}.$$

But

$$a_{n+1} > b;$$

hence

$$b-\sum_{i=1}^{n} b_i < a.$$

Comments.

1) The construction of the sequence (b_p) is done by an archaic induction, less explicit than the induction applied by Ibn al-Haytham.[5]

2) The terms b_p are not defined in a unique manner, but each b_p is to be taken in the defined interval starting with b_{p-1}:

$$b > b_1 \geq k\, b,$$

$$b - b_1 > b_2 \geq k\,(b - b_1),$$

$$\ldots$$

$$b-\sum_{i=1}^{p-1} b_i > b_p \geq k\,(b-\sum_{i=1}^{p-1} b_i).$$

[5] Cf. Vol. II.

The propositions that follow are dedicated to the study of the volume of a parabolic dome described by a parabolic section *ABC* rotated about the diameter *BC*, *AC* being the ordinate associated to *BC*.

Propositions 32–35 invoke a division of the diameter *BC* into *n* segments proportional to the consecutive odd numbers 1, 3 … $2n-1$. The abscissae of the points of this division are then proportional to the squares of consecutive integers and, by the equation of the parabola, the ordinates associated with them are proportional to consecutive integers, properties established by the author in p_{16}. Therefore, two sequences of segments satisfy the hypotheses of Propositions 13, 21 and 29. The construction of such segments was used in p_{17} and p_{18}.

To each partition of *BC* into *n* segments there correspond *n* circles inscribed on the parabolic dome. The smallest is the closest to the vertex; let r_1 be its radius and s_1 its area. The largest is the circle at the base described by *A*; let $r_A = r_n$ be its radius and s_A its area.

2.3.2.7. *Calculation of the volumes of paraboloids*

Proposition 32. — *Let the diameter* BC *be divided by the* n *points* $E_0 = B$, $E_1, E_2 \dots E_n = C$ *such that* $\frac{E_{p-1}E_p}{E_0E_1} = \frac{2p-1}{1}$ *and let there be on the arc* BA *the points* $D_0 = B, D_1 \dots D_n = A$ *that are associated to them. Let* S_n *be the solid produced by the rotation about* BC *of the polygon* $BD_1D_2 \dots AC$ *and* v_s *its volume. Then*

$$v_s + \frac{2}{3}BC \cdot \frac{s_1}{4} = \frac{1}{2}BC \cdot s_A.$$

By hypothesis,

$$\frac{E_{p-1}E_p}{E_0E_1} = \frac{2p-1}{1}, \qquad 1 \le p \le n,$$

and we deduce

$$\frac{E_pD_p}{E_1D_1} = \frac{p}{1} = \frac{2p}{2} \qquad \text{(Fig. 2.3.9 below).}$$

First case (Fig. II.2.32a, p. 313, and Fig. 2.3.9 below): The diameter *BC* is the axis of the parabola; then $E_pD_p \perp BC$, and E_pD_p is the radius r_p of the circle described by D_p. The two sequences $(E_{p-1}E_p)$ and (E_pD_p) satisfy the hypotheses of Proposition 13, and we thus have

$$(1)\quad \frac{1}{3} BE_1 \cdot s_1 + \frac{1}{3}\pi\left[\sum_{p=2}^{n} E_{p-1} E_p \left(E_{p-1} D_{p-1}^2 + E_{p-1} D_{p-1} \cdot E_p D_p + E_p D_p^2\right)\right] + \frac{2}{3} BC \cdot \frac{s_1}{4} = \frac{1}{2} BC \cdot s_A.$$

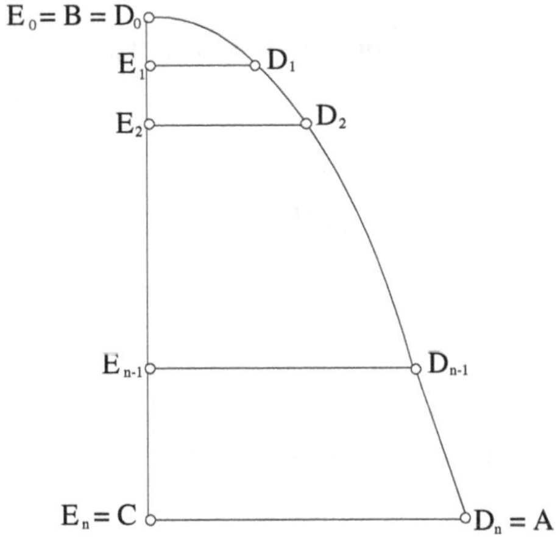

Fig. 2.3.9

Using the established expressions in Proposition 15 for the volumes of the cone of revolution described by the right-angled triangle BE_1D_1 and the frusta of the cone described by the rectangular trapezoids $E_{p-1}E_pD_pD_{p-1}$, volumes for which the sum is the volume v_s of the solid S_n, we have

$$v_s + \frac{2}{3}BC \cdot \frac{s_1}{4} = \frac{1}{2} BC \cdot s_A,$$

which we may write as

$$v_s + \frac{2}{3}\pi BC \,.\left(\frac{r_1}{2}\right)^2 = \frac{1}{2}\pi \,.\, BC \,.\, r_A^2.$$

Second and third cases (Fig. II.2.32b, c, p. 313, and Fig. 2.3.10 below): we drop from the points $D_1\ D_2 \ldots D_n = A$, the perpendiculars D_1F_1, $D_2F_2 \ldots$ AF_n onto BC, and we have for $1 \leq p \leq n$

$$\frac{E_1D_1}{F_1D_1} = \frac{E_pD_p}{F_pD_p}.$$

Hence

$$\frac{F_pD_p}{F_1D_1} = \frac{p}{1};$$

the sequences $(E_{p-1}\,E_p)$ and $(F_p\,D_p)$ satisfy the conditions of Proposition 13. Replacing $E_1\,D_1$, $E_{p-1}\,D_{p-1}$, $E_p\,D_p$ from (1) respectively by $F_1\,D_1$, $F_{p-1}\,D_{p-1}$, $F_p\,D_p$, we recover the expressions for volumes described by the triangle BE_1D_1 and by the trapezoids $E_{p-1}\,E_p\,D_p\,D_{p-1}$: expressions established in Propositions 16 and 17. The sum of these volumes is v_s and we have as in the first case

$$v_s + \frac{2}{3}BC \cdot \frac{s_1}{4} = \frac{1}{2}\,BC \cdot s_A.$$

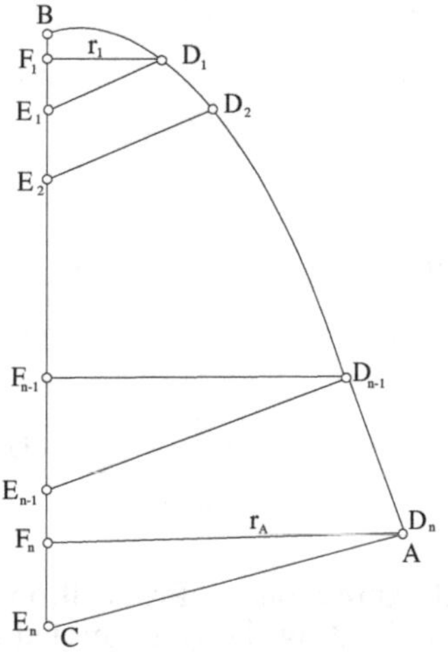

Fig. 2.3.10

Comments.

1) In the second case, the solid described by the triangle *BED* is a hollow cone and the solids described by the trapezoids are all frusta of hollow cones. But in the third case, the type of the solids generated depends on the angles *EBD*, *GD′F* and *GF′F*, if we denote by *D′* and *F′* the points of intersection of the respective segments *FD* and *AF* with the diameter *BC*.

The triangle *DBE* produces:

a solid rhombus if $E\hat{B}D < \frac{\pi}{2}$ (Fig. II.2.32c, p. 313)

a hollow cone if $E\hat{B}D > \frac{\pi}{2}$ (Fig. 2.3.11 below)

a cone if $E\hat{B}D = \frac{\pi}{2}$.

The trapezoid *EDFG* produces:

a frustum of a solid rhombus if $G\hat{D}'F < \frac{\pi}{2}$ (Fig. II.2.32c, p. 313) a frustum of a hollow cone if $G\hat{D}'F \geq \frac{\pi}{2}$ (Fig. 2.3.11 below).

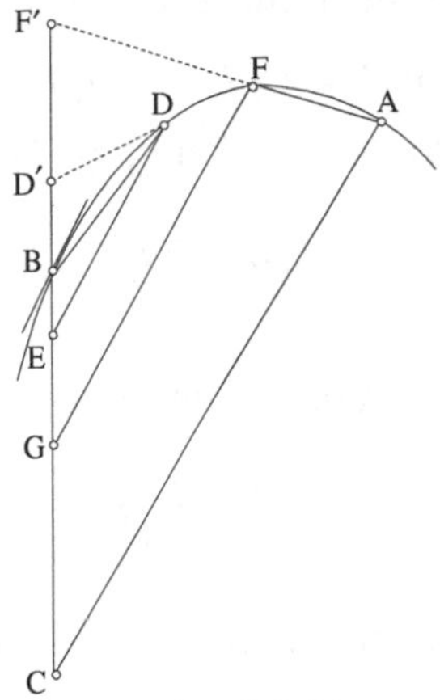

Fig. 2.3.11

In the case of Fig. II.2.32c, we have a solid rhombus and two frusta of solid rhombuses; in the case of Fig. 2.3.11, we have a hollow cone, a frustum of a hollow cone and a frustum of a solid rhombus, since $GF'F < \frac{\pi}{2}$.

2) We may note here, as we have already seen in the case of the parabola, that the subdivision of the diameter according to the odd numbers adopted by Thābit arranges the ordinates in an arithmetic progression such that the integration is accomplished according to ordinates instead of abscissae: in our terms the volume

$$\int_0^{BC} \pi y^2\, dx = \int_0^{r_A} \pi y^2 \cdot \frac{ydy}{p} = \frac{\pi}{4p}\, r_A^4 = \frac{\pi}{2}\, BC \cdot r_A^2,$$

writing the equation for the parabola as $y^2 = 2px$. Hence

$$r_A^2 = 2p \cdot BC \quad \text{and} \quad dx = \frac{ydy}{p}.$$

Propositions 33 and 34. — *It is possible to inscribe in every parabolic dome of revolution of volume* v *a piecewise conic solid* S_n *whose volume* v_s *satisfies* $v - v_s < \varepsilon$, *with* ε *being a given known volume.*

Proposition 33. — The *parabolic dome under consideration has the same axis as that of the parabola.*

Let v_1 be the volume of the cone ABC. If $v - v_1 < \varepsilon$, the problem is solved.

If $v - v_1 \geq \varepsilon$, we consider the division of AC into two equal parts, denoted $e_1 = (I_0, I_1, I_2)$, with $I_0 = C$, $I_2 = A$, $I_0 I_1 = \frac{1}{2} I_0 I_2$, and we associate it on the diameter BC with the division $d_1 = (E_0, E_1, E_2)$, with $E_0 = B$ and $E_2 = C$, satisfying[6]

$$\frac{E_0E_1}{1} = \frac{E_1E_2}{3},$$

and on the arc AC of the parabola with the points $F_0 = B$, F_1 and $F_2 = A$. We then have

$$\frac{E_1F_1}{1} = \frac{E_2F_2}{2}$$

and we apply Proposition 21 to obtain

$$BE_1 \cdot E_1F_1^2 + E_1E_2\left[E_1F_1^2 + E_2F_2^2 + E_1F_1 \cdot E_2F_2\right] - BC \cdot AC^2 > BE_1 \cdot AC^2.$$

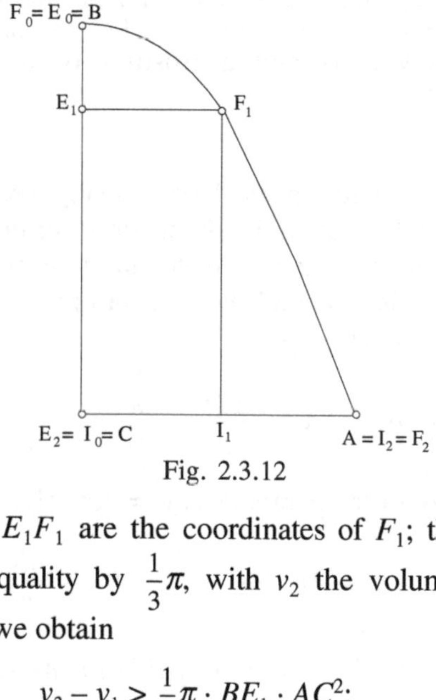

Fig. 2.3.12

But by p_{16}, BE_1 and E_1F_1 are the coordinates of F_1; thus on multiplying both sides of the inequality by $\frac{1}{3}\pi$, with v_2 the volume of the solid S_2 described by AF_1BC, we obtain

$$v_2 - v_1 > \frac{1}{3}\pi \cdot BE_1 \cdot AC^2;$$

[6] We write the ratio in this way in order to simplify notation. Thābit would have written $\frac{E_1E_2}{E_0E_1} = \frac{3}{1}$.

and using the tangent to the point F_1 and the property of the subtangent, Proposition 18, as well as Proposition 19, we show that

$$\frac{1}{3}\pi \cdot BE_1 \cdot AC^2 > \frac{1}{3}(v - v_1).$$

Hence

$$v_2 - v_1 > \frac{1}{3}(v - v_1),$$

from which we deduce

$$v - v_2 < \frac{2}{3}(v - v_1).$$

- If $v - v_2 < \varepsilon$, the solid S_2 solves the problem.

- If $v - v_2 \geq \varepsilon$, we repeat the process by dividing AC into 2^2 equal parts according to the subdivision $e_2 = (I_0, I_1, I_2, I_3, I_4)$, with $I_0 = C$, $I_4 = A$ and

$$\frac{I_0I_1}{1} = \frac{I_0I_2}{2} = \frac{I_0I_3}{3} = \frac{I_0I_4}{4};$$

on BC we associate it with the subdivision d_2 $(E_0, E_1, E_2, E_3, E_4)$, with $E_0 = B$ and $E_4 = C$, satisfying

$$\frac{E_0E_1}{1} = \frac{E_1E_2}{2} = \frac{E_2E_3}{5} = \frac{E_3E_4}{7},$$

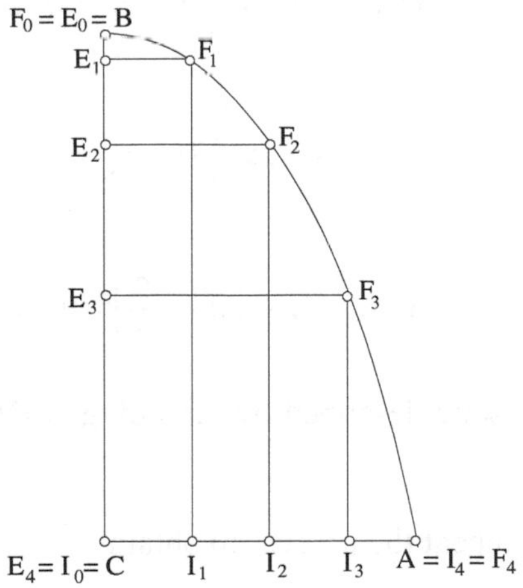

Fig. 2.3.13

and on the arc AB with the points $F_0 = B$, F_1, F_2, F_3, $F_4 = A$, we then have

$$\frac{E_1F_1}{2} = \frac{E_2F_2}{4} = \frac{E_3F_3}{6} = \frac{E_4F_4}{8}.$$

So, by Proposition 29, we get

$$E_2E_3\left(E_2F_2^2 + E_2F_2 \cdot E_3F_3 + E_3F_3^2\right) + E_3E_4\left(E_3F_3^2 + E_3F_3.E_4F_4 + E_4F_4^2\right)$$
$$-E_2E_4\left(E_2F_2^2 + E_2F_2 \cdot E_4F_4 + E_4F_4^2\right) > E_0E_1\left(E_4F_4^2 - E_2F_2^2\right).$$

Hence we deduce

$$v(E_2F_2F_3E_3) + v(E_3F_3F_4E_4) - v(E_2F_2F_4E_4) > \frac{1}{3}\pi E_0E_1\left(E_4F_4^2 - E_2F_2^2\right).$$

The left side is the volume of the torus described by the triangle $A\ F_3\ F_2$ and we show using the tangents to F_1 and F_3[7] and Propositions 18–20 that

$$\frac{1}{3}\pi\, E_0\, E_1\left(E_4\, F_4^2 - E_2\, F_2^2\right) > \frac{1}{3}v.\ \text{sg.}\ (AF_3F_2);$$

thus

$$v \cdot \text{tr.}\ (AF_3F_2) > \frac{1}{3}\ v.\ \text{sg.}\ (AF_3F_2).$$

Likewise,

$$v \cdot \text{tr.}\ (F_2F_1B) > \frac{1}{3}\ v \cdot \text{sg.}\ (F_2F_1B).$$

Hence by addition

$$v_3 - v_2 > \frac{1}{3}\ (v - v_2),$$

and hence

$$v - v_3 < \frac{2}{3}\ (v - v_2) < \left(\frac{2}{3}\right)^2 (v - v_1).$$

If $v - v_3 < \varepsilon$, the solid described by the polygon $BF_1F_2F_3AC$ solves the problem.

If $v - v_3 \geq \varepsilon$, we repeat the process to obtain

[7] *i.e.* S and O in the text.

$$v - v_4 < \frac{2}{3}(v - v_3) < \left(\frac{2}{3}\right)^3 (v - v_1),$$

...

$$v - v_n < \frac{2}{3}(v - v_{n-1}) < \left(\frac{2}{3}\right)^{n-1} (v - v_1).$$

The sequence so obtained is decreasing, and we can find n such that

$$\left(\frac{2}{3}\right)^{n-1} (v - v_1) < \varepsilon$$

and *a fortiori* $v - v_n < \varepsilon$.

The solid S_n corresponding to the division of AC into 2^n equal parts solves the problem.

Comments.

1) Thābit does not use the decreasing sequence $\left(\frac{2}{3}\right)^p$, but relies upon its reasoning in Proposition 31; yet we have seen that the proof of this last general case is made using a sequence $(1 - k)^p$ with $(1 - k) < 1$.

2) To show that the tangent at O and that at S meet the diameter FG at the same point R (Fig. II.2.33b, p. 320), it suffices to show that $OO' = SS'$, O' and S' being the respective midpoints of FB and FA. This follows from Proposition 18:

$$AQ = SS' = \frac{1}{2}(CP - KP) = BU,$$

as

$$KP = 5\ BU \text{ and } PC = 7\ BU$$

and, on the other hand,

$$OO' = I'B = BU.$$

Proposition 34. — *The axis BC of the dome is an arbitrary diameter of the parabola* (see Fig. II.2.34a , b, pp. 324, 325).

By successively dividing AC into 2, 2^2, … , 2^n equal parts, we construct as in Proposition 33 the solids S_1, S_2, …, S_n inscribed in the dome. As in the second and third cases of Proposition 32, we drop perpendiculars from the vertices of the obtained polygons to the axis of the paraboloid. Thābit rehashes the reasoning in referring to each stage of the preceding proposition. Thus, in every case, we can find n such that $v - v_n < \varepsilon$.

Proposition 35. — *In every parabolic dome* ABC *with vertex* B *(regular or not), and axis* BD, *we can inscribe a solid whose volume* v_S *is less than half the volume* V *of the cylinder whose base is the circle of diameter* AC *and whose height is equal to* BD, *and differs by a quantity less than a given volume* ε.

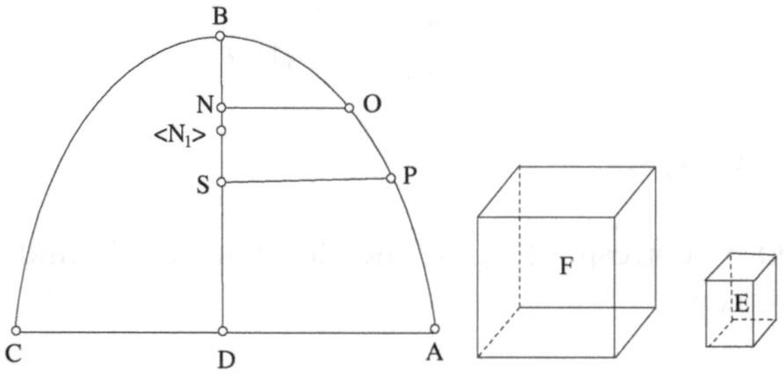

Fig. 2.3.14

We thus want to determine a solid whose volume v_S satisfies

$$0 < \frac{V}{2} - v_S < \varepsilon.$$

Now we have shown in Proposition 32 that if the axis BD is partitioned into n segments proportional to 1, 3, 5, …, $2n - 1$, we can associate with this partition a solid S_n whose volume v_S satisfies

$$\frac{V}{2} - v_S = \frac{2}{3}\pi\left(\frac{r_1}{2}\right)^2 . BD,$$

with r_1 the radius of the circle closest to the vertex B. We thus have to show that we can determine a partition of BD by which we have

(1) $$\frac{2}{3}\pi\left(\frac{r_1}{2}\right)^2 \cdot BD < \varepsilon \Leftrightarrow \pi\, r_1^2 \cdot BD < 6\varepsilon.$$

Thābit's approach comprises two parts:

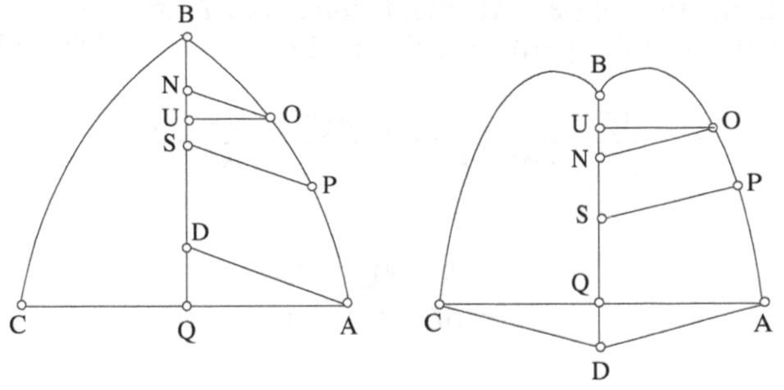

Fig. 2.3.15

a) With given volumes V and ε, to make use of (1) in part b) of this proposition, we have to set $\eta = 6\varepsilon$. We define the volume F by

$$\frac{V}{F} = \frac{F}{\eta}.$$

We then have

$$\left(\frac{V}{F}\right)^2 = \frac{V}{\eta}.$$

In the three cases, AD is an ordinate; we consider an ordinate ON satisfying

$$\frac{AD}{ON} > \frac{V}{F}.$$

We then have

$$\frac{AD^2}{ON^2} > \frac{V}{\eta}.$$

• First case: $AC = 2AD$, $ON \perp BD$. The ratio of volumes of the right cylinders of radii AD and ON and of height BD is

$$\frac{\pi \cdot BD \cdot AD^2}{\pi \cdot BD \cdot ON^2} = \frac{AD^2}{ON^2} > \frac{V}{\eta}.$$

But in this case we have

$$V = \pi \cdot BD \cdot AD^2;$$

hence

$$\pi \cdot BD \cdot ON^2 < \eta.$$

• Second and third cases: AC meets the axis BD at point Q, $AC = 2AQ$. We produce from O the perpendicular to BD, so $OU \parallel AQ$ and $ON \parallel AD$, hence

$$\frac{AQ}{OU} = \frac{AD}{ON} > \frac{V}{F} \text{ and } \frac{AQ^2}{OU^2} = \frac{AD^2}{ON^2} > \frac{V}{\eta}.$$

Then

$$\frac{\pi \cdot BD \cdot AQ^2}{\pi \cdot BD \cdot OU^2} > \frac{V}{\eta}.$$

But, in the two cases,

$$V = \pi \cdot BD \cdot AQ^2;$$

hence

$$\pi \cdot BD \cdot OU^2 < \eta.$$

If we designate by r_N the radius of the circle described by the point O ($r_N = ON$ or $r_N = OU$), we thus have in the three cases

$$(2) \qquad \pi BD \cdot r_N^2 < \eta.$$

Comment. — The inequality $\frac{AD^2}{ON^2} > \frac{V}{\eta}$ does not define a unique point N, but it is satisfied by all the points of a segment. By the equation of the parabola, we have $\frac{AD^2}{ON^2} = \frac{BD}{BN}$, and N thus satisfies $\frac{BN}{BD} < \frac{\eta}{V}$ or $BN < \frac{\eta \cdot BD}{V}$.

Let N_1 be the point defined by $BN_1 = \frac{\eta \cdot BD}{V}$; then all points N on the segment BN_1 verify the inequality.

b) In order for a solid S_n associated to a partition of BD into segments proportional to consecutive odd numbers to solve the problem, it suffices that the first point of the partition should be a point of the preceding segment BN_1.

Let N be this point and let $APOBD$ be the polygon corresponding to the partition of BD. We then have $r_1 = r_N$; thus, on the one hand,

$$\frac{V}{2} - v_S = \frac{2}{3}\pi\left(\frac{r_N}{2}\right)^2 \cdot BD \qquad \text{by Proposition 32;}$$

on the other hand,

$$\pi BD \,.\, r_N^2 < \eta \quad \text{with } \eta = 6\,\varepsilon \quad \text{by (2).}$$

But

$$\frac{2}{3} \cdot \left(\frac{r_N}{2}\right)^2 = \frac{r_N^2}{6},$$

and thus

$$\frac{V}{2} - v_S < \varepsilon.$$

Comment. — The result of Proposition 35 could have been directly obtained from Proposition 32 without using part a) of Proposition 35. The solid S_n corresponds to a partition of AD into n equal segments; we thus have $r_1 = \frac{AD}{n}$. The problem is to find n so that r_1 satisfies (1), *i.e.*

$$\left(\frac{r_1}{2}\right)^2 < \frac{3\varepsilon}{2\pi \,.\, BD} \Leftrightarrow \frac{1}{n^2} < \frac{6\varepsilon}{\pi \,.\, BD \,.\, AD^2} \Leftrightarrow n^2 > \frac{\pi \,.\, BD \,.\, AD^2}{6\varepsilon} \Leftrightarrow n^2 > \frac{V}{6\varepsilon};$$

V and ε are given.

We may thus ask why Thābit did not follow this path. This style of writing was perhaps too arithmetical for the ninth-century mathematician.

Proposition 36. — *The volume* v *of every parabolic dome* ABC *with axis* BD *is half the volume* V *of the cylinder of height* h *and whose base is a circle of diameter* AC,

$$\mathrm{v} = \frac{1}{2}\,\mathrm{V} = \frac{1}{2}\,\pi\,\mathrm{h} \cdot \frac{\mathrm{AC}^2}{4}.$$ [Fig. II.2.36, p. 331]

Thābit proceeds by *reductio ad absurdum.*

• Suppose $v > \frac{V}{2}$ and let $v = \frac{V}{2} + \varepsilon$.

By Propositions 33 and 34, it is possible to inscribe in the dome a solid of revolution of volume v_S such that $v - v_S < \varepsilon$, so $v_S + \varepsilon > v$. Thus $v_S + \varepsilon > \frac{V}{2} + \varepsilon$, and hence $v_S > \frac{V}{2}$, which is absurd since in Proposition 35 we have shown that $v_S < \frac{V}{2}$.

• Suppose $v < \frac{V}{2}$ and let $\frac{V}{2} = v + \varepsilon$.

By Proposition 35, it is possible to inscribe in the dome a solid of volume v_S such that $\frac{V}{2} - v_S < \varepsilon$ or $v_S + \varepsilon > \frac{V}{2}$. Then $v_S + \varepsilon > v + \varepsilon$, and hence $v_S > v$, which is absurd since the solid is inside the dome. We thus have

$$v = \frac{V}{2}.$$

(1) Here, Thābit refers to Proposition 35, but in Proposition 32 we have shown that

$$v_S = \frac{V}{2} - \frac{2}{3}\pi h r_1^2;$$

thus

$$v_S < \frac{V}{2};$$

and in Proposition 35 we have used Proposition 32 to show that for a given ε we can find a solid whose volume v_S satisfies

$$\frac{V}{2} - v_S < \varepsilon.$$

The inequality $v_S < \frac{V}{2}$ does not require a *reductio ad absurdum*, but Thābit wanted, in all evidence, to keep to the strict apagogical rhetorical method.

2.3.2.8. *Parallel between the treatise on the area of the parabola and the treatise on the volume of the paraboloid*

In the two treatises, Thābit uses a subdivision of the diameter of a parabolic section into segments proportional to consecutive odd numbers. The points of the parabola corresponding to this subdivision are then abscissae proportional to the squares of integers and ordinates proportional to consecutive integers.

These points determine:

in the plane:	in space:
a polygon inscribed in the parabola and decomposed into trapezoids	a solid of revolution inscribed in the paraboloid and decomposed into conic-type solids
s area of the parabola S area of the associated parallelogram s_i area of a trapezoid	v volume of the paraboloid V volume of the associated cylinder v_i volume of a conic solid

Thābit shows that, given $\varepsilon > 0$, we can find N such that for all $n > N$, we have

$\frac{2}{3}S - \sum_{i=1}^{n} s_i < \varepsilon$ (Propositions 17 and 19)	$\frac{V}{2} - \sum_{i=1}^{n} v_i < \varepsilon$ (Propositions 32 and 35)
$s - \sum_{i=1}^{n} s_i < \varepsilon$ (Proposition 18)	$v - \sum_{i=1}^{n} v_i < \varepsilon$ (Propositions 33 and 34).

In other words, he has thus shown that

$\frac{2}{3}S =$ upper bound $\sum_{i=1}^{n} s_i$	$\frac{V}{2} =$ upper bound $\sum_{i=1}^{n} v_i$
$s =$ upper bound $\sum_{i=1}^{n} s_i$	$v =$ upper bound $\sum_{i=1}^{n} v_i$

By a *reductio ad absurdum,* he then shows in each case the uniqueness of the upper bound:

$s = \frac{2}{3}S$ (Proposition 20)	$v = \frac{V}{2}$ (Proposition 36)

2.3.3. *Translated text*

Thābit ibn Qurra

On the Measurement of the Paraboloids

In the name of God, the Merciful, the Compassionate

THĀBIT IBN QURRA

On the Measurement of the Paraboloids

<Definitions>

The solid figures which I call *paraboloidal* are of two sorts: one is obtained by the rotation of a segment of a parabola about a straight line, I call this sort the *paraboloid of revolution*; the other is obtained by the rotation of a straight line about the perimeter of a segment of a parabola.

Among the paraboloids of revolution, there are two genera which comprise five species. The first of the two genera is that surrounded by half of a portion of the parabola, when its diameter is fixed and one of the two parts of the line of the parabola is rotated at the same time as one of the two halves of its base, which is adjacent to it, from an arbitrary position up to where it returns to its original position; I call this genus *the parabolic dome*. By the expression *half of a portion of the parabola* I mean that which has been limited by the diameter of that portion and one of the two halves of the line of the parabola which are on opposite sides <of that diameter> and half of the base of the portion. The other genus is that surrounded by a portion of the parabola when its base is fixed and the line of the parabola is rotated around it from an arbitrary position up to where it returns to its original position. I call this genus *the parabolic sphere*. I call the vertex of the portion about which one rotates the half to generate the parabolic dome *the vertex of the dome*. I call the two extremities of the base of the portion which one rotates to generate the parabolic sphere *the two poles of the <parabolic> sphere*.

The parabolic dome is of three species. The first is obtained by rotating a half of the portion of the parabola whose diameters are the axes;[1] I call this species *the dome with a regular vertex* because of the regularity of its vertex in its emergence in relation to that which surrounds it. The second

[1] In the case where the diameter chosen as axis of rotation is the axis of the parabola.

species is obtained by rotating the most distant half from the axis, among the two halves of the portion of the parabola whose diameters are not the axes;[2] I call this species *the dome with a pointed vertex* because of the great elevation of the height of its vertex and of its emergence in relation to that which surrounds it. The third species is obtained if we rotate half of the portion of the parabola whose diameters are not axes;[3] I call this type *the dome with a sunken vertex* because of the depression of its vertex in relation to that which surrounds it.

The parabolic sphere is of two species. The first is obtained by rotating a portion of the parabola whose diameters are axes; I call this species *a <parabolic sphere> like a melon*, as the shapes of these figures are rounded[4] and similar to the shapes of some kinds of melon. The other species is obtained by rotating a portion of the parabola whose diameters are not axes. I call it *a <parabolic sphere> like an egg* because of the thinness of one of its extremities and of the thickness of the other extremity.

If, in an obtuse-angled triangle, one of the sides that enclose the obtuse angle is fixed and if we rotate the two sides which remain, then I call the figure thus generated *a hollow cone of revolution*. If one of the two sides that enclose an accute angle of a triangle is fixed and if we rotate the two sides which remain of the triangle, then the figure thus generated is called *a solid rhombus*.

If a plane parallel to the base of a cone of revolution cuts it, then I call the portion of cone located between that plane and the base of the cone *a frustum of a cone of revolution.*[5] If we take away a hollow cone of revolution from another hollow cone of revolution, such that the angle of the two generating triangles for the two cones, which is at the vertex of the two cones, is common to the two triangles and such that the two straight lines of the two triangles, intercepted by that angle are parallel, then I call the portion which remains a *frustum of a hollow cone of revolution.*[6] I call the homologue of this in the solid rhombus *a frustum of a solid rhombus.*[7]

If an arbitrary figure and a straight line are in the same plane such that the straight line is outside of that figure, if we fix the straight line and if we rotate about it the plane with the figure which it contains from an arbitrary position up to where it returns to its original position, then I call the solid

[2] In the case where the diameter chosen is not the axis.
[3] See the previous note.
[4] Lit.: in form of a dome.
[5] Lit.: the residue of the cone of revolution.
[6] Lit.: the residue of the hollow cone of revolution.
[7] Lit.: the residue of the solid rhombus.

bounded by this figure which is in the plane *a torus*. If this figure is triangular, I call the solid a *triangular torus*; if it is squared, I call it a *square torus*, and so on in the same way for the others.

– **1** – Given two successive square numbers, the difference between them is equal to twice the side of the smaller of them, increased by one.

Let *AB* and *C* be two successive square numbers, let the side of *AB* be the number *DE* and let *F* be the side of *C*; let *GB* be equal to *C*.

I say that AG *is equal to twice* F, *increased by one.*

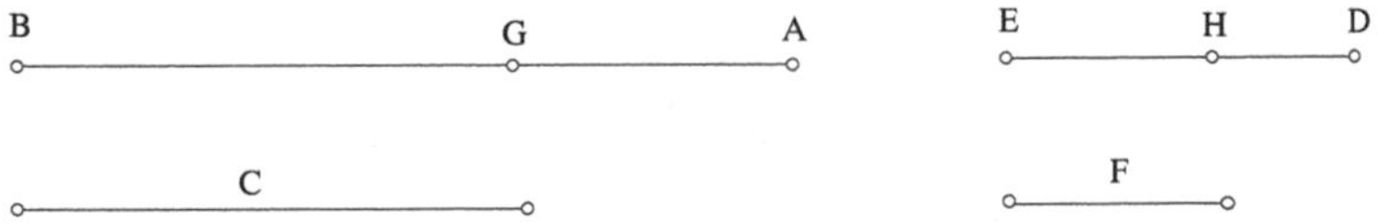

Fig. II.2.1

Proof: The two squares *AB* and *C* are successive; therefore the difference between their sides is one, as if it were not thus, there would be a number between them and its square would be between the two square numbers *AB* and *C*; this is not possible, as the two numbers *AB* and *C* are two successive squares.

If we suppose *HE* to be equal to *F*, then *DH* will be one and the square of the number *DE* will be equal to the sum[8] of the squares obtained from *DH* and *HE* and the double-product of *DH* and *HE*. Regarding the square of *HE*, it is *C* as *HE* is equal to *F*. Regarding the square of *DE*, it is *AB* and the difference between *AB* and *C* is the number *AG*. The double-product of *DH* and *HE*, plus the square of *DH*, is equal to the number *AG*. The double-product of *DH* and *HE* is twice *HE*, as *DH* is one; regarding the square of *DH*, it is one, thus the number *AG* is equal to twice *HE*, increased by one. But *HE* is equal to *F*; thus the number *AG* is equal to twice *F*, increased by one. That is what we wanted to prove.

< **2** > If an odd square number is increased by one, then the sum is equal to twice its side, plus the quadruple of the sum of the successive odd <numbers> beginning with one and which are less than the side.

Let *A* be an odd square number, let *B* be its side and let the successive odd <numbers> beginning with one and which are less than *B* be the odd numbers *C*, *D*, *E*.

[8] We sometimes add 'sum' for the purposes of the translation.

I say that if one is added to the number A, *the sum will be equal to twice the number* B, *plus the quadruple of the sum of the <numbers>* C, D, E.

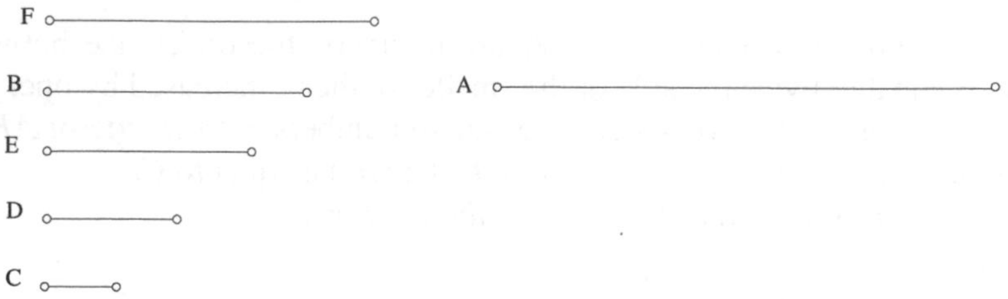

Fig. II.2.2

Proof: The number *B* is odd, as it is the side of the number *A* which is odd. If one is added to it, the sum will be even. Let the number *F* be this sum and let the number *G* be the square of the number *F*. The number *G* is thus the quadruple of the square of half of the number *F*. But the square of half of the number *F* is equal to the sum of the numbers *C*, *D*, *E*, *B*, according to what has been shown in proposition four of our treatise *On the Measurement of the Parabola*; thus the number *G* is the quadruple of the sum of the numbers *C*, *D*, *E*, *B*; this is why the number *G* from which one subtracts twice the number *B* is equal to the quadruple of the sum of the numbers *C*, *D*, *E*, plus twice the number *B*. But the number *A* is less than the number *G* by twice *B*, which is the side of the number *A*, increased by one, as the two numbers *A* and *G* are successive squares; indeed, the difference between the two numbers *B* and *F*, which are their sides, is one. If one is thus added to the number *A*, the sum will be equal to twice the number *B*, plus the quadruple of the sum of the numbers *C*, *D*, *E*. That is what we wanted to prove.

< **3** > If we add to an odd cubic number its side, then the sum is equal to the double-product of the side of the cube with itself and twice the sum of the successive odd <numbers> beginning with one and which are less than it.

Let *A* be an odd cubic number, *B* its side and *C*, *D*, *E* the successive odd numbers beginning with one and which are less than *B*.

I say that the sum of the numbers A *and* B *is equal to the double-product of the number* B *with itself and twice the sum of the numbers* C, D, E.

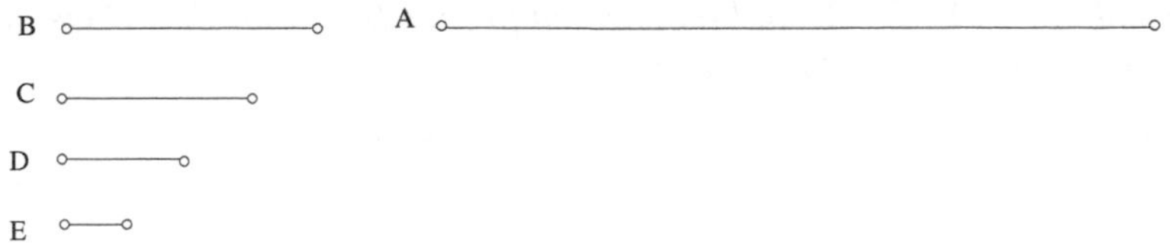

Fig. II.2.3

Proof: The number *B* is odd, as it is the side of the cube *A* which is odd; this is why the square of the number *B* is odd. Thus, if one is added to it, the sum will be <even and> equal to twice *B*, plus the quadruple of the sum of *C*, *D*, *E*;[9] this is why the product of the number *B* and the square of the number *B* – square increased by one – is equal to the sum of the product of the number *B* and its double and the quadruple of the sum of *C*, *D*, *E*. The product of the number *B* and the square of the number *B* – square increased by one – is consequently twice the sum of the product of the number *B* with itself and twice the sum of the numbers *C*, *D*, *E*. But the product of the number *B* and the square of the number *B* is the cube *A* and the product of the number *B* and one is equal to the number *B*; thus the cube *A*, plus the number *B*, is equal to twice the product of the number *B* with itself and twice the sum of the numbers *C*, *D*, *E*. That is what we wanted to prove.

– **4** – Given successive odd cubic numbers beginning with one and, in equal number, other numbers, namely squares of the successive square numbers which are associated with them and beginning with one, if we then add to each of the cubic numbers its side, the sum is equal to twice the difference between the squares of that the one of the square numbers which is associated with it and of that which preceeds it, if there is a number that preceeds it; otherwise it is equal only to its double.[10]

Let *A*, *B*, *C*, *D* be successive odd cubic numbers beginning with one and let, in equal number, *E*, *F*, *G*, *H* be other numbers, namely squares of the successive square numbers which are associated with them and beginning with one, let *I* be the side of the cube *A*, *K* the side of *B*, *L* the side of *C* and *M* the side of *D*.

I say that if we add to each of the numbers A, B, C, D *its side, then the sum is equal to twice the difference between its associated number, among*

[9] By Proposition 2.

[10] That is to say, to twice the square of the square number which is associated with it.

the numbers E, F, G, H, *and that which preceeds it, and that if we add* A *and* I, *their sum is equal to twice* E.

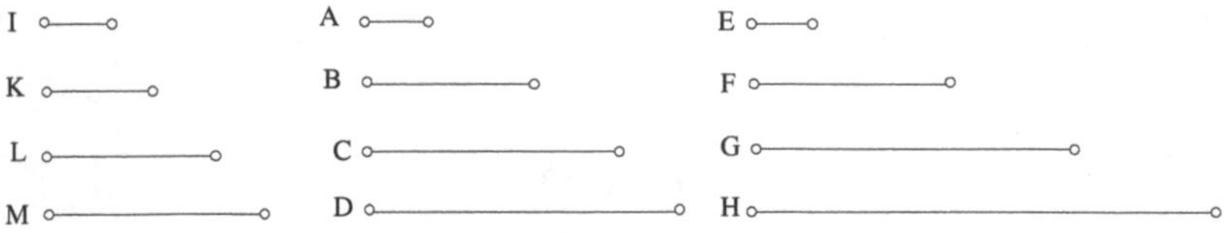

Fig. II.2.4

Proof: If we suppose that the square numbers which are the sides of the numbers *E*, *F*, *G*, *H* are, in succession, the numbers *N*, *S*, *O*, *P*, then the numbers *N*, *S*, *O*, *P* are successive squares beginning with one. The differences[11] between the successive square numbers beginning with one are the successive odd numbers beginning with three, according to what has been shown in proposition three of our treatise *On the Measurement of the Parabola*. But the successive odd numbers beginning with three are the numbers *K*, *L*, *M*, as they are the sides of the cubes *B*, *C* and *D* which are the cubes of the successive odd numbers beginning with the first of the odd cubic numbers;[12] thus the difference between the two numbers *N* and *S* is the number *K*, the difference between the two numbers *S* and *O* is the number *L* and the difference between the two numbers *O* and *P* is the number *M*. The sum of the double-product of the number *K* and *N*, the square of the number *K* and the square of *N* is equal to the square of the number *S*. This is why the difference between the square of the number *S* and the square of *N* is equal to the sum of the double-product of the number *K* and *N*, and the square of the number *K*. But the square of *N* is *E* and the square of the number *S* is *F*; therefore the difference between *E* and *F* is equal to the sum of the double-product of *K* and *N*, and the square of the number *K*. But *N* is equal to the sum of the odd numbers less than the number *K*, as has been shown in proposition three of our treatise *On the Measurement of the Parabola.*[13] The difference between *E* and *F* is thus equal to the sum of the products of the number *K*, once with itself and twice the sum of the odd numbers which are less than it; it is thus equal to the sum of the products of the number *K* with itself and twice the sum of the odd numbers which are less than it. But the two numbers *B* and *K*, if we add them up, are equal to twice the sum of the products of the number *K*

[11] Lit.: the difference. What matters here is the sequence of the differences between the successive square numbers taken two by two.

[12] Other than one.

[13] See Propositions 3 and 4.

with itself and twice the sum of the odd numbers which are less than it,[14] as the number *B* is an odd cube and its side is the number *K*; the two numbers *B* and *K*, if they are added, are thus equal to twice the difference between the two square numbers *E* and *F*. Likewise, we also show that the two numbers *C* and *L*, if they are added, are equal to twice the difference between the two numbers *F* and *G*, and that the two numbers *D* and *M*, if they are added, are equal to twice the difference between the two numbers *G* and *H*. It is clear that *A*, plus *I*, is twice *E*. That is what we wanted to prove.

– **5** – If we add all the successive odd cubic numbers beginning with one and if we add to them their sides, then the sum is equal to twice the square of the number that is equal to the sum of the sides.

Let *A*, *B*, *C*, *D* be successive odd cubic numbers beginning with one, *E*, *F*, *G*, *H* their sides and the number *I* the sum of the sides.

I say that if the numbers A, B, C, D, E, F, G, H *are added, the sum is equal to twice the square of the number* I.

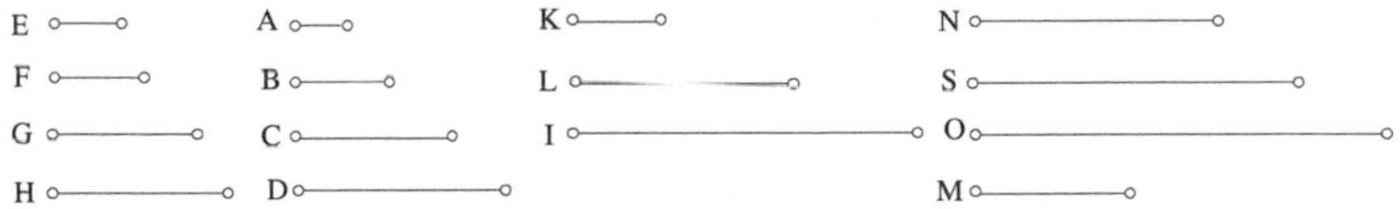

Fig. II.2.5

Proof: If we let the number *K* be equal to the sum of *E* and *F* and if we let the number *L* be equal to the sum of *E*, *F* and *G* – with the number *I* being equal to the sum of *E*, *F*, *G*, *H* – then the numbers *E*, *K*, *L* and *I* will be successive squares, beginning with one; this was shown in the third proposition of our treatise *On the Measurement of the Parabola*, as the numbers *F*, *G*, *H* are successive odd numbers beginning with three. If we let the squares of the numbers *E*, *K*, *L*, *I*, be the numbers *M*, *N*, *S*, *O*, then the numbers *M*, *N*, *S*, *O* are the squares of the successive square numbers beginning with one. But the numbers *A*, *B*, *C*, *D* are successive odd cubes beginning with one, and their sides are *E*, *F*, *G*, *H*. If it is thus, if then *A* and *E* are added, the sum is equal to twice *M* and if *B* and *F* are added, the sum is equal to twice the excess of *N* over *M*;[15] thus the sum of the numbers *A*, *B*, *E*, *F* is twice the number *N*. Likewise, if the two numbers *C* and *G* are also added, the sum is twice the excess of the number *S* over the number *N*. If the numbers *A*, *B*, *C*, *E*, *F*, *G* are added, the sum is equal to

[14] By Proposition 3.
[15] By Proposition 4.

twice the number *S*. By this example, it has been also shown that the sum of the two numbers *D* and *H* is twice the excess of the number *O* over the number *S*. The numbers *A*, *B*, *C*, *D*, *E*, *F*, *G*, *H*, if they are added, are equal to twice the number *O*. But the number *O* is the square of the number *I*; thus the numbers *A*, *B*, *C*, *D*, *E*, *F*, *G*, *H*, if they are added, are equal to twice the square of the number *I*. That is what we wanted to prove.

– **6** – Given successive odd numbers beginning with one, then the sum of the products of each of them and the triple of its square, increased by three, is equal to six times the square of the number equal to the sum of these odd numbers.

Let *A*, *B*, *C* be successive odd numbers beginning with one, the number *D* their sum and the numbers *E*, *F*, *G* their successive squares.

I say that the sum of the product of A *and the triple of* E, *increased by three, the product of* B *and the triple of* F, *increased by three, and the product of* C *and the triple of* G, *increased by three, is equal to six times the square of the number* D.

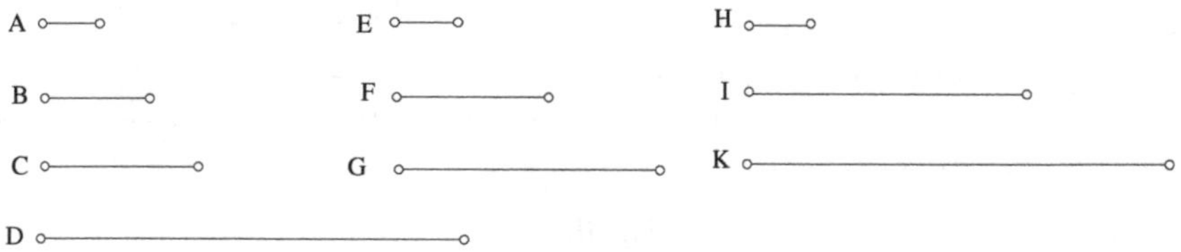

Fig. II.2.6

Proof: If we let *H* be the product of *A* and *E*, *I* the product of *B* and *F* and *K* the product of *C* and *G*, then the numbers *H*, *I*, *K* are successive odd cubic numbers beginning with one, their sides are *A*, *B*, *C* and the sum of these sides is *D*; thus the sum of the numbers *A*, *B*, *C*, *H*, *I*, *K* is equal to twice the square of the number *D*;[16] this is why the triple of the sum of the numbers *A*, *B*, *C*, *H*, *I*, *K* is equal to six times the square of the number *D*. The triple of *H* is equal to the product of *A* and the triple of *E*, the triple of *I* is equal to the product of *B* and the triple of *F* and the triple of *K* is equal to the product of *C* and the triple of *G*; thus the triple of the sum of *A*, *B*, *C* and the sum of the product of *A* and the triple of *E*, the product of *B* and the triple of *F* and the product of *C* and the triple of *G* is equal to six times the square of the number *D*. But the sum of the products of each <of the numbers> *A*, *B*, *C* and three is equal to the triple of the sum of the numbers *A*, *B*, *C*; thus the sum of the product of *A* and the triple of *E*, to which we

[16] By Proposition 5.

add three, the product of *B* and the triple of *F*, to which we add three, and the product of *C* and the triple of *G*, to which we add three, is equal to six times the square of the number *D*. That is what we wanted to prove.

– **7** – If one is added to the planar number obtained from the multiplication of two successive even numbers by each other, then the sum obtained is equal to the square of the odd number that is between the two even numbers.

Let *A* be a planar number and let its sides be two successive even numbers *BC* and *D*, and *EF* the odd number that is between the two even numbers.

I say that if one is added to the number A, *the sum will be equal to the square of the number* EF.

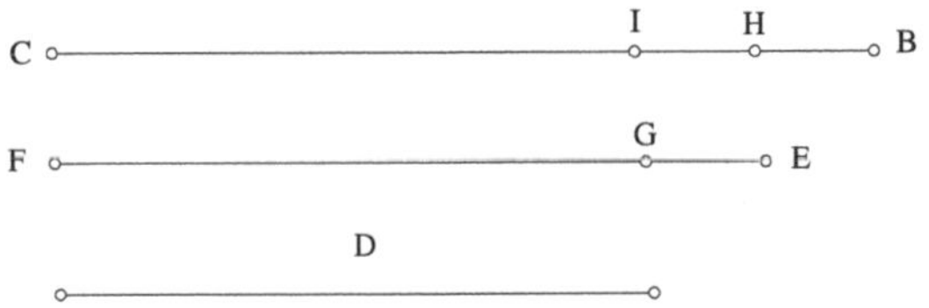

Fig. II.2.7

Proof: The numbers *D*, *EF* and *BC* are successive; thus the difference between each of them and that which follows it is one. If we let *FG* be equal to *D*, *CH* equal to *EF* and if we also let *CI* be equal to *D*, each <of the numbers> *BH*, *HI*, *EG* will be equal to one. But the product of *BC* and *D* is equal to the product of *BI* and *IC*, plus the square of *IC*. But the product of *BI* and *IC* is equal to the double-product of *HI* and *IC*, as *HI* is half of *BI*, the product of *BC* and *D* is thus equal to the double-product of *HI* and *IC*, plus the square of *IC*. Regarding *HI*, it is equal to *EG* and *IC* is equal to *GF*; thus the product of *BC* and *D* is equal to the double-product of *EG* and *GF*, plus the square of *GF*. If we add one on both sides, which is the square of *EG*, then the product of *BC* and *D*, to which one is added, is equal to the sum of the double-product of *EG* and *GF* and of the two squares obtained from *GF* and *EG*. Yet, this is equal to the square of the number *EF*; thus the product of *BC* and *D*, to which one is added, is equal to the square of the number *EF*. But the product of *BC* and *D* is equal to the number *A*. If one is added to the number *A*, then the sum will be equal to the square of *EF*. That is what we wanted to prove.

– **8** – Consider successive odd numbers beginning with three and, in equal number, the planar numbers which are associated with them, obtained from the multiplication of the successive even numbers beginning with two, each by that which follows it; then if we add the products of each of the successive odd <numbers> and the triple of its associate, among the planar numbers, increased by six, to the product of one and six, the sum will be equal to six times the square of the number equal to the sum of the odd numbers, including unity.

Let *A*, *B*, *C* be the successive odd numbers beginning with three; let *D*, *E*, *F* be the planar numbers, in equal number, which are associated with them, obtained from the multiplication of the successive even numbers beginning with two, each by that which follows it; let *G* be unity, let *H* be six and let *I* be the number equal to the sum of the numbers *G*, *A*, *B*, *C*.

I say that the product of G *and* H, *the product of* A *and the triple of* D, *increased by* H, *the product of* B *and the triple of* E, *increased by* H, *and the product of* C *and the triple of* F, *increased by* H, *have a sum equal to six times the square of the number* I.

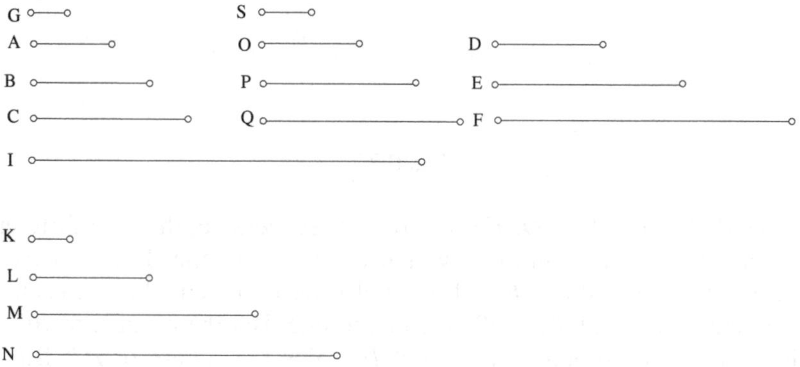

Fig. II.2.8

Proof: If we let the squares of *G*, *A*, *B*, *C* be the numbers *K*, *L*, *M*, *N*, with the numbers *G*, *A*, *B*, *C* being the successive odd <numbers> beginning with one, and their squares being *K*, *L*, *M*, *N*, then the sum of the product of *G* and the triple of *K*, increased by three, the product of *A* and the triple of *L*, increased by three, the product of *B* and the triple of *M*, increased by three, and the product of *C* and the triple of *N*, increased by three, is equal to six times the square of the number *I*.[17] If we let the successive even numbers beginning with two be the numbers *S*, *O*, *P*, *Q*, then the odd number *A* will be between the two numbers *S* and *O*, the odd number *B* will be between the two numbers *O* and *P*, and the odd number *C* will be between the two numbers *P* and *Q*. But the product of the number *S* and the number *O* is the planar number *D*; if one is added to the number

[17] By Proposition 6.

D, the sum will be equal to the square of the number *A*, which is the number *L*.[18] Likewise, we also show that if one is added to the number *E*, the sum will be equal to the square of the number *B*, which is the number *M*, and that if one is added to the number *F*, the sum will be equal to the square of the number *C*, which is equal to the number *N*. This is why, if we add three to the triple of each of the numbers *D*, *E*, *F*, the sum will be equal to the triple of its homologue[19] among the numbers *L*, *M*, *N*. If we add the number three on both sides, if we then add six to the triple of each of the numbers *D*, *E*, *F*, the sum will be equal to the triple of its homologue among the numbers *L*, *M*, *N*, increased by three. The sum of the product of *A* and the triple of *L*, increased by three, the product of *B* and the triple of *M*, increased by three, and the product of *C* and the triple of *N*, increased by three, is equal to the sum of the product of *A* and the triple of *D*, increased by six, the product of *B* and the triple of *E*, increased by six, and the product of *C* and the triple of *F*, increased by six.

If we let the product of *G* and the triple of *K*, increased by three, which is equal to its product and *H*, on both sides, then the sum of the product of *G* and the triple of *K*, increased by three, the product of *A* and the triple of *L*, increased by three, the product of *B* and the triple of *M*, increased by three, and the product of *C* and the triple of *N*, increased by three, is equal to the sum of the product of *G* and *H*, the product of *A* and the triple of *D*, increased by six, the product of *B* and the triple of *E*, increased by six, and the product of *C* and the triple of *F*, increased by six. Yet, we have shown[20] that the sum of the product of *G* and the triple of *K*, increased by three, the product of *A* and the triple of *L*, increased by three, the product of *B* and the triple of *M*, increased by three, and the product of *C* and the triple of *N*, increased by three, is equal to six times the square of the number *I*. The sum of the product of the number *G* and *H*, the product of *A* and the triple of *D*, increased by *H* – which is six – the product of *B* and the triple of *E*, increased by *H*, and the product of *C* and the triple of *F*, increased by *H*, is equal to six times the square of the number *I*. That is what we wanted to prove.

< **9** > For every pair of successive even numbers, the sum of their squares and the planar number obtained from their product by each other is equal to the triple of their product by each other, increased by four.

[18] By Proposition 7.

[19] That is to say, the sum associated with each of the numbers is the triple of the homologue of that number: $3D + 3 = 3L$.

[20] In Proposition 6.

Let *A* and *BC* be two successive even numbers, let the number *D* be the square of *A*, the number *E* the square of *BC* and let the number *F* be the product of *A* and *BC*.

I say that the sum of the numbers D, F, E *is equal to the triple of the number* F, *increased by four.*

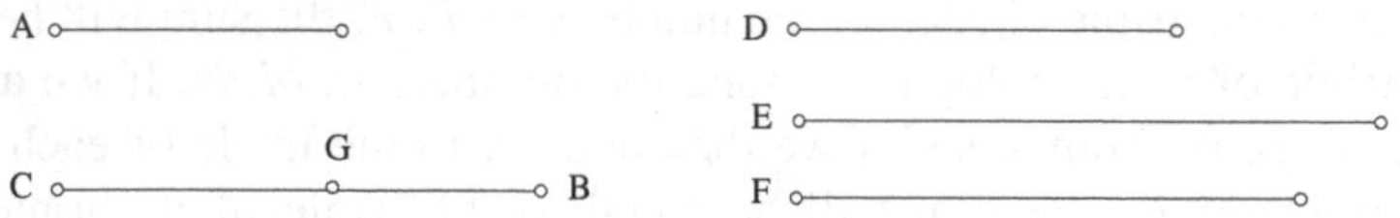

Fig. II.2.9

Proof: If we let *CG* be equal to *A*, then the two squares obtained from *BC* and *CG* have a sum equal to the double-product of *BC* and *CG*, plus the square of *BG*. If we let the product of *BC* and *CG* on both sides, then the two squares obtained from *BC* and *CG* and the product of *BC* and *CG* have a sum equal to the triple of the product of *BC* and *CG*, plus the square of *BG*. But *CG* is equal to *A*, thus the sum of the two squares obtained from *A* and *BC* – which are *D* and *E* – and the product of *A* and *BC* – which is *F* – is equal to the triple of the product of *A* and *BC* – which is the triple of *F* – increased by the square of *BG*. But the square of *BG* is four as *BG* is two; indeed, it is the difference between two successive even numbers. The sum of the numbers *D*, *F*, *E* is equal to the triple of the number *F*, increased by four. That is what we wanted to prove.

– **10** – Given successive odd numbers beginning with three and, in equal number, the planar numbers which are associated with them, obtained by the multiplication of the successive even numbers beginning with two, each by its successor, then the sum of the products of each of these odd numbers and its associate among the planar numbers and the squares of the two sides of that planar number, increased by two, and the product of one and the number six, is equal to six times the square of the number equal to the sum of the odd numbers, including the unit.

Let *A*, *B*, *C* be successive odd numbers beginning with three and, in equal number, the planar numbers *D*, *E*, *F* which are associated with them, obtained by the multiplication of the successive even numbers beginning with two, each by its successor, their sides <being> the successive even numbers *G*, *H*, *I*, *K* beginning with two and whose squares are the numbers *L*, *M*, *N*, *S*; let the unit be *O*, let <the number> six be *P* and let the number *U* be equal to the sum of the numbers *O*, *A*, *B*, *C*.

I say that the sum of the product of O *and* P, *the product of* A *and the sum of the numbers* D, L, M, *plus two, the product of* B *and the sum of the numbers* E, M, N, *plus two, and the product of* C *and the sum of the numbers* F, N, S, *plus two, is equal to six times the square of the number* U.

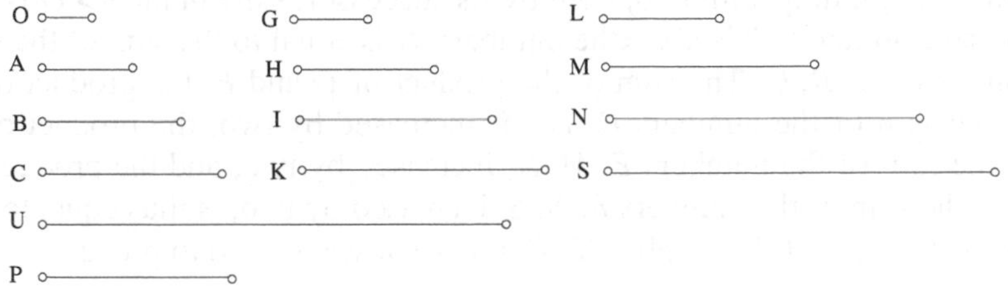

Fig. II.2.10

Proof: The two numbers *G* and *H* are successive even numbers, their squares are the two numbers *L* and *M*, their product by each other is the number *D*, the sum of the numbers *D*, *L*, *M* is thus equal to the triple of the number *D*, increased by four.[21] If we let the number two on both sides, the sum of the numbers *D*, *L*, *M*, plus two, is equal to the triple of the number *D*, increased by six. Likewise, we also show that the sum of the numbers *E*, *M*, *N*, plus two, is equal to the triple of the number *E*, increased by six, and that the sum of the numbers *F*, *N*, *S*, plus two, is equal to the triple of the number *F*, increased by six. The sum of the product of *A* and the sum of the numbers *D*, *L*, *M*, increased by two, the product of *B* and the sum of the numbers *E*, *M*, *N*, increased by two, and the product of *C* and the sum of the numbers *F*, *N*, *S*, increased by two, is equal to the sum of the product of *A* and the triple of *D*, increased by six, the product of *B* and the triple of *E*, increased by six, and the product of *C* and the triple of *F*, increased by six. If we let the product of *O* and *P* on both sides, then the sum of the product of *O* and *P*, the product of *A* and the sum of the numbers *D*, *L*, *M*, increased by two, the product of *B* and the sum of the numbers *E*, *M*, *N*, increased by two, and the product of *C* and the sum of the numbers *F*, *N*, *S*, increased by two, is equal to the sum of the product of *O* and *P*, the product of *A* and the triple of *D*, increased by six, the product of *B* and the triple of *E*, increased by six, and the product of *C* and the triple of *F*, increased by six. But the sum of the product of *O* and *P*, the product of *A* and the triple of *D*, increased by six, the product of *B* and the triple of *E*, increased by six, and the product of *C* and the triple of *F*, increased by six, is equal to six times the square of the number *U*.[22] Indeed, these numbers that we have mentioned are such that, on the one hand, the numbers *A*, *B*, *C* are successive odd numbers beginning with three and, on the other, the

[21] By Proposition 9.
[22] By Proposition 8.

numbers *D*, *E*, *F* are the planar numbers, in equal number, which are associated with them, obtained by the multiplication of the successive even numbers beginning with two, each by its successor; <the number> *O* is the unit, <the number> *P* is six, <the number> *U* is equal to the sum of the odd numbers *O*, *A*, *B*, *C*. The sum of the product of *O* and *P*, the product of *A* and the sum of the numbers *D*, *L*, *M*, increased by two, the product of *B* and the sum of the numbers *E*, *M*, *N*, increased by two, and the product of *C* and the sum of the numbers *F*, *N*, *S*, increased by two, is thus equal to six times the square of the number *U*. That is what we wanted to prove.

– **11** – Given successive odd numbers beginning with one and, in equal number, successive even numbers beginning with two which are associated with them, if we multiply each of the odd <numbers> by the square of its associate, among the even <numbers>, and if its associate is preceeded by an even <number>, we also multiply it by the square of that even <number> and by the planar number obtained from the multiplication of that associate by <the even number> that preceeds it; if we add all of those, if we take its third and if we add to it two thirds of the number equal to the sum of the odd numbers, then the result will be equal to half of the product of the number equal to the sum of the odd <numbers> and the square of the greater even <number>.

Let *A*, *B*, *C*, *D* be successive odd numbers beginning with one and, in equal number, let the successive even numbers beginning with two which are associated with them be *E*, *F*, *G*, *H*; let the planar <number> of *E* times *F* be the number *I*, and the planar <number> of *F* times *G* be the number *K*, and the planar number of *G* times *H* be the number *L*, and let the number *M* be equal to the sum of the numbers *A*, *B*, *C*, *D*.

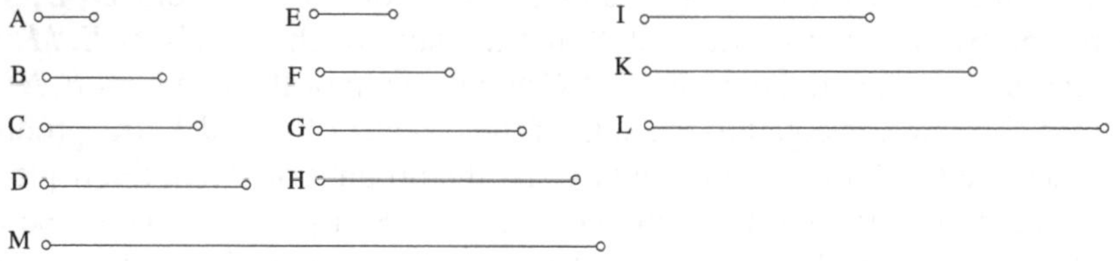

Fig. II.2.11

I say that if we add the product of A *and the square of the number* E, *the product of* B *and the sum of the squares of the two numbers* E, F, *and the number* I, *the product of* C *and the sum of the squares of the two numbers* F, G, *and the number* K, *and the product of* D *and the sum of the squares of the two numbers* G, H, *and the number* L; *if we take one third of the sum and if we add to it two thirds of the number* M, *the sum will be equal to half of the product of the number* M *and the square of the number* H.

Proof: The numbers *B*, *C*, *D* are successive odd numbers beginning with three, the numbers *I*, *K*, *L* are in equal number[23] and are associated with them, they are the planar <numbers> obtained by the multiplication of the successive even numbers beginning with two, each by its successor, and the number *M* is equal to the sum of the numbers *B*, *C*, *D*, plus the unit which is *A*. If we add the product of the number *B* and the planar number *I*, and the squares of the numbers *E*, *F* which are its sides, and two, the product of the number *C* and the planar number *K*, and the squares of the two numbers *F*, *G* which are its sides, and two, and the product of the number *D* and the planar number *L*, and the squares of the two numbers *G*, *H* which are its sides, and two, if we add to this the product of one and six, the sum will be equal to six times the square of the number *M*.[24] The product of one and six is equal to the product of *A* and the square of the number *E*, plus two. The product of each of the <numbers> *A*, *B*, *C*, *D* and two is equal to twice the sum of the numbers *A*, *B*, *C*, *D*, which is the number *M*. If we add the product of *A* and the square of the number *E*, the product of *B* and the squares of the two numbers *E*, *F* and the number *I*, the product of *C* and the squares of the two numbers *F*, *G* and the number *K*, and the product of *D* and the squares of the two numbers *G*, *H* and the number *L*; if we add to this twice the number *M*, the sum will be equal to six times the square of the number *M*. Likewise, the numbers *A*, *B*, *C*, *D* are successive odd numbers beginning with one; the numbers *E*, *F*, *G*, *H*, in equal number, are their associates and are successive even <numbers> beginning with two; thus each of the numbers *E*, *F*, *G*, *H* exceeds its associate among the numbers *A*, *B*, *C*, *D* by the unit. If we thus add one to the number *D*, we obtain the number *H*; if we take the square of its half, it will be equal to the sum of the numbers *A*, *B*, *C*, *D*, which is the number *M*, according to what has been shown in proposition four of our treatise *On the Measurement of the Parabola*. The square of the number *M* is thus equal to the product of the number *M* and the square of half of the number *H*; thus the sum of the product of *A* and the square of the number *E*, the product of *B* and the squares of the two numbers *E*, *F* and the number *I*, the product of *C* and the squares of the two numbers *F*, *G* and the number *K*, the product of *D* and the squares of the two numbers *G*, *H* and the number *L*, plus twice the number *M*, is equal to six times the square of the number *M*; it is thus equal to six times the product of the number *M* and the square of half of the number *H*. But six times the product of the number *M* and the square of

[23] 'in equal number' concerns the even numbers from which *I*, *K*, *L* are derived.
[24] By Proposition 10.

half of the number H is equal to one and a half times the product of the number M and the square of the number H. The sum of the product of A and the square of the number E, the product of B and the sum of the squares of the two numbers E, F and the number I, the product of C and the sum of the squares of the two numbers F, G and the number K, and the product of D and the sum of the squares of the two numbers G, H and the number L, plus twice the number M, is equal to one and a half times the product of the number M and the square of the number H. From this, one shows that one third of the sum of the product of A and the square of the number E, the product of B and the sum of the squares of the two numbers E, F and the number I, the product of C and the sum of the squares of the two numbers F, G and the number K, the product of D and the sum of the squares of the numbers G, H and the number L, plus two thirds of the number M, is equal to half of the product of the number M and the square of the number H. That is what we wanted to prove.

– **12** – Consider straight lines following the ratios of the successive odd numbers beginning with one and, in equal number, and other straight lines, which are associated with them, following the ratios of the successive even numbers beginning with two, and such that the smallest of the straight lines which are following the ratios of the odd <numbers> are half of the smallest of the straight lines which are following the ratios of the even <numbers>; if we multiply each of the straight lines that are following the ratios of the odd numbers by the square of its associate – among the straight lines which are following the ratios of the even numbers – and if there is another straight line before its associate, we multiply it also by the square of that straight line and by the product of its associate and the straight line that preceeds it; if we add the solids thus obtained and if we take one third of the sum to which we add two thirds of the solid formed as the product of the straight line equal to the sum of the straight lines – which are following the ratios of the odd numbers – and the square of half of the smallest of the straight lines which are following the ratios of the even numbers, then the result is equal to half of the solid formed as the product of the straight line that is equal to the sum of the straight lines – which are following the ratios of the odd numbers – and the square of the greater of the straight lines which are following the ratios of the even numbers.

Let the straight lines A, B, C following the ratios of the successive odd numbers beginning with one and let, in equal number, the straight lines which are associated with them following the ratios of the successive even numbers beginning with two, be the straight lines D, E, F; let the straight line A be half of the straight line D and let the straight line G be equal to the sum of the straight lines A, B, C.

I say that if we add up the solids formed as the product of the straight line A *and the square of the straight line* D, *as the product of* B *and the sum of the squares of the straight lines* D, E *and the product of* D *and* E, *and as the product of* C *and the sum of the squares of the straight lines* E, F *and the product of* E *and* F ; *and if we take one third of this sum, to which are added two thirds of the solid formed as the product of the straight line* G *and the square of half of the straight line* D, *the result is equal to half of the solid formed as the product of the straight line* G *and the square of the straight line* F.

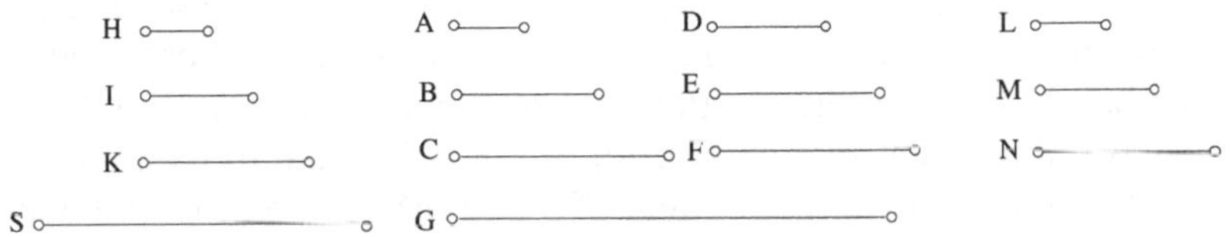

Fig. II.2.12

Proof: If we let the odd numbers whose ratios are equal to the ratios of the straight lines A, B, C be the numbers H, I, K, and the even numbers whose ratios are equal to the ratios of the straight lines D, E, F be the numbers L, M, N, then the ratio of A to D is equal to the ratio of H to L, as it is its half.[25] The ratio of each of the straight lines A, B, C to each of the straight lines D, E, F is thus equal to the ratio of the homologue of that straight line – among the numbers H, I, K – to the homologue of the other straight line, among the numbers L, M, N. The ratio of the solid formed as the product of A and the square of the straight line D to the cube obtained from the straight line D is thus equal to the ratio of the product of H and the square of the number L to the cube obtained from L. Likewise, one also shows that the ratio of the solids formed as the product of the straight line B and the square of the straight line D, and the square of the straight line E and the product of D and E to the cube obtained from the straight line E, is equal to the ratio of the sum of the products of the number I and the squares of the numbers L, M and the product of L and M to the cube obtained from M, and that the ratio of the solids formed as the products of the straight line C and the squares of the two straight lines E, F and the product of E and F to the cube obtained from the straight line F is equal to the ratio of the sum of the products of the number K and the squares of the

[25] By hypothesis $A = \frac{D}{2}$.

two numbers *M*, *N* and the product of *M* and *N* to the cube obtained from the number *N*. But the ratio of each of the cubes of the straight lines *D*, *E*, *F* to the cube of the straight line *F* is equal to the ratio of its associate – among the cubes of the numbers *L*, *M*, *N* – to the cube of the number *N*. Thus the ratio of the solids formed as the products[26] *of the straight line *B* and the squares of the two straight lines *D*, *E* and the product of *D* and *E* to the cube of the straight line *F* is equal to the ratio of the product of the number *I* and the squares of the two numbers *L*, *M* and the product of *L* and *M* to the cube obtained from the number *N*. And the ratio of the solids formed as the products of the straight line *C* and the squares of the two straight lines *E*, *F* and the product of *E* and *F* to the cube of the straight line *F* is equal to the ratio of the product of the number *K* and the squares of the two numbers *M*, *N* and the product of *M* and *N* to the cube obtained from the number *N*. But the ratio of the cube of the straight line *F* to the solid formed as the product of *G* and the square of *F* is equal to the ratio of the cube of the number *N* to the product of *S* and the square of the number *N*. Thus the ratio of the solid formed as the product of the straight line *A* and the square of the straight line *D* to the solid formed as the product of the straight line *G* and the square of the straight line *F* is equal to the ratio of the product of the number *H* and the square of *L* to the product of the number *S* and the square of the number *N*; the ratio of the solids formed as the products of the straight line *B* and the squares of the two straight lines *D*, *E* and the product of *D* and *E* to the solid formed as the product of the straight line *G* and the square of the straight line *F* is equal to the ratio of the product of the number *I* and the squares of the two numbers *L*, *M* and the product of *L* and *M*, to the product of the number *S* and the square of the number *N*; the ratio of the solids formed as the products of the straight line *C* and the squares of the straight lines *E*, *F* and the product of *E* and *F* to the solid formed as the product of the straight line *G* and the square of the straight line *F* is equal to the ratio of the product of the number *K* and the squares of the numbers *M*, *N* and the product of *M* and *N* to the product of the <number> *S* and the square of the number *N*. Thus the ratio of one third of the solids formed as the product of the straight line *A* and the square of the straight line *D* and the product of the straight line *B* and the squares of the straight lines *D*, *E* and the product of *D* and *E*, and the product of the straight line *C* and the squares of the straight lines *E*, *F* and the product of *E* and *F*, if they are added, to the solid formed as the product of the straight line *G* and the square of the straight line *F* is equal to the ratio of one third of the product of *H* and the square of the number *L*, the

[26] *... * The paragraph between the two asterisks renders the Arabic text reconstituted by us.

product of *I* and the squares of the numbers *L*, *M* and the product of *L* and *M*, and the product of *K* and the squares of the numbers *M*, *N* and the product of *M* and *N*, if they are added, to the product of *S* and the square of *N*. But the ratio of the solid formed* as the product of the square of the straight line *A* and the straight line *G* to the solid formed as the product of the straight line *G* and the square of the straight line *F* is equal to the ratio of the number *S* to the product of the number *S* and the square of the number *N*. This is why the ratio of two thirds of the solid formed as the product of the straight line *G* and the square of half of the straight line *D* to the solid formed as the product of the straight line *G* and the square of the straight line *F* is equal to the ratio of two thirds of the number *S* to the product of the number *S* and the square of the number *N*. Yet, we have shown that the ratio of one third of the solids formed as the product of the straight line *A* and the square of the straight line *D*, as the product of the straight line *B* and the squares of the straight lines *D*, *E* and the product of *D* and *E*, and as the product of the straight line *C* and the squares of the straight lines *E*, *F* and the product of *E* and *F*, if they are added, to the solid formed as the product of the straight line *G* and the square of the straight line *F*, is equal to the ratio of one third of the product of *H* and the square of the number *L*, of the product of *I* and the squares of the numbers *L*, *M* and the product of *L* and *M*, and of the product of *K* and the squares of the numbers *M*, *N* and the product of *M* and *N*,[27] *if they are added, to the product of *S* and the square of *N*. The ratio of one third of the solids formed as the product of the straight line *A* and the square of the straight line *D*, as the product of the straight line *B* and the squares of the straight lines *D*, *E* and the product of *D* and *E*, and as the product of the straight line *C* and the squares of the straight lines *E*, *F* and the product of *E* and *F*, if we add them up, and if we add to them two thirds of the product of the straight line *G* and the square of half of the straight line *D*, to the solid formed as the product of the straight line *G* and the square of the straight line *F* is equal to the ratio of one third of the sum of the product of *H* and the square of the number *L*, of the product of *I* and the two squares of the numbers *L*, *M* and the product of *L* and *M*, and of the product of *K* and the two squares of the numbers *M*, *N* and the product of *M* and *N*,* if they are added, and if we add two thirds of the number *S*, to the product of the number *S* and the square of the number *N*. But one third of the sum of the product of *H* and the square of the number *L*, of the product of *I* and the squares of the numbers *L*, *M* and the product of *L* and *M*, and of the product of *K* and the squares of the numbers *M*, *N* and the product of *M* and *N*, if they are added

[27] *...* The paragraph between the two asterisks renders the Arabic text reconstituted by us.

and if we add to them two thirds of the number *S*, is equal to half of the product of the number *S* and the square of the number *N*,[28] as the numbers *H*, *I*, *K* are successive odd numbers beginning with one and whose sum is equal to the number *S* and the numbers *L*, *M*, *N* are successive even numbers beginning with two. Thus one third of the solids formed as the product of *A* and the square of the straight line *D*, as the product of *B* and the squares of the straight lines *D*, *E* and the product of *D* and *E*, and as the product of *C* and the squares of the straight lines *E*, *F* and the product of *E* and *F*, if we add them up and if we add to them two thirds of the product of the straight line *G* and the square of half of the straight line *D*, is equal to half of the solid formed as the product of the straight line *G* and the square of the straight line *F*. That is what we wanted to prove.

– **13** – Consider straight lines following the ratios of the successive odd numbers beginning with one and, in equal number, other straight lines which are associated with them following the ratios of the successive even numbers beginning with two, and if the smallest of the straight lines which are following the ratios of the odd numbers is not half of the smallest of the straight lines which are following the ratios of the even <numbers> and if we multiply each of the straight lines which are following the ratios of the odd numbers by the square of its associate, among the straight lines that are following the ratios of the even <numbers>, and if there is another straight line which preceeds its associate, we multiply it also by the square of that straight line and by the product of its associate and the straight line that preceeds it; if we add the solids thus formed and if we take one third of the sum and if we add to it two thirds of the solid formed as the product of the straight line which is equal to the sum of the straight lines that are following the ratios of the odd numbers, and the square of half of the smallest of the straight lines which are following the ratios of the even numbers, then the sum is equal to half of the solid formed as the product of the straight line, which is equal to the sum of the straight lines that are following the ratios of the odd numbers, and the square of the greater of the straight lines that are following the ratios of the even numbers.

Let the straight lines *A*, *B*, *C* following the ratios of the successive odd numbers beginning with one and let, in equal number, the straight lines *D*, *E*, *F* which are associated with them following the ratios of the successive even numbers beginning with two; let the straight line *A* not be half of the straight line *D* and let the straight line *G* be equal to the sum of the straight lines *A*, *B*, *C*.

[28] By Proposition 11.

I say that the solids formed as the product of the straight line A *and the square of the straight line* D, *as the product of* B *and the squares of the straight lines* D, E *and the product of* D *and* E *and as the product of* C *and the squares of the straight lines* E, F *and the product of* E *and* F, *if we add them up, and if we take one third of the sum, to which we add two thirds of the solid formed as the product of the straight line* G *and the square of half of the straight line* D, *then the sum will be equal to half of the solid formed as the product of the straight line* G *and the square of the straight line* F.

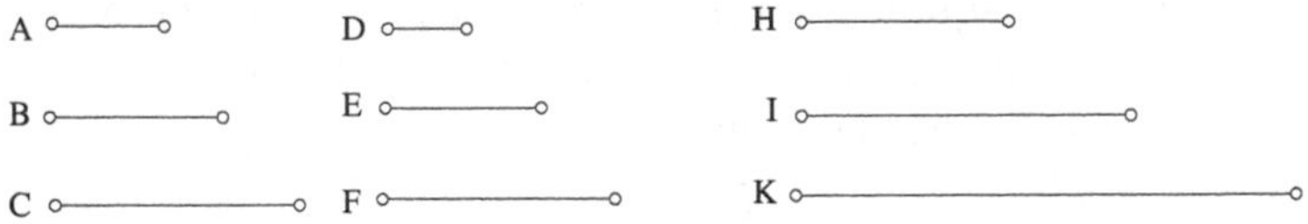

Fig. II.2.13

Proof: If we let the straight line *H* be twice the straight line *A* and if we let the ratios of the straight lines *H*, *I*, *K*, to each other, taken in succession, be equal to the ratios of the straight lines *D*, *E*, *F*, to each other, taken in succession, then the ratio of the solid formed as the product of the straight line *A* and the square of the straight line *D* to the solid formed as the product of the straight line *A* and the square of the straight line *H* is equal to the ratio of the square of the straight line *D* to the square of the straight line *H*, and the ratio of the solid formed as the product of the straight line *B* and the square of the straight line *D* to the solid formed as the product of the straight line *B* and the square of the straight line *H* is equal to the ratio of the product of *D* and *E* to the product of *H* and *I*. But the ratio of the product of *D* and *E* to the product of *H* and *I* is equal to the ratio of the solid formed as the product of the straight line *B* and the product of *D* and *E* to the solid formed as the product of the straight line *B* and the product of *H* and *I*; thus the ratio of the solid formed as the product of the straight line *B* and the square of the straight line *D* to the solid formed as the product of *B* and the square of the straight line *H*, which is equal to the ratio of the square of the straight line *D* to the square of the straight line *H*, is equal to the ratio of the solid formed as the product of the straight line *B* and the product of *D* and *E* to the solid formed as the product of the straight line *B* and the product of the straight line *H* and *I*. We also show, in the same way, that each of the ratios of the solids formed as the product of *B* and the square of the straight line *E* and as the product of *C* and the squares of the straight lines *E*, *F* and the product of *E* and *F*, to its homologue – among the solids formed as the product of *B* and the square of the straight line *I* and as the product of *C* and the squares of the straight lines *I*, *K* and the product of *I* and *K* – is equal to the ratio of the square of the straight line *D* to the square of the straight line *H*. It is, likewise, the ratio of one third of the first solids to one third of the second solids. It is, likewise, the ratio of two thirds of the solid formed as the product of the straight line *G* and the square of half of the straight line *D* to two thirds of the solid formed as the product of the straight line *G* and the square of half of the straight line *H*. If we add them up, the ratio of one third of the solids formed as the product of *A* and the square of the straight line *D*, as the product of *B* and the squares of the straight lines *D*, *E* and the product of *D* and *E*, and as the product of *C* and the squares of the straight lines *E*, *F* and the product of *E* and *F*, if we add them up, plus two thirds of the solid formed as the product of the straight line *G* and the square of half of the straight line *D*, to one third of the solids formed as the product of *A* and the square of the straight line *H*, as the product of *B* and the squares of the straight lines *H*, *I* and the product of *H* and *I*, and as the product of *C* and the squares of the straight lines *I*, *K* and the product of *I* and *K*, if we add them up, plus two thirds of the solid formed as the product of the straight line *G* and the square of half of the straight line *H*, is equal to the ratio of the square of the straight line *D* to the square of the straight line *H*, which is equal to the ratio of the square of the straight line *F* to the square of the straight line *K*. But the ratio of the square of the straight line *F* to the square of the straight line *K* is equal to the ratio of the solid formed as the product of the straight line *G* and the square of the straight line *F* to the solid formed as the product of the straight line *G* and the square of the straight line *K*; thus the ratio of one third of the solids formed as the product of *A* and the square of the straight line *D*, as the product of *B* and the squares of the straight lines *D*, *E* and the product of *D* and *E*, and as the product of *C* and the squares of the straight lines *E*,[29] **F* and the product of *E* and *F*, if we add them up, plus two thirds of the solid formed as the product of the straight line *G* and the square of half of the straight line *D*, to one third of the solids formed as the product of *A* and the square of the straight line *H*, as the product of *B* and the squares of the straight lines *H*, *I* and the product of *H* and *I*, and as the product of *C* and the squares of the straight lines *I*, *K* and the product of *I* and *K*, if we add them up, plus two thirds of the solid formed as the product of the straight line *G* and the square of half of the straight line *H*, is equal to the ratio of half of the solid formed as the product of the straight line *G* and the square of the straight line *F* to half of the solid formed as the product of the straight line *G* and the square of the straight line *K*. But one third of the solids formed as the product of *A* and the square of the straight line *H*, as the product of *B* and the squares of the straight lines *H*, *I* and the

[29] *… * The paragraph between the two asterisks renders the Arabic text reconstituted by us.

product of *H* and *I*, and as the product of* *C* and the squares of the straight lines *I*, *K* and the product of *I* and *K*, if we add them up, plus two thirds of the solid formed as the product of the straight line *G* and the square of half of the straight line *H*, is equal to half of the solid formed as the product of the straight line *G* and the square of the straight line *K*. The solids formed as the product of the straight line *A* and the square of the straight line *D*, as the product of the straight line *B* and the squares of the straight lines *D*, *E* and the product of *D* and *E*, and as the product of *C* and the squares of the straight lines *E*, *F* and the product of *E* and *F*, if we add them up and if we take one third of it, to which we add two thirds of the solid formed as the product of the straight line *G* and the square of half of the straight line *D*, the sum will be equal to half of the solid formed as the product of the straight line *G* and the square of the straight line *F*. That is what we wanted to prove.

– **14** – Given five magnitudes such that the ratio of the first to the second is equal to the ratio of the third to the fourth and is equal to the ratio of the fourth to the fifth, and if the first is less than the second, then the product of the first and the excess of the fifth over the third is equal to the product of the excess of the second over the first and the sum of the third and of the fourth.

Let *A*, *BC*, *D*, *E* and *FG* be five magnitudes such that the ratio of *A* to *BC* is equal to the ratio of *D* to *E* and is equal to the ratio of *E* to *FG* and such that *A* is less than *BC*, let *HC* be equal to *A* and *IG* equal to *D*.

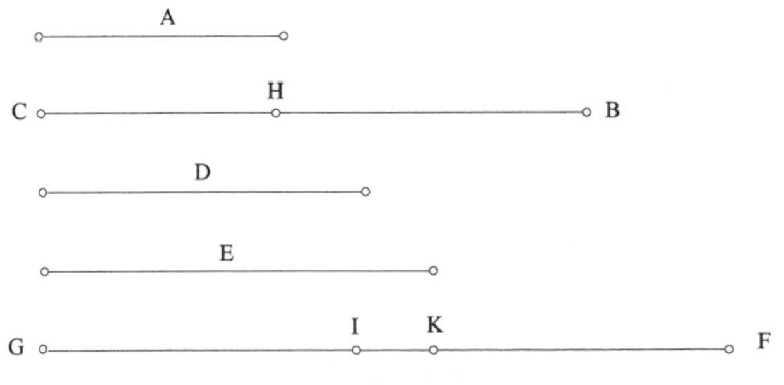

Fig. II.2.14

I say that the product of A *and* FI *is equal to the product of* BH *and the sum of the two magnitudes* D *and* E.

Proof: The ratio of *A* to *BC* is equal to the ratio of *E* to *FG*; but *HC* is equal to the magnitude *A*; thus the ratio of *HC* to *CB* is equal to the ratio of *E* to *FG*. If we set *GK* equal to *E*, the ratio of *HC* to *CB* is equal to the ratio of *KG* to *GF*. If we separate, the ratio of *CH* to *HB* is equal to the ratio of

GK to *KF*. But the ratio of *D* to *E* is also equal to the ratio of *E* to *GF*, the magnitude *D* is equal to *IG* and the magnitude *E* is equal to *GK*; thus the ratio of *IG* to *GK* is equal to the ratio of *GK* to *GF*. If we separate, the ratio of *GI* to *IK* is equal to the ratio of *GK* to *KF*. If we add them up, the ratio of the sum of *GI* and *GK* to the sum of *IK* and *KF*, which is equal to *IF*, is equal to the ratio of *GK* to *KF* and which is equal to the ratio of *CH* to *HB*; thus the ratio of the sum of *GI* and *GK* to *IF* is equal to the ratio of *CH* to *HB*. But *GI* is equal to *D* and *GK* is equal to *E* and *CH* is equal to *A*; thus the ratio of *A* to *HB* is equal to the ratio of the sum of *D* and *E* to *IF*; thus the product of *A* and *IF* is equal to the product of *BH* and the sum of the two magnitudes *D* and *E*. That is what we wanted to prove.

– **15** – The volume of every frustum of a cone of revolution is equal to one third of the product of its height and the sum of three circles, where the first is its upper circle, the other its base circle, and the third, a circle whose square of diameter is equal to the product of the diameter of the upper circle of the frustum of a cone and the diameter of its base circle.

Let there be a frustum of a cone of revolution whose base <circle> is *ABC* and whose upper circle is *DEF*, and let the square of the diameter of another circle, which is the circle *GHI*, be equal to the product of the diameter of the circle *ABC* and the diameter of the circle *DEF*.

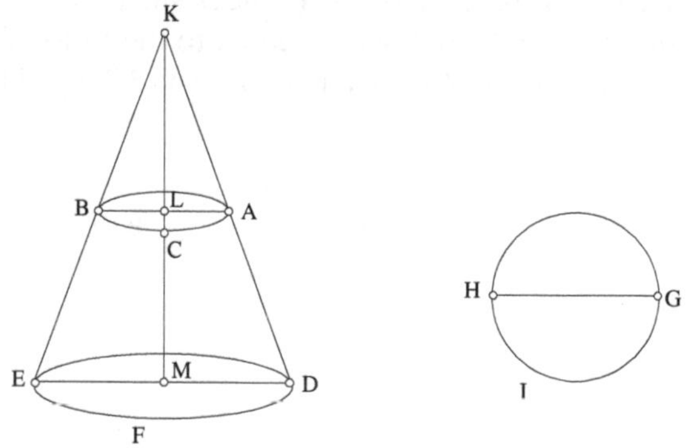

Fig. II.2.15

I say that the volume of the frustum of a cone ABC *is equal to one third of the product of its height and the sum of the three circles* ABC, DEF *and* GHI.

Proof: If we let the point *K* be the vertex point of the two cones where the first is subtracted from the other leaving the frustum of a cone as the remainder, if we let their axis be *KLM* and if we make the plane *DAKBE* pass through the axis *KLM*, then the section *DAKBE* is a triangle and *AB*, which is the intersection of this plane and the plane of the circle *ABC*, is a diameter of the circle *ABC* and *DE*, which is the intersection of this plane and the plane of the circle *DEF*, is a diameter of the circle *DEF*. If we let *GH* be the diameter of the circle *GHI*, then the product of *AB* and *DE* is equal to the square of the straight line *GH*, thus the ratio of *AB* to *GH* is equal to the ratio of *GH* to *DE* and the ratio of the square of the straight line *AB* to the square of the straight line *GH* is consequently equal to the ratio of the square of the straight line *GH* to the square of the straight line *DE* and is equal to the ratio of *AB* to *DE*. But the ratio of *AB* to *DE* is equal to the ratio of *KL* to *KM*, as the straight lines *AB* and *DE* are parallel; indeed, they are the two intersections of the plane *KDE* and the planes of the circles *ABC* and *DEF*, which are parallel. Thus the ratio of the square of the straight line *AB* to the square of the straight line *GH* is equal to the ratio of the square of the straight line *GH* to the square of the straight line *DE* and is equal to the ratio of the straight line *KL* to the straight line *KM*. On the one hand, the ratio of the square of the straight line *AB* to the square of the straight line *GH* is equal to the ratio of the circle *ABC* to the circle *GHI* and, on the other, the ratio of the square of the straight line *GH* to the square of the straight line *DE* is equal to the ratio of the circle *GHI* to the circle *DEF*; thus the ratio of *KL* to *KM* is equal to the ratio of the circle *ABC* to the circle *GHI* and is equal to the ratio of the circle *GHI* to the circle *DEF*. The product of *KL* and the excess of the circle *DEF* over the circle *ABC* is thus equal to the product of *LM* and the sum of the circles *ABC* and *GHI*.[30] If we let the product of *LM* and the circle *DEF* on both sides, the product of *KL* and the excess of the circle *DEF* over the circle *ABC*, plus the product of *LM* and the circle *DEF*, is equal to the product of *LM* and the sum of the three circles *GHI*, *DEF* and *ABC*. But the product of *KL* and the excess of the circle *DEF* over the circle *ABC*, plus the product of *LM* and the circle *DEF*, is equal to the triple of the volume of the frustum of a cone *ABCDEF*. The product of *LM* and the sum of the three circles *ABC*, *DEF*, *GHI* is thus equal to the triple of the volume of the frustum of a cone *ABCDEF*. One third of the product of *LM* and the sum of the three circles *ABC*, *DEF*, *GHI* is thus equal to the volume of the frustum of a cone *ABCDEF*. That is what we wanted to prove.

– **16** – The volume of every frustum of a hollow cone of revolution[31] is equal to one third of the product of its axis and the sum of three circles, where the first is its upper circle, the other its base circle and the third a

[30] By Proposition 14.
[31] Definition given in the introduction to this treatise, *supra*, p. 262.

circle whose square of diameter is equal to the product of the diameter of the first of the two circles and the diameter of the other.

Let there be a frustum of a hollow cone of revolution whose upper circle is *ABC*, whose base circle is *DEF* and whose axis is *GH*;[32] let the square of the diameter of another circle, which is the circle *IKL*, be equal to the product of the diameter of the circle *ABC* and the diameter of the circle *DEF*.

I say that the volume of the frustum of a hollow cone AGBEHD *is equal to one third of the product of* GH *and the sum of the three circles* ABC, DEF, IKL.

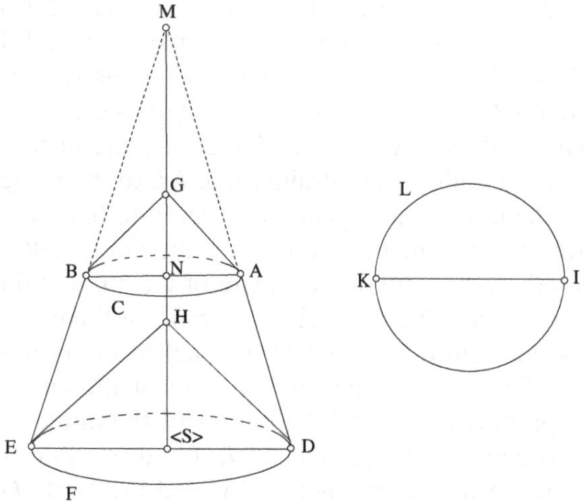

Fig. II.2.16

Proof: If we let the vertex point of the two hollow cones of revolution, where the first is subtracted from the other leaving the frustum of a cone as the remainder, be the point *M*, and their axis *MGNH*, and if we produce the plane *DAMBE* on the axis *MGNH* and if we join the straight line *DE*, then *DAMBE* is a triangle; *AB*, which is the intersection of this plane and the plane of the circle *ABC*, is a diameter of the circle *ABC*, and *DE*, which is the intersection of this plane and the plane of the circle *DEF*, is a diameter of the circle *DEF*. One third of the product of *MN* and the circle *ABC* is the volume of the cone of revolution whose base is the circle *ABC* and whose vertex is the point *M*. One third of the product of *GN* and the circle *ABC* is the volume of the cone of revolution whose base is the circle *ABC* and whose vertex is the point *G*. One third of the product of *MG* and the circle

[32] Moreover, by the definition given in the introduction, we have *GA* || *HD*.

ABC is thus the volume of the hollow cone of revolution *AMBG*. In the same way, we also show that one third of the product of *MH* and the circle *DEF* is the volume of the hollow cone of revolution *DMEH*. One third of the product of *GH* and the circle *DEF*, plus one third of the product of *MG* and the difference between the two circles *DEF* and *ABC*, is the volume of the frustum of a hollow cone of revolution *AGBEHD*. If we also let the diameter of the circle *IKL* be the straight line *IK*, then the product of *AB* and *DE* is equal to the square of the straight line *IK*; thus the ratio of *AB* to *IK* is equal to the ratio of *IK* to *DE*. The ratio of the square of the straight line *AB* to the square of the straight line *IK* is consequently equal to the ratio of the square of *IK* to the square of the straight line *DE* and is equal to the ratio of *AB* to *DE*. But the ratio of *AB* to *DE* is equal to the ratio of *AM* to *MD*, which is equal to the ratio of *GM* to *MH*.[33] The ratio of the square of the straight line *AB* to the square of the straight line *IK* is thus equal to the ratio of the square of the straight line *IK* to the square of the straight line *DE* and is equal to the ratio of *GM* to *MH*. But, on the one hand, the ratio of the square of the straight line *AB* to the square of the straight line *IK* is equal to the ratio of the circle *ABC* to the circle *IKL*, and, on the other, the ratio of the square of the straight line *IK* to the square of the straight line *DE* is equal to the ratio of the circle *IKL* to the circle *DEF*. Thus the ratio of *GM* to *MH* is equal to the ratio of the circle *ABC* to the circle *IKL* and is equal to the ratio of the circle *IKL* to the circle *DEF*. The product of *GM* and the excess of the circle *DEF* over the circle *ABC* is thus equal to the product of *GH* and the sum of the circles *ABC* and *IKL*. If we let the product of *GH* and the circle *DEF* on both sides, then the product of *MG* and the excess of the circle *DEF* over the circle *ABC*, plus the product of *GH* and the circle *DEF*, is equal to the product of *GH* and the sum of the three circles *ABC*, *DEF*, *IKL*. Thus one third of the product of *GH* and the circle *DEF*, plus one third of the product of *MG* and the excess of the circle *DEF* over the circle *ABC*, is equal to one third of the product of *GH* and the sum of the three circles *ABC*, *DEF* and *IKL*. Yet, we have shown that one third of the product of *GH* and the circle *DEF*, plus one third of the product of *MG* and the excess of the circle *DEF* over the circle *ABC*, is the volume of the frustum of a hollow cone *AGBEHD*. The volume of the frustum of a hollow cone *AGBEHD* is thus equal to one third of the product of *GH* and the sum of the three circles *ABC*, *DEF* and *IKL*. That is what we wanted to prove.

Moreover, we have shown that the volume of every hollow cone of revolution is equal to one third of the product of its axis and its base circle.

[33] This assumes that *AG* is parallel to *HD*, which has been indicated in the definitions.

– **17** – The volume of every frustum of a solid rhombus[34] is equal to one third of the product of its axis and the sum of three circles, where the first is its upper circle, the other its base circle and the third a circle whose square of diameter is equal to the product of the diameter of the first of these circles and the diameter of the other.

Let there be a frustum of a solid rhombus whose upper circle is *ABC*, whose base circle is *DEF* and whose axis is *GH*; let the square of the diameter of another circle – the circle *IKL* – be equal to the product of the diameter of the circle *ABC* and the diameter of the circle *DEF*.

I say that the volume of the frustum of a solid rhombus AGBEHD *is equal to one third of the product of* GH *and the sum of the three circles* ABC, DEF *and* IKL.

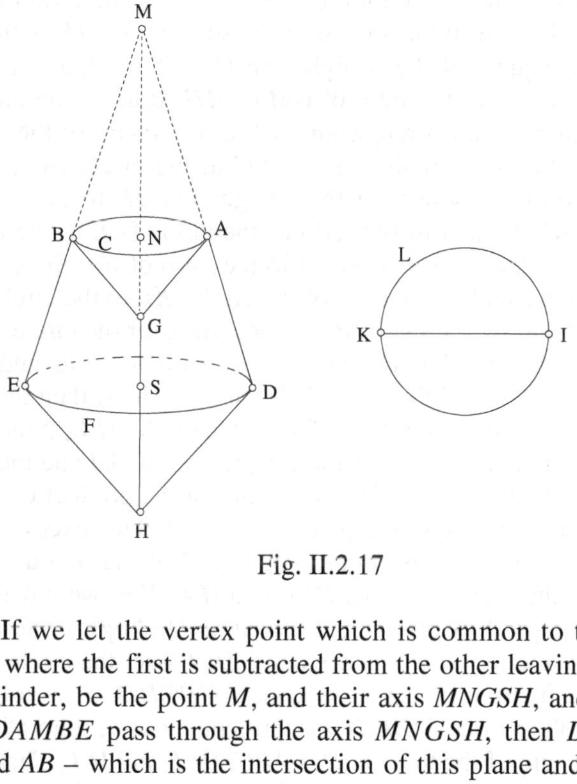

Fig. II.2.17

Proof: If we let the vertex point which is common to the two solid rhombuses, where the first is subtracted from the other leaving the frustum as the remainder, be the point *M*, and their axis *MNGSH*, and if we make the plane *DAMBE* pass through the axis *MNGSH*, then *DAMBE* is a triangle, and *AB* – which is the intersection of this plane and the plane of the circle *ABC* – is a diameter of the circle *ABC*, and *DE* – which is the intersection of this plane and the plane of the circle *DEF* – is the diameter of the circle *DEF*. One third of the product of *MN* and the circle *ABC* is the volume of the cone of revolution whose base is the circle *ABC* and whose

[34] See the definition in the introduction to this treatise, *supra*, p. 262.

vertex is the point *M*. One third of the product of *GN* and the circle *ABC* is the volume of the cone of revolution whose base is the circle *ABC* and whose vertex is the point *G*. One third of the product of *MG* and the circle *ABC* is thus the volume of the solid rhombus *AMBG*. In the same manner, one also shows that one third of the product of *MH* and the circle *DEF* is equal to the volume of the solid rhombus *DMEH*; thus one third of the product of *GH* and the circle *DEF*, plus one third of the product of *MG* and the difference between the two circles *DEF* and *ABC*, is the volume of the frustum of a solid rhombus *AGBEHD*. We show, as we have shown in the previous proposition, that one third of the product of *GH* and the circle *DEF*, plus one third of the product of *MG* and the excess of the circle *DEF* over the circle *ABC*, is equal to one third of the product of *GH* and the sum of the three circles *ABC*, *DEF* and *IKL*. Thus the volume of the frustum of the solid rhombus *AGBEHD* is equal to one third of the product of *GH* and the sum of the three circles *ABC*, *DEF* and *IKL*. That is what we wanted to prove.

Moreover, we have shown that the volume of every solid rhombus is equal to one third of the product of its axis and its base circle.

– **18** – If we mark on the line of a portion of a parabola three points in the one of the halves of the portion, and if we produce parallel straight lines from these points to the diameter of the section such that the excesses of the parallel straight lines over each other are equal, and if we produce from the intermediate point, among the three points, a straight line tangent to the section and if we produce from the two points which remain two straight lines parallel to the diameter of the portion until they meet the tangent straight line, then these two straight lines are equal and each of them is equal to half of the difference between the two straight lines separated by the parallel straight lines on the diameter of the portion.

Let *AB* be a portion of a parabola, of diameter *CD*. Let us mark on the line of the parabola the three points *A*, *E*, *F* in one of the halves of the portion. Let us produce the parallel straight lines *AG*, *EH* and *FI* from these points to the diameter. Let the excess of *AG* over *EH* be equal to the excess of *EH* over *FI*. Let a straight line *KEL* tangent to the parabola *AB* pass through the point *E* and let one produce from the points *A* and *F* two straight lines *AK* and *FL*, parallel to the straight line *CD* and let them meet the straight line *KEL* at the points *K* and *L*.

I say that the two straight lines AK *and* FL *are equal and that each of them is equal to half of the difference between the straight lines* GH *and* HI.

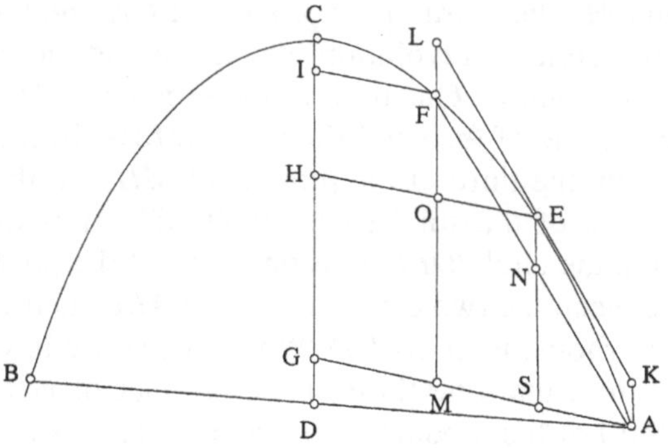

Fig. II.2.18

Proof: If we extend the straight line *LF* up to the point *M* and if we produce from the point *E* a straight line *ENS* parallel to the straight line *CD*, then the straight lines *FI* and *EH* are equal to the straight lines *MG* and *SG*, thus the straight lines *AS* and *SM* are equal, as the excess of *AG* over *EH* is equal to the excess of *EH* over *FI*. If we join the straight line *ANF*, then the ratio of *AS* to *SM* is equal to the ratio of *AN* to *NF*; thus the straight line *AN* is equal to the straight line *NF*, and the straight line *ENS* is the first of the diameters of the section according to what has been shown in proposition forty-six of the first book of the work by Apollonius on the *Conics*.[35] Yet *AF* has been divided into two halves. But Apollonius has shown in proposition five of the second book of his work on the *Conics*[36] that if it is thus, then the straight line *AF* is parallel to the straight line tangent to the section at the point *E*; thus the straight line *AF* is parallel to the straight line *KL*. But the straight lines *AK*, *NE* and *FL* are parallel; they are consequently equal. But the straight line *AN* is equal to the straight line *NF*; thus the straight line *AN* is half of the straight line *AF*; this is why the straight line *SN* is equal to half of the straight line *FM*. But the straight line *FM* is equal to the straight line *IG*; thus the straight line *NS* is equal to half of the straight line *IG*. But the straight line *EN* is the difference between the straight lines *ES* and *NS*; thus the straight line *EN* is equal to the difference between the straight line *ES* and half of the straight line *IG*. But the straight line *ES* is equal to the straight line *HG*; thus the straight line *EN* is equal to the difference between *HG* and half of the straight line *IG*. But the difference between *HG* and half of the straight line *IG* is equal to half of the difference between the straight lines *GH* and *HI*; thus the

[35] Lit.: the conic.
[36] Lit.: the conic.

straight line *EN* is equal to half of the difference between the straight lines *GH* and *HI*. Yet, we have shown that each of the straight lines *AK* and *FL* is equal to the straight line *EN*; thus the straight lines *AK* and *FL* are equal and each of them is equal to half of the difference between the straight lines *GH* and *HI*. That is what we wanted to prove.

From that, it is clear that if any of the three points is the vertex of the portion, such as the point *F* which is the vertex of the portion whose diameter is *FM*, and if *AM* is twice *EO*, then the straight lines *AK* and *FL* are equal and each of them is equal to half of the difference between of the straight lines *MO* and *OF*.

< **19** > Consider two parallelograms having a same base, in the same direction and between two parallel straight lines; if we fix the straight line parallel to their base and if we rotate all of the other sides, then the two solids generated by the rotation of the parallelograms are equal.

Let *ABCD*, *AEFD* be two parallelograms on the same base *AD*, in the same direction and between the two parallel straight lines *AD* and *BF*.

I say that if we fix the straight line BF *and if we rotate the other sides of the two parallelograms, then the solid generated by the rotation of the parallelogram* ABCD *is equal to the solid generated by the rotation of the parallelogram* AEFD.

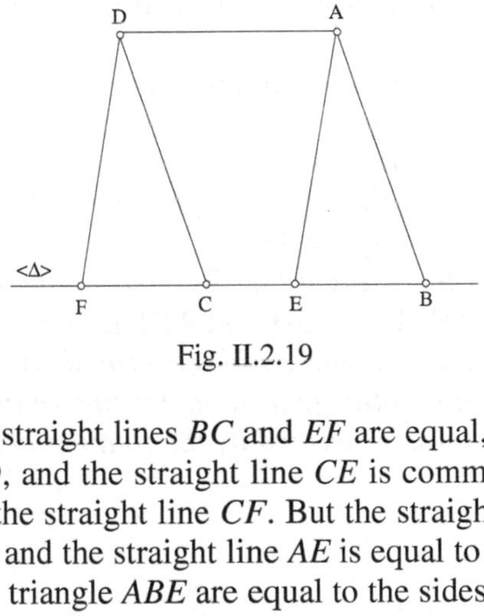

Fig. II.2.19

Proof: The two straight lines *BC* and *EF* are equal, as they are equal to the straight line *AD*, and the straight line *CE* is common; thus the straight line *BE* is equal to the straight line *CF*. But the straight line *AB* is equal to the straight line *CD* and the straight line *AE* is equal to the straight line *DF*; thus the sides of the triangle *ABE* are equal to the sides of the triangle *DCF* and their angles are equal. The generated solid – when we fix *BE* and rotate the two remaining sides of the triangle *ABE* – is thus equal to the solid generated when we fix *CF* and rotate the two remaining sides of the triangle *DCF*. If we subtract the first of these two solids – let that be generated when we fix *BE* and rotate the two remaining sides of the triangle *ABE* – from the solid generated when we fix the straight line *BF* and rotate all of the other sides of the figure *ABFD*, there remains the solid generated when we fix the straight line *EF* and rotate all of the other sides of the surface *AEFD*. If we subtract the other solid, among the two equal solids that we have mentioned – which is generated when we fix the straight line *CF* and rotate the two remaining sides of the triangle *DCF* – from the same solid generated when we fix the straight line *BF* and rotate the other sides of the figure *ABFD*, there remains the solid generated when we fix the straight line *BC* and rotate all of the other sides of the surface *ABCD*. The solid generated when we fix the straight line *BC* and rotate all of the other sides of the surface *ABCD* is thus equal to the solid generated when we fix the straight line *EF* and rotate all of the other sides of the surface *AEFD*. That is what we wanted to prove.

– **20** – If two parallelograms are in the same plane, having two equal bases, and in the same direction, such that their bases are on the <same> straight line, if two straight lines join the extremities of the two parallel straight lines to their bases in order to make a third parallelogram, if we fix the straight line over which the bases of the first two parallelograms are and if we rotate all of the other sides of the three surfaces, according to their form,[37] then the difference between the two solids generated by the rotation of the first two surfaces is equal to the torus[38] generated by the rotation of the third surface.

Let *ABCD* and *EFGH* be two parallelograms in the same plane, having two equal bases, *BC* and *FG*, over the same straight line; let the two surfaces be in the same direction. Let the extremities of the straight lines *AD* and *EH* be joined to the two straight lines *AE* and *DH* and let us produce from that a parallelogram *ADHE*.

I say that if we fix the straight line BG *and rotate all of the other sides of the three surfaces* ABCD, EFGH *and* ADHE, *according to their form,*[39] *then the difference between the solid generated by the rotation of the surface* ABCD *and the solid generated by the rotation of the surface* EFGH *is equal to the torus generated by the rotation of the surface* ADHE.

[37] That is to say, without deforming themselves.
[38] See the definition in the introduction to this treatise, *supra*, p. 262.
[39] That is to say, without deforming themselves.

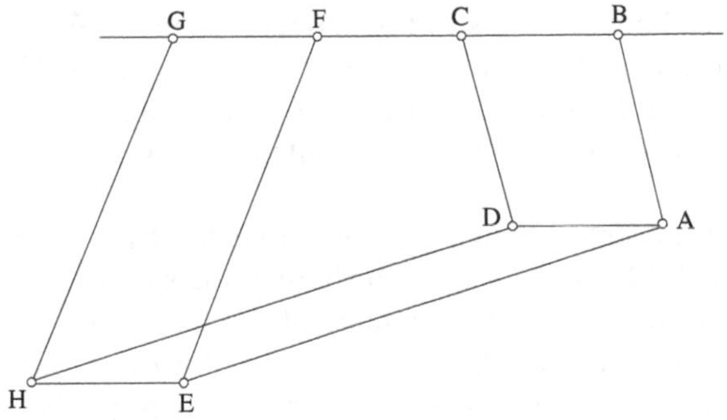

Fig. II.2.20

Proof: The two straight lines *BC* and *FG* are equal and the straight line *CF* is common; thus the two straight lines *BF* and *CG* are equal; the two parallel straight lines *AB* and *CD* are equal as they join the extremities of two parallel straight lines; likewise, the two straight lines *EF* and *HG* and the two straight lines *EA* and *HD* are equal. The sides of the figure *ABFE* are thus equal to the sides of the figure *DCGH* and their angles are equal, since the sides of the three figures are parallel. The solid generated when we fix the straight line *BF* and rotate all of the other sides of the figure *ABFE* is thus equal to the solid generated when we fix the straight line *CG* and rotate all of the other sides of the figure *DCGH*. If we subtract the first of these two solids – which is generated when we fix the straight line *BF* and rotate all of the other sides of the figure *ABFE* – from the solid generated when we fix the straight line *BG* and rotate all of the other sides of the figure *ABGHE*, there remains the solid generated when we fix the straight line *FG* and rotate all of the other sides of the surface *EFGH*. If we subtract the other solid, among the two equal solids that we have mentioned – which is generated when we fix the straight line *CG* and rotate all of the other sides of the figure *DCGH* – from the same solid generated when we fix the straight line *BG* and rotate all of the other sides of the figure *ABGHE*, there remains the solid generated when we fix the straight line *BC* and rotate all of the other sides of the figure *EABCDH*, according to their form. The solid generated when we fix the straight line *FG* and rotate all of the other sides of the surface *EFGH* is thus equal to the solid generated when we fix the straight line *BC* and rotate all of the other sides of the figure *EABCDH*, according to their form. If we remove from the two solids the solid generated when we fix the straight line *BC* and rotate all of the other sides of the surface *ABCD*, there remains the torus that the parallelogram *ADHE* generates when we fix the straight line *BG* and rotate all of the other sides of the three surfaces *ABCD*, *EFGH* and *EADH*, equal to the difference between the solid generated, when we fix the straight line *FG* and rotate all of the other sides of the surface *EFGH* and the solid generated when we fix the straight line *BC* and rotate all of the other sides of the surface *ABCD*. That is what we wanted to prove.

– **21** – If four straight lines are such that the first is one third of the second and the third is half of the fourth, then the sum of the solids formed as the product of the first and the square of the third, and as the product of the second and the sum of the square of the third and the square of the fourth, and the surface obtained from the product of the third and the fourth, from which we subtract the solid formed as the product of the sum of the first and the second and the square of the fourth, gives a remainder greater than the solid formed as the product of the first and the square of the fourth.

Let *A*, *B*, *C*, *D* be four straight lines; let *A* be one third of *B* and let *C* be half of *D*.

I say that the sum of the solids formed as the products of A *and the square of the straight line* C, *and as the product of* B *and the sum of the squares of the straight lines* C, D *and the product of* C *and* D, *from which we subtract the solid formed as the product of the sum of the straight lines* A, B *and the square of the straight line* D, *gives a remainder greater than the solid formed as the product of* A *and the square of the straight line* D.

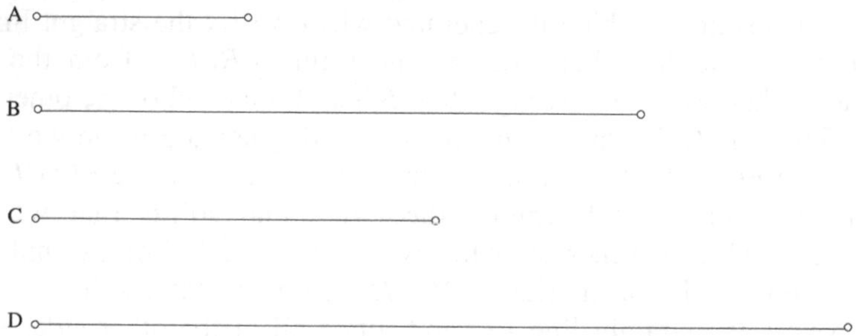

Fig. II.2.21

Proof: The straight line *C* is half of the straight line *D*; thus its square is one quarter of the square of *D*. The straight line *A* is also one quarter of the sum of the straight lines *A* and *B*; thus the ratio of the square of the straight line *C* to the square of the straight line *D* is equal to the ratio of the straight line *A* to the sum of the straight lines *A* and *B*. This is why the solid formed as the product of *A* and the square of the straight line *D* is equal to the solid formed as the product of the sum of the straight lines *A*, *B* and the square of the straight line *C*. Likewise, the ratio of the product of *C* and *D* to the square of the straight line *D* is equal to the ratio of *C* to *D*. But the ratio of *C* to *D* is greater than the ratio of *A* to *B*; thus the ratio of the product of *C* and *D* to the square of the straight line *D* is greater than the ratio of *A* to *B*. This is why the product of *B* and the product of *C* and *D* is greater than the product of *A* and the square of the straight line *D*. But we have shown that the product of *A* and the square of the straight line *D* is equal to the product of the sum of the straight lines *A* and *B* and the square of the straight line *C*. The product of *B* and the product of *C* and *D*, plus the product of the sum of the straight lines *A* and *B* and the square of the straight line *C*, is thus greater than the double-product of *A* and the square of the straight line *D*. We set the product of *B* and the square of the straight line *D* on both sides; then the product of *B* and the product of *C* and *D*, and the product of the sum of the straight lines *A*, *B* and the square of the straight line *C* and the product of *B* and the square of the straight line *D* have a sum greater than the double-product of *A* and the square of the straight line *D*, plus the product of *B* and the square of the straight line *D*. But the double-product of *A* and the square of the straight line *D*, plus the product of *B* and the square of the straight line *D*, is equal to the product of the sum of the straight lines *A*, *B* and the square of the straight line *D*, plus the product of *A* and the square of the straight line *D*. The product of *B* and the product of *C* and *D* and the squares of the straight lines *C* and *D*, plus the product of *A* and the square of the straight line *C*, have a sum greater than the product of the sum of the straight lines *A*, *B* and the square of the straight line *D*, plus the product of *A* and the square of the straight line *D*. We remove on both sides the product of the sum of the straight lines *A*, *B* and the square of the straight line *D*, the remainder – which is the solids formed as the product of *A* and the square of the straight line *C*, as the product of *B* and the sum of the squares of the straight lines *C*, *D*, and as the product of *C* and *D*, from which we subtract the solid formed as the product of the sum of the straight lines *A*, *B* and the square of the straight line *D* – is greater than the solid formed as the product of *A* and the square of the straight line *D*. That is what we wanted to prove.

– **22** – If three numbers are successive, then the product of the greater and the middle is equal to the square of the smaller, increased by the smaller and by twice the middle.

Let *AB*, *CD* and *E* be three successive numbers, where *AB* is the greatest.

I say that the product of AB *and* CD *is equal to the square of* E, *increased by the number* E *and by twice the number* CD.

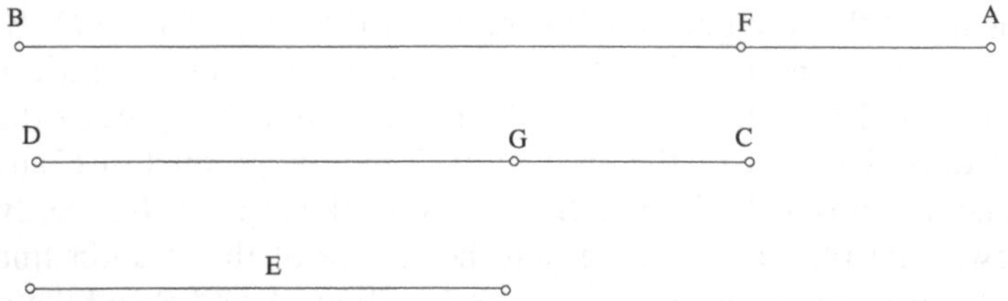

Fig. II.2.22

Proof: If we let *BF* be equal to *CD*, then *AF* is one and the product of *AB* and *BF* is equal to the product of *AF* and *FB*, plus the square of the number *BF*. But, on the one hand, the product of *AF* and *FB* is *FB* as *AF* is one and, on the other, the square of the number *FB* is equal to the square of the number *CD*; thus the product of *AB* and *CD* is equal to the square of the number *CD*, increased by the number *CD*. Likewise, if we let *DG* be equal to *E*, *CG* is one and the square of the number *CD* is equal to the product of *CD* and *DG* and *CG*. But the product of *CD* and *DG* is equal to the square of *DG*, increased by the product of *DG* and *GC*; thus the square of the number *CD* is equal to the product of *CG* and *CD* and *DG*, plus the square of *DG*. But the product of *CG* and *CD* and *DG* is equal to *CD* and *DG*. The square of the number *CD* is thus equal to the square of *DG*, increased by the sum of *CD* and of *DG*. But *DG* is equal to *E*; thus the square of the number *CD* is equal to the square of *E*, plus the numbers *CD* and *E*. But we have shown that the product of *AB* and *CD* is equal to the square of the number *CD*, increased by the number *CD*; thus the product of *AB* and *CD* is equal to the square of *E*, increased by the number *E* and by twice the number *CD*. That is what we wanted to prove.

– **23** – If three numbers are successive, then the sum of the square of the greatest and of the square of the smallest is equal to the product of the sum of the greatest and of the smallest, and the middle, increased by two.

Let *AB*, *CD* and *E* be three successive numbers and let *AB* be the greatest.

I say that the sum of the squares of the numbers AB *and* E *is equal to the product of the sum of* AB, E *and* CD, *increased by two.*

Proof: If we let *BF* be equal to *CD*, then *AF* is one and the square of the number *AB* is equal to the product of *AB* and *BF* and *AF*; the square of the number *AB* is thus equal to the product of *AB* and *BF*, plus the product of *AB* and one. Yet, on the one hand, *BF* is equal to *CD* and, on the other, the product of *AB* and one is equal to *AB*. The square of the number *AB* is thus equal to the product of *AB* and *CD*, plus the number *AB*.

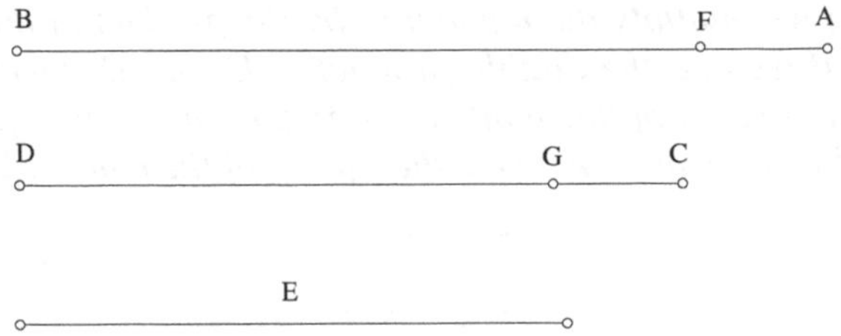

Fig. II.2.23

Likewise, if we let *DG* be equal to *E*, then *CG* is one and the product of *CD* and *DG* is equal to the square of *DG*, plus the product of *DG* and *GC*. Regarding *GC*, it is the unit. Yet, *DG* is equal to *E* and its square is equal to the square of *E*; thus the product of *CD* and *E* is equal to the square of *E*, plus the product of the unit and *E*, which is equal to *E*. But we have shown that the square of the number *AB* is equal to the product of *AB* and *CD*, plus the number *AB*; thus if we add the squares of the numbers *AB* and *E*, the sum will be equal to the product of *AB* and of *E* and *CD*, plus the excess of the number *AB* over *E*. But the excess of the number *AB* over *E* is two as the excesses of the numbers *AB*, *CD* and *E*, over each other, taken in succession, are always one. If we add the squares of the numbers *AB* and *E*, the sum will be equal to the product of *AB* and of *E* and *CD*, increased by two. That is what we wanted to prove.

– **24** – Consider more than two successive numbers beginning with one and, in equal number, successive odd numbers beginning with one which are associated with them; if we take among the successive numbers, three successive numbers, whichever these three <numbers> may be, if we multiply the odd number associated with the middle number of these three <numbers> by the product of the sum of the smallest of the numbers, among them, and of the greatest number and the middle number, and if we add to the result the double-product of the middle and the greater, then the sum is greater than the product of this odd number associated with the middle number and the sum of the square of the small <number> and the square of the great <number>, increased by twice the square of the small <number>.

Let *A*, *B*, *C*, *D* be more than two successive numbers beginning with one, and in equal number; let *E*, *F*, *G*, *H*, be successive odd numbers beginning with one and which are associated with them. Let us take among the numbers *A*, *B*, *C*, *D* three successive numbers, whichever these three <numbers> may be, let the numbers be *B*, *C*, *D*.

I say that if we multiply the number G *by the product of the sum of* B, D *and* C, *and if we add the double-product of* C *and* D, *then the sum is greater than the product of the number* G *and the sum of the squares of the numbers* B *and* D, *increased by twice the square of the number* B.

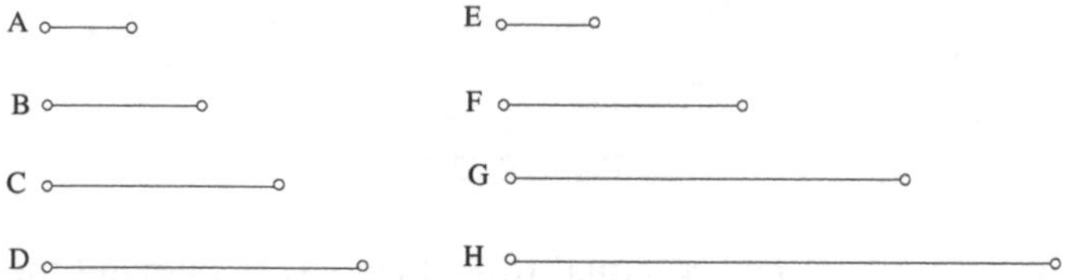

Fig. II.2.24

Proof: The numbers *A*, *B*, *C*, *D* are successive beginning with one; if we take numbers in equal number to that of the numbers *A*, *B*, *C*, *D* and such that each of them is twice its homologue among the numbers *A*, *B*, *C*, *D*, then the chosen numbers are successive even numbers beginning with two and each of them exceeds by a unit its homologue among the successive odd numbers beginning with one, which are *E*, *F*, *G*, *H*. Thus twice the number *C* is greater than the number *G*; this is why twice the number *C*, increased by the number *B*, is much greater than the number *G*. If we set the square of the number *B* on both sides, then the sum of the number *B*, twice the number *C* and the square of the number *B* is greater than the square of the number *B*, plus the number *G*. But the sum of the number *B*, twice the number *C* and the square of the number *B* is equal to the product of *C* and *D*, since the numbers *B*, *C*, *D* are successive. The product of *C* and *D* is thus greater than the square of the number *B*, increased by the number *G*. This is why twice the product of *C* and *D* is greater than twice the square of the number *B*, plus twice the number *G*. But twice the number *G* is the product of the number *G* and two, thus the double-product of *C* and *D* is greater than twice the square of the number *B*, plus the product of the number *G* and two. If we set the product of the number *G* and the product of the sum of the two <numbers> *B*, *D* and *C* on both sides, then the product of the number *G* and the product of the sum of *B*, *D* and *C* and the double-product of *C* and *D* have a sum greater than the product of the number *G* and the sum of two and the product of the sum of *B*, *D* and *C*, plus twice the square of the number *B*. But the product of the sum of *B*, *D* and *C*, plus two, is equal to the sum of the squares of the numbers *B* and *D*, as the numbers *B*, *C*, *D* are successive. The product of the number *G* and the product of the sum of *B*, *D* and *C*, plus the double-product of *C* and *D* is thus greater than the product of the number *G* and the sum of the squares of the numbers *B* and *D*, plus twice the square of the number *B*. That is what we wanted to prove.

– **25** – Consider more than two successive numbers beginning with one and, in equal number, successive odd numbers beginning with one and which are associated with them; if we take, among the successive numbers, three successive numbers, whichever these three <numbers> may be, if we multiply the odd number associated with the middle number, among these three numbers, by the sum of the square of the smallest of them, the square of the middle and the product of the smallest and the middle, if we add to this product the product of the odd number associated with the greatest of the three <numbers> and the sum of the square of the greatest number, the square of the middle number and the product of the middle and the greatest, then the sum obtained is greater than the product of the sum of the odd number associated with the middle number and the odd number associated with the great <number> with the sum of the squares of the three numbers.

Let *A*, *B*, *C*, *D* be more than two successive numbers beginning with one and, in equal number, let *E*, *F*, *G*, *H*, be successive odd numbers beginning with one, which are associated with them; let us take, among the numbers *A*, *B*, *C*, *D*, three successive numbers, whichever these three <numbers> may be, let the numbers be *B*, *C*, *D*.

I say that if we multiply the number G *by the sum of the squares of the numbers* B, C *and by the product of* B *and* C, *and if we add to the result the product of* H *and the sum of the squares of the numbers* C, D, *and the product of* C *and* D, *the result is greater than the product of the sum of the numbers* G, H *and the sum of the squares of the numbers* B, C, D.

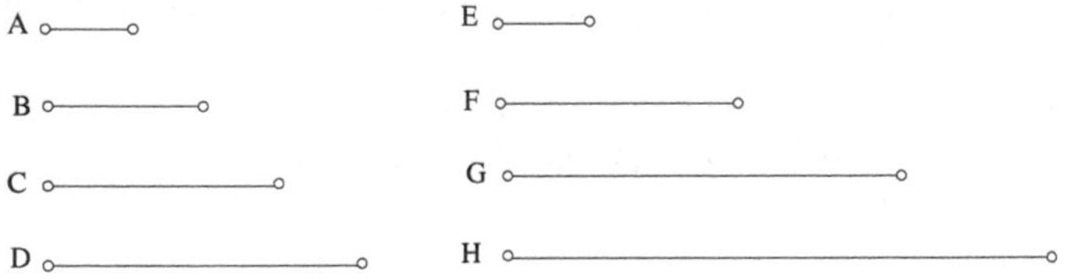

Fig. II.2.25

Proof: The numbers *A*, *B*, *C*, *D* are successive beginning with one and the numbers *E*, *F*, *G*, *H* are successive odd numbers beginning with one; thus the product of the number *G* and the product of the sum of *B*, *D* and *C*, if we add to it twice the product of *C* and *D*, is greater than the product of the number *G* and the squares of the numbers *B* and *D*, increased by twice the square of the number *B*. But, on the one hand, the double-product of *C* and *D* is equal to the product of two and the product of *C* and *D* and, on the other, twice the square of the number *B* is equal to the product of two and the square of the number *B*. The product of the number *G* and the product of *B* and *C* plus the product of the number *G*, increased by two, and the product of *C* and *D* is greater than the product of the number *G* and the square of the number *D*, plus the product of the number *G*, plus two, and the square of the number *B*. But the number *H* is equal to the number *G*, plus two, as the numbers *G* and *H* are successive odd numbers, thus the product of the number *G* and the product of *B* and *C*, plus the product of the number *H* and the product of *C* and *D*, is greater than the product of the number *G* and the square of the number *D*, plus the product of the number *H* and the square of the number *B*. If we set the product of the number *G* and the square of the number *B*, plus the product of the number *H* and the square of the number *D* on both sides, the product of the number *G* and the sum of the square of the number *B* and of the product of *B* and *C*, plus the product of the number *H* and the sum of the product of *C* and *D*, and the square of the number *D*, will be greater than the product of the sum of the number *G* and the number *H* and the sum of the squares of the numbers *B* and *D*. If we set the product of the sum of the numbers *G*, *H* and the square of the number *C* on both sides, the product of the number *G* and the sum of the squares of the numbers *B*, *C* and of the product of *B* and *C*, plus the product of the number *H* and the sum of the squares of the numbers *C*, *D* and the product of *C* and *D*, will be greater than the product of the sum of the numbers *G*, *H* and the sum of the squares of the numbers *B*, *C*, *D*. That is what we wanted to prove.

– **26** – Let there be three successive numbers, if one is added to the product of the smallest and the greatest, the sum will be equal to the square of the middle number.

Let *AB*, *C*, *D*, be three successive numbers.

I say that the product of AB *and* D, *increased by one, is equal to the square of the number* C.

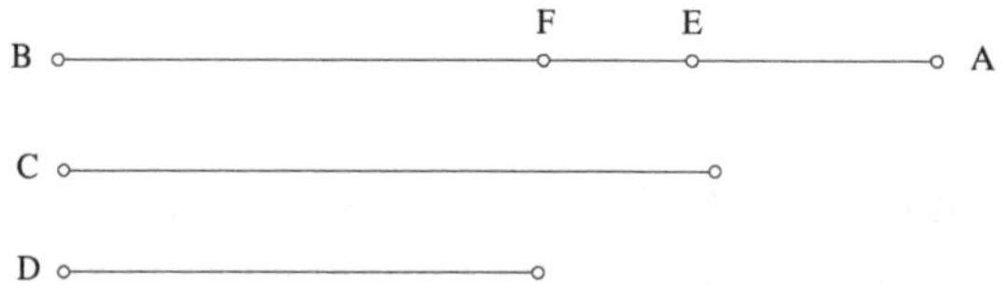

Fig. II.2.26

Proof: If we let *BE* be equal to *C* and *BF* equal to *D*, then each <of the numbers> *AE* and *EF* is equal to one; but the square of the number *BE* is equal to the sum of the squares of *BF* and *FE*, plus the double-product of *BF* and *FE*. But the double-product of *BF* and *FE* is equal to the product of *BF* and *FA*; thus the square of the number *BE* is equal to the sum of the squares of *BF* and *FE*, plus the product of *BF* and *FA*. On the one hand, the product of *BF* and *FA*, plus the square of *BF*, is equal to the product of *AB* and *BF* and, on the other, the square of *FE* is equal to one; thus the square of the number *BE* is equal to the product of *AB* and *BF*, plus one. Yet, on the one hand, *BF* is equal to *D* and, on the other, *BE* is equal to *C*; thus the product of *AB* and *D*, increased by one, is equal to the square of the number *C*. That is what we wanted to prove.

– **27** – Consider more than two successive numbers beginning with one and, in equal number, successive odd numbers beginning with one which are associated with them; if we take, among the successive numbers, three numbers which follow each other, whichever these three <numbers> may be, if we multiply the odd number associated with the middle number among the three <numbers>, by the sum of the square of the smallest <number> among them, the square of the middle <number> and the product of the smallest and the middle, if we add to the result the product of the odd number associated with the greatest of the three <numbers> and the sum of the square of the greatest number, the square of the middle number and the product of the greatest and the middle, if we subtract from the result the product of the sum of the two odd numbers associated with the middle number and with the greatest number and the sum of the square of the smallest <number>, the square of the greatest <number> and the product of the smallest and the greatest, then the remainder will be greater than the difference between the square of the greatest <number> and the square of the smallest.

Let *A*, *B*, *C*, *D*, be more than two successive numbers beginning with one and, in equal number, let the numbers *E*, *F*, *G*, *H* be successive odd numbers beginning with one, which are associated with them; let us take, among them, three numbers which follow each other, whichever these three <numbers> may be; let the numbers be *B*, *C*, *D*.

I say that if we multiply the number G *by the sum of the squares of the numbers* B, C *and the product of* B *and* C, *if we add to the result the product of the number* H *and the sum of the squares of the numbers* C, D, *and the product of* C *and* D, *and if we subtract from the result the product of the sum of the numbers* G, H *and the sum of the squares of the numbers* B, D, *and the product of* B *and* D, *then the remainder is greater than the difference between the squares of the numbers* D *and* B.

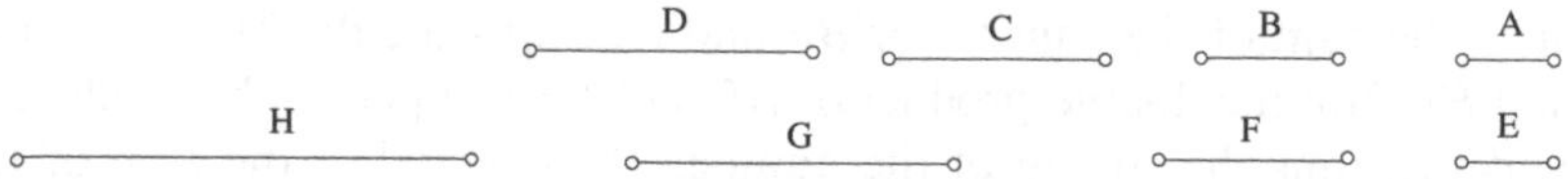

Fig. II.2.27

Proof: The numbers *A*, *B*, *C*, *D* are successive beginning with one and the numbers *E*, *F*, *G*, *H* are successive odd numbers beginning with one; if we multiply the number *G* by the sum of the squares of the numbers *B* and *C* and the product of *B* and *C*, and if we add to the result the product of *H* and the sum of the squares of the numbers *C* and *D* and the product of *C* and *D*, the result is greater than the product of the sum of *G*, *H* and the sum of the squares of the numbers *B*, *C*, *D*.[40] But the square of the number *C* is equal to the product of *B* and *D*, increased by one, as the numbers *B*, *C*, *D* are successive; thus the product of the number *G* and the sum of the squares of the numbers *B*, *C*, and the product of *B* and *C*, increased by the product of the number *H* and the sum of the squares of the numbers *C*, *D* and the product of *C* and *D*, is greater than the product of the sum of the numbers *G*, *H* and the sum of the squares of the numbers *B*, *D* and the product of *B* and *D*, plus the product of the sum of the numbers *G*, *H* and the unit. If we commonly subtract from the two sums the product of the sum of the numbers *G*, *H* and the sum of the squares of the numbers *B*, *D* and the product of *B* and *D*, then the product of the number *G* and the sum of the squares of the two numbers *B*, *C* and the product of *B* and *C*, if we add to it the product of the number *H* and the sum of the squares of the numbers *C*, *D* and the product of *C* and *D*, and if we subtract from the sum the product of the sum of the numbers *G*, *H* and the sum of the squares of the numbers *B*, *D* and the product of *B* and *D*, then the result is greater than the product of the sum of the two numbers *G*, *H* and the unit, which is equal to the sum of *G*, *H*. But the sum of the numbers *G* and *H* is equal to the difference of the squares of the numbers *D* and *B*, as is proven in proposition three of our treatise *On the Measurement of the Parabola*. The product of the number *G* and the sum of the squares of the numbers *B*, *C* and the product of *B* and *C*, if we add to it the product of the number *H* and the sum of the squares of the numbers *C*, *D* and the product of *C* and *D*, and if we subtract from the result the product of the sum of the numbers *G*, *H* and the sum of the squares of the numbers *B*, *D* and the product of *B* and *D*, the remainder is thus greater than the difference of the squares of the numbers *D* and *B*. That is what we wanted to prove.

[40] By Proposition 25.

– **28** – Consider more than two straight lines following the ratios of the successive numbers beginning with one and, in equal number, straight lines following the ratios of the successive odd numbers beginning with one and which are associated with them; if the smallest of the straight lines which are following the ratios of the successive numbers is equal to the smallest of the straight lines which are following the ratios of the odd numbers, if we take three straight lines which follow each other, among the straight lines which are following the ratios of the successive numbers, whichever these three <straight lines> may be, if we multiply the straight line associated with the middle straight line among the three <straight lines>, by the sum of the square of the smallest straight line among them, the square of the middle straight line and their product with each other, if we add to the result the product of the straight line associated with the greatest straight line of the three and the sum of the square of the greatest straight line, the square of the middle straight line and their product with each other and if we subtract from the result the product of the sum of the two straight lines associated with the middle and the greatest and the sum of the square of the smallest straight line, the square of the greatest straight line and their product with each other, then the remainder is greater than the product of the smallest of the straight lines which are following the ratios of the odd numbers and the difference between the square of the greatest of the three straight lines and the square of the smallest.

Let A, B, C, D be straight lines following the ratios of the successive numbers E, F, G, H beginning with one; let, in equal number, I, K, L, M be the straight lines which are associated with them, and which are following the ratios of the successive odd numbers N, S, O, P, beginning with one. Let one take three straight lines which follow each other, among the straight lines A, B, C, D, whichever these three <straight lines> may be, namely B, C, D and let A be equal to I.

I say that if, to the product of the straight line L *and the sum of the squares of the straight lines* B, C *and the product of* B *and* C, *we add the product of the straight line* M *and the sum of the squares of the straight lines* C, D *and the product of* C *and* D *and if we subtract from the sum the product of the sum of the straight lines* L, M *and the sum of the squares of the straight lines* B, D *and the product of* B *and* D, *the remainder is greater than the product of the straight line* I *and the difference between the squares of the straight lines* B *and* D.

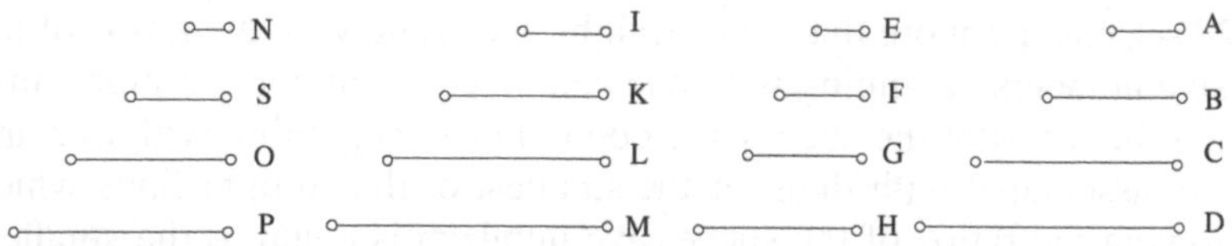

Fig. II.2.28

Proof: The ratios of the straight lines *A*, *B*, *C*, *D* to each other are equal to the ratios of the numbers *E*, *F*, *G*, *H* to each other and the ratios of the straight lines *I*, *K*, *L*, *M* to each other are equal to the ratios of the numbers *N*, *S*, *O*, *P* to each other. The ratio of *A* to *I* is equal to the ratio of *E* to *N*, as *A* is equal to *I*; thus the ratio of each of the straight lines *A*, *B*, *C*, *D* to each of the straight lines *I*, *K*, *L*, *M* is equal to the ratio of its homologue, among the numbers *E*, *F*, *G*, *H*, to the homologue of the other among the numbers *N*, *S*, *O*, *P*. This is why the ratio of the product of the straight line *L* and the sum of the squares of the straight lines *B* and *C* and the product of *B* and *C* to the cube of the straight line *C* is equal to the ratio of the product of the number *O* and the sum of the squares of the numbers *F* and *G* and the product of *F* and *G*, to the cube of the number *G*. This is also why the ratio of the product of the straight line *M* and the sum of the squares of the straight lines *C* and *D* and the product of *C* and *D* to the cube of the straight line *C* is equal to the ratio of the product of the number *P* and the sum of the squares of the numbers *G* and *H* and the product of *G* and *H* to the cube of the number *G*. If we add them up, the ratio of the product of the straight line *L* and the sum of the squares of the straight lines *B* and *C* and the product of *B* and *C*, to which we add the product of the straight line *M* and the sum of the squares of the straight lines *C* and *D* and the product of *C* and *D* to the cube of the straight line *C*, is equal to the ratio of the product of the number *O* and the sum of the squares of the numbers *F* and *G* and the product of *F* and *G*, to which one adds the product of the number *P* and the sum of the squares of the numbers *G* and *H* and the product of *G* and *H* to the cube of the number *G*.

Likewise, we show that the ratio of the product of the sum of the straight lines *L*, *M* and the sum of the squares of the straight lines *B* and *D* and the product of *B* and *D* to the cube of the straight line *D* is equal to the ratio of the product of the sum of the numbers *O* and *P* and the sum of the squares of the numbers *F* and *H* and the product of *F* and *H* to the cube of the number *H*. But the ratio of the cube of the straight line *D* to the cube of the straight line *C* is equal to the ratio of the cube of the number *H* to the cube of the number *G*. By the ratio of equality, the ratio of the product of the sum of the straight lines *L*, *M* and the sum of the squares of the straight lines B, D and the product of B and D to the cube of the straight line C is equal to the ratio of the product of the sum of the numbers O and P and the sum of the squares of the numbers F, H and the product of F and H to the cube of the number G. But we have shown that the ratio of the product of the straight line L and the sum of the squares of the straight lines B, C and the product of B and C, if we add to it the product of the straight line M and the sum of the squares of the straight lines C, D and the product of C and D, to the cube of the straight line C, is equal to the ratio of the product of the number O and the sum of the squares of the numbers F and G and the product of F and G, if we add to it the product of the number P and the sum of the squares of the numbers G, H and the product of G and H to the cube of the number G. Thus the ratio of the excess of the product of the straight line L and the sum of the squares of the straight lines B, C and the product of B and C, if we add to it the product of the straight line M and the sum of the squares of the straight lines C, D and the product of C and D over the product of the sum of the straight lines L, M and the sum of the squares of the straight lines B, D and the product of B and D, to the cube of the straight line C, is equal to the ratio of the excess of the product of the number O and the sum of the squares of the numbers F, G and the product of F and G if we add to it the product of the number P and the sum of the squares of the numbers G, H and the product of G and H over the product of the sum of the numbers O, P and the sum of the squares of the numbers F, H and the product of F and H, to the cube of the number G. But the ratio of the cube of the straight line C to the product of I and the difference between the squares of the straight lines B, D is equal to the ratio of the cube of the number G to the product of N and the difference of the squares of the numbers F and H, since the ratio of the base which is the square of the straight line C to the base which is the difference between the squares of the straight lines B and D is equal to the ratio of the base, which is the square of the number G, to the base which is the difference between the squares of the numbers F and H, and since the ratio of the height which is the straight line C, to the height which is the straight line I, is equal to the ratio of the height which is the number G, to the height which is <the number> N. By the ratio of equality, the ratio of the product of the straight line L and the sum of the squares of the straight lines B, C and the product of B and C, if we add to it the product of the straight line M and the sum of the squares of the straight lines C, D and the product of C and D, and if we subtract from the result the product of the sum of the straight lines L, M and the sum of the squares of the straight lines B, D and the product of B and D, to the product of the straight line I and the difference of the squares of the straight lines B and D, is thus equal to the ratio of the product of the number O and the sum of the squares of the numbers F and G and the product of F and G, if we add to it the product of the number P and the sum of the squares of the numbers G, H and the product of G and H and if we subtract from the result the product of the sum of the numbers O, P and the sum of the squares of the numbers F, H and the product of F and H to the product of N and the difference between the squares of the numbers F and H. But if, to the product of the number O and the sum of the squares of the numbers F, G and the product of F and G, we add the product of the number P and the sum of the squares of the numbers G, H and the product of G and H and if we subtract from the sum the product of the sum of the numbers O, P and the sum of the squares of the numbers F, H and the product of F and H, the remainder is greater than the product of N and the difference of the squares of the numbers F and H, as N is the unit and its product by the difference between the squares of the numbers F and H is equal to the difference between these two squares. If, thus, to the product of the straight line L and the sum of the squares of the straight lines B, C and the product of B and C, we add the product of the straight line M and the sum of the squares of the straight lines C, D and the product of C and D and if we subtract from it the product of the sum of the straight lines L, M and the sum of the squares of the straight lines B, D and the product of B and D, the remainder is greater than the product of I and the difference between the squares of the straight lines B and D. That is what we wanted to prove.

– **29** – Consider more than two straight lines following the ratios of the successive numbers beginning with one and, in equal number, straight lines following the ratios of the successive odd numbers beginning with one, which are associated with them, and if the smallest of the straight lines which are following the ratios of the successive numbers is not equal to the smallest of the straight lines which are following the ratios of the odd numbers, if we take three straight lines which follow each other among the straight lines which are following the ratios of the successive numbers, whichever these three <straight lines> may be, if we multiply the straight line associated with the middle straight line, among the three <straight lines>, by the sum of the square of the smallest among them, and the square of the middle straight line and their product with each other, and if we add to the sum the product of the straight line associated with the greatest straight line of the three and the sum of the square of the greatest straight line, the square of the middle straight line and their product by each other and if we subtract from the sum the product of the sum of the two associated straight lines, the middle and the greatest, and the sum of the square of the smallest straight line, the square of the greatest straight line and their product with each other, then the remainder is greater than the product of the smallest of the straight lines which are following the ratios of the odd numbers and the difference between the squares of the greatest of the three straight lines and the smallest among them.

Let *A*, *B*, *C*, *D* be straight lines following the ratios of the successive numbers beginning with one; let *E*, *F*, *G*, *H* be in equal number, straight lines which are associated with them and which are following the ratios of the successive odd numbers beginning with one and such that *A* is not equal to *E*. Let us take three straight lines which follow each other among the straight lines *A*, *B*, *C*, *D*, whichever these three <straight lines> may be, let them be *B*, *C*, *D*.

I say that if, to the product of the straight line G *and the sum of the squares of the straight lines* B, C *and the product of* B *and* C, *we add the product of <the straight line>* H *and the sum of the squares of the straight lines* C, D *and the product of* C *and* D, *and if we subtract from it the product of the sum of the straight lines* G, H *and the sum of the squares of the straight lines* B, D *and the product of* B *and* D, *the remainder is greater than the product of the straight line* E *and the difference between the squares of the straight lines* B *and* D.

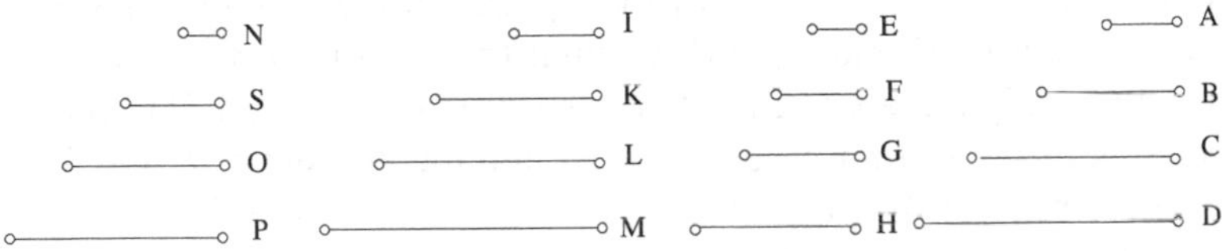

Fig. II.2.29

Proof: If we let the straight line *I* be equal to the straight line *E* and if we let the ratios of the straight lines *I*, *K*, *L*, *M* to each other, taken in succession, be equal to the ratios of the straight lines *A*, *B*, *C*, *D*, to each other, taken in succession, then the ratios of the straight lines *I*, *K*, *L*, *M*, to each other, taken in succession, are equal to the ratios of the successive numbers beginning with one, and the straight line *E* is equal to the straight line *I*. If, to the product of the straight line *G* and the sum of the squares of the straight lines *K* and *L* and the product of *K* and *L*, we add the product of the straight line *H* and the sum of the squares of the straight lines *L*, *M* and the product of *L* and *M* and if, from the result, we subtract the product of the sum of the straight lines *G*, *H* and the sum of the squares of the straight lines *K*, *M* and the product of *K* and *M*, the remainder is greater than the product of the straight line *E* and the difference between the squares of the straight lines *K* and *M*.[41] But the ratio of the difference between the squares of the straight lines *B* and *D* to the difference between the squares of the straight lines *K* and *M* is equal to the ratio of the square of the straight line *B* to the square of the straight line *K*, as the ratios of the straight lines *I*, *K*, *L*, *M* to each other, taken in succession, are equal to the ratios of the straight lines *A*, *B*, *C*, *D* to each other, taken in succession. Thus the ratio of the product of the straight line *E* and the difference between the squares of the straight lines *B* and *D* to its product and the difference between the squares of the straight lines *K* and *M* is equal to the ratio of the square of the straight line *B* to the square of the straight line *K*. But the ratio of the square of the straight line *B* to the square of the straight line *K* is equal to the ratio of the square of the straight line *C* to the square of the straight line *L*; it is equal to the ratio of the square of the straight line *D* to the square of the straight line *M*; it is equal to the ratio of the product of *B* and *C* to the product of *K* and *L*; it is equal to the ratio of the product of *C* and *D* to the product of *L* and *M* and it is equal to the ratio of the product of *B* and *D* to the product of *K* and *M*. Thus the ratio of the product of *E* and the difference between the squares of the straight lines *B* and *D* to its product and the difference between the squares of the straight lines *K* and *M* is equal to the ratio of the sum of the squares of the straight lines *B* and *C* and the product of *B* and *C* to the sum of the squares of the straight lines *K* and *L* and the product of *K* and *L*, and is equal to the ratio of the sum of the squares of the straight lines *C* and *D* and the product of *C* and *D* to the sum of the squares of the straight lines *L* and *M* and the product of *L* and *M* and is equal to the ratio of the sum of the squares of the straight lines *B* and *D* and the product of *B* and *D* to the sum of the squares of the straight lines *K* and *M* and the product of *K* and *M*. But if a straight line is multiplied by two surfaces, then the ratio of the solid formed as its multiplication by the one to the solid formed as its multiplication by the other is equal to the ratio of the first of the surfaces to the second. Thus the ratio of the product of *E* and the difference between the squares of the straight lines *B* and *D* to its product and the difference of the squares of the straight lines *K* and *M* is equal to the ratio of the product of the straight line *G* and the sum of the squares of the straight lines *B*, *C* and the product of *B* and *C* to the product of the straight line *G* also and the sum of the squares of the straight lines *K*, *L* and the product of *K* and *L*, and is equal to the ratio of the product of the straight line *H* and the sum of the squares of the straight lines *C*, *D* and the product of *C* and *D* to the product of the straight line *H* and, also, the sum of the squares of the straight lines *L*, *M* and the product of *L* and *M*, and is equal to the ratio of the product of the sum of the straight lines *G*, *H* and

[41] By Proposition 28.

the sum of the squares of the straight lines B, D and the product of B and D to the product of the sum of the two straight lines G, H and, also, the sum of the squares of the straight lines K and M and the product of K and M. If we thus take the product of G and the sum of the squares of the straight lines B and C and the product of B and C, if we then add to it the product of H and the sum of the squares of the straight lines C, D and the product of C and D and if we subtract from the sum the product of the sum of the two straight lines G, H and the sum of the squares of the straight lines B, D and the product of B and D, then the ratio of the remainder to that which remains – if we take the product of G and the sum of the straight lines K and L and the product of K and L, if we add to it the product of H and the sum of the squares of the straight lines L, M and the product of L and M and if we subtract from the sum the product of the sum of the two straight lines G and H and the sum of the squares of the straight lines K, M and the product of K and M – is equal to the ratio of the product of E and the difference between the squares of the straight lines B and D to the product of E and the difference between the squares of the straight lines K and M. But if we permute, they will be also proportional. But we have shown that if, to the product of G and the sum of the squares of the straight lines K, L and the product of K and L, we add the product of H and the sum of the squares of the straight lines L, M and the product of L and M and if we subtract from the sum the product of the sum of the straight lines G, H and the sum of the squares of the straight lines K, M and the product of K and M, that which remains is greater than the product of E and the difference between the squares of the straight lines K and M. If, to the product of G and the sum of the squares of the straight lines B, C and the product of B and C, we add the product of H and the sum of the squares of the straight lines C, D and the product of C and D and if we subtract from the sum the product of the sum of the straight lines G, H and the sum of the squares of the straight lines B, D and the product of B and D, the remainder will be greater than the product of E and the difference between the squares of the straight lines B and D. That is what we wanted to prove.

– **30** – If three magnitudes are such that each has a ratio with each of its two associates and such that the first is the smallest and the third the greatest, we may then find successive magnitudes following the ratio of the first to the second beginning with the first and ending to a magnitude greater than the third.

Let the three magnitudes, such that each has a ratio with each of its two associates, be the magnitudes A, BC, DE let the smallest be A and the greatest DE.

I say that we can find successive magnitudes following the ratio of A *to* BC *beginning with* A *and ending to a magnitude greater than* DE.

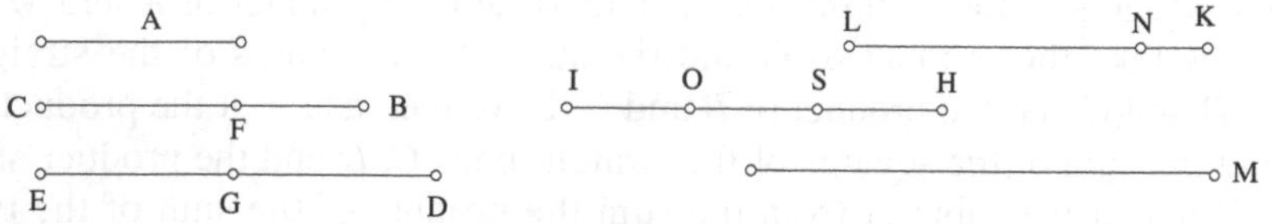

Fig. II.2.30

Proof: If we let the excess of the magnitude *BC* over the magnitude *A* be equal to the magnitude *BF* and the excess of the magnitude *DE* over the magnitude *A* be equal to the magnitude *DG*, the magnitude *BF* has a ratio to the magnitude *DG*; it is then possible by sufficiently multiplying *BF* by itself, that it exceeds *DG*. If we suppose that this multiple which exceeds the magnitude *DG* is of the magnitude *HI* and if we let the ratio of *KL* to *BC* be equal to the ratio of *BC* to *A*, and equal to the ratio of *M* to *KL* and if we continue to proceed thus up to where the number of the magnitudes *BC*, *KL* and *M* is equal to the number of times *BF* is in the magnitude *HI*; if we divide *HI* into that many times *BF*, which is to say into parts *HS*, *SO*, *OI*, if we let *KN* be equal to the excess of *KL* over *BC* – the ratio of *A* to *BC* being equal to the ratio of *BC* to *KL* – if we subtract the two smallest from the two greatest, then the ratio of the remainder, which is *BF*, to the remainder, which is *KN*, is equal to the ratio of *A* to *BC*. But the magnitude *A* is smaller than the magnitude *BC*, thus the magnitude *KN*, which is the excess of *KL* over *BC*, is greater than the magnitude *BF* which is the excess of *BC* over *A*. Likewise, we also show that the excess of *M* over *KL* is greater than the excess of *KL* over *BC* and much greater than the excess of *BC* over *A*. Thus, on the one hand, *BF*, which is the excess of *BC* over *A*, is equal to *HS* and, on the other, each of the excesses of *KL* over *BC* and *M* over *KL* is greater than *SO* and *OI* respectively. But the number of the excesses is equal to the number of parts of the straight line *HI*. If we add them up, then the excess of *M* over *A* will be greater than *HI*. But *HI* is greater than *DG*; thus the excess of *M* over *A* is much greater than *DG* and the magnitude *A* is equal to the magnitude *GE*; thus the excess of the magnitude *M* over the magnitude *A*, plus the magnitude *A*, is greater than the magnitude *DE*. But the excess of the magnitude *M* over the magnitude *A*, plus the magnitude *A*, is equal to the magnitude *M*; thus the magnitude *M* is greater than the magnitude *DE*. The magnitudes *A*, *BC*, *KL*, *M* are successive following the ratio of *A* to *BC*. That is what we wanted to prove.

– **31** – If two magnitudes are such that the one is smaller than the other and two other magnitudes such that the first[42] is smaller than the greater of the first two magnitudes, if we subtract from the greater of the first two magnitudes, a magnitude whose ratio to it is not less than the ratio of the smallest of the two other magnitudes to the greater, if we subtract from the remainder a magnitude whose ratio to it is also not less than the ratio of the smaller of the two latter magnitudes to the greater, and if we then continue to proceed thus with that which remains, then there will remain of the greater magnitude, a magnitude smaller than the smaller magnitude.

Let *AB* and *CD* be two magnitudes, such that *AB* is greater than *CD*, and let *E* and *FG* be two other magnitudes such that *E* is smaller than *FG*, and *FG* smaller than *AB*.

I say that if we subtract from AB *a magnitude whose ratio to it is not less than the ratio of* E *to* FG *and if we subtract from the remainder a magnitude whose ratio to it is also not less than this ratio, and if we then continue to proceed thus with that which remains, then there will remain of* AB *a magnitude smaller than the magnitude* CD.

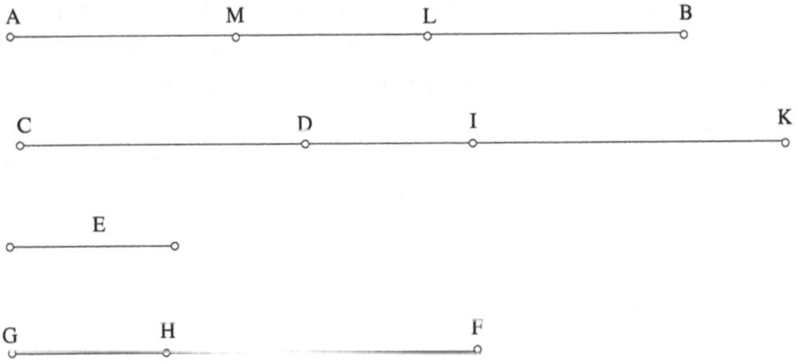

Fig. II.2.31

Proof: If we separate from *FG*, *HG* equal to *E*, and if we let the ratio of *ID* to *DC* be equal to the ratio of *GH* to *HF*, then either *CI* will be greater than *AB* or it will not be greater that it. If *CI* is greater that it, *then[43] the ratio of *AB* to *CD* is smaller than the ratio of *CI* to *CD*. But the ratio of *CI* to *CD* is equal to the ratio of *FG* to *FH*; thus the ratio of *AB* to *CD* is smaller than the ratio of *FG* to *FH*. If we let the ratio of *BL* to *AB* be equal to the ratio of *GH* to *GF*, then the ratio of *AB* to *AL* is equal to the ratio of *FG* to *FH*; thus the ratio of *AB* to *CD* is smaller than the ratio of *AB* to *AL*; thus *AL* is smaller than *CD*. That is what we wanted.*

[42] It consists of the greater, as we see in the example.

[43] *…* The paragraph between the two asterisks renders the Arabic text reconstituted by us.

Otherwise the magnitudes *AB*, *CI* and *CD* have a ratio to each other and the greatest of which is *AB* and the smallest is *CD*; we can thus find successive magnitudes following the ratio of *CD* to *CI*, beginning with *CD* and ending to a magnitude greater than *AB*.[44] If we let *CD*, *CI*, *CK* be these magnitudes, then the ratio of *ID* to *DC* is equal to the ratio of *KI* to *IC* and equal to the ratio of *GH* to *HF*. If we suppose that the ratio of *BL* to *BA* is not smaller than the ratio of *E* to *FG* and likewise, for the ratio of *LM* to *AL*, and if we continue to proceed thus up to where the number of parts *BL*, *LM*, *MA* is equal to the number of the straight lines *KI*, *ID*, *DC*, then the ratio of *BL* to *BA* will not be smaller than the ratio of *E* to *FG*; but *E* is equal to *GH*. If we separate, then the ratio of *BL* to *AL* will not be smaller than the ratio of *GH* to *HF*. But the ratio of *GH* to *HF* is equal to the ratio of *KI* to *IC*; thus the ratio of *BL* to *AL* is not smaller than the ratio of *KI* to *IC*. Likewise, we also show that the ratio of *LM* to *MA* is not smaller than the ratio of *ID* to *DC*. From that, one shows that the ratio of *BM* to *MA* is not smaller than the ratio of *KD* to *DC*.[45] If we compose <the ratios>, then the ratio of *BA* to *AM* is not smaller than the ratio of *KC* to *CD*. If we permute, then the ratio of *BA* to *KC* is not smaller than the ratio of *AM* to *CD*; it is thus either equal to it or it is greater. If it is equal to it – *BA* being smaller than *KC* – then *AM* which remains of *AB* is smaller than *CD*. That is what we wanted.

If it is greater than it, the ratio will be equal to the ratio of a magnitude greater than *AM* to the magnitude *CD*; let this magnitude be *AN*; *AN* will thus be smaller than *CD*, as its ratio to *CD* is equal to the ratio of *AM* to *CK*; thus the magnitude *AM* which remains of *AB* is much smaller than *CD*. That is what we wanted to prove.

– **32** – If we produce in a portion of a parabola its diameter and if we extend in one of its two halves ordinate straight lines to this diameter such that the ratios of the parts of the diameter separated by the ordinate straight lines, to each other, taken in succession, are equal to the ratios of the successive odd numbers beginning with one, to each other, and such that the smallest of these parts is on the side of the vertex of the section, if straight lines join the extremities of the ordinate straight lines which are on the same side and the vertex of the parabola also to the extremity of the smallest of the drawn ordinates, generating in the section a polygonal figure[46] inscribed in half of the portion of the parabola, if we fix the diameter of this portion of a parabola and if we rotate all of the other sides

[44] By Proposition 30.

[45] The author does not give any explication; see the commentary.

[46] Lit.: a figure of rectilinear sides.

of the figure which is in its half from an arbitrary position up to where it returns to its original position, then the solid enclosed by this figure is less than half of the cylinder whose base is the base circle of this figure if its base is a circle, or the base of its lower part if its lower part is the surface of a cone of revolution whose height is equal to the diameter of this portion of the parabola, by two thirds of the solid formed as the product of the diameter of the portion and the circle whose diameter is the perpendicular dropped from the extremity located over the line of the section from one of the two extremities of the ordinates produced in the section over the diameter of the section.

Let *ABC* be half of a portion of a parabola, *BC* its diameter and, in this half of the portion, let *DE*, *FG* and *AC* be ordinate straight lines to the diameter *BC*. Let the ratios of the straight lines *BE*, *EG*, *GC* to each other, taken in succession, be equal to the ratios of the successive odd numbers *H*, *I*, *K* beginning with one, and let *BE* be the smallest of the straight lines; join the straight lines *AF*, *FD* and *DB*.

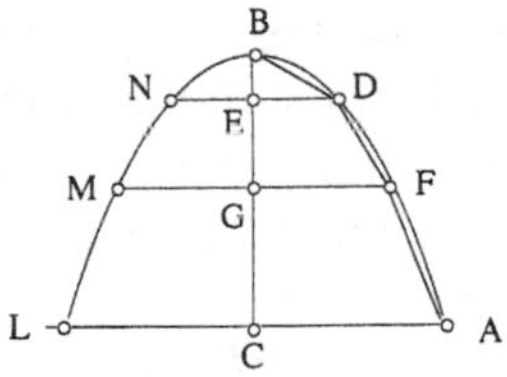

Fig. II.2.32a

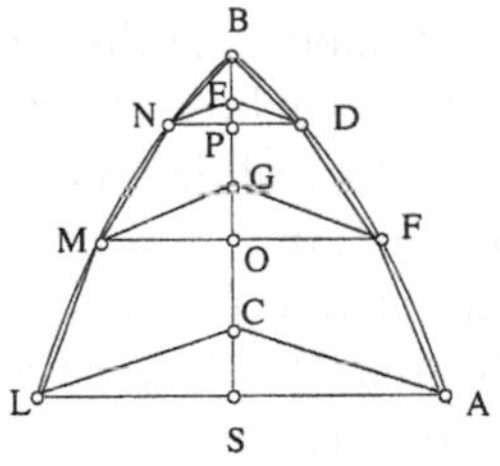

Fig. II.2.32b

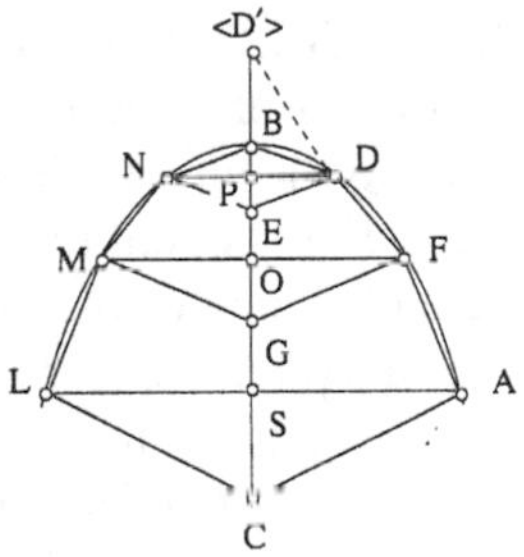

Fig. II.2.32c

I say that if we fix the straight line BC *and if we rotate all of the other sides of the figure* CAFDB, *from an arbitrary position up to its original position,*[47] *then the solid enclosed by that figure is less than half of the cylinder whose base is the base circle of this solid, if its lower part is a circle, or the base circle of its lower part if its lower part is the surface of a cone of revolution, and whose height is equal to the straight line* BC, *by two thirds of the solid formed as the product of* BC *and the circle whose diameter is the perpendicular produced from the point* D *to the diameter* BC.

Proof: The ratio of the square of the straight line *FM* to the circle of diameter *FM* is equal to the ratio of the square of the straight line *AL* to the

[47] Lit.: they have started.

circle of diameter *AL* and is equal to the ratio of the product of *FM* and *AL* to the circle whose square of the diameter is equal to the product of *FM* and *AL* and is equal to the ratio of the square of the straight line *DN* to the circle whose diameter is *DN* and is equal to the ratio of the product of *FM* and *DN* to the circle whose square of the diameter is equal to the product of *FM* and *DN*. Thus one third of the solids formed as the product of *BE* and the circle of diameter *DN*, plus the product of *EG* and the sum of the two circles whose diameters are the straight lines *DN*, *FM* and the circle whose square of the diameter is equal to the product of *DN* and *FM*, plus the product of *GC* and the sum of the circles whose diameters are the straight lines *FM*, *AL* and the circle whose square of the diameter is equal to the product of *FM* and *AL*, and two thirds of the solid formed as the product of *BC* and the circle whose diameter is *DE*, have a sum equal to half of the solid formed as the product of *BC* and the circle whose diameter is *AL*.

Yet, on the one hand, one third of the solid formed as the product of *BE* and the circle of diameter *DN* is equal to the volume of the cone of revolution whose base is the circle of diameter *DN* and whose height is the straight line *BE*.[48] On the other hand, one third of the solid formed as the product of *EG* and the two circles whose diameters are the straight lines *DN*, *FM* and the circle whose square of the diameter is equal to the product of *DN* and *FM* is equal to the frustum of a cone of revolution whose base is the circle of diameter the straight line *FM*, and whose upper surface is the circle of diameter the straight line *DN*. Regarding one third of the product of *GC* and the two circles of diameters the straight lines *FM*, *AL* and the circle whose square of the diameter is equal to the product of *FM* and *AL*, it is equal to the frustum of a cone of revolution whose base is the circle of diameter *AL* and whose upper surface is the circle of diameter *FM*. And the solid that we mentioned from the cone and the two frusta of cones of revolution, if we add them up, is equal to the solid generated when we fix the straight line *BC* and rotate all of the other sides of the figure *CAFDB*. This solid, plus two thirds of the solid formed as the product of *BC* and the circle of diameter the straight line *DE*, is equal to half of the solid whose base is the circle of diameter *AL* and whose height is *BC*. But this solid is the cylinder whose base is the circle of diameter *AL* and whose height is *BC*. The solid generated, when we fix the straight line *BC* and rotate all of the other sides of the figure *CAFDB*, from an arbitrary position up to its original position, is less than half of the cylinder whose base is the circle of diameter *AL* and whose height is the straight line *BC*, by two thirds of the

[48] The author first treats the case where the diameter *BC* is the axis of the parabola (Fig. II.2.32a).

solid formed as the product of *BC* and the circle of diameter *DE* which is perpendicular to *BC*.

Likewise, if we do not suppose the ordinate straight lines to be perpendicular to the diameter *BC* – let the perpendiculars produced from the points *A*, *F*, *D* to the diameter[49] be the perpendiculars *AS*, *FO* and *DP* as in the second and the third case of figure – and if we join the straight lines *LS*, *MO*, *NP*, then the lines *ASL*, *FOM*, *DPN* are straight lines and the angles *DPE* and *FOG* are equal as they are right angles. But the straight line *DE* is parallel to the straight line *FG*; thus the angle *DEP* is equal to the angle *FGO* and there remains the angle *PDE* of the triangle *EDP* equal to the angle *GFO* of the triangle *OFG*. The two triangles *EDP* and *OFG* are thus similar; this is why the ratio of *DE* to *FG* is equal to the ratio of *DP* to *FO*. In the same way, we also show that the ratio of *FG* to *AC* is equal to the ratio of *FO* to *AS*. But the ratios of the straight lines *DE*, *FG* and *AC* to each other, are equal to the ratios of the successive even numbers beginning with two. But the straight line *DN* is twice the straight line *DP*, the straight line *FM* is twice the straight line *FO* and the straight line *AL* is twice the straight line *AS*; thus the ratios of the straight lines *DN*, *FM*, *AL*, to each other, taken in succession, are equal to the ratios of the successive even numbers beginning with two, and the ratios of the straight lines *BE*, *EG*, *GC*, to each other, taken in succession, are equal to the ratios of the successive odd numbers beginning with one. One third of the sum of the product of *BE* and the square of the straight line *DN*, the product of *EG* and the squares of the straight lines *DN*, *FM* and the product of *DN* and *FM*, and the product of *GC* and the squares of the straight lines *FM*, *AL*, and the product of *FM* and *AL*, plus two thirds of the product of *BC* and the square of half of the straight line *DN* which is equal to *DP*, is equal to half of the product of *BC* and the square of the straight line *AL*. But the ratio of the square of the straight line *DN* to the circle of diameter *DN* is equal to the ratio of the square of the straight line *FM* to the circle of diameter *FM*; it is equal to the ratio of the product of *DN* and *FM* to the circle whose square of the diameter is equal to the product of *DN* and *FM*; it is equal to the ratio of the square of the straight line *AL* to the circle of diameter *AL*; it is equal to the ratio of the product of *FM* and *AL* to the circle whose square of the diameter is equal to the product of *FM* and *AL*, and it is equal to the ratio of the square of the straight line *DP* to the circle of diameter *DP*. One third of the sum of the solids formed as the product of *BE* and the circle of diameter *DN*, the product of *EG* and the circles of diameters *DN*, *FM* and the circle whose square of the diameter is equal to the product of *DN* and *FM*, and the product of *GC* and the two circles of diameters the straight

[49] Lit.: to the axis.

lines *FM*, *AL* and the circle whose square of the diameter is equal to the product of *FM* and *AL*, plus two thirds of the product of *BC* and the circle of diameter *DP*, is equal to half of the solid formed as the product of *BC* and the circle of diameter *AL*. Yet, on the one hand, one third of the product of *BE* and the circle of diameter *DN* is equal to the volume of the hollow cone *DBNE* in the second case of figure, and also to the volume of the solid *DBNE*, in the third case of figure which is either a solid rhombus or a hollow cone. On the other hand, one third of the product of *EG* and the two circles whose diameters are *DN*, *FM* and the circle whose square of the diameter is equal to the product of *DN* and *FM* is equal to the volume of a frustum of a hollow cone *DENMGF* in the second case of figure, and also to the volume of the solid *DENMGF* in the third case of figure, which is either a frustum of a solid rhombus or a frustum of a hollow cone. *Regarding[50] one third of the product of *GC* and the two circles whose diameters are *FM*, *AL* and the circle whose square of the diameter is equal to the product of *FM* and *AL*, it is equal to the volume of a frustum of a hollow cone *FGMLCA* in the second case of figure, and also to the volume of the solid *FGMLCA* in the third case of figure, which is either a frustum of a solid rhombus or a frustum of a hollow cone.* Regarding the product of *BC* and the circle of diameter *AL*, it is equal to the cylinder whose base is the circle of diameter *AL* and whose height is *BC*. Thus the solid *DBNE*, which is a hollow cone in the second case of figure and either a solid rhombus or a hollow cone in the third case of figure, and the two solids *DENMGF* and *FGMLCA*, which are the frusta of hollow cones in the second case of figure and either two frusta of solid rhombuses, or two frusta of hollow cones, or the first a frustum of a hollow cone and the other a frustum of a solid rhombus, in the third case of figure, with two thirds of the solid formed as the product of *BC* and the circle of diameter *DP*, have a sum equal to half of the cylinder whose base is the circle of diameter *AL* and whose height is *BC*. But the hollow cone of revolution that we mentioned, plus the two frusta of hollow cones, have a sum equal to the solid generated by the rotation of the figure *CAFDB*, in the second case of figure, if the fixed straight line is *BC*. It is likewise for the solid rhombus or the hollow cone in the third case of figure with, in this case, the two frusta of solid rhombuses or the two hollow cones, or the two solids where the first is a solid rhombus and the other is a hollow cone. The solid generated, if we fix the straight line *BC* and rotate all of the other sides of the figure *CAFDB* from an arbitrary position up to where it returns to that position, is thus less than half of the cylinder whose base is the circle of diameter *AL*

50 *...* The paragraph between the two asterisks renders the Arabic text reconstituted by us.

and whose height is the straight line *BC*, by two thirds of the solid formed as the product of *BC* and the circle of diameter *DP* which is the perpendicular dropped from the point *D* on the diameter *BC*. That is what we wanted to prove.

– **33** – Consider a known parabolic dome of regular vertex and a known solid; then it is possible to describe on the lateral surface of the dome, circles parallel to the base of that surface such that, when we produce ordinates from their circumferences to the axis of the dome, they divide it into parts such that the ratios of the ones to the others, taken in succession, are equal to the ratios of the successive odd numbers beginning with one and such that the smallest is on the side of the vertex of the dome. If we join the surfaces between the circumferences of the circles described over the dome and another surface between the circumference of the smallest circle and the vertex point of the dome, it generates in the dome an inscribed solid such that the excess of the dome over this solid is smaller than the known solid.

Let a parabolic dome be known, of regular vertex, with *AB* half of the portion of the parabola which one has rotated, and which has generated it; let *BC* be its axis, which is the axis of the parabola; let *BD* be that half of the portion, from the other side, where it has turned above the plane in which it was at the start, and let *E* be the known solid.

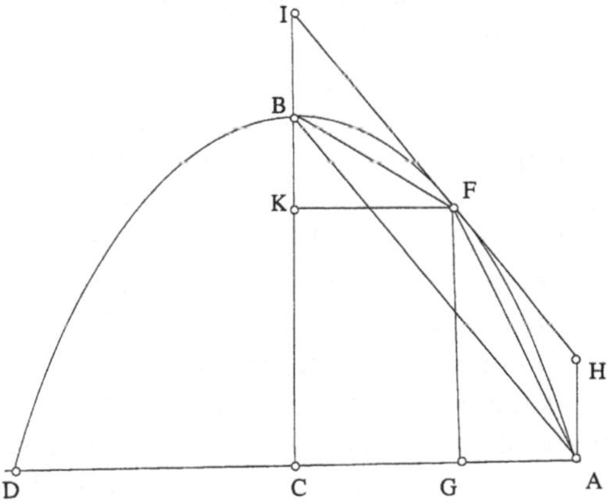

Fig. II.2.33a

I say that it is possible to describe on the surface of the dome ABD *circles parallel to the base circle of its surface such that, when we produce, from the circumferences of these circles, ordinates to the axis* BC, *which divide* BC *into parts whose ratios to each other, taken in succession, are equal to the ratios of the successive odd numbers beginning with the unit, and if we join the surfaces between the circumferences of these circles and another surface between the vertex of the dome and the circumference of the smallest of these circles, an inscribed solid is generated in the dome such that the excess of the dome over this solid is smaller than the solid* E.

Proof: If we join the straight line AB and if we let the ordinate produced from the point A to the axis be the straight line AC, then the torus generated by the rotation of the portion AFB, if we fix the straight line BC and if we rotate the half ABC of the section, is either smaller than the solid E or it is not thus. If it is smaller, then that is what we want. Otherwise, if we divide the straight line AC into two halves at the point G, if we produce from the point G a straight line GF parallel to the straight line BC, <if we join the straight lines AF and FB>, if we make a tangent HFI pass through the point F to the section which meets the axis at the point I, and if we produce from the point A a straight line AH parallel to the straight line BC, then the parallelogram $HIBA$ is circumscribed about the portion AFB of the parabola. The solid generated by the parallelogram $HIBA$, if we fix the straight line IC and rotate all of the other sides of the parallelogram $HIBA$ at the same time as the surface of half of the section, is greater than the torus generated by AFB. But the solid generated by the rotation of the parallelogram $HIBA$ is equal to the product of BK and the circle whose semi-diameter is the straight line AC.[51] Likewise, BK, KC, FK and AC are four straight lines such that BK is one third of KC and FK is half of AC.[52] If we add the solids formed as the product of BK and the square of the straight line FK and as the product of KC and the squares of the straight lines FK, AC and the product of FK and AC, and if from the sum, we subtract the solid formed as the product of BC and the square of the straight line AC, the remainder is greater than the solid formed as the product of BK and the square of the straight line AC.[53] But the ratio of the solids formed as the product of BK and the square of the straight line FK, as the product of KC and the squares of the straight lines FK, AC and the product of FK and AC and as the product of BC and the square of the straight line AC, to the solids formed as the product of BK and the circle of semi-diameter FK, as the product of KC and the circles whose two semi-diameters are FK, AC and the circle whose square of the semi-diameter is equal to the product of FK and AC, and as the product of BC and the circle of semi-diameter AC, each to its homologue, is equal to the ratio of the solid formed as the product of BK and the square of AC to the solid formed as the product of

[51] As $BK = BI$, property of the sub-tangent, and $BI = AH$.
[52] See the mathematical commentary.
[53] By Proposition 21.

BK and the circle of semi-diameter *AC*. The solids formed as the product of *BK* and the circle of semi-diameter *FK*, and as the product of *KC* and the two circles whose semi-diameters are *FK*, *AC* and the circle whose square of the semi-diameter is equal to the product of *FK* and *AC*, if we add them up and if we subtract from them the solid formed as the product of *BC* and the circle of semi-diameter *AC* – on the one hand, the solid formed as the product of *BK* and the circle of semi-diameter *FK* being the triple of the cone of revolution generated by the rotation of *FBK*; on the other, the solid formed as the product of *KC* and the two circles whose semi-diameters are *FK*, *AC* and the circle whose square of the semi-diameter is equal to the product of *FK* and *AC* being the triple of the frustum of a cone of revolution generated by the rotation of the trapezium *FKCA*; and the solid formed as the product of *BC* and the circle of semi-diameter *AC* being the triple of the cone of revolution generated by the rotation of the triangle *ABC* – if we thus subtract the triple of the entire cone generated by the rotation of the triangle *ABC*, from the triple of the entire solid generated by the rotation of the trapezium *AFBC*,[54] then the remainder will be greater than the product of *BK* and the circle of semi-diameter *AC*. But, on the one hand, that which remains of the triple of the solid generated by the rotation of the trapezium *AFBC*, if we subtract from it the triple of the cone generated by the rotation of the triangle *ABC*, is equal to the triple of the torus generated by the rotation of the triangle *AFB*, if the fixed straight line is *BC*; on the other, we have shown that the product of *BK* and the circle of semi-diameter *AC* is equal to the solid generated by the rotation of the parallelogram *HIBA*. The torus generated by the rotation of the triangle *AFB*, if the fixed straight line is *BC*, is thus greater than one third of the solid generated by the rotation of the parallelogram *HIBA*. Yet, we have shown that this solid is greater than the torus generated by the rotation of the portion *AFB* of the parabola, if the fixed straight line is *BC*; thus the torus *generated[55] by the rotation of the triangle *AFB*, if the fixed straight line is *BC*, is much greater than one third of the torus generated by the rotation of the portion *AFB* of the parabola if the fixed straight line is *BC*; that which thus remains of the dome *ABD*, after having subtracted the figure generated by the rotation of the quadrilateral *AFBC*, is composed of the two tori generated by the rotation of the two portions *AF* and *FB* of the parabola, if the fixed straight line is *BC*; this remainder, is either smaller than the solid *E*, or it is not. If it is smaller than it, that is what we wanted.

[54] The polygon *AFBC* is formed from a triangle and a trapezium.

[55] *...* The paragraph between the two asterisks renders the Arabic text reconstituted by us.

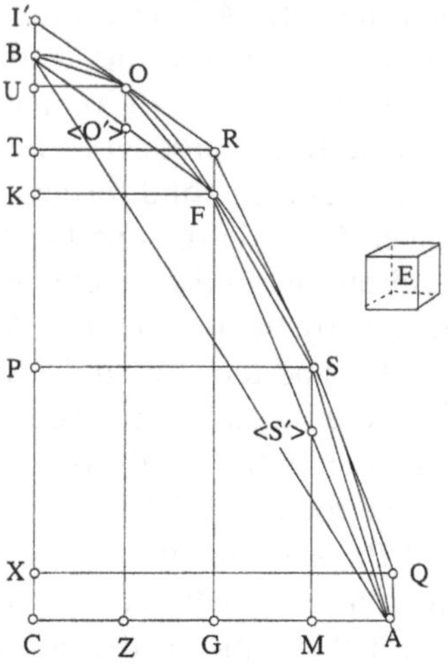

Fig. II.2.33b

Otherwise, if we divide the two straight lines *AG* and *GC* always into two halves, at the points *M* and *Z*, and if we produce from these two points the two straight lines *MS* and *ZO* parallel to the axis and if we make two straight lines *QSR* and *ROI′*,[56] tangent to the parabola, pass through the points *S* and *O*, and if we produce from the points *S* and *O* two parallel straight lines to the straight line *AC* and which meet the axis *BC* at the points *P* and *U* and if we produce from the points *Q* and *R* the perpendiculars *QX* and *RT* to the axis, then the parallelogram *AQRF* is circumscribed about the portion *ASF* of the parabola and the parallelogram *FRI′B* is circumscribed about the portion *FOB* of the parabola. The torus generated by the rotation of the parallelograms *AQRF* and *FRI′B* is equal to the solid* generated by the rotation of *AQXC* because the two parallelograms *AQXC* and *FRTK* have equal bases *CX* and *KT*, as they are equal to the two straight lines *AQ* and *FR*, bases which are on a single straight line *BC*, the two parallelograms being on the same side.[57] If we fix the straight line *BC* and if we rotate the three parallelograms *FRTK*, *AQXC* and *AQRF* according to their shape, then the torus generated by the rotation of the parallelogram *AQRF* is equal to the difference between the two solids generated by the rotation of the two parallelograms *AQXC* and *FRTK*.[58] But, on the one hand, the solid generated by the rotation of the

[56] See the mathematical commentary.
[57] By Propositions 19 and 20.
[58] See previous note 57.

parallelogram *AQXC* is equal to the product of *AQ* and the circle of semi-diameter *AC*. On the other hand, the solid generated by the rotation of the parallelogram *FRTK* is equal to the product of *FR* and the circle of semi-diameter *FK*. Thus the torus generated by the rotation of the parallelogram *AQRF*, if the fixed straight line is *BC*, is equal to the product of *BU* and the difference between the two circles whose semi-diameters are *FK* and *AC*, as we have shown that *BU* is equal to each of the straight lines *AQ* and *FR*.[59] And also the ratios of the doubles of the straight lines *OU*, *FK*, *SP* and *AC*, to each other, taken in succession, are equal to the ratios of the successive even numbers beginning with two. But the number of the straight lines *BU*, *UK*, *KP* and *PC* is the same as the number of those, and their ratios to each other, taken in succession, are equal to the ratios of the successive odd numbers beginning with one. If we add the solids formed as the product of *KP* and the squares of the straight lines *FK*, *SP* and the product of *FK* and *SP*, and as the product of *PC* and the squares of the straight lines *SP*, *AC* and the product of *SP* and *AC*, and if we subtract from the sum the solids formed as the product of *KC* and the squares of the straight lines *FK*, *AC* and the product of *FK* and *AC*, then the remainder is greater than the solid formed as the product of *BU* and the difference between the squares of the straight lines *FK* and *AC*.[60] But the ratio of the square of the straight line *FK* to the circle of semi-diameter *FK* is equal to the ratio of the square of the straight line *SP* to the circle of semi-diameter *SP* and is equal to the ratio of the product of *FK* and *SP* to the circle whose square of the semi-diameter is equal to this product and equal to the ratio of the square of the straight line *AC* to the circle of semi-diameter *AC* and is equal to the ratio of the product of *SP* and *AC* to the circle whose square of the semi-diameter is equal to this product, and is equal to the ratio of the difference between the squares of the straight lines *FK* and *AC* to the difference between the two circles whose semi-diameters are the straight lines *FK* and *AC*. Thus, if we add the product of *KP* and the two circles whose semi-diameters are *FK*, *SP* and the circle whose square of the semi-diameter is equal to the product of *FK* and *SP*, and the product of *PC* and the two circles whose semi-diameters are *SP*, *AC* and the circle whose square of the semi-diameter is equal to the product of *SP* and *AC*, and if we subtract from the sum the product of *KC* and the two circles whose semi-diameters are *FK*, *AC* and the circle whose square of the semi-diameter is equal to the product of *FK* and *AC*, then the remainder is greater than the product of *BU* and the difference between the two circles whose semi-diameters are the straight lines *FK* and *AC*. But, on the one hand, the

[59] By Proposition 18.
[60] By Proposition 29.

product of *KP* and the two circles whose semi-diameters are the straight lines *FK*, *SP* and the circle whose square of the semi-diameter is equal to the product of *FK* and *SP*, plus the product of *PC* and the two circles whose semi-diameters are the straight lines *SP*, *AC* and the circle whose square of the semi-diameter is equal to the product of *SP* and *AC*, is the triple of the solid generated by the rotation of the figure *ASFKC*, if the fixed straight line is *KC*, as it is composed of the two frusta of cones of revolution *FKPS* and *SPCA*. On the other hand, the product of *KC* and the two circles of semi-diameters *FK*, *AC* and the circle whose square of the semi-diameter is equal to the product of *FK* and *AC*, is the triple of the frustum of a cone of revolution generated by the rotation of the trapezium *AFKC*, if the fixed straight line is *KC*. The excess of the triple of the solid generated by the rotation of the figure *ASFKC*, if the fixed straight line is *KC*, over the triple of the solid generated by the rotation of the trapezium *AFKC*, if the fixed straight line is *KC*, is thus greater than the product of *BU* and the difference between the two circles whose semi-diameters are the straight lines *FK* and *AC*. On the one hand, the excess of the triple of the solid generated by the rotation of the figure *ASFKC*, if the fixed straight line is *BC*, over the triple of the solid generated by the rotation of the trapezium *AFKC*, if the fixed straight line is *BC*, is equal to the triple of the torus generated by the rotation of the triangle *ASF*, if the fixed straight line is *BC*. And on the other hand, we have shown that the product of *BU* and the difference between the two circles whose semi-diameters are the straight lines *FK* and *AC* is equal to the torus generated by the rotation of the parallelogram *AQRF*, if the fixed straight line is *BC*. The torus generated by the rotation of the triangle *ASF*, if the fixed straight line is *BC*, is thus greater than one third of the torus generated by the rotation of the parallelogram *AQRF*, if the fixed straight line is *BC*, and the torus generated by the rotation of the parallelogram *AQRF*, if the fixed straight line is *BC*, is greater than the torus generated by the rotation of the portion *ASF* of the parabola, if the fixed straight line is *BC*, as the straight line *QSR* is tangent to the section. The torus generated by the rotation of the triangle *ASF*, if the fixed straight line is *BC*, is thus greater than one third of the torus generated by the rotation of the portion *ASF* of the parabola, if the fixed straight line is *BC*. Yet, we have shown that the torus generated by the rotation of the triangle *FOB*, if the fixed straight line is *BC*, is greater than one third of the torus generated by the rotation of the portion *FOB* of the parabola, if the fixed straight line is *BK*. The two tori generated by the rotation of the triangles *ASF* and *FOB*, if the fixed straight line is *BC*, are greater than one third of the two tori generated by the rotation of the two portions *ASF* and *FOB* of the parabola, if the fixed straight line is *BC*. The portion which remains of the dome *ABD* after having subtracted the solid generated by the rotation of the figure *ASFOBC*, if the fixed straight line is *BC*, composed of the tori generated by the rotation of the portions *AS*, *SF*, *FO* and *OB*, if the fixed straight line is *BC*, is either smaller than the solid *E* or it is not thus. If it is smaller than it, that is what we wanted. Otherwise, it is necessary, when we continue to proceed in the same way as numerous times, that we lead to tori, remaining from the dome, which are less than the solid *E*, as if two magnitudes are such that the one is greater than the other and if we subtract from the greater of the two a magnitude whose ratio to that is greater than a given ratio, and of the remainder, a magnitude whose ratio to that is greater than this ratio, and if we continue to proceed thus, it is necessary that we lead from the greater, to a thing which remains from it, smaller than the smallest.[61] Let the tori generated by the rotation of the portions *AS*, *SF*, *FO* and *OB* of the parabola, if the fixed straight line is *BC*, be that which remains of the dome and let them be smaller than the solid *E*; it is then possible to construct oin the lateral surface of the dome *ABD* of regular vertex circles parallel to the circle of the base of its surface. If ordinate straight lines are produced from the circumferences of these circles to the axis *BC*, they divide it into arbitrary parts such that their ratios, to each other, taken in succession, are equal to the ratios of the successive odd numbers beginning with one, and if we join the surfaces between these circles and another surface between the vertex of the dome and the smaller circle, an inscribed solid is generated in the dome, such that the excess of the dome over this solid is smaller than the solid *E*; <these circles> are like the circles whose semi-diameters are *OU*, *FK*, *SP* and *AC*. That is what we wanted to prove.

– **34** – Consider a known parabolic dome, with a pointed vertex or with a sunken vertex, and a known solid; then it is possible to describe on the lateral surface of the dome circles parallel to the base of that surface, such that when ordinates are produced from their circumferences to the axis of the dome, they divide it into parts such that the ratios of the ones to the others, taken in succession, are equal to the ratios of the successive odd numbers beginning with one and such that the smallest is on the side of the vertex of the dome, and if we join the surfaces between the circumferences of the circles described over the dome and another surface between the circumference of the smallest of these circles and the vertex of the dome, an inscribed solid is generated in the dome such that the excess of the dome over this solid is smaller than the known solid.

[61] By Proposition 31.

Let there be a known parabolic dome with a pointed vertex or with a sunken vertex, let *AB* be half of the portion where one has rotated the base, let *BC* be its axis which is the diameter of the section, let *BD* be half of the portion on the other side, when it has rotated above the plane in which it was first, and let *E* be the known solid.

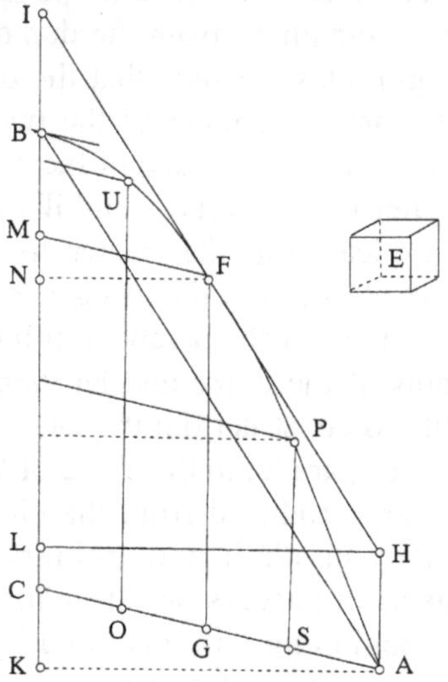

Fig. II.2.34a

I say that we may describe on the surface ABD *circles parallel to the base circle of its surface such that when ordinates are produced from the circumferences of these circles to the axis* BC, *they divide* BC *into parts whose ratios to each other, taken in succession, are equal to the ratios of the successive odd numbers beginning with one, and if we join the surfaces between the circumferences of these circles and the surface between the vertex of the dome and the circumference of the smaller circle, an inscribed solid is generated in the dome, such that the excess of the dome over this solid is smaller than the solid* E.

Proof: If we join the straight line *AB* and if we let the ordinate produced from the point *A* to the axis be the straight line *AC*, then the torus generated by the rotation of the portion *AFB* of the parabola, when we fix the straight line *BC* and rotate half *CAB* of the parabola, is either smaller than the solid *E*, or it is not thus. If it is smaller than it, that is what we wanted.

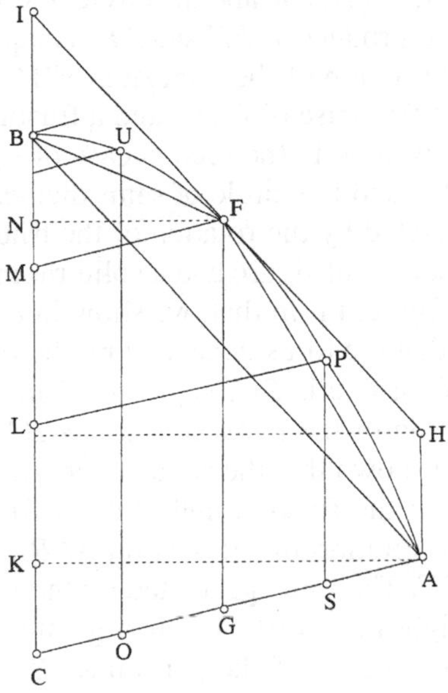

Fig. II.2.34b

Otherwise, if we divide the straight line AC into two halves at the point G, if we produce from the point G a straight line parallel to the straight line BC, which is the straight line GF, if we join the straight lines AF and FB, if we make a straight line HFI pass through the point F, tangent to the parabola, which meets the diameter at the point I, and if we produce from the point A a straight line parallel to the diameter BC, which is AH, we then show as we have shown in the previous proposition, that the solid generated by the surface $HIBA$, if we fix the straight line IC and rotate the plane of the half AB of the portion, is greater than the torus generated by the portion AFB of the parabola, and that the surface $HIBA$ is a parallelogram. If we produce two perpendiculars AK and HL from the points A and H to the straight line BC, then the surface $HLKA$ is a parallelogram and its base is the same as the base of the surface $HIBA$, which is AH, and they are in the same direction and between the parallel straight lines AH and CI. * [62]If we produce from the point F an ordinate FM, then FM and AC are two parallel straight lines; if we produce from the point F the perpendicular FN to the axis BC, then FN and AK are two parallel straight lines. The solid formed as the product of MC and the two

[62] *...* The paragraph between the two asterisks renders the Arabic text reconstituted by us.

circles* of semi-diameters *FN*, *AK* and the circle whose square of the semi-diameter is equal to the product of *FN* and *AK*, is equal to the triple of the solid generated by the rotation of the trapezium *FMCA*, which is a frustum of a hollow cone in the first case of figure and a frustum of a solid rhombus or a frustum of a hollow cone in the second case of figure. But, on the one hand, the product of *BC* and the circle of semi-diameter *AK* is equal to the triple of the solid generated by the rotation of the triangle *ABC*, which is a hollow cone in the first case of figure and a solid rhombus or a hollow cone in the second case of figure. From that, we show in the same way as in the previous proposition, that the torus generated by the rotation of the triangle *AFB*, if the fixed straight line is *BC*, is greater than one third of the solid generated by the rotation of the parallelogram *HIBA*, if the fixed straight line is *IC*. But we have shown that the solid generated by the rotation of the parallelogram *HIBA*, if the fixed straight line is *BC*, is greater than the torus generated by the rotation of the section *AFB* of the parabola, if the fixed straight line is *BC*. The torus generated by the rotation of the triangle *AFB*, if the fixed straight line is *BC*, is much greater than one third of the torus generated by the rotation of the portion *AFB* of the parabola, if the fixed straight line is *BC*. After having subtracted the figure generated by the rotation of the trapezium *AFBC*,[63] that which remains of the dome *ABD* is composed of the two tori generated by the rotation of the two portions *AF* and *FB* of the parabola, if the fixed straight line is *BC*; the remainder is either smaller than the solid *E*, or it is not thus. If it is smaller than it, that is what we wanted; otherwise, if we divide the straight lines *AG* and *GC* always in halves at the points *S* and *O*, if we produce from these the straight lines *SP* and *OU* parallel to the diameter and if we join the straight lines *AP*, *PF*, *FU* and *UB*, we show as we have shown in the previous proposition that the ratios of the ordinates produced from the points *A*, *P*, *F*, *U* to the perpendiculars produced from these points to the diameter are equal. If we pursue this in an analogous way to what we have followed in the previous proposition, we show that the tori generated by the rotation of the triangles *APF* and *FUB*, if the fixed straight line is *BC*, are greater than one third of the tori generated by the rotation of the portions *APF* and *FUB* of the parabola, if the fixed straight line is *BC*, as the way in this proposition and in that which preceeds is a same way, except that, here, we use the perpendiculars[64] in place of the ordinates and the straight lines which are parallel to them, and in place of the cone of revolution, the hollow cone of revolution, in the first case of figure and the solid rhombus or the hollow cone of revolution, in the second case of figure, and in place

[63] The polygon *AFBC* is formed from a triangle and a trapezium.
[64] The straight lines perpendicular to the axis.

of the frustum of a cone of revolution, the frustum of a hollow cone of revolution in the first case of figure, and the frustum of a solid rhombus or the frustum of a hollow cone in the second case of figure. We show from this that, when we proceed in the same manner as numerous times, it is necessary that we reach tori which remain of the dome *ABD* less than the solid *E*. We reach the tori generated by the rotation of the portions *AP*, *PF*, *FU* and *UB* of the parabola, if the fixed straight line is *BC*; then it is possible to construct over the lateral surface of the dome *ABD* circles parallel to the circle of the base of its surface. If ordinates are produced from their circumferences to the axis *BC*, they divide it into parts such that their ratios of the ones to the others, taken in succession, are equal to the ratios of successive odd numbers beginning with one, and if we join the surfaces between the circumferences of these circles and another surface between the circumference of the smallest of these circles and the vertex of the dome, an inscribed solid is generated in the dome *ABD*, such that the excess of the dome over this solid is smaller than the solid *E*; these circles are the circles that the points *A*, *P*, *F*, *U* describe during their rotation. That is what we wanted to prove.

– **35** – Consider a known parabolic dome and a known solid; then it is possible to construct in the dome an inscribed solid figure which is less than half of the cylinder, whose base is the circle which is the base of the dome, if the dome is of regular vertex, or the circle which is the lower base of the dome if it is not of regular vertex, and whose height is equal to the axis of the dome, of a magnitude smaller than the known solid.

Let the known parabolic dome be the dome *ABC* and let *AB* be half of the section by which it has been generated; let *BD* be the axis of the dome and *E* the known solid.

I say that we can construct in the dome ABC *an inscribed solid which is less than half of the cylinder whose base is the base circle of the dome* ABC *if it is of regular vertex, or the circle of its lower base if it is not of regular vertex, and whose height is equal to the straight line* BD, *of a magnitude smaller than the solid* E.

Proof: If we produce from the point *A* the straight line *AC* to the point *C*, *AC* will be perpendicular to the axis. If we let the solid *F* be in proportion with the solid[65] and with the solid *E*,[66] and the ratio of *AD* to *ON*[67] greater than the ratio of the solid whose base is the circle of diameter *AC* and whose height is *BD* to the solid *F* – the dome *ABC* is either of

[65] *i.e.* the solid whose base is the circle of diameter *AC* and whose height is *BD*.

[66] He means that the cylinder and the solids *F* and *E* are in continuous proportion.

[67] In the three cases of figure, *ON* is an ordinate.

regular vertex or it is not – so if it is of regular vertex, *ON* is perpendicular to the axis *BD* and *AD* will be half of *AC* as in the first case of figure, and the ratio of the circle of semi-diameter *AD* to the circle of semi-diameter *ON* is greater than the ratio of the solid whose base is the circle of semi-diameter *AD* and whose height is *BD*, to the solid *F*, repeated twice.[68] On the one hand, the ratio of the circle whose semi-diameter is *AD* to the circle whose semi-diameter is *ON* is equal to the ratio of the solid whose base is the circle of semi-diameter *AD* and whose height is *BD*, to the solid whose base is the circle of semi-diameter *ON* and whose height is *BD*. On the other hand, the ratio of the solid whose base is the circle of diameter *AC* and whose height is *BD* to the solid *F*, repeated twice, is equal to the ratio of the solid whose base is the circle of diameter *AC* and whose height is *BD*, to the solid *E*, as the solid whose base is the circle of diameter *AC* and whose height is *BD*, the solid *F* and the solid *E* are in proportion. The ratio of the solid whose base is the circle of semi-diameter *AD* and whose height is *BD* to the solid whose base is the circle of semi-diameter *ON* and whose height is *BD*, is thus greater than the ratio of the solid whose base is the circle of diameter *AC* and whose height is *BD* to the solid *E*. But the solid whose base is the circle of semi-diameter *AD* and whose height is *BD* is the solid whose base is the circle of diameter *AC* and whose height is *BD*; the solid whose base is the circle of semi-diameter *ON* and whose height is *BD* is thus smaller than the solid *E*.

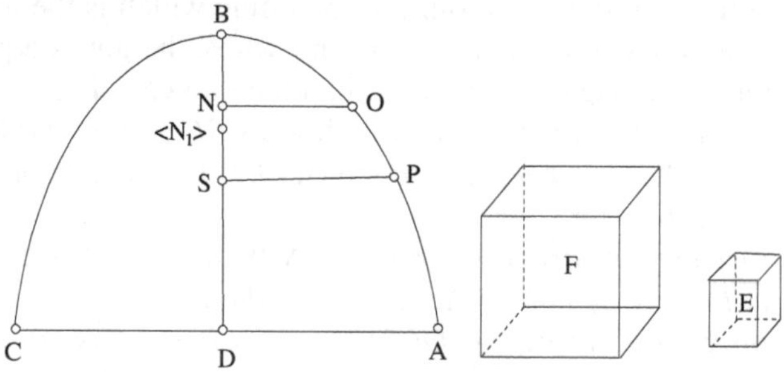

Fig. II.2.35a

If the dome *BC* is of pointed vertex or of sunken vertex, if we produce from the point *O* the perpendicular *OU* to the axis *BD*, as in the second case of figure and the third case of figure, and if the perpendicular dropped from the point *A* onto *BD* is the perpendicular *AQ* which is half of *AC*, then the straight lines *AQ* and *OU* are parallel. But the straight lines *AD* and *ON*

[68] That is to say to the square of the ratio.

are also parallel, as they are ordinates. The triangle *ADQ* is thus similar to the triangle *OUN* and, consequently, the ratio of *AD* to *ON* is equal to the ratio of *AQ* to *OU*. Yet, we have shown that the ratio of *AD* to *ON* is greater than the ratio of the solid whose base is the circle of diameter *AC* and whose height is *BD*, to the solid *F*; thus the ratio of *AQ* to *OU* is greater than the ratio of the solid whose base is the circle of diameter *AC* and whose height is *BD* to the solid *F*. But the ratio of *AQ* to *OU*, repeated twice, is equal to the ratio of the circle of semi-diameter *AQ* to the circle of semi-diameter *OU*; thus the ratio of *AQ* to *OU*, repeated twice, is greater than the ratio of the solid whose base is the circle of diameter *AC* and whose height is *BD*, to the solid *F*, repeated twice. On the one hand, the ratio of the circle whose semi-diameter is *AQ* to the circle whose semi-diameter is *OU* is equal to the ratio of the solid whose base is the circle of semi-diameter *AQ* and whose height is *BD* to the solid whose base is the circle of semi-diameter *OU* and whose height is *BD*; and on the other, the ratio of the solid whose base is the circle of diameter *AC* and whose height is *BD* to the solid *F*, repeated twice, is equal to the ratio of the solid whose base is the circle of diameter *AC* and whose height is *BD* to the solid *E*. Thus the ratio of the solid whose base is the circle of semi-diameter *AQ* and whose height is *BD*, to the solid whose base is the circle of semi-diameter *OU* and whose height is *BD*, is greater than the ratio of the solid whose base is the circle of diameter *AC* and whose height is *BD* to the solid *E*. But the solid whose base is the circle of semi-diameter *AQ* and whose height is *BD* is the solid whose base is the circle of diameter *AC* and whose height is *BD*, thus the solid whose base is the circle of semi-diameter *OU* and whose height is *BD* is smaller than the solid *E*.

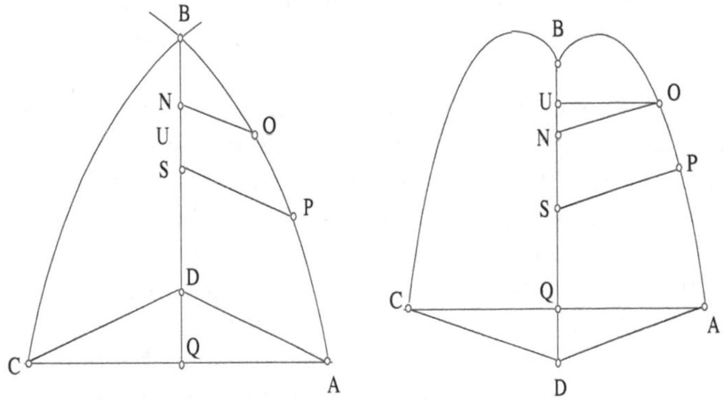

Fig. II.2.35b

Yet, we have shown in the first case of figure – that of the dome of regular vertex – that the solid whose base is the circle of semi-diameter *ON* and whose height is *BD* is smaller than the solid *E*. Thus in the three cases of figure, the solid whose base is the circle of semi-diameter the perpendicular dropped from the point *O* on the axis *BD* and whose height is *BD* is smaller than the solid *E*. But the solid generated by the rotation of the polygon *APOBD*, if the fixed straight line is *BD*, in the three cases of figure, is less than half of the cylinder whose base is the circle of diameter *AC* and whose height is *BD*, by two thirds of the solid whose base is the circle of diameter the perpendicular dropped from the point *O* on the axis *BD* and whose height is *BD*,[69] as the ratios of the parts *BN*, *NS* and *SD* to each other, taken in succession, are equal to the ratios of the successive odd numbers beginning with one. Thus the solid generated by the rotation of the polygon *APOBD*, if the fixed straight line is *BD*, and inscribed in the dome, is less than half of the cylinder whose base is the circle of diameter *AC* and whose height is *BD*, of a magnitude smaller than the solid *E*, and the circle, whose diameter is *AC*, is in the first case of figure, the base of the dome, and in the second and third cases of figure, it is the lower base. That is what we wanted to prove.

– **36** – The volume of every parabolic dome is equal to half of the volume of the cylinder whose base is the base circle of the dome, if the dome is of regular vertex, or the lower base of the circle if it is not of regular vertex, and whose height is equal to the axis of the dome.

Let *ABC* be a parabolic dome, let its axis be *BD*, and the diameter of its base or of the lower base, the straight line *AC*.

I say that the volume of the dome ABC *is equal to half of the volume of the cylinder whose base is the circle of diameter* AC *and whose height is* BD.

Proof: If the dome *ABC* is not equal to half of the cylinder that we mentioned, then either it is greater than half, or it is smaller than it. First let it be greater than half, if it were possible, and let its excess over the half be equal to the solid *E*; it is possible to describe on the lateral surface of the dome *ABC* circles parallel to the base of that surface, such that when ordinates are produced from their circumferences to the diameter, they divide it into parts such that their ratios to each other, taken in succession, are equal to the ratios of the successive odd numbers beginning with one, and such that the smallest circle is on the side of the vertex of the dome; if

[69] We may associate with the point *O* the polygon *APOBD* defined as in Proposition 32. We do not know if Thābit omitted to give this justification or if there is something missing from the manuscript. See the mathematical commentary.

we join the surfaces between the circumferences of the circles and another surface between the smallest of the circles and the vertex of the dome, then the excess of the dome *ABC* over the figure generated in the dome is smaller than the solid *E*. If we suppose that the figure generated in the dome is the solid figure *APGHBIKLCD*, then the solid figure *APGHBIKLCD*, increased by the solid *E*, is greater than the dome *ABC*. But the dome *ABC* is equal to the semi-cylinder whose base is the circle of diameter *AC* and whose height is *BD*, increased by the solid *E*. The solid figure *APGHBIKLCD*, increased by the solid *E*, is thus greater than half of the cylinder whose base is the circle of diameter *AC* and whose height is *BD*, increased by the solid *E*. If we remove that which is common, which is the solid *E*, there remains the figure *APGHBIKLCD* greater than half of the cylinder whose base is the circle of diameter *AC* and whose height is *BD*. Yet, it has been shown in the previous propositions that it is smaller than its half; this is contradictory. The dome *ABC* is thus not greater than half of the cylinder whose base is the circle of diameter *AC* and whose height is *BD*.

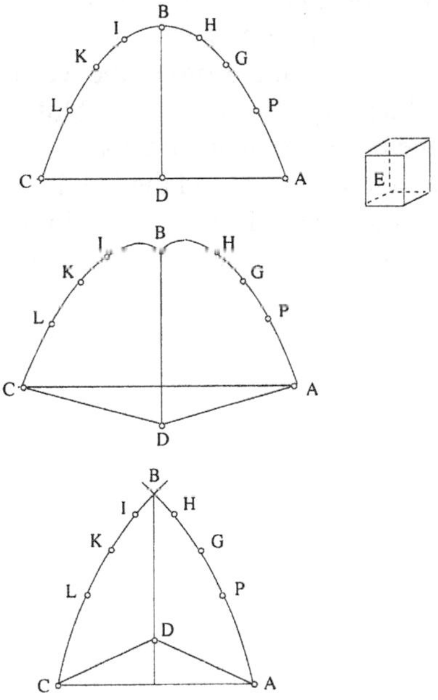

Fig. II.2.36

I say that the dome ABC *is not smaller than half of the cylinder that we mentioned.*

If it were possible, let it be less than its half by the magnitude of the solid *E*. It is thus possible to construct in the dome *ABC*, a solid figure inscribed in it and such that it is less than half of the cylinder that we mentioned, by a magnitude smaller than the solid *E*. Let this figure be the solid figure *APGHBIKLCD*; thus the solid figure *APGHBIKLCD*, plus the solid *E*, is greater than half of the cylinder whose base is the circle of diameter *AC* and whose height is *BD*. But the dome *ABC*, plus the solid *E*, is equal to half of the cylinder whose base is the circle of diameter *AC* and whose height is *BD*. Thus, the solid figure *APGHBIKLCD*, plus the solid *E*, is greater than the dome *ABC*, plus the solid *E*. But if we remove that which is common, which is the solid *E*, there remains the solid figure *APGHBIKLCD* greater than the dome *ABC*; it is thus greater than the dome and it is inscribed in it; this is contradictory. The dome *ABC* is thus not smaller than half of the cylinder whose base is the circle of diameter *AC* and whose height is *BD*. But we have shown that it is not greater than its half, it is consequently equal to its half. That is what we wanted to prove.

The treatise of Thābit ibn Qurra on the measurement of paraboloids
is completed.
Thanks be to God, Lord of the worlds.
May the blessing of God be upon Muḥammad
the prophets' seal, and his own.
Written by Aḥmad ibn Muḥammad ibn ʿAbd al-Jalīl at Shīrāz,
the night of Saturday, eight days left to go in Rabīʿ al-awwal,
the year three hundred fifty-eight.

2.4. ON THE SECTIONS OF THE CYLINDER AND ITS LATERAL SURFACE

2.4.1. *Introduction*

Not only has the *Treatise on the Sections of the Cylinder and its Lateral Surface*, like the two preceding treatises, made its mark on the history of infinitesimal mathematics, but it is also one of the most important texts on geometry. Even more so, as it touched on the study of geometrical point-wise transformations, it steered research into geometry in a new direction, and by this fruitful action it influenced algebra as well. The influence of this treatise may be detected in the work of Ibrāhīm ibn Sinān, of Ibn Sahl, of Ibn al Haytham and of Sharaf al-Dīn al-Ṭūsī, among others.

This feature is not the only difference between the first two treatises of Ibn Qurra and the one under consideration here: in the field of infinitesimal mathematics itself, Thābit forged here a new path, more geometric, which owed nothing to either arithmetic lemmata or integral summation. To this was added a divergence in historical terms: in *The Measurement of the Parabola*, just as in *The Measurement of the Paraboloid*, Thābit had no predecessors. Unaware of the works by Archimedes on this subject, he conceived a general work that was entirely innovative. In the introduction to the *Treatise on the Sections of the Cylinder*, by way of contrast, Thābit made reference himself to a study by al-Ḥasan ibn Mūsā, his elder and without a doubt his master, saying that with this book Thābit was adding his name to a tradition that had never stopped being theirs, the tradition of the Banū Mūsā.

Unfortunately, this book by al-Ḥasan ibn Mūsā is lost. In order to understand the role it played in the beginnings of Thābit's research, every bit as much as, later on, in its contribution to the work of the mathematician of the Islamic West, Ibn al-Samḥ, only a few indirect statements are available to us. The first comes to us from the author's own brothers, Muḥammad and Aḥmad, which we have referred to previously.[1] They inform us that al-Ḥasan, without any real knowledge of the *Conics* of Apollonius – he had a faulty copy of it which he could neither understand nor translate – studied the ellipse, its properties as a plane section of a cylinder, as well as the different types of elliptical sections. Thābit himself recalls that al-Ḥasan ibn Mūsā calculated the area of an ellipse. This was therefore very much in the field Thābit ibn Qurra was to make his own. But we also know, thanks to another witness, the tenth-century mathematician al-Sijzī, that al-Ḥasan ibn Mūsā worked by the bifocal method in order to study this 'elongated circular' figure. One should expect then, if one trusts

[1] See Chapter I: Banū Mūsā.

in the practice of the mathematicians of the time, which, furthermore, was marked by a conformity with the requirement for rigour, that one part of al-Ḥasan's book was dedicated to establishing that the figure obtained by the use of the bifocal method is the same as that generated by the section, and that it proves, in particular, the fundamental relationship – the *symptoma* – which one could come to know through a simple glance at the first book of the *Conics*. This hypothesis is far from being arbitrary: it tallies with the statement of his two brothers, Muḥammad and Aḥmad, according to which al-Ḥasan conceived a theory of the ellipse and of elliptical sections following a different path from that of Apollonius; it sheds light for us, on the other hand, on the researches of Ibn al-Samḥ, who is an excellent witness to this method, as we shall see further on.

If now we come to the treatise by Thābit on the sections of the cylinder, we may note that there is nowhere any question of use of the bifocal method. Of the various subjects of the Banū Mūsā, Thābit thus kept with only a part. What might appear to be a restrictive choice makes sense if one recollects another difference with al-Ḥasan ibn Mūsā: Thābit ibn Qurra, in contrast to the latter, had excellent knowledge of the *Conics* of Apollonius. He even translated the last three of the seven books that have survived in Greek. He therefore had at his disposal from the outset the text of Apollonius and al-Ḥasan's book, and it was with the methods of the first that he stepped into the tracks of the second. From Apollonius, he took up a project and various applications; from al-Ḥasan, he obtained an account of the *Conics*, from it the ellipse, and the powerful means for studying them. This was a hitherto unknown situation, which saw the project become transformed and develop, and the means evolve and bend in a manner other than the way in which they were employed in their original field. And it is true – and here lies the main point of Thābit's book – that the project became one of elaborating a theory of the cylinder and of its plane sections analogous to that of the cone and its sections. As for the means, they were enriched by the projections and point-wise transformations. To elaborate the theory of the cylinder and of its plane sections by drawing inspiration from the model of the *Conics* was the transformed project Thābit undertook to achieve, by applying – perhaps in this respect he followed al-Ḥasan ibn Mūsā, but in doing so went very much further – projections and point-wise transformations. Let us explain a little about these essential features of the *Treatise on the Sections of the Cylinder*, which have until this point remained hidden in the shadows.

Thābit ibn Qurra considered, and he was the first to take a step in this direction, the cylindrical surface as a conic surface, and the cylinder as a cone whose vertex would be projected to infinity in a given direction.

Indeed, he replaced straight lines passing through a point and planes passing through a point, in the case of the cone, with parallel straight lines and planes parallel to a straight line, or containing this straight line, in the case of the cylinder. He began by defining the cylindrical surface then the cylinder, as Apollonius in the *Conics* had first defined the conic surface then the cone. It was also the order as found in Apollonius that he followed for his definitions: axis, generating line, base, right or oblique cylinder.

Thābit defined the height of a cylinder as extended from the centre of its base. Even if an analogous definition did not occur in the work of his predecessor, the role, in the work of Thābit, of the plane containing the axis and the height (thus perpendicular to the base) and in Apollonius of the plane passing through the axis and perpendicular to the base, is evident in both authors, from Proposition 5 in Apollonius and Thābit's Proposition 9. This plane, which we call the principal plane, is a plane of symmetry for the cone and for the cylinder, hence its importance.

Thābit did not give, as we understand it, definitions for a diameter, for two conjugate diameters, or one for the axes of a curve, as one finds at the beginning of the *Conics*. On the other hand, he does give a definition for two opposite generating lines, which do not make an obvious appearance in Apollonius.

Confirmation of the similarity in the approach of the two authors is obtained when one examines the first propositions in the book of Thābit. Propositions 1, 2, 3, 4, 8, 9, 10 and 11 correspond respectively to Propositions 1, 2, 3, 4, 5, 9 and 13 in Apollonius. Let us quickly examine these correlations, and note to begin with that Propositions 5 and 6 in Thābit, which demonstrate a necessary and sufficient condition whereby the section of the cylinder through a plane parallel to its axis or containing it is a rectangle, and Proposition 7, which defines the cylindrical projection, have no equivalent in Apollonius. Conversely, we find no clear sign in Thābit of propositions corresponding to Propositions 6–8 in Apollonius, which concern the parabola or the hyperbola. Let us note next of all that the similarity in the first four propositions is so clear that it is not worth detaining ourselves over it.[2] In Propositions 9 in Thābit and 5 in Apollonius a correspondence occurs in the methods they employ. Indeed, the method used to study a section by means of a plane antiparallel to the plane of the base is the same: the method is based on a characteristic property of the circle that we translate algebraically as $y^2 = x(d - x)$, d being its diameter

[2] See the first four propositions: *Apollonius Pergaeus*, ed. I.L. Heiberg, Stuttgart, 1974, vol. 1; *Apollonius: Les Coniques,* tome 1.1: *Livre I*, commentaire historique et mathématique, édition et traduction du texte arabe par R. Rashed, Berlin/New York, 2008.

(with the tangent at one of its extremities, it defines the system of axes). In Propositions 8 and 10 Thābit makes recourse to the cylindrical projection, and in Propositions 10 and 11 the methods differ. In Proposition 10 Thābit shows that the section being studied is an ellipse or a circle, and in Proposition 11 he shows that it cannot be a circle, whereas Apollonius begins by showing in Proposition 9 that it is not a circle, in order then, in Proposition 13, to characterize the ellipse by the *symptoma*, by which he will deduce a characteristic property in Proposition 21: it is to this latter that Thābit makes recourse in his Propositions 10 and 11, while establishing that the plane section obtained is no other than the ellipse defined by Apollonius. Henceforward he makes references to Apollonius in terms of the properties of the ellipse: conjugate diameters, smaller and larger diameter, etc.

If Thābit therefore found in the *Conics* of Apollonius a model for elaborating his theory of the cylinder, he would develop, for the needs of the latter, the study of geometric transformations. This is the second quality of the *Treatise on the Sections of the Cylinder* to be emphasized here.

As a matter of fact, he made recourse, in Propositions 7 and 8 of the treatise, to the cylindrical projection p of one plane upon another, which is parallel to it, whilst in Proposition 10 he moved on to the cylindrical projection of one plane upon any sort of plane whatsoever. In the second part of this last proposition, Thābit set down two cylindrical projections. In Proposition 12, he showed that two ellipses with the same centre I, whose axes (a, b) and (a', b') are respectively collinear and satisfy $a'/a = b'/b$, correspond to each other in terms of a homothety $h(I, a'/a)$, seen as composed of two cylindrical projections and of the homothety between the base circles.

Thābit reminds us that for p and h the ratio of any two segments is equal to the ratio of their homologues.

Proposition 13 defined the correspondence through orthogonal affinity between the ellipse and each of the circles having its axes for diameters. Only the ratios between the two segments, perpendicular to the axis of affinity, or parallel to it, are retained. In Proposition 14, Thābit showed that the ratio between the areas of two homologous polygons in an affinity f is equal to the ratio $\frac{a}{b}$ of the affinity, and showed that one may move from an ellipse to the equivalent circle by a transformation $h \circ f$, h being a homothety of ratio $\sqrt{\frac{b}{a}}$. He thus defined a transformation in which two homologous areas are equal, a transformation he made use of in Propositions 15–17 to obtain a circle segment equivalent to a segment of an ellipse. In these three cases a simple geometric construct was defined of $h \circ f$. In Propositions 24 and 26, the result, first of all established for two homothetic ellipses, is then extended to two similar ellipses. Thābit defined the displacement that allowed the movement from homothety to similarity. Moreover, Proposition 9 introduces orthogonal symmetry in relation to a plane: it transformed the base circle in the section through an antiparallel plane.

The possibilities inherent in these transformations are too numerous, and their role too fundamental in the development of the book, for us to see them as simple circumstantial facts. What is more, they would outlive Thābit, as we have said, in this field as in others. It was all of these means, in any case, that enabled Thābit to pursue the elaboration of the theory of the cylinder and of its sections.

Let us come now to the commentary on the propositions of Thābit, in order to trace the implementation of these means in the concrete progress of his project, and mark their renewal in this context of Archimedean methods. We shall therefore compare as often as necessary the approach used by Thābit with that used by Archimedes, hoping to obtain by that comparative test a better perception of Thābit's contribution. We shall bear in mind at all times that the latter was not familiar with Archimedes' *On Conoids and Spheroids*.

Let us start with a reminder of our explicit definitions:

D_1 cylinder axis
D_2 cylinder 'side' or generating line
D_3 cylinder's lateral surface
D_4 opposite sides
D_5 cylinder's height
D_6 right cylinder (when the height is equal to the axis)
D_7 oblique cylinder (when the height is different from the axis).

2.4.2. *Mathematical commentary*

2.4.2.1. *Plane sections of the cylinder*

Proposition 1. — *Every generating line is parallel to the axis.*

By definition a generating line and the axis are coplanar and the base circles have the same radius. The result is immediate by application of Euclid I.33.

Proposition 2. — *The only straight lines lying on the lateral surface of a cylinder are the generating lines.*

Thābit used a *reductio ad absurdum* based on the property that a straight line and a circle have at most two common points.

Proposition 3. — *If a plane containing the axis or parallel to the axis cuts the lateral surface of the cylinder, then the intersection is formed of two straight lines. If the plane does not contain the axis and is not parallel to it, the intersection does not include any straight lines.*

The proof makes use of Propositions 1 and 2 and is based on the uniqueness of a parallel to a given straight line drawn through a given point.

Proposition 4. — *If a plane containing the axis or parallel to the axis cuts a cylinder, the section is a parallelogram.*

This result is derived immediately from Proposition 3 by using Euclid XI.16.

In the case of a right cylinder, the section is a rectangle.

Proposition 5. — *For a plane passing through the axis of an oblique cylinder to cut it producing a rectangle, it must be and has only to be perpendicular to the principal plane.*

Proposition 6. — *For a plane parallel to the axis of an oblique cylinder to cut it producing a rectangle, it must be and has only to be perpendicular to the principal plane.*

Propositions 5 and 6 are the immediate consequences of Proposition 4 and demonstrate their proofs by using the properties of perpendicular planes and those of straight lines perpendicular to a plane (Euclid XI.18, 19).

Proposition 7. — *Given two parallel planes* (P) *and* (P′), $A \in (P)$, $E \in (P')$, *and a figure* F *in the plane* (P), *the straight lines parallel to* AE *passing through the points of* F *cut* (P′) *and the points of intersection belong to a figure* F′ *similar to and equal to* F.

Proposition 7 is thus the study of the cylindrical projection in a parallel direction to AE with a figure F in a plane (P) above a parallel plane (P').

Although the transformation under study here is also a translation of vector AE, the continuation shows that Thābit was looking to set down the characteristics of cylindrical projection even when planes (P) and (P') are not parallel, as we shall see in Proposition 10.

The proof is by *reductio ad absurdum*.

Figures F and F' are therefore isometric.

Proposition 8. — *The section of the lateral surface of a cylinder through a plane parallel to its base is a circle equal to the base circle and centred on the axis.*

More generally, the sections of the lateral surface of a cylinder through two parallel planes cutting all the generating lines are isometric figures.

Proposition 8 is an application of Proposition 7.

Proposition 9. — Antiparallel sections

1. Definition of antiparallel planes: for a cone or an oblique cylinder with circular bases on axis GH and with height GI, the plane of base P and a plane P' not parallel to P are said to be antiparallel if:

1) P' is perpendicular to the plane GHI, the plane of symmetry for the cone or cylinder;

2) the intersections of the plane (GHI) with the planes P and P' are antiparallel straight lines, in other words they make equal angles with the straight line GH.

2. In the case of the oblique cylinder, the bisecting plane of planes P and P', which is perpendicular to GH and is therefore a plane of right section for the cylinder, will be the plane of a symmetry transforming the base circle into the antiparallel circle lying in P'. We shall see that this plane gives a minimal section (Propositions 18 and 19).

3. The intersection of the lateral surface of the cylinder with a plane antiparallel to the plane of the base is a circle centred on the axis of the cylinder and equal to the base circle or a portion of such a circle.

This intersection is called a section of 'contrary position τομὴ ὑπεναντία – مخالف الوضع' by Thābit as by Apollonius in the case of the cone (Proposition I.5).

The method followed by Thābit is, furthermore, the one used by Apollonius, in making use of the characteristic property of the circle $MN^2 = NO \cdot NS$; that is to say, its equation in relation to a system of axes defined by a diameter and the tangent to one of its extremities is the equation $y^2 = x(d - x)$, where d is the diameter.

In order to gain a better understanding of certain aspects of Thābit's approach in Propositions 8 and 9, let us consider the figure in the plane of symmetry of the cylinder. The parallelogram $ABED$ is the orthogonal projection of the cylinder on to this plane, and segments AB and DE the projections of the base circles.

In Proposition 8, Thābit defined a family of circles of which AB is a member.

If A' is the point of AD such that $BA = BA'$, we have $B\hat{A}A' = A\hat{A}'B$; thus $A'B$ is the projection of an antiparallel plane. In Proposition 9, Thābit thus defined a family of circles of which $A'B$ is a member.

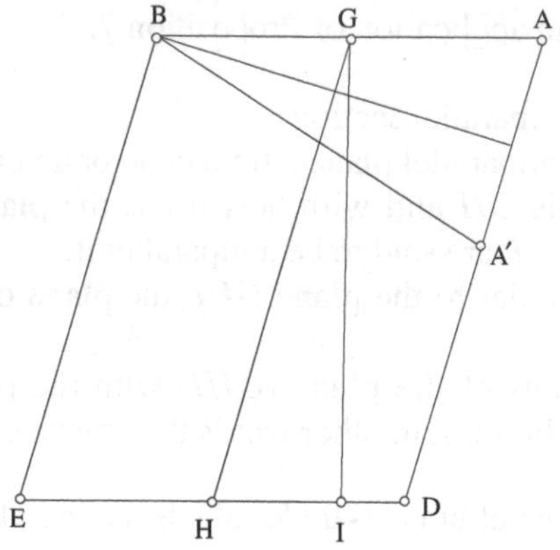

Fig. 2.4.1

The bisector of ABA' is the orthogonal projection of a plane perpendicular to the axis and is a member of a family of planes. One plane from this family will be a plane of symmetry for the figure formed by a circle from the first family and a circle from the second.

Proposition 10. — *The cylindrical projection of a circle* (ABC) *with centre* D *on a plane* (P) *not parallel to the plane of the circle is a circle or an ellipse.*

Let us make p the cylindrical projection being considered.

a) The plane (P) passes through D. It cuts the plane (ABC) following diameter AB. Let $DC \perp AB$. For every point E of the circle projected orthogonally in H on to AB, we have $EH^2 = HA \cdot HB$, $y^2 = x\,(2a - x)$, if $AB = 2a$.

If $F = p\,(E)$, triangle FEH is defined up to a similarity; if $G = p\,(C)$, we have $\frac{EH}{FH} = \frac{DC}{DG} = k$; hence $k^2\,FH^2 = HA \cdot HB$. In the plane (P), FH is an ordinate y' relative to AB, $FH \parallel GD$, and for every point F such that $F = p\,(E)$, we have

$$k^2 y'^2 = x\,(2a - x), \qquad 0 \le x \le 2a.$$

The totality of points F is thus a circle or an ellipse.

b) The plane (P) does not pass through D. Let $(P') \parallel P$ and passing through D, with p the projection of (ABC) onto (P), p' the projection of (P') onto (P), and p'' the projection of (ABC) onto (P'), the three projections being made in a parallel direction to the same straight line. According to

Proposition 7, p' is a displacement, Thābit uses here, as we can see, the composition of transformations

$$p = p' \circ p''.$$

The figure obtained in (P) is thus equal to the figure obtained in (P'), and is a circle or an ellipse.

Let us observe that the thirteenth-century mathematician Ibn Abī Jarrāda in the course of his commentary on this proposition made a study of the cylindrical projection of an ellipse in order to provide a more important version of it (cf. Supplementary note [3]).

Proposition 11. — *Let there be an oblique cylinder* (**C**) *with bases* ABC *and* DEF *and a plane* (P) *that is neither parallel nor antiparallel to* (ABC), *that does not contain the axis and is not parallel to it. If, furthermore,* (P) ∩ (ABC) = Ø *and* (P) ∩ (DEF) = Ø, *then* (P) ∩ (**C**) *is an ellipse.*

Thābit distinguished two cases according to whether the intersection of plane P with the principal plane is parallel to the bases or is not.

According to Proposition 10, we know that (P) ∩ (**C**) is a circle or an ellipse, Thābit showed by a *reductio ad absurdum*, based on the uniqueness of the perpendicular drawn from a point to a straight line, that (P) ∩ (**C**) cannot be a circle.

From Propositions 8, 9 and 11 it is clear that the only circles lying on an oblique cylinder are situated in planes parallel to the planes of the bases or antiparallel to those planes.

2.4.2.2. *Area of an ellipse and elliptical sections*

Proposition 12. — *The plane sections of two cylinders with circular bases having the same axis and the same height are homothetic, the centre of homothety being their common centre lying on the axis and the ratio of homothety being the ratio of the diameters of the base circles.*

Thābit takes as the property characterizing two ellipses with similar axes $(2a, 2b)$ and $(2a', 2b')$ the equality $\frac{a'}{a} = \frac{b'}{b}$.

This equality is not set forth in Apollonius, *Conics* VI.12, but is a consequence of it [cf. Supplementary notes].

If d and d' are the diameters of the base circles, and δ and δ' any two collinear diameters in the large and the small ellipse, we have

$$\frac{\delta'}{\delta} = \frac{d'}{d}$$

whatever the diameters being considered; hence

$$\frac{d'}{d} = \frac{2a'}{2a} = \frac{2b'}{2b} = \frac{\delta'}{\delta}.$$

The homothety $h\left(I, \frac{a'}{a}\right)$, I being the common centre of the two ellipses, has been defined from equalities in ratios resulting uniquely from the cylindrical projection in a parallel direction to the cylinder's axis.

Proposition 13. — *Let there be an ellipse with major axis* AC = 2a *and with minor axis* 2b *and a circle of diameter* AC. *For every perpendicular to* AC *cutting the circle, the ellipse and the axis respectively in* G, H and I *we have* $\frac{\mathrm{GI}}{\mathrm{HI}} = \frac{\mathrm{b}}{\mathrm{a}}$.

The proof uses the characteristic property of the circle and that of the ellipse in relation to AC. We have

$$y^2 = x\ (2a - x),$$

$$y'^2 = \frac{cx}{2a}(2a - x),$$

c being the *latus rectum* relative to the axis AC; hence

$$\frac{y'^2}{y^2} = \frac{c}{2a} = \frac{b^2}{a^2}.$$

Thābit thus defined an orthogonal affinity with axis AC and with ratio $\frac{b}{a} <$ 1 in which the ellipse is the image of the circle of diameter AC, an affinity which is a contraction.

In the same way, the ellipse is the image of the circle having as diameter its minor axis in an orthogonal affinity with ratio $\frac{a}{b} > 1$, which is a dilatation.

Let there be in an orthogonal reference system, with $b < a$:

$$(C_1) = \{(x, y),\quad x^2 + y^2 = a^2\ \},$$
$$(C_2) = \{(x, y),\quad x^2 + y^2 = b^2\ \},$$
$$(E)\ = \{(X, Y),\quad \frac{X^2}{a^2} + \frac{Y^2}{b^2} = 1\ \}.$$

1) $(E) = \varphi\,((\mathrm{C}_1))$, with $\varphi : (x, y) \rightarrow (X, Y)$, $\begin{cases} X = x, \\ Y = \frac{b}{a}y; \end{cases}$

φ is a contraction.

2) $(E) = \Psi((C_2))$, with $\Psi: (x, y) \to (X, Y)$, $\begin{cases} X = \frac{a}{b}x, \\ Y = y; \end{cases}$

Ψ is a dilatation.

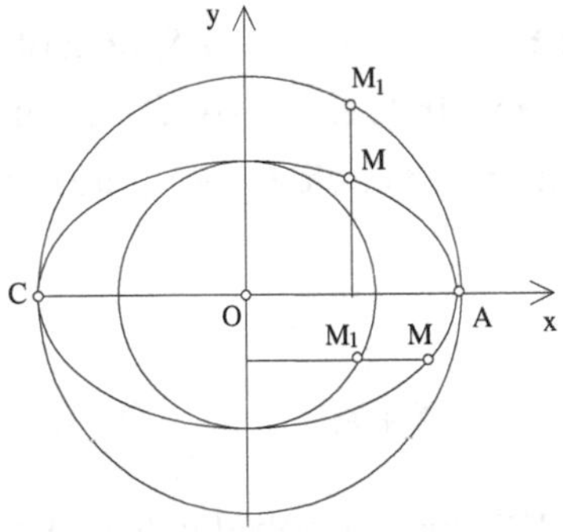

Fig. 2.4.2

If one calls principal diameters the major axis and the minor axis of an ellipse or the diameter of a circle, Proposition 13 is equivalent to the following:

Two closed conics, ellipse or circle, that have in common a principal diameter $2a$, and for second diameter $2b$ and $2b'$, can be derived from each other by an orthogonal affinity

$$\begin{pmatrix} x' \\ y' \end{pmatrix} = A \begin{pmatrix} x \\ y \end{pmatrix}, \qquad \text{with } A = \begin{pmatrix} 1 & 0 \\ 0 & \frac{b}{b'} \end{pmatrix},$$

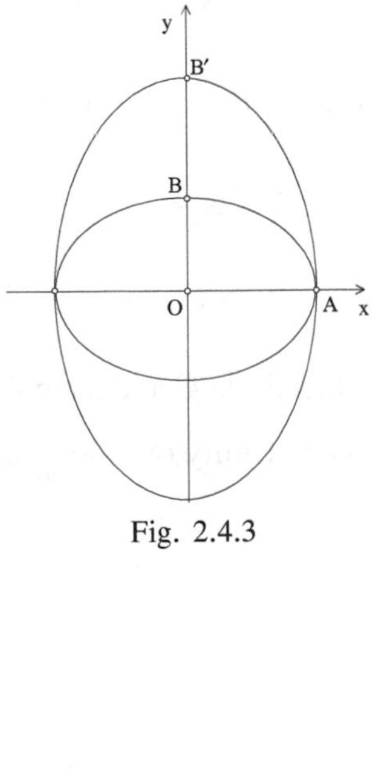

Fig. 2.4.3

the coordinates being related to the same orthogonal basis with $\overline{OA} = a$, $\overline{OB} = b$, $\overline{OB'} = b'$, $\overline{OB'} \neq a$ or $\overline{OB'} = a$.

Proposition 14. — *If* S *is the area of the ellipse* **E** *with axes* 2a *and* 2b *and* Σ *that of the circle* E *with radius* $\mathrm{r} = \sqrt{\mathrm{ab}}$, *then* S = Σ.

Notation:

S area of the ellipse **E** → S_n area of P_n inscribed in **E**,

Σ area of the equivalent circle *E* → Σ_n area of Π_n inscribed in *E*,

S′ area of the circumscribed circle **C** → S'_n area of P'_n inscribed in **C**.

Thābit's proof:

a) If $S > \Sigma$, then

(1) $$S = \Sigma + \varepsilon.$$

Let P_n be a polygon of 2^{n+1} sides inscribed in the ellipse **E** and derived from P_{n-1}, the number of vertices of which are doubled in cutting the ellipse by diameters that pass through the middles of the sides of P_{n-1}. The first polygon P_1 is the rhombus defined by the vertices of the ellipse. If S_n is the area of P_n, we have successively

$$S_1 > \frac{1}{2} S \Rightarrow S - S_1 < \frac{1}{2} S$$

$$S_2 - S_1 > \frac{1}{2}(S - S_1) \Rightarrow S - S_2 < \frac{1}{2^2} S$$

$$\ldots$$

$$S_n - S_{n-1} > \frac{1}{2}(S - S_{n-1}) \Rightarrow S - S_n < \frac{1}{2^n} S \cdot$$

So for ε defined by (1), there exists $n \in \mathbf{N}$ such that $\frac{1}{2^n} S < \varepsilon$; hence

$$S - S_n < \varepsilon,$$

$$S_n > \Sigma.$$

We may therefore consider the circle **C** and the polygon P'_n derived from **E** and from P_n by the orthogonal affinity of ratio $\frac{a}{b}$ and let S'_n be the area of P'_n and S' the area of **C**:

$$\frac{S_n}{S'_n} = \frac{b}{a} = \frac{ab}{a^2} = \frac{\Sigma}{S'};$$

but $S_n > \Sigma$, hence $S'_n > S'$, which is impossible.

b) If $S < \Sigma$, we have

$$\frac{S}{S'} < \frac{\Sigma}{S'};$$

hence

(2)
$$\frac{\Sigma}{S'} = \frac{S}{S' - \varepsilon'}.$$

Taking the circle **C** and the preceding polygons P'_n, we have successively:

$$S' - S'_1 < \frac{1}{2} S',$$
$$S' - S'_2 < \frac{1}{2^2} S',$$
$$\ldots$$
$$S' - S'_n < \frac{1}{2^n} S'.$$

So for ε' defined by (2), there exists $n \in \mathbf{N}$ such that $\frac{1}{2^n} S' < \varepsilon'$, therefore

(3)
$$S' - S'n < \varepsilon'.$$

If P_n is the polygon inscribed in **E** corresponding to P'_n by the orthogonal affinity of ratio $\frac{b}{a}$, then

$$\frac{S_n}{S'_n} = \frac{\Sigma}{S'} = \frac{S}{S' - \varepsilon'};$$

but by (3)

$$S'_n > S' - \varepsilon',$$

hence $S_n > S$, which is absurd.

From a) and b) we therefore deduce $S = \Sigma$.

Comments

We move from the ellipse **E** to the circle **C** by the orthogonal dilatation f of ratio $k_1 = \frac{a}{b}$ and from the circle **C** of radius a to the circle E of radius r

such that $r^2 = ab$ by a homothety h of ratio $k_2 = \frac{r}{a} = \frac{\sqrt{ab}}{a} = \sqrt{\frac{b}{a}}$. Thus $E = h \circ f(\mathbf{E})$, the transformation $h \circ f$ retaining the areas since $k_1 \cdot k_2^2 = 1$.

The object of Proposition 14 is precisely to show this property in the case of the ellipse **E**.

With the foregoing notations Thābit made use of $\frac{\Sigma}{S'} = \frac{b}{a} = k_2^2$ and showed that $\frac{S_n}{S'_n} = \frac{b}{a} = \frac{1}{k_1}$ for any value of n; hence

$$S = \Sigma \Leftrightarrow \frac{S}{S'} = \frac{\Sigma}{S'} \Leftrightarrow \frac{S}{S'} = \frac{S_n}{S'_n}.$$

Thābit's method thus corresponds to the following two stages:

a) $$\frac{S_n}{S'_n} < \frac{S}{S'}, \qquad \text{so } \frac{S_n}{S'_n} = \frac{S - \varepsilon_1}{S'} \qquad (1).$$

We can show that

$$\exists\ P_n \subset \mathbf{E} \text{ such that } S - \varepsilon_1 < S_n < S.$$

Now

$$f(P_n) = P'_n \subset \mathbf{C} \text{ proves (1);}$$

hence

$$S'_n > S',$$

which is impossible.

b) $$\frac{S_n}{S'_n} > \frac{S}{S'}, \qquad \text{so } \frac{S_n}{S'_n} = \frac{S}{S' - \varepsilon_2} \qquad (2).$$

We can show that $\exists\ P'_n \subset \mathbf{C}$ such that $S' - \varepsilon_2 < S'_n < S'$. Now $f^{1}(P'_n) = P_n \subset \mathbf{E}$ proves (2); hence

$$S_n > S,$$

which is impossible.

We have thus proven that

$$\frac{S}{S'} = \frac{S_n}{S'_n}.$$

Thus moving away from the property of the orthogonal affinity, which expresses that the ratio of the areas S'_n and S_n of the two homologous polygons P_n and P'_n is, for any value n, equal to the ratio $\frac{a}{b}$ of the affinity, Thābit deduced from it that the same holds for area S of the ellipse **E** and area S' of **C**. This amounts to saying that the ratio is retained when reaching the limit

$$\forall n \qquad \frac{S_n}{S'_n} = \frac{b}{a}$$

and

$$\frac{S}{S'} = \frac{\lim S_n}{\lim S'_n} = \lim \frac{S_n}{S'_n} = \frac{b}{a}.$$

We may observe that Luca Valerio took this type of assertion as the basis of his method.[3] This method did not involve the use of integral sums.

It only remains to say that this same result had been obtained by Archimedes in *On Conoids and Spheroids*, Proposition 4. But this book was unknown to the mathematicians of the time, including Thābit. To compare the approach of the former with that followed by the latter is doubly advantageous: we would be in a position to form a better understanding of the contribution of the ninth-century mathematician, and also to apprehend better what knowledge there was of the Archimedean *corpus* at this time.

Proposition 4 of *On Conoids and Spheroids*[4] may be rewritten, if one makes use of the notations of Thābit's Proposition 14:

The ratio of an area S *of an ellipse* **E** *of major axis* 2a *and minor axis* 2b *to the area* S′ *of a circle* **C** *of diameter* 2a *is* $\frac{\mathrm{S}}{\mathrm{S'}} = \frac{\mathrm{b}}{\mathrm{a}}$.

Archimedes immediately brought it back to the statement of a proposition equivalent to Thābit's Proposition 14. He defined the circle Φ of area Σ, such that $\frac{\Sigma}{S'} = \frac{b}{a}$, and wrote, 'I say that Φ is equivalent to **E**', in other words that $\Sigma = S$. The circle Φ is none other than Thābit's circle E.

α) $\Sigma > S$.

Let Π_n be a regular polygon of 2^{n+1} sides inscribed in E with area Σ_n such that $\Sigma_n > S$. So if φ_1 is the similarity of ratio $\sqrt{\frac{a}{b}}$ and φ_2 the orthogonal affinity of ratio $\frac{b}{a}$, we have

[3] *De Centro Gravitatis Solidorum Libri Tres*, Bologna, 1661, Book II, Propositions I–III, pp. 69–75.

[4] Archimedes, *On Conoids and Spheroids*, text established and translated by Charles Mugler, Collection des Universités de France, Paris, 1970, vol. I, pp. 166–9.

φ_1: $E \to \mathbf{C}$,
$\Pi_n \to P'_n$, polygon inscribed in **C**;
φ_2: $\mathbf{C} \to \mathbf{E}$,
$P'_n \to P_n$, polygon inscribed in **E**.

We thus have

$$\frac{S'_n}{\Sigma_n} = \frac{a}{b} \text{ and } \frac{S_n}{S'_n} = \frac{b}{a};$$

hence

$$S_n = \Sigma_n,$$

which is impossible because $S_n < S$ and we put $\Sigma_n > S$.

Let us note that Archimedes did not prove that the correspondence φ_2 is an orthogonal affinity: he employed $\frac{b}{a}$ without justification.

β) $\Sigma < S$.

Let P_n be a regular polygon of 2^{n+1} sides inscribed in **E**, such that $S_n > \Sigma$.

φ_2^{-1}: $\mathbf{E} \to \mathbf{C}$,
$P_n \to P'_n$;
φ_1^{-1}: $\mathbf{C} \to E$,
$P'_n \to \Pi_n$.

We have

$$\frac{S'_n}{S_n} = \frac{a}{b} \text{ and } \frac{\Sigma_n}{S'_n} = \frac{b}{a};$$

hence

$$\Sigma_n = S_n,$$

which is impossible because $\Sigma_n < \Sigma'$, and we put $S_n > \Sigma'$.

From α) and β) we may deduce $S = \Sigma$.

Thābit took the two parts of his proof in reverse order to those in Archimedes.

• $\Sigma < S$ (a for Thābit, β for Archimedes).

Thābit related in detail the explanation of the construction of the polygons P_n and made use of Apollonius I.17 in order to introduce the coefficient $\frac{1}{2}$ so as to apply Euclid's Proposition X.1 to explain the existence of n such that $S - S_n < \varepsilon$, with $\varepsilon = S - \Sigma$. In β Archimedes did not explain the construction of P_n: he reckoned it was obtained as in α from Π_n, but he gave no explanation here of the existence of n such that $S_n > \Sigma$.

Thābit then passed from P_n to P'_n by the orthogonal dilatation $f = \varphi_2^{-1}$, which he characterized in Proposition 13 and which Archimedes employed without justification. Both of them established that $\frac{S_n}{S'_n} = \frac{b}{a}$, by breaking down the polygons into trapeziums and triangles.

Thābit did not involve Π_n inscribed in E.

• $\Sigma > S$ (b for Thābit, α for Archimedes).

Archimedes began with Π_n, of area Σ_n inscribed in E such that $\Sigma > \Sigma_n > S$.

He had already used the existence of such a polygon in *The Measurement of a Circle*, Proposition 1, and had considered this existence as 'evident' in Proposition 6 of *The Sphere and the Cylinder*, and 'conveyed in the *Elements*'. Thābit began directly with P'_n and justified as in a) the existence of n such that $S' - S'_n < \varepsilon'$. Both of them used as previously $\frac{S_n}{S'_n} = \frac{b}{a}$. The two authors used the following as postulate: of two plane surfaces one of which surrounds the other, the surrounded surface is the smaller.

Comments on Archimedes — Archimedes made use of the right cylinder and the isosceles cone in several propositions of *The Sphere and the Cylinder* (Propositions 7, 10, 11 and 12) and amongst the lemmata that precede Proposition 17, Lemma 5 clearly shows that the cones considered are isosceles. At no point in his text is there any question of the oblique cylinder or the scalene cone. The author holds with Euclid's Definitions XI.21 and 28.

In his treatise on *Conoids and Spheroids* no definition is given for the three conics.

From Proposition 4 Archimedes deduced two propositions – 5 and 6 – and a corollary.

Proposition 5. — *If* E *is an ellipse with axes* 2a *and* 2b *and* C *a circle of diameter* d = 2r, *we have*

$$\frac{S(E)}{S(C)} = \frac{4ab}{d^2} = \frac{ab}{r^2}.$$

Proposition 6. — *If* E *is an ellipse with axes* 2a *and* 2b *and* E′ *an ellipse with axes* 2a′ *and* 2b′, *we have*

$$\frac{S(E)}{S(E')} = \frac{ab}{a'b'}.$$

Corollary. If E *and* E′ *are similar,*

$$\frac{S(E)}{S(E')} = \frac{a^2}{a'^2} = \frac{b^2}{b'^2}.$$

Thābit's Propositions 21 and 22, which he deduced from Proposition 14, are particular cases of Archimedes' Proposition 5. If S_m is the area of a minimal ellipse, S_M that of the maximal ellipse and S that of the circle of radius r, the base of the cylinder, we have the following:

Proposition 21.

$$\frac{S_m}{S} = \frac{b_m}{r} \qquad \text{(since } a_m = r\text{)}.$$

Proposition 22.

$$\frac{S_M}{S} = \frac{a_M}{r} \qquad \text{(since } b_M = r\text{)}.$$

Proposition 23 is a corollary of Propositions 21 and 22, but is also a consequence of Proposition 6.

Proposition 23.

$$\frac{S_m}{S_M} = \frac{b_m}{a_M}.$$

Proposition 27, which Thābit derived from Proposition 14, is none other than Archimedes' Proposition 6 and its corollary.

Proposition 15. — *Let* **E** *be an ellipse of major axis* EB = 2a, *with minor axis* 2b *and* E *the equivalent circle with radius* $r = \sqrt{ab}$. *A chord* AC ⊥ BE *and a chord* IL *of the circle separate in* **E** *and in* E *respectively the segments* ABC *of area* S_1 *and* IKL *of area* Σ_1, *if* $\frac{AC}{b} = \frac{IL}{\sqrt{ab}}$, *so* $S_1 = \Sigma_1$.

If **C** is the circle of diameter *EB*, by orthogonal affinity *f* with axis *EB* and with ratio $\frac{a}{b}$, Thābit associated with segment *ABC* a segment *TBV*. He constructed in segment *ABC* a polygon P_n by the procedure indicated in Proposition 14 and by *f* associated with it a polygon P'_n. He showed that segments *TBV* and *IKL* correspond in a homothety of ratio $\sqrt{\frac{b}{a}}$; therefore

(1) $$(IKL) = h \circ f((ABC)).$$

The proof is identical then to that of Proposition 14.

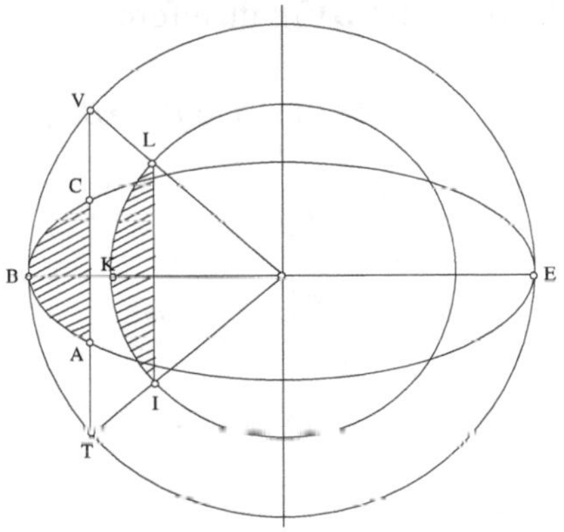

Fig. 2.4.4

Comment. — From (1) and knowing *ABC*, we may deduce a simple geometric construction of *IKL* if we suppose that *E* and **E** are concentric.

We have in fact

$$\frac{AC}{IL} = \sqrt{\frac{b}{a}} \text{ and } \frac{IL}{TV} = \sqrt{\frac{b}{a}};$$

hence

$$IL^2 = AC \cdot TV.$$

The chord *IL* of the circle *E* is the geometrical mean of the chords *AC* of the ellipse and *TV* of the circle, chords taken by a same perpendicular to axis *BE*, hence a simple construction of *IL*.

Proposition 16. — *Let* **E** *be an ellipse with minor axis* EB = 2b *and with major axis* 2a *and circle* E *equivalent to* **E**. *A chord* AC ⊥ BE *and a chord* IL *of the circle separate in* **E** *and in* E *segments* (ABC) *and* (IKL) *with respective areas* S_2 *and* Σ_2. *If*

$$\frac{AC}{a} = \frac{IL}{\sqrt{ab}},$$

then

$$S_2 = \Sigma_2.$$

If **C′** is the circle of diameter *EB*, it is the image of **E** in the affinity *f′* of axis *EB* and with ratio $\frac{b}{a}$; Thābit then constructed $(UBV) = f'((ABC))$.

The circle *E* is derived from **C′** in a homothety *h′* of ratio $\sqrt{\frac{a}{b}}$, and Thābit showed that $(IKL) = h'((UBV))$; therefore

$$(2) \qquad (IKL) = h' \circ f'((ABC)).$$

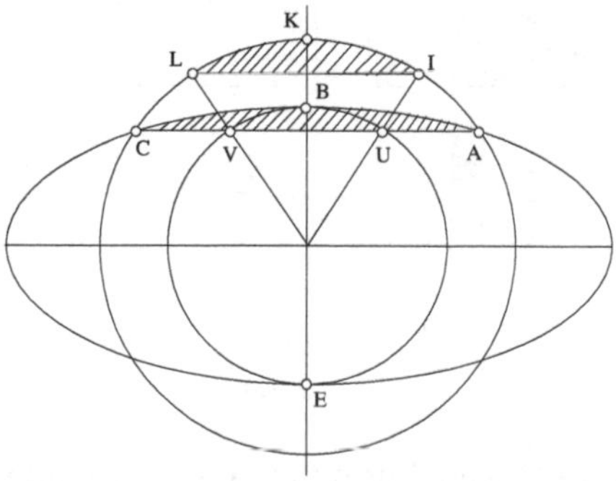

Fig. 2.4.5

Polygons P_n and P'_n are defined as before and the proof is identical then to that of Proposition 14.

Comment. — From (2) we may deduce a geometric construction of *IKL*.
As in Proposition 15, we can write

$$\frac{AC}{IL} = \sqrt{\frac{a}{b}},$$

and on the other hand we have

$$\frac{IL}{UV} = \sqrt{\frac{a}{b}};$$

hence

$$IL^2 = AC \cdot UV.$$

Proposition 17. — *Let* **E** *be an ellipse, one of the axes of which is* DE *and* E *the equivalent circle of which* NO *is a diameter. From two points* A *and* C *of the ellipse are dropped the perpendiculars* AI *and* CK *onto* DE *and from the points* L *and* M *of the circle perpendiculars* LP *and* MU *onto* NO. *If the position of* LM *in relation to* NO *is the same as that of* AC *in relation to* DE, *if the position of points* P *and* U *in relation to the centre of the circle is the same as that of points* I *and* K *in relation to the centre of the ellipse, and if,* 2a *being the major axis and* 2b *the minor axis, we have*

$$\frac{\mathrm{LP}}{\sqrt{\mathrm{ab}}} = \frac{\mathrm{AI}}{\mathrm{b}} \quad \textit{and} \quad \frac{\mathrm{MU}}{\sqrt{\mathrm{ab}}} = \frac{\mathrm{CK}}{\mathrm{b}} \qquad (\textit{when } \mathrm{DE} = 2\mathrm{a})$$

or

$$\frac{\mathrm{LP}}{\sqrt{\mathrm{ab}}} = \frac{\mathrm{AI}}{\mathrm{a}} \quad \textit{and} \quad \frac{\mathrm{MU}}{\sqrt{\mathrm{ab}}} = \frac{\mathrm{CK}}{\mathrm{a}} \qquad (\textit{when } \mathrm{DE} = 2\mathrm{b})$$

then the two segments separated by AC *in the ellipse and by* LM *in the circle are both equivalent to each other.*

Notation: areas of a segment S_{sg},
triangle S_{tr},
trapezium S_{tp},
area of **E** or of E: S.

The method used by Thābit consists of establishing the areas of the segment of an ellipse and the segment of a circle considered here, with the aid of sums or differences of respectively equal areas.

Thābit distinguished eight cases. Let us extend the perpendiculars AI and CK as far as Q and R and the perpendiculars LP and MU as far as V and T. Let S be the area of the ellipse and of the circle.

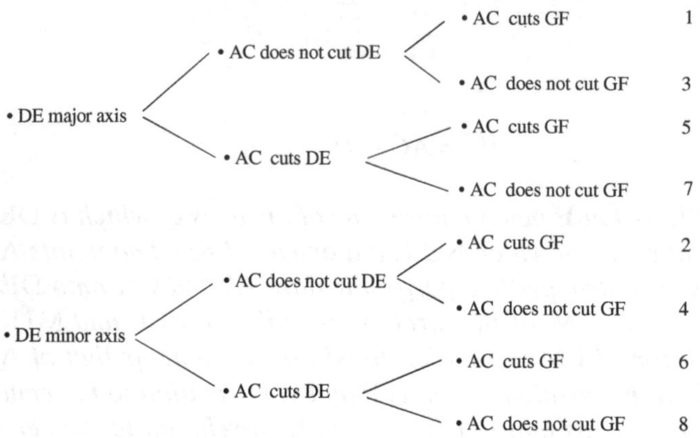

1) $S_{sg}(ABC) < \frac{1}{2}\, S$ and $S_{sg}(LM) < \frac{1}{2}\, S$.

According to Propositions 15 and 16, we have in all cases

$$S_{sg}(ADQ) = S_{sg}(LNV) \text{ and } S_{sg}(CDR) = S_{sg}(MNT).$$

a) In figures 1, 2, 3, 4 we have from the hypotheses

$$S_{tp}(AQRC) = S_{tp}(LVTM).$$

Likewise, we have

$$S_{sg}(ABC) = \frac{1}{2}\, [S_{sg}(CDR) - S_{sg}(ADQ) - S_{tp}(AQRC)],$$

$$S_{sg}(LM) = \frac{1}{2}\, [S_{sg}(MNT) - S_{sg}(LNV) - S_{tp}(LVTM)];$$

hence

$$S_{sg}(ABC) = S_{sg}(LM).$$

b) For figures 5, 6, 7, 8 we have

$$S_{sg}(ABC) = S_{sg}(ADQ) + S_{sg}(QC) + S_{tr}(AQC),$$

$$S_{sg}(LVM) = S_{sg}(LNV) + S_{sg}(VM) + S_{tr}(LVM).$$

According to Propositions 15 and 16, we have

$$S_{sg}(ADQ) = S_{sg}(LNV),$$

by a) we have

$$S_{sg}(QC) = S_{sg}(VM),$$

and by the hypotheses we have

$$S_{tr}(AQC) = S_{tr}(LVM).$$

Therefore

$$S_{sg}(ABC) = S_{sg}(LVM).$$

2) $S_{sg}(ABC) > \frac{1}{2}\, S$ and $S_{sg}(LVM) > \frac{1}{2}\, S$.

By 1) we know that

$$S - S_{sg}(ABC) = S - S_{sg}(LVM);$$

hence

$$S_{sg}(ABC) = S_{sg}(LVM).$$

3) If $S_{sg}(ABC) = \frac{1}{2}\, S$, then $S_{sg}(LVM) = \frac{1}{2}\, S = S_{sg}(ABC)$.

If we relate to the same reference system (Ox, Oy), ellipse **E**, circle **C** having for diameter the major axis and circle E equivalent to **E**, O being their common centre, we saw in Propositions 14 and 15 that $E = h \circ f(\mathbf{E})$.

$$f \;:\; \mathbf{E} \;\rightarrow \mathbf{C},$$
$$(x, y) \rightarrow (x', y') = \left(x, \frac{a}{b} y\right),$$
$$h \;:\; \mathbf{C} \;\rightarrow E,$$
$$(x', y') \rightarrow (X, Y) = \left(\sqrt{\frac{b}{a}} x', \sqrt{\frac{b}{a}} y'\right).$$

Therefore

$$\frac{Y}{\sqrt{ab}} = \frac{y}{b};$$

this is the relation given by Thābit in the case where points A and C are projected on to the major axis. Proposition 17 thus proves that if M_1 and M_2 are two points of the ellipse, and M''_1 and M''_2 their images by $h \circ f$, then

$$S_{sg}(M_1M_2) = S_{sg}(M''_1M''_2).$$

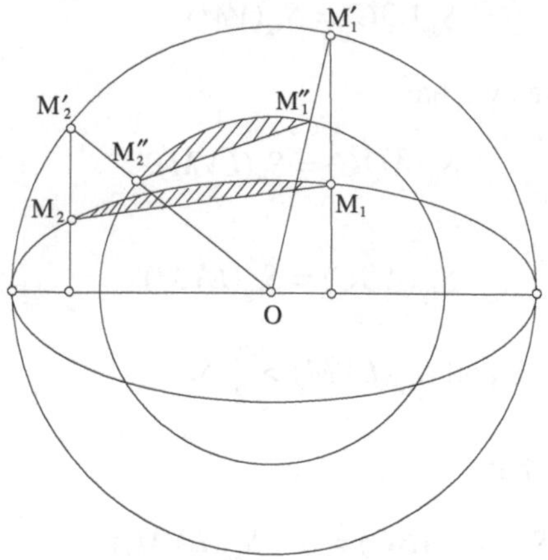

Fig. 2.4.6

A geometric construction has been deduced from it.

2.4.2.3. *Concerning the maximal section of the cylinder and concerning its minimal sections*

Proposition 18. — *The section of an oblique cylinder with axis* IK *with height* IL *by a plane* (P) *perpendicular to* IK *is an ellipse with axes* 2a *and* 2b, a > b, *such that* 2a = d, *the diameter of the base circle, and*

$$\frac{b}{a} = \frac{IL}{IK}.$$

By Proposition 11, we know that the section is an ellipse **E**. The plane (Q) which passes through IK and is perpendicular to the principal plane cuts the plane (P) following a diameter of the ellipse. This diameter is

1) equal to the diameter of the base circle,
2) the largest diameter of **E**.

Therefore $2a = d$.

We have $(P) \cap (Q) \perp (P) \cap (IKL)$; therefore the minor axis is in (IKL). By the similarity of two right-angled triangles, we can show that

$$\frac{2b}{d} = \frac{IL}{IK}.$$

Comments. — The reasoning employs Proposition 5 and the properties of straight lines and perpendicular planes.

The principal plane IKL is a plane of symmetry at one and the same time for the cylinder and for the plane (P); it is therefore a plane of symmetry for **E**, and therefore contains one of the axes of the ellipse.

The plane (P) is called a plane of right section. The cosine of the angle of (P) with the base plane is

$$\frac{IL}{IK} = \cos K\hat{I}L.$$

Proposition 19. — *Let* $\mathbf{E}_m$ *be the ellipse obtained in a plane of right section* (P) *and* **E** *an ellipse in any plane* (Q). If $2a_m$, $2b_m$, S_m *and* 2a, 2b, S *are respectively the major axis, the minor axis and the area of* $\mathbf{E}_m$ *and of* **E**, *we have*

$$a \geq a_m,\ b_m \leq b \leq a_m \ \textit{and}\ S \geq S_m.$$

1) $(P) \parallel (Q)$, so by Proposition 8: $a = a_m$, $b = b_m$, $S = S_m$.

2) $(P) \nparallel (Q)$. Let $d = 2r$ be the diameter of the base circle; then by Proposition 18, we have $a_m = r$.

If $a = r$,	then $a = a_m$ and $b < a_m$.
If $b = r$,	then $b = a_m$, $b > b_m$ and $a > a_m$.
If $a \neq r$ and $b \neq r$,	then $a > r > b$, hence $a > a_m > b$.

We therefore have in all cases $a \geq a_m$ and $b \leq a_m$.

The plane containing the axis of the cylinder and the minor axis of **E** contains a diameter δ of $\mathbf{E}_m$, $2a_m \geq \delta \geq 2b_m$, but $2b \geq \delta$; therefore $b \geq b_m$. We have in all cases $a_m \geq b \geq b_m$. We have deduced from it $a_m \cdot b_m \leq a \cdot b$; hence $S \geq S_m$.

Every ellipse obtained in a plane of right section is called a minimal ellipse.

Proposition 20. — *Let* AE *be the longer diagonal in the intersection of a cylinder* **C** *with its principal plane* GHI. *Let* (P) *be the plane containing* AE, *such that* (P) $\perp$ (GHI), *so* (P) $\cap$ (**C**) *is an ellipse* $\mathbf{E}_M$. If $2a_M$, $2b_M$, S_M *and* 2a, 2b, S *are respectively the major axis, the minor axis and the surface of* $\mathbf{E}_M$ *and of any ellipse* **E** *situated on the cylinder, then*

$$a_M \geq a,\quad b_M \geq b \quad \textit{and} \quad S_M \geq S.$$

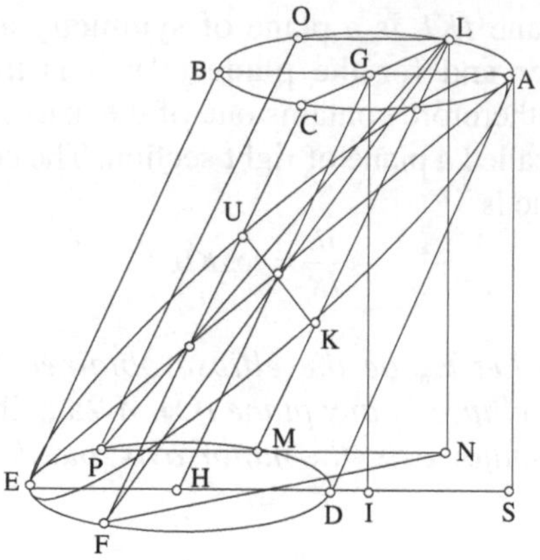

Fig. 2.4.7

1) Thābit showed that

a) AE is the largest of the segments joining two points situated on two opposite generating lines.

b) AE is the largest of the segments joining two points of any two generating lines.

AE is then the largest of the major axes of the ellipses of the cylinder, and $AE = 2a_M$, therefore $a \leq a_M$.

2) The minor axis of $\mathbf{E}_M$ is perpendicular to AE and is in (P); it is therefore perpendicular to the principal plane, so $2b_M = d$, the diameter of the base circle; but by Proposition 19, $2b \leq d = 2r$, so $b \leq b_M$. We may deduce from it $S \leq S_M$.

The ellipse $\mathbf{E}_M$ is called a maximal ellipse, it is unique in a given cylinder.

Taking into account Propositions 19 and 20, we have

$$a_m = b_M = r,$$
$$a_m \leq a \leq a_M,$$
$$b_m \leq b \leq b_M,$$
$$S_m \leq S \leq S_M.$$

Proposition 21. — *If* GH *and* GI *are the axis and the height of an oblique cylinder,* (ABC) *its base circle of diameter* d = 2r *and* $\mathrm{S_m}$ *the area of a minimal section, we have*

$$\frac{\mathrm{S_m}}{\mathrm{S(ABC)}} = \frac{\mathrm{b_m}}{\mathrm{r}} = \frac{\mathrm{b_m}}{\mathrm{a_m}} = \frac{\mathrm{GI}}{\mathrm{GH}}.$$

This result is derived immediately from Propositions 14, 18 and 19.

Proposition 22. — *If* S_M *is the area of the maximal section, then*

$$\frac{S_M}{S(ABC)} = \frac{a_M}{r} = \frac{a_M}{b_M}.$$

This result is derived from Propositions 14 and 20.

Proposition 23. — *Corollary of Propositions 21 and 22*:

$$\frac{S_m}{S_M} = \frac{b_m}{a_M}.$$

Proposition 24. — *If we have two similar ellipses* **E** *and* **E′** *with the same centre such that their major axes, just as their minor axes, are collinear, then the tangent at any point on the small ellipse determines in the large one a chord the point of contact of which is the centre.*

The similarity of two ellipses can be characterized by the equality $\frac{2a}{c} = \frac{2a'}{c'}$ (Apollonius, *Conics* VI.12), c and c' being the *latera recta* relative to axes $2a$ and $2a'$, or by the equality of the ratios of the axes $\frac{a}{b} = \frac{a'}{b'}$. By using these two equalities and Apollonius, *Conics* I.13, Thābit showed that for every half-straight line produced from the centre I and cutting the small ellipse at N and the large ellipse at L, we have

$$\text{(1)} \qquad \frac{IN}{IL} = \frac{a'}{a}.$$

This property and Proposition I.50 in the *Conics* of Apollonius permit us to conclude.

The equality (1) defines the homothety $h\left(I, \frac{a'}{a}\right)$ in which $\mathbf{E'} = h(\mathbf{E})$.

In the last paragraph, Thābit stated a result that concerns the general case of the similarity. Let $\mathbf{E''}$ be an ellipse equal to $\mathbf{E'}$, and let k be the displacement, translation or rotation, such that $\mathbf{E''} = k(\mathbf{E'})$; then $\mathbf{E''} = k \circ h(\mathbf{E})$, $k \circ h$ is a similarity. The displacement k retains the angles; therefore two homologous diameters of **E** and **E″** make equal angles with two homologous axes, as Thābit defined them.

The homothety $h\left(I, \frac{a'}{a}\right)$ has been defined here by equalities of ratios, equalities obtained starting with metric relations, in opposition to what was established in Proposition 12.

Proposition 25. — *Given two homothetic ellipses of the same centre I, construct a polygon inscribed in the large ellipse whose sides do not touch the small ellipse and admitting* I *as the centre of symmetry.*

Let AC and EG be their major axes, BD and FH their minor axes, $AC > EG$, and $BD > FH$.

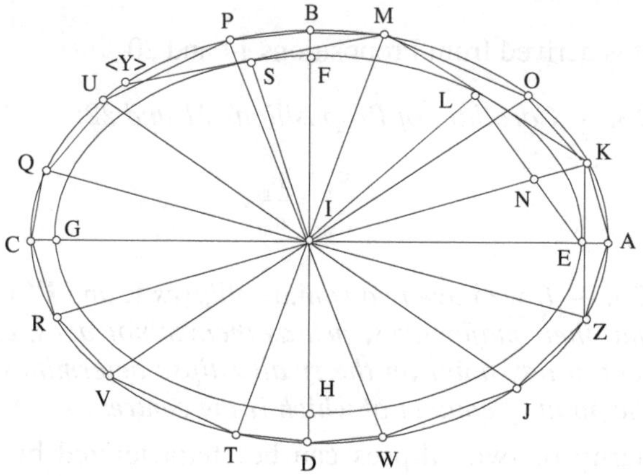

Fig. 2.4.8

Let EK be the tangent at E and let $\alpha_1 = K\hat{I}E$. We then draw the tangent KLM, then the tangent MSY and repeat until obtaining a tangent that cuts segment IB. We put $\alpha_2 = M\hat{I}E$, $\alpha_3 = Y\hat{I}E$ to the nth tangent corresponding to α_n.

Thābit showed by employing Proposition 24 and *Conics* II.29 and V.11 from Apollonius that

$$\alpha_2 > 3\alpha_1, \quad \alpha_3 > 5\alpha_1 \ldots \quad \alpha_n > (2n-1)\alpha_1;$$

and he admits (by virtue of the axiom of Eudoxus–Archimedes) the existence of n such that $(2n-1)\alpha_1 > \frac{\pi}{2}$; hence $\alpha_n > \frac{\pi}{2}$. He thus obtained the desired tangent.

The vertices of the polygon on arc AB of the large ellipse are in turn any A, K, O on arc KM defined by the second tangent, M a point on the following arc, and so on until the extremity of the $(n-1)$th tangent, and in the end the point B.

The other vertices of the polygon are obtained:

1) by symmetry with relation to axis BD;

2) by symmetry with relation to I.

We also obtain a polygon of $8(n-1)$ sides. The existence of such a polygon will make an appearance in Propositions 26, 31 and 32.

Proposition 26. — *The ratio of the perimeters of two similar ellipses is equal to the ratio of similarity.*

The reasoning is achieved using two homothetic ellipses of the same centre k, $\mathbf{E}_1$ and $\mathbf{E}_2$, of which the major axis, the minor axis and the perimeter are respectively $(2a_1, 2b_1, p_1)$ and $(2a_2, 2b_2, p_2)$. We assume $a_1 < a_2$; hence $b_1 < b_2$ and $p_1 < p_2$. We want to show that

$$\frac{p_1}{p_2} = \frac{a_1}{a_2}.$$

a) Let us assume $\frac{p_1}{p_2} > \frac{a_1}{a_2}$.

So there exists an a_3 such that $\frac{p_1}{p_2} = \frac{a_3}{a_2}$, $a_1 < a_3 < a_2$.

Let f be the homothety $\left(K, \frac{a_1}{a_2}\right)$ and g the homothety $\left(K, \frac{a_3}{a_2}\right)$; we have $\mathbf{E}_1 = f(\mathbf{E}_2)$ and may consider ellipse $\mathbf{E}_3 = g(\mathbf{E}_2)$. After Proposition 25, we know how to construct a polygon P_n of $8(n-1)$ sides inscribed in $\mathbf{E}_3$ and without common points with $\mathbf{E}_1$, let p'_3 be its perimeter; we therefore have $p_1 < p'_3 < p_2$. Cf. the postulate of Archimedes on the lengths of convex curves [*The Sphere and the Cylinder*, Postulate 2].[5]

If $\mathbf{P}_n = g^{-1}(P_n)$, $\mathbf{P}_n$ is inscribed in $\mathbf{E}_2$, let p'_2 be its perimeter; we have $\frac{p'_3}{p'_2} = \frac{a_3}{a_2}$, and therefore $\frac{p'_3}{p'_2} = \frac{p_1}{p_2}$; this is absurd because $p'_3 > p_1$ and $p'_2 < p_2$.

We therefore have $\frac{p_1}{p_2} \leq \frac{a_1}{a_2}$.

b) Let us suppose $\frac{p_1}{p_2} < \frac{a_1}{a_2}$.

Then there exists a'_3 such that $\frac{p_1}{p_2} = \frac{a_1}{a'_3}$, $a'_3 > a_2 > a_1$.

Let h be the homothety $\left(K, \frac{a'_3}{a_1}\right)$; we can construct the ellipse $\mathbf{E}'_3 = h(\mathbf{E}_1)$ and in $\mathbf{E}'_3$ a polygon P'_n without common points with $\mathbf{E}_2$, and we may

[5] Archimedes, *On the Sphere and the Cylinder*, ed. and transl. by Charles Mugler, vol. I, pp. 10–11.

deduce from it $\mathbf{P'}_n = h^{-1}(P'_n)$ inscribed in $\mathbf{E}_1$. If p'_3 and p'_1 are their respective perimeters, we have

$$\frac{p'_1}{p'_3} = \frac{a_1}{a'_3} = \frac{p_1}{p_2},$$

which is impossible since $p'_3 > p_2$ and $p'_1 < p_1$. We therefore have $\frac{p_1}{p_2} \geq \frac{a_1}{a_2}$.

From a) and b) we derive $\frac{p_1}{p_2} = \frac{a_1}{a_2}$.

The result established for the homothetic ellipses $\mathbf{E}_1$ and $\mathbf{E}_2$ is still valid if we replace $\mathbf{E}_1$ with an ellipse $\mathbf{E'}_1$ derived from $\mathbf{E}_1$ by a displacement; this extends the result to two similar ellipses.

In this proposition, starting with the fact that the ratio of the two perimeters of two similar polygons is equal to the ratio of similarity, Thābit proved that the same applies for the ratio of the perimeters of two similar ellipses.

Thābit's method is, on the one hand, based on an infinitesimal argument that revisits Proposition 25 in knowing that we can always interpose between two homothetic ellipses in relation to their common centre a polygon that is inscribed in the larger one and does not touch the small one, and this whatever the ratio of homothety even if it is very close to 1, and, on the other hand, based on an apagogic argument, namely an upper bounding or a lower bounding.

Why did Thābit not calculate the perimeter?

In Proposition 14 for determining the area of the ellipse $\mathbf{E}$, Thābit moved from $\mathbf{E}$ to the equivalent circle E by making up two transformations, orthogonal affinity and homothety, for which we know the ratio of the homologous areas.

One cannot compare the perimeter of the ellipse with that of its great circle by starting with regular polygons $\mathbf{P'}_n$ of perimeters p'_n inscribed in the circle and with their homologues[6] $\mathbf{P}_n$ of perimeters p_n inscribed in the ellipse, since the ratio of two homologous segments in the affinity in question is not constant $\frac{p_n}{p'_n} \neq \frac{b}{a}$, and the affinity does not retain the ratio of the lengths and thus cannot be of service as in the case of the areas.

Let us note, however, that this is the first time that the length of the ellipse has been considered, or more generally that of a curve, apart from the circle.

[6] $\mathbf{P}_n$ that are not regular.

Proposition 27.

a) *The ratio of the areas* S_1 *and* S_2 *of two ellipses with axes* $(2a_1, 2b_1)$ *and* $(2a_2, 2b_2)$ *is*

$$\frac{S_1}{S_2} = \frac{a_1 b_1}{a_2 b_2}.$$

b) *If the ellipses are similar and if* δ_1 *and* δ_2 *are two homologous diameters,*

$$\frac{S_1}{S_2} = \frac{a_1^2}{a_2^2} = \frac{b_1^2}{b_2^2} = \frac{\delta_1^2}{\delta_2^2}.$$

a) is a corollary of Proposition 14,
b) a corollary of a), using the ratio of similarity.

2.4.2.4. *Concerning the lateral area of the cylinder and the lateral area of portions of the cylinder lying between the plane sections touching all sides*

Proposition 28. — *Two generating lines of a right or oblique cylinder are opposite if and only if they pass through the extremities of a diameter of either elliptical or circular section.*

Following the definition, two generating lines Δ and Δ' are said to be opposite if they have originated from the extremities of a diameter of one of the bases. They are thus in a plane the same as the one containing the axis – hence the result, which we can write in the form: for two generating lines Δ and Δ' to be opposite, it must be and is sufficient that a segment joining any point on Δ to any point on Δ' touches the axis.

Proposition 29. — *In every right or oblique cylinder, the sum of the given segments on two generating lines opposed by two planes that do not cut into the interior of the cylinder and that touch all the generating lines is constant and equal to twice the given segment on the axis. One of the planes can be the plane of one of the bases.*

The proof makes use of the property of the segment joining the midpoints of the two non-parallel sides of a trapezium, which Thābit demonstrated starting with the property of the segment that joins the midpoints of two sides of a triangle.

Proposition 30. — *Let there be an oblique cylinder, a minimal section* $\mathbf{E}_m$ *and any section whatsoever, ellipse or circle, that does not touch* $\mathbf{E}_m$*. If in* $\mathbf{E}_m$ *one inscribes a polygon* P *whose vertices are diametrically opposed to each other, then the generating lines passing through the vertices of* P *determine a prismatic surface the lateral area of which is* $\Sigma = \frac{1}{2}\mathrm{p}(\ell + \mathrm{L})$*, if* p *is the perimeter of polygon* P, and ℓ *and* L *the segments lying on the two opposite generating lines between the two sections.*

The proof makes use of Proposition 29 and the area of the trapezium. The result remains true if the two sections are tangential.

Proposition 31. — *The lateral area* Σ *of a portion of an oblique cylinder included between two right sections is*

$$\Sigma = \mathrm{p} \cdot \ell,$$

where p *is the perimeter of a minimal ellipse and* ℓ *the length of the segment of generating line between the two sections.*

Let $\mathbf{E}$ be one of the sections, K its centre and $2a$ its major axis.

1) If $\Sigma < p \cdot \ell$, a length g exists, $g < p$, such that $\Sigma = g \cdot \ell$.

Let there be h such that $g < h < p$. An area ε exists such that $\Sigma + \varepsilon = h \cdot \ell$; hence $\varepsilon = \ell(h - g)$.

We may construct the ellipse $\mathbf{E}_1 = \varphi(\mathbf{E})$, φ being the homothety of centre K with ratio $\frac{a_1}{a}$ such that $1 > \frac{a_1}{a} > \frac{h}{p}$. Its perimeter p_1 is such that $\frac{p_1}{p} = \frac{a_1}{a}$ following Proposition 26; hence $\frac{p_1}{p} > \frac{h}{p}$ and consequently $p_1 > h$.

Let P_n be a polygon inscribed in $\mathbf{E}$ and without contact with $\mathbf{E}_1$, P'_n its projection onto the other base and p_n their perimeter. If Σ_n is the lateral area of the prismatic surface with bases P_n and P'_n, we have $\Sigma_n = p_n \cdot \ell$; but $p_n > p_1 > h$, and hence $\Sigma_n > h\ell$:

$$\Sigma_n > \Sigma + \varepsilon. \tag{1}$$

a) If $\frac{\varepsilon}{2} \geq s$, the areas s and s' of the two bases, which are minimal ellipses, being equal, we have $\varepsilon \geq s + s'$; hence $\Sigma_n > \Sigma + s + s'$.

The lateral area of the prism inscribed in the cylinder would be larger than the total area of the cylinder, which is absurd.

b) If $\frac{\varepsilon}{2} < s$, we furthermore impose on a_1 the condition $\frac{a_1^2}{a^2} > \frac{s - \frac{\varepsilon}{2}}{s}$, but, s_1 being the area of $\mathbf{E}_1$, $\frac{s_1}{s} = \frac{a_1^2}{a^2}$; hence $s - s_1 < \frac{\varepsilon}{2}$.

If s_n is the area of P_n, s'_n that of P'_n, we have

$$s_n = s'_n,\ \ s > s_n > s_1,\ s - s_n < \frac{\varepsilon}{2} \text{ and } \varepsilon > (s - s_n) + (s' - s'_n).$$

From (1) we may deduce $\Sigma_n > \Sigma + (s - s_n) + (s' - s'_n)$, which is absurd.

From a) and b) we deduce $\Sigma \geq p \cdot \ell$.

2) If $\Sigma > p \cdot \ell$, there exists a length g, $g > p$ such that $\Sigma = g \cdot \ell$.

Let there be h, $p < h < g$ and let ε be an area such that $\Sigma = h \cdot \ell + \varepsilon$.

Let $\mathbf{E}_1 = \varphi(\mathbf{E})$, φ being the homothety of centre K with ratio $\frac{a_1}{a}$ such that

$$\frac{a_1}{a} < \frac{h}{p} \text{ and } \frac{a_1^2}{a^2} < \frac{s + \frac{\varepsilon}{2}}{s}.$$

If p_1 in the perimeter of $\mathbf{E}_1$, we have $\frac{p_1}{p} = \frac{a_1}{a}$, hence $p_1 < h$.

We inscribe in $\mathbf{E}_1$ a polygon P_n, without common points with $\mathbf{E}$. With the notation of the first part, we have $\Sigma_n = p_n \cdot \ell$; but $h > p_1 > p_n$, and hence $\Sigma_n < h \cdot \ell$ and consequently

$$\Sigma > \Sigma_n + \varepsilon. \tag{2}$$

But

$$\frac{s_1}{s} = \frac{a_1^2}{a^2};$$

hence

$$s_1 < s + \frac{\varepsilon}{2}.$$

Now

$$s_1 - s > s_n - s;$$

hence

$$s_n - s < \frac{\varepsilon}{2}.$$

We know that

$$\Sigma_n + (s_n - s) + (s'_n - s') > \Sigma;$$

hence

$$\Sigma_n + \varepsilon > \Sigma,$$

which is the opposite of (2). We therefore have $\Sigma \leq p \cdot \ell$.

From (1) and (2) we may deduce $\Sigma = p \cdot \ell$.

Let us note that the only areas of curved surfaces considered until this point were those of the right cylinder, the right cone and the sphere (Archimedes, *The Sphere and the Cylinder*). Thābit was the first to study the area of the oblique cylinder, which we shall express by means of an elliptic integral.

Let us pursue the comparison between the results and the methods created by Archimedes to obtain them, and Thābit's Proposition 31 as well as his own methods. To proceed with this comparison, let us first recall the propositions of Archimedes set out in *The Sphere and the Cylinder*, one of the Arabic translations of which had been revised by Thābit himself. This concerns successively Propositions 11, 12 and 13 in Archimedes.

Proposition 11. — *The area σ of a portion of the lateral surface of a right cylinder contained between two generating lines is larger than the area* s *of the rectangle defined by the latter lines,* $\sigma >$ s.

Let AA' and BB' be the given generating lines and EE' any generating line in the portion under consideration:

area $(AA'BB') = s$,

area $(AEE'A') = s_1$,

area $(BEE'B') = s_2$.

We have $AB < AE + EB$; hence $s < s_1 + s_2$.

Let us put $s_1 + s_2 = s + h$.

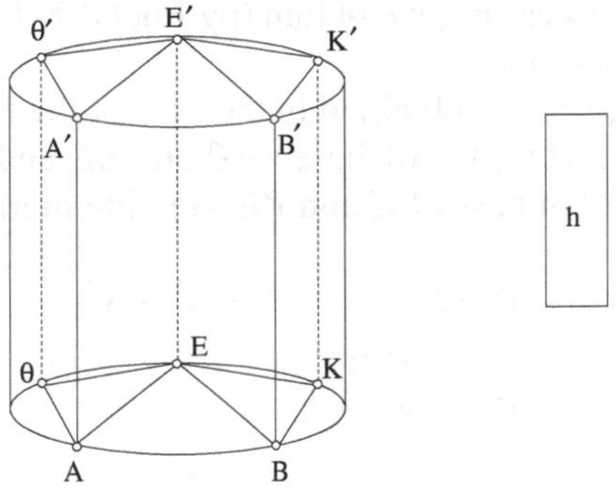

Fig. 2.4.9

1) Let us suppose $h > [S_{sg}(AE) + S_{sg}(EB)] + [S_{sg}(A'E') + S_{sg}(E'B')]$.

We know by postulate 4 of *The Sphere and the Cylinder* that

$$\sigma + [S_{sg}(AE) + S_{sg}(EB)] + [S_{sg}(A'E') + S_{sg}(E'B')] > s_1 + s_2.$$

We therefore have $\sigma + h > s + h$; hence

$$\sigma > s.$$

2) Let us suppose $h < [S_{sg}(AE) + S_{sg}(EB)] + [S_{sg}(A'E') + S_{sg}(E'B')]$.

Let θ and K be the midpoints of arcs AE and EB, and $\theta\theta'$ and KK' the generating lines coming from these points.[7]

$$S_{tr}(A\theta E) > \frac{1}{2} S_{sg}(AE);$$

therefore

$$S_{sg}(AE) - S_{tr}(A\theta E) < \frac{1}{2} S_{sg}(AE);$$

that is to say

$$S_{sg}(A\theta) + S_{sg}(\theta E) < \frac{1}{2} S_{sg}(AE),$$
$$S_{sg}(EK) + S_{sg}(KB) < \frac{1}{2} S_{sg}(EB),$$
$$S_{sg}(A'\theta') + S_{sg}(\theta'E') < \frac{1}{2} S_{sg}(A'E'),$$
$$S_{sg}(E'K') + S_{sg}(K'B') < \frac{1}{2} S_{sg}(E'B').$$

[7] Cf. Thābit's Proposition 14.

By reiterating if necessary, we obtain (by Euclid X.1) a sum of segments whose area z is smaller than h.

Let us assume this result obtained in the case of the figure, let s'_1 and s''_1 be the areas of the rectangles of bases $A\theta$ and θE and let s'_2 and s''_2 the areas of the rectangles of bases EK and KB. Bearing in mind Postulate 4,

$$\sigma + z > s'_1 + s''_1 + s'_2 + s''_2$$
$$> s_1 + s_2,$$
$$\sigma + z > s + h.$$

But $z < h$; hence

$$\sigma > s.$$

Corollary of Proposition 11. — *If* Σ *is the lateral area of a cylinder and* Σ_n *the lateral area of a prism inscribed in the cylinder, then, whatever this prism,* $\Sigma > \Sigma_n$.

Proposition 12. — *Let* AC *be an arc of the base circle of a right cylinder; the tangents at* A *and* C *intersect at* H. *The area* σ *of the part of the lateral surface of the cylinder contained between the generating lines* AA′ *and* CC′ *is smaller than the sum of the areas of the base rectangles* AH *and* HC *and the height of which is equal to* AA′, $\sigma < s_1 + s_2$.

Let B be a point on arc AC; the tangent at B cuts HA and HC respectively at E and Z. We have $EH + HZ > EZ$; hence

$$AH + HC > AE + EZ + ZC.$$

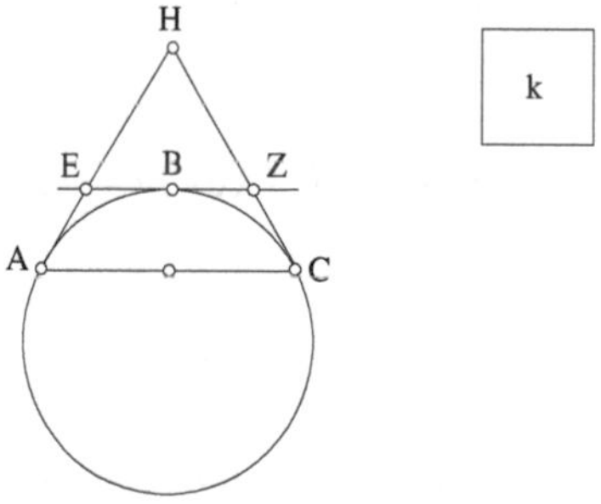

Fig. 2.4.10

Let s'_1, s'_2, s'_3 be the areas of the base rectangles AE, EZ, ZC and the height is equal to AA'. We have $s_1 + s_2 > s'_1 + s'_2 + s'_3$.

Let k be the area such that $s_1 + s_2 = s'_1 + s'_2 + s'_3 + k$.

From Postulate 4, we have, whilst taking into consideration equal figures, trapeziums or segments, in the two bases,

$$s'_1 + s'_2 + s'_3 + 2\, S_{tp}\,(AEZC) > \sigma + 2S_{sg}\,(ABC),$$
$$s'_1 + s'_2 + s'_3 + 2\,[S_{tp}\,(AEZC) - S_{sg}\,(ABC)] > \sigma\,.$$

1) If $\frac{k}{2} \geq [S_{tp}\,(AEZC) - S_{sg}\,(ABC)]$, then $s'_1 + s'_2 + s'_3 + k > \sigma$; hence $s_1 + s_2 > \sigma$.

2) If $\frac{k}{2} < [S_{tp}\,(AEZC) - S_{sg}\,(ABC)]$.

We take a point θ on arc AB and a point K on arc BC; we draw the tangents at θ and K and repeat until the sum of the differences of the areas between each trapezium obtained and the segment that is associated with it becomes smaller than $\frac{k}{2}$. We have then $s'_1 + s'_2 + s'_3 + k > \sigma$; hence $s_1 + s_2 > \sigma$.

Corollary of Proposition 12. — *If* Σ *is the lateral area of a cylinder and* Σ_n *the lateral area of a prism circumscribed around the cylinder, then whatever this prism* $\Sigma < \Sigma_n$.

Proposition 13. — *The lateral area of a right cylinder is equal to the area of a circle whose radius* r *is the geometrical mean between the generating line* ℓ *of the cylinder and the diameter* d *of its base:*

$$r^2 = \ell \cdot d.$$

Let A be the base circle, of diameter $d = CD$, and let B be the circle of radius r such that $r^2 = \ell \cdot d$.

Let Σ be the lateral area of the cylinder and S that of circle B; we want to prove that $\Sigma = S$.

1) Let us suppose $S < \Sigma$.

From Proposition 5 in Archimedes, we can construct two polygons, P_n circumscribed around B and Q_n inscribed in B, with respective areas s_n and s'_n such that $\frac{s_n}{s'_n} < \frac{\Sigma}{S}$. And let R_n be circumscribed around A, R_n similar to P_n. Let σ_n be the area of R_n and p_n its perimeter. Let us put $p_n = KD = ZL$ and $EZ = \ell$.

The prism of base R_n circumscribed around the cylinder has for lateral area

$$\Sigma_n = \ell \cdot p_n = EZ \cdot ZL.$$

Let T be the middle of CD and P such that $ZP = 2ZE$; then

$$\Sigma_n = S_{\text{tr}}\,(LZP).$$

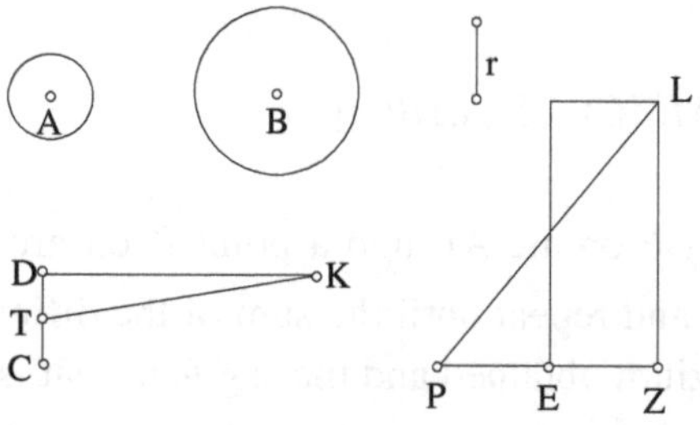

Fig. 2.4.11

On the other hand,

$$\frac{\sigma_n}{s_n} = \frac{TD^2}{r^2} = \frac{TD^2}{2TD.\,EZ} = \frac{TD}{PZ}.$$

But

$$\frac{S_{\text{tr}}(KTD)}{S_{\text{tr}}(PLZ)} = \frac{TD}{PZ} \text{ and } \sigma_n = S_{\text{tr}}\,(KTD);$$

hence

$$s_n = S_{\text{tr}}\,(PLZ),$$

and consequently

$$\Sigma_n = s_n.$$

But by hypothesis

$$\frac{s_n}{s'_n} < \frac{\Sigma}{S};$$

hence

$$\frac{\Sigma_n}{s'_n} < \frac{\Sigma}{S},$$

which is absurd since $\Sigma_n > \Sigma$ and $s'_n < S$. We therefore have $S \geq \Sigma$.

2) Let us suppose $S > \Sigma$.

From Proposition 5 in Archimedes, we can construct P_n circumscribed around B and Q_n inscribed in B such that $\frac{s_n}{s'_n} < \frac{S}{\Sigma}$. Then let there be R'_n inscribed in A and similar to Q_n; let σ'_n and p'_n be respectively its area and its perimeter.

Let us put again as in 1) $KD = ZL = p'_n$. We have

$$\sigma'_n < S_{\text{tr}}(KTD) \text{ and } \frac{\sigma'_n}{s'_n} = \frac{TD^2}{r^2} = \frac{TD}{PZ} = \frac{S_{\text{tr}}(KTD)}{S_{\text{tr}}(ZPL)};$$

hence

$$s'_n < S_{\text{tr}}(ZPL).$$

Let Σ'_n be the lateral area of the prism of base R'_n, inscribed in the cylinder. We have $\Sigma'_n = p'_n \cdot \ell = EZ \cdot ZL = S_{\text{tr}}(LZP)$; hence $\Sigma'_n > s'_n$, and consequently $s'_n < \Sigma$. But we have put

$$\frac{s_n}{s'_n} < \frac{S}{\Sigma},$$

which is impossible since $s_n > S$ and $s'_n < \Sigma$. We therefore have $S \leq \Sigma$.

From 1) and 2) we deduce $S = \Sigma$.

Thābit's Proposition 31 is, as we have seen, a stage towards the determination of the lateral area of an oblique cylinder with circular bases and of the lateral area of the whole portion of an oblique cylinder contained between two parallel planes or not.

In Thābit's Proposition 31, which concerns a portion of an oblique cylinder contained between two planes of right section, this portion is a right cylinder with elliptical base. The proposition is therefore more general than the one in Archimedes, which treats the right circular cylinder of revolution.

In Proposition 13 in Archimedes he proved that $\Sigma = \pi \cdot r^2$, with $r^2 = d \cdot \ell$; therefore $\Sigma = \pi d \cdot \ell$, and πd is the perimeter p of the base circle. Hence $\Sigma = p \cdot \ell$, which is the form of the result of Proposition 31 in Thābit in which p is the perimeter of the right section. On the other hand, the form $\Sigma = p \cdot \ell$ is the one which proceeds logically from the expression of the lateral area of a right prism, $\Sigma_n = p_n \cdot \ell$, to which both authors refer, and which extends, furthermore, to the lateral area of the oblique prism if p_n is the perimeter of a right section of the prism; $\Sigma = p \cdot \ell$ will extend to the general case of the oblique cylinder.

The definitions and postulates given by Archimedes at the beginning of the treatise of the *Sphere and the Cylinder*, concerning the concavity of surfaces and the order of size of the areas of two surfaces one of which surrounds the other in the conditions specified in Postulate 4, are employed by both authors.

Propositions 11 and 12 in Archimedes and their respective corollaries are lemmata for Proposition 13. They make use of postulate 4, in Proposition 11 for a prism inscribed in the cylinder, and in Proposition 12 for a circumscribed prism. Additionally, Proposition 11 in Archimedes uses Euclid X.1.

In Proposition 13, Archimedes used Proposition 5 in both parts of the *reductio ad absurdum*, in order to deduce from it on the one hand a prism inscribed in the cylinder and on the other hand a circumscribed prism.

In the first part of Proposition 31, Thābit began with a prism inscribed in the cylinder and showed that the hypothesis $\Sigma < p \cdot \ell$ and Postulate 2 (the lengths of convex curves) are contradictory. In the second part, he began with a prism that surrounds the cylinder, without any contact with it; Postulate 4 applied again and is in contradiction with the hypothesis $\Sigma > p \cdot \ell$.

The approaches are different. Archimedes based his reasoning on circle B equivalent to the lateral surface, and on the inscribed polygon and the circumscribed polygon associated with circle B. From this he deduced by similarity a polygon inscribed in the given circle A or a polygon circumscribing this circle.

Thābit used Proposition 25 to construct directly in the first part of his reasoning a polygon inscribed in $\mathbf{E}$ and without common point with the homothetic ellipse $\mathbf{E}_1$, and in part 2) a polygon inscribed in the ellipse $\mathbf{E}_1$ homothetic to $\mathbf{E}$, a polygon that surrounds $\mathbf{E}$, without any contact. He used Propositions 26 and 27, which give the ratio of the perimeters and the ratio of the areas of similar ellipses. His approach is more natural and leads to a proof distinctly more easy to follow.

To apply Euclid's Proposition X.1, Archimedes turned to a property of the segments of a circle, which allowed him to make apparent the coefficient $\frac{1}{2}$; by iteration he obtained $\frac{1}{2^n}$. We have seen that Thābit made use of the same procedure in Proposition 14, in applying it to the segments of an ellipse in the first part and to the segments of a circle in the second.

In Proposition 31, Thābit used 'the principle of continuity' in **R** (if $g < p$, there exists $h \in \,]\, g, p\, [$).

Proposition 32. — *The lateral area* Σ *of a portion of an oblique cylinder with circular bases contained between a right section of perimeter* p *and any section at all is*

$$\Sigma = \frac{1}{2}\, \mathrm{p}(\ell + \mathrm{L}),$$

if ℓ *and* L *are the lengths of the portions of two opposite generating lines contained between the two sections.*

1) Let us suppose $\Sigma < \frac{1}{2}\, p(\ell + L)$.

Let g be a length $g < p$ such that $\Sigma = \frac{1}{2}\, g(\ell + L)$; let h be a length and ε an area such that

$$(1) \qquad g < h < p \text{ and } \varepsilon = \frac{1}{2}\,(\ell + L)\,(h - g).$$

Let G and d be the centre and the diameter of circle **C**, the base of the cylinder, let **C′** be the circle homothetic with **C** in the homothety $\left(G, \frac{d'}{d}\right)$ such that $1 > \frac{d'}{d} > \frac{h}{p}$. The cylinder, of base **C′** and with the same axis GH as the given cylinder, cuts the plane Π with right section following an ellipse **E′** homothetic with the minimal ellipse **E**. Let p' be the perimeter of **E′**; we have $\frac{p'}{p} = \frac{d'}{d} > \frac{h}{p}$, and hence $p' > h$.

Let P_n be a polygon of $2n$ sides inscribed in **E** whose vertices are two by two diametrically opposite and that is exterior to the ellipse **E′**; let p_n be the perimeter of the polygon. We have

$$p_n > p' > h.$$

To polygon P_n we join a frustum of a prism. Its lateral area is

$$\Sigma_n = \frac{1}{2}\, p_n(\ell + L) > \frac{1}{2}\, h(\ell + L),$$

but from (1)

$$\frac{1}{2}\,(\ell + L)\,h = \varepsilon + \frac{1}{2}\,(\ell + L)\,g;$$

hence

$$(2) \qquad \Sigma_n > \Sigma + \varepsilon.$$

Let s be the area of the minimal ellipse **E** and s_1 the area of the second section $\mathbf{E}_1$; we have $s_1 > s$.

a) If $\frac{1}{2}\,\varepsilon \geq s_1$, then $\frac{1}{2}\varepsilon > s$.

From (2) we derive

$$\Sigma_n > \Sigma + s + s_1,$$

which is absurd.

b) If $\frac{1}{2}\,\varepsilon < s_1$, we place on d' a supplementary condition

$$\frac{d'^2}{d^2} > \frac{s_1 - \frac{1}{2}\varepsilon}{s_1}.$$

Let $\mathbf{E'}_1$ be the homothetic ellipse of $\mathbf{E}_1$. We have

$$\frac{s_1'}{s_1} = \frac{d'^2}{d^2} = \frac{s'}{s} > \frac{s_1 - \frac{1}{2}\varepsilon}{s_1}.$$

Hence

$$s_1' > s_1 - \frac{1}{2}\varepsilon$$

and

$$\frac{s_1 - s_1'}{s_1} = \frac{s - s'}{s};$$

hence

$$s - s' < s_1 - s'_1 < \frac{1}{2}\,\varepsilon,$$

and so

$$\Sigma + (s - s') + (s_1 - s_1') < \Sigma + \varepsilon.$$

Now, from Postulate 4

$$\Sigma + (s - s') + (s_1 - s'_1) > \Sigma_n,$$

$$\Sigma + \varepsilon > \Sigma_n,$$

which is impossible by (2); therefore

(3) $$\Sigma \geq \frac{1}{2}\, p \cdot (\ell + L).$$

2) Let us show that we cannot have $\Sigma > \frac{1}{2}\, p\ (\ell + L)$. Let LS be the larger segment of the generating line between the sections in question. The planes of right section passing through L and S determine a cylindrical surface the bases of which are the minimal ellipses **E** and $\mathbf{E}_2$; by Proposition 31, its area is $\Sigma_2 = p \cdot L$, if L is the length of the larger segment of the generating line.

The cylindrical surface between $\mathbf{E}_1$ and $\mathbf{E}_2$ has, after the first part, an area equal to

$$\Sigma_1 \geq \frac{1}{2}\, p \cdot (\ell - L).$$

But

$$\Sigma_1 + \Sigma = \Sigma_2;$$

hence

(4) $$\Sigma \leq \frac{1}{2}\, p \cdot (\ell + L).$$

From (3) and (4) we deduce

$$\Sigma = \frac{1}{2}\, p \cdot (\ell + L).$$

Comments

1) In the first part of the proof of Proposition 32, the method is the same as in Proposition 31. The prism from Proposition 31 with lateral area $\Sigma_n = p_n \cdot \ell$ that leads to $\Sigma = p \cdot \ell$ for the cylinder is replaced in Proposition 32 by a frustum of a prism with lateral area $\Sigma_n = \frac{1}{2}\, p_n\ (\ell + L)$, which leads for the frustum of a cylinder to $\Sigma = \frac{1}{2}\, p\ (\ell + L)$.

2) Instead of treating the second part with a *reductio ad absurdum* of the same type as that in the first part, Thābit showed that by assuming the conclusion of the first part to be true, $\Sigma \geq \frac{1}{2}\, p \cdot (\ell + L)$, we end up, by proceeding with the sum or difference of areas, with $\Sigma \leq \frac{1}{2}\, p \cdot (\ell + L)$.

The five propositions that follow are corollaries of Proposition 32. The notations remain the same: p is the perimeter of a right section, L and ℓ are the lengths of the segments defined on two opposite generating lines by the planes P and Q of the two sections in question, and Σ is the lateral area of the portion of cylinder contained between P and Q.

Proposition 33. — *If* P *and* Q *are any planes,*

$$\Sigma = \frac{1}{2}\ \mathrm{p} \cdot (\ell + \mathrm{L}).$$

Thābit brought in a plane of right section and proceeded by the difference of two surfaces that satisfied the conditions of Proposition 32.

Proposition 34. — *If* P *and* Q *are two parallel planes, then* $\mathrm{L} = \ell$ *and* $\Sigma = \mathrm{p} \cdot \mathrm{L}$.

If P *and* Q *are the base planes of the cylinder,* L *is then the length of the generating lines and* Σ *the lateral area of the cylinder itself.*

Comment. — If P and Q are planes of right section, we find the result of Proposition 31, we have a right cylinder with elliptical base.

Proposition 35. — *Particular case of Proposition 33.*

If the sections of the cylinder through planes P *and* Q *are tangent at a point, we have* $\Sigma = \frac{1}{2}\ \mathrm{p} \cdot \mathrm{L}$, L *being the length of the segment of the generating line opposite to the segment of no length.*

Propositions 36 and 37 are notes using Proposition 29. If ℓ_m and L_M are the lenghs of the shortest and longest segments of the generating line and L_1 the length of the segment defined on the axis of the cylinder by planes P and Q, we have:

Proposition 36.

$$\Sigma = \frac{1}{2}\ \mathrm{p} \cdot (\ell_m + \mathrm{L}_\mathrm{M}).$$

Proposition 37.

$$\Sigma = \mathrm{p} \cdot \mathrm{L}_1.$$

By their very nature, problems of rectification or calculation of the area of curved surfaces do not reduce directly to quadratures. We understand therefore that Thābit did not use integral sums in this treatise. As we have seen, the principal means implemented in the course of his research are:

- point-wise transformations,
- Archimedes' postulate 2 on convexity,
- the postulate of Eudoxus–Archimedes and Euclid X.1,
- the construction of a polygon inscribed in an ellipse and not touching a smaller homothetic ellipse.

2.4.3. *Translated text*

Thābit ibn Qurra

On the Sections of the Cylinder and its Lateral Surface

In the name of God, the Merciful, the Compassionate

THE BOOK OF THĀBIT IBN QURRA AL-ḤARRĀNĪ

On the Sections of the Cylinder and its Lateral Surface

The chapters of this treatise

At the beginning of this treatise the species of sections of the right cylinder and of the oblique cylinder are characterized; these sections have parallel sides and are surfaces with parallel sides or circles or portions of circles and the greater part of these cylinder sections are of the species called *ellipse* belonging to the conic sections, or *portion of an ellipse*.

We shall proceed by speaking of the area of a section of a cylinder, which has been determined by Abū Muḥammad al-Ḥasan ibn Mūsā – may God be pleased with him – and which is the ellipse belonging to the conic sections, and of the area of the species of portions of that section.

We shall continue by speaking of the sections of the cylinder having the greatest area, having the smallest area, having the longest diameter, having the shortest diameter, of their ratios to one another and of the ratios of their axes to one another.

The remainder of this treatise concerns the area of the lateral surface of the right cylinder and of the oblique cylinder, and the area of what is lying on the lateral surface of each of them between the sections meeting their sides.

This is the start of the treatise.

<Definitions>

If we have two equal circles in two parallel planes, if their centres are joined together by a straight line and their circumferences by another straight line – these two straight lines being in the same plane – if we fix the two circles and the straight line joining the two centres and if we rotate the second straight line on the circumferences of the two circles from a position on one of these latter until it returns to the original position – both this line and the straight line joining the two centres being in the same plane during the entire rotation – then the solid defined by this straight line and the two parallel circles is called a *cylinder*.

The straight line joining the centres of the two circles is called the *axis of the cylinder*.

The straight line joining the circumferences of the two circles and which is rotated, whatever its position, is called the *side of the cylinder*.

The two parallel circles we have mentioned are called *the two bases of the cylinder*.

The surface in which the side of the cylinder lies is called the *lateral surface of the cylinder*.

Let us call two of the sides of the cylinder, which are between the extremities of two of the diameters of its bases, *two opposite sides out of the cylinder's sides*.

Let us call the perpendicular dropped from the centre of one of the two bases of the cylinder on to the surface of the other base, *the height*[1] *of the cylinder*.

If the axis of the cylinder is its height, then that cylinder is called a *right cylinder*; if its axis is not its height, the cylinder is called an *oblique cylinder*.

The introduction of the treatise is finished.

<I. *Plane sections of the cylinder*>

– **1** – Every side of a cylinder is parallel to its axis and to all its other sides.

Let there be a cylinder, whose two bases are <the circles> *ABC* and *DEF* with centres *G* and *H*, and whose axis is *GH*; let *AD* be one of the sides of the cylinder.

I say that AD *is parallel to axis* GH *and to each of the sides of the cylinder.*

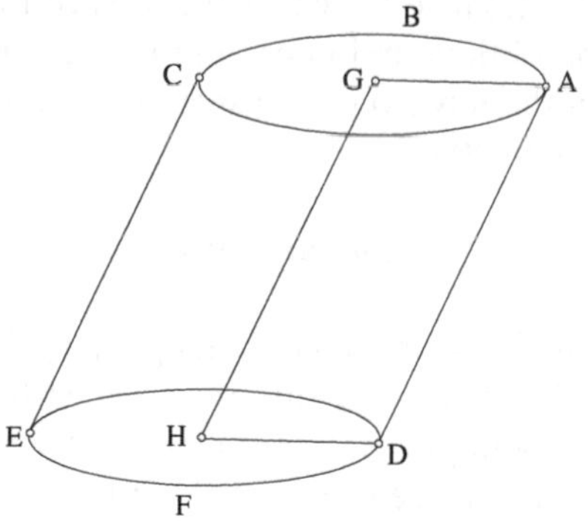

Fig. II.3.1

Proof: The straight line *AD* is one of the sides of the cylinder; it is therefore in the same plane as the axis *GH*, this plane cutting the planes of the two circles

[1] Literally: perpendicular. We shall translate it here as 'height'.

ABC and *DEF*. If we make the two intersections of this plane and the planes of circles *ABC* and *DEF* the two lines *GA* and *HD*, then the two lines *GA* and *HD* are straight lines because they are the two intersections of plane *GADH* with the planes of the two circles *ABC* and *DEF*; they are parallel since the planes of the two circles *ABC* and *DEF* are parallel, they are the two halves of the diameters of circles *ABC* and *DFE* because the centres of these two circles are the two points *G* and *H*, and they are equal because these two circles are equal. The two straight lines *AD* and *GH* which connect their extremities are therefore parallel.[2] Thus each of the sides of the cylinder is parallel to its axis. It can also be shown from this that the straight line *AD* is equal to each of the other sides of the cylinder. That is what we wanted to prove.

– **2** – Every straight line lying on the lateral surface of the cylinder is one of its sides or a portion of one of its sides.

Let there be a cylinder, whose two bases are *ABC* and *DEF*, and let there be a straight line on the lateral surface of the cylinder which is *GH*.

I say that GH *is one of the sides of the cylinder or a portion of one of its sides.*

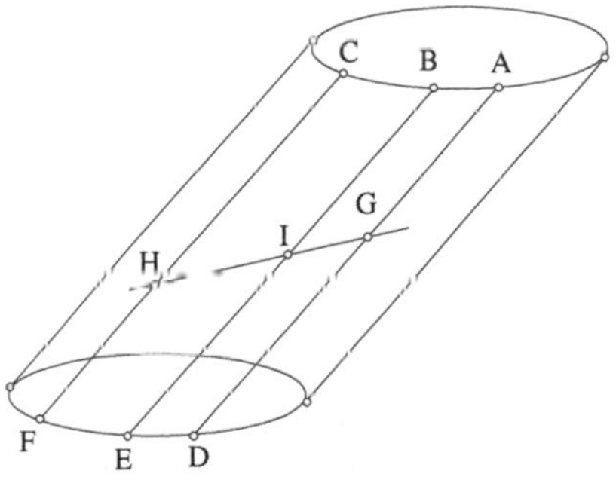

Fig. II.3.2

Proof: If we mark on the straight line *GH* any three points, let them be *G*, *I* and *H*, then these points are on the lateral surface of the cylinder since the whole of the straight line *GH* is on its lateral surface. If it was possible that the line *GH* was not one of the sides of the cylinder or a portion of one of its sides, then if we make the sides of the cylinder which pass through the points *G*, *I* and *H* the straight lines *AD*, *BE* and *CF*, none of them is superposed on the line *GH*; these straight lines are parallel, the line *GIH* is a straight line and cuts them; they are therefore in the same plane. That is why the three points *D*, *E* and *F* are in that plane, but they are also in the plane of circle *DFE*; accordingly they are at the intersection of these two planes. But every intersection of two planes is a straight line; therefore only one straight line passes through points *D*, *E* and *F*

[2] If we assume *A* and *D* are on the same side of *GH*; cf. Euclid, *Elements* I.33.

and meets the circumference of circle *DEF* at three points; that is impossible. The straight line *GH* is therefore one of the sides of the cylinder or a portion of one of its sides. That is what we wanted to prove.

– **3** – Every plane cutting a cylinder and passing through its axis or parallel to that axis cuts its lateral surface following two straight lines; if this plane does not pass through the axis and is not parallel to it, it will not cut the lateral surface of this cylinder following a straight line.

Let there be a cylinder whose bases are *ABC* and *DEF* with centres *G* and *H* and whose axis is *GH*; let any plane cut this cylinder.

I say that if this plane passes through the axis GH *or is parallel to it, then it cuts the lateral surface of the cylinder* ABCDEF *following two straight lines; but if it does not do so, then it doesn't cut the lateral surface of this cylinder following a straight line.*

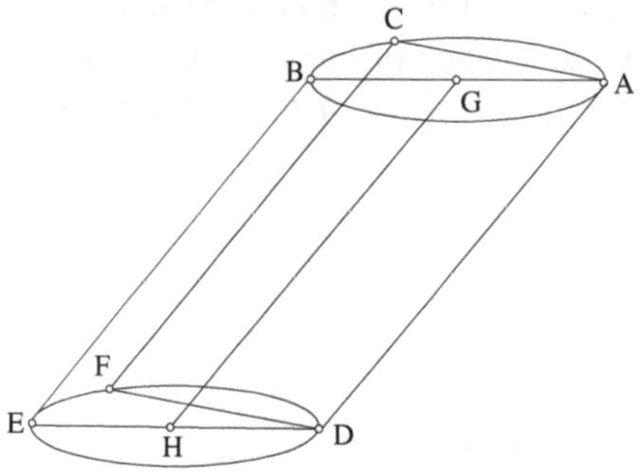

Fig. II.3.3

Proof: If the plane cutting the cylinder passes through axis *GH*, then it cuts the lateral surface of the cylinder following two lines. If we make the intersections of this plane and the lateral surface of the cylinder the two lines *AD* and *BE*, and if we draw the straight line *AGB*, then the two straight lines *AGB* and *GH* are in the plane cutting the cylinder. The line *AD* is therefore the intersection of the plane containing the two straight lines *AGB* and *GH* and the lateral surface of the cylinder, and it passes through point *A*. But the side of the cylinder drawn from point *A* is in the same plane as *GH*, and the straight line *AGB* which cuts them is likewise in this plane; the side of the cylinder drawn from point *A* is therefore in the plane containing the two straight lines *AGB* and *GH* and it is also on the lateral surface of the cylinder; accordingly it is their intersection passing through point *A*. But we have shown that the line *AD* is also their intersection passing through point *A*; therefore the line *AD* is one of the sides of the cylinder; consequently it is a straight line. In the same way we can also show that the line *EB* is a straight line.

Similarly, if the plane cutting the cylinder is parallel to axis *GH*, if we make the line *AD* an intersection[3] of the latter with the lateral surface of the cylinder and if we make the plane *ABE* passing through axis *GH* and through a point on the line *AD*, then it will pass through the whole of line *AD*, which is an intersection of the secant plane and the lateral surface of the cylinder, and it will cut the lateral surface of the cylinder with any straight line passing through a point on the line *AD*[4] and will cut the secant plane with another straight line passing through this point on the line *AD*. If we set its section onto the lateral surface of the cylinder following the straight line *AI* passing through point *A* and its section onto the plane cutting the cylinder following the straight line *AK* also passing through point *A*, then the straight line *AI* is one of the sides of the cylinder or a portion of one of its sides. In fact, it is a straight line; it is therefore parallel to axis *GH*. But the straight line *AK* is in the same plane as the straight line *GH* and is parallel to it, because if it was not parallel to it, it would have met it since it is in the same plane as it, and if it had met it, the axis *GH* would have cut the secant plane to the cylinder, since *AK* is in that plane; that is not possible, because the secant plane to the cylinder is parallel to axis *GH*. The straight line *AK* is thus parallel to axis *GH*. Now, we have shown that the straight line *AI* is also parallel to axis *GH*; therefore the two straight lines *AI* and *AK* are parallel, but they met at point *A*, which is not possible. The plane *ABE* therefore passes through the line *AD* and line *AD* is an intersection of the plane *ADFC* with the lateral surface of the cylinder, and is therefore a straight line.

In the same way, the secant plane to the cylinder cuts the lateral surface of the cylinder following another straight line. If, in fact, it was not cutting it following the straight line *AD* alone, it would be a tangent to the cylinder without cutting it, because *AD* is a straight line. If, therefore, it cuts it, it cuts its lateral surface following another straight line than *AD*, as the plane *ACFD* cuts it following the line *CF*. We can show as we did previously that the line *CF* is also a straight line.

Furthermore, if the secant plane to the cylinder does not pass through axis *GH* and is not parallel to it, and if we make the line *AD* an intersection of this plane and of a portion of the lateral surface of the cylinder, it will not be a straight line. If that was possible, let the line *AD* be a straight line; it would therefore be one of the sides of the cylinder or a portion of one of its sides and would hence be parallel to axis *GH*, and would then be in the plane *GHDA* with it. But the straight line *GH* meets the secant plane to the cylinder, and meets it accordingly at the intersection of this plane and the plane *GHDA*. But their intersection is the straight line *AD*; the straight line *GH* thus meets the straight line *AD*. But we have shown that it is parallel to it, which is contradictory. The line *AD* is therefore not a straight line. That is what we wanted to prove.

[3] *AD* is a part of the intersection.

[4] *i.e.* point *A*.

– **4** – If a plane cuts a cylinder by passing through its axis or in parallel to it, then the section generated in the cylinder is a parallelogram.

Let there be a cylinder whose two bases are *ABC* and *DEF* with centres *G* and *H*, and whose axis is *GH*. Let the cylinder be cut by a plane passing through axis *GH*, as in the first case of figure, or by a plane parallel to axis *GH*, as in the second case of figure. Let this plane generate in the cylinder the section *ABED*.

I say that ABED *is a parallelogram.*

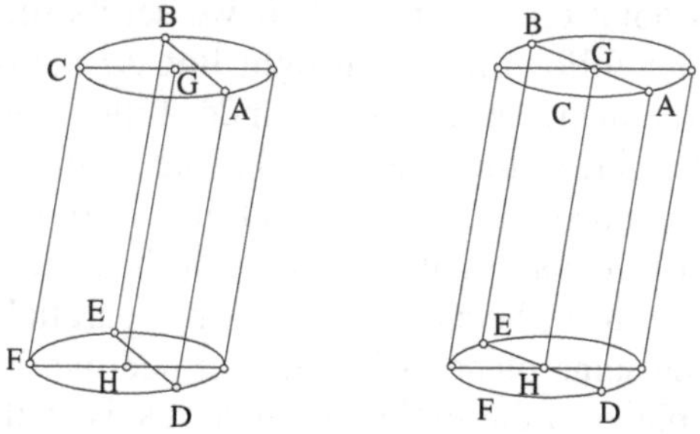

Fig. II.3.4

Proof: The two lines *AB* and *DE* are both straight lines in the two cases of figure, because they are the two intersections of the plane *ABED* with the planes of the two circles *ABC* and *DEF*, and they are parallel because the planes of these two circles are parallel. But the two lines *AD* and *BE* joining their extremities are straight lines because they are the two intersections of the plane *ABED* – which passes through axis *GH* or is parallel to it – and the lateral surface of the cylinder; they are thus two of the sides of the cylinder; that is why they are parallel. The section[5] *ABED* is therefore a parallelogram. That is what we wanted to prove.

And it is clear from what we have said that if a plane cuts a right cylinder and passes through its axis or is parallel to that axis, then the section generated in the cylinder is a rectangle.

– **5** – If a plane cuts an oblique cylinder and if it passes through its axis perpendicularly to the plane which passes through its height and through its axis, then the section which it produces in the cylinder is a rectangle, and the sections generated by all the other planes which pass through the axis are not rectangles.

Let there be an oblique cylinder whose bases are *ABC* and *DEF* with centres *G* and *H*, whose axis is *GH* and whose height is *GI*. Let a plane passing through axis *GH* cut the cylinder, let it be the plane *ABED*, and let the plane passing through axis *GH* and through the height *GI* cut the plane *ABED* perpendicularly.

[5] Literally: surface. Henceforth, in this context, we will translate it as 'section'.

I say that the section ABED *is a rectangle and that the sections generated by all the other planes which pass through axis* GH *are not rectangles.*

Proof: The straight line *GI* is perpendicular to the plane of circle *DEF*; therefore all the planes which pass through it are perpendicular to the plane of circle *DEF*. The latter is likewise perpendicular to all these planes; therefore the plane of circle *DEF* is perpendicular to the plane which passes through the two straight lines *GH* and *GI*. But the plane *ABED* is also perpendicular to the latter; therefore the intersection of these two planes, which is the straight line *DE*, is perpendicular to the plane which passes through the two straight lines *GH* and *GI*, and is accordingly perpendicular to all the straight lines drawn from point *H* in that plane. But one of these straight lines is the straight line *HG*; therefore the straight line *EH* is perpendicular to the straight line *HG* and the straight line *AD* is parallel to the straight line *GH*; therefore angle *ADH* is a right angle. But the section *ABED* is a parallelogram, and is consequently a rectangle.

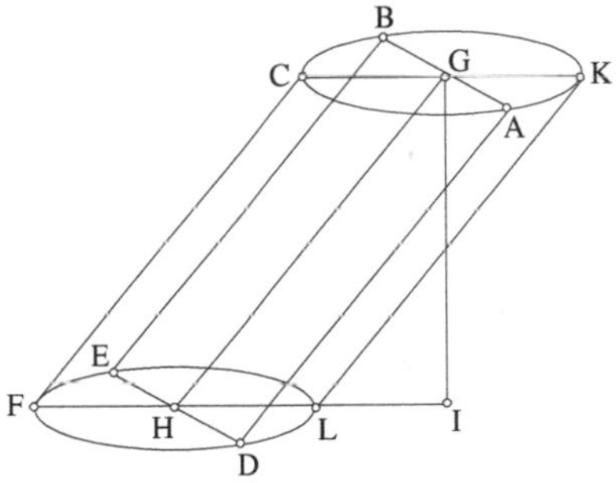

Fig. II.3.5

I say that none of the sections generated by the planes which cut the cylinder and which pass through its axis GH, *other than the section* ABED, *is a rectangle.*

If it can be otherwise, let the section *CKLF*[6] also be a rectangle <whose plane> passes through axis *GH*; the angle *CFL* is therefore a right angle and the straight line *GH* is parallel to the straight line *CF*; therefore angle *FHG* is a right angle. Now we have shown that angle *DHG* is a right angle; therefore axis *GH* is perpendicular to the plane containing the two straight lines *DH* and *FH*, which is <the plane> of circle *DEF*, the axis is therefore perpendicular to it. But the cylinder is oblique; this is impossible. Therefore *CKLF* is not a rectangle, and no other section generated by a plane which passes through axis *GH*, except for the plane *ABED*, is one. That is what we wanted to prove.

[6] This section *CKLF* is not represented in the figure in the text.

– **6** – If a plane cuts an oblique cylinder and if it is parallel to the rectangle[7] which passes through its axis, then the section generated in the cylinder is a rectangle, and there is not, amongst the sections parallel to the remaining planes which pass through its axis, any rectangular section.

Let there be an oblique cylinder, whose two bases are *ABC* and *DEF* with centres *G* and *H*, and whose axis is *GH*. On the rectangular section which passes through axis *GH*, we have *ABED*, and on the section parallel to the plane *ABED*, we have *ICFK*.

I say that the section ICFK *is a rectangle and that there is not, amongst the sections parallel to the remaining planes which pass through the axis, any rectangular section.*

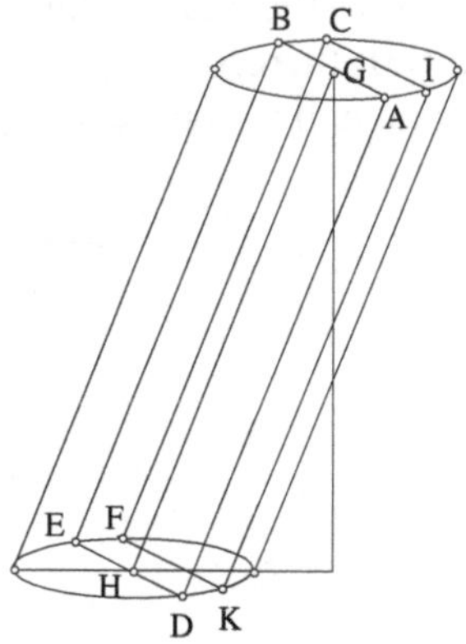

Fig. II.3.6

Proof: The two straight lines *AD* and *IK* are parallel, because they are two sides of the cylinder. But the <plane of> circle *ABCI* cut two parallel planes, that is planes *ABED* and *ICFK*; therefore the two intersections of the former and these latter – which are *AB* and *CI* – are parallel; therefore the two straight lines *AB* and *AD* are parallel to the two straight lines *IC* and *IK*, each one to its homologue. The angle *DAB* formed by the two straight lines *AB* and *AD* is therefore equal to the angle formed by the two straight lines *IC* and *IK*. But angle *DAB* is a right angle, so angle *KIC* is a right angle. But the section *ICFK* is a parallelogram, and is consequently a rectangle.

In the same way, we can make the section *ABED* one of the sections which passes through axis *GH* without being a rectangle and so that the section *ICFK* is parallel to it.

I say that the latter is not a rectangle.

Proof: We can show, as we have shown previously, that angle *DAB* is equal to angle *KIC*. But angle *DAB* is not a right angle; therefore angle *KIC* is not a

[7] Literally: to the perpendicular plane; the plane studied in Proposition 5.

right angle, so the section *ICFK* is not a rectangle. That is what we wanted to prove.

It is clear from what we have said that if a plane cuts a right cylinder and if it is parallel to its axis, then the section generated in the cylinder is a rectangle.[8]

– **7** – If we have two parallel planes each containing a figure and if a point on the circumference or on the perimeter[9] of one of the two figures is joined with a straight line to another point on the perimeter of the second figure, so that each straight line drawn from a point on the perimeter of the first figure in parallel to the first straight line drawn falls to a point on the perimeter of the second figure, then the two figures are similar and equal.

Let there be two figures in two planes, on the one we have *ABCD* and on the other we have *EFGH*, and let there be between the circumference or the perimeter of figure *ABCD* and the circumference or perimeter of figure *EFGH* a straight line, which is *AE*. Let every straight line drawn from a point on the perimeter of figure *ABCD* in parallel to the straight line *AE*, fall on a point on the perimeter of figure *EFGH*.

I say that the two figures ABCD *and* EFGH *are similar and equal.*

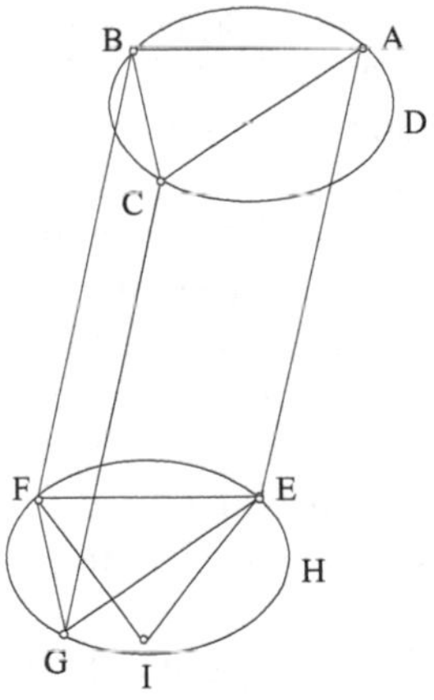

Fig. II.3.7

Proof: If we mark on the circumference or the perimeter of figure *ABCD*, the point *B*, whatever its position, and if we draw from this point a straight line parallel to the straight line *AE*, it falls on a point on the perimeter of figure *EFGH*. If we suppose that it falls at the point *F*, the two straight lines *AE* and *BF* are parallel, and are therefore in the same plane. But the two planes *ABCD* and *EFGH* are parallel; if therefore the plane containing the straight lines *AE*

[8] See Supplementary note [2] at the end of the volume.

[9] *Al-khaṭṭ al-muḥīṭ* is translated as 'circumference' and *al-khuṭūṭ al-muḥīṭa* as 'perimeter'.

and *BF* cuts them, then the intersections of these two latter and the former are parallel. But these two intersections are the straight line which joins the two points *A* and *B* and the straight line which joins the two points *E* and *F*; the two straight lines *AB* and *EF* are therefore parallel. But the two straight lines *AE* and *BF* are likewise parallel; therefore the two straight lines *AB* and *EF* are equal. If we therefore superpose figure *ABCD* on to figure *EFGH*, if we superpose point *A* of the former on to point *E* of the other figure and if we superpose the straight line *AB* on to the straight line *EF* so that point *B* falls on point *F*, then the rest of the figure falls on the rest of the other figure and is superposed on to it. In fact the perimeter of figure *ABCD* is superposed on to the perimeter of figure *EFGH*, because if it was possible that it was not superposed on to it, then if we suppose that the point *C* of figure *ABCD* is superposed on to a point which is not on the perimeter of figure *EFGH*, as the point *I*, and if we draw the straight lines *CA*, *CB*, *IE* and *IF*, then the straight line *AC* will be superposed on to the straight line *EI* and the straight line *CB* on to the straight line *IF*. But the straight line drawn from point *C* in parallel to the straight line *AE* falls on a point on the perimeter of figure *EFGH*; if we set this point the point *G* and if we draw the two straight lines *GE* and *GF*, we can show as we have shown previously that the straight line *CA* is equal to the straight line *GE* and that the straight line *CB* is equal to the straight line *GF*. But points *A*, *B* and *C* are superposed on to points *E*, *F* and *I*, the straight line *AC* is superposed on to the straight line *EI* and the straight line *CB* is superposed on to the straight line *IF*; therefore the two straight lines *EI* and *IF* are equal to the two straight lines *EG* and *GF*, each one to its homologue. Now, they have come from the points of origin of the straight lines *EG* and *GF*, on the straight line *EF* and in their direction, and they met at another point other than *G*, which is impossible. The whole perimeter of figure *ABCD* falls on the whole perimeter of figure *EFGH* and is thus superposed on to it. Consequently, the two figures *ABCD* and *EFGH* are similar and equal. That is what we wanted to prove.

– **8** – If a plane cuts a cylinder in parallel to both its bases, then the section generated in the cylinder is a circle whose centre is the point at which the plane cuts the axis.

Let there be a cylinder, whose bases are *ABC* and *DEF*, with centres *G* and *H* and whose axis is *GH*. Let a plane cut the cylinder in parallel to <the planes of> the two circles *ABC* and *DEF*, let the section generated be the surface *IKL*, let this section cut the axis at point *M*.

I say that IKL *is a circle, with centre at point* M.

Proof: If we make the line which bounds the section generated in the cylinder, the line *IKL*, then the two figures *ABC* and *IKL* are in two parallel planes; and if we draw from a point on the circumference of circle *ABC* one of the sides of the cylinder like the straight line *AID*, then every straight line drawn from a point on the circumference of circle *ABC* in parallel to the straight line

AID is one of the sides of the cylinder. Each of these straight lines therefore falls on a point on the line which bounds the section *IKL*. Consequently, the two figures *ABC* and *IKL* are similar and equal. But figure *ABC* is a circle, therefore figure *IKL* is a circle. And by the same way followed in the previous Proposition, we can show that the centre of this circle is the point *M*. That is what we wanted to prove.

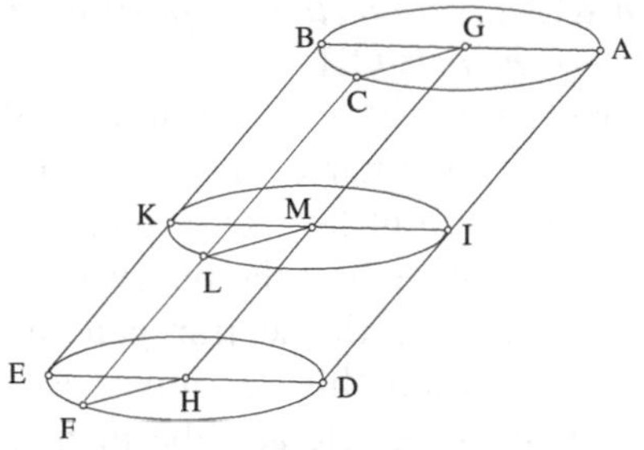

Fig. II.3.8

Furthermore, we can likewise show that the circle we have mentioned is equal to each of the cylinder's two bases. In the same manner as well we can show that if we have two parallel planes which cut all the sides of the cylinder, then they generate in the latter two similar and equal sections; and when a point on one is placed on its homologue on the other, *i.e.* the one through which passes the side which passes through the first, then it is possible to put the whole of the section on to the whole of the section, and the one will be superposed on to the other without it being either larger or smaller than it.

– **9** – If a plane cuts an oblique cylinder and passes through its axis and through its height, and if another plane perpendicular to the plane mentioned cuts the cylinder, so that the intersection of the two planes we have mentioned meets the two sides of the section generated by the first plane, which are two of the sides of the cylinder – whether inside the cylinder or outside – and forms with each of them an angle equal to the angle which is on the same side, amongst the two angles formed by this side and by one of the two sides remaining in this plane,[10] then the section generated in the cylinder, from the second of the two planes we have mentioned, is a circle or a portion of a circle, whose centre is the point at which it meets the axis. Let us call this circle an antiparallel[11] section.

Let there be an oblique cylinder whose bases are *ABC* and *DEF*, with centres *G* and *H*, and whose axis is *GH*; and let *GI* be the perpendicular dropped from point *G* to the plane of the circle *DEF*. Let the section generated by the

[10] *i.e.* the diameter of one or other of the base circles.

[11] Literally: section of contrary position, an expression found in the *Conics* of Apollonius, Book I.5 (subcontrary; ὑπεναντία).

plane which passes through the two straight lines *GH* and *GI* in the cylinder be the parallelogram *ABED*. Let another plane perpendicular to the plane *ABED* cut the cylinder; it therefore meets the two straight lines *AD* and *BE* either inside of the cylinder, or outside. Let the section generated by this plane in the cylinder be section *KLM*, and let the intersection of this plane and plane *ABED* be the straight line *KL*. Let the two angles *AKL* and *KAB* be equal.

I say that the section KLM *is a circle or a portion of a circle, whose centre is the point at which it meets the axis* GH.

Proof: The straight line *KL* is either secant to one of the two straight lines *AB* and *DE*, or it is not secant to either of them. If it is not secant to either of them, if we then mark on the straight line *KL* a point *N*, whatever its position, and if we make a plane parallel to each of the planes of the two circles *ABC* and *DEF* pass through this point, so that the section generated by this plane in the cylinder is the section *SMO*, then this section is a circle. If we make the intersection <of the plane> of this circle and the plane *ABED* the straight line *SO*, then *SO* is a diameter of circle *SMO*, because its centre is on the axis *GH*. But the straight line *GI* is perpendicular to the plane *DEF*, and is accordingly perpendicular to the plane *ABC* which is parallel to it; therefore every plane passing through the perpendicular *GI* is perpendicular to the planes of the two circles *ABC* and *DEF*; therefore the plane *ABED* is perpendicular to the planes of the two circles *ABC* and *DEF*. These two planes are likewise perpendicular to plane *ABED*, and in the same way as in plane *SMO*. But the plane *KLM* is also perpendicular to plane *ABED*; if we therefore make the intersection of these two planes the straight line *NM*, it will be perpendicular to plane *ABED*, and is accordingly perpendicular to each of the two straight lines *KL* and *SO*, because they are in plane *ABED*. Now, we have shown that the straight line *SO* is a diameter of circle *MSO*, so the product[12] of *SN* and *NO* is equal to the square of straight line *NM*. But angle *NSD* is equal to angle *BAD*, because the two straight lines *AB* and *SO* are parallel – since they are the two intersections of plane *ABED* with the two parallel planes *ABC* and *SMO*. Now, we have made angle *BAD* equal to angle *AKL*; therefore angle *NSD* is equal to angle *AKL*, and the triangle *SNK* is consequently isosceles. From that, we can likewise show that triangle *ONL* is isosceles; therefore the product of *SN* and *NO* is equal to the product of *KN* and *NL*. But we have shown that the product of *SN* and *NO* is equal to the square of the straight line *NM*; therefore the product of *KN* and *NL* is equal to the square of the straight line *NM*.

In the same way, we can likewise show that for every perpendicular falling from a point on the line which bounds section *KML* on to the straight line *KL*, its square is equal to the product of one of the two parts into which the straight line *KL* is divided, and the other part. Section *KML* is therefore a circle of diameter *KNL*.

[12] Literally: the surface obtained by multiplication, henceforth we shall translate this expression as 'product'.

I say that the centre of circle KML *is the point at which axis* GH *cuts plane* KML.

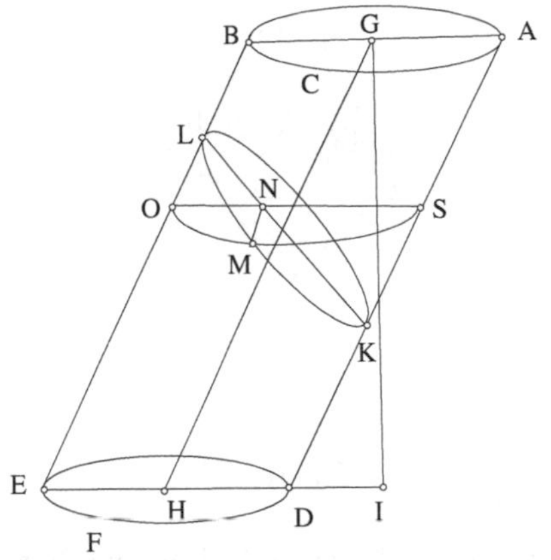

Fig. II.3.9

Proof: If we suppose that this point is the point *N*, if, in plane *ABED*, we make a straight line parallel to *AB*, let it be *SNO*, pass through it, and if we show, as we have shown previously, that the two triangles *KNS* and *ONL* are isosceles and that the straight line *SN* is equal to the straight line *NO*, then the straight line *KN* is equal to the straight line *NL*. But we have shown that the straight line *KNL* is a diameter of circle *KML*, its centre is accordingly point *N*.

With the same procedure, we can likewise show that if the straight line *KL* is secant to the straight line *AB*, or to the straight line *DE*, or to both of them, then the section generated in the cylinder from it is a portion of a circle, with as centre the point at which it meets the axis. That is what we wanted to prove.

Let us call the circle we have referred to an *antiparallel section*.

It then becomes clear that the antiparallel section is equal to each of the two bases of the cylinder and that all the antiparallel sections which cut the cylinder are parallel with each other.

– **10** – If we have a circle in any plane whatsoever, if we draw straight lines from its circumference to another plane, and if each of the straight lines drawn is parallel to the others, then they fall in the other plane at points through which a single line passes which surrounds an ellipse or a circle.

Let there be a circle *ABC* with centre *D*.

I say that if straight lines are drawn from the circumference of circle ABC *to a plane other than its own, and if each of them is parallel to the others, then all of them fall at points through which a single line passes which surrounds an ellipse or a circle.*

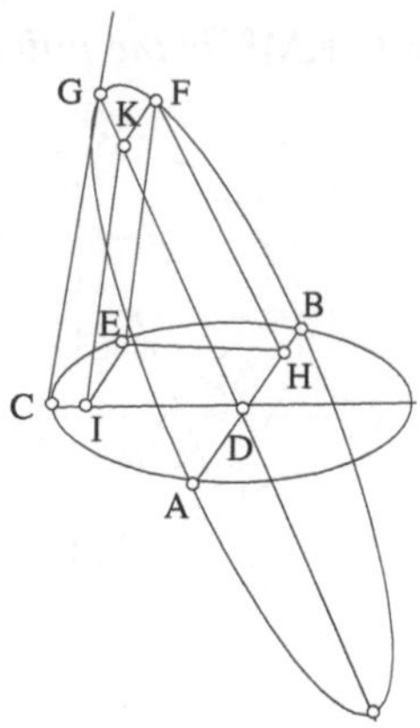

Fig. II.3.10

Proof: The plane on which the straight lines we have mentioned fall either passes through the centre of circle *ABC*, which is the point *D*, or does not pass through it. If we make it first of all as passing though *D*, then it cuts the plane of circle *ABC* and their intersection is a straight line which passes through point *D*. If we make this intersection the straight line *ADB*, if we draw from point *D* in the plane of circle *ABC* the straight line *DC* perpendicular to the straight line *ADB*, if we mark on the circumference of circle *ABC* any point at all, amongst the points from which the parallel straight lines we have mentioned begin, *i.e.* point *E*, and if we draw from the latter the parallel straight line which falls on the plane on which all the parallel straight lines fall, it will be the straight line *EF* and will fall at point *F* on this plane. If we draw from point *C* to this plane likewise the straight line *CG* parallel to the straight line *EF*, if we draw from point *E* the straight line *EH* perpendicular to *BD*, and if we join the two points *D* and *G* by the straight line *DG* and thc two points *F* and *H* by the straight line *FH*, then the two straight lines *FH* and *DG* are in the plane *AFGB* on which <all> the parallel straight lines fall, because they join the points, amongst the points which are in this plane. If we draw from point *E* in the plane of circle *ABC* the straight line *EI* parallel to the straight line *DH*, then the surface *EHDI* is a parallelogram. In fact, the straight line *EH* is parallel to the straight line *ID* because they are two perpendiculars to *BD*. The two straight lines *EI* and *HD* which join their extremities are parallel; therefore the two straight lines *EH* and *DI* are equal; the same applies to the two straight lines *HD* and *EI*. Similarly, if we draw from point *I*, in the plane of triangle *DCG*, the straight line *IK* parallel to the straight line *CG*, and if we join the two points *F* and *K* by the straight line *FK*, then the straight line which joins them is in the plane *AFGD* because the two straight lines *HF* and *DG* are in this plane; it is likewise in the same plane as all the straight lines *FE*, *EI* and *IK*, because the two straight lines *EF* and *IK* are parallel, since they are parallel to the straight line *CG*; the straight line *FK* is therefore the intersection of the plane in which are points *F*, *H*, *D* and *G*, and of the plane in which are points *F*, *E*, *I* and *K*. If we draw from point *F* a straight line parallel to one of the two straight lines *EI* and *HD*, then it will be parallel to the other one because they are parallel, and it will be in the same plane with each of them; accordingly it will be in the plane in which are points *F*, *H*, *D* and *G*, and likewise in the plane in which are points *F*, *E*, *I* and *K*; consequently it will be the intersection of these two planes. But we have shown that the intersection is the straight line *FK*; therefore the straight line *FK* is parallel to the straight line *EI*. But the straight line *EF* is parallel to the straight line *IK*, and accordingly is equal to it. We have likewise shown that the straight line *EI* is equal to the straight line *DH*; now it is also equal to the straight line *FK*; therefore the straight line *DH* is equal to the straight line *FK* and is parallel to it. The two straight lines *FH* and *KD* which join their extremities are then equal and parallel. But we have shown that the straight line *DI* is equal to the straight line *EH* and that the straight line *EF* is equal to the straight line *IK*, therefore the sides of the triangle *HEF* are equal to the sides of the triangle *DIK*. But triangle *DIK* is similar to triangle *DCG* because the straight line *IK* is parallel to the straight line *CG*; therefore the triangle *FEH* is similar to the triangle *GCD*; therefore the ratio of the square of the straight line *EH* to the square of the straight line *HF* is equal to the ratio of the square of the straight line *CD* to the square of the straight line *DG*. But the square of the straight line *EH* is equal to the product obtained from *AH* and *HB*, because *AB* is the diameter of circle *ABC* and the straight line *EH* is perpendicular to it; the square of the straight line *CD* is also equal to the product obtained from *AD* and *DB*; therefore the ratio of the product obtained from *AH* and *HB* to the square of the straight line *HF* is equal to the ratio of the product of *AD* and *DB* to the square of the straight line *DG*. The two points *F* and *G* are therefore on the perimeter of an ellipse with centre *D* and with *BA* as one of its diameters, and the straight lines ordinatewise to this diameter meet it at an angle like *ADG*, or on <the circumference> of a circle having this property, according to what has been shown from the reciprocal of Proposition 21 in Book I of the work by Apollonius on the *Conics*.

In the same way, we can likewise show that all the straight lines drawn from the circumference of circle *ABC* in parallel to the straight line *EF* fall on the perimeter of the ellipse or of the circle on which the straight line *EF* fell, let it be *AFGB*.

If likewise we make the plane on which the parallel straight lines fall, a plane which does not pass through point *D* – which is the centre of circle *ABC* – and if we draw a plane which passes through point *D* and which is parallel to the plane on which the parallel straight lines fall, like the plane *AFGB*, it can be shown, as we have shown previously, that the parallel straight lines we have mentioned cut plane *AFGB* at points through which there passes a single line which surrounds an ellipse with centre *D* and with *AB* as one of its diameters, or a circle having the same property. If they are extended until they fall on to the other plane parallel to plane *AFGB*, they fall from the former on to the points through which there passes a line which surrounds an ellipse or a circle and such that this ellipse or this circle is equal to the ellipse or to the circle on which they fall in plane *AFGB*. That is what we wanted to prove.

Furthermore, it has been shown that the centre of the ellipse or of the circle on which the parallel straight lines fall is the position on which there falls the straight line parallel to those straight lines drawn from the centre of the first circle.[13]

– **11**[14] – If a plane cuts a cylinder, without being parallel either to its bases or to its axis, without passing through the axis and without the section generated by it in the oblique cylinder being an antiparallel section, or a portion of an antiparallel section, then it is an ellipse or a portion of an ellipse. If the plane does not cut the two bases of the cylinder or just the one of them, it is an ellipse; if it cuts one of the two bases, it is a portion of an ellipse limited by a straight line and by a part of the perimeter of the ellipse; if it cuts the two bases at the same time, then it is a portion of an ellipse limited by two parallel straight lines and two parts of the perimeter of the ellipse; the centre of this ellipse is the point on which the cylinder's axis falls.

Let there be a cylinder whose bases are *ABC* and *DEF* with centres *G* and *H*, and whose axis is *GH*; let it be cut by a plane which is not parallel either to its bases, or to its axis and which does not pass through its axis, let this plane generate in it the section *IKL*, such that this section is not – if the cylinder is oblique – either an antiparallel section, or a portion of an antiparallel section.

Let the plane *IKL*, first of all, not be secant to the two bases of the cylinder nor to one of them.

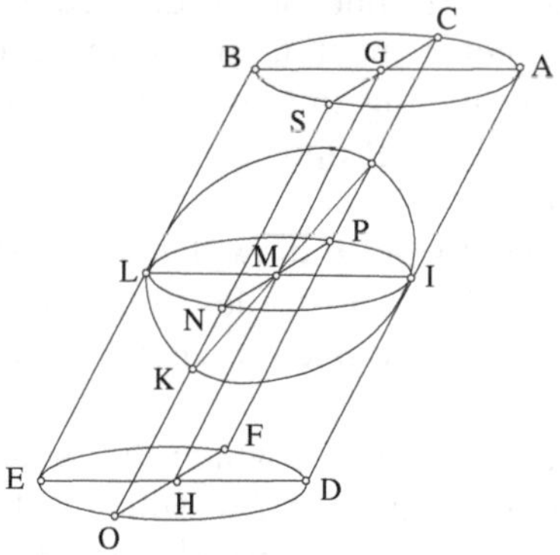

Fig. II.3.11a*

[13] See Supplementary note [3].
[14] See Supplementary note [4].
* There is only one figure in the manuscript; we shall divide it into two.

I say that the section IKL *is an ellipse and that its centre is the point at which it cuts the axis* GH.

Proof: At every point on the perimeter of the section *IKL* there falls one of the sides of the cylinder drawn from the circumference of circle *ABC*. But each of these straight lines we have called to mind, that is the sides of the cylinder, is parallel to the others and to the straight line *GH*. Through all these points marked on the perimeter of section *IKL* there passes a single line which surrounds an ellipse or a circle, whose centre is on the straight line *GH*; section *IKL* is therefore either an ellipse or a circle.

I say that it is an ellipse.

If it is not possible that it is so, let it be a circle with centre the point *M* on the straight line *GH*. If we draw a plane which passes through the straight line *GH* and through the height of the cylinder, dropped from point *G* on the plane *DEF*, then the section generated by this plane in the cylinder will be a parallelogram. If we make it parallelogram *ABED*, if we make the intersection of this plane and plane *IKL* the straight line *IL*, and if we pass through point *M* a plane parallel to each of the bases *ABC* and *DEF*, then this plane generates a circle in the cylinder; this circle is not the section *IKL*, because the plane *IKL* is not parallel to the two bases of the cylinder. The intersection of this circle parallel with plane *ABED* is either the straight line *IL* or a straight line other than *IL*.

If first of all we make their intersection the straight line *IL*, such that the circle parallel to the two bases is circle *INL*, and if we pass through axis *GH* a plane which cuts the plane *ABED* perpendicularly such that the section generated by this plane in the cylinder is *CSOF*, then the surface *CSOF* will be a rectangle. If we make the intersection of this plane and the plane of circle *INL* the straight line *NMP*, if we make the point in section *IKL* through which the straight line *SNO* passes the point *K*, and if we join the two points *K* and *M* with the straight line *MK*, then the plane *CSOF* cuts three parallel planes which are the planes *ABC*, *INL* and *DEF*; therefore the intersections of the former and these latter are parallel. If we make them the straight lines *SC*, *NM* and *FO*, being given the right angle *CSO* since the surface *CSOF* is a rectangle, then angle *MNK* is a right angle. Point *M* is likewise the centre of the circle *INL*, therefore the straight line *IM* is equal to the straight line *NM*; point *M* is also the centre of section *IKL*, so if the section *IKL* was a circle then the straight line *IM* would be equal to the straight line *MK*. But the straight line *IM* is equal to the straight line *MN*; therefore the straight line *MN* would be equal to the straight line *MK*; that is why angle *MNK* in triangle *NMK* would be equal to angle *MKN* in this same triangle. But we have shown that angle *MNK* is a right angle; therefore angle *MKN* is likewise a right angle. Now, they are both in the same triangle, which is impossible. Section *IKL* is therefore not a circle.

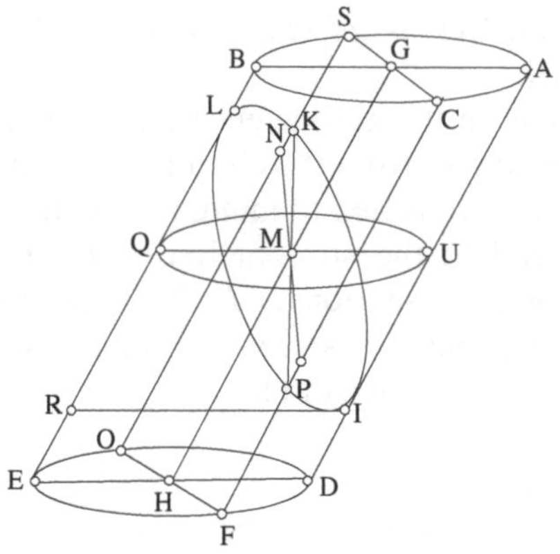

Fig. II.3.11b

In the same way, if we make the intersection of the plane *ABED* and <the plane> of the circle parallel to the two bases of the cylinder a straight line other than the straight line *IL*, let it be the straight line *UQ*, then it is clear that this straight line passes through point *M* which is the centre of this circle, and that the intersection <of the planes> of this circle and circle *IKL* also passes through point *M*; accordingly it is a common diameter of these two circles, and will then be equal to the straight line *IL* – since it is a diameter of circle *IKL* – and also equal to the straight line *UQ* – since it is a diameter of the circle parallel to the two bases of the cylinder. The straight line *IL* is therefore equal to the straight line *UQ*. If we draw from point *I* a straight line *IR*[15] parallel to the straight line *UQ*, then the straight line *IR* will also be equal to the straight line *UQ*. The straight line *IL* is therefore equal to the straight line *IR*, angle *ILR* will therefore be equal to angle *LRI*. But the straight line *UQ* is parallel to the straight line *DE* since they are the intersections <of the planes> of circle *DEF* and of the circle of diameter *UQ* which is parallel to it with plane *ABED*, and the straight line *IR* is also parallel to the straight line *UQ*; therefore the two straight lines *DE* and *IR* are parallel, angle *DEL* is then equal to angle *IRL*. But angle *IRL* is equal, as we have shown, to angle *ILR*; therefore angle *DEL* is equal to angle *ILR*. If it is thus, then the two angles *LID* and *IDE* are equal, the antiparallel section therefore passes through the straight line *IL*; but section *IKL* which passes through the straight line *IL* is not the antiparallel section. If we therefore make the antiparallel section another circle which passes through the straight line *IL*, that is the circle *INL*, such that the intersection of its <plane> with plane *CSOF* is the straight line *NMP*, then each of the straight lines *NM* and *KM* will be equal to the straight line *IM*; therefore the two straight lines *NM* and *KM* are equal and the straight line *MN* is the semi-diameter of the antiparallel section; accordingly it is equal to the straight line *GS*, which is the semi-diameter of

[15] See Supplementary note [5].

circle *ABC*. But the straight line *GS* is perpendicular to the straight line *NS* and the two straight lines *SG* and *MN* are between two parallel lines; therefore the straight line *MN* is likewise perpendicular to *SO* and consequently angle *MNO* is a right angle. But we have shown that the straight line *MN* is equal to the straight line *MK*, therefore angle *MNK* is equal to angle *MKN*. But angle *MNK* is a right angle, so angle *MKN* is also a right angle; then in triangle *MNK* there are two right angles, which is impossible. Section *IKL* is therefore not a circle, and is consequently an ellipse with its centre at point *M*.

In the same way, if we make the plane secant to the cylinder, secant to the two bases or to one of them, then if this plane is extended and if the lateral surface of the cylinder is also extended by extending its sides, this plane cuts the lateral surface of the extended cylinder and generates an ellipse, and what there is in the cylinder *ABED* is a portion of the ellipse. When the plane is secant to only one of the two bases of the cylinder, then this portion is limited by a straight line and a part of the perimeter of the ellipse. But when the plane is secant to the two bases at the same time, then this portion is limited by two parts of the perimeter of the ellipse and two parallel straight lines, since the planes of the two bases are parallel and are cut by the section plane; therefore their intersections with the latter are two parallel straight lines. That is what we wanted to prove.

<II. *Area of the ellipse and of portions of the ellipse*>

– **12**[16] – If we have two cylinders such that the two base circles of one are in the planes of the two base circles of the other and the centres of one pair are the centres of the others, and if the same plane cuts the two cylinders at once by cutting their sides in these latter,[17] then the two sections generated in the two cylinders are similar and the ratios of their diameters one to another, each to its homologue, are equal to the ratio of the diameter of the base circle of the first cylinder to the diameter of the base circle of the other.

Let there be two cylinders of which the two base circles of the one are *ABC* and *DEF* and the two base circles of the other *GHI* and *KLM*. Let the two circles *ABC* and *GHI* be in the same plane, and let point *N* be their common centre. Let the two circles *DEF* and *KLM* likewise be in the same plane, let point *S* be their common centre, and let *NS* be the axis of the two cylinders. Let the two cylinders be cut by a plane which cuts their sides in these latter, and generates in cylinder *ABCDEF* the section *OPU* and in cylinder *GHIKLM* the section *QRV*.

I say that the two sections OPU *and* QRV *are similar and that the ratio of each of the diameters of section* OPU *to its homologue, amongst the diameters*

[16] See Supplementary note [6].
[17] See Supplementary note [7].

of section QRV, *is equal to the ratio of the diameter of circle* ABC *to the diameter of circle* GHI.

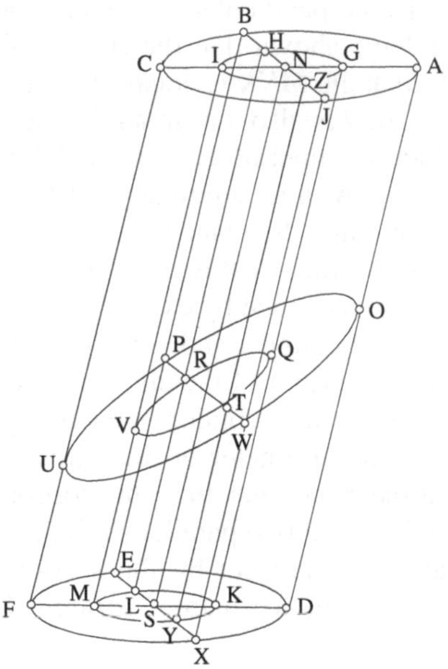

Fig. II.3.12

Proof: If we cut the two cylinders at the same time by a plane which passes through their axis, which is *NS*, it generates in the two cylinders two parallelograms. If we make the two surfaces *ADFC* and *GKMI*, then the sides of these two surfaces are parallel. If we make the intersection of these two parallelograms with the first plane which cuts the two cylinders the straight line *OQVU*, then thc ratio of *OU* to *QV* is equal to the ratio of *AC* to *GI* and is equal to the ratio of *DF* to *KM* because the straight lines *AOD*, *GQK*, *IVM* and *CUF* are parallel and the two straight lines *OU* and *QV* are two of the diameters of the two sections *OPU* and *QRV* because they pass through the position where these two sections cut axis *NS*, which is the centre of these two sections.

In the same way, we can also show that from each of the diameters of section *OPU* is separated, in section *QRV*, one of the diameters of *QRV*, as what is separated from diameter *OU* is diameter *QV*.

If we cut the two cylinders by a plane which passes through another diameter of section *OPU*, whatever this diameter, like diameter *PRTW*, such that it produces in the base circles of the two cylinders the intersections *BJ*, *HZ*, *EX* and *LY*, then the ratio of *PW* to *RT* is equal to the ratio of *BJ* to *HZ* and is equal to the ratio of *EX* to *LY*. But we have shown that the ratio of *OU* to *QV* is equal to the ratio of *AC* to *GI* and is equal to the ratio of *DF* to *KM*. But the straight lines *AC*, *DF*, *BJ* and *EX* are equal because they are diameters of the two circles *ABC* and *DEF*, and the straight lines *GI*, *HZ*, *KM* and *LY* are also equal because they are diameters of the two circles *GHI* and *KLM*; therefore the ratio of diameter *PW* to diameter *RT* is equal to the ratio of diameter *OU* to diameter *QV*. In the same way, we can show that all the diameters of the two sections *OPU* and *QRV* are in the same situation. If this is the case, then the two sections *OPU* and *QRV* are either both circles – and what we wanted is then proved – or they are not so – and then what separates from the largest of the diameters of section *OPU* to the inside of section *QRV* is the largest of the diameters of section *QRV* and what separates from the smallest of the diameters of section *OPU*, to the inside of section *QRV*, is the smallest of the diameters of section *QRV*. The largest diameter of every section is its largest axis, and its smallest diameter is its smallest axis.[18] Therefore the ratio of the largest of the two axes of section *OPU* to the largest of the two axes of section *QRV* is equal to the ratio of the smallest of the two axes of section *OPU* to the smallest of the two axes of section *QRV*. But if we permute, the ratio of the largest of the two axes of section *OPU* to the smallest axis is equal to the ratio of the largest of the two axes of section *QRV* to the smallest axis; then the two sections *OPU* and *QRV* are similar according to what has been shown in Proposition 12 of Book VI of the work of Apollonius on the *Conics*.[19] It has also been shown that the ratio of each diameter of section *OPU* to its homologue, amongst the diameters of section *QRV*, is equal to the ratio of the diameter of circle *ABC* to the diameter of circle *GHI*. That is what we wanted to prove.

– **13** – If we have an ellipse and if we construct on its larger axis a semi-circle, then the perpendiculars drawn from the arc of this semi-circle to the largest axis of the ellipse have equal ratios with their parts inside the ellipse.

Let there be an ellipse *ABCD* and its larger axis be *AC*, let there be a semi-circle *AEC* on *AC*. Let us draw from the arc *AEC* to axis *AC* the perpendiculars *EBF*, *GHI* and *KLM*.

I say that the ratios of EF *to* FB, *of* GI *to* IH *and of* KM *to* ML *are equal ratios.*

Proof: The ratio of the product obtained from *AF* and *FC* to the square of the straight line *FB* is equal to the ratio of axis *AC* to the *latus rectum*, according to what has been shown in Proposition 21 of Book I of the work of Apollonius on the *Conics*. But the product obtained from *AF* and *FC* is equal to the square of the straight line *EF*, therefore the ratio of the square of the straight line *EF* to the square of the straight line *FB* is equal to the ratio of axis *AC* to its *latus rectum*.

[18] See Supplementary note [8].
[19] See Supplementary note [9].

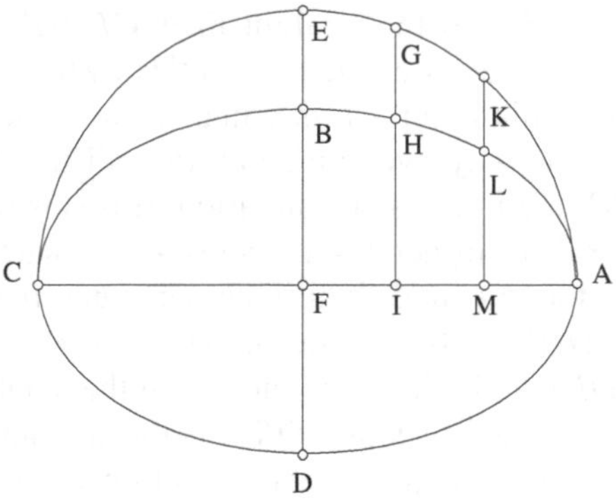

Fig. II.3.13

In the same way, we can show as well that the ratio of the square of the straight line *GI* to the square of the straight line *IH* and the ratio of the square of the straight line *KM* to the square of the straight line *ML* are, each of them, equal to the ratio of axis *AC* to its *latus rectum*. Therefore the ratios of *EF* to *FB*, of *GI* to *IH* and of *KM* to *ML* are equal ratios because the ratios of their squares are equal. That is what we wanted to prove.

It can also be shown by the same procedure that there necessarily follows for the small axis the analogue of what we have said for the large axis.

– **14**[20] – The area of every ellipse is equal to the area of a circle the square of whose diameter is equal to the surface obtained from multiplying one of the two axes of this section by the other.

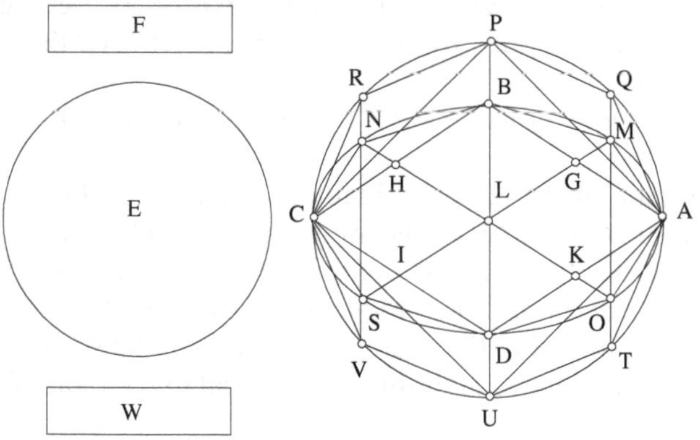

Fig. II.3.14

[20] See Supplementary note [10].

Let there be an ellipse *ABCD*, its large axis *AC*, its small axis *BD*; let there be a circle *E* such that the square of its diameter is equal to the product obtained from *AC* and *BD*.

I say that the area of the section ABCD *is equal to the area of circle* E.

Proof: If the area of the section *ABCD* is not equal to the area of circle *E*, then it will either be larger than it or smaller than it.

Let the area of section *ABCD* first of all be larger than the area of circle *E*, if that is possible, and let the excess over it be equal to the surface *F*. If we draw the straight lines *AB*, *BC*, *CD* and *DA*, then either <the sum> of the portions *AB*, *BC*, *CD* and *DA* of the section is smaller than surface *F* or it is not.

If it is smaller than it, that is what we want; if not, then we can divide the straight lines *AB*, *BC*, *CD* and *DA* into two halves at points *G*, *H*, *I* and *K*;[21] if we make the centre of the section point *L*, if we draw the straight lines *LG*, *LH*, *LI* and *LK* which we extend to points *M*, *N*, *S* and *O* on the perimeter of the section and if we draw the straight lines *AM*, *MB*, *BN*, *NC*, *CS*, *SD*, *DO* and *OA*, then the triangles *AMB*, *BNC*, *CSD* and *DOA* are <respectively> larger than the halves of portions *AB*, *BC*, *CD* and *DA* of the section, because if straight lines were drawn tangentially to the section at points *M*, *N*, *S* and *O*, they would be parallel to the straight lines *AB*, *BC*, *CD* and *DA* according to Proposition 17 of Book I of the work of Apollonius on the *Conics*. If <the sum> of the portions *AM*, *MB*, *BN*, *NC*, *CS*, *SD*, *DO* and *OA* of the section is smaller than the surface *F*, then that is what we want; if not, if we continue to proceed as we have done previously, we shall of necessity arrive at portions which subtract from the section less than surface *F*. Let us then make these portions which subtract from the section less than surface *F*, the portions *AM*, *MB*, *BN*, *NC*, *CS*, *SD*, *DO* and *OA*, we get the polygon *AMBNCSDO* larger than circle *E*. If we describe on the straight line *AC* a circle such that *AC* is one of its diameters, namely circle *APCU*, and if we draw the two straight lines *MO* and *NS* so that they cut the axis *AC* at right angles, if we extend them to the circle *APCU* at points *Q*, *R*, *V* and *T*, if we likewise extend the straight line *BD* to points *P* and *U*, if we draw the straight lines *AQ*, *QP*, *PR*, *RC*, *CV*, *VU*, *UT* and *TA*, then the ratio of triangle *AMO* to triangle *AQT* is equal to the ratio of the base *MO* to base *QT*, and the ratio of the surface *MBDO* to surface *QPUT* is equal to the ratio of the sum of the two straight lines *MO* and *BD* to the sum of the two straight lines *QT* and *PU*, because these two surfaces have equal height. Similarly, the ratio of the surface *BNSD* to surface *PRVU* is also equal to the ratio of the sum of the two straight lines *BD* and *NS* to the sum of the two straight lines *PU* and *RV*, and the ratio of triangle *NCS* to triangle *RCV* is equal to the ratio of *NS* to *RV* and the ratios of *MO*, *BD* and *NS* to *QT*, *PU* and *RV*, each to its homologue, are equal ratios because the ratios of their halves are equal; therefore the ratio of the entire polygon *AMBNCSDO* to the entire polygon *AQPRCVUT* is equal to the ratio of

[21] See Supplementary note [11].

BD to *UP*. But the ratio of *BD* to *UP* is equal to the ratio of the product obtained from *AC* and *BD* to the product obtained from *AC* and *PU* which is equal to the square of the straight line *PU*; therefore the ratio of polygon *AMBNCSDO* to polygon *AQPRCVUT* is equal to the ratio of the product obtained from *AC* and *BD* to the square of the straight line *PU*. But the product obtained from *AC* and *BD* is equal to the square of the diameter of circle *E*, the ratio of polygon *AMBNCSDO* to polygon *AQPRCVUT* is therefore equal to the ratio of the square of the diameter of circle *E* to the square of the straight line *PU* which is the diameter of the circle *APCU*. But the ratio of the square of the diameter of circle *E* to the square of the diameter of circle *APCU* is equal to the ratio of circle *E* to circle *APCU*, the ratio of polygon *AMBNCSDO* to polygon *AQPRCVUT* is therefore equal to the ratio of circle *E* to circle *APCU*. But polygon *AMBNCSDO* is larger than circle *E*; therefore polygon *AQPRCVUT* is larger than circle *APCU*; now, this last one is circumscribed by it, which is impossible. In consequence, the area *ABCD* is not larger than the area of circle *E*.

I say likewise that it isn't smaller than it. If that was possible, then let the area of section *ABCD* be smaller than the area of circle *E*. The ratio of circle *E* to circle *APCU* shall be equal to the ratio of section *ABCD* to a surface smaller than circle *APCU*. If we make it equal to the ratio of section *ABCD* to surface *F*, if we make the excess of circle *APCU* over surface *F* equal to the surface *W* and if we draw the straight lines *AP*, *PC*, *CU* and *UA*, then either <the sum> of the portions *AP*, *PC*, *CU* and *UA* of the circle is smaller than surface *W* or it is not. If it is smaller than it, that is what we wanted; otherwise, if we divide the arcs *AP*, *PC*, *CU* and *UA* into two halves at points *Q*, *R*, *V* and *T* and if we draw the straight lines *AQ*, *QP*, *PR*, *RC*, *CV*, *VU*, *UT* and *TA*, then the triangles *AQP*, *PRC*, *CVU* and *UTA* of the circle are <respectively> larger than the halves of the portions *AQP*, *PRC*, *CVU* and *UTA*. If therefore <the sum> of the portions *AQP*, *PRC*, *CVU* and *UTA* of the circle is smaller than surface *W*, that is what we want; if not, we can proceed exactly as we have done previously, and of necessity we shall arrive at portions which subtract from circle *APCU* less than surface *W*. If we make the portions which subtract less than surface *W*, the portions *AQP*, *PRC*, *CVU* and *UTA*, there remains the polygon *AQPRCVUT* larger than surface *F* and the surface *F* smaller than it. If we draw the straight lines *QT*, *PU* and *RV* which cut the perimeter of section *ABCD* at points *M*, *B*, *N*, *S*, *D* and *O* and if we draw the straight lines *AM*, *MB*, *BN*, *NC*, *CS*, *SD*, *DO* and *OA*, it can be shown as we have shown before that the ratio of the polygon *AMBNCSDO* to polygon *AQPRCVUT* is equal to the ratio of circle *E* to circle *APCU*. But we have made the ratio of circle *E* to circle *APCU* equal to the ratio of section *ABCD* to surface *F*. The ratio of polygon *AMBNCSDO* to polygon *AQPRCVUT* is therefore equal to the ratio of section *ABCD* to surface *F*. But polygon *AQPRCVUT* is larger than surface *F*; therefore polygon *AMBNCSDO* is larger than section *ABCD*; now, the section is circumscribed by it; that is impossible. The area of section *ABCD* is therefore not smaller than the area of circle *E*. But we have shown that it is not larger than it, therefore it is equal to it. That is what we wanted to prove.

It is clear from what we know that every ellipse is in proportion between the two circles constructed on its axes.[22]

– **15**[23] – Every portion of an ellipse whose diameter is perpendicular to the base, such that this diameter is a portion of the large axis, has an area equal to the area of a portion of the circle equal to the whole ellipse, such that the ratio of its chord to the diameter of this circle is equal to the ratio of the base of the portion of ellipse to the smaller of the two axes of the ellipse, on the understanding that, if the portion of the ellipse is smaller than half the ellipse, the portion of the circle is smaller than half the circle and that, if the portion of the ellipse is not smaller than half the ellipse, the portion of the circle is not smaller than half the circle.

Let there be a portion of an ellipse *ABC* whose base is *AC* and whose diameter is *BD*; let *BD* be perpendicular to *AC* and let *BD* also be a portion of the larger of the two axes of the ellipse. Let *ABCE* be the whole ellipse whose large axis is *BE* and small axis *FG*, and let there be the circle *HIKL* with diameter *HK*, equal to the ellipse. Let the ratio of the chord *IL* to diameter *HK* be equal to the ratio of *AC* to *FG*. If the portion *ABC* of the ellipse is smaller than half of it, let the portion *IKL* of the circle be smaller than half the circle; if portion *ABC* of the ellipse is not smaller than half the ellipse, then portion *IKL* of the circle is not smaller than half the circle.

I say that the area of portion ABC *of the ellipse is equal to the area of portion* IKL *of the circle.*

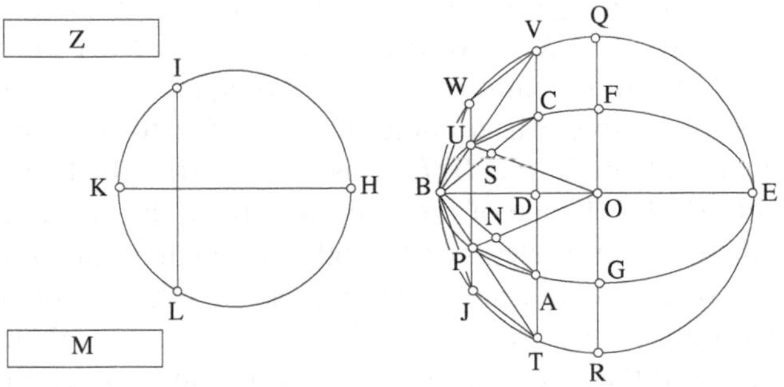

Fig. II.3.15

Proof: If the area of portion *ABC* of the ellipse is not equal to the area of portion *IKL* of the circle, then it is either larger than it or smaller than it.

[22] See Supplementary note [12].
[23] See Supplementary note [13].

Let the area of the portion *ABC* of the ellipse first of all be larger than the area of the portion *IKL* of the circle, if that is possible; let its excess over it be equal to the surface *M*. If we draw the two straight lines *AB* and *BC*, then <the sum> of the two portions *AB* and *BC* of the ellipse is either smaller than surface *M* or it will not be so. If it is smaller than it, that is what we wanted; otherwise, if we divide the two straight lines *AB* and *BC* into two halves at points *N* and *S*, if we make the centre of the ellipse the point *O*, if we draw the straight lines *ON* and *OS*, if we extend them to points *P* and *U* on the perimeter of the ellipse, if we draw the straight lines *AP*, *PB*, *BU* and *UC*, then the two triangles *APB* and *BUC* are <respectively> larger than half of the two portions *AB* and *BC* of the ellipse because if two tangents to the ellipse were drawn to points *P* and *U*, they would be parallel to the two straight lines *AB* and *BC*, according to what has been shown in Proposition 17 of Book I of the *Conics*. If <the sum> of the portions *AP*, *PB*, *BU* and *UC* of the ellipse is smaller than surface *M*, that is what we wanted; otherwise, if we continue to proceed as we have done previously, of necessity, we shall arrive at the portions which subtract from portion *ABC* less than surface *M*; let them be the portions *AP*, *PB*, *BU* and *UC*, the circle portion *IKL* then becomes smaller than the polygon *APBUC*.

If we describe on the straight line *BE* a circle such that *BE* is a diameter of it, let it be the circle *BQER*, if we extend the two straight lines *FG* and *CA* to the points *Q*, *R*, *V* and *T*; if we join the straight line *PU* between the two points *U* and *P*, if we extend it to circle *BQER* to points *W* and *J* and if we draw the straight lines *TJ*, *JB*, *BW* and *WV*, it can be shown from that, as we have shown in the previous proposition, that the ratio of the polygon *APBUC* to polygon *TJBWV* is equal to the ratio of *CA* to *TV*, which is equal to the ratio of *FG* to *QR*, and that the ratio of the first polygon to the second is equal to the ratio of the circle *HIKL* to circle *BQER*.[24] In the same way, the ratio of *CA* to *TV* is equal to the ratio of *FG* to *QR*. If we permute, the ratio of *CA* to *FG* is equal to the ratio of *TV* to *QR*. But the ratio of *AC* to *FG* is equal to the ratio of *IL* to *HK*, therefore the ratio of *VT* to *QR* is equal to the ratio of *IL* to *HK*. The straight line *HK* is the diameter of the circle *HIKL*; as for the straight line *QR*, it is the diameter of circle *BQER*. If one of the two portions of the two circles *TBV* and *IKL* is smaller than a semi-circle, then the other is smaller than a semi-circle. If the portion is not smaller than a semi-circle, then the other is not smaller than a semi-circle, accordingly they are similar. The ratio of each of them to the other is equal to the ratio of the circle of which it is a portion, to the circle of which the other is a portion. The ratio of the portion of circle *IKL* to the portion of circle *TBV* is equal to the ratio of circle *HIKL* to circle *BQER*. But we have shown that the ratio of circle *HIKL* to circle *BQER* is equal to the ratio of the polygon *APBUC* to polygon *TJBWV*; therefore the ratio of the portion of circle *IKL* to the portion of circle *TBV* is equal to the ratio of polygon *APBUC* to

24 $\frac{FG}{QR}=\frac{FG}{EB}=\frac{FG\cdot EB}{EB^2}=\frac{KH^2}{EB^2}$.

polygon *TJBWV*. But the portion of circle *IKL* is smaller than polygon *APBUC*; therefore the portion of circle *TBV* is smaller than polygon *TJBWV*; that is impossible because the circle is circumscribed by it. The area of the portion *ABC* of the ellipse is therefore not larger than the area of the portion of circle *IKL*.

I say as well that it is not smaller than it. If that was possible, then let the area of portion *ABC* of the ellipse be smaller than the area of the portion *IKL* of the circle; the ratio of the portion of circle *IKL* to the portion of circle *TBV* is therefore equal to the ratio of the portion *ABC* of the ellipse to a surface smaller than the portion of circle *TBV*. If we make this ratio equal to the ratio of the portion *ABC* of the ellipse to a surface *M*, if we make the excess of the portion of circle *TBV* over surface *M* equal to the surface *Z* and if we draw the two straight lines *TB* and *BV*, then either the two portions *TB* and *BV* of the circle have <a sum> smaller than surface *Z* or it is not thus. If <their sum> is smaller than it, that is what we wanted; otherwise, if we divide the two arcs *TB* and *BV* into two halves at the two points *J* and *W* and if we draw the straight lines *TJ*, *JB*, *BW* and *WV*, then the two triangles *TJB* and *WBV* are <respectively> larger than half of the two portions of circle *TJB* and *WBV*. If <the sum> of the portions *TJ*, *JB*, *BW* and *WV* of the circle is smaller than the surface *Z*, that is what we wanted; otherwise, if we continue to proceed as we have done previously, of necessity we shall arrive at the portions which subtract from the portion of circle *TBV*, less than surface *Z*. If we make the portions which subtract less than surface *Z*, the portions *TJ*, *JB*, *BW* and *WV*, what is left is the polygon *TJBWV* larger than the surface *M*, and surface *M* is smaller than it. If we draw the straight line *WJ* which then cuts the perimeter of the portion *ABC* of the ellipse at points *U* and *P*, if we draw the straight lines *AP*, *PB*, *BU* and *UC* and if we follow an analogous method to what we followed previously, it can be shown as we have shown before, that the ratio of the portion of circle *IKL* to the portion of circle *TBV* is equal to the ratio of the polygon *APBUC* to polygon *TJBWV*. But the ratio of the portion of circle *IKL* to the portion of circle *TBV* is equal to the ratio of portion *ABC* of the ellipse to surface *M*. The ratio of portion *ABC* of the ellipse to surface *M* is therefore equal to the ratio of polygon *APBUC* to polygon *TJBWV*. The portion *ABC* of the ellipse is therefore smaller than polygon *APBUC*; that is impossible, because it is circumscribed by it. The area of portion *ABC* of the ellipse is therefore not smaller than the area of portion *IKL* of the circle; now we have shown that it is not larger than it, consequently it is equal to it. That is what we wanted to prove.

It then becomes clear that the area of portion *ABC* of the ellipse is equal to the area of a portion of circle *HIKL* such that the ratio of its axis to the diameter *HK* is equal to the ratio of the diameter of portion *ABC*, that is *BD*, to *BE* which is the largest axis; in fact, *BD* is also the axis of the arc *TBV*, which is similar to the arc *IKL*.[25]

– **16** – Every portion of an ellipse whose diameter is perpendicular to its base – this diameter being a portion of the small axis – is such that its area is equal to the area of a portion of the circle equal to the whole ellipse, a portion such that the ratio of its chord to the diameter of the circle is equal to the ratio of the base of the portion of ellipse to the larger of the two axes of the ellipse on the understanding that, if the portion of the ellipse is smaller than half the ellipse, the portion of circle is smaller than half the circle, and that if the portion of the ellipse is not smaller than half the ellipse, the portion of circle is not smaller than half the circle.

Let there be a portion of ellipse *ABC* whose base is *AC* and diameter *BD*; let *BD* be perpendicular to *AC*, let *BD* likewise be a portion of the smaller of the two axes of the ellipse, let the whole ellipse be *ABCE*, its small axis *BE* and its large axis *FG* and let the circle equal to the ellipse be *HIKL* and its diameter *HK*. Let the ratio of the chord *IL* to diameter *HK* be equal to the ratio of *AC* to *FG*. If the portion *ABC* of the ellipse is smaller than half of it, then the portion of circle *IKL* is smaller than half the circle, and if the portion *ABC* of the ellipse is not smaller than half of it, then the portion of circle *IKL* is not smaller than half of it.

I say that the area of the portion ABC *of the ellipse is equal to the area of the portion* IKL *of the circle.*

Proof: If the area of the portion *ABC* of the ellipse was not equal to the area of the portion of circle *IKL*, then it would either be larger than it or smaller.

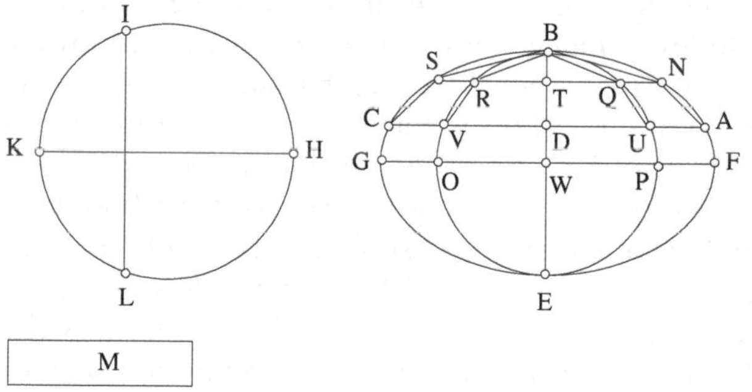

Fig. II.3.16

Let the area of the portion *ABC* of the ellipse first of all be larger than the area of the portion of circle *IKL*, if that is possible; let its excess over it be equal to the surface *M*. If we follow an analogous method to what we followed in the previous proposition to construct in the portion *ABC* of the ellipse a polygon

[25] See the mathematical commentary.

larger than the portion of circle *IKL*, then this polygon is the polygon *ANBSC*, we can construct on the straight line *BE* a circle such that *BE* is one of its diameters; this circle is *BOEP*. If we join the straight line *NQRS* between the points *N* and *S*, then the ratio of *QT* to *TN* is equal to the ratio of *UD* to *DA* and is equal to the ratio of *PW* to *WF*. We can show, as we have shown in the two previous propositions, that the ratio of polygon *ANBSC* to polygon *UQBRV* is equal to the ratio of the portion of circle *IKL* to the portion of circle *UBV* which is similar to it. But polygon *ANBSC* is larger than the portion of circle *IKL*, therefore polygon *UQBRV* is larger than the portion of circle *UBV*; this is impossible because the circle is circumscribed by it. Therefore the area of the portion *ABC* of the ellipse is not larger than the area of the portion *IKL* of the circle.

By an analogous method to what we followed in the previous proposition, we can show that it is not smaller than it; accordingly it is equal to it. That is what we wanted to prove.

It then becomes clear that the area of the portion *ABC* of the ellipse is equal to the area of a portion of circle *HIKL*, such that the ratio of its axis to the diameter *HK* is equal to the ratio of the diameter of portion *ABC*, that is *BD*, to *BE* which is the small axis; in fact, *BD* is also the axis of arc *UBV* which is similar to the arc *IKL*.[26]

– **17** – The area of every portion of ellipse, whatever that portion of ellipse is, is equal to the area of a portion of the circle equal to that ellipse, a portion such that if there are drawn from the two extremities of its base two perpendiculars to one of the diameters of the circle and if there are drawn from the two extremities of the base of the portion of ellipse two perpendiculars to one of the axes of the ellipse, then the ratio of each of the two perpendiculars falling on this axis to the other axis is equal to the ratio of its homologue, amongst the two perpendiculars falling on the diameter of the circle, to the diameter of the circle. The two perpendiculars falling on the axis of the ellipse both fall on to it on the same side and let the two perpendiculars falling on the diameter of the circle both likewise fall on to it on the same side, or the two perpendiculars falling on the axis of the ellipse fall on to it on two opposite sides and the two perpendiculars falling on the diameter of the circle also fall on to it on two opposite sides and the centre of the ellipse be between the ends of the two perpendiculars falling on its axis and the centre of the circle is between the feet of the two perpendiculars falling on its diameter, or the centre of the ellipse is not between the feet of the two perpendiculars falling on its axis and the centre of the circle is not between the ends of the two perpendiculars falling on its diameter and the portion of the ellipse is smaller than half the ellipse and the portion of the circle is smaller than half the circle or the portion of the

[26] See the mathematical commentary.

ellipse is not smaller than half of it and the portion of the circle is not smaller than half of it.

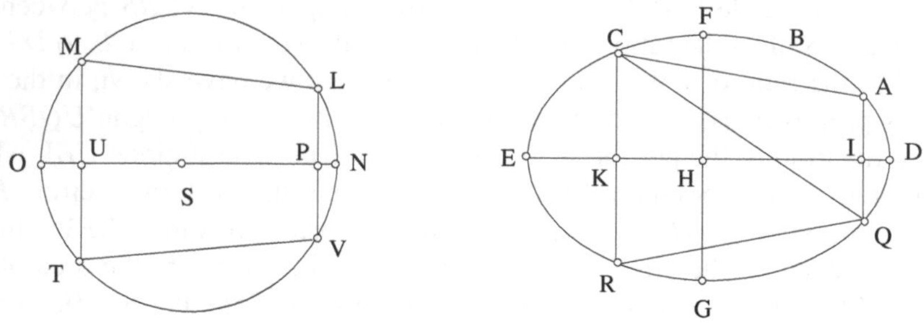

Fig. II.3.17.1

Let there be a portion of ellipse *ABC* with base *AC*; let it first of all be smaller than half the ellipse. Let the ellipse be *ABCD* and let *DE* be the large axis and *FG* the small axis in the first, third, fifth and seventh cases of figure. As for the second, fourth, sixth and eighth cases of figure, let us have the contrary, *i.e.* let the large axis be *FG* and the small axis *DE*. Let point *H* be the centre of the ellipse; let there be drawn from two points *A* and *C* two perpendiculars to axis *DE* in all the cases of figure of the ellipse, let them be *AI* and *CK*. Let the circle equal to the ellipse *ABCD* be the circle *LMN* with centre *S*. Let there be a portion of base *LM* smaller than half of it, let there be drawn from two points *L* and *M* to one of the diameters of the circle, namely the diameter *NO*, two perpendiculars, let them be *LP* and *MU*. Let the ratio of each of the two perpendiculars *AI* and *CK* to axis *FG* in all the cases of figure of the ellipse be equal to the ratio of its homologue, amongst the two perpendiculars *LP* and *MU*, to diameter *NO* in all the cases of figure of the circle. Let the two perpendiculars *AI* and *CK* fall either both on the same side[27] on axis *DE* and let the two perpendiculars *LP* and *MU* likewise both fall on the same side of diameter *NO* as in the first, second, third and fourth cases of figure, or they fall on two different sides of axis *DE* and the two perpendiculars *PL* and *MU* likewise on two different sides of diameter *NO* as in the remaining cases of figure. Let there be centre *H*, either between the two perpendiculars *AI* and *CK* and centre *S* between the perpendiculars *LP* and *MU* as in the first, second, fifth and sixth cases of figure, or not between the two perpendiculars *AI* and *CK* and centre *S* not between the two perpendiculars *LP* and *MU* as in the third, fourth, seventh and eighth cases of figure.

Then I say that the area of the portion ABC *of the ellipse is equal to the area of the portion* LM *of the circle.*

[27] Literally: on the same side of the two sides of the axis.

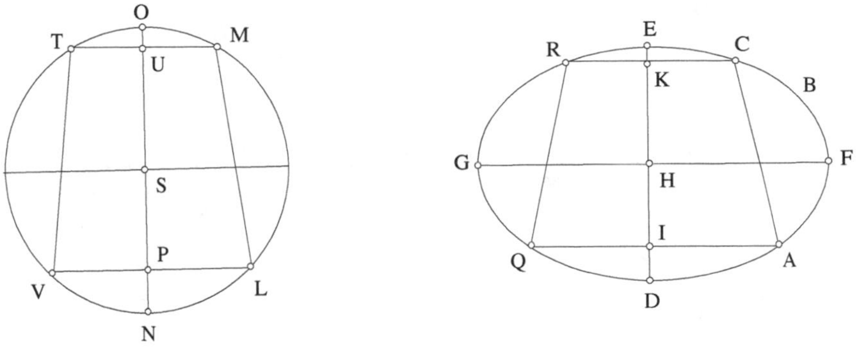

Fig. II.3.17.2

Proof: If we draw the perpendiculars *AI*, *CK*, *LP* and *MU* in all the cases of figure to the points *Q*, *R*, *V* and *T* and if we draw first of all in the first, second, third and fourth cases of figure the straight line *QA*, then the ratio of *AQ*, which is the base of the portion *ADQ* of the ellipse, to axis *FG* is equal to the ratio of the chord *LV* to diameter *NO*. But the straight line *DI* is a diameter of portion *ADQ* of the ellipse and is a portion of axis *DE*, portion *ADQ* of the ellipse is smaller than half of it and portion *LV* of the circle is likewise smaller than half the circle, therefore the area of portion *ADQ* of the ellipse is equal to the area of portion *LNV* of the circle.

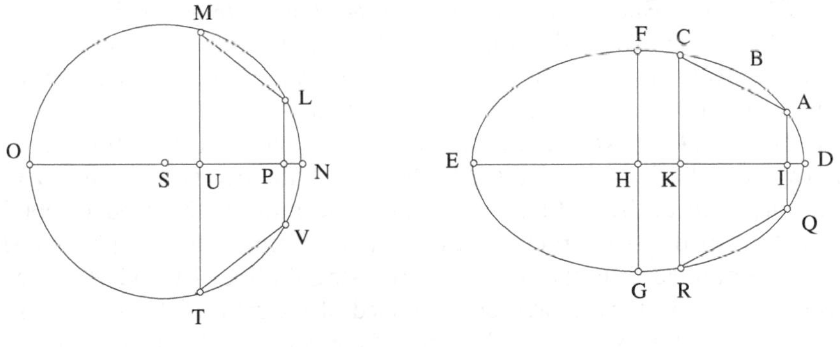

Fig. II.3.17.3

In the same manner, we can show that the area of portion *CDR* of the ellipse – which is, in the first and second cases of figure, larger than half of it and, in the third and fourth cases of figure, smaller than half of it – is equal to the area of the portion *MNT* of the circle, given that it is also larger than half the circle in the first and second cases of figure and smaller than half of it in the third and fourth cases of figure; therefore, there remains the area of portion *AIQRKC* of the ellipse equal to the area of portion *LPVTUM* of the circle.

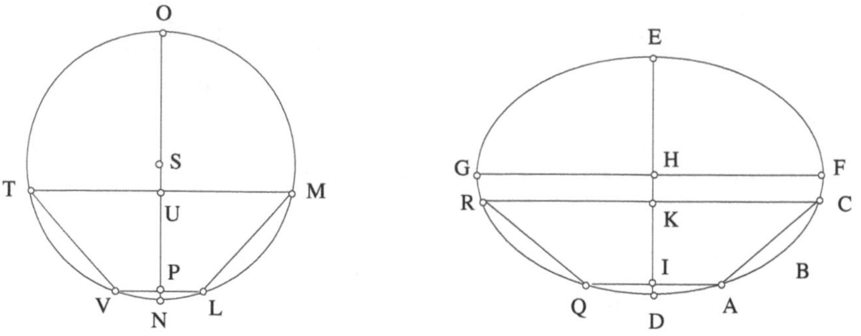

Fig. II.3.17.4

Moreover, the ratio of *DI* to *DE* is equal to the ratio of *NP* to *NO* and the ratio of *DK* also to *DE* is equal to the ratio of *NU* to *NO*; there remains the ratio of *IK* to *DE* equal to the ratio of *PU* to *NO*. If we permute, the ratio of *IK* to *PU* is equal to the ratio of *DE* to *NO*. We also have the ratio of *AQ*, given that it is the double of *AI*, to *FG*, equal to the ratio of *LV*, given that it is the double of *LP*, to *NO*; and the ratio of *CR*, given that it is the double of *CK*, to *FG*, equal to the ratio of *MT*, given that it is the double of *MU*, to *NO*. If we add up, the ratio of the sum of the two straight lines *AQ* and *CR* to *FG* is equal to the ratio of the sum of *LV* and *MT* to *NO*. If we permute, the ratio of the sum of the two straight lines *AQ* and *CR* to the sum of the two straight lines *LV* and *MT* is equal to the ratio of *FG* to *NO*. But we have shown that the ratio of *IK* to *PU* is equal to the ratio of *DE* to *NO*. The ratio compounded of the ratio of the sum of the two straight lines *AQ* and *CR* to the sum of the two straight lines *LV* and *MT* and of the ratio of *IK* to *PU* is equal to the ratio compounded of the ratio of *FG* to *NO* and the ratio of *DE* to *NO*. As for the ratio compounded of the ratio of the sum of the two straight lines *AQ* and *CR* to the sum of the two straight lines *LV* and *MT* and the ratio of *IK* to *PU*, it is equal to the ratio of the product obtained from the sum of the two straight lines *AQ* and *CR* times the straight line *IK* to the product obtained from the sum of the two straight lines *LV* and *MT* times the straight line *PU*. As for the ratio compounded of the ratio of *FG* to *NO* and the ratio of *DE* to *NO*, it is equal to the ratio of the product obtained from *FG* and *DE* to the square of the straight line *NO*. But the product obtained from *FG* and *DE* is equal to the square of the straight line *NO*, the product obtained from the sum of the two straight lines *AQ* and *CR* times the straight line *IK* is therefore equal to the product obtained from the sum of the two straight lines *LV* and *MT* times the straight line *PU*. But half of the product obtained from the sum of the two straight lines *AQ* and *CR* times the straight line *IK* is the surface of the trapezium *AQRC* and half of the product obtained from the sum of the two straight lines *LV* and *MT* times the straight line *PU* is the surface of the trapezium *LVTM*. The surface of the trapezium *AQRC* is therefore equal to the surface of the trapezium *LVTM*. But we have shown that the area of the portion

QIABCKR of the ellipse is equal to the area of the portion *VPLMUT* of the circle; therefore there remains the area of portions *ABC* and *QR* of the ellipse, if we add them together, equal to the area of portions *LM* and *VT* of the circle, if we add them together. But the two portions *LM* and *VT* of the circle are equal and the two portions *ABC* and *QR* of the ellipse are also equal according to what has been shown in Proposition 8 of Book VI of the work of Apollonius on the *Conics*, therefore the area of portion *ABC* of the ellipse is equal to the area of portion *LM* of the circle.

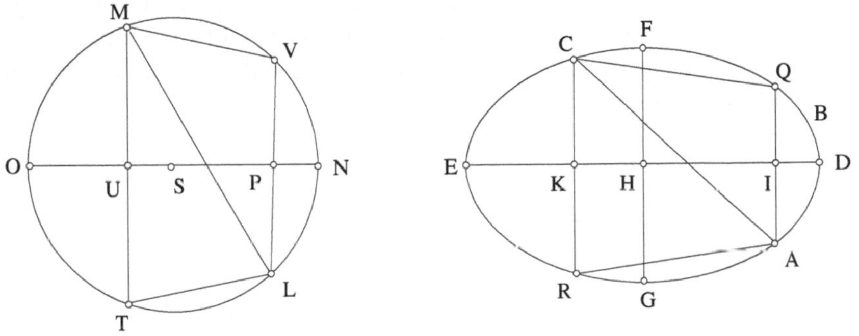

Fig. II.3.17.5

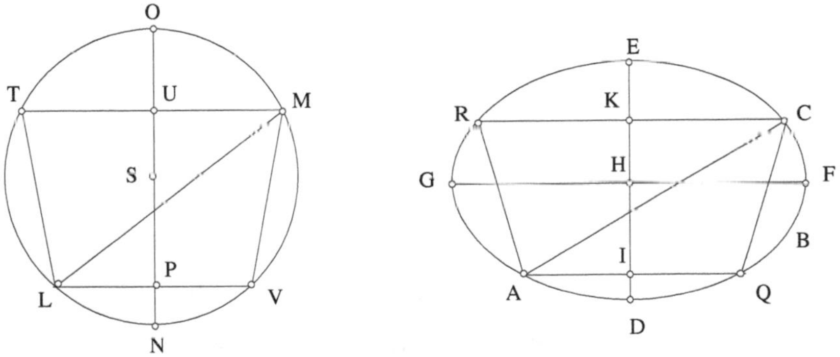

Fig. II.3.17.6

Similarly, we are going to speak of the fifth, sixth, seventh and eighth cases of figure. We can draw the straight lines *QC*, *AR*, *LT* and *VM* and we can show, as we have shown before, that the area of the portion *ADQ* of the ellipse is equal to the area of the portion *LNV* of the circle and that the area of the trapezium *QCRA* is equal to the area of trapezium *LVMT* and that the area of the portion *QC* of the ellipse is equal to the area of the portion *VM* of the circle. We also have the ratio of *AQ*, given that it is the double of *AI*, to *FG*, equal to the ratio of *LV*, given that it is the double of *LP*, to *NO*, and the ratio of *FG* to *CR*, given that it is the double of *CK*, is equal to the ratio of *NO* to *MT*, given that it is the double of *MU*. By the ratio of equality (*ex aequali*), the ratio of *AQ* to *CR* is equal to the ratio of *LV* to *MT*. As for the ratio of *AQ* to *CR*, it is equal to the ratio of the triangle *ACQ* to triangle *ACR*, because the heights of these two triangles are equal; in fact, the height of each of them is equal to *IK*. As for the ratio of *LV* to *MT*, it is equal to the ratio of the triangle *LVM* to triangle *LTM*, because the heights of these triangles are equal since the height of each of them is equal to *PU*. Therefore the ratio of triangle *AQC* to triangle *ARC* is equal to the ratio of triangle *LVM* to triangle *LTM*. But we have shown that the trapeziums *AQCR* and *LVMT* are equal; therefore triangle *AQC* is equal to triangle *LVM*. Now, we have shown that the area of portions *ADQ* and *QC* of the ellipse is equal to the area of portions *LNV* and *VM* of the circle; therefore the area of the whole portion *AQC* of the ellipse is equal to the area of the whole portion *LVM* of the circle.

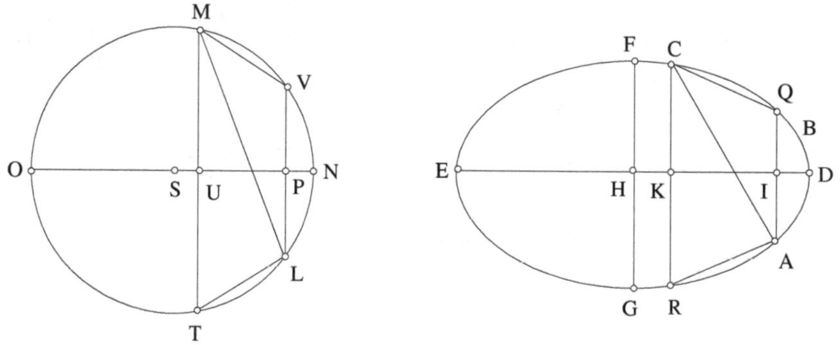

Fig. II.3.17.7

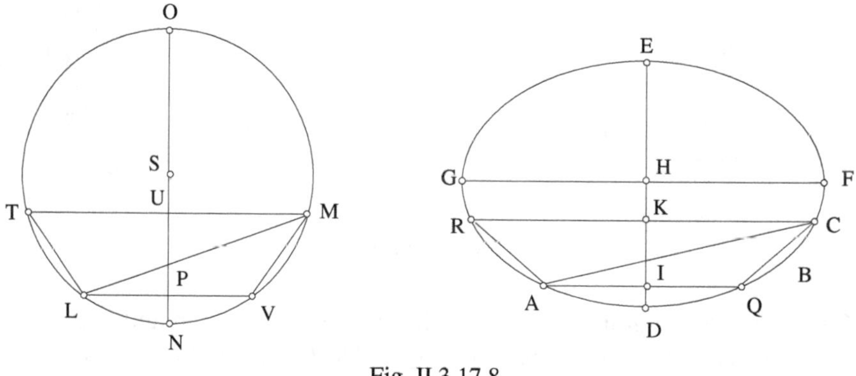

Fig. II.3.17.8

Similarly, if the portion of the ellipse is larger than half the ellipse, like portion *AEC* and if the portion of the circle is larger than half the circle, like *LOM*, then the area of portion *AEC* of the ellipse is equal to the area of portion *LOM* of the circle, because the area of the whole ellipse is equal to the area of the whole circle and the area of portion *ABC*, which is smaller than half the ellipse, is equal to the area of portion *LM* of the circle, which is smaller than half the circle; the result therefore is that the area of portion *AEC* of the ellipse is equal to the area of portion *LOM* of the circle. If the portion of the ellipse is equal to half the ellipse and if the portion of the circle is equal to half the circle, the equality of their areas is obvious. That is what we wanted to prove.

It then becomes clear that if the ratio of *DI* to axis *DE* is equal to the ratio of *NP* to diameter *NO* and if the ratio of *IK* to axis *DE* is equal to the ratio of *PU* to diameter *NO*, then the area of portion *ABC* of the ellipse is equal to the area of portion *LM* of the circle and <the area of> portion *AEC* of the ellipse is equal to <the area of> portion *LOM* of the circle.

<III. *On the maximal section of the cylinder and on its minimal sections*>

– **18** – If a plane cuts an oblique cylinder and if the axis of the cylinder meets this plane, whether in the cylinder or outside it, such that it is perpendicular to it, then the section generated in the cylinder is an ellipse whose large axis is equal to the diameter of each of the bases of the cylinder and whose small axis is a straight line such that its ratio to the diameter of each of the bases of the cylinder is equal to the ratio of the height of the cylinder to its axis, or a portion of ellipse having the property we have set out.

Let there be an oblique cylinder with bases *ABCD* and *EFGH* and with axis *IK*, let us draw from point *I* the height of the cylinder, let it be *IL*. Let a plane cut the cylinder and be met by the straight line *IK* either inside the cylinder or outside it, such that this straight line is perpendicular to this plane and that the plane generates in the cylinder the section *MNSO*. If the plane *MNSO* meets all the sides of the cylinder on the inside of the latter, then the section is an ellipse. This has been shown because the plane is not parallel to the two bases of the cylinder, and is not an antiparallel section.

I say that its large axis is equal to the diameter of each of the circles ABCD *and* EFGH *and that its small axis is a straight line whose ratio to the diameter of each of the circles* ABCD *and* EFGH *is equal to the ratio of* IL *to* IK.

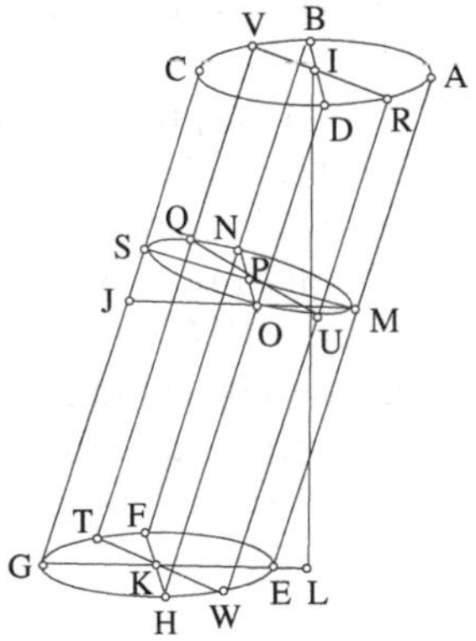

Fig. II.3.18

Proof: If we cut the cylinder by a plane which passes through the two straight lines *IK* and *IL*, that is the plane *ACGE*, and by another plane which cuts this plane perpendicularly and passes through axis *IK*, that is the plane *BDHF*, then the surface *BDHF* is a rectangle. If we make the intersection of this plane and the plane *MNSO* the straight line *NO*, the straight line *IPK* will cut the straight line *NO* perpendicularly because the straight line *IP* is perpendicular to plane *MNSO*, and the straight line *NO* is likewise perpendicular to the straight lines *BNF* and *DOH* since they are parallel to the axis *IPK*, given that they are two of the sides of the cylinder. The two straight lines *BN* and *DO* are perpendicular to the straight line *BD* because surface *BDHF* is a rectangle; the straight line *NO* is therefore equal to the straight line *BD* which is the diameter of the circle *ABCD* and to the straight line *FH* which is the diameter of circle *EFGH*. <The straight line> *NO* is also one of the diameters of the section *MNSO*, because it passes through its centre which is the point *P*. If we draw another of its diameters, whatever this diameter is, that is *UQ*, and if we draw with the two straight lines *UQ* and *IP* a plane *RVTW* which cuts the cylinder, then the section generated is a parallelogram not a rectangle. But the straight line *UQ* is perpendicular to the straight line *IK*, because *IP* is perpendicular to the plane *MNSO*, the straight line *UQ* is accordingly perpendicular to the two parallel straight lines *RW* and *VT*, the straight line *RV* is therefore longer than the straight line *UQ*; but the straight line *RV* is one of the diameters of circle *ABCD*, now every diameter of circle *ABCD* is equal to the straight line *NO*, therefore the straight line *NO* is longer than the straight line *UQ*. In the same way, it can also be shown that the straight line *NO* is the longest of all the diameters of the ellipse *MNSO*, and is consequently its longest axis, because the longest axis is the longest of the ellipse's diameters, according to what has been shown in Proposition 11 of Book V of the work of Apollonius on the *Conics*. And we have shown that *NO* is equal to the diameter of each of the two circles *ABCD* and *EFGH*.

I say likewise that the ratio of the small axis of the ellipse MNS *to the diameter of each of the two circles* ABCD *and* EFGH *is equal to the ratio of* IL *to* IK.

Proof: The plane *BDHF* cuts the plane *ACGE* perpendicularly; if we therefore make the intersection of plane *ACGE* with plane *MNSO* the straight line *MS*, then it is perpendicular to *IPK*. Therefore, in the two planes *ACGE* and *BDHF*, two perpendiculars have been drawn to *IPK*, which is their intersection, that is *PN* and *PM*. Therefore angle *NPM* is a right angle, the straight line *MS* cuts perpendicularly the straight line *NO* which is the largest axis and passes through the point *P* which is the centre of the ellipse; therefore the straight line *MS* is the smaller of the two axes of the ellipse *MNSO*, according to what has been shown in Proposition 15 of Book I of the work of Apollonius on the *Conics*.[28]

If we draw from point *M* a straight line parallel to the straight lines *AC* and *EG*, that is *MJ*, then the external angle *MJS* is equal to the internal angle *EGS* which is opposite to it. In the same way, angle *EGS* is also equal to angle *IKE* because the straight line *IK* is parallel to the straight line *GS*; therefore angle *MJS*, which is one of the angles of triangle *MSJ*, is equal to angle *IKL*, which is one of the angles of triangle *LKI*, and the two angles *MSJ* and *ILK*, which are angles of these triangles, are also equal because they are two right angles; there remains the angle *JMS* of triangle *MSJ* equal to the angle *KIL* of triangle *LIK*. The two triangles *MSJ* and *IKL* are therefore similar. The ratio of *MS* to *MJ* is accordingly equal to the ratio of *LI* to *IK*. But the straight line *MJ* is equal to the straight line *AC* which is one of the diameters of circle *ABCD*, since this straight line and the straight line *AC* are parallel and are between two parallel straight lines, therefore the ratio of *MS* to the diameter of circle *ABCD*, which is equal to the diameter of circle *EFGH*, is equal to the ratio of *LI* to the perpendicular *IK*. Now we have shown that *MS* is the smaller of the two axes of ellipse *MNSO*; therefore the ratio of the smaller of the two axes of ellipse *MNSO* to the diameter of each of the circles *ABCD* and *EFGH* is equal to the ratio of *LI* to *IK*.

If the section *MNSO* does not cut all the sides of the cylinder, then if the cylinder is extended along its sides in both directions and if the plane *MNSO* is drawn until it cuts all its sides, it can be shown according to what we have said before, that *MNSO* is a portion of an ellipse having the property we have mentioned before. That is what we wanted to prove.

– **19** – If a plane cuts an oblique cylinder and if the axis of the cylinder meets this plane whether in the cylinder or outside it, such that it is perpendicular to it, then the ellipse generated by this plane in the cylinder, or a part of which is in the cylinder, is such that amongst the large axes of the ellipses of this cylinder, there is none which is smaller than its large axis and which, amongst their small axes, there is none which is larger than its large axis which is equal to the diameter of each of the bases of the cylinder, nor smaller than its small axis, and that none of the sections of this cylinder which meet its sides in the latter is not smaller than this ellipse. Let us call this ellipse the minimal section of the cylinder.

Let there be an oblique cylinder whose bases are *ABC* and *DEF* and whose axis is *GH*, let it be cut by a plane met by the straight line *GH*, whether in the cylinder or outside it, and such that *GH* is perpendicular to this plane; let this plane generate in the cylinder the ellipse *IKL* or a portion of it.

I say that, amongst the large axes of the ellipses of this cylinder, there is no axis smaller than the large axis of ellipse IKL *and that, amongst the small axes,*

[28] See Supplementary note [14].

no axis is larger than its large axis, which is equal to the diameter of each of the bases ABC *and* DEF, *nor smaller than its small axis; and that none of the sections of this cylinder which meet its sides in the latter is smaller than the section* IKL.

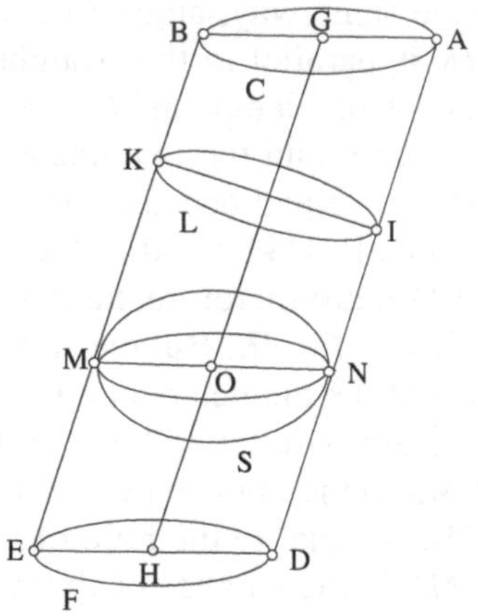

Fig. II.3.19

Proof: Every ellipse, amongst the ellipses of cylinder *ABED*, with the exception of ellipse *IKL*, is either parallel to ellipse *IKL*, or not parallel to it. If it is parallel to it, it is similar to it and equal to it and the two axes of the one are equal to the two axes of the other, we therefore have with respect to it and with respect to its axes what we have set out.

If it is not parallel to it, then if we make it the section *MNS*, its centre will be the point *O* at which section *MNS* cuts axis *GH*. But if we cut the cylinder by a plane which passes through point *O* and which is parallel to the bases of the cylinder, it generates in the cylinder a circle and the intersection of this circle and section *MNS* will be one of the diameters of this circle because it passes through point *O* which is the centre of the latter, and it is equal to the diameter of each of the circles *ABC* and *DEF*. If we make this intersection the straight line *MON*, then the straight line *MON* is one of the diameters of section *MNS*, because it passes through its centre, and the straight line *MN* is equal to the diameter of each of the circles *ABC* and *DEF*; the straight line *MN* is therefore either the large axis of section *MNS* or its small axis or one of its other diameters. If the diameter *MN* is the large axis of section *MNS*, it is clear that its large axis is not smaller than the large axis of section *IKL* because we have shown that the large axis of section *IKL* is equal to the diameter of each of the circles *ABC* and *DEF*. It has also been shown that the small axis of section *MNS* is not larger than the large axis of section *IKL*; on the contrary it is smaller than it because it is smaller than the straight line *MN*. If the straight line *MN* is the smallest of the axes of section *MNS*, then its large axis is not smaller than the large axis of section *IKL*; on the contrary it is larger than it because it is larger than the straight line *MN*. It is likewise clear that the small axis of section *MNS*, which is *MN*, is not larger than the large axis of section *IKL* because it is equal to it, both being equal to the diameter *AB*. If diameter *MN* is not one of the axes of section *MNS* it is therefore smaller than its large axis and larger than its small axis according to what has been shown in Proposition 11 of Book V of the work of Apollonius on the *Conics*. If *MN* is smaller than the largest of the axes of section *MNS*, and if it is equal to the largest of the axes of section *IKL*, then the largest of the axes of section *IKL* is smaller than the largest of the axes of section *MNS*. The largest of the axes of section *MNS* is therefore not smaller than the largest of the axes of section *IKL*. If the straight line *MN* is larger than the smallest of the axes of section *MNS*, then the largest of the axes of section *IKL* will be larger than the smallest of the axes of section *MNS*. The smallest of the axes of section *MNS* is therefore not larger than the largest of the axes of section *IKL*.

I say likewise that the smallest of the axes of section MNS *is not smaller than the smallest of the axes of section* IKL.

Proof: If we make the smallest of the axes of section *MNS*, the straight line *MN*, and if we cut the cylinder with the plane which passes through the two straight lines *GH* and *MN*, it generates in the cylinder a parallelogram. If we make this plane the plane *ABED* and if we make the intersection of plane *ABED* and plane *IKL* the straight line *KI*, the straight lines *AD* and *BE* are two sides of the cylinder, they are therefore parallel to its axis, which is *GH*. But the axis *GH* is perpendicular to plane *IKL*; therefore each of the straight lines *AD* and *BE* is perpendicular to plane *IKL*, and each of them is therefore perpendicular to every straight line drawn from one of its points[29] in plane *IKL*; therefore each of the straight lines *AD* and *BE* is perpendicular to *IK* and the straight line *IK* is perpendicular to them; therefore there is no other straight line which might be drawn between them, which meets them and which is smaller than the straight line *IK*; the straight line *MN* is therefore not smaller than the straight line *IK*. If the straight line *IK* is the smallest of the axes of section *IKL*, it has also been shown that the small axis of section *MNS* is not smaller than it. If the straight line *IK* is not the smallest of the axes of section *IKL*, then its small axis is smaller than the straight line *IK*, because the straight line *IK* is one of its diameters and the small axis, in every ellipse, is smaller than all its other diameters, according to what has been shown in Proposition 11 of Book V of the work of Apollonius on the *Conics*. The straight line *MN* which is the smallest of the axes of section *MNS* is not smaller than the smallest of the axes of section *IKL*.

It has also been shown, from what we have said, that amongst the sections of this cylinder there is not a section smaller than *IKL* because, amongst the large axes of these sections, no axis is smaller than its large axis and, amongst

[29] This point can only be the point *I* on *AD* and the point *K* on *BE*.

the small axes of these sections, no axis is smaller than its small axis. That is what we wanted to prove.

Let the section *IKL* be called the minimal section of the cylinder.

It then becomes clear that, amongst the small axes of the ellipses of the cylinder, there is no axis larger than the diameter of the circle of one of the two bases of the cylinder and that in the cylinder no straight line can be drawn which cuts its axis and whose extremities end at its lateral surface, and which is smaller than the small axis of section *IKL*.

– **20**[30] – If a plane cuts an oblique cylinder by passing through its axis and through its height and if another plane perpendicular to this plane passes through the largest of the two diagonals, then the ellipse generated in the cylinder by this last plane is such that its large axis is larger than the axes of the other ellipses generated in this cylinder, its small axis is a straight line such that no other of their small axes is larger than it, and its surface is larger than all the surfaces of the other sections of this cylinder which meet in the latter's sides. Let us call this section the maximal section of the cylinder.

Let there be an oblique cylinder whose bases are *ABC* and *DEF* with centres *G* and *H*, and whose axis is *GH* and height *GI*. Let this cylinder be cut by the plane which passes through the two straight lines *GH* and *GI* and which generates in the cylinder the parallelogram *ABED*; let us draw the straight line *AE*. The larger of the two diagonals of the parallelogram *ABED* is the straight line *AE*. Let another plane passing through the straight line *AE* and perpendicular to the parallelogram *ABED* likewise cut the cylinder, let it generate in the cylinder the ellipse *AKE*.

I say that the large axis of section AKE *is larger than the axis of every ellipse generated in this cylinder and that, amongst the small axes of these ellipses, none is larger than its small axis and that its surface is larger than the surfaces of all the other sections of this cylinder which meet in the latter's sides.*

Proof: The straight line *AE* is the largest of the straight lines drawn in the parallelogram *ABED* because it is the largest diagonal.

I say that it is the largest of the straight lines drawn in all the sections of the cylinder which pass through its axis. In fact, if we make any one of these sections, the surface *LCFM*, it will be a parallelogram and if we make the larger of its two diagonals the straight line *LF*, *LF* will be the largest of the straight lines drawn in the parallelogram *LCFM*. If we draw from the point *L* a perpendicular to the plane in which there is the circle *DEF*, let it be the perpendicular *LN*, and if we join the two points *N* and *M* with the straight line *MN*, the straight line *MN* will not be on the extension of the straight line *MF* because if it was on its extension, plane *LCFM* would be the plane which passes through the axis of the cylinder and its height. Now this plane is not thus;

[30] See Supplementary note [15].

therefore the straight line *MN* is not on the extension of the straight line *MF*. If we join the two points *N* and *F* with the straight line *NF*, it will be smaller than the sum of the two straight lines *NM* and *MF*. In the same way, if we draw from point *A* a perpendicular to the straight line *DE*, let it be the perpendicular *AS*, then *AS* will be perpendicular to the plane of circle *DEF* because it is parallel to the perpendicular *GI* and, for that reason, the straight line *AS* will be equal to the straight line *LN* because they are parallel and they are between two parallel planes. But the straight line *AD* is equal to the straight line *LM* because they are two of the sides of the cylinder, there remains the side *SD* of the right-angled triangle *ASD* equal to side *NM* of the right-angled triangle *LNM*. But the straight line *DE* is equal to the straight line *MF* because they are two diameters of circle *DEF*; therefore the sum of the two straight lines *NM* and *MF* is equal to the straight line *SE* and the sum of the two straight lines *NM* and *MF* is larger than the straight line *NF*; therefore the straight linc *SE* is larger than the straight line *NF*. But the perpendicular *AS* is equal to the perpendicular *LN* and the straight line *AE* which is intercepted by the right angle is larger than the straight line *LF* which is intercepted by the right angle; therefore the straight line *AE* is the largest of the straight lines drawn in one of the sections of the cylinder which pass through its axis.[31]

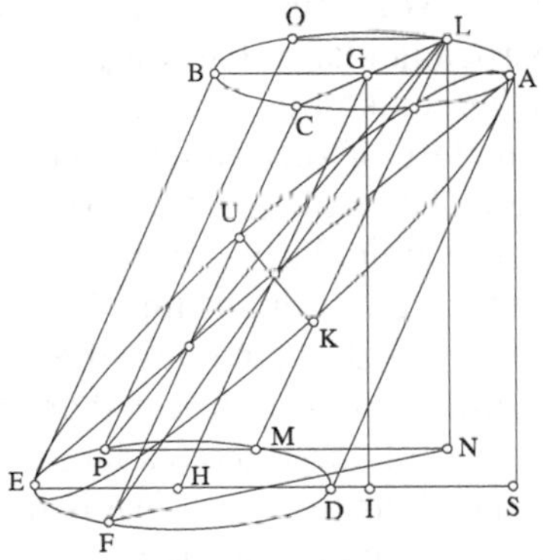

Fig. II.3.20

I say likewise that the straight line *AE* is the largest of the straight lines drawn in every section amongst the sections parallel to the axis of the cylinder. In fact, if we make one of these sections the surface *LOPM*, it will be a parallelogram, and if we make the larger of its two diagonals the straight line *LP*, *LP* will be the largest of the straight lines drawn in the parallelogram *LOPM*. If we proceed as we have done previously, it becomes clear that the straight line *NM* is equal to the straight line *SD*. But the straight line *DE* is larger than the straight line *MP*, because *DE* is a diameter of the circle, and the straight line which joins the two points *N* and *P*, let it be the straight line *NMP*

[31] In this context it is clear that this means the planes of the sections.

or another straight line,[32] is smaller than the straight line SE. But the perpendicular AS is equal to the perpendicular LN; therefore the straight line AE is larger than the straight line LP; it is consequently the largest of the straight lines drawn in one of the sections of the cylinder parallel to its axis. But we have shown that it is the largest of the straight lines drawn in the sections which pass through the axis; it is consequently the largest of the straight lines drawn in the cylinder, because either each of these straight lines is in the same plane as the axis GH, or it is possible that a plane parallel to the straight line GH passes through it. If it is thus, it is clear that the straight line AE is the largest of the diameters of section AKE and that it is larger than every diameter amongst the diameters of all the sections of the cylinder which meet the sides of the cylinder in the latter. Therefore the straight line AE is the large axis of section AKE, according to what has been shown in Proposition 11 of Book V of the work of Apollonius on the *Conics*, and it is larger than the axes of all the ellipses of the cylinder and than all the diameters of the circles which are in the latter.

I say likewise that the small axis of section AKE is the largest of the small axes of all the sections.

It is in fact equal to the diameter of each of the bases of the cylinder, because if a plane cuts the cylinder by passing through axis GH and is such that it is perpendicular to plane $ABED$, then the intersection of this plane and the plane of section AKE is perpendicular to the straight line GH and is equal to the diameter of circle ABC. If we make this intersection the straight line KU, then KU will be one of the diameters of section AKE, because it passes through its centre. But the straight line KU cuts plane $ABED$ perpendicularly; accordingly it cuts the straight line AE perpendicularly; but the straight line AE is the large axis of section AKE; therefore the straight line KU is its small axis and it is equal to the diameter of circle ABC. Amongst the small axes of the ellipses which there are in the cylinder, nonc is larger than the small axis of section AKE. But we have shown that its large axis is larger than their large axes; its surface is therefore larger than their surfaces. That is what we wanted to prove.

Let us call section AKE the maximal section.

It then becomes clear that the largest of the axes of the maximal section of the cylinder is the largest of the straight lines drawn in this cylinder, and that the smallest of its axes is equal to the diameter of each of the bases of the cylinder and that it is also equal to the largest of the axes of its minimal section.

– **21** – For every oblique cylinder, the ratio of each of its minimal sections to each of the two circles of its bases is equal to the ratio of each of the straight lines which are such that no straight line smaller than it is drawn in this cylinder between two of its sides and passing through its axis, to the diameter of each of

[32] The points M, N and P are aligned if $MP \parallel ED$, they are not aligned generally.

its two bases, and it is also equal to the ratio of the height of this cylinder to its axis.

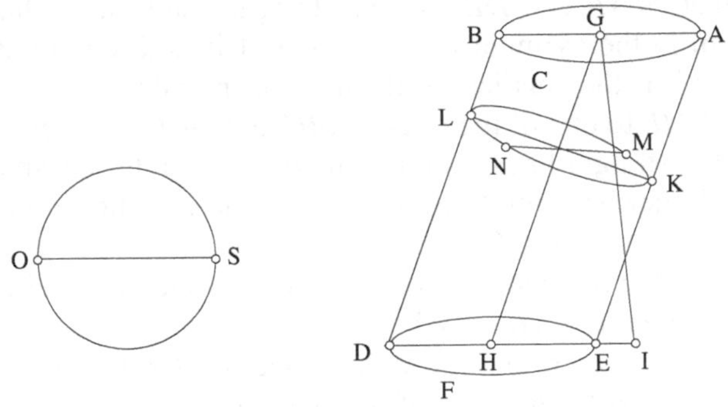

Fig. II.3.21

Let there be an oblique cylinder whose bases are *ABC* and *DEF*, with diameters *AB* and *DE*, and whose axis is *GH*. Let us draw from point *G* the height of the cylinder, which is *GI*, and let there be one of the minimal sections of the cylinder, *KL*.

I say that the ratio of section KL *to each of the circles* ABC *and* DEF *is equal to the ratio of each of the straight lines which are such that no straight line smaller than it is drawn in this cylinder between two of its sides and passing through its axis, to each of the diameters* AB *and* DE, *and that it is also equal to the ratio of* GI *to* GH.

Proof: If we make the small axis of section *KL* the straight line *KL*, and its large axis the straight line *MN*, if we make the square obtained from the straight line *SO* equal to the product obtained from *KL* and *MN* and if we describe on the straight line *SO* a circle such that *SO* is one of its diameters, then circle *SO* will be equal to section *KML*. The ratio of circle *SO* to one[33] of the two circles *ABC* and *DEF* is equal to the ratio of the square of diameter *SO* to one of the squares of the diameters *AB* and *DE*, the ratio of section *KML* to one of the circles *ABC* and *DEF* is therefore equal to the ratio of the square of diameter *SO* to one of the squares of the diameters *AB* and *DE*. But the square of diameter *SO* is equal to the product obtained from *KL* and *MN*; therefore the ratio of section *KML* to one of the circles *ABC* and *DEF* is equal to the ratio of the product obtained from *KL* and *MN* to one of the squares of the diameters *AB* and *DE*. But *MN* is the largest of the axes of the minimal section *KML*; consequently it is equal to each of the diameters *AB* and *DE*; the ratio of section *KMLN* to one of the circles *ABC* and *DEF* is therefore equal to the ratio of the product obtained from *KL* and one of the diameters *AB* and *DE* to the square of one of the diameters *AB* and *DE*. But the ratio of the product obtained from *KL* and one of the diameters *AB* and *DE* to the square of one of the diameters *AB* and *DE* is

[33] Literally: to each one. In this context, 'each one' will be translated by 'one'.

equal to the ratio of *KL* to one of the diameters *AB* and *DE*; therefore the ratio of section *KML* to one of the circles *ABC* and *DEF* is equal to the ratio of *KL* to one of the diameters *AB* and *DE*. But no straight line smaller than the straight line *KL* is drawn in the cylinder between two of its sides and passing through the axis, because it is the smallest of the axes of the minimal section *KML*. The ratio of section *KML* to one of the circles *ABC* and *DEF* is therefore equal to the ratio of one of the straight lines such that no other straight line smaller than it is drawn in the cylinder between two of its sides and passing through its axis, to one of the diameters *AB* and *DE*.

I say likewise that the ratio of section KML *to one of the circles* ABC *and* DEF *is equal to the ratio of* GI *to* GH.

Proof: We have shown that the ratio of section *KML* to one of the circles *ABC* and *DEF* is equal to the ratio of *KL*, which is the smallest of the axes of the minimal section *KML*, to one of the diameters *AB* and *DE*. But the ratio of the smallest of the axes of the minimal section of the cylinder to one of the diameters *AB* and *DE* is equal to the ratio of *GI* to *GH*; therefore the ratio of section *KML* to one of the circles *ABC* and *DEF* is equal to the ratio of *GI* to *GH*. That is what we wanted to prove.

– **22** – For every oblique cylinder, the ratio of its maximal section to one of the circles of its bases is equal to the ratio of the largest straight line drawn in the cylinder to the diameter of one of the circles of its bases.

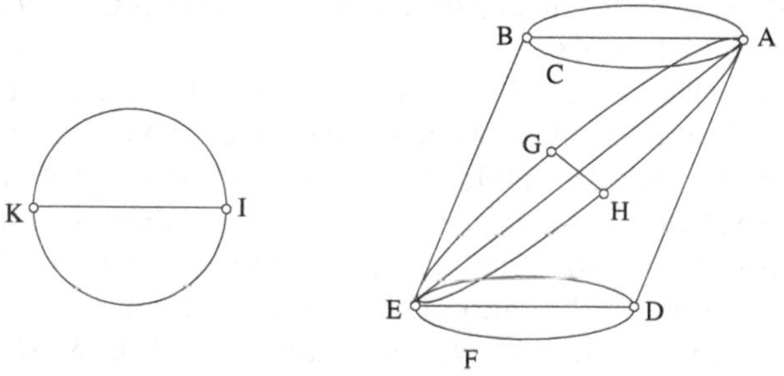

Fig. II.3.22

Let there be an oblique cylinder whose bases are *ABC* and *DEF* and the diameters of these bases *AB* and *DE*. Let there be amongst the sections of the cylinder the maximal section *AGEH*.

I say that the ratio of section AGEH *to one of the circles* ABC *and* DEF *is equal to the ratio of the largest straight line drawn in the cylinder* ABCDEF *to one of the straight lines* AB *and* DE.

Proof: If we make the largest of the axes of section *AGEH* the straight line *AE*, and the smallest of its axes *GH*, and if we make the square obtained from *IK* equal to the product obtained from *AE* and *GH*, and if we describe on the straight line *IK* a circle such that *IK* is one of its diameters, circle *IK* will be equal to section *AEGH*, and the ratio of circle *IK* to one of the circles *ABC* and *DEF* is equal to the ratio of the square of the diameter *IK* to one of the squares of the diameters *AB* and *DE*. The ratio of section *AGEH* to one of the circles *ABC* and *DEF* is therefore equal to the ratio of the square of the diameter *IK* to one of the squares of the diameters *AB* and *DE*. But the square of diameter *IK* is equal to the product of *AE* and *GH*; therefore the ratio of section *AGEH* to one of the circles *ABC* and *DEF* is equal to the ratio of the product obtained from *AE* and *GH* to one of the squares of the diameters *AB* and *DE*. As for the straight line *AE*, it is the largest straight line drawn in the cylinder *ABCDEF*. As for the straight line *GH*, it is equal to one of the diameters *AB* and *DE*; therefore the ratio of section *AGEH* to one of the circles *ABC* and *DEF* is equal to the ratio of the product obtained from the largest straight line drawn in the cylinder *ABCDEF* and one of the diameters *AB* and *DE*, to one of the squares of diameters *AB* and *DE*, which is equal to the ratio of the largest straight line drawn in the cylinder *ABCDEF* to one of the diameters *AB* and *DE*. The ratio of section *AGEH* to one of the circles *ABC* and *DEF* is therefore equal to the ratio of the largest straight line drawn in the cylinder *ABCDEF* to one of the diameters *AB* and *ED*. That is what we wanted to prove.

– **23** – For every oblique cylinder, the ratio of one of its minimal sections to its maximal section is equal to the ratio of one of the straight lines such that no straight line smaller than it is drawn in this cylinder between two of its sides and passing through its axis, to the largest straight line drawn in this cylinder.

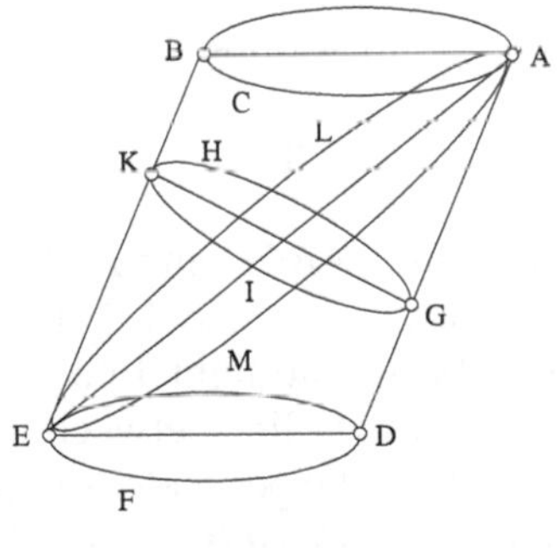

Fig. II.3.23

Let there be an oblique cylinder whose bases are *ABC* and *DEF*, *GHIK* one of its minimal sections and *ALEM* its maximal section.

I say that the ratio of section GHIK *to section* ALEM *is equal to the ratio of one of the straight lines such that no straight line smaller than it is drawn in the cylinder* ABCDEF *between two of its sides and passing through its axis, to the largest straight line drawn in this cylinder.*

Proof: The ratio of section *GHIK* to circle *ABC* is equal to the ratio of one of the straight lines such that no straight line smaller than it is drawn in the cylinder *ABCDEF* between two of its sides and passing through its axis, to the diameter *AB*. But the ratio of circle *ABC* to section *ALEM* is equal to the ratio of diameter *AB* to the largest straight line drawn in the cylinder *ABCDEF*. By the ratio of equality, we have the ratio of section *GHIK* to section *ALEM* equal to the ratio of one of the straight lines such that no straight line smaller than it is drawn in the cylinder *ABCDFE* between two of its sides and passing through its axis, to the largest straight line drawn in the cylinder *ABCDEF*. That is what we wanted to prove.[34]

– **24** – If we have in the same plane two similar ellipses such that their centre is common and that the larger of the two axes of one is a portion of the larger of the two axes of the other, and if there is drawn between them a straight line tangent to the smallest and such that its extremities end at the perimeter of the largest, then the point of contact divides this straight line into two halves.

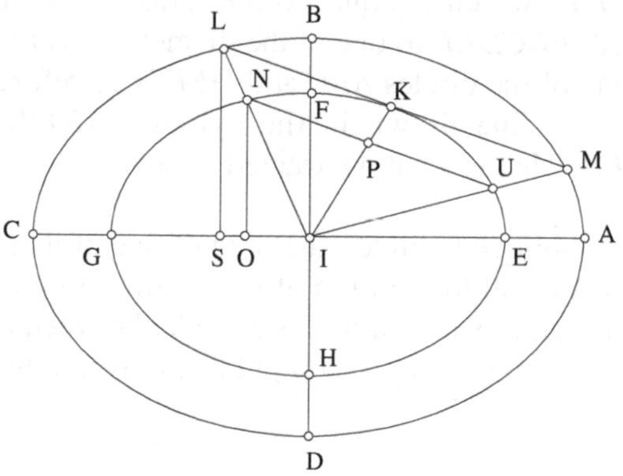

Fig. II.3.24

Let there be two similar ellipses *ABCD* and *EFGH*, let *I* be their centre and let the large axis of *ABCD* be the straight line *AC*, its small axis the straight line *BD*, the large axis of *EFGH* be the straight line *EG*, and its small axis the straight line *FH*. Let there be a straight line between the two sections tangent to *EFGH* and whose extremities end at the perimeter of section *ABCD*.

I say that the point of contact divides this straight line into two halves.

Proof: The tangent straight line is either tangent to section *EFGH* at one of the points *E*, *F*, *G* and *H*, or is tangent to it at another point than these points. If it is tangent to it at one of the points *E*, *F*, *G* and *H*, it is clear that <this point> divides it into two halves because it is one of the ordinate straight lines since it cuts the axis perpendicularly, according to what has been shown in Propositions 13 and 15 of the first book of the work of Apollonius on the *Conics*. If it is not tangent to it at one of these points, then if we make its contact with it point *K* and if we make the tangent straight line *LKM*, if we join the two points *K* and *I*

[34] See Supplementary note [16].

by the straight line *KPI*, if we join the two points *L* and *I* by the straight line *LNI* and if we draw from points *L* and *N* two perpendiculars *LS* and *NO* to *AC*, then the ratio of the product obtained from *AS* and *SC* to the square of the straight line *LS* is equal to the ratio of *AC* to the *latus rectum*, according to what has been shown in Proposition 21 of Book I of the work of Apollonius on the *Conics*. That is why likewise the ratio of the product obtained from *EO* and *OG* to the square of the straight line *NO* is equal to the ratio of *EG* to its *latus rectum*. But the ratio of *AC* to its *latus rectum* is equal to the ratio of *EG* to its *latus rectum*, because the two sections *ABCD* and *EFGH* are similar;[35] therefore the ratio of the product obtained from *AS* and *SC* to the square of the straight line *LS* is equal to the ratio of the product obtained from *EO* and *OG* to the square of the straight line *NO*. If we permute, the ratio of the product obtained from *AS* and *SC* to the product obtained from *EO* and *OG* is equal to the ratio of the square of the straight line *LS* to the square of the straight line *NO*. But the ratio of the square of the straight line *LS* to the square of the straight line *NO* is equal to the ratio of the square of the straight line *SI* to the square of the straight line *IO*. The ratio of the product obtained from *AS* and *SC*, plus the square of the straight line *SI*, to the product obtained from *EO* and *OG*, plus the square of the straight line *OI*, is therefore equal to the ratio of the square of the straight line *LS* to the square of the straight line *NO*. As for the product obtained from *AS* and *SC*, plus the square of the straight line *SI*, it is equal to the square of the straight line *AI*. As for the product of *EO* and *OG* plus the square of the straight line *OI*, it is equal to the square of the straight line *EI*. The ratio of the square of the straight line *LS* to the square of the straight line *NO* is therefore equal to the ratio of the square of the straight line *AI* to the square of the straight line *EI*. The ratio of *LS* to *NO* is consequently equal to the ratio of *AI* to *IE*. But the ratio of *LS* to *NO* is also equal to the ratio of *LI* to *IN*, because the straight line *LS* is parallel to the straight line *NO*; therefore the ratio of *AI* to *EI* is equal to the ratio of *LI* to *IN*. It will be the same for all the straight lines drawn from point *I* to section *ABCD*. If we therefore join the two points *I* and *M* by the straight line *IUM*, the ratio of *MI* to *IU* is also equal to the ratio of *AI* to *IE*, which we have shown is equal to the ratio of *LI* to *IN*; the ratio of *MI* to *IU* is therefore equal to the ratio of *LI* to *IN*. If we therefore join the two points *U* and *N* by the straight line *UN*, the straight line *UN* will be parallel to the straight line *LM*. But the straight line *LM* is tangent to section *EFGH*; if we therefore draw a diameter from point *K*, the straight line *UN* will be an ordinate to it, according to what has been shown in Proposition 50 of the first book[36] of the work of Apollonius on the *Conics*. If we therefore join the two points *I* and *K* by the straight line *IPK*, the ratio of *NP* to *PU* is equal to the ratio of *LK* to *KM*, because the two straight lines *NU* and *LM* are parallel and the straight line *NP* is equal to the

[35] Apollonius, VI.12.

[36] *i.e.* Proposition 47 in the Heiberg edition of the *Conics* of Apollonius, that is, the version of Eutocius.

straight line *PU*, since *NU* is an ordinate to diameter *IK*, therefore the straight line *LK* is equal to the straight line *KM*. That is what we wanted to prove.

Furthermore, it has also been shown that the ratios of the diameters of similar ellipses one to another – each one to its homologue which describes with its axis an angle equal to the angle which its associate describes with its <homologous> axis – are equal to the ratios of their axes, one to another, each of the axes to its homologue.[37]

– **25** – We want to show how to construct in the larger of two similar, unequal ellipses, which are in the same plane, which have a common centre and which are such that the large axis of the one is a portion of the large axis of the other, a polygon inscribed in the largest section, surrounding the smallest and such that the sides of this polygon are not tangent to the smallest section and such that if its opposite vertices[38] are joined by straight lines, they are diameters of the largest section.[39]

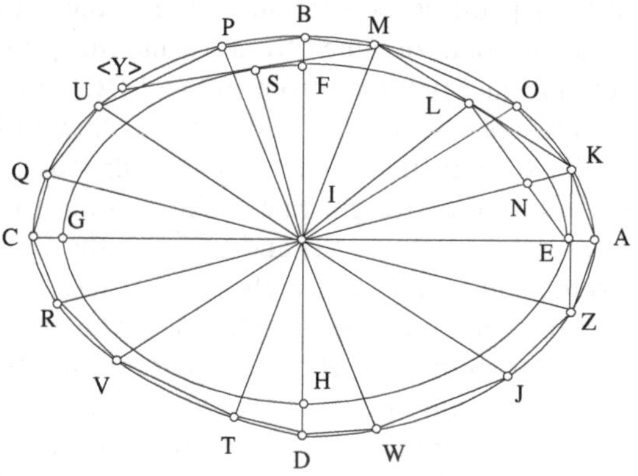

Fig. II.3.25

Let there be two similar and unequal sections in the same plane which are *ABCD* and *EFGH*, let point *I* be the centre for both of them, let the large axis of the largest section be the straight line *AC*, its small axis the straight line *BD* and the large axis of the other section *EG* the small axis *FH*. We want to show how to construct a polygon inscribed in section *ABCD* and which surrounds section *EFGH*, without its sides being tangent to it and such that if straight lines join the opposite vertices, they are diameters of section *ABCD*.

Let us draw from point *E* a perpendicular to *EG*, which is *EK*, the straight line *EK* becomes tangent to section *EFGH*, according to what has been shown in Proposition 17 of Book I of the work by Apollonius on the *Conics*. Let us

[37] See commentary.
[38] Literally: angles.
[39] See Supplementary note [17].

draw from point *K* another straight line tangent to section *EFGH*, which is the straight line *KL*; let us extend it so that it meets the perimeter of section *ABCD*; let it meet it at point *M*. If the straight line *KLM* met the straight line *BD*, that is what we wanted; otherwise we can draw from point *M* likewise a straight line tangent to section *EFGH* and we can proceed as we have done before.

I say that if we continue to proceed in this way, one of the straight lines which we draw tangent to section EFGH *will of necessity meet the straight line* BD *at a point which is not outside section* ABCD.

Proof: If we join the straight line *EL* between two points of contact, if we divide it into two halves at point *N* and if we draw from point *K* a straight line to point *N*, let it be the straight line *KN*, straight line *KN* will be a portion of one of the diameters of section *ABCD*, according to what has been shown in Proposition 29 of Book II of the work by Apollonius on the *Conics*. If the straight line *KN* is extended, it ends at point *I* which is the centre of the two sections, *i.e.* the straight line *KNI*, and if we draw from point *I* the straight *IL*, the straight line *LN* is equal to the straight line *NE*; but the straight line *IL* is smaller than the straight line *IE*, according to what has been shown in Proposition 11 of Book V of the work by Apollonius on the *Conics*. The ratio of *LN* to *NE* is therefore larger than the ratio of *LI* to *IE*. The angle *LIN* is therefore larger than the angle *NIE* because the straight line which divides angle *LIE* into two halves divides *LE* in an equal ratio <to the ratio> of *LI* to *IE*.

Similarly, the straight line *KLM* is tangent to section *EFGH* and its extremities ended at section *ABCD*, therefore the straight line *ML* is equal to the straight line *LK*. If we draw the straight line *MI*, it will be smaller than the straight line *IK*, according to what has been shown in Proposition 11 of Book V of the work by Apollonius on the *Conics*. The ratio of *ML* to *LK* is therefore larger than the ratio of *MI* to *IK*, therefore angle *MIL* is larger than angle *LIK*. But angle *LIK* is larger than angle *KIE*; therefore angle *MIL* is much larger than angle *KIE*. Then angle *LIE* is larger than double angle *KIE* and angle *MIE* is larger than triple this angle.

Similarly, we can also show for all angles generated between the straight line *EI* and the straight lines which we draw from point *I* – if we follow the preceding method in relation to straight lines tangent to section *EFGH* – that they all of them exceed, if taken successively, angle *KIE*. These angles of necessity therefore arrive at an angle which is their junction and which will be larger than angle *BIE*. If they arrive at this limit, the last tangent straight line which is drawn will of necessity meet the straight line *IB*; let this tangent straight line which meets *BI*, without going beyond section *ABCD*, be the straight line *MS*. Let us draw the straight lines *AK* and *MB*, let us mark on the portion *KM* of the section a point *O* with any position at all and let us draw from this point the straight lines *OK* and *OM*. The straight lines *AK*, *KO*, *OM* and *MB* are between the two sections and do not meet section *EFGH* because the straight lines *EK*, *KM* and *MS* are tangent to it. If we draw in portion *BC* of the section chords equal to the chords *BM*, *MO*, *OK* and *KA*, in order and in succession – as regards the chord *BP*, it is like chord *BM*; as regards the chord *PU*, it is like chord *MO*, and the same for the other chords – then there will be found in portion *BC* of the section equal chords – and equal in number – to those found in portion *AB*, because if half of section *DAB* is reversed and if it is placed on half of section *BCD*, it is positioned over it and superposed on to the other entirely, then point *A* is placed over point *C*, according to what has been shown in Proposition 4 of Book VI of the work of Apollonius on the *Conics*. If we extend the straight line *KI* to point *R* and if we draw the straight line *CR*, the two straight lines *KI* and *IA* are equal to the straight lines *RI* and *IC* because the centre *I* divides the diameters *AC* and *KR* into two halves, according to what has been shown in Proposition 30 of the first Book of the work of Apollonius on the *Conics*, and the two opposite angles *KIA* and *RIC* are equal, the base *AK* is therefore equal to the base *CR*. But if half of section *ABC* is placed on to half of section *CDA* such that point *A* of the former is put on to point *C* of the latter, it is superposed completely on to the other, according to what has been shown in Proposition 4 of Book VI of the work of Apollonius on the *Conics*. The straight line *AK* will be placed on the straight line *CR*; therefore the straight line *CR* is not tangent to section *EFGH* because half of section *EFG* is likewise entirely superposed on to half of section *GHE* if it is placed on to it. In the same way, if we likewise draw from the points *O*, *M*, *P*, *U* and *Q* diameters of the section, and if we join the extremities of the diameters drawn by the straight lines *RV*, *VT*, *TD*, *DW*, *WJ*, *JZ* and *ZA*, there will thus have been constructed in section *ABCD* a polygon inscribed in section *ABCD* and which surrounds section *EFGH* without touching it, namely the polygon *AKOMBPUQCRVTDWJZ*, such that the straight lines which join its opposite vertices[40] are diameters of section *ABCD*. That is what we wanted to prove.

It is likewise clear according to what we have said that if a polygon is constructed in an ellipse and if the straight lines which join its vertices[41] are diameters of this section, then the opposite sides are equal.

– **26** – The ratios of the perimeters of similar ellipses, one to another, are equal to the ratios of their axes, one to another, each axis to its homologue.

Let there be two similar ellipses *ABCD* and *EFGH*, let their large axes be *AC* and *EG* and let their small axes be *BD* and *FH*.

I say that the ratio of the perimeter of section ABCD *to the perimeter of section* EFGH *is equal to the ratio of axis* AC *to axis* EG *and is equal to the ratio of axis* BD *to axis* FH.

[40] Literally: angles.
[41] Literally: angles.

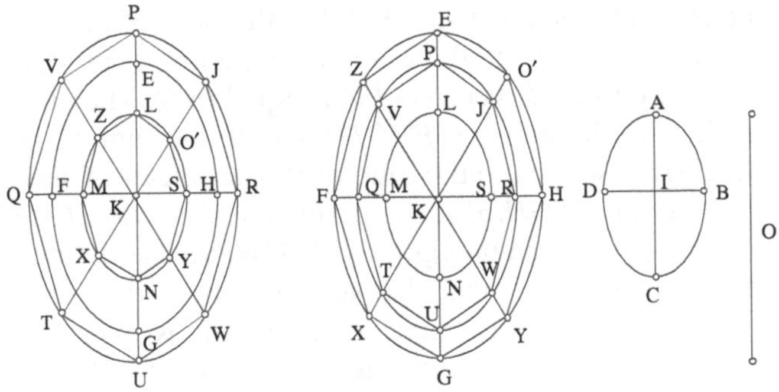

Fig. II.3.26

Proof: If we make the smaller of the two sections the section *ABCD* with centre the point *I* and the centre of section *EFGH* the point *K*, and if we place section *ABCD* in the plane of section *EFGH*, centre *I* of the former on centre *K* and the large axis, which is *AC*, on what it covers of the large axis *EG*, then its small axis will be placed on a part of its small axis; let the whole section be placed in the position of section *LMNS* and let its large axis be *LN* and its small axis *MS*.

If it was possible that the ratio of the perimeter of section *ABCD* to the perimeter of section *EFGH* was not equal to the ratio of *AC* to *EG*, then the ratio of the perimeter of section *LMNS* to the perimeter of section *EFGH* would not be equal to the ratio of *LN* to *EG*; it would therefore be either larger than it or smaller.

If we suppose first of all that it is larger than it, if that is possible, and if we make the ratio of the straight line *O* to the straight line *EG* equal to the ratio of the perimeter of section *LMNS* to the perimeter of section *EFGH*, the straight line *O* will be larger than the straight line *LN*; but it is clear that it is smaller than the straight line *EG*. If we make each of the straight lines *KP* and *KU* in the first case of figure equal to half of straight line *O*, if we construct on the straight line *PU* an ellipse similar to each of the sections *EFGH* and *LMNS*, *PU* will be its large axis and *QR* its small axis, let it be the section *PQUR*, if we construct a polygon inscribed in section *PQUR* and which surrounds section *LMNS*, let the polygon be *PVQTUWRJ*, if we draw the straight lines *KV*, *KT*, *KW* and *KJ* which we extend to the points *Z*, *X*, *Y* and *O′* and if we draw the straight lines *EZ*, *ZF*, *FX*, *XG*, *GY*, *YH*, *HO′* and *O′E*, then the ratio of *KP* to *KE* is equal to the ratio of *KV* to *KZ*; therefore the two straight lines *PV* and *EZ* are parallel and the ratio of *PV* to *EZ* is equal to the ratio of *KP* to *KE*. In the same way, we can show that the ratios of the sides which are left in the polygon *PVQTUWRJ* to their homologues amongst the sides of polygon *EZFXGYHO′* are equal to the ratio of *KP* to *KE*. The ratio of the sum of the sides of the polygon *PVQTUWRJ* to the sum of the sides of the polygon *EZFXGYHO′* is equal to the ratio of the straight line *KP* to the straight line *KE*; consequently it is equal to the ratio of the straight line *PU* to the straight line *EG*; but the ratio of the straight line *PU* to the straight line *EG* is equal to the ratio of the perimeter of section *LMNS* to the perimeter of section *EFGH* because *PU* is equal to *O*. Therefore the ratio of the sum of the sides of the polygon *PVQTUWRJ* to the sum of the sides of the polygon *EZFXGYHO′* is equal to the ratio of the perimeter of section *LMNS* to the perimeter of section *EFGH*. But the perimeter of section *LMNS* is smaller than the sum of the sides of the polygon *PVQTUWRJ*. The perimeter of section *EFGH* is therefore smaller than the sum of the sides of the polygon *EZFXGYHO′*; now, it surrounds them, which is impossible. The ratio of the perimeter of section *LMNS* to the perimeter of section *EFGH* is not larger than the ratio of axis *LN* to axis *EG*.

I say that it is not smaller than it.

If it was possible that it was smaller than it, then it would be equal to the ratio of axis *LN* to a straight line *O*. The straight line *O* will therefore be larger than axis *EG*. If we make each of the straight lines *KP* and *KU*, in the second example, equal to half of the straight line *O*, if we construct on the straight line *PU* an ellipse such that *PU* is its large axis and it is similar to each of the sections *EFGH* and *LMNS*, namely the section *PQUR*, if we construct a polygon *PVQTUWRJ* which surrounds section *EFGH* and which is inscribed within section *PQUR*, and if we draw the straight lines *KZV*, *KXT*, *KYW* and *KO′J* and the straight lines *LZ*, *ZM*, *MX*, *XN*, *NY*, *YS*, *SO′* and *O′L*, then we can show as we have shown before that the ratio of the sum of the sides of the polygon *LZMXNYSO′* to the sum of the sides of the polygon *PVQTUWRJ* is equal to the ratio of *KL* to *KP*, which is equal to the ratio of *LN* to *PU*. But the ratio of *LN* to *PU* is equal to the ratio of the perimeter of section *LMNS* to the perimeter of section *EFGH*. The ratio of the sum of the sides of the polygon *LZMXNYSO′* to the sum of the sides of the polygon *PVQTUWRJ* is equal to the ratio of the perimeter of section *LMNS* to the perimeter of section *EFGH*. But the sum of the sides of the polygon *LZMXNYSO′* is smaller than the perimeter of section *LMNS*; therefore the sum of the sides of the polygon *PVQTUWRJ* is smaller than the perimeter of section *EFGH*; now, these sides surround it, which is impossible. The ratio of the perimeter of section *LMNS* to the perimeter of section *EFGH* is therefore not smaller than the ratio of *LN* to *EG*, which is equal to the ratio of *AC* to *EG*. Now we have shown that it is not larger than it, it is consequently equal to it. That is what we wanted to prove.

– **27** – The ratios of the ellipses one to another are compounded of ratios of their axes one to another. And if these ellipses are similar, then their ratios one to another are equal to the ratios of the squares of their diameters, one to another: the square of each of these diameters to the square of its homologue.

Let there be two ellipses *ABCD* and *EFGH*; the straight line *AC* is the large axis of section *ABCD*, the straight line *BD* is its small axis, the straight line *EG* is the large axis of section *EFGH* and the straight line *FH* is its small axis.

I say that the ratio of section ABCD *to section* EFGH *is compounded of the ratio of* AC *to* EG *and the ratio of* BD *to* FH. *If the two sections* ABCD *and* EFGH *are similar, then the ratio of section* ABCD *to section* EFGH *is equal to the ratio of the square of each of the diameters of section* ABCD *to the square of its homologue amongst the diameters of section* EFGH.

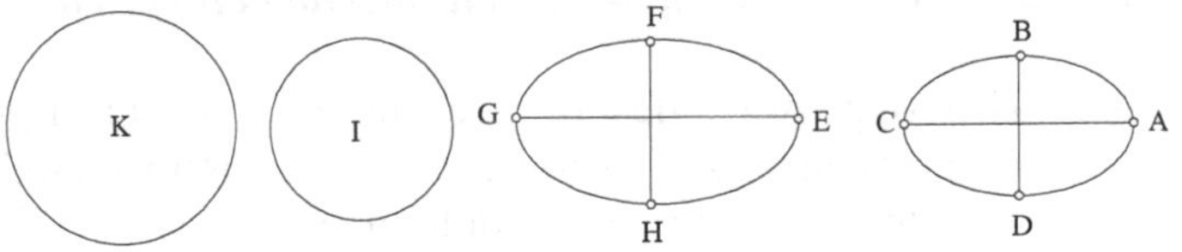

Fig. II.3.27

Proof: If we make the square of the diameter of a circle *I* equal to the product of *AC* and *BD* and if we make the square of the diameter of a circle *K* equal to the product of *EG* and *FH*, circle *I* will be equal to section *ABCD* and circle *K* will be equal to section *EFGH*. The ratio of section *ABCD* to section *EFGH* is therefore equal to the ratio of circle *I* to circle *K*. But the ratio of circle *I* to circle *K* is equal to the ratio of the square of the diameter of circle *I* to the square of the diameter of circle *K*. But the square of the diameter of circle *I* is equal to the product of *AC* and *BD* and the square of the diameter of circle *K* is equal to the product of *EG* and *FH*. The ratio of section *ABCD* to section *EFGH* is therefore equal to the ratio of the product of *AC* and *BD* to the product of *EG* and *FH*. This ratio is compounded of the ratio of *AC* to *EG* and the ratio of *BD* to *FH*. The ratio of section *ABCD* to section *EFGH* is therefore compounded of the ratio of *AC* to *EG* and the ratio of *BD* to *FH*.

Furthermore, if the two sections *ABCD* and *EFGH* are similar, then the ratio of *AC* to *BD* is equal to the ratio of *EG* to *FH*. If we permute, the ratio of *AC* to *EG* will be equal to the ratio of *BD* to *FH*. The ratio compounded of the ratio of *AC* to *EG* and the ratio of *BD* to *FH* is equal to the ratio of *AC* to *EG* repeated twice, which is equal to the ratio of the square of the straight line *AC* to the square of the straight line *EG* and is equal to the ratio of the square of the straight line *BD* to the square of the straight line *FH*. The ratio compounded of the ratio of *AC* to *EG* and the ratio of *BD* to *FH* is equal to the ratio of the square of the straight line *AC* to the square of the straight line *EG* and is equal to the ratio of the square of the straight line *BD* to the square of the straight line *FH*. Now we have shown that the ratio of section *ABCD* to section *EFGH* is equal to the ratio compounded of the ratio of *AC* to *EG* and the ratio of *BD* to *FH*. The ratio of section *ABCD* to section *EFGH* is therefore equal to the ratio of the square of the straight line *AC* to the square of the straight line *EG* and is equal to the ratio of the square of the straight line *BD* to the square of the straight line *FH* and it is also equal to the ratio of the square of every diameter of section *ABCD* amongst the remaining diameters to the square of its homologue amongst the diameters of section *EFGH*, because the ratios of the diameters of section $ABCD$ to their homologues amongst the diameters of section $EFGH$ are equal ratios. That is what we wanted to prove.

<IV. *On the lateral area of the cylinder and the lateral area of portions of the cylinder located between plane sections meeting all the sides*>

– **28** – Any two opposite sides of a cylinder pass through the two extremities of one of the diameters of every section – through which they pass – amongst the sections of this cylinder, which meets its sides. Two of the cylinder's sides, which pass through the two extremities of one of the diameters of one of its sections which meet its sides, are two opposite sides amongst the cylinder's sides.

Let there be a cylinder whose bases are ABC and DEF, with centres G and H, and whose axis is GH. Let there be in the cylinder one of its sections which meet its sides, namely IKL; let the two sides of the cylinder, namely AID and BKE, pass through it.

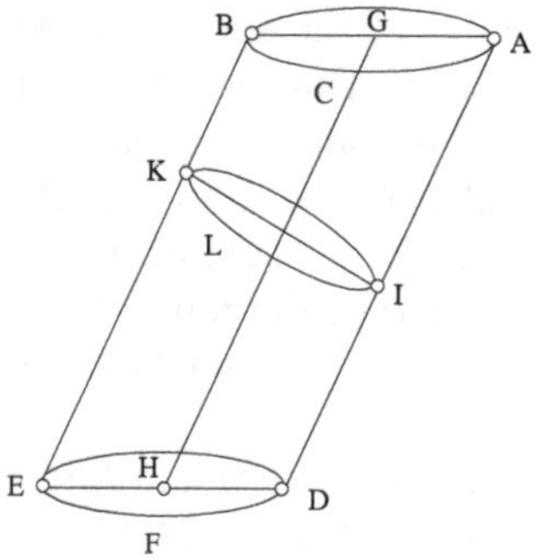

Fig. II.3.28

I say that if the two sides AID *and* BKE *are two opposite sides amongst the sides of the cylinder, then they pass through the extremities of one of the diameters of section* IKL. *And if they pass through the extremities of one of the diameters of section* IKL, *then they are two opposite sides amongst the sides of the cylinder.*

Let us first of all make the two straight lines AID and BKE two opposite sides amongst the sides of the cylinder.

I say that they pass through the extremities of one of the diameters of section IKL.

Proof: If we join the extremities of the straight lines AID and BKE by the two straight lines AB and DE, the two straight lines AB and DE are two of the diameters of circles ABC and DEF; they therefore pass through points G and H. The straight lines AID and BKE are two of the sides of the cylinder; they are therefore in the same plane because they are parallel; that is, why the two straight lines which join their extremities are in this plane, they are the straight lines *AGB* and *DHE*; these two straight lines being in this plane, then the straight line *GH* which links them is likewise in this plane, that is, plane *ADEB*. If we join the points *I* and *K* at which plane *IKL* cuts the sides *AID* and *BKE*, with a straight line *IK*, this straight line is in this plane and cuts the axis *GH* at the point of intersection of this axis and section *IKL*, which is the centre of section *IKL*. The straight line *IK* accordingly passes through the centre of section *IKL*; it is therefore one of its diameters; the sides *AID* and *BKE* therefore pass through the extremities of one of the diameters of section *IKL*, which is *IK*.

Similarly, let us make the two sides *AID* and *BKE* passing through the extremities of one of the diameters of section *IKL*, which is diameter *IK*.

I say that AID *and* BKE *are two opposite sides of the cylinder.*

Proof: The straight line *IK* is one of the diameters of section *IKL*; it therefore passes through its centre, which is the point at which plane *IKL* cuts axis *GH*. The straight line *IK* therefore cuts axis *GH* and meets the straight line *AID*. But the straight lines *AID* and *GH* are in the same plane; the straight line *IK* is therefore together with them in this plane and this plane is the one in which are the straight lines *GH* and *IK*. In the same way, we can also show that the straight line *BKE* is in this plane; the three straight lines *AID*, *GH* and *BKE* are therefore in the same plane. The intersections of this plane and the two planes *ABC* and *DEF* are two straight lines one of which passes through the three points *A*, *G* and *B*, which is the straight line *AGB*, while the other passes through the three points *D*, *H* and *E*, which is the straight line *DHE*. But the two straight lines *AGB* and *DHE* are two of the diameters of the circles *ABC* and *DEF* because they pass through their centres; therefore the two straight lines *AID* and *BKE* are two opposite sides of the cylinder. That is what we wanted to prove.

It is clear from what we have said that, if we have in any cylinder at all sections, in whatever number, amongst those which meets its sides, if diameters are drawn in these sections and if all these diameters are drawn from one only of the sides of the cylinder, then the other extremities of these diameters all end at the opposite side to the first side of the cylinder, from which these diameters have been drawn.

– **29** – For every cylinder, <the sum> of the portions[42] located on any two opposite sides amongst the sides of the cylinder, between two of its sections which do not intersect but which meet the sides of the cylinder, or between one of these sections and one of the bases of the cylinder, if this section does not cut it, is equal to the sum of the portions located between these sections on any two other opposite sides, amongst the sides of the cylinder, and is also equal to twice the portion located between them on the cylinder's axis.

[42] Literally: that which is; we use 'portion' for this expression, or an equivalent term.

Let there be a cylinder whose bases are *ABC* and *DEF* with centres *G* and *H* and whose axis is *GH*, let there be two sections in this cylinder, which do not intersect and which are *IKL* and *MNU*; let the straight lines *AIMD* and *BKNE* be any two opposite sides amongst the sides of the cylinder and let the straight lines *CLSF* and *OPUQ* likewise be two other opposite sides of the cylinder, whatever these two sides are.

I say that the sum of the two straight lines IM *and* KN *which are located between the two sections* IKL *and* MNS *on the opposite sides* AD *and* BE *is equal to the sum of the two straight lines* LS *and* PU *located between the two sections mentioned on the two opposite sides* CF *and* OQ *and is also equal to the double of* RV *which is located on axis* GH *between the two sections, and that the sum of the two straight lines* AI *and* BK *which are located between section* IKL *and the base* ABC *on the two opposite sides* AD *and* BE *is equal to the sum of the two straight lines* CL *and* OP *which are likewise located between section* IKL *and the base* ABC *on the two opposite sides* CF *and* OQ *and is also equal to the double of* GR *which is located on axis* GH *between* ABC *and* IKL.

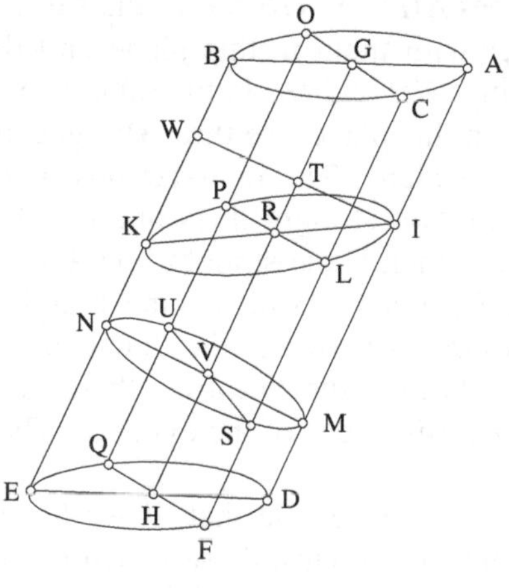

Fig. II.3.29

Proof: The two sections *IKL* and *MNS* are either parallel or not parallel. If they are parallel, then all the straight lines located between them and which are parts of sides of the cylinder and of the cylinder's axis are equal, because they are parallel and between two parallel planes. If the two straight lines *IM* and *KN* are added together, the sum will be equal to the two straight lines *LS* and *PU*, if they are added together, and also equal to twice the straight line *RV*. If the two sections *IKL* and *MNS* are not parallel, then if we draw, from points *I* and *M* which are both on side *AIMD* of the cylinder, two diameters of the sections *IKL* and *MNS*, they end at the side opposite to side *AIMD* which is *BKNE*, and the two diameters we have imagined, diameters *IRK* and *MVN*, of necessity pass through the centres of the sections, which are the two points at which they cut the axis, namely *R* and *V*. The straight lines *IM* and *KN* are therefore in the same plane because they are parallel, and the two diameters *IK* and *MN* which join them are in this plane. If we draw in this plane, from point *I*, a straight line parallel to the straight line *MN*, like the straight line *ITW*, the straight line *RT* is parallel to the straight line *KW* which is a side of triangle *KIW*. The ratio of side *KI* of triangle *KIW* to the straight line *IR* is equal to the ratio of *KW* to *RT*; but *KI* is the double of *IR* because point *R* is the centre of section *IKL*; therefore the straight line *KW* is the double of the straight line *RT*. But the straight lines *IM*, *WN* and *TV* are equal because they are parallel and between two parallel straight lines; the double of the straight line *TV* is therefore equal to the sum of the straight lines *IM* and *WN*. Now we have shown that the straight line *KW* is also the double of the straight line *RT*; therefore the sum of the straight lines *IM* and *KN* which are between the two sections *IKL* and *MNS*, on opposite sides of the cylinder, *AD* and *BE*, is equal to twice the straight line *RV* which is between these two sections on the axis of the cylinder. In the same way, we can show that the sum of the two straight lines *LS* and *PU* which are between these two sections on opposite sides of the cylinder, *CF* and *OQ*, is equal to twice *RV*. The sum of the straight lines *IM* and *KN* is therefore equal to the sum of the two straight lines *LS* and *PU* and is also equal to twice the straight line *RV*. Similarly, we can also show that the sum of the two straight lines *AI* and *BK* is equal to the sum of the two straight lines *CL* and *OP* and is also equal to twice the straight line *RG*. That is what we wanted to prove.

It is clear from what we have said that, if the portion located between the two sections *IKL* and *MNS* on one of the opposite sides *AD* and *BE* is the smallest straight line located on one of the sides of the cylinder between these two sections or if the two sections are tangent at <a point> on this side, then the portion located between them on the other opposite side which passes through the other two extremities of the two diameters of the two sections drawn from this first side is the largest straight line located on one of the sides of the cylinder located between these two sections. If the portion located on the first side we have mentioned, between the two sections, is the largest of the straight lines located on one of the cylinder's sides between the two sections, then the portion located between them on the opposite side is the smallest of the straight lines located between them on one of the cylinder's sides or otherwise the two sections are tangent at <a point> on this opposite side. It will be the same as well if one of the bases of the cylinder is substituted for one of the sections.

– **30** – If we have two of the sections of an oblique cylinder which meet its sides in the latter without cutting; if one of them is a minimal section, if there is constructed in this minimal section a polygon inscribed in this section and such that any two opposite sides of the sides of the polygon are between the extremities of two diameters of the section, if portions of the cylinder's sides located between the two sections and passing through the vertices[43] of the polygon are drawn and if straight lines then join their extremities which are in the other section, then the area of the sum of the trapeziums generated between these two sections the bases of which are the sides of the polygon constructed in the minimal section, is equal to half the product of the sum of the portions located on any two opposite sides amongst the sides of the cylinder, between the two sections, and the sum of the sides of the polygon constructed in the minimal section. The case will be the same if in place of one of these sections we have one of the bases of the cylinder, the other being a minimal section.

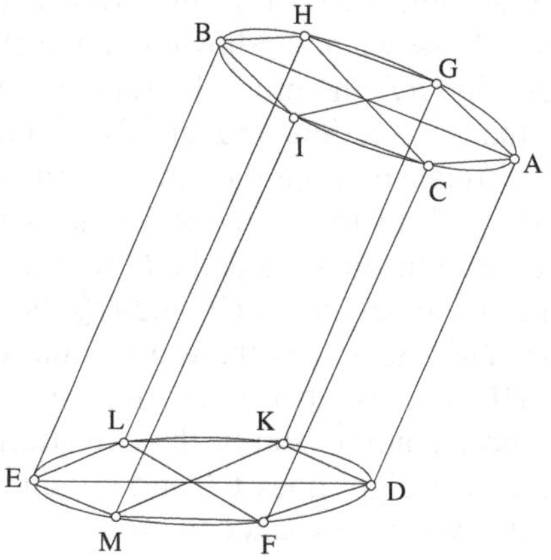

Fig. II.3.30

Let two of the sections of the cylinder which meet its sides in the latter – or otherwise one of these two sections and one of the bases of the cylinder – be *ABC* and *DEF*, without cutting. Let section *ABC* be one of the minimal sections, let there be constructed in section *ABC* the inscribed polygon *AGHBIC*, let any two opposite sides, amongst its sides, be between the extremities of two diameters of section *ABC*, let the portions located on the sides of the cylinder between the two sections *ABC* and *DEF* which pass through the points *A*, *G*, *H*, *B*, *I* and *C* be the straight lines *AD*, *GK*, *HL*, *BE*, *IM* and *CF*, let the straight lines *DK*, *KL*, *LE*, *EM*, *MF* and *FD* join their extremities in section *DEF*.

I say that the area of the sum of the trapeziums ADKG, GKLH, HLEB, BEMI, IMFC *and* CFDA *is equal to half the product of the sum of the portions located between the two sections* ABC *and* DEF, *on any two opposite sides amongst the sides of the cylinder, and the sum of the straight lines* AG, GH, HB, BI, IC *and* CA.

Proof: The straight lines *AD*, *GK*, *HL*, *BE*, *IM* and *CF* are portions of sides of the cylinder, they are parallel and parallel to the axis of the cylinder and the cylinder's axis cuts the plane *ABC* perpendicularly, because *ABC* is one of the

[43] Literally: angles.

minimal sections. The straight lines *AD*, *GK*, *HL*, *BE*, *IM* and *CF* are likewise perpendicular to the plane of section *ABC*; they are therefore perpendicular to all the straight lines drawn from their extremities in this plane; that is why the straight lines we have mentioned are perpendicular to the sides of the polygon *AGHBIC* and surround them forming right angles. The area of the trapezium *ADKG* is therefore equal to half the product of the sum of the two straight lines *AD* and *GK* times the straight line *AG*. We can also show in the same way that the area of the trapezium *BEIM* opposite to the trapezium we have mentioned is equal to half the product of the sum of the straight lines *BE* and *IM* times the straight line *BI*. But side *BI* of the polygon *AGHBIC* is equal to side *AG* of the latter, because it is opposite to it following the diameter;[44] the area of the sum of the two trapeziums *AGKD* and *BEMI* is therefore equal to half the product of <the sum> of the four straight lines *AD*, *GK*, *BE* and *IM* times the straight line *AG*. But the two straight lines *AD* and *GK* are opposite to the two straight lines *BE* and *IM* following two diameters of *ABC*, because the two sides *AG* and *BI*, amongst the sides of polygon *AGHBIC*, join the extremities of two diameters of section *ABC*. The two straight lines *BE* and *IM* are therefore two portions of two sides of the cylinder opposite to the sides *AD* and *GK*; the sum of the two straight lines *AD* and *BE* is therefore equal to the sum of the two straight lines *GK* and *IM*. Now, we have shown that the area of the sum of the two trapeziums *ADKG* and *BEMI* is equal to half the product <of the sum> of the four straight lines *AD*, *GK*, *BE* and *IM* times the straight line *AG*, the product <of the sum> of the two straight lines *AD* and *BE* times the straight line *AG* is therefore equal to the area <of the sum> of the two trapeziums *ADKG* and *BEMI*. But we have shown that the straight line *AG* is equal to the straight line *BI*, therefore the product of the sum of the two straight lines *AD* and *BE* which are two portions of two opposite sides amongst the sides of the cylinder, and the sum of the two straight lines *AG* and *BI*, is equal to twice the area <of the sum> of the two trapeziums *ADKG* and *BEMI* and the area <of the sum> of these two trapeziums is equal to half of what we have said. In the same way, we can also show that for any two opposite trapeziums amongst the trapeziums located between the two sections *ABC* and *DEF*, the area of their sum is equal to half the product <of the sum> of the portions of two opposite sides amongst the sides of the cylinder, located between these two sections, and <the sum> of two sides of these two trapeziums, which are in section *ABC*. But the sum of the portions of any two opposite sides amongst the sides of the cylinder, located between the two sections *ABC* and *DEF*, is constant.[45] The area of the sum of the trapeziums *ADKG*, *GKLH*, *HLEB*, *BEMI*, *IMFC* and *CFDA* is equal to half the product <of the sum> of the portions of any two opposite sides amongst the sides of the cylinder, located between sections *ABC* and *DEF*, and the sum of the sides of polygon *AGHBIC*.

[44] The extremities of *AG* and *BI* are diametrically opposite.

[45] Literally: is the same thing.

It may sometimes be that a few surfaces limited by sections *ABC* and *DEF* are triangular; this takes place if the two sections are tangent.[46] The method of the proof in this case is like the method we have mentioned previously. It will be the same if *DEF* is one of the bases of the cylinder. That is what we wanted to prove.

– **31**[47] – For every portion of the lateral surface of an oblique cylinder, located between two of the minimal sections of this cylinder which meet its sides in the latter, its area is then equal to the product of the portion of one of the sides of the cylinder, located between these two sections, whatever this side may be, and the perimeter of one of the two minimal sections, whatever it is.

Let there be a portion of the lateral surface of an oblique cylinder located between two of the sections of the cylinder which meet its sides in the latter, which are *ABC* and *DEF*; let these sections be two of the minimal sections of the cylinder, let the straight line *AD* be located between them on one of the sides of the cylinder.

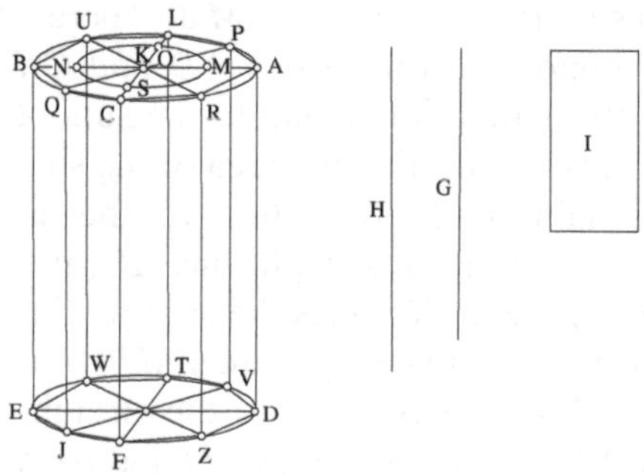

Fig. II.3.31a

I say that the area of the portion of the lateral surface of the cylinder located between the two sections ABC *and* DEF *is equal to the product of* AD *and the perimeter of section* ABC.

Proof: If the area of the portion of the lateral surface of the cylinder located between the two sections *ABC* and *DEF* is not equal to the product of *AD* and the perimeter of section *ABC*, then it is either smaller than it, or larger than it.

Let the area of the portion of the lateral surface of the cylinder which is located between the two sections *ABC* and *DEF* first of all be smaller than the product of the straight line *AD* and the perimeter of section *ABC*, if that is possible. The area of the portion of the lateral surface of the cylinder located between the two sections *ABC* and *DEF* will be equal to the product of the straight line *AD* and a line smaller than the perimeter of section *ABC*. If we

[46] Two trapeziums are replaced by two triangles having a common apex, which is the point of contact.

[47] See Supplementary note [18].

make this line the straight line *G* and if we make the straight line *H* larger than it and smaller than the perimeter of section *ABC*, then the product of the straight line *AD* and the straight line *H* will be larger than the area of the portion of the lateral surface of the cylinder located between the two sections *ABC* and *DEF*; let its excess over it be equal to the surface *I*. Let the centre of section *ABC* be the point *K*, its large axis *AB* and its small axis *CL*. Half of the surface *I* is either not smaller than section *ABC*, or is smaller than it.

If half of the surface *I* is not smaller than section *ABC*, we can separate from the straight line *KA* a straight line such that its ratio to the straight line *KA* is larger than the ratio of the straight line *H* to the perimeter of section *ABC*, which is the straight line *KM* in the first case of figure.[48] And if half of the surface *I* is smaller than section *ABC*, we can make, in this case of figure, the ratio of *KM* to *KA* larger than the ratio we have mentioned, and we can also make the ratio of the square of the straight line *KM* to the square of the straight line *KA* larger than the ratio of the excess of section *ABC* over half of surface *I* to section *ABC*. If we make, in both cases at once, each of the ratios of *KN* to *KB*, of *KS* to *KC*, of *KO* to *KL* equal to the ratio of *KM* to *KA* and if we imagine an ellipse such that its large axis is *MN* and its small axis *SO*, which is the ellipse *MSNO*, then the section *MSNO* will be similar to section *ABC* because the ratio of axis *MN* to axis *AB* is equal to the ratio of axis *SO* to axis *CL*. The ratio of the perimeter of section *MSNO* to the perimeter of section *ABC* is equal to the ratio of *MN* to *AB* which is larger than the ratio of the straight line *H* to the perimeter of section *ABC*. The ratio of the perimeter of section *MSNO* to the perimeter of section *ABC* is therefore larger than the ratio of the straight line *H* to the perimeter of section *ABC*, and that is why the perimeter of section *MSNO* is larger than the straight line *H*. If we construct in section *ABC* a polygon inscribed in the latter and which surrounds section *MSNO* without its sides being tangent to it, let the polygon be *APLUBQCR*, and if we draw from points *P*, *L*, *U*, *B*, *Q*, *C* and *R* portions of the sides of the cylinder which pass through these points and are located between the two sections, let them be the straight lines *PV*, *LT*, *UW*, *BE*, *QJ*, *CF* and *RZ*, then these straight lines are parallel to the axis of the cylinder and are perpendicular to each of the planes of the two sections *ABC* and *DEF*, because these sections are two of the minimal sections, for which we have shown that the axis of the cylinder is perpendicular to them. If we draw the straight lines *DV*, *VT*, *TW*, *WE*, *EJ*, *JF*, *FZ* and *ZD*, the straight lines we have mentioned before, which are portions of the sides of the cylinder, are perpendicular to these straight lines and to the sides of the polygon *APLUBQCR* and the surfaces generated between the two sections *ABC* and *DEF* from all the straight lines we have mentioned, are rectangles; and if they are added together, their area will be equal to the product of the straight line *AD* and the sum of the sides of the polygon *APLUBQCR*, because the portions of the sides of the

[48] See Supplementary note [19].

cylinder located between the two sections *ABC* and *DEF* are all equal to the straight line *AD*. But the sum of the sides of the polygon *APLUBQCR* is larger than the perimeter of section *MSNO*, which, we have shown, is larger than the straight line *H*. The sum of the surfaces we have mentioned, located between the two sections *ABC* and *DEF*, is therefore much larger than the product of the straight line *AD* and the straight line *H*. Now we have shown that the product of the straight line *AD* and the straight line *H* is larger than the area of the portion of the lateral surface of the cylinder located between the two sections *ABC* and *DEF*, and we have made its excess over it equal to the surface *I*. The sum of the surfaces we have mentioned, located between the two sections *ABC* and *DEF*, is therefore much larger than the portion of the lateral surface of the cylinder located between these two sections, and its excess over it is larger than surface *I*. Surface *I*, plus the portion of the lateral surface of the cylinder located between the two sections *ABC* and *DEF* is therefore smaller than the sum of the surfaces we have mentioned, located between these two sections. Half of the surface *I*, either is not smaller than section *ABC* or is smaller than it.

If it is not smaller than it, then it is not smaller than section *DEF* because these two sections are equal, given that they are minimal sections. The whole surface *I* is therefore not smaller than the sum of the two sections *ABC* and *DEF*. Now, we have shown that surface *I*, plus the portion of the lateral surface of the cylinder located between the two sections *ABC* and *DEF*, is smaller than <the sum> of the surfaces with parallel sides located between these two sections. The sum of these two sections and the portion of the lateral surface of the cylinder located between them is smaller than <the sum> of the surfaces with parallel sides we have mentioned, located between these two sections; this is impossible, because what surrounds is not smaller than what is surrounded. The portion of the lateral surface of the cylinder located between the two sections *ABC* and *DEF* is therefore not smaller than the product of the straight line *AD* and the perimeter of section *ABC*.

If half of the surface *I* is smaller than section *ABC*, then the ratio of the square of the straight line *KM* to the square of the straight line *KA* is larger than the ratio of what section *ABC* exceeds half of surface *I* by, to the section *ABC*, because we have made it thus in this case. But the ratio of the square of the straight line *KM* to the square of the straight line *KA* is equal to the ratio of the square of the straight line *MN* to the square of the straight line *AB*; therefore the ratio of the square of the straight line *MN* to the square of the straight line *AB* is larger than the ratio of what section *ABCR* exceeds half of surface *I* by, to section *ABC*. But the ratio of the square of the axis *MN* to the square of the axis *AB* is equal to the ratio of section *MSNO* to section *ABC*, because these two sections are similar. The ratio of section *MSNO* to section *ABC* is therefore larger than the ratio of what section *ABC* exceeds half of section *I* by, to section *ABC*. If we inverse, the ratio of section *ABC* to the difference surrounded by the two perimeters of sections *ABC* and *MSNO* and located between them, which is the difference between these two sections, is also larger than the ratio of section *ABC* to half of surface *I*. The surface of the figure surrounded by the perimeters of sections *ABC* and *MSNO* and located between them is smaller than half of surface *I*. But the surface of this figure we have mentioned – the one surrounded by the perimeters of sections *ABC* and *MSNO* and located between them – is larger than its portions delimited by the straight lines *AP*, *PL*, *LU*, *UB*, *BQ*, *QC*, *CR* and *RA* and the curved lines of which these straight lines we have mentioned are chords. If these portions we mentioned are added together, the ones which are delimited by the curved lines and their chords, <their sum> is much smaller than half of surface *I*. But these portions we mentioned are equal to the homologous portions of section *DEF*, because if section *DEF* is placed over section *ABC*, it will be superposed on to it and each of its points will be placed on to its homologue in section *ABC*, at which there ends the side of the cylinder which passes through the first point. If the portions delimited by the curved lines and their chords in section *ABC* and their homologues in section *DEF* are added together, <their sum> will be smaller than surface *I*. But <the sum of> all these portions we have mentioned and the portions of the lateral surface of the cylinder which are between them – whose sum is the portion of the lateral surface of the cylinder located between the two sections *ABC* and *DEF* – is larger than the sum of the surfaces with parallel sides which are located between these two sections because it surrounds it. Surface *I*, plus the portion of the lateral surface of the cylinder located between the two sections *ABC* and *DEF*, is much larger than the sum of the surfaces with parallel sides located between these two sections. Now, we have shown that it is smaller than it; this is contradictory. The portion of the lateral surface of the cylinder located between the two surfaces *ABC* and *DEF* is therefore not smaller than the product of the straight line *AD* and the perimeter of section *ABC*, given that half of surface *I* is smaller than section *ABC*. Now, we have shown that it is not smaller than it if half of surface *I* is not smaller than section *ABC*; therefore the portion of the lateral surface of the cylinder located between the two sections *ABC* and *DEF* is not smaller than the product of the straight line *AD* and the perimeter of section *ABC*.

I say likewise that it is not larger than it.

If that was possible, then it would be larger than it; we have then the area of the portion of the lateral surface of the cylinder located between the two sections *ABC* and *DEF* equal to the product of the straight line *AD* and a line larger than the perimeter of section *ABC*. If we make this line the straight line *G*, and if we make the straight line *H* smaller than it and larger than the perimeter of section *ABC*, the product of the straight line *AD* and the straight line *H* will be smaller than the area of the portion of the lateral surface of the cylinder located between the two sections *ABC* and *DEF*. If we make their difference equal to the surface *I*, if we make the ratio of the straight line *KM*, in the second example, to the straight line *KA* which is smaller than it, smaller than the ratio of the straight line *H* to the perimeter of section *ABC*, if we make the ratio of its square to its square likewise smaller than the ratio of section *ABC*, plus half of surface *I*, to section *ABC*; if we make each of the ratios of *KN* to *KB*, *KS* to *KC*, and *KO* to *KL* equal to the ratio of *KM* to *KA* and if we construct outside the section *ABC* an ellipse such that its large axis is *MN* and its small axis *SO*, let it be the section *MSNO*, then section *MSNO* will also be similar to section *ABC* and the ratio of its perimeter to the perimeter of section *ABC* will be equal to the ratio of *MN* to *AB* which is smaller than the ratio of *H* to the perimeter of section *ABC*. The ratio of the perimeter of section *MSNO* to the perimeter of section *ABC* is therefore smaller than the ratio of the straight line *H* to the perimeter of section *ABC* as well, and it is for that reason that the perimeter of section *MSNO* is smaller than the straight line *H*. If we construct in the plane of section *ABC* a polygon inscribed in section *MSNO*, and which surrounds section *ABC*, without its sides meeting it, namely the polygon *MPOUNQSR*, and if we draw from the vertices of the angles[49] of this polygon perpendiculars to its plane which end at the plane where section *DEF* is, let the straight lines be *MV*, *PT*, *OW*, *UJ*, *NZ*, *QX*, *SY* and *RO′*, these straight lines will be parallel to the axis of the cylinder and to its sides and equal to the straight line *AD*, because the sections *ABC* and *DEF* are two of the minimal sections.

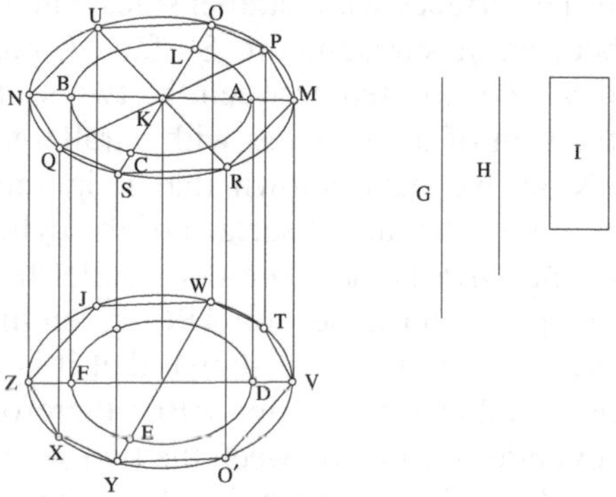

Fig. II.3.31b*

If we draw the straight lines *VT*, *TW*, *WJ*, *JZ*, *ZX*, *XY*, *YO′* and *O′V*, surfaces with parallel sides outside the lateral surface of the cylinder are generated from them. It is clear from what we have said, as likewise we have shown, that the area of the sum of these surfaces is equal to the product of the straight line *AD* and the sum of the sides of polygon *MPOUNQSR*. But the sum of the sides of the polygon mentioned is smaller than the perimeter of section *MSNO* which, we have shown, is smaller than the straight line *H*. The sum of the surfaces with parallel sides we have mentioned is much smaller than the product of the

[49] Literally: from the points of the angles.

* This figure is not in the manuscript.

straight line *AD* and the straight line *H*. But we have shown that the product of the straight line *AD* and the straight line *H* is smaller than the area of the portion of the lateral surface of the cylinder located between sections *ABC* and *DEF* and we have made their difference equal to surface *I*; the sum of the surfaces with parallel sides we have mentioned is therefore much smaller than the portion of the lateral surface of the cylinder located between sections *ABC* and *DEF* and their difference is larger than surface *I*. Surface *I*, plus the sum of the surfaces with parallel sides we have mentioned, have <a sum> smaller than the portion of the lateral surface of the cylinder located between sections *ABC* and *DEF*. Furthermore, section *MSNO* is similar to section *ABC*; therefore its ratio to the latter is equal to the ratio of the square of axis *MN* to the square of axis *AB*, which is equal to the ratio of the square of the straight line *KM* to the square of the straight line *KA*. But we have made the ratio of the square of the straight line *KM* to the square of the straight line *KA* smaller than the ratio of section *ABC*, plus half of section *I*, to section *ABC*; therefore the ratio of section *MSNO* to section *ABC* is smaller than the ratio of section *ABC*, plus half of section *I*, to section *ABC*. If we separate, we have the ratio of what section *MSNO* exceeds section *ABC* by – which is the figure delimited by the perimeters of sections *ABC* and *MSNO* and located between them – to section *ABC*, smaller than the ratio of half of surface *I* to section *ABC*. The surface of the figure delimited by the perimeters of sections *ABC* and *MSNO* and located between them, is therefore smaller than half of surface *I*. But the surface of this figure we have mentioned, which is delimited by the perimeters of sections *ABC* and *MSNO* and located between them, is larger than the surface of the figure delimited by the sides of polygon *MPOUNQSR*, located between them and section *ABC*. The surface of this figure we have mentioned – located between the sides of polygon *MPOUNQSR* and section *ABC* – is much smaller than half of surface *I*. But this surface we have mentioned is equal to its homologue located around section *DEF* and which is located between section *DEF* and the sides of the polygon *VTWJZXYO′* because if this polygon and section *DEF* are placed according to their shape over polygon *MPOUNQSR* and section *ABC*, they will be superposed and each of their points will be placed on its homologue in the two others. The sum of the surfaces of the two figures we have mentioned one of which is around section *ABC* and the other around section *DEF*, is smaller than surface *I*. But the sum of these two surfaces and the surfaces with parallel sides whose bases are the sides of polygon *MPOUNQSR* is larger than the portion of the lateral surface located between the two sections *ABC* and *DEF*, because it surrounds it. The sum of surface *I* and the surfaces with parallel sides whose bases are the sides which surround polygon *MPOUNQSR* is much larger than the portion of the lateral surface of the cylinder located between the two sections *ABC* and *DEF*. Now, we have shown that it is smaller than it; this is contradictory. The area of the portion of the lateral surface of the cylinder located between the two sections *ABC* and *DEF* is therefore not larger than the product of the straight line *AD* and the perimeter of section *ABC*.

But we have shown that it is not smaller than it; it is consequently equal to it. That is what we wanted to prove.

– **32**[50] – Every portion of the lateral surface of an oblique cylinder located between two of its sections which do not intersect and which meet, in the cylinder, all its sides and of which one is one of the minimal sections of the cylinder and the other one of the other sections of the cylinder, or located between one of the bases of the cylinder and one of the minimal sections which do not cut it, has an area equal to half the product <of the sum> of the portions of any two opposite sides of the cylinder, located between the two sections or between the section and the base, and the perimeter of any one of the minimal sections.

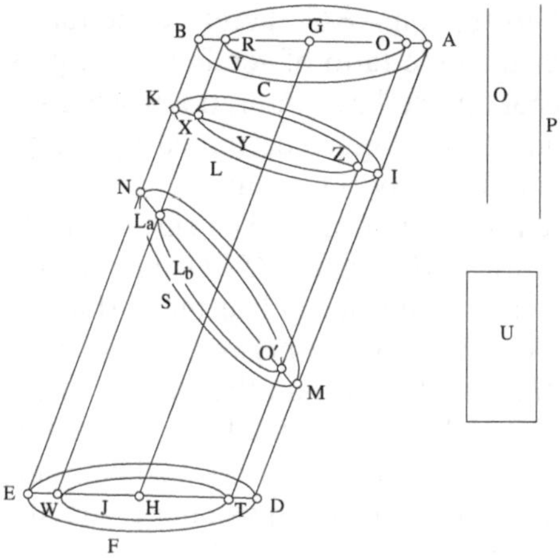

Fig. II.3.32a

Let there be an oblique cylinder with bases *ABC* and *DEF*, with centres *G* and *H*, and let two of the sections of the cylinder which meet its sides in the latter, either being tangent at a point, or without being tangent and without intersecting; namely the sections *IKL MNS*. Then from these two sections, let section *IKL* be the one which is minimal and let the two straight lines *AIMD* and *BKNE* be any two opposite sides of the cylinder.

I say that the area of the portion of the lateral surface of the cylinder which is located between the sections IKL *and* MNS *is equal to half the product of the sum of the two straight lines* IM *and* KN *times the perimeter of section* IKL, *and that the area of the portion of the lateral surface of the cylinder located between section* IKL *and the base* ABC – *if it does not cut the base – is equal to half the*

[50] See Supplementary note [20].

product of the sum of the two straight lines AI *and* BK *times the perimeter of section* IKL.

Proof: If the area of the portion of the lateral surface of the cylinder located between the two sections *IKL* and *MNS* is not equal to half the product[51] we have set out, then it is either smaller than half or larger than half.

Let it first of all be smaller than half, if that was possible; the area of the portion of the lateral surface of the cylinder located between the two sections *IKL* and *MNS* will be equal to half the product of the sum of the two straight lines *IM* and *KN* times a line smaller than the perimeter of section *IKL*. If we make this line the straight line *O*, and if we make a straight line *P* larger than *O* and smaller than the perimeter of section *IKL*, the half-product of the sum of the straight lines *IM* and *KN* times the straight line *P* is larger than the area of the portion of the lateral surface of the cylinder located between the two sections *IKL* and *MNS*; if we make what this half-product exceeds this area by equal to the surface *U*, then half of surface *U* either is not smaller than section *MNS* or is smaller than it.

If this half is not smaller than it, we can describe around centres *G* and *H* equal circles, smaller than the two circles *ABC* and *DEF* and such that the ratio of the diameter of each of them to the diameter of each of the circles *ABC* and *DEF* is larger than the ratio of the straight line *P* to the perimeter of section *IKL*. If half of surface *U* is smaller than section *MNS*, we can make the ratio of the diameter of each of the circles we have mentioned to the diameter of each of the circles *ABC* and *DEF* larger than the ratio we have mentioned and we can likewise make the ratio of the square of the diameter of each of them to the square of the diameter of each of the circles *ABC* and *DEF* larger than the ratio of the excess of section *MNS* over half of surface *U* to section *MNS*; let the two circles we have described be the two circles *QRV* and *TWJ*. If we imagine, taking these two cases into account at the same time, a cylinder inside the first cylinder, the circles *QRV* and *TWJ* will be its bases and the two sections generated in this small cylinder by the planes of sections *IKL* and *MNS* – which are the sections *ZXY* and $O'L_aL_b$ – will be similar to sections *IKL* and *MNS* – each to its homologue – and the ratio of the perimeter of section *ZXY* to the perimeter of section *IKL* will be equal to the ratio of each of the diameters of section *ZXY* to the homologous diameter in section *IKL*; this ratio is equal to the ratio of the diameter of circle *QRV* to the diameter of circle *ABC*, which we have made larger than the ratio of the straight line *P* to the perimeter of section *IKL*. The ratio of the perimeter of section *ZXY* to the perimeter of section *IKL* is larger than the ratio of the straight line *P* to the perimeter of section *IKL* as well. And it is for that reason that the perimeter of section *ZXY* will be larger than the straight line *P*. If we imagine in the plane of section *IKL* a polygon inscribed in section *IKL* and which surrounds section *ZXY* without its sides touching it in

[51] Literally: of that which.

such a way that the straight lines joining its opposite vertices[52] are diameters of section *IKL* and if we imagine that portions of the sides of the large cylinder have been drawn from the vertices[53] of this polygon in such a way that their extremities end at section *MNS* and that straight lines join its extremities which are in section *MNS*, then polygonal surfaces are generated from that between the lateral surface of the large cylinder and the lateral surface of the small cylinder and in section *MNS* is generated a polygon inscribed in section *MNS* and which surrounds section $O'L_aL_b$. The area of the sum of the surfaces located between the two sections *IKL* and *MNS* is equal to the half-product of the sum of the two straight lines *IM* and *KN* times the sum of the sides of the polygon we have imagined in section *IKL*. But the sum of the sides of this polygon is larger than the perimeter of section *ZXY* which, we have shown, is larger than the straight line *P*. The sum <of the areas> of the polygonal surfaces we have mentioned, located between sections *IKL* and *MNS*, is much larger than the half-product of the sum of the straight lines *IM* and *KN* times the straight line *P*. But we have shown that the half-product of the sum of the two straight lines *IM* and *KN* times the straight line *P* is larger than the area of the portion of the lateral surface of the large cylinder delimited by sections *IKL* and *MNS*. If we make what this product exceeds the area by equal to the surface *U*, then the sum of the surfaces mentioned located between the two sections *IKL* and *MNS*, is much larger than the portion of the lateral surface of the large cylinder, delimited by sections *IKL* and *MNS*, and its excess over it is larger than surface *U*. Surface *U*, plus the portion of the lateral surface of the large cylinder located between the two sections *IKL* and *MNS*, is therefore smaller than the surfaces we have mentioned, located between these two sections. But half of surface *U* either is smaller than section *MNS* or is not smaller than it. If it is not smaller than it, then it is not smaller than section *IKL*, because section *IKL* is a minimal section and section *MNS* is not a minimal section; therefore the whole surface *U* is not smaller than the sum of sections *MNS* and *IKL*. But we have shown that surface *U*, plus the portion of the lateral surface of the large cylinder located between the two sections *IKL* and *MNS*, is smaller than the sum of the surfaces located between these two sections and located between the two lateral surfaces of the two cylinders. The sum of sections *IKL* and *MNS* and the portion of the lateral surface of the large cylinder located between them is smaller than the sum of the surfaces located between the two sections; that is impossible, because what surrounds cannot be smaller than what is surrounded. The portion of the lateral surface of the large cylinder located between sections *IKL* and *MNS* is therefore not smaller than the half-product of the sum of the straight lines *IM* and *KN* times the perimeter of section *IKL*.

[52] Literally: angles.
[53] Literally: angles.

If half of surface *U* is smaller than section *MNS*, then the ratio of the square of the diameter of circle *QRV* to the square of the diameter of circle *ABC* is larger than the ratio of what section *MNS* exceeds half of surface *U* by to section *MNS*, because we have assumed thus in this case. But the planes *MNS* and *IKL* cut the two cylinders the bases of one of which are the circles *ABC* and *DEF* and the bases of the other are the circles *QRV* and *TWJ*, and generated in the large cylinder the sections *MNS* and *IKL* and in the small cylinder the sections $O'L_aL_b$ and *ZXY*; the sections $O'L_aL_b$ and *ZXY* are therefore similar to sections *MNS* and *IKL*, each section to its homologue, and the ratio of the square of each of the diameters of one to the square of the homologous diameter of its associate which is similar to it is equal to the ratio of the square of the diameter of circle *QRV* to the square of the diameter of circle *ABC*. The ratio of the square of each of the diameters of sections $O'L_aL_b$ and *ZXY* to the square of the homologous diameter of the section *MNS* or *IKL* which is similar to it is larger than the ratio of what section *MNS* exceeds half of surface *U* by to section *MNS*. But the ratio of the square of each of the diameters of sections $O'L_aL_b$ and *ZXY* to the square of the homologous diameter of the section *MNS* or *IKL* which is similar to it is equal to the ratio of each of the two first sections to its homologue between the two last, because the two first sections are similar to the two last sections, each to its homologue. The ratio of each of the sections $O'L_aL_b$ and *ZXY* to its homologue, which is similar to it, between sections *MNS* and *IKL*, is therefore larger than the ratio of what section *MNS* exceeds half of surface *U* by to section *MNS*. If we inverse, then the ratio of section *MNS* to its excess over section $O'L_aL_b$ – which is the surface located between the perimeters of these two sections – and the ratio of section *IKL* to its excess over section *ZXY* – which is the surface located between the perimeters of these two sections – are, each of them, larger than the ratio of section *MNS* to half of surface *U*. As for the surface located between the perimeters of the two sections *MNS* and $O'L_aL_b$, it is smaller than half of surface *U* as is shown from what we have just said. As for the surface located between the perimeters of the two sections *IKL* and *ZXY*, it has been shown from what we have said that the ratio of section *IKL* to this surface is larger than the ratio of section *MNS* to half of surface *U*. But section *IKL* is smaller than section *MNS*, given that section *IKL* is one of the minimal sections; the surface located between the perimeters of the two sections *IKL* and *ZXY* will therefore be, as well, much smaller than half of surface *U*. If it is thus, then the sum of the two surfaces one of which is located between the perimeters of the two sections *IKL* and *ZXY* and the other located between the perimeters of the two sections *MNS* and $O'L_aL_b$, is smaller than surface *U*. But the sum of these two surfaces we have mentioned is larger than <the sum of> their portions which are delimited and contained by the sides of the two polygons one of which is the one whose sides we have imagined between the perimeters of sections *IKL* and *ZXY* and the other of which is the one whose sides have been generated between the perimeters of the two sections

MNS and $O'L_aL_b$, owing to the fact that we have drawn the straight lines between the extremities of the portions of the sides of the large cylinder which are in section *MNS*. Each of the portions we have mentioned is surrounded by one of the sides of the two polygons we have described and the curved line subtended by this side. The sum of these portions we have mentioned, which are surrounded by the curved lines and their chords, is much smaller than surface *U*. But these portions which we have mentioned, if they are added to the portions of the lateral surface of the large cylinder located between them and whose sum is the portion of the lateral surface of the large cylinder located between sections *IKL* and *MNS*, are larger than the <plane> surfaces located between these two sections, placed between the lateral surface of the large cylinder and the lateral surface of the small cylinder because the portions surround them. <The sum of> surface *U* and the portion of the lateral surface of the large cylinder located between sections *IKL* and *MNS* is larger than the sum of the <plane> surfaces located between these two sections and placed between the large cylinder and the small cylinder. But we have shown that it is smaller than it, it is therefore larger than it and smaller than it; that is contradictory. The portion of the lateral surface of the large cylinder located between sections *IKL* and *MNS* is therefore not smaller than half the product of the sum of the straight lines *IM* and *KN* times the perimeter of section *IKL*, if half of surface *U* is smaller than section *MNS*. But we have shown that the lateral surface is not smaller than half of this product. If it is not thus, the portion of the lateral surface of the large cylinder located between sections *IKL* and *MNS* is therefore not smaller than half the product of the sum of the straight lines *IM* and *KN* times the perimeter of section *IKL*.

I say likewise that it is not larger than half of it.

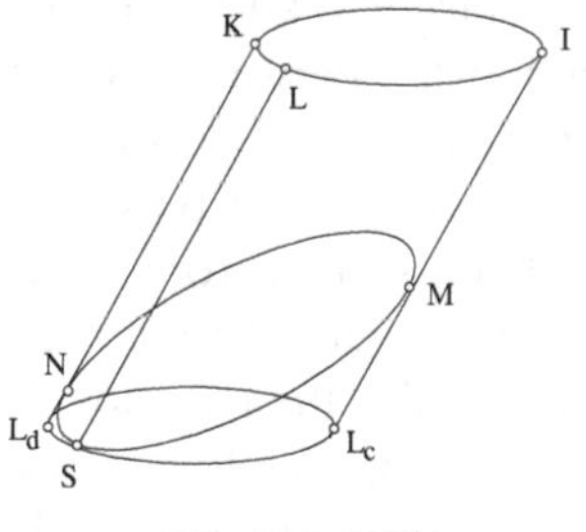

Fig. II.3.32b*

In fact, if we make the largest of the sides of the large cylinder located between the two sections *IKL* and *MNS*, the straight line *LS*, and if we pass through the point *S* a plane parallel to the plane of section *IKL* which cuts the large cylinder either as it is, or once the cylinder is extended along its sides, and which generates in the cylinder the section SL_cL_d, then the portions of the sides

* This figure is not in the manuscript.

of the large cylinder, one of which is the straight line *LS*, placed between the two sections *IKL* and SL_cL_d are equal, because they are parallel and between two parallel planes; the sections *MNS* and SL_cL_d are then tangent only at point *S* because the straight line *LS* is the longest of the sides of the large cylinder located between sections *MNS* and *IKL* and the section SL_cL_d is one of the minimal sections. We can show from that, as we have shown before, that the area of the portion of the lateral surface of the large cylinder located between sections *MNS* and SL_cL_d is not smaller than the half-product of <the sum> of the portions of any two opposite sides, amongst the sides of the large cylinder, located between sections *MNS* and SL_cL_d times the perimeter of section SL_cL_d which is equal to the perimeter of section *IKL*. But the area of the whole portion of the lateral surface of the large cylinder located between sections SL_cL_d and *IKL* is equal to the half-product <of the sum> of the portions of these same two opposite sides we have mentioned amongst the sides of the large cylinder, located between these two sections times the perimeter of section *IKL* because the sections SL_cL_d and *IKL* are minimal sections and also the sum of the portions of two opposite sides, amongst the sides of the large cylinder, is double the straight line *LS*. The remainder – that is the area of the portion of the lateral surface of the large cylinder located between the two sections *IKL* and *MNS* – is not larger than the half-product <of the sum> of the portions of these two opposite sides we have mentioned, amongst the sides of the large cylinder, located between the two sections times the perimeter of section *IKL*. But the portions of these two opposite sides we have mentioned, amongst the sides of the large cylinder, located between sections *IKL* and *MNS*, are either the two straight lines *IM* and *KN*, or two straight lines such that their sum is equal to the sum of the other two. The area of the portion of the lateral surface of the large cylinder located between sections *IKL* and *MNS* is therefore not larger than the half-product of the sum of the two straight lines *IM* and *KN* times the perimeter of section *IKL*. But we have shown that it is not smaller than it; consequently it is equal to the half-product.

In the same way, we can also show that the area of the portion of the lateral surface of the large cylinder located between section *IKL* and the base *ABC* is equal to half the product of the sum of the two straight lines *AI* and *BK* times the perimeter of section *IKL*. That is what we wanted to prove.

– **33** – The lateral surface of every oblique cylinder and every portion of this surface located between two of the sections of the cylinder which meet its sides in the latter, without intersecting and without one of them being a minimal section, or located between one of these sections and one of the bases of the cylinder without cutting it, are such that the area of the lateral surface and the area of the portion are equal to the half-product <of the sum> of the two portions of any two opposite sides of the cylinder, located between the plane of the highest <section> and the plane of its base, times the perimeter of one of the minimal sections of the cylinder, whatever that section is.

Let there be an oblique cylinder whose bases are *ABC* and *DEF*; let two sections, which meet its sides in it, cut it, namely the sections be *GHI* and *KLM*. Let the two straight lines *AGKD* and *BHLE* be two of the opposite sides of the cylinder.

I say that the area of the lateral surface of the cylinder ABCDEF *is equal to the half-product of the sum of the two straight lines* AD *and* BE *times the perimeter of any one of the minimal sections of the cylinder, that the area of the portion of the lateral surface of the cylinder located between sections* GHI *and* KLM *is equal to the half-product of the sum of the two straight lines* GK *and* HL *times the perimeter of any one of the minimal sections of the cylinder, and that the area of the portion of the lateral surface of the cylinder located between section* GHI *and base* ABC *is equal to the half-product of the sum of the two straight lines* AG *and* BH *times the perimeter of any one of the minimal sections of the cylinder.*

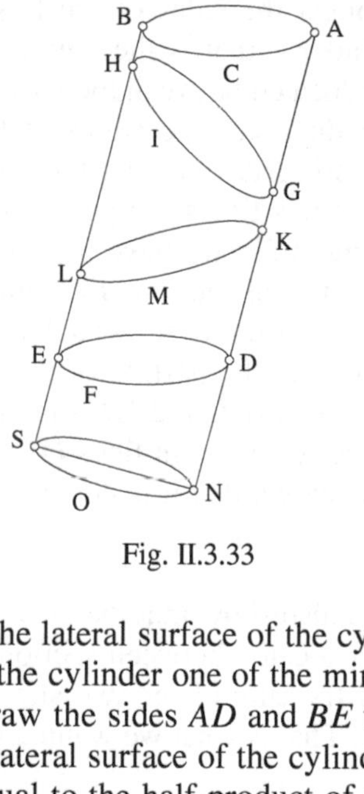

Fig. II.3.33

Proof: If we produce the lateral surface of the cylinder along its sides, if we imagine on the outside of the cylinder one of the minimal sections, which is the section *NSO*, and if we draw the sides *AD* and *BE* to points *N* and *S*, then the area of the portion of the lateral surface of the cylinder located between section *NSO* and circle *ABC* is equal to the half-product of the sum of the two straight lines *AN* and *BS* times the perimeter of section *NSO* because the straight lines *AD* and *BE* are two of the opposite sides of the cylinder. That is also why the area of the portion of the lateral surface of the cylinder located between section *NSO* and circle *DEF* is equal to the half-product of the sum of the two straight lines *DN* and *ES* times the perimeter of section *NSO*. The area of the lateral surface of the cylinder which is left between the two circles *ABC* and *DEF*, which are the bases of the cylinder, is equal to the half-product of the sum of the two straight lines *AD* and *BE* times the perimeter of section *NSO*, which is one of the minimal sections. In the same way, the area of the portion of the lateral surface of the cylinder located between sections *NSO* and *GHI* is equal to the half-product of the sum of the straight lines *GN* and *HS* times the perimeter of section *NSO*. The area of the portion of the lateral surface of the cylinder located between sections *NSO* and *KLM* is also equal to the half-product of the sum of the straight lines *KN* and *LS* times the perimeter of section *NSO*. There remains the area of the portion of the lateral surface of the cylinder located between sections *GHI* and *KLM* equal to the half-product of the sum of the straight lines *GK* and *HL* times the perimeter of section *NSO*, which is one of the minimal sections. In the same way, we can also show that the area of the portion of the lateral surface of the cylinder located between section *GHI* and base *ABC* is equal to the half-product of the sum of the straight lines *AG* and *BH* times the perimeter of section *NSO*, which is one of the minimal sections; now, all the minimal sections are equal. That is what we wanted to prove.

– **34** – The lateral surface of every oblique cylinder and of every portion of the latter located between two parallel planes amongst those which meet its sides, in the cylinder, are such that the area of each of them is equal to the product of the portion of any one of the sides of the cylinder, located between its upper plane and the plane of its base, times the perimeter of any one of the minimal sections.

Let there be an oblique cylinder whose bases are *ABC* and *DEF*, let there be a portion of the latter located between two parallel planes *GHI* and *KLM* which meet its sides in the cylinder, let the straight line *AGKD* be one of the sides of the cylinder.

I say that the area of the lateral surface of the cylinder ABCDEF *is equal to the product of the straight line* AD *and the perimeter of any one of the minimal sections of the cylinder, and that the area of the portion of the lateral surface of the cylinder located between the two planes* GHI *and* KLM *is equal to the product of the straight line* GK *and the perimeter of any one of the minimal sections of the cylinder.*

Proof: If we draw in the lateral surface of the cylinder the side opposite to side *AD* of the cylinder, let it be the straight line *BHLE*, then the area of the lateral surface of the cylinder *ABCDEF* is equal to the half-product of the sum of the straight lines *AD* and *BE* – given that they are two of the opposite sides of the cylinder – and the perimeter of any one of the minimal sections of the cylinder. But the two straight lines *AD* and *BE* are equal, because they are parallel and are between two parallel planes, the area of the lateral surface of the cylinder *ABCDEF* is therefore equal to the product of the straight line *AD* and the perimeter of any one of the minimal sections of the cylinder. Similarly, the area of the portion located between the two planes *GHI* and *KLM* is equal to the half-product of the sum of the straight lines *GK* and *HL* – given that they are two of the opposite sides of the cylinder – and the perimeter of any one of the minimal sections. But the two straight lines *GK* and *HL* are equal because they are parallel between two parallel planes; the area of the portion of the lateral surface of the cylinder located between planes *GHI* and *KLM* is therefore equal to the product of the straight line *GK* and the perimeter of any one of the minimal sections of the cylinder.

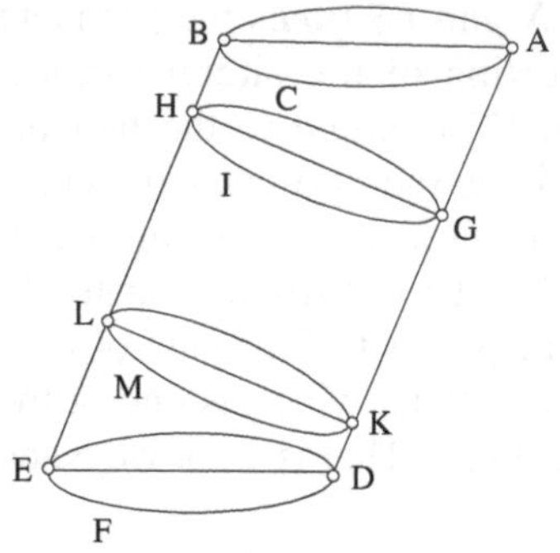

Fig. II.3.34

In the same way, we can also show that if the plane *GHI* is parallel to the <plane of> circle *ABC*, then the area of the portion of the lateral surface of the cylinder located between this plane and circle *ABC* is equal to the product of the straight line *AG* and the perimeter of any one of the minimal sections. That is what we wanted to prove.

– **35** – Every portion of the lateral surface of an oblique cylinder located between two of the sections of the cylinder which meet its sides in the latter, and which are tangent at a single point, or every portion located between one of these sections and one of the bases of the cylinder, if the section is tangent to it at a single point, is such that its area is equal to the half-product of the largest side of the cylinder which is between the two sections, or between the section and the base, and the perimeter of any one of the minimal sections of the cylinder.

Let there be a portion of the lateral surface of an oblique cylinder located between two of the sections of the cylinder, or between a section and a base, let them be *ABC* and *ADE*. Let *ABC* and *ADE* meet the sides of the cylinder in the latter and be tangent at a single point, namely the point *A*. Let the largest of the sides of the cylinder located between *ABC* and *ADE* be the straight line *BD*.

I say that the area of the portion of the lateral surface of the cylinder located between ABC *and* ADE *is equal to the half-product of the straight line* BD *and the perimeter of any one of the minimal sections of the cylinder.*

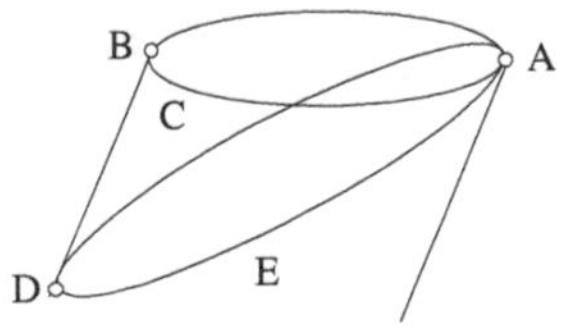

Fig. II.3.35

Proof: The area of the portion of the lateral surface of the cylinder located between *ABC* and *ADE* is equal to the half-product <of the sum> of the portions of any two opposite sides of the cylinder located between these two sections, and the perimeter of any one of the minimal sections of the cylinder. But the side of the cylinder opposite to side *BD* is the one which passes through point *A*, which is the point of contact; nothing of this side lies between *ABC* and *ADE*. The area of the portion of the lateral surface of the cylinder located between *ABC* and *ADE* is therefore equal to the half-product of *BD* and the perimeter of any one of the minimal sections of the cylinder. That is what we wanted to prove.[54]

– **36** – Every portion of the lateral surface of an oblique cylinder located between two of the sections of the cylinder which meet its sides in it and which do not intersect in the cylinder and are not parallel, or located between one of these sections and one of the bases of the cylinder which do not intersect in the cylinder, is such that its area is equal to the half-product of the sum of the largest straight line of one of the sides of the cylinder, located between the upper plane of the portion and its lower plane, and of the smallest straight line of one of the sides of the cylinder located between these planes as well, and the perimeter of any one of the minimal sections of the cylinder.

Let there be a portion of the lateral surface of an oblique cylinder located between *ABC* and *DEF*, and let *ABC* and *DEF* be two of the sections of the cylinder which meet its sides in it, or let them be one of its sections and one of the bases of the cylinder, and let these two sections be not parallel and not intersect in the cylinder. Let the largest portion of one of the sides of the cylinder located between *ABC* and *DEF* be the straight line *AD* and let the smallest portion of one of the sides of the cylinder located between these latter be the straight line *BE*.

I say that the area of the portion of the lateral surface of the cylinder located between ABC *and* DEF *is equal to the half-product of the sum of the straight lines* AD *and* BE *and the perimeter of any one of the minimal sections of the cylinder.*

[54] See Supplementary note [21].

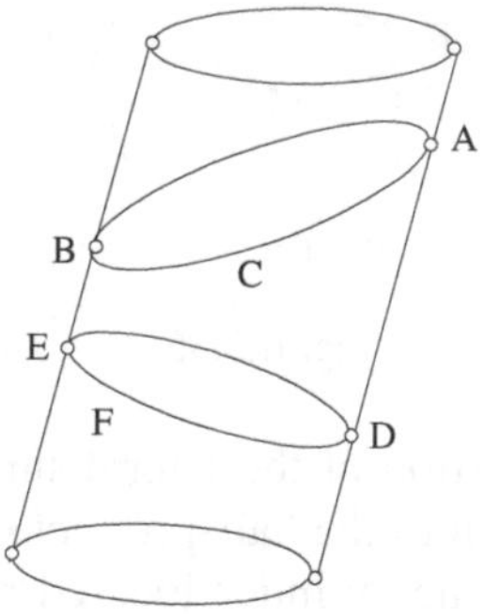

Fig. II.3.36

Proof: The straight line *AD* is the largest portion of one of the sides of the cylinder located between *ABC* and *DEF*; therefore the portion of the cylinder side opposite to the straight line *AD* located between *ABC* and *DEF* is the smallest portion of one of the sides of the cylinder located between *ABC* and *DEF*. But the smallest portion of one of the sides of the cylinder located between *ABC* and *DEF* is the straight line *BE*. Therefore the straight line *BE* is the portion of one of the sides of the cylinder, opposite to the side of which *AD* is the portion located between *ABC* and *DEF*. But the area of the portion of the lateral surface of the cylinder located between *ABC* and *DEF* is equal to the half-product <of the sum> of the portions of any two of the opposite sides of the cylinder, located between these two sections, and the perimeter of any one of its minimal sections. The area of the portion of the lateral surface of the cylinder located between *ABC* and *DEF* is therefore equal to the half-product of the sum of the two straight lines *AD* and *BE* and the perimeter of any one of the minimal sections of the cylinder. That is what we wanted to prove.

– **37** – The area of the lateral surface of every oblique cylinder and the area of every portion of the latter located between two sections of the cylinder which meet all its sides in it, but without intersecting in the cylinder, or located between one of these two sections and one of the bases of the cylinder, is equal to the product of the part of the axis of the cylinder located between the upper plane of each of these portions and the plane of its base, and the perimeter of any one of the minimal sections of the cylinder.

Let there be the lateral surface of a cylinder or a portion of the latter located between *ABC* and *DEF*. Let *ABC* and *DEF* meet all the sides of the cylinder in it, without intersecting within the cylinder; let the part of the axis of the cylinder located between them be *GH*.

I say that the area of the portion of the lateral surface of the cylinder located between ABC *and* DEF *is equal to the product of* GH *and the perimeter of any one of the minimal sections of the cylinder.*[55]

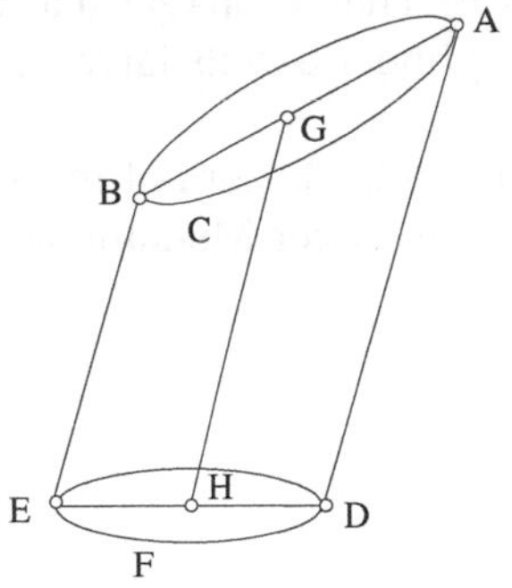

Fig. II.3.37

Proof: *ABC* and *DEF* either are parallel or are not. If they are parallel and if we make the portions of any two opposite sides of the cylinder located between *ABC* and *DEF* the two straight lines *AD* and *BE*, then the area of the portion of the lateral surface of the cylinder located between *ABC* and *DEF* is equal to the product of *AD* and the perimeter of any one of the minimal sections of the cylinder. But the straight line *AD* is equal to the straight line *GH* because they are parallel and between two parallel planes. The area of the portion of the lateral surface of the cylinder located between *ABC* and *DEF* is therefore equal to the product of the straight line *GH* and the perimeter of any one of the minimal sections of the cylinder.

But if *ABC* and *DEF* are not parallel and if they are tangent at a single point or if they do not intersect, then the area of the portion of the lateral surface of the cylinder located between them is equal to the half-product of the sum of the two straight lines *AD* and *BE* and the perimeter of any one of the minimal sections of the cylinder. But the straight line *GH* is equal to half the sum of the straight lines *AD* and *BE*; therefore the area of the portion of the lateral surface of the cylinder located between *ABC* and *DEF* is equal to the product of the straight line *GH* and the perimeter of any one of the minimal sections of the cylinder. That is what we wanted to prove.

You should know that the situation with the circles parallel to the bases of the right cylinder in the latter is like the situation with the minimal section in the oblique cylinder and that everything we have explained for the oblique cylinder concerning the area of its lateral surface and of the portions of its lateral surface must be identical for the right cylinder. When, in place of its minimal sections,

[55] See Supplementary note [22].

we make the circles parallel to the bases of the right cylinder, the method of the proof in both cases is the same method.[56]

The Book of Thābit ibn Qurra al-Ḥarrānī
on the sections of the cylinder and its lateral surface is completed.

Thanks be to God, Lord of the worlds.
Blessing upon His messenger Muḥammad and all his own.

[56] See Supplementary note [23].

CHAPTER III

IBN SINĀN, CRITIQUE OF AL-MĀHĀNĪ: THE AREA OF THE PARABOLA

3.1. INTRODUCTION

3.1.1. *Ibrāhīm ibn Sinān: 'heir' and 'critic'*

Ibrāhīm ibn Sinān ibn Thābit ibn Qurra was born in Baghdad in 296/909, where he died following an illness 37 years later, in 335/946.[1] He was an 'heir' in the strict sense, but in fact, as we shall see, in all senses of the word, for he was also a mathematician of genius: all the signs forecast great works. Despite the shortness of his life, Ibrāhīm ibn Sinān was by no means disappointed.

It is enough to read his full name, to conjure up the high reputation enjoyed by his parents and their relations, the Ṣabi'ūn, to be convinced that who we have here is very much an heir. We have but hardly finished with his grandfather: Thābit ibn Qurra encouraged his son, Ibrāhīm's father, to pursue his medical studies: Sinān excelled to such a degree in this art that he became doctor to three successive caliphs – al Muqtadir, al-Qāhir and al-

[1] Al-Nadīm, *Kitāb al-fihrist*, ed. R. Tajaddud, Tehran, 1971, p. 332; al-Qifṭī, *Ta'rīkh al-ḥukamā'*, ed. J. Lippert, Leipzig, 1903, pp. 57–9; Ibn Abī Uṣaybi'a: 'His birth was the year two hundred and ninety-six and his death on the Sunday mid-Muḥarram the year three hundred and thirty-five in Baghdad. The cause of his death was a liver tumour' (*'Uyūn al-anbā' fī ṭabaqāt al-aṭibbā'*, ed. A. Müller, 3 vols, Cairo/Königsberg, 1882–84, vol. I, p. 226, 29–32; ed. N. Riḍā, Beirut, 1965, p. 307, 14–17). On the other hand, the manuscript of the autobiography of Ibn Sinān is part of Collection 2519 of Khuda Bakhsh – *vide infra*. The pages of this text are scattered. A. S. Saidan has noted this fact and given the order of the pages. Cf. 'Rasā'il of al-Bīrūnī and Ibn Sinān', *Islamic Culture*, 34, 1960, pp. 173–5. In 1981 G. Saliba put out a critical edition of this text, with the title 'Risālat Ibrāhīm ibn Sinān ibn Thābit ibn Qurra fī al-Ma'ānī allatī istakhrajahā fī al-handasa wa-al-nujūm', *Studia Arabica & Islamica*, Festschrift for Iḥsān 'Abbās, ed. Wadād al-Qāḍī, American University of Beirut, 1981, pp. 195–203. A. S. Saidan in turn put out his edition of this text, in *The Works of Ibrāhīm ibn Sinān*, Kuwait, 1983, pp. 23–30. See our edition and French translation in R. Rashed and H. Bellosta, *Ibrāhīm ibn Sinān. Logique et géométrie au Xe siècle*, Leiden, 2000.

Rāḍī – and, according to al-Qifṭī, 'foremost of doctors'. A great doctor, Sinān was also a geometer: his name is associated with several writings in mathematics, one of which, dedicated to the Buyid king ʿAḍud al-Dawla, concerned inscribed and circumscribed polygons. His son, Ibrāhīm's brother, Thābit ibn Sinān, followed the example of his father, whose place he filled for caliph al-Rāḍī; he was at the same time in charge of the hospital in Baghdad. Thābit ibn Sinān was also a historian whose annals are still well known.[2] The cousin of Thābit and Ibrāhīm was no other than the famous man of letters Hilāl ibn al-Muḥassin al-Ṣābi'. These few names, these few titles, amongst others, make their contribution to assembling the scene of an intellectual and social aristocracy, the members of which were the moving forces behind the currents of power, but so too the high circles of science and of medicine. It was there that Ibrāhīm saw the light, it was there that he grew up, before being the object of a momentary persecution he later referred to.[3]

Ibrāhīm ibn Sinān was also the heir to a history. He belonged to a privileged generation, the fourth down from the Banū Mūsā. The translation of the chief mathematical texts had, in the main, been done, and the great traditions of research were already well established: that of the algebraists, which, born with al-Khwārizmī, was carried on with Abū Kāmil; that of the geometers, al-Jawharī, al-Nayrīzī, etc., which followed the Euclidean project; the tradition of the Banū Mūsā, finally, which, thanks to mathematicians like Thābit ibn Qurra, had already amassed a considerable body of results, had developed new methods and elaborated further theories: so much knowledge which had allowed their successors to see further and in greater depth. It was in this tradition that Ibrāhīm ibn Sinān took his place right away; namely, where deliberately an Archimedean geometry, as a geometry of measurement, was combined with a geometry that was concerned with the properties of positions, as the geometry of Apollonius. Benefiting from the works of scholars in this tradition, and, in the absolute first place, from those of his grandfather, Thābit ibn Qurra, Ibrāhīm ibn Sinān would develop the study of geometric transformations and of their applications to conic sections, as well as to the calculation of the area of a portion of parabola. He would take their works on sundials deeper to fashion the theory of a whole class of these instruments. There were, finally, the inquiries of his predeces-

[2] Al-Qifṭī, *Ta'rīkh al-ḥukamā'*, p. 195.

[3] Autobiography of Ibn Sinān, in R. Rashed and H. Bellosta, *Ibrāhīm ibn Sinān. Logique et géométrie au Xe siècle*, pp. 6–8. See also the introduction to his book on *The Movements of the Sun* (*Fī ḥarakāt al-shams)*, ed. A. S. Saidan, in *The Works of Ibrāhīm ibn Sinān*, p. 275.

sors into analysis and synthesis that had prompted him to write the first treatise on this subject worthy of the name.

At the same time we can catch a glimpse of his situation functioning as a connecting hinge and its eventual impact: with a keen and searching gaze this heir opened up several fields of the mathematics of the future – so many privileged training-grounds for his more distinguished successors, like Ibn al-Haytham a half century later, whose works cannot be properly understood without Ibn Sinān's research. As a matter of fact it was following on from the latter, but also in opposition to him, that Ibn Haytham put together his magisterial treatise *On the Lines of the Hours*.[4] The same applies to his treatise, no less magisterial, on *Analysis and Synthesis*.[5]

One would hope for a copious amount of information on the life and work of a mathematician of such broad interests, and of one of those precocious creative forces whom fate has carried off before their time. But we have learned not to be surprised any more: Ibn Sahl, Sharaf al-Dīn al-Ṭūsī, to take only these examples, have made us familiar with having only a paucity of information. The situation with Ibrāhīm ibn Sinān is at all events better: al-Nadīm devoted an account to him, which was originally to have been longer; Ibn Abī Uṣaybiʿa a few lines, like al-Nadīm. Al-Qifṭī did no more, but he did have in his hands a succinct autobiography of Ibn Sinān, which on occasion he made clumsy attempts at summarizing. Much more fortunately, this autobiography has come down to us.

Ibn Sinān makes it clear that he drafted it after his 25th year, after 934. Concerning his life itself, he remains more discreet. He refers rather vaguely to a time of persecution,[6] without specifying either the period, or the reasons, even whether it is right to assume that this persecution was connected to his political set. He announces his desire in this autobiography to make a checklist of his writings until that time, the reasons that have pushed him to write them down, the aims that are his own, so that neither works that are not his own might be attributed to him nor one of his writings might be claimed by someone else. These last have all come down to us, with the exception of one important book, of his subject matter even, on *Tangent Circles*. Apart from this, al-Nadīm, in his bio-bibliography, cites under Ibn Sinān's name two titles that the latter did not mention in his autobiography: a *Commentary on the First Book of the* Conics and the

[4] R. Rashed, *Les mathématiques infinitésimales du IXe au XIe siècle*. Vol. II: *Ibn al Haytham*, London, 1993, pp. 491–4.

[5] R. Rashed, 'La philosophie mathématique d'Ibn al Haytham. I: L'analyse et la synthèse', *M.I.D.E.O.*, 20, 1991, pp. 31–231, and *Les Mathématiques infinitésimales du IXe au XIe siècle*, vol. IV: *Méthodes géométriques, transformations ponctuelles et philosophie des mathématiques*, London, 2002.

[6] Cf. Note 3.

Intentions (aghrāḍ) of the Book of the Almagest.[7] Lastly, there exists a *Treatise on the Astrolabe* that carries as its author's name Ibn Sinān, but which no known list counts and whose authenticity has still not been established. The works cited by al-Nadīm have not come down to us: does that mean they are books written after his autobiography, or simply that they are apocryphal works? On this question, no one is in any position to give an answer.

From the autobiography of Ibn Sinān stands out what has not stopped being repeated, since al-Nadīm at least: this was a gifted and precocious mathematician. According to his own assertions, he began his studies at 15; by 16 or 17 years old he had put together a first version of his book *On Shadow Instruments*, which he was to revise at 25. In this book he wrote:

> I have set forth everything which concerns dials. In fact, I have brought together all the constructions of dials with a plane surface under one single construction common to them all, which I have demonstrated, in addition to other things which I have shown [...].[8]

A year after – at the age of 18 – he was discussing and forming criticisms of Ptolemy's views on *The Determination of the Anomalies of Saturn, Mars and Jupiter*, in a treatise which he completed six years later, at the age of 24. In geometry, Ibn Sinān wrote some treatises: *On the Tangent Circles*, *On Analysis and Synthesis*, *On Chosen Problems*, *On the Measurement of the Parabola*, and *On the Drawing of the Three Conic Sections*. Written before his 25th year, all these writings had been revised by him before this same date.

This autobiography allows us besides to arrange Ibn Sinān's writings in their relations to one another, and to separate out the standard features that govern them. For each one of them, Ibn Sinān was set on making explicit the purpose he was addressing, its range just as much as its place in the totality of his work. As regards the standard features he complied with, we can but be struck by the demands of 'criticism': criticism is founded in positive and recognized value, which is practised systematically and in all directions. It is brought to bear on the works of the ancients, such as those of Ptolemy, but the writings of the moderns do not escape it, those of al-Māhānī, for example. On the other hand, at this period, as we shall see quite particularly with Ibn Sinān, rigour is not the only criterion putting its restraint on a proof, but elegance must equally be sought out; there are so many things to respond to in renewing one's research. From the beginning of his mathematical career Ibn Sinān, moreover, applied himself to the theo-

[7] Al-Nadīm, *al-Fihrist*, p. 332.

[8] Autobiography of Ibn Sinān, in R. Rashed and H. Bellosta, *Ibrāhīm ibn Sinān. Logique et géométrie au Xe siècle*, p. 8.

retical problems of proofs, and a good part of his work refers to what we might call a theory of proofs. Thus is explained, in part at least, the interest he always held for the subject of analysis and synthesis. In other respects, the desire for simplicity and elegance are reasons he explicitly considered as being sufficient for taking up again the proof, already correct, of a proposition.

It is this context that sheds light on the only text by him written on infinitesimal mathematics, by isolating the features of the work and giving their sense to Ibn Sinān's themes. Let us read what he wrote concerning this treatise on *The Measurement of the Parabola*:

> I have written a work on the measurement of the parabola, in one book. My grandfather had determined the measurement of this section. Certain contemporary geometers have informed me that there is on this subject a work by al-Māhānī, which they presented to me, and easier than that of my grandfather. I did not like it that there was a work by al-Māhānī more advanced than my grandfather's, without there being amongst us one which surpassed him in its work. My grandfather had determined it in 20 propositions. He preceded it with numerous arithmetic lemmata, included in the twenty propositions. The question of the measurement of the section appeared clearly to him through the method of the absurd. Al-Māhānī also preceded what he proved with arithmetic lemmata. He then demonstrated what he wanted by the method of the absurd, in five or six propositions that are lengthy. I then proved it in three geometrical propositions, without preceding them by any arithmetic lemma. I showed the measurement of this same section by the method of direct proof, and I had no need for the method of the absurd.[9]

An expression of the pride of an heir and of the assurance of a scholar outside the common, these remarks are also the reflection of the standards of Ibn Sinān the mathematician: brevity, competence and elegance. Fruitful and creative, these standards are equally at work in the very heart of Ibn Sinān's work: he took up his own editing to refine the lines of his proof.

3.1.2. *The two versions of* The Measurement of the Parabola*: Texts and translations*

In the introduction to his treatise on *The Measurement of a Portion of the Parabola*, Ibn Sinān writes:

> Some time ago, I wrote a book on the area of this section. Later, I made a number of changes to one of the propositions. This corrected copy and the

[9] Autobiography of Ibn Sinān, in R. Rashed and H. Bellosta, *Ibrāhīm ibn Sinān. Logique et géométrie au Xe siècle*, p. 18.

> older copy have now been lost and I therefore need to repeat my earlier work in this book.[10]

Having this confirmation from Ibn Sinān himself, modern historians and biographers have assumed that all the surviving manuscripts of this treatise derive from one single version, the last one. This is not at all the case.

In his autobiography, we learn that Ibn Sinān submitted his own works for critical evaluation and all had been revised before he reached the age of twenty-five, that is before 321/934. He writes: 'Those of my books that I had not already corrected as I was writing them, were corrected by me before I reached twenty-five years of age'.[11] That is to say that by 321/934 there already existed two versions of this text on the measurement of the parabola; the first either in its original or reworked form, but already lost, and the final version that was intended to replace it. It is this lost version that has been rediscovered, making it possible to do that which was inconceivable in the past; trace the development of Ibn Sinān's thought and mathematical techniques.

The study of the manuscript traditions has made it possible, not only to recover the lost text of Ibn Sinān, but also to demonstrate that its existence was known to earlier copyists. In this regard, one result of an examination of manuscript 4832 in the Aya Sofya collection in Istanbul should be noted.[12] This manuscript is derived from the same source as that copied by Muṣṭafā Ṣidqī in 1159/1746–1747, that is, manuscript Riyāḍa 40 in the Dār al-Kutub in Cairo. The following appears in the margin of the first copy of the Ibn Sinān treatise (fol. 78^v):

> He wrote the following in his introduction: Abū Isḥāq Ibrāhīm ibn Sinān ibn Thābit originally wrote this book, he then mentioned that he had lost it, and finally he wrote another book and mentioned this copy in the introduction to this new treatise.

One can read this citation word for word in the margin of fol. 182^v, written by Muṣṭafā Ṣidqī in his own hand, and in a version of Ibn Sinān on the measurement of the parabola. In other words, they have both transcribed a comment already noted by their common predecessor, which very likely dates it to before the fifth century of the Hegira, and certainly before the sixth century, as we have shown. It is therefore clear that this predecessor knew that the text he was copying was that lost by Ibn Sinān. The critical transcription of this text, and its analysis, reveals the rest; that it is definitely the first version, and not some degenerate version of the second.

[10] *Vide infra*, p. 495.
[11] Autobiography of Ibn Sinān, p. 18.
[12] See Chapter II, section 2.1.3.

All the manuscripts known to us and available to us at the present time therefore fall into one of two groups: two transmit the lost version and three transmit the final version. The first two have already been cited and described: Aya Sofya 4832, fols 76^{v}–79^{r}, referred to as A, and Dār al-Kutub, Riyāḍa 40, fols 182^{v}–186^{v}, referred to as Q. It should also be noted that the text by Ibn Sinān comes to us in the manuscript Damascus 5648, fols 159–165. This manuscript is a recent copy of the Cairo manuscript, Riyāḍa 40, and no other.

The second and final version of the treatise by Ibn Sinān therefore comes down to us through the following manuscripts:

1) Manuscript 2457 in the Bibliothèque Nationale de Paris, copied by al-Sijzī in 358/967–969 in Shīrāz, fols 134^{v}–136^{r}. We have previously made reference to this famous collection.[13] We should, however, draw attention to one notable feature of this text. After he had translated the Ibn Sinān treatise, al-Sijzī compared it with another manuscript and marked all the variations in another colour. The copy is written in black ink, while the variations from the other manuscript are written in red ink. Apart from a single word (*annahu*) on fol. 135^{v} written in black and added in the margin almost certainly during the copying process, all the other corrections are in red ink. Al-Sijzī also uses red ink for the end of the treatise and for the colophon, in which he declares explicitly that he has compared the copy with another source. There are around 40 marginal corrections in red ink, together with another 40 or so notes in the same colour above or below words in the text. He has also added a number of diacritical marks in red ink. In all, 11 phrases and 10 words are taken from the second source manuscript. We refer to this manuscript as manuscript B, the source being copied as x_1 and the other copy as x_2.

2) Manuscript 461 in the India Office (Loth 767, fols 191–197), which we have described elsewhere.[14] This manuscript, copied in 1198/1784 from a source manuscript also located in India in *nasta'līq*, contains no additions or marginal annotations. We refer to this manuscript as manuscript L.

3) The third manuscript forms part of collection 2519 in the Khuda Bakhsh library in Patna, India (referred as Kh).[15] This important collection includes 42 treatises by Archimedes, al-Qūhī, Ibn 'Irāq, al-Nayrīzī, and others. The manuscript is a collection of 327 sheets (32 lines per page,

[13] See Chapter II, Section 2.1.3.

[14] R. Rashed, *Sharaf al-Dīn al-Ṭūsī, Œuvres mathématiques. Algèbre et géométrie au XIIe siècle*, Paris, 1986, vol. I, pp. XLII–XLIII.

[15] Corresponds to No. 2468 in the *Catalogue of the Arabic and Persian Manuscripts in the Oriental Public Library at Bankipore*, volume XXII (Arabic MSS.) Science, prepared by Maulavi Abdul Hamid, Patna, 1937, pp. 60–92.

dimension 24/15, and 20/12.5 for the text). It was copied in 631–632 of the Hegira, *i.e.* 1234–1235, in Mosul, in *naskhī*. Ibn Sinān's text occupies folios 132^r–134^v, and contains no additions or marginal annotations.[16]

An analysis of errors and other incidents results in the *stemma* proposed below:

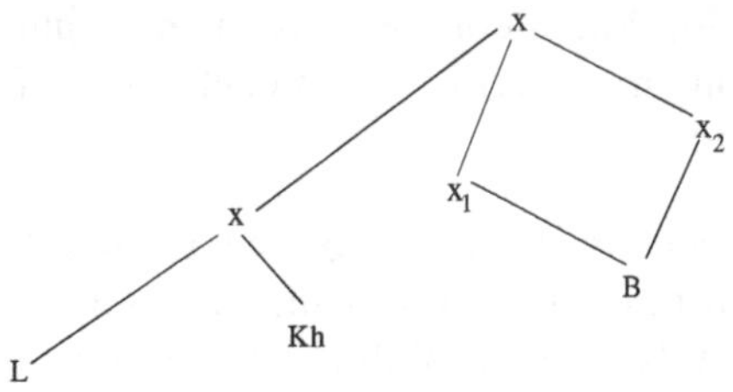

To conclude our discussion of the various editions and translations of the two versions by Ibn Sinān: As we have said, the distinction between them has never been made by biobibliographers and historians. No edition of the first version exists in any form, and the second has not been the subject of a critical edition until now. Two non-critical editions of the manuscript Kh have been published. The first dates from 1947: *Rasā'il Ibn-I-Sinān*, edited and published by Osmania Oriental Publications Bureau, Hyderabad–Deccan, 1948. The second is by A. S. Saidan, *The Works of Ibrāhīm ibn Sinān*, Kuwait, 1983, pp. 57–65. Translations include that by H. Suter: 'Abhandlung über die Ausmessung der Parabel von Ibrāhīm b. Sinān b. Thābit', in *Vierteljahrsschrift der Naturforschenden Geselschaft in Zürich*, Herausgegeben von Hans Schinz, 63, Zürich, 1918, pp. 214–28. This translation has been made from manuscript B alone.

We have shown the existence of the two versions in *Les Mathématiques infinitésimales*, vol. I, pp. 695–735, to which the reader is referred.

3.2. MATHEMATICAL COMMENTARY

To follow the evolution of Ibn Sinān's thought on the measure of the parabola, we shall simultaneously examine the two versions of his treatise so as to compare them. The first – the older – consists of three propositions. These are all also found in the newer version, which includes a further corollary to the last proposition.

[16] For a detailed comparison of the manuscripts and the history of the manuscript tradition, see *Les mathématiques infinitésimales*, vol. I, pp. 680–1.

Proposition 1. — *Let* $A = (A_0, A_1, \ldots, A_n)$ *and* $B = (B_0, B_1, \ldots, B_n)$ *be two convex polygons. We project the points* $A_1, A_2, \ldots, A_{n-1}$ *on* A_0A_n *in parallel to* $A_{n-1}A_n$ *to the points* $A'_1, A'_2, \ldots, A'_{n-1} = A_n$ *and the points* $B_1, B_2, \ldots, B_{n-1}$ *on* B_0B_n *in parallel to* $B_{n-1}B_n$ *to the points* $B'_1, B'_2, \ldots, B'_{n-1} = B_n$. If we have

$$\frac{A_0A'_1}{B_0B'_1} = \ldots = \frac{A'_{n-2}A_n}{B'_{n-2}B_n} = \lambda,$$

and

$$\frac{A_1A'_1}{B_1B'_1} = \ \ldots \ = \frac{A_{n-1}A_n}{B_{n-1}B_n} = \mu,$$

then

$$\frac{\text{tr.}(A_0, A_{n-1}, A_n)}{\text{p.}(A_0, A_1, \ldots, A_n)} = \frac{\text{tr.}(B_0, B_{n-1}, B_n)}{\text{p.}(B_0, B_1, \ldots, B_n)}.$$

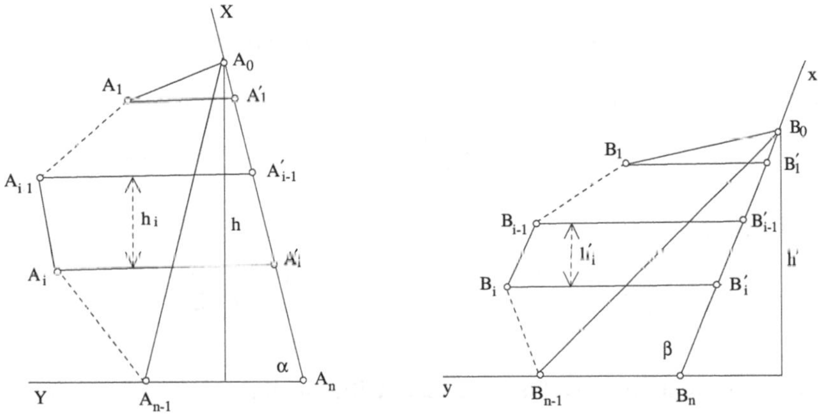

Fig. 3.1

Let h and h' be the respective heights of the triangles (A_0, A_{n-1}, A_n) and (B_0, B_{n-1}, B_n); h_1 and h'_1 those of the triangles (A_0, A_1, A'_1) and (B_0, B_1, B'_1), h_i and h'_i those of the trapeziums $(A_{i-1}, A'_{i-1}, A'_i, A_i)$ and $(B_{i-1}, B'_{i-1}, B'_i, B_i)$ for $2 \le i \le n-1$. We have

$$s = \text{tr.}\ (A_0, A_{n-1}, A_n) = \tfrac{1}{2}\, h \cdot A_{n-1}A_n,$$
$$s_1 = \text{tr.}\ (A_0, A_1, A'_1) = \tfrac{1}{2}\, h_1 \cdot A_1A'_1,$$
$$s_i = \text{tp.}\ (A_{i-1}, A'_{i-1}, A'_i, A_i) = \tfrac{1}{2}\, h_i \cdot (A_{i-1}A'_{i-1} + A_iA'_i),$$

$$S = \text{p. } (A_0, A_1, ..., A_n) = \sum_{i=1}^{n-1} s_i,$$

$$s' = \text{tr. } (B_0, B_{n-1}, B_n) = \tfrac{1}{2} h' \cdot B_{n-1}B_n,$$

$$s'_1 = \text{tr. } (B_0, B_1, B'_1) = \tfrac{1}{2} h'_1 \cdot B_1B'_1,$$

$$s'_i = \text{tp. } (B_{i-1}, B'_{i-1}, B'_i, B_i) = \tfrac{1}{2} h'_i . (B_{i-1}B'_{i-1} + B_iB'_i),$$

$$S' = \text{p. } (B_0, B_1, ..., B_n) = \sum_{i=1}^{n-1} s'_i.$$

But, by hypothesis, we have on the one hand,

$$\frac{A_1A'_1}{B_1B'_1} = \frac{A_{i-1}A'_{i-1} + A_iA'_i}{B_{i-1}B'_{i-1} + B_iB'_i} = \mu \qquad (2 \le i \le n-1),$$

and on the other hand

$$h_i = A'_{i-1}\, A'_i \sin \alpha, \quad h'_i = B'_{i-1}\, B'_i \sin \beta;$$

hence

$$\frac{h}{h'} = \frac{h_i}{h'_i} = \lambda \frac{\sin\alpha}{\sin\beta} \qquad (1 \le i \le n-1).$$

We then deduce

$$\frac{s}{s'} = \frac{s_1}{s'_1} = \ \dots \ = \frac{s_i}{s'_i} = \ \dots \ = \frac{s_{n-1}}{s'_{n-1}} = \frac{\sum_{i=1}^{n-1} s_i}{\sum_{i=1}^{n-1} s'_i} = \frac{S}{S'} = \lambda\mu \frac{\sin\alpha}{\sin\beta},$$

and hence arrive at the conclusion given by Ibn Sinān:

$$\frac{s}{S} = \frac{s'}{S'}.$$

Comparison of the two versions

In the first version, containing four figures, Ibn Sinān explains in detail the construction of the two considered polygons starting with two similar groups such that $(A_0, A'_1, \dots, A'_i, \dots, A_n)$ and $(B_0, B'_1, \dots, B'_i, \dots, B_n)$. This detailed construction does not appear in the second version, which contains only a single figure. The reasoning in the two versions is based on the supposition that $\alpha \neq \frac{\pi}{2}$ and $\beta \neq \frac{\pi}{2}$. It remains valid, however, whether α and β are right angles or not.

In the older version, Ibn Sinān considers, to end the proposition, the particular cases where $\alpha = \frac{\pi}{2}$ or $\beta = \frac{\pi}{2}$ and explains that the segments A_0A_n and B_0B_n and their parts $A'_{i-1}A'_i$ and $B'_{i-1}B'_i$ replace in this case the respective heights h and h', h_i and h'_i of the triangles and trapeziums considered.

In the later version, the calculations are much more rapid than in the first. For example the equations for which

$$\frac{h}{h_i} = \frac{A_0A_n}{A'_{i-1}A'_i}$$

are directly deduced by way of the parallelism of the segments given in the the later version, whereas they are obtained in the first with the help of similar triangles.

In the two versions, Ibn Sinān uses the hypotheses in the form

$$\frac{a_1}{a_2} = \frac{b_1}{b_2}, \dots, \frac{a_{i-1}}{a_i} = \frac{b_{i-1}}{b_i}, \dots, \frac{a_{n-1}}{a_n} = \frac{b_{n-1}}{b_n},$$

with $a_i = A'_{i-1}A'_i$, $b_i = B'_{i-1}B'_i$ for $1 \leq i < n-1$, and $(A'_0 - A_0)$ $(B'_0 = B_0)$. In the later version, Ibn Sinān deduces without justification that

$$\frac{a_1}{a_{n-1}} = \frac{b_1}{b_{n-1}},$$

whereas in the first, he obtains this equation by constructing the given ratios step by step. There, the difference must be purely formal.

Let us note that in the two versions, the conclusions pertaining to areas are obtained in the form

$$\frac{s}{s_1} = \frac{s'}{s'_1}, \quad \frac{s}{s_2} = \frac{s'}{s'_2}, \dots, \frac{s}{s_i} = \frac{s'}{s'_i}, \dots, \frac{s}{s_{n-1}} = \frac{s'}{s_{n-1}}.$$

In the later version, he immediately deduces

$$\frac{s}{s'} = \frac{s_1}{s'_1} = \dots = \frac{s_i}{s'_i} = \dots = \frac{s_{n-1}}{s'_{n-1}} = \frac{\sum_{i=1}^{n-1} s_i}{\sum_{i=1}^{n-1} s'_i}.$$

In the first version, Ibn Sinān transforms firstly by a permutation the ratios in the first expression.

Lastly, in the two versions, Ibn Sinān proceeds with the help of the same pointwise transformation T defined in the statement of that proposition and in which the polygon $(B_0, B_1, ..., B_n)$ has as its image the polygon $(A_0, A_1, \ldots, A_n)$. This transformation, we show, is an *affine mapping*.

Let us take as a system of reference $x\,B_n\,y$ and $X\,A_n\,Y$, with $A_0 \in A_n X$ and $B_0 \in B_n x$, $A_{n-1} \in A_n Y$ and $B_{n-1} \in B_n y$, and consider A_0, B_0, A_{n-1} and B_{n-1} as unit points on the axes. Then

$$B_0\,(x_0 = 1\,;\,y_0 = 0), \qquad B_{n-1}\,(x_{n-1} = 0\,;\,y_{n-1} = 1),$$
$$A_0\,(X_0 = 1\,;\,Y_0 = 0), \qquad A_{n-1}\,(X_{n-1} = 0\,;\,Y_{n-1} = 1).$$

For each point $B'_i\,(x_i\,;\,0)$ on $B_n x$ and its homologue $A'_i\,(X_i\,;\,0)$ on $A_n X$, we have by hypothesis

$$\frac{B_n B'_i}{B_n B_0} = \frac{A_n A'_i}{A_n A_0}.$$

Hence

$$\frac{x_i}{x_0} = \frac{X_i}{X_0};$$

thus $X_i = x_i$.

For each point $B_i\,(x_i, y_i)$ and its homologue $A_i\,(X_i, Y_i)$, we likewise have by hypothesis

$$\frac{B'_i B_i}{B_n B_{n-1}} = \frac{A'_i A_i}{A_n A_{n-1}}.$$

Hence

$$\frac{y_i}{y_{n-1}} = \frac{Y_i}{Y_{n-1}};$$

thus $Y_i = y_i$.

Thus, the points B_i for $(0 \leq i \leq n)$ with respect to the system of reference $x\,B_n\,y$ have the same coordinates as the respective images A_i considered with respect to the system of reference $X\,A_n\,Y$. They are thus homologous under the transformation T defined with respect to the two systems of reference $x\,B_n\,y$ and $X\,A_n\,Y$, with the two systems of reference being able to be as here in the same plane, or in two different planes. The function T is thus an arbitrary affine bijection and the ratio k of an arbitrary area to the homologous area is independent of the area chosen. In the example treated by Ibn Sinān, k is determined by the data

$$k = \frac{s}{s'} = \frac{s_i}{s'_i} = \frac{S}{S'} = \lambda\mu\frac{\sin\alpha}{\sin\beta}.$$

In the particular case where $\alpha = \beta$, we have $k = \lambda\mu$ and the ratio k is thus the product of the ratios λ and μ of two affinities (dilatations or contractions). One can thus consider the transformation T as the product of two affinities, oblique or orthogonal depending on whether α is or is not a right angle, and of a displacement (and likewise of an isometry). In the particular case where $\alpha = \beta$ and $\lambda = \mu$, we have $k = \lambda^2$, and the transformation T is then a similarity of ratio λ.

Proposition 2. — *The ratio of the areas of the two portions of a parabola sections is equal to the ratio of the areas of the two triangles which are associated with them.*

Let ABC and DEG be two portions of a parabola, S and S' their respective areas, and S_1 and S'_1 the areas of the triangles $\mathbf{P}_1$ and $\mathbf{P'}_1$ associated with these sections. We want to show that

$$\frac{S'}{S} = \frac{S'_1}{S_1}.$$

Ibn Sinān's proof is by a *reductio ad absurdum* that relies on the following lemma:

LEMMA: *If* M *is the vertex associated with an arbitrary chord* BC *of a parabola, then*

$$\text{tr. (BMC)} > \tfrac{1}{2}\,\text{port. (BMC)}.$$

The tangent at M is parallel to BC; it meets the diameter BH at O and the parallel to BH through C at S. We have

$$\text{tr. }(BMC) = \tfrac{1}{2}\,\text{area }(BOSC).$$

But

$$\text{area }(BOSC) > \text{port. }(BMC);$$

hence the result.

We then suppose

$$\frac{S'_1}{S_1} \neq \frac{S'}{S}.$$

1) If $\frac{S_1'}{S_1} > \frac{S'}{S}$, there exists an area J such that $\frac{S_1'}{S_1} = \frac{S'}{J}$; we then have $J < S$ and we let $S - J = \varepsilon$. We thus have

$$S - S_1 \leq \varepsilon \text{ or } S - S_1 > \varepsilon.$$

If $S - S_1 \leq \varepsilon$, then $S - S_1 \leq S - J$; hence $S_1 \geq J$, which is impossible, as $\frac{S_1'}{S_1} = \frac{S'}{J}$ and $S_1' < S'$.

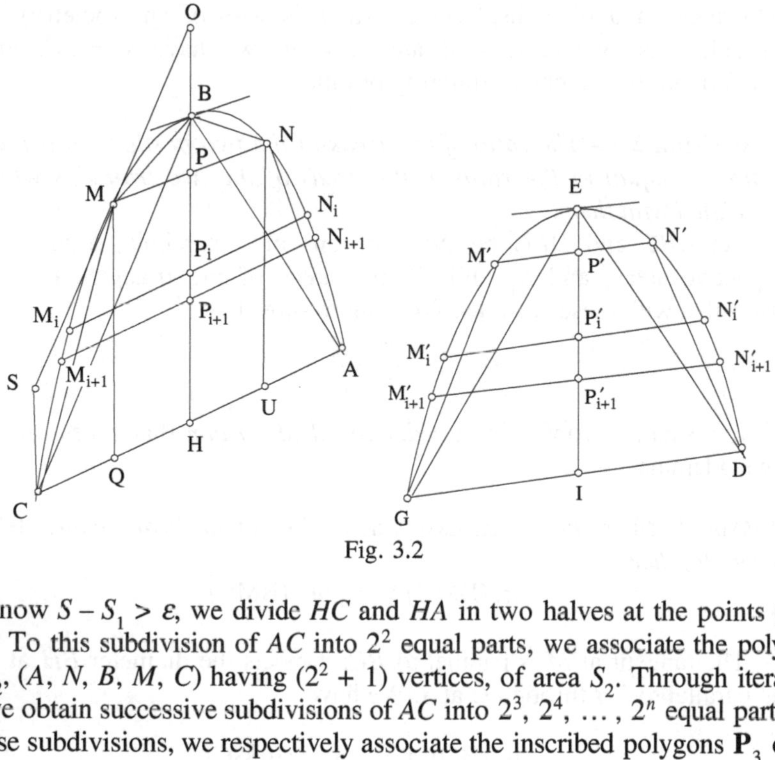

Fig. 3.2

If now $S - S_1 > \varepsilon$, we divide HC and HA in two halves at the points Q and U. To this subdivision of AC into 2^2 equal parts, we associate the polygon $\mathbf{P}_2$, (A, N, B, M, C) having $(2^2 + 1)$ vertices, of area S_2. Through iteration, we obtain successive subdivisions of AC into $2^3, 2^4, \ldots, 2^n$ equal parts. To these subdivisions, we respectively associate the inscribed polygons $\mathbf{P}_3$ of area $S_3, \ldots, \mathbf{P}_n$ of area S_n. The polygon $\mathbf{P}_n$ has $(2^n + 1)$ vertices. By the lemma, we have

$$S - S_1 < \tfrac{1}{2} S,$$
$$S - S_2 < \tfrac{1}{2}(S - S_1) < \frac{1}{2^2} S,$$
$$\ldots$$
$$S - S_n < \tfrac{1}{2}(S - S_{n-1}) < \frac{1}{2^n} S.$$

Thus, for this ε given $\exists N \in \mathbf{N}^*$; $\forall n \geq N$, we have $S - S_n < \varepsilon$, that is to say

$$S - S_n < S - J;$$

hence

$$S_n > J.$$

Let $\mathbf{P}_n$ be the polygon corresponding to this number n. The vertices M_i and N_i for $(0 \leq i \leq 2^{n-1})$ and with $M_0 = N_0 = B$, $M_2{}^{n-1} = C$, $N_2{}^{n-1} = A$, have pairwise equal ordinate; there thus corresponds to them a single abscissa on BH. If a is the *latus rectum* of the parabola, the vertices M_i (x_i, y_i) of $\mathbf{P}_n$ satisfy the equation $y_i^2 = a\,x_i$.

To these vertices there is associated a subdivision along the diameter BH by the points P_i of abscissa x_i, with $x_0 = 0$ and $x_{2^{n-1}} = BH$. To this subdivision of BH, we associate on the diameter EI of the second section a similar subdivision by the points P'_i of abscissa x'_i. We then have

$$\text{(1)} \qquad \frac{x_i}{x'_i} = \frac{BH}{EI} = \lambda.$$

The points M'_i and N'_i of common abscissa x'_i of the parabola DEG define a polygon $\mathbf{P}'_n$ of area S'_n. If a' is the *latus rectum* of DEG, the coordinates (x'_i, y'_i) of M'_i satisfy the equation $y'^2 = a'x'_i$. We have

$$\frac{y_i^2}{y_i'^2} = \frac{ax_i}{a'x'_i} = \frac{a}{a'}\lambda,$$

whence

$$\text{(2)} \qquad \frac{y_i}{y'_i} = \sqrt{\frac{a}{a'}}\lambda = \mu.$$

By (1) and (2), the polygons $(H, B, \ldots M_i, \ldots, C)$ and $(H, B, \ldots, N_i, \ldots, A)$ and their respective homologues $(I, E, \ldots, M'_i, \ldots, G)$ and $(I, E, \ldots, N'_i, \ldots, D)$ satisfying the hypotheses of Proposition 1, we deduce

$$\frac{S'_n}{S_n} = \frac{S'_1}{S_1},$$

whence

$$\frac{S'_n}{S_n} = \frac{S'}{J},$$

which is impossible, as $S_n > J$ and $S'_n < S'$.

2) If $\frac{S'_1}{S_1} < \frac{S'}{S}$, there exists an area J' such that $\frac{S'_1}{S_1} = \frac{J'}{S}$ with $J' < S'$. We show in the same manner that this is impossible.

From cases 1) and 2), we conclude that

$$\frac{S'_1}{S_1} = \frac{S'}{S}.$$

Comparison of the two versions

1. The difference between the two versions concerns the uniqueness of the abscissa. In fact, in the older version (*vide infra* pp. 488–9 and n. 3), to the two points M and L, which have equal ordinates by construction, Ibn Sinān associates two different abscissae IT and IV. To the points T and V taken to be distinct on IG he associates the distinct points R and X on EH and then separately considers the polygons (G, I, L, C) and (E, H, Q, A) on the one hand and (G, I, M, D) and (E, H, J, B) on the other hand, which satisfy the hypotheses of Proposition 1. From this, he deduces the conclusion for the polygons (C, L, I, M, D) and (A, Q, H, I, B).

The fact of having given different abscissae to the two points M and L which have equal ordinates does not, however, have an effect on the rigour of the reasoning. The uniqueness of the abscissa results from Proposition I.20 of Apollonius's *Conics*. Ibn Sinān did not think about this in the course of his first composition in spite of his great familiarity with the *Conics*.

In the later version, on the other hand, Ibn Sinān proves that the points M and N, which have by hypothesis equal ordinates $HQ = HU$, have a same and single abscissa BP. He then directly considers the polygons (A, N, B, M, C) and (D, X, E, T, G) without indicating that these are the polygons (H, B, M, C) and (I, E, T, G) on the one hand, and (H, B, N, A) and (I, E, X, D) on the other, which satisfy the hypotheses of Proposition 1.

Yet, in using the affine mapping T deduced in Proposition 1, we have directly the correspondance between the polygons (A, N, B, M, C) and (D, X, E, T, G) and we can conclude without separating these polygons into two parts. It seems likely that it was this very idea that impelled Ibn Sinān to produce a new edition of his treatise.

2. Proposition 2, like those that will follow, bears upon parabolic sections. The vertex of a portion of a parabola is the extremity of the diameter

conjugate to the chord which is the base of the portion.[17] This vertex and this chord determine the triangle associated with the portion of the parabola. Now, this triangle plays here an important role.

In fact, in the statement of the proposition in both versions Ibn Sinān relates each portion of a parabola to the triangle 'for which the base is its base and the vertex its vertex'.

In the statement of Proposition 3, as in that of Proposition 4 of the newer version, Ibn Sinān relates the portion of the parabola 'to the triangle on the same base and with the same height', whereas in Proposition 3 from the older version, he relates the portion of the parabola to the parallelogram 'which has for base its base and for height its height'. Yet in each of these propositions, the height of the triangles or of the parallelogram considered does not appear. Ibn Sinān uses, in fact, the segment joining the vertex to the middle of the base. This segment only becomes the height in the particular case where the diameter considered is the axis of the parabola; something Ibn Sinān knew perfectly well.

Proposition 3. — *The area of a portion of a parabola is four thirds the area of the triangle which is associated with it.*

Let ABC be a portion of a parabola with base AC and diameter BD; let S_p be its area and S_T that of the triangle ABC. We have

$$S_p = \tfrac{4}{3}\, S_T.$$

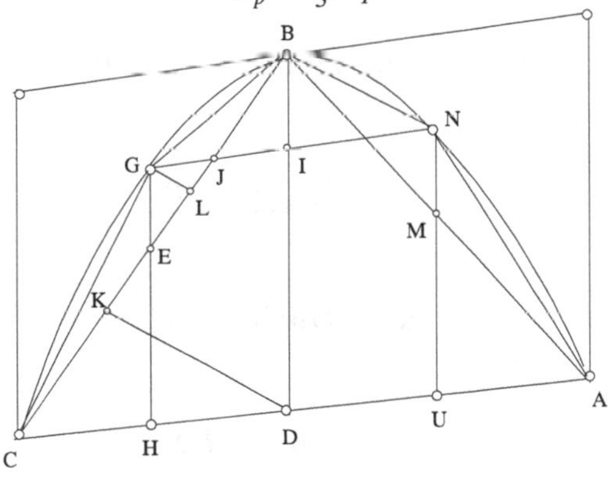

Fig. 3.3

[17] *The Conics*, Book I, definition of diameter (R. Rashed, *Apollonius: Les Coniques,* tome 1.1: *Livre I*, commentaire historique et mathématique, édition et traduction du texte arabe, Berlin/New York, 2008, p. 254); the word 'vertex' appears in order to designate the extremity of the diameter.

Let E and M be the respective midpoints of BC and BA, GH and NU the corresponding diameters, and H and U the respective midpoints of DC and DA; thus GN is parallel to AC and cuts BD at I and BC at J.

Let $DK \perp BC$ and $GL \perp BC$. We have

$$\frac{IJ}{CD} = \frac{BI}{BD},$$

and on the other hand

$$\frac{BI}{BD} = \frac{IG^2}{CD^2},$$

whence $IG^2 = IJ \cdot CD$. But $DC = 2\ HD = 2GI$, whence $GI = 2IJ$ and as a result $GJ = \frac{1}{2}\ GI = \frac{1}{4}\ DC$.

The right-angled triangles DKC and GLJ are similar as $C\hat{D}K = L\hat{G}J$ (acute angles with parallel sides); thus

$$\frac{GL}{DK} = \frac{GJ}{DC},$$

whence

$$GL = \tfrac{1}{4}\ DK.$$

The triangles DBC and GBC have the same base BC, whence

$$\text{tr. }(GBC) = \tfrac{1}{4}\ \text{tr. }(DBC) = \tfrac{1}{8}\ \text{tr. }(ABC) = \tfrac{1}{8}\ S_T.$$

But, by proposition 2,

$$\frac{\text{tr. }(GBC)}{S_T} = \frac{\text{area }(GBC)}{S_p};$$

thus

$$\text{area }(GBC) = \tfrac{1}{8}\ S_p.$$

Likewise

$$\text{area }(NBA) = \tfrac{1}{8}\ S_p,$$

whence

$$\text{area }(GBC) + \text{area }(NBA) = \tfrac{1}{4}\ S_p$$

and

$$S_T = \tfrac{3}{4}\ S_p,$$

whence the result

$$S_p = \frac{4}{3} S_T.$$

We now analytically treat Ibn Sinān's result in the case where BD is the axis of the parabola. In an orthonormal system of reference $B(0,0)$, $A(x_0, y_0 = \sqrt{ax_0})$, $C(x_0, -\sqrt{ax_0})$, we have

$$S_p = 2\int_0^{x_0} \sqrt{ax}\, dx = \frac{4}{3}\, x_0\sqrt{a\, x_0}\,;$$

yet

$$S_T = x_0\, \sqrt{ax_0}\,,$$

whence

$$S_p = \frac{4}{3} S_T.$$

Comparison of the two versions

The two proofs are very close to one another; however, in the older version, to show that $HO = \frac{1}{4} DP$ (corresponding to $GL = \frac{1}{4} DK$ in the newer version), Ibn Sinān makes appeal to the properties of the tangent at a point, that is to say to Book I of Apollonius's *Conics*,[18] and the proof is markedly longer than that of the newer version which is summarised here.

In the figure in the older version, we find as before two different abscissae for two points H and I for which the ordinates are equal. The result is two distinct points, in place of a single one, for the intersections of the tangents to H and I with the diameter.[19] But as before, this does not enter into the reasoning.

Let us note ultimately that, in the older version, Ibn Sinān ends by giving the ratio $\frac{2}{3}$ of the parabola to the associated parallelogram.

The following proposition is a corollary to Proposition 3 and is not found in the older version, but only in the newer.

Proposition 4. — *Let* ACE *and* BCD *be two portions of the same parabola. If the bases* AE *and* BD *are parallel and if they meet the*

[18] R. Rashed, *Apollonius: Les Coniques,* tome 1.1: *Livre I*, pp. 318 and 320.

[19] It is clear that, by their construction, the points I and H have equal ordinates, $DC/2$ and $DB/2$, and thus have the same abscissa; we must therefore have $X = R$ (cf. Fig. 3.3.). Moreover, the vertex A being the midpoint of the subtangent, we must also have $K = L$.

diameter CH *associated with them at the points* H *and* G, *then their respective areas* S_1 and S_2 *satisfy*

$$\frac{S_1}{S_2} = \frac{CH}{CG} \cdot \sqrt{\frac{CH}{CG}}.$$

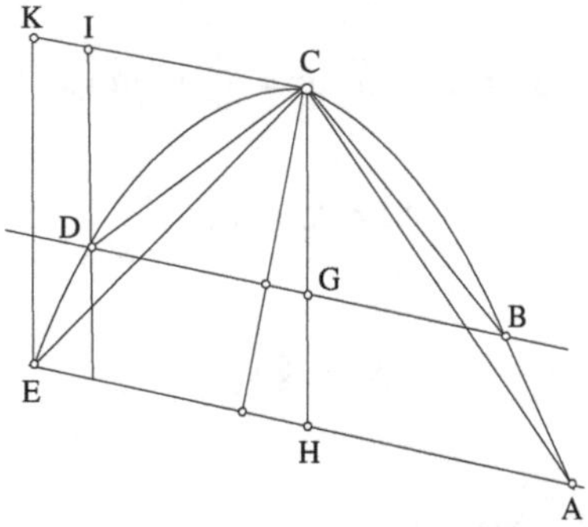

Fig. 3.4

In fact, by Proposition 3, we have

$$S_1 = \tfrac{4}{3} \text{ tr. } (ACE) = \tfrac{4}{3} \text{ area } (HCKE),$$
$$S_2 = \tfrac{4}{3} \text{ tr. } (BCD) = \tfrac{4}{3} \text{ area } (GCID);$$

hence

$$\frac{S_1}{S_2} = \frac{\text{area } (HCKE)}{\text{area } (GCID)} = \frac{CH \cdot HE}{CG \cdot GD},$$

as $I\hat{D}G = K\hat{E}H$. But

$$\frac{HE^2}{GD^2} = \frac{CH}{CG};$$

hence

$$\frac{S_1}{S_2} = \frac{CH}{CG} \cdot \sqrt{\frac{CH}{CG}}.$$

Comment: If we put $CH = x_1$, $CG = x_2$, we get

$$\frac{S_1}{S_2} = \left(\frac{x_1}{x_2}\right)^{\frac{3}{2}}.$$

If we call h_1 and h_2 the distances from C to the chords AE and BD, then h_1 and h_2 are the heights drawn from C in the triangles ACE and BCD and we have

$$\frac{S_1}{S_2} = \left(\frac{h_1}{h_2}\right)^{\frac{3}{2}},$$

as

$$\frac{h_1}{h_2} = \frac{x_1}{x_2}.$$

The affine transformation T defined by Ibn Sinān in Proposition 1 and characterized by the two ratios λ and μ associates with a portion ABC of the parabola of diameter BD and whose *latus rectum* relative to the diameter is a, a portion $A'B'C'$ of the parabola of diameter $B'D'$ such that $B'D' = \lambda \cdot BD$ and of base $A'C'$ such that $A'C' = \mu \cdot AC$. The *latus rectum* a' relative to the diameter $B'D'$ is then

$$a' = a \cdot \frac{\mu^2}{\lambda}.$$

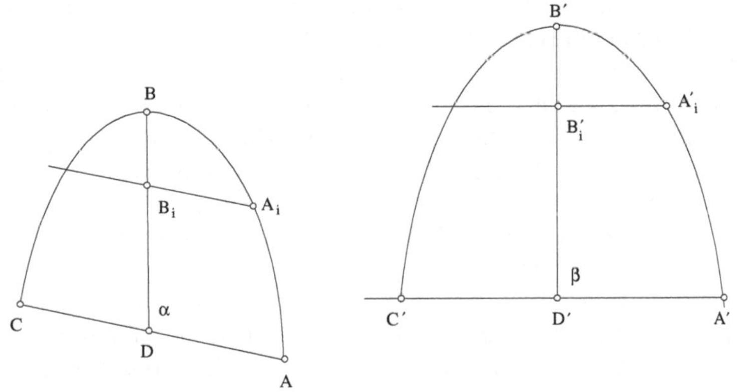

Fig. 3.5

In fact, we have

$$B'B'_i = \lambda \cdot BB_i \text{ or } x'_i = \lambda\, x_i,$$
$$B'A'_i = \mu \cdot B_iA_i \text{ or } y'_i = \mu y_i.$$

By hypothesis, we have

$$y_i^2 = ax_i.$$

Hence

$$\frac{y_i'^2}{\mu^2} = a\frac{x_i'}{\lambda},$$

and these would be the angles α and β.

Reciprocally, if two parabolic sections ABC and $A'B'C'$ are given, there exists a transformation T such that $(A'B'C') = T\,(ABC)$.

The affine transformation T becomes a similarity of ratio λ if $\alpha = \beta$ and $\lambda = \mu$; we then have $a' = a\lambda$, $D'B' = \lambda \cdot DB$, $A'C' = \lambda \cdot AC$.

Reciprocally, if two parabolic sections (ABC) and $(A'B'C')$ satisfy

$$\alpha = \beta,\ D'B' = \lambda \cdot DB,\ a' = \lambda a,$$

then they correspond by a similarity of ratio

$$\lambda\ (D'B' = \lambda \cdot DB \Rightarrow A'C' = \lambda \cdot AC).$$

Thus in the course of his treatise Ibn Sinān introduces in the study of the area of the parabola the notion of an affine transformation at the same time as that of infinitesimal procedures. The different stages of Ibn Sinān's approach in this treatise are thus articulated in the following manner:

• In Proposition 1, he shows that the affine transformation T conserves the ratio of areas in the case of triangles and polygons.

• He then shows in Proposition 2 that the same holds when one turns to the ratio of the area of a portion of a parabola to that of its associated triangle and to the ratio of their homologues. The subjacency property is in fact the conservation of ratios of areas (even curvilinear) by every affine transformation. The mathematical perspective of the epoch, however, did not lead him to consider general classes of curves and Ibn Sinān states this property only for polygons and parabolic sections.

For this, he uses Proposition X.1 of the *Elements* or, if one prefers, the Lemma of Archimedes, to show that it is possible to inscribe in the portion of the parabola a polygon whose area differs as little as one likes from that of the parabola.

• Having shown this, the calculation of the ratio of the area of a portion of a parabola to that of the associated triangle did not require any further infinitesimal treatment, but only the fact that the ratio does not depend on the portion considered (as was precisely established by Ibn Sinān).

This strategy of Ibn Sinān's, based on the combination of affine transformations and infinitesimal methods succeeded in reducing the number of lemmas to just two.

3.3. *Translated texts*

Ibrāhīm ibn Sinān

3.3.1. *On the Measurement of the Parabola*

3.3.2. *On the Measurement of a Portion of the Parabola*

In the name of God, the Merciful, the Compassionate

BOOK OF IBRĀHĪM IBN SINĀN

On the Measurement of the Parabola[1]

– **1** – Consider two straight lines *AB* and *CD*, and if they are divided into an arbitrary number of parts at the points *E*, *G*, *H* and *I* such that the ratios of the straight lines *BG*, *GE*, *EA* are equal to the ratios of the straight lines *DI*, *IH*, *CH*, if the parallel lines *BN*, *GL*, *EK* and the parallel lines *DS*, *IM*, *HJ* are drawn also such that the ratio of *BN* to *GL* is equal to the ratio of *DS* to *MI* and the ratio of *GL* to *EK* equal to the ratio of *IM* to *HJ*, and if the straight lines *AN*, *AK*, *LK*, *LN*, *CS*, *CJ*, *JM*, *SM*, are joined, then the ratio of the triangle *BNA* to the triangle *DSC* is equal to the ratio of the polygon *AKLNB* to the polygon *CJMSD*.

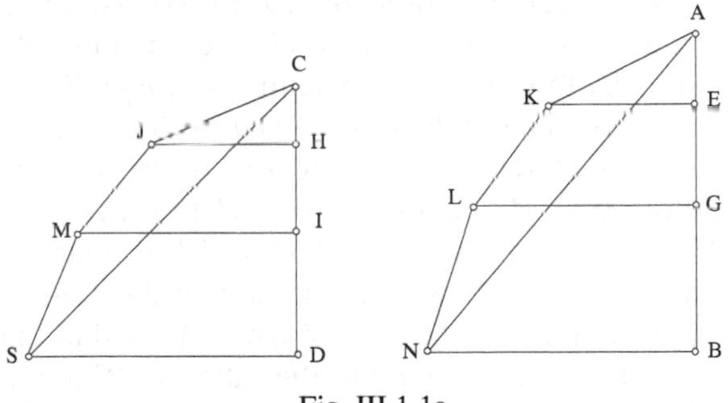

Fig. III.1.1a

Proof: The straight lines *BN*, *GL* and *EK* are either perpendicular to the straight line *BA* or are not thus. If they are perpendicular, we use the straight lines *BG*, *GE* and *EA* as the perpendiculars to the parallel straight lines, otherwise we produce from the point *A* a perpendicular *AO* to *NB*, on

[1] We find in the margin of the manuscripts: 'He wrote the following in his introduction: Abū Isḥāq Ibrāhīm ibn Sinān ibn Thābit originally wrote this book, he then mentioned that he had lost it, and finally he wrote another book and mentioned this copy in the introduction to this new treatise.'

which it lies at *O*, from the point *E* the perpendicular *EX* to the straight line *GL*, and from the point *G* the perpendicular *GQ* to the straight line *NB*. Likewise, if the straight lines *HJ*, *IM*, *DS* are perpendicular to the straight line *CD*, we use the straight lines *DI*, *IH*, *HC* as the perpendiculars to the parallel straight lines, otherwise we produce from the points *C*, *H*, *I* the homologous perpendiculars to those produced in the other figure: *CF* on *DS*, *HT* on *MI*, *IV* on *SD*. We extend *KE* to *AO*; it then meets it at the <point> *R* and it is perpendicular to it because it is perpendicular to a straight line parallel to it. Likewise, we extend *HJ* to *CF*; it then meets it at the point *Z* and it will also be perpendicular to the straight line *CF*. The triangle *ANB* is half of the product of *AO* and *NB* and the triangle *AEK* is half of the product of *AR* and *EK*; thus the ratio of the triangle *ANB* to the triangle *AEK* is equal to the ratio of half the product of *AO* and *NB* to half the product of *AR* and *EK* and is equal to the ratio of the product of *AO* and *NB* to the product of *AR* and *EK*. But this ratio is compounded of the ratio of *AO* to *AR* and of the ratio of *BN* to *EK*, and the ratio of *AO* to *AR* is equal to the ratio of *BA* to *EA* as *ER* is parallel to *BO*; thus the ratio of the triangle *ABN* to the triangle *AEK* is compounded of the ratio of *BN* to *EK* and of the ratio of *BA* to *AE*. In the same way, we also show that the ratio of the triangle *CSD* to the triangle *CHJ* is compounded of the ratio of *DS* to *HJ* and of the ratio of *DC* to *CH*. But since the ratio of *BG* to *GE* is equal to the ratio of *DI* to *IH*, the ratio of *BE* to *EG* is, by composition (*componendo*), equal to the ratio of *DH* to *HI*. But the ratio of *GE* to *EA* is equal to the ratio of *IH* to *CH*. By the equality (*ex aequali*), it follows that the ratio of *BE* to *EA* is equal to the ratio of *DH* to *CH*; by composition, the ratio of *BA* to *AE* is equal to the ratio of *DC* to *CH*, the ratio of *BN* to *GL* is equal to the ratio of *DS* to *MI* and the ratio of *GL* to *EK* is equal to the ratio of *MI* to *HJ*. By the equality, the ratio of *BN* to *EK* is equal to the ratio of *DS* to *HJ*. But since we have shown that thc ratio of the triangle *ABN* to the triangle *AEK* is compounded of the ratio of *BA* to *AE* and of the ratio of *BN* to *EK*, and that these ratios are equal to the ratio of *DC* to *CH* and to the ratio of *DS* to *HJ*, as we have shown, the ratio of the triangle *ABN* to the triangle *AEK* is compounded of the ratio of *DC* to *CH* and of the ratio of *DS* to *HJ*, and of these two ratios is compounded a ratio equal to the ratio of the triangle *CSD* to the triangle *CHJ* as we have shown. Thus the ratio of the triangle *ABN* to the triangle *AEK* is equal to the ratio of the triangle *CSD* to the triangle *CHJ*. By permutation, the ratio of the triangle *ABN* to the triangle *CSD* is thus equal to the ratio of the triangle *AEK* to the triangle *CHJ*.

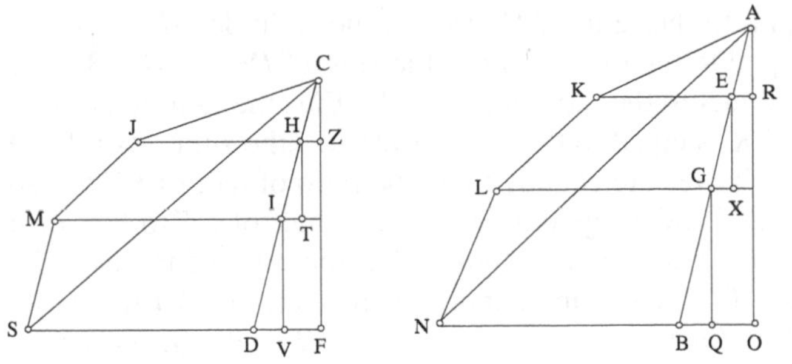

Fig. III.1.1b

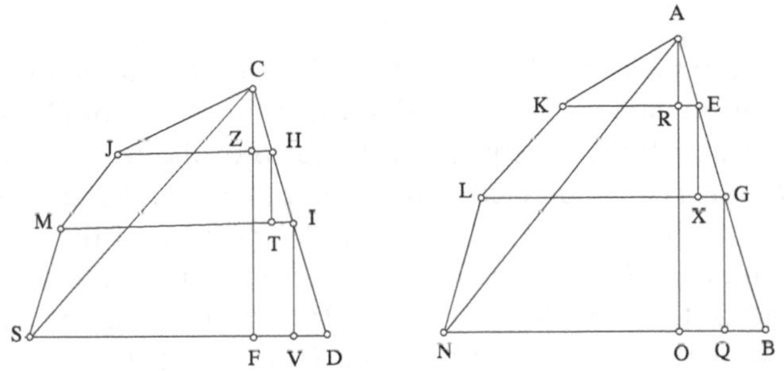

Fg. III.1.1c

Moreover, the trapezium *EKLG* contains the two parallel straight lines *GL* and *EK*; it is thus equal to the product of the half-sum of the two straight lines *GL*, *EK* and *EX*, which is perpendicular to them. Thus the ratio of the triangle *ABN* to the trapezium *EKLG* is equal to the ratio of the product of the perpendicular *AO* and half of the straight line *BN* to the product of *EX* and the half-<sum> of the two straight lines *GL* and *EK*. The ratio of the triangle *ABN* to the trapezium *EGLK* is consequently compounded of the ratio of *AO* to *EX* and of the ratio of half of the straight line *BN* to the half-<sum> of the straight lines *EK* and *GL*. But the ratio of *AO* to *EX* is equal to the ratio of *AB* to *EG* from the fact that the triangle *EXG* is similar to the triangle *AOB*, since the straight line *AO* is parallel to the straight line *EX*, as they are perpendicular to two parallel straight lines. But the straight line *BO* is parallel to the straight line *GX* and the straight line *AB* is on the extension of the straight line *EG*; thus the ratio of the triangle *ABN* to the trapezium *EGLK* is compounded of the ratio of *AB* to *EG* and of the ratio of half of *BN* to the half-sum of *GL* and *EK*. In the same way, we show that the ratio of the triangle *CDS* to the trapezium *HJMI* is compounded of the ratio of *CD* to *HI* and of the ratio of half of *DS* to the half-<sum> of the straight lines *IM* and *HJ*. But since the ratio of *GL* to *EK* is equal to the ratio of *IM* to *HJ*, the ratio de *GL* to the sum of *LG* and *EK* is equal to the ratio of *IM* to the sum of *IM* and *HJ*. But the ratio of *BN* to *GL* is equal to the ratio of *DS* to *IM*; thus the ratio of *BN* to <the sum of> *GL* and *EK* is equal to the ratio of *DS* to <the sum of> *IM* and *HJ*; the ratios of their halves are equally thus: the ratio of half of *BN* to half of <the sum of> *GL* and *EK* is equal to the ratio of half of *DS* to half of <the sum of> *IM* and *HJ*. Yet, we have shown that the ratio of the triangle *ABN* to the trapezium *EGLK* is compounded of the ratio of *AB* to *EG* and of the ratio of half of *BN* to half of <the sum of> *KE* and *LG*. As for the ratio of *AB* to *EG*, it is equal to the ratio of *DC* to *IH*, as the ratio of *AB* to *AE* is equal to the ratio of *DC* to *CH*, as we have shown, and the ratio of *EA* to *EG* is equal to the ratio of *CH* to *HI*; thus, by the equality, the ratio of *AB* to *EG* is equal to the ratio of *CD* to *HI*; as for the ratio of half of *BN* to the half-<sum> of *EK* and *GL*, it is equal to the ratio of half of *DS* to the half-<sum> of *HJ* and *IM*. Consequently, the ratio of the triangle *ABN* to the trapezium *EGLK* is compounded of the ratio of *CD* to *HI* and of the ratio of half of *DS* to the half-<sum> of *HJ* and *IM*. The ratio of the triangle *CSD* to the trapezium *HJMI* is compounded of these two ratios, as we have said; thus the ratio of the triangle *ABN* to the trapezium *EGLK* is equal to the ratio of the triangle *CDS* to the trapezium *HIMJ*. By permutation, the ratio of the triangle *ABN* to the triangle *CSD* is equal to the ratio of the trapezium *EGLK* to the trapezium *HJMI*. Likewise, we show that it is also equal to the ratio of the trapezium *BNLG* to the trapezium *SDIM*; yet the ratio is also equal to the ratio of the triangle *AEK* to the triangle *CHJ*. The ratio of each polygon to its associated one is thus equal to the ratio of all to all. The ratio of the triangle *ABN* to the triangle *CSD* is thus equal to the ratio of the sum of the triangle *AEK*, the trapezium *EGLK* and the trapezium *GBNL* to the sum of the triangle *CHJ*, the trapezium *HJMI* and the trapezium *SMID*; this is equal to the ratio of the polygon *AKLNB* to the polygon *CJMSD*. That is what we wanted to prove.

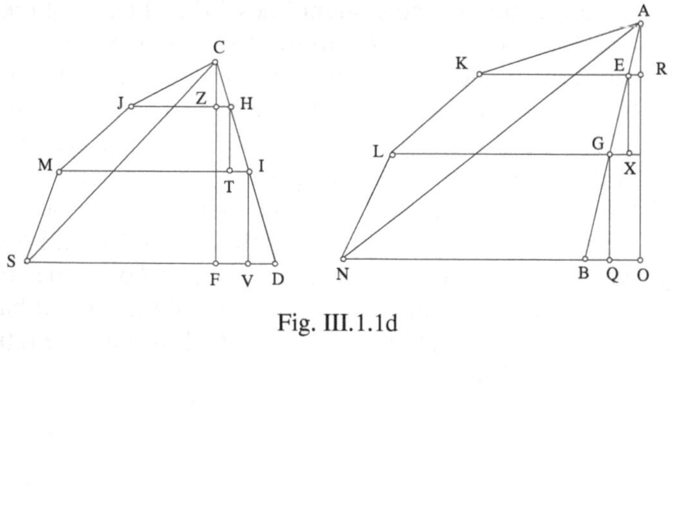

Fig. III.1.1d

We must show this whether only the angle *B* is a right angle or each of the two angles *B* and *D* is, so the proof of this case is similar to that one, as we use the ratio between the straight line *AB* and each of its parts in place of using the ratio between the perpendicular *AO* and the perpendicular *EX* or the perpendicular *GQ*. Likewise, we use in place of the product of *AO* and *BN*, the product of *AB* and *BN* and in place of the product of *EX* and half <the sum> of *KE* and *LG*, the product of *EG* and half <the sum> of *KE* and *LG*; likewise for the polygon *CJMSD*.

– **2** – For two portions of a parabola, the ratio of the one to the other is equal to the ratio of the triangle whose base is the base of the first and whose vertex is its vertex, to the triangle whose base is the base of the other and whose vertex is its vertex.

Let *ABCD* be a parabola; we cut it into two portions *AB* and *CD*, we divide the straight lines *AB* and *CD* in two halves at the points *E* and *G* and we make two diameters *EH* and *GI* pass through these points, which meet the parabola at the points *H* and *I*. We join *AH*, *HB*, *CI* and *ID*.

I say that the ratio of the portion AHB *of the parabola to the portion* CID *of the parabola is equal to the ratio of the triangle* AHB *to the triangle* CID.

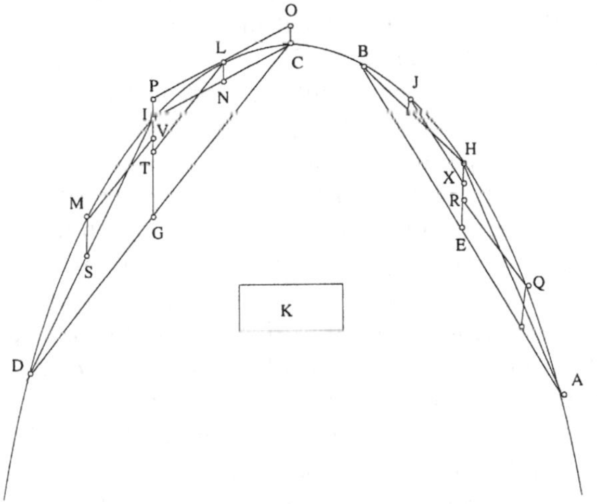

Fig. III.1.2

If it is not thus, let the ratio of the triangle *AHB* to the triangle *CID* be equal to the ratio of the portion *AHB* of the parabola to a smaller surface than the portion *CID* of the parabola, let that surface be *K*. <The sum> of the two portions bounded by the straight line *CI* and the portion *CI* of the line of the parabola, and by the straight line *DI* and the corresponding portion *ID* of the line of the parabola, is either greater than the excess of the portion *CID* of the parabola over the surface *K*, or is not greater than this excess.

Let it be first not greater than this excess; there remains the triangle *CDI* which is not smaller than the surface *K*; therefore the ratio of the portion *AHB* of the parabola to the surface *K* is not smaller than the ratio of the portion *AHB* to the triangle *CDI*. But the ratio of the portion *AHB* to the surface *K* is equal to the ratio of the triangle *AHB* to the triangle *CDI*; thus the ratio of the triangle *AHB* to the triangle *CDI* is not smaller than the ratio of the portion *AHB* to the triangle *CDI*; this is impossible, as the triangle *ABH* is smaller than the portion *AHB* of the parabola.

Now, let the sum of the two portions *CLI* and *IMD* of the parabola be greater than the excess of the portion *CID* of the parabola over the surface *K* which exceeds the triangle *CDI*. We divide the two straight lines *CI* and *ID* in two halves at the points *N* and *S*, making two diameters parallel to the straight line *GI* pass through these points, since the diameters of this parabola are parallel; these diameters are *NL* and *SM*. We join the straight lines *CL*, *LI*, *IM* and *MD*, and we make a straight line parallel to the straight line *CNI* pass through the point *L*; it is thus tangent to the parabola, as was shown in the book of *Conics*,[2] let the straight line be *OLP*; we produce to it the diameter *GI* that it meets at *P*, and from the point *C* a diameter parallel to the diameter *GI*, let it be *CO*. The surface *COPI* is a parallelogram circumscribed about the portion *CIL* of the parabola; it is thus greater than it, its half is thus greater than its half. The triangle *CLI*, which is half of the parallelogram *COPI*, is thus greater than half of the portion *CLI* of the parabola. Likewise, we show that the triangle *IMD* is greater than half the portion of the parabola in which it is inscribed. If we proceed in the same manner on the portions of the parabola which are on the straight lines *CL*, *LI*, *IM* and *MD*, that is, we separate from each of them <a surface> greater than its half, we then obtain a remainder of the portion *CID* smaller than the excess of the portion *CID* over the surface *K*. Let the remainder be the portions of a parabola *CL*, *LI*, *IM*, *MD*, so the polygon *CDMIL* will be greater than the surface *K*. But since the diameter *IG* cut the straight line *CD* in two halves, the straight line *CD* is an ordinate. We then produce from the points *L* and *M* two straight lines parallel to it, that is, ordinates to the diameter *GI*, let them be *LT* and *MV*,

[2] Apollonius, I.17.

which meet the diameter at the points *T* and *V*.[3] We divide the straight line *EH* which is on the diameter *EH* following the ratios of the parts of the straight line *IG* at the points *X* and *R* so that the ratio of the straight line *HX* to *EH* should be equal to the ratio of *IV* to *IG* and that the ratio of *RH* to *EH* should be equal to the ratio of *TI* to *IG*. We produce from the points *X* and *R* two ordinates from the diameter *EH*, that is, parallel to the straight line *AB*, as the straight line *AB* is also an ordinate since the diameter *EH* cuts it in two halves. Let us produce the two straight lines *XJ* and *RQ* in two different directions; they fall on the parabola at the points *J* and *Q*; we join *AQ*, *QH*, *HJ*, *JB*. But since the diameter of the parabola is *EH* and since it had been cut by the ordinates which are *AE* and *QR*, the ratio of the square of *AE* to the square of *QR* is equal to the ratio of *EH* to the straight line *RH*; this did part of what was shown in the book of *Conics*.[4] Likewise, the ratio of the square of *DG*[5] to the square of *LT* will be equal to the ratio of the straight line *GI* to the straight line *IT*. But since the ratio of the straight line *EH* to the straight line *RH* is equal to the ratio of the straight line *GI* to the straight line *IT*, the ratio of the square of the straight line *AE* to the square of the straight line *RQ* is equal to the ratio of the square of the straight line *GD* to the square of *LT*. But these straight lines are also proportional in length. Yet it has been shown in the previous proposition that if one has two straight lines *EH* and *GI* and that the straight line *EH* has been divided at the point *R* and the straight line *GI* at the point *T*, such that the ratio of *ER* to *RH* is equal to the ratio of *GT* to *TI*, if we produce two parallel straight lines *AE* and *RQ* and two parallel straight lines *GD* and *LT*, such that the ratio of *AE* to *RQ* is equal to the ratio of *GD* to *LT*, and if we join the straight lines, then the ratio of the triangle *AEH* to the triangle *CIG*[6] is equal to the ratio of the polygon *AEHQ* to the polygon *CLIG*. In the same manner, we show that the ratio of the triangle *EBH* to the triangle *GID* is equal to the ratio of the polygon *EHJB* to the polygon *IMDG*. But the ratio of the triangle *EAH* to the triangle *ICG* is equal to the ratio of the triangle *EBH* to the triangle *GID*. Indeed, since the straight line *AE* is equal to the straight line *EB*, the triangle *AEH* must be equal to the triangle *EBH*; in the same way, the triangle *GCI* is equal to the triangle

[3] In this version, Ibn Sinān does not show that the points *T* and *V* coincide with a single point in the middle of the straight line *LM*. To the points *T* and *V* on *PG* he associates the points *X* and *R* on *EH*. He then separates each of the triangles *ABH* and *CDI* and each of the polygons *ABJHQ* and *CDMIL* into two parts in order to apply the result of proposition 1.

[4] Apollonius, I.20.

[5] We know that $DG = GC$.

[6] See the previous note.

GID. That is why the ratio of the polygon *HJBE* to the polygon *GDMI* is equal to the ratio of the polygon *AEHQ* to the polygon *CLIG*. The ratio of the polygon *AEHQ* to the polygon *CLIG* is equal to the ratio of the polygon *ABJHQ* to the polygon *CDMIL*. But the ratio of the polygon *AEHQ* to the polygon *CLIG* is equal to the ratio of the triangle *EHA* to the triangle *CGI*, and is equal to the ratio of the multiples of these triangles. The ratio of the triangle *ABH* to the triangle *DIC* is thus equal to the ratio of the polygon *BAQHJ* to the polygon *DCLIM*. But we have stated that the ratio of the portion *AHB* of the parabola to the surface *K* is equal to the ratio of the triangle *AHB* to the triangle *DIC* and we have shown that the surface *K* is smaller than the polygon *DCLIM*; thus the ratio of the portion *AHB* of the parabola to the surface *K* is greater than its ratio to the polygon *DCLIM*. But its ratio to the surface *K* is equal to the ratio of the triangle *HAB* to the triangle *DIC*, as we have stated; thus the ratio of the triangle *HAB* to the triangle *DIC* is greater than the ratio of the portion *AHB* to the polygon *DCLIM*. But the ratio of the portion *AHB* to the polygon *DCLIM* is greater than the ratio of the polygon *ABJHQ* to the polygon *DCLIM*; thus the ratio of the triangle *AHB* to the triangle *DIC* is much even greater than the ratio of the polygon *ABJHQ* to the polygon *DCLIM*. But we have shown that the ratio of the triangle *ABH* to the triangle *DCI* is equal to the ratio of the polygon *ABJHQ* to the polygon *DCLIM*, which is impossible. It is thus not possible that the ratio of the triangle *ABH* to the triangle *DCI* should be equal to the ratio of the portion *ABH* to a smaller figure than the portion *DIC*.

Were it possible for there to be a surface greater than it, then the ratio of the triangle *DIC* to the triangle *ABH* would be equal to the ratio of the portion *DIC* to a smaller surface than the portion *ABH*. This is contradictory and not possible.

The ratio of the triangle *ABH* to the triangle *CID* is thus not equal to the ratio of the portion *ABH* to a surface which is neither smaller nor larger than the portion *ICD*, the ratio of the triangle *ABH* to the triangle *ICD* is therefore equal to the ratio of the portion bounded by the straight line *AB* and by a section *AB* of the line of the parabola to the portion bounded by the straight line *CD* and the line *CD* of the parabola. In the same way, for two arbitrary portions of a parabola, the ratio of the one to the other is equal to the ratio of the triangle of the same base and vertex to the triangle in the other. That is what we wanted to prove.

– **3** – Every portion of a parabola is equal to two thirds the parallelogram with the same base and the same height, and is equal to one and one third times the triangle of the same base and vertex.

Let *TAV* be a parabola, and let it be cut by an arbitrary straight line, that is straight line *BC*, which separates the portion *BAC*. Let the straight line *BC* be divided into two halves at the <point> *D*. Let us produce from the point *D* a diameter *DA* for the portion. We join *AB* and *AC* and make a straight line parallel to the straight line *BC* passing through point *A*, that is the straight line *NAS*, and produce through the points *B* and *C* two diameters parallel to the diameter *AD*, which are *BN* and *CS*.

I say that the ratio of the portion BAC *of the parabola – on the one hand – to the parallelogram* NBCS *is equal to the ratio of four to six and – on the other hand – to the triangle* ABC *is equal to the ratio of four to three.*

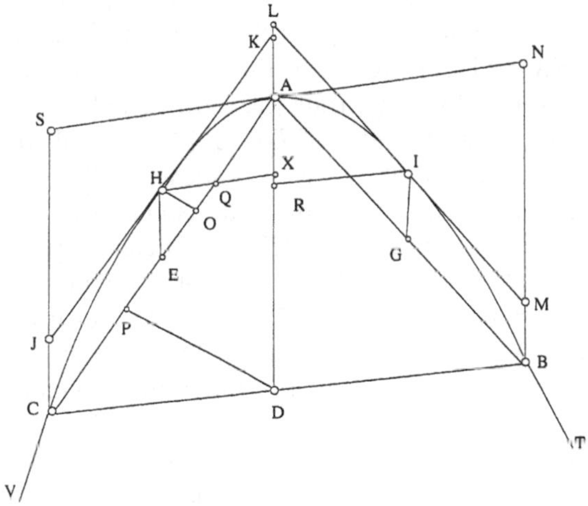

Fig. III.1.3

Proof: We divide each of the straight lines *AC* and *AB* into two halves at the points *E* and *G* and we make to pass through these two points two diameters which cut the parabola; that which passes through the point *G* cuts it at *I*, and the other cuts it at *H*. We produce from the points *I* and *H* two straight lines *IL* and *HK* tangent to the parabola; they meet the diameter *AD* at the points *K* and *L*. We extend *LI* so that it meets *BN* at the <point> *M* and the straight line *KH* so that it meets *SC* at the <point> *J*. We produce from the point *H* the ordinate straight line *HX* to the diameter *AD* and likewise the straight line *IR*.[7] We produce as well from the point *H* the perpendicular *HO* to *AC* and from the point *D* the perpendicular *DP* to *AC*, so that the straight line *HX* meets the straight line *AC* at <the point> *Q*. Given that the straight line *EH* is a diameter which cuts the straight line *CA* into two halves, then *CA* is an ordinate. But the straight line *HK* has been produced tangent to the parabola at the extremity of the diameter, it is thus parallel to the ordinates.[8] The straight line *AC* is thus parallel to the straight line *KJ* and, moreover, the straight line *KH* is tangent to the parabola. One has produced from the point of contact to the diameter *AD* an ordinate which is *HX* and the diameter *AD* met the straight line tangent at *K*; the straight line *KA* is thus equal to the straight line *AX*[9] and the ratio of *AK* to *AX* is equal to the ratio of *HQ* to *QX* since *AQ* is parallel to *KH*; thus the straight line *HQ* is equal to the straight line *QX*. Likewise, since the straight lines *HX* and *CD* are the ordinates of the diameter *AD*, *CD* is parallel to the straight line *XQ*; thus the ratio of *DA* to *AX* is equal to the ratio of *DC* to *XQ*. But the ratio of *DA* to *AX* is equal to the ratio of the square of *DC* to the square of *XH*, as was shown in the *Conics*.[10] The ratio of *DC* to *XQ* is thus equal to the ratio of the square of *DC* to the square of *XH*; for this reason, the straight line *HX* is in mean proportion between the two straight lines *DC* and *XQ*. The product of *DC* and *XQ* is thus equal to the square of *XH*, but the square of *XH* is equal to four times the product of *XQ* and *QH* since *XQ* is equal to *HQ*, as we have shown. Consequently, the product of *DC* and *XQ* is equal to four times the product of *HQ* and *XQ*; the straight line *HQ* is thus one quarter the straight line *DC*. But since the straight line *DC* is parallel to the straight line *HQ*, the straight line *DP* is a perpendicular, the straight line *HO* is a perpendicular and the straight line *QO* is the extension of the straight line *OP*, the triangle *DPC* is similar to the triangle *HOQ*; thus the ratio of *HO* to *PD* is equal to the ratio of *HQ* to *DC*. But *HQ* is one quarter of *CD*; thus *HO* is one quarter of *DP*. But the ratio of *HO* to *DP* is equal to the ratio of the product of the perpendicular *HO* and *AC* to the product of the perpendicular *DP* and the straight line *AC*. But this ratio is the ratio of the triangle *AHC* to the triangle *ADC*; the triangle *AHC* is thus one quarter of the triangle *ADC*, and it is thus one eighth of the triangle *ABC*, as the triangle *ABC* is the double of the triangle

[7] It is clear that, by their construction, the points *I* and *H* have equal ordinates, $\frac{DC}{2}$ and $\frac{DB}{2}$; they thus have the same abscissa, and we must thus have $X = R$. Moreover, the vertex *A* being the midpoint of the sub-tangent, we must also have $K = L$.

[8] Book I.17 (tangent at the vertex).

[9] I.33 and 35 (sub-tangent).

[10] Book I.20.

ADC, since the straight line *BC* is twice the straight line *CD*. Yet, it has been shown in the previous proposition that for two portions of a parabola, the ratio of the one to the other is equal to the ratio of the triangle whose base is the base of the first and whose vertex is its vertex, to the triangle which is its homologue in the other. Thus the portion *AHC* of the parabola is one eighth of the portion *BAC* of the parabola. In the same way, the portion *BIA* of the parabola is one eighth of the portion *BAC* of the parabola; the sum of the two portions is thus one quarter of the portion *BAC* and the triangle *BAC* which remains is three quarters of the portion *BAC*. The portion is thus equal to one and one third times the triangle. But the parallelogram *BNSC* is the double of the triangle *BAC*; thus the ratio of the portion *BAC* to the triangle *BAC* is equal to the ratio of four to three and, to the parallelogram *BNSC*, is equal to the ratio of four to six. That is what we wanted to prove.

Ibrāhīm ibn Sinān's book on the measurement of the parabola is completed.

In the name of God, the Merciful, the Compassionate

BOOK OF IBRĀHĪM IBN SINĀN IBN THĀBIT

On the Measurement of a Portion of the Parabola

Some time ago, I wrote a book on the area of this section. Later, I made a number of changes to one of the propositions. This corrected copy and the older copy have now been lost and I therefore need to repeat my earlier work in this book. If a copy where the terms differ from those of that copy, comes down, or if, in one of its parts, which contains a notion which differs from some of the notions from that copy, then it is the one of the two copies which I have evoked. My forebear Thābit ibn Qurra, as well as al-Māhānī, have composed writings on this subject.

– **1** – Consider a polygon *ABCDE* and a polygon *GHIJK* as well, if the straight lines *BL*, *CM*, *HN*, *IS* are drawn parallel to the straight line *DE* and to the straight line *JK*, such that the ratios of the straight lines *AL*, *LM*, *ME* are following the ratios of the straight lines *GN*, *NS*, *SK* and the ratios of the straight lines *BL*, *CM* and *DE* are following the ratios of the straight lines *HN*, *IS* and *JK*, and if *AD* and *JG* are joined, then the ratio of the triangle *ADE* to the triangle *JKG* is equal to the ratio of the polygon *ABCDE* to the polygon *GHIJK*.

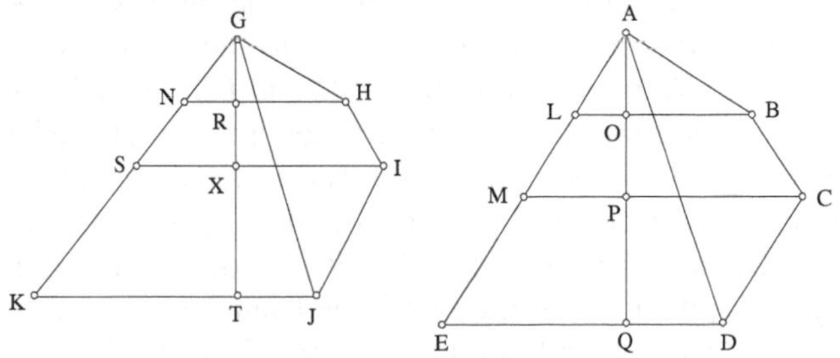

Fig. III.2.1

Proof: Let us produce the perpendicular *AOPQ* to the parallel straight lines *BL*, *CM* and *DE* and the perpendicular *GRXT* to the parallel straight lines *HN*, *IS* and *JK*, so the ratio of the triangle *ADE* to the trapezium *CDEM* is equal to the ratio of the product of *AQ* and half of *DE* to the product of *PQ* and half the sum[1] of *DE* and *CM*; in fact, their area is equal to the product of the straight lines that we have mentioned, the ones with the others. Consequently, the ratio of the triangle *ADE* to the trapezium *CMED* is compounded of the ratio of *AQ* to *QP* and the ratio of half of *DE* to half the sum of *DE* and *CM*.

Likewise, we show that the ratio of the triangle *GJK* to the trapezium *JKSI* is compounded of the ratio of *GT* to *TX* and of the ratio of half of *JK* to half the sum of *JK* and *IS*.

On the one hand, the ratio of *AQ* to *QP* is equal to the ratio of *AE* to *EM* by reason of the parallelism of the two straight lines *DE* and *CM* and is equal to the ratio of *GK* to *KS* – as we assumed from the start that the ratios of these straight lines are equal – and is equal to the ratio of *GT* to *TX*. On the other hand, the ratio of half of *DE* to half the sum of *DE* and *CM* is equal to the ratio of *DE* to the sum of *DE* and *CM*, and this ratio is equal to the ratio of *JK* to the sum of *JK* and *IS*, as they have been supposed thus by separation, and this ratio is equal to the ratio of half of *JK* to half the sum of *JK* and *IS*, the ratio of half of *DE* to half the sum of *DE* and *CM* is consequently equal to the ratio of half of *JK* to half the sum of *JK* and *IS*. The ratios from which a ratio equal to the ratio of the triangle *ADE* to the trapezium *CDEM* is compounded are consequently equal to the ratios from which a ratio equal to the ratio of the triangle *GJK* to the trapezium *JKSI* is compounded. That is why the ratio of the triangle *ADE* to the trapezium *DEMC* is equal to the ratio of the triangle *GJK* to the trapezium *JKSI*. Likewise, the ratio of the triangle *ADE* to the trapezium *BCML* is equal to the ratio of the triangle *GJK* to the trapezium *HNSI*; in fact, from the sides of the rectangles which are equal to them, a same ratio is compounded, as when we say: the ratio of *AQ* to *OP* is equal to the ratio of *GT* to *RX* and the ratio of half of *DE* to half the sum of *CM* and *BL* is equal to the ratio of half of *JK* to half the sum of *HN* and *IS*. Likewise, the ratio of the triangle *ADE* to the triangle *GJK* is equal to the ratio of the triangle *ABL* to the triangle *GHN*, as the ratio of the perpendicular *AQ* to *OA* is equal to the ratio of *GT* to *GR* and the ratio of *DE* to *BL* is equal to the ratio of *JK* to *HN*. Consequently, the ratio of the two large triangles is equal to the ratios of the trapeziums, each to its homologue. If we thus add them up, the ratio of the trapezium *CMED* to the trapezium *ISKJ* will be equal to the ratio of the polygon *ABCDE* to the polygon *GHIJK* and will be equal to the ratio of

[1] We add 'sum', throughout the text, for the sake of the translation.

the triangle *ADE* to the triangle *GJK*. Consequently, it has been shown by the proof what we sought.

– **2** – That having been proven, we show that for two portions of parabola, the ratio of the one to the other is equal to the ratio of the triangle whose base is its base[2] and whose vertex is its vertex, to the triangle constructed in the other, in the same way.

Let *ABC* be a portion of the parabola and *DEG* be a portion of the parabola[3] whose bases are *AC* and *DG*. Let us divide them into two halves at *H* and at *I*. Let *BH* and *EI* be the diameters of the two portions; let us join *ABC* and *DEG*.

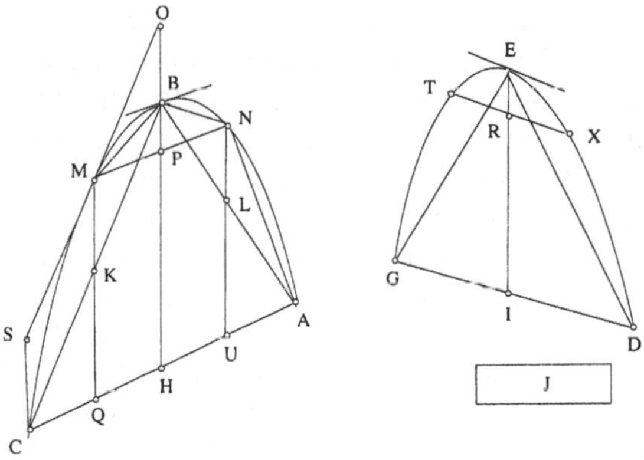

Fig. III.2.2

I say that what we mentioned is true. If it is false, let the ratio of the triangle *DEG* to the triangle *ABC* be equal to the ratio of the portion *DEG* to a surface smaller than the portion *ABC*, which is the surface *J*. We divide *BC* into two halves at the <point> *K* and *AB* into two halves at the <point> *L* and we produce *KM* and *LN*, two diameters parallel to the diameter *BH*, which fall on the points *M* and *N* of the parabola. We join *AN*, *NB*, *BM* and *MC*, so each of the triangles *ANB* and *BMC* is greater than half of the section in which it is inscribed; in fact, if we produce a straight line tangent to the parabola at the point *M*, as the straight line *SMO*, it will be parallel to the straight line *BKC* which is an ordinate to the diameter *MK*. If we produce the diameter *CS*, it will be parallel to the straight line *BH*. Let *HB* meet *MO* at *O*, so the triangle *BCM* is half of the

[2] The base of the first.

[3] In the three manuscripts, we have a figure with two different parabolas; the reasoning is valid for two sections of the same parabola.

parallelogram *BOCS*; yet the parallelogram is greater than the portion *BMCK*; its half, *i.e.* the triangle *BMC*, is greater than half of the portion.

We continue to divide the straight lines *AN*, *NB*, *MB*, *CM* and their homologues into two halves, to produce diameters through the midpoints and to join the straight lines that form triangles that are greater than half the portions in which they are inscribed, until there is a remainder smaller than the excess of the portion *ABC* over the surface *J*. Let the magnitude that remains be the portions *AN*, *NB*, *BM*, *MC*; thus the surface *AHCMBN* is greater than the surface *J*. The ratio of the triangle *DEG* to the triangle *ABC* is consequently equal to the ratio of the portion *DEG* to a surface smaller than the surface *ANBMCH*. We join *MN*; it meets the diameter *OH* at *P*; it will be its ordinate. In fact, we make it so that the diameter *MK* meets *HC* at *Q* and that the diameter *NL* meets *AH* at *U*. Since *AL* is equal to *LB* and that the diameter *LU* is parallel to the diameter *BH*, *AU* will be equal to *UH*. Likewise, *HQ* will be equal to *QC*. But *AH* is equal to *CH*, thus *HU* is equal to *HQ*; thus the straight line dropped ordinatewise from *M* to the diameter *BH* falls on the diameter *BH* and will be equal to *HQ*. Likewise, the produced ordinate from the point *N* is equal to *HQ*; thus the produced ordinate from *N* is equal to that produced from *M*; they thus fall on a single point, let <the point> be *P*.[4] We divide *EI* according to the ratio of *BP* to *BH* at the point *R*; we produce the ordinate *XRT* parallel to *DG* and we join *DX*, *XE*, *ET* and *TG*. Since the ratio of *HB* to *BP* is equal to the ratio of *EI* to *ER*, the ratio of the square of *DG* to the square of *TX* is equal to the ratio of the square of *AC* to the square of *MN*. In fact, Apollonius showed in the book on the *Conics* that the ratio of the square of the ordinates, in the parabola, is equal to the ratio of that which they separate from the diameter for which they are the ordinates. Consequently, the ratios of the straight lines *DG*, *XT* and *AC*, *MN* – in length – are equal. Consequently, the two straight lines *EI* and *BH* are divided at two points *R* and *P* in equal ratios, the parallels *DG* and *XT* are drawn and likewise *AC* and *MN*; the ratio of *DG* to *XT* is thus equal to the ratio of *AC* to *MN*.

The ratio of the triangle *DEG* to the triangle *ABC* is consequently equal to the ratio of the polygon *DXETG* to the polygon *ANBMC*, as we have shown in the first proposition; yet the ratio of the portion *DEG* to a surface smaller than *ANBMC* is equal to the ratio of the triangle *DEG* to the triangle *ABC*; consequently, the ratio of the polygon *DXETG* to the polygon *ANBMC* is equal to the ratio of the portion *DEG* to a surface smaller than the surface *ANBMC*; this is impossible, from an evident impossibility and a manifest absurdity, which cannot be, as the portion

[4] This is a consequence of Apollonius's *Conics* I.20, whose statement Ibn Sinān recalls in the paragraph that follows.

DEG is greater than *DXETG*. The ratio of the triangle *DEG* to the triangle *ABC* is thus not equal to the ratio of the portion *DEG* to a surface smaller than the portion *ABC*.

If this were possible, let it be equal to a surface larger than it; consequently, the ratio of the triangle *ABC* to the triangle *DEG* would be equal to the ratio of the portion *ABC* to a surface smaller than the portion *DEG*. We show that this is impossible, as was shown previously, which is the inverse of what we treat now. The ratio of the triangle *DEG* to the triangle *ABC* is consequently equal to the ratio of the portion *DEG* to the portion *ABC*. That is what we wanted to prove.

– **3** – *I say that the ratio of every portion of a parabola to the triangle of the same base and same height is equal to the ratio of four to three.*

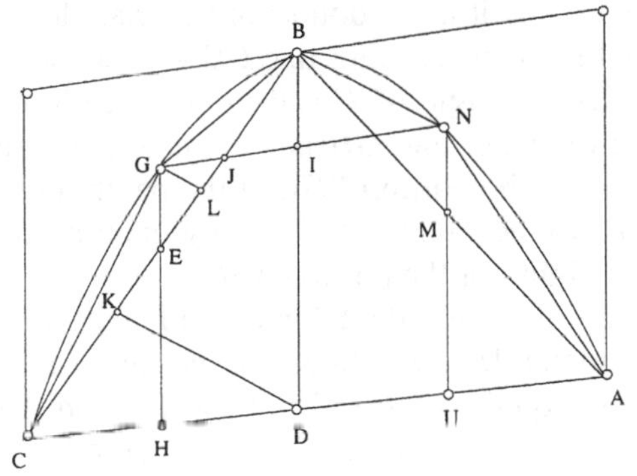

Fig. III.2.3

Proof: We consider the portion *ABC* of base *AC* whose midpoint is *D*, and with diameter *BD*; we draw the two straight lines *AB* and *BC*, we divide *BC* into two halves at the <point> *E* and we draw *GEH* parallel to *BD* which meets the parabola at *G*; we join *BG* and *GC*, we produce an ordinate straight line *GJI* which meets the diameter *BD* at *I* and the straight line *BC* at *J*. Since the ratio of *DC* to *IJ* is equal to the ratio of *DB* to *BI*, which is equal to the ratio of the square of *DC* to the square of *IG*, as had been shown for the ordinates in the book of *Conics*,[5] the straight line *IG* will be in mean proportion between *DC* and *IJ*, as the ratio of *DC* to *IJ* is equal to the ratio of the square of *DC* to the square of *IG*, as we have shown. But since *BE* is equal to *EC* and since the diameter *EH* is parallel to the diameter *BD*, *DH* is equal to *HC*; consequently, *DC* is twice *IG*, as it is

[5] Apollonius, I.20.

twice *DH* which is equal to *IG*, as *GIDH* is a parallelogram by reason of parallelism of the ordinate straight lines and of parallelism of the diameters in the parabola. But the ratio of *DC* to *IG* is equal to the ratio of *GI* to *IJ*, thus *GI* is the double of *IJ*; consequently, *IJ* is equal to *JG*; *DC* – which is the double of *GI* – is hence the quadruple of *JG*.

If we draw the perpendicular *DK* to *BC* and the perpendicular *GL* to *BC*, they will be parallel; but *DC* is parallel to *GJ* and the straight line *BC* fell on both of them; thus the angle *DKC* is equal to the angle *GLJ*, as *DKL* is equal to *GLK*, alternate angles, and the angle *DCK* is equal to the angle *GJL*, alternate angles; thus the two triangles *GJL* and *DKC* are similar, the ratio of *DC* to *GJ* is then equal to the ratio of *DK* to *GL*. Consequently, since *DC* is the quadruple of *GJ*, *DK* will be the quadruple of *GL*. The product of *DK* and half of *BC*, *i.e.* the triangle *BCD*, is consequently the quadruple of the product of *GL* and half of *BC*, *i.e.* the triangle *BGC*. The triangle *ABC* – given that it is the double of the triangle *BDC* as the straight line *AC* is the double of the straight line *CD* – is consequently eight times the triangle *BGC*. The triangle *BGC* is then one eighth of the triangle *ABC*. But since *BD* is a diameter and *GH* is a diameter, the ratio of the portion *ABC* of the parabola to the portion *BGC* of the parabola is equal to the ratio of the triangle *ABC* to the triangle *BGC*; consequently, the portion *BGC* of the parabola is one eighth of the portion *ABC*.

In the same way, if we divide *AB* into two halves at <the point> *M* and if we draw the diameter *MN*, we show that the ratio of the triangle *ABC* to the triangle *ANB* is equal to the ratio of the portion *ABC* to the portion *ANB*, we also show that the triangle *ANB* is one eighth of the triangle *ABC*; consequently, the portion *ANB* is one eighth of the portion *ABC*.

The sum of the two portions *ANB* and *BGC* is consequently one quarter of the portion *ABC*. If we set the portion *ABC* four, the sum of the two portions *ANB* and *BGC* will be one, and there remains the triangle *ABC*, three. Thus the ratio of the portion *ABC* to the triangle *ABC* is equal to the ratio of four to three. The ratio of every portion of a parabola to the triangle of the same base and same height is, consequently, equal to the ratio of four to three. That is what we wanted to prove.

– **4** – *I say that if two portions of a parabola have parallel bases, then the ratio of one to the other is equal to the ratio of the height of the one to the height of the other, multiplied by a ratio such that if it is multiplied by itself, it will be equal to the ratio of the height of the one to the height of the other.*

Let *ABCDE* be a portion of the parabola, with *AE* parallel to *BD* and *CGH* the diameter which cuts the straight lines *AE* and *BD* into two halves.

We draw a straight line parallel to *AE* and *BD*, which is *CI*;[6] we produce two straight lines *DI* and *EK* parallel to *CH*. The parallelogram *GDIC* is thus equal to the triangle whose base is *BD* and vertex is *C*, as *BD* is twice *DG*. Likewise, the parallelogram *HEKC* is equal to the triangle of base *AE* and vertex *C*. That is why the ratio of the portion *ACE* to the portion *BCD* is equal to the ratio of the parallelogram *KCHE* to the parallelogram *ICGD*. But this ratio – by the equality of the angles of these two parallelograms – is equal to the ratio of *HC* to *GC*, multiplied by the ratio of *HE* to *GD*. The ratio of the portion *ACE* to the portion *BCD* is thus equal to the ratio of *HC* to *GC*, multiplied by the ratio of *HE* to *GD*. Yet, it is clear that the ratio of *HE* to *GD*, if it is multiplied by itself, will be equal to the ratio of the square of *HE* to the square of *GD*, which is equal to the ratio of *CH* to *CG*. Consequently, the ratio of *HE* to *GD*, if it is multiplied by itself, is equal to the ratio of *HC* to *CG*. The ratio of the portion *ACE* to the portion *BCD* is consequently equal to the ratio of *HC* to *CG*, multiplied by a ratio such that if it is multiplied by itself, it is equal to the ratio of *CH* to *CG*.

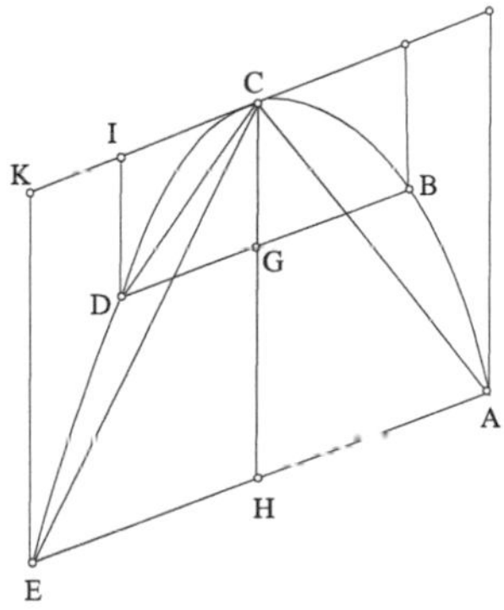

Fig. III.2.4

In the same way, we show that the same holds for two arbitrary portions[7] of a parabola; that is what we intended to prove.

Ibrāhīm ibn Sinān ibn Thābit's book
on the measurement of the parabola is completed.

[6] *CI* is thus the tangent at *C* to the parabola.
[7] to parallel bases.

CHAPTER IV

ABŪ JAʿFAR AL-KHĀZIN: ISOPERIMETRICS AND ISEPIPHANICS*

4.1. INTRODUCTION

4.1.1. *Al-Khāzin: his name, life and works*

For the historian of mathematics, the work of al-Khāzin counts in the results it incorporates and the domains it encompasses, but above all in its very significance in its own right. Algebra, geometry, number theory and astronomy are equally chapters in which al-Khāzin was inventive. But he represents more than any other of his generation a stream of the research of his era, that of mathematicians who had learnt to marry the Greek geometric heritage and the algebraic legacy of the ninth century, and so to advance the frontiers of the former in offering new extensions to the latter. If one is to believe the testimony of the expert on the material, al-Khayyām, al-Khāzin is the first to have successfully applied the conics to the solution of a cubic equation,[1] thereby openning the possibility of that chapter in the theory of algebraic equations (founded later by al-Khayyām). He is one of the first, along with al-Khujandī, who had conceived of the whole of Diophantine analysis.[2] In astronomy, his work is likewise, in the view of his critics, such as Ibn ʿIrāq[3] and al-Bīrūnī, among the most distinguished contributions of its time.

* Isoperimetric: as having equal perimeters. Isepiphanic: as having equal surface areas.

[1] R. Rashed and B. Vahabzadeh, *Al-Khayyām mathématicien*, Paris, 1999; English version (without the Arabic texts): *Omar Khayyam. The Mathematician*, Persian Heritage Series no. 40, New York, 2000.

[2] R. Rashed, 'L'analyse diophantienne au Xe siècle: l'exemple d'al-Khāzin', *Revue d'histoire des sciences*, 32, 1979, pp. 193–222.

[3] Ibn ʿIrāq, 'Taṣḥīḥ zīj al-ṣafā'iḥ', in *Rasā'il Mutafarriqa fī al-hay'a*, Hyderabad, 1948. It suffices to read the words of Ibn ʿIrāq to understand the prestige of al-Khāzin at the time:

وإن كان بعض الناس يعظم أن يستدرك على مثل أبي جعفر في تأليفاته سهو وقع له ...

Despite the range of his work, often pioneering, despite the recognition he received from his contemporaries and successors, the bibliographic sources are nearly mute on his life and works. Recent uncertainties surrounding his name have even resulted in presenting him in terms of a split persona: he has been despoiled of an important swath of his titles, which have been seen to glorify another author who never even existed. Let us start with him.

The name most cited currently is Abū Jaʿfar al-Khāzin. It is under this name that al-Nadīm called him in a brief entry (one line),[4] and incidentally cited him twice more.[5] It is also under this name that he is mentioned by al-Qifṭī,[6] and that his successors referred to him – Ibn ʿIrāq, al-Bīrūnī, al-Khayyām, among others. One notes, however, three interesting variants. The first is that of the contemporary historian of al-Khāzin, Abū Naṣr al-ʿUtbī, who tells us a story is relegated from ʿAbū al-Ḥusayn Jaʿfar ibn Muḥammad al-Khāzin'.[7] The second variant, less important, is from the hand of al-Nadīm, who one time adds al-Khurāsānī – from Khurāsān,[8] to indicate his residence. The third, later because it is due to al-Samaw'al, and thus from the twelth century, gives ʿAbū Jaʿfar Muḥammad ibn al-Ḥusayn al-Khāzin'.[9] Yet this name, noted by al-Samaw'al, is that found in several books that have come to us from al-Khāzin.[10] The name reported by Abū Naṣr al-ʿUtbī is in fact the same, to two close inversions. But, whereas past mathematicians and historians have never entertained any doubt about the person's identity, one thought oneself able to subscribe, after F. Woepcke[11] to the existence of two mathematicians: Abū Jaʿfar al-Khāzin and Abū

See also the mentions made by al-Bīrūnī of al-Khāzin in *Taḥdīd nihāyāt al-amākin* (*vide infra*, Note 15).

[4] Al-Nadīm, *Kitāb al-Fihrist*, ed. R. Tajaddud, Tehran, 1971, p. 341.

[5] *Ibid*, pp. 153 and 311.

[6] Al-Qifṭī, *Ta'rīkh al-ḥukamā'*, ed. J. Lippert, Leipzig, 1903, p. 396.

[7] *Tārīkh Abī Naṣr al-ʿUtbī*, in the margin of *Sharḥ al-Yamīnī al-musammā bi-al-Fatḥ al-Wahbī al-Manīnī,* Cairo 1286/1870, vol. I, p. 56.

[8] Al-Nadīm, *al-Fihrist*, p. 325.

[9] Al-Samaw'al, *Fī kashf ʿuwār al-munajjimīn*, MS Leiden 98.

[10] Cf. *Mukhtaṣar mustakhraj min kitāb al-Makhrūṭāt bi-iṣlāḥ Abī Jaʿfar Muḥammad ibn al-Ḥusayn al-Khāzin*, MS Oxford, Bodleian, Huntington 539. Cf. another copy of this text, with the same authorial name, MS Alger, B.N. 1446, fol. 125^r. One has, as well, a commentary on Euclid's book X by 'Abū Jaʿfar Muḥammad ibn al-Ḥusayn al-Khāzin', MS Istanbul, Feyzullah 1359, fol. 245^r; as well as the version Tunis B.N. 16167, fol. 65^v. Finally, the tract we consider here is also under this name.

[11] F. Woepcke, 'Recherches sur plusieurs ouvrages de Léonard de Pise', *Atti Nuovi Lincei*, 14, 1861, pp. 301–24.

Ja'far Muḥammad ibn al-Ḥusayn. A. Anbouba has recently been able to show that it pertains well to the same and single person.[12]

This confusion for once dissipated and the figure's identity established, we are not however better informed about his dates or his works. Let us address ourselves to historians as well as mathematicians: already al-Nadīm[13] tells us that al-Kindī's student, the *littérateur* and philosopher Abū Zayd al-Balkhī, addressed to al-Khāzin his commentary on Aristotle's *De Cœlo*. Yet we know that al-Balkhī[14] died in 322/934. This gives us a first reference point: al-Khāzin had to have been born at the end of the third century of the Hegira, or thereabouts.

On the other hand, according to al-Bīrūnī, al-Khāzin attended the observation by mathematician and astronomer al-Harawī 'of the altitude of the sun at noon on the 12th Wednesday of Rabī' the second, the year three hundred forty-eight of the Hegira',[15] which would indicate he was still active in 959 at least. Better yet, the historian al-'Utbī[16] reports a story told by al-Khāzin, on the subject of the arrival of Sebüktijīn in Bukhārā in the period of Samanide Manṣūr ibn Nūḥ, which is to say around the middle of the third century of the Hegira. In the seventeenth century, the commentator of the story of al-'Utbī, al-Manīnī, wrote that al-Khāzin was one of the ministers of the Samanides,[17] which we cannot confirm, but which suggests that al-Khāzin maintained ties with those who were in power. This version fits with the data we have from the historian Ibn al-Athīr, and from the *littérateur* al-Ṭawḥīdī. According to the former,[18] al-Khāzin was the envoy of the leader – 'Alī ibn Muḥtāj – of the expedition of Prince Samanide Nūḥ ibn Naṣr in 342/953 against the Būyid Rukn al-Dawla, to negotiate a halt to the combat. Here is what he wrote: '[...] and the envoy was Abū Ja'far al-Khāzin, author of the *Zīj al-Ṣafā'iḥ*, and who was learned in mathematics'. This testimony shows clearly that al-Khāzin was at this time a mature man, enjoying both the confidence of the Prince and a solid scientific reputation.

[12] A. Anbouba, 'L'algèbre arabe aux IXe et Xe siècles: Aperçu général', *Journal for the History of Arabic Science*, 2, 1978, pp. 66–100.

[13] Al-Nadīm, *al-Fihrist*, pp. 153 and 311.

[14] Yāqūt, *Kitāb irshād al-arīb ilā ma'rifat al-adīb* (*Mu'jam al-udabā'*), ed. D. S. Margoliouth, London, 1926, vol. VII, pp. 141, 150–1.

[15] Al-Bīrūnī, 'Kitāb taḥdīd nihāyāt al-amākin li-taṣḥīḥ masāfāt al-masākin', edited by P. Bulgakov and revised by Imām Ibrāhīm Aḥmad, in *Majallat Ma'had al-Makhṭūṭāt,* 8, 1962, p. 98.

[16] Al-'Utbī, vol. I, p. 56.

[17] *Ibid.*

[18] Ibn al-Athīr, *Al-Kāmil fī al-tā'rīkh* photographed edn., Beirut, 1979, from that of Carolus Johannes Tornberg, Leiden, 1862, under the title *Ibn-El-Athiri Chronicon quod perfectissimum inscribitur*, vol. 8 (see *The events of the year*, 342).

Al-Tawḥīdī confirms this portrait of al-Khāzin on all counts when he later depicts him in the concurrent state, that of Būyids: he is in the court of Rukn al-Dawla, protégé of the famous minister Ibn al-ʿAmīd.[19] Recall that the passage from one court to another was common practice and admitted amongst the scholars and writers, resulting originally from a competition between the courts that was favorable to the development of the arts and sciences (one thinks of the voyages of al-Mutanabbī). In brief, born, it seems, at the start of the tenth century, al-Khāzin is always living in the sixties. A renowned and famous scholar, it must also be that he was a dignitary so that his name would thus be retained in literature and history. There ends the knowledge we have of his life and works.

4.1.2. *The treatises of al-Khāzin on isoperimeters and isepiphanics*

Of the works by al-Khāzin in the field of infinitesimal mathematics, we know only of the single treatise translated here. However, as the title indicates, it is a part of a commentary on the first book of the *Almagest* by al-Khāzin. One phrase reads, 'We have copied from the commentary by Abū Jaʿfar Muḥammad ibn al-Ḥusayn al-Khāzin on the first book of the *Almagest*...' Two references by al-Bīrūnī[20] confirm the existence of this commentary, together with its extension. It was not as short as the surviving text. This text was not, therefore, an independent treatise intended as an examination of the isoperimetric problem alone, but rather the contribution of a mathematician to the proof of a proposition stated, but not proved, by Ptolemy. Its scope was limited, in contrast as we shall see, to the ambitious nature of the text by Ibn al-Haytham.

This work by al-Khāzin is only known to have survived, for the moment at least, in the form of a single manuscript, part of collection 4821 (8), fols 47v–68v in the Bibliothèque Nationale de Paris.[21] Unlike the other texts in this collection, all written in the same hand, al-Khāzin's text is not dated. However, the colophons in these manuscripts leave the date in no

[19] Al-Tawḥīdī, *Mathālib al-wazīrayn al-Ṣāḥib ibn ʿAbbād wa-Ibn al-ʿAmīd*, ed. Muḥammad al-Ṭanjī, Beirut, 1991, p. 346. For further information about al-Khāzin, see the articles dedicated to him in the *Dictionary of Scientific Biography* by Y. Dold-Samplonius, New York, 1973, t. VII, pp. 334–5, and in *EI*², IV pp. 1215–6 by J. Samsó.

[20] Al-Bīrūnī, *al-Qānūn al-Masʿūdī*, ed. Osmania Oriental Publications Bureau, 3 vols, Hyderabad, 1954–1956, vol. II, p. 653; *Taḥdīd nihāyāt al-amākin*, p. 95.

[21] See G. Vajda, *Index général des manuscrits arabes musulmans de la Bibliothèque Nationale de Paris*, Paris, 1953, together with the supplements and corrections added by the late author to his work, held in the Bibliothèque Nationale.

doubt. Al-Khāzin's treatise was copied in 544/1149, either in Ḥamadān or in Asadabad, by Ḥusayn ibn Muḥammad ibn ʿAlī. The collection consists of 86 paper pages, 230 × 150 mm, with 18 lines per page. The foliation is more recent, and the collection was in Istanbul in the fifteenth century and was probably still there at the end of the seventeenth century, before it came to the Bibliothèque Nationale.

This unique manuscript is a very careful copy, written in perfectly clear *naskhī*. All additions and erasures are in the hand of the copyist, and were almost certainly made at the time of the transcription. There is no evidence that the copyist compared the finished copy with the source. This treatise was originally edited and translated into English by R. Lorch.[22] The improvements – some 20 in number – that we have been able to make to this excellent work are not, in themselves, sufficient to justify a new edition. But we include it here simply as part of a project to bring all the contributions to this topic that are known to us together in a single book.

4.2. MATHEMATICAL COMMENTARY

4.2.1. *Introduction*

To show that, for regions of the plane having a given perimeter, the disc has the greatest area; and that, for solids having the same total area, it is the sphere that has the greatest volume: this 'extremal' undertaking interested the mathematicians as much as the astronomers. The latter needed it to establish the sphericity of the heavens and the earth's surface, while the former were enmeshed in the task presumably to satisfy the latter. The question of isoperimetrics and isepiphanics seems in every case, over a long period of its history, tied to this cosmological perspective: it is that perspective which assured it permanence and fecundity for centuries. The detailed history of this question will be retraced in the last volume of this book, but for now we must set down some names and titles. The first: Zenodorus, the successor of Archimedes, and his lost writing on *Isoperimetric Figures*. Fortunately, Theon of Alexandria cites it in his *Commentary on the First Book of the Almagest*,[23] on the topic of a famous formula of Ptolemy's: 'Because, amongst different figures having the same perimeter, those which have more sides are largest, amongst plane figures it is the circle which is

[22] R. Lorch, 'Abū Jaʿfar al-Khāzin on isoperimetry and the Archimedian tradition', *Zeitschrift für Geschichte der Arabisch-Islamischen Wissenschaften* 3, 1986, pp. 150–229.

[23] A. Rome, *Commentaires de Pappus et de Théon d'Alexandrie sur l'Almageste*, text edited and annotated, vol. II: Théon d'Alexandrie, *Commentaire sur les livres 1 et 2 de l'Almageste*, Vatican, 1936, pp. 355 sqq.

largest, and amongst solids, the sphere, and the sky is the largest of the bodies'.[24] Commentators on the *Almagest*, already since Theon, could no longer put forward a single formula in silence, without giving its proof. Other mathematicians were interested in this problem, including Hero of Alexandria, and Pappus, in the fifth book of the *Collection*.[25] But what is important here is that Theon's text, as well as the *Almagest*, were known by mathematicians and astronomers from Baghdad in the ninth century, and that they fueled a new tradition of research, which started with al-Kindī. He claimed to have treated this problem 'in <his> book on spheres';[26] whereas the thirteenth century bibliographer Ibn Abī Uṣaybiʿa attributes to him *The Sphere is the Largest Solid Figure*.[27]

In this tradition will register, and under very different titles, Ibn Hūd, Jābir ibn Aflaḥ ..., and above all al-Khāzin and Ibn al-Haytham, who are the principal figures who we know of for now. The reading and analysis of these last two contributions will reveal the great distance between two mathematicians who nonetheless formed part of a single and unified tradition. While the former develops the past, the latter, in accomplishing it, graces the banks of the future. But to understand, if only partially, the sense of this sibylline affirmation, we begin by analyzing al-Khāzin's text. Al-Khāzin works from the citation of Ptolemy, which he proposes to establish

[24] J.L. Heiberg, *Claudii Ptolemaei opera quae exstant omnia. I. Syntaxis mathematica*, Leipzig, 1898, p. 13, lines 16–19. Here is the Arabic translation made in 212/827 by al-Ḥajjāj (MS Leiden 680, fols 3^v–4^r):

ومن أجل أن الأشكال الكثيرة الأضلاع التي تكون في دوائر متساوية أكثرها زوايا أعظمها عظماً، تكون الدائرة أعظم الأشكال البسيطة وتكون الكرة أعظم الأشكال المجسمة، فالسماء أعظم مما سواها من الأجسام.

[25] Cf. note 1 and also the translation of P. Ver Eecke: Pappus d'Alexandrie, *La Collection Mathématique*, Paris/Bruges, 1933, t. I, pp. 239 sqq.

[26] In his book *Fī al-ṣināʿat al-ʿuẓmā*, al-Kindī writes: 'Just as the largest of the figures in the circle having equal sides is that which has the most angles, and the largest of the solid figures having equal planar faces is the sphere as we explained in our book *On spheres*, the sky is thus greater than all other bodies, and it is spherical as it must have the largest shape.' Here is the Arabic text which we have established on the basis of the manuscript from Istanbul, Aya Sofya 4860, fol. 59^v:

وأيضاً، لأن أعظمَ الأشكال التي في الدائرة المتساوية الأضلاع أكثرُها زوايا، وأعظمَ الأشكال المجسمة المعتدلة المتساوية السطوح الكرةُ كما أوضحنا ذلك في كتابنا في الأكر، تكون السماء إذاً هي أعظمَ مما سواها من الأجسام كريّةً، لأنه ينبغي أن يكون لها الشكل الأعظم.

[27] Ibn Abī Uṣaybiʿa, *'Uyūn al-anbā' fī ṭabaqāt al-aṭibbā'*, ed. A. Müller, 3 vols, Cairo/Königsberg, 1882–84, vol. I, p. 210, 18; ed. N. Riḍā, Beirut, 1965, p. 289, 27–28.

not with the aid of calculation (*ḥisāb*) but by means of geometry. The guiding idea, which seems perfectly conscious for al-Khāzin is that, of all the convex figures of a given type (triangle, rhombus, parallelogram, ...), the most symmetric achieves an *extremum* for a certain magnitude (area, ratio of area, perimeter,...). One procedes in the following manner: one fixes a parameter and varies the figure by way of making it more symmetric with respect to a certain straight line. Thus, in fixing the perimeter of a parallelogram, one transforms this parallelogram into a rhombus by making it symmetric with respect to a diagonal; the area increases in the process.

As for the treatise, it is divided into two parts, one dedicated to isoperimetrics and the other to isepiphanics, both also depending on unstated notions and undeclared axioms. Amongst these notions is that of convexity: the polygons and polyhedra considered in this treatise are convex. Among other axioms, one notably has the following:

A_1 If a convex polygon is inscribed in a circle, then its perimeter is less than that of the circle.

A_2 If a convex polygon is circumscribed about a circle, then its perimeter is greater than that of the circle.

A_3 If a convex polyhedron is inscribed in a sphere, then its area is less than that of the sphere.

A_4 If a convex polyhedron is circumscribed about a sphere, then its area is greater than that of the sphere.

One will remark that from A_1 and A_2, al-Khāzin deduces the results relating to areas in Lemma 8; and that from A_3 and A_4 he deduces the results relating to volumes in Proposition 19.

Let us consider the two parts of the treatise in succession.

4.2.2. *Isoperimetrics*

Al-Khāzin required eight lemmas and one proposition to establish the isoperimetric theorem. The first four lemmas are related to isosceles and equilateral triangles, and show that the area of an equilateral triangle is greater than that of every isosceles triangle of the same perimeter. The fifth shows that the area of an equilateral triangle is greater than that of every triangle of the same perimeter. In the course of that demonstration, al-Khāzin established a result already proven by Zenodorus and by Pappus, to wit: 'Among the isoperimetric figures with an equal number of sides, the largest is that which is equilateral and equiangular.' In Lemma 6, he compares the parallelogram to a square of the same perimeter. In Lemma 7, al-Khāzin takes the example of a regular pentagon, deduces from it an irregular pentagon having the same perimeter, and shows that the second has a smaller area than the first. Finally, in Lemma 8, he passes to convex polygons admitting an inscribed circle and a circumscribed circle.

All is now in place to establish the isoperimetric property of regular polygons, before ultimately passing to the theorem for the circle. We shall follow step by step al-Khāzin's deliberately progressive path in reasoning.

Lemma 1. — *Let* ABC *be an equilateral triangle and* ADE *an isosceles triangle* (D *and* E *on the segment* BC). *One has*

$$AB - AD < AD - BE$$

and

$$(AB - AD) + (AD - BE) = AB - BE = BD$$

and

$$BE \cdot 9AB < (AB + BE + AE)^2.$$[28]

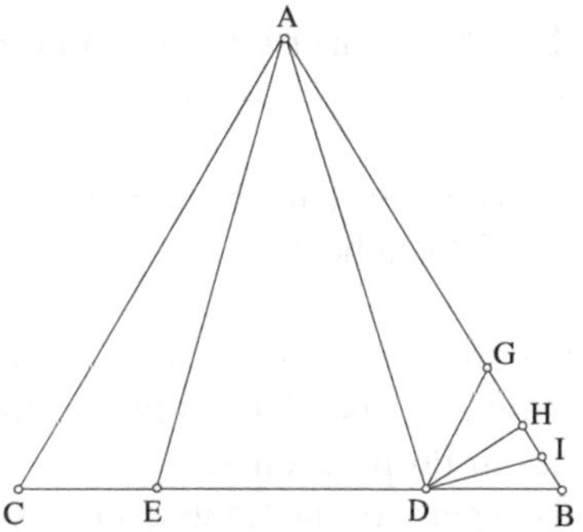

Fig. 4.2.1

Proof: The point D being between B and C, we have $AD < AB$ (as angle ADB is obtuse) and $AD > DC$ (as $A\hat{C}D > C\hat{A}D$); thus $AD > BE$.

If $DG \parallel AC$, then $AG = CD = BE$ and $GB = DB$, and if $DH \perp AB$, we have $BH = HG$ and $AH < AD$.

Let I be a point on AB such that $AI = AD$, then $AH < AI < AB$, I is between B and H, and thus $BI < IG$. Moreover,

[28] Al-Khāzin does not give this result in the statement of the lemma, but it is the object of his proof and he will use it in Lemma 2.

$$BI = AB - AD \text{ and } GI = AI - AG = AD - BE;$$

thus

$$AB - AD < AD - BE$$

and

$$(AB - AD) + (AD - BE) = BD.$$

We can write

$$AB - BE + AB - AD = BG + BI = 2\,AB - (AD + BE)$$

and

$$AB - BE + AD - BE = BD + IG = (AB + AD) - 2BE.$$

Then

$$2AB - (AD + BE) < (AB + AD) - 2BE;$$

and as a result

$$3AB - (AB + BE + AD) < (AB + BE + AD) - 3BE.$$

Dividing the left side by $AB + BE + AD$ and the right by $3BE$ (we know that $AB + BE + AD > 3BE$), it follows that

$$\frac{3AB}{AB+BE+AD} < \frac{AB+BE+AD}{3BE};$$

and using the fact that $AE = AD$, we have

$$BE \cdot 9AB < (AB + BE + AE)^2.$$

Comment. — The figure, on the one hand, and the reasoning, on the other, made by al-Khāzin suppose that D and E lie on the segment BC. If D and E are on the line BC, but on opposite sides of the segment ($DE > BC$), we have $AD > AB$ as angle ABD is obtuse and $BE > AE$, as $E\hat{A}B > A\hat{B}E$; thus $BE > AD$. The parallel to AC taken through D cuts AB at G; therefore $AG = CD = BE$ and $GB = GD$.

Let H and I be on BG such that $DH \perp BG$ and $AI = AD$; we have

$$AB < AH < AI < AG;$$

moreover,

$$BI = AD - AB \text{ and } GI = AG - AD = BE - AD.$$

We have $BI > IG$; thus

$$AD - AB > BE - AD$$

and

$$(AD - AB) + (BE - AD) = BG = BD = BE - AB.$$

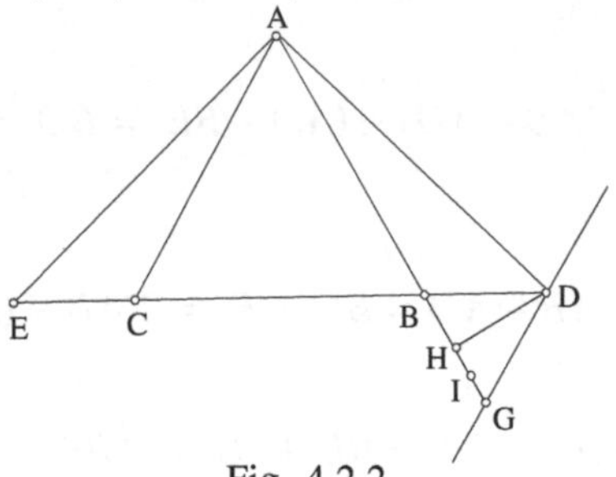

Fig. 4.2.2

From this we deduce

$$BG + BI = (BE - AB) + (AD - AB) = (BE + AD) - 2AB,$$
$$BD + IG = (BE - AB) + (BE - AD) = 2BE - (AB + AD),$$

so we have

$$(BE + AD) - 2AB > 2BE - (AB + AD);$$

hence

$$(AB + BE + AD) - 3AB > 3BE - (AB + BE + AD).$$

Divide the left side by $3AB$ and the right by $AB + BE + AD$ (we know that $3AB < AB + BE + AD$); it follows that

$$\frac{AB + BE + AD}{3AB} > \frac{3EB}{AB + BE + AD};$$

and as a result, taking account of the fact $AD = AE$,

$$BE \cdot 9AB < (AB + BE + AE)^2.$$

Hence, whatever the case of the figure,

$$BE \cdot 9AB < [\text{per. } (ABE)]^2.$$

Lemma 2. — Under the same conditions,

$$\frac{[\text{per. } (ADE)]^2}{[\text{per. } (ABC)]^2} > \frac{\text{area } (ADE)}{\text{area } (ABC)}.$$

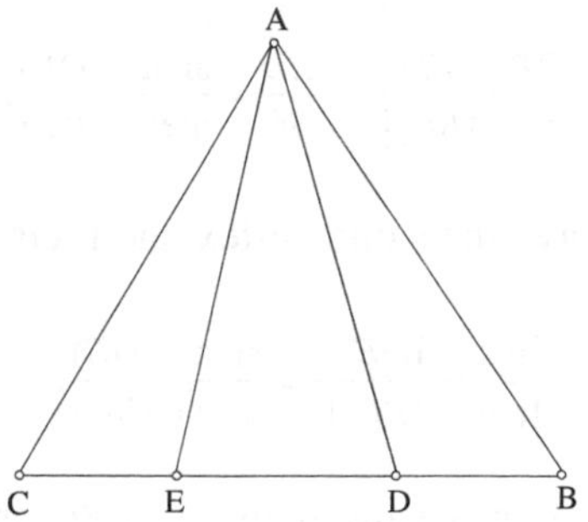

Fig. 4.2.3

Proof: By Lemma 1, we have

$$[\text{per. }(ABE)]^2 > BE \cdot 9BC,$$

$$[\text{per. }(ADC)]^2 > DC \cdot 9BC.$$

Adding the respective sides, we obtain

$$[\text{per. }(ABE)]^2 + [\text{per. }(ADC)]^2 > 9BC \cdot (BE + DC) > 9BC^2 + 9BC \cdot ED,$$

$$(1) \qquad [\text{per. }(ABE)]^2 + [\text{per. }(ADC)]^2 > [\text{per. }(ABC)]^2 + 9BC \cdot ED.$$

But

$$\text{per. }(ABE) + \text{per. }(ADC) = \text{per. }(ABC) + \text{per. }(ADE),$$

with

$$\text{per. }(ABE) = \text{per. }(ADC) \text{ and per. }(ABC) \neq \text{per. }(ADE).$$

From this, we deduce

$$(2) \qquad [\text{per. }(ABC)]^2 + [\text{per. }(ADE)]^2 > [\text{per. }(ABE)]^2 + [\text{per. }(ADC)]^2.$$ [29]

As a result of (1) and (2), we obtain

$$[\text{per. }(ADE)]^2 > 9BC \cdot ED$$

and

$$\frac{[\text{per. }(ADE)]^2}{[\text{per. }(ABC)]^2} > \frac{9BC \,.\, ED}{[\text{per. }(ABC)]^2}.$$

[29] Let the four numbers a, b, a' and b' be such that $a = b$, $a' \neq b'$ and $a + b = a' + b'$. We have $2a = a' + b'$ and $4a^2 = a'^2 + b'^2 + 2a'b'$; but $a'^2 + b'^2 > 2a'b'$, as $(a' - b')^2 > 0$; hence $4a^2 < 2(a'^2 + b'^2)$ and as a result $2a^2 < a'^2 + b'^2$; hence $a^2 + b^2 < a'^2 + b'^2$.

But

$$\frac{9BC \cdot ED}{[\text{per. }(ABC)]^2} = \frac{ED}{BC} = \frac{\text{area }(ADE)}{\text{area }(ABC)},$$

as these two triangles have the same vertex and their bases fall on the same line. We then have

$$\frac{[\text{per. }(ADE)]^2}{[\text{per. }(ABC)]^2} > \frac{\text{area }(ADE)}{\text{area }(ABC)}.$$

Comment. — The reasoning is valid with $GE < BC$ or $GE > BC$.

Lemma 3. — *If* ABC *is an isosceles triangle with vertex* A *and* G *is a point on parallel to* BC *through* A, *then*

$$GB + GC > AB + AC.$$

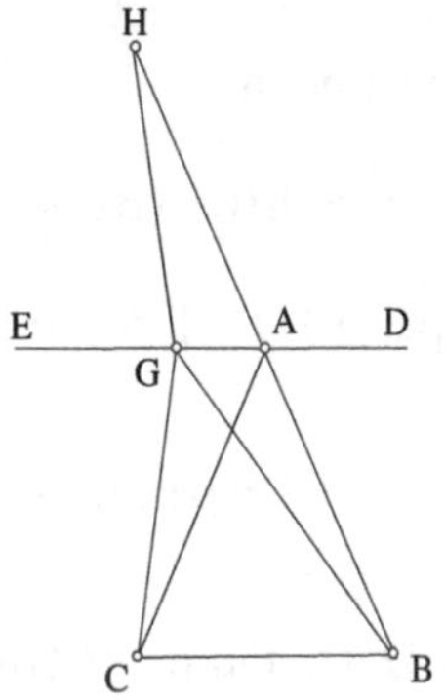

Fig. 4.2.4

We extend BA by a length equal to AH, so the triangles HAG and CAG are equal and we have $GH = GC$. From this, we deduce

$$GB + GC = GB + GH > BH;$$

thus

$$GB + GC > AB + AC.$$

Lemma 4. — *If an equilateral triangle* ABC *and an isosceles triangle* DEG (DE = DG) *have the same perimeter, then*

$$\text{area (ABC)} > \text{area (DEG)}.$$

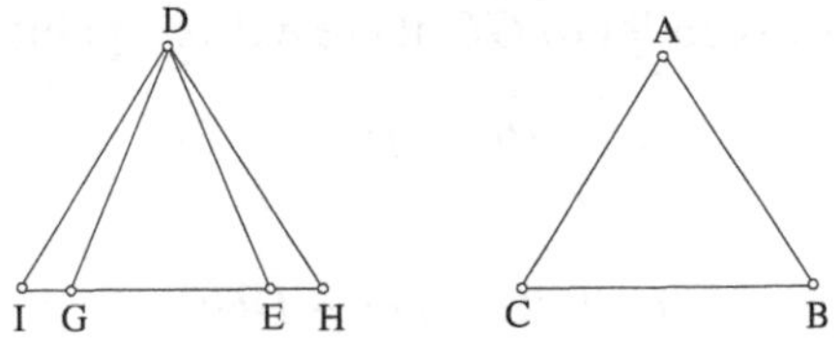

Fig. 4.2.5

Proof: Let H and I be two points on the line DE such that DHI is equilateral. We have by Lemma 2:

$$\frac{[\text{pér. }(DGE)]^2}{[\text{pér. }(DHI)]^2} > \frac{\text{aire }(DGE)}{\text{aire }(DHI)}.$$

But

$$\text{per. }(DGE) = \text{per. }(ABC);$$

moreover, ABC and DHI being equilateral,

$$\frac{[\text{pér. }(ABC)]^2}{[\text{pér. }(DHI)]^2} = \frac{\text{aire }(ABC)}{\text{aire }(DHI)}.$$

Hence

$$\frac{\text{area }(ABC)}{\text{area }(DHI)} > \frac{\text{area }(DGE)}{\text{area }(DHI)};$$

thus

$$\text{area }(ABC) > \text{area }(DGE).$$

Lemma 5. — *If an equilateral triangle* ABC *and an arbitrary triangle* DEG *have the same perimeter, then*

$$\text{area (ABC)} > \text{area (DEG)}.$$

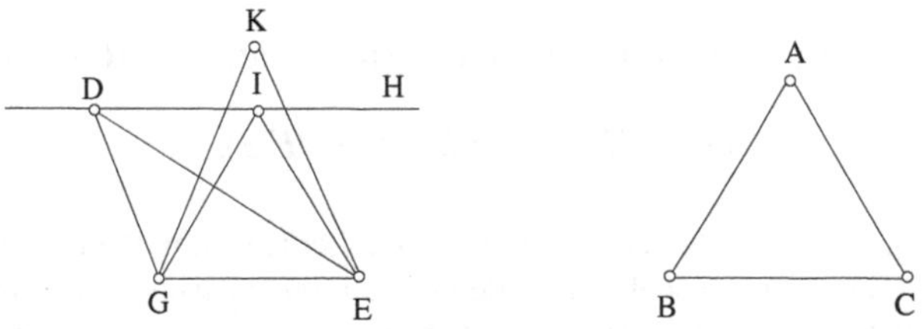

Fig. 4.2.6

Proof: On the line *DH* parallel to *GE*, there exists a point *I* such that

$$IG = IE,$$

and by Lemma 3

$$IE + IG < DE + DG;$$

thus

$$\text{per.}\ (IEG) < \text{per.}\ (DEG).$$

But

$$\text{area}\ (IEG) = \text{area}\ (DGE).$$

Let *K* be such that $KG = KE$ and per.(KGE) = per.(DGE) = per. (ABC); we then have

$$\text{area}\ (KGE) > \text{area}\ (DGE);$$ [30]

and by Lemma 4

$$\text{area}\ (ABC) > \text{area}\ (KGE).$$

As a result

$$\text{area}\ (ABC) > \text{area}\ (DEG).$$

Lemma 6. — We take up the arbitrary triangle *DEG* and the equilateral *ABC* having the same perimeter, and we complete the parallelogram *DEGI* and the rhombus *ABCH*.

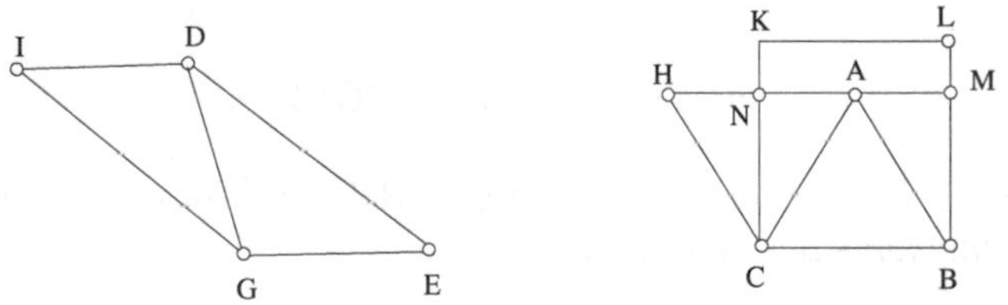

Fig. 4.2.7

If per.(DEG) = per.(ABC), we have seen that area (ABC) > area (DEG). But

$$\text{area}\ (DEGI) = 2\ \text{area}\ (DEG),$$

[30] Al-Khāzin has thus demonstrated without explicitly stating the following result: if an arbitrary triangle and an isosceles triangle have the same perimeter and an equal base, then the area of the isosceles triangle is greater than that of the arbitrary triangle.

Note nevertheless that the area of an arbitrary triangle is not less than that of every isosceles triangle having the same perimeter (for there exist isosceles triangles of a given perimeter for which the area is next to nothing).

$$\text{area}\,(ABCH) = 2\ \text{area}\,(ABC);$$

thus

$$\text{area}\,(ABCH) > \text{area}\,(DEGI).$$

But in general

$$\text{per.}\,(ABCH) \neq \text{per.}(DEGI);$$

they are only equal if we assume $DG = AC$.[31]

[31] The parallelogram and the rhombus constructed by al-Khāzin do not in general have the same perimeter.

$$\text{per.}(ABCH) = \text{per.}(EDGI) \Leftrightarrow 2AC = ED + EG;$$

but by hypothesis

$$3AC = ED + EG + DG.$$

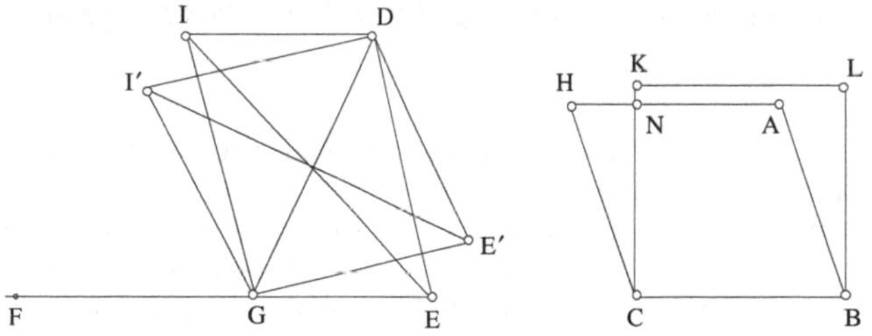

Fig. 4.2.8

It is thus necessary to have the supplementary condition $AC = DG$. We can take up the reasoning without making use of the equilateral triangle ABC.

Starting with a parallelogram $DEGI$, let us construct a rhombus of the same perimeter. The diagonal DG separates the figure into two triangles of equal area:

$$\text{area}\,(DGI) = \text{area}\,(DGE).$$

If we construct the points I' and E' on the perpendicular bisector of DG such that

$$I'D = I'G = E'D = E'G = \frac{1}{2}\,(DE + EG) = \frac{1}{2}\,EF,$$

the rhombus $DE'GI'$ is then of the same perimeter as $DEGI$. But, by the note from Lemma 5, area $(DE'G) >$ area (DEG); thus

$$\text{area}\,(DE'GI') > \text{area}\,(DEGI).$$

So let $ABCH$ be a rhombus equal to $DE'GI'$ and $BCKL$, the square constructed on BC:

$$\text{area}\,(ABCH) = BC \cdot NC,$$
$$\text{area}\,(BCKL) = BC \cdot KC.$$

Thus the area of the square is greater than that of every rhombus of the same perimeter, which is itself greater than that of every parallelogram of the same perimeter.

Let the segment $CK \perp BC$ be such that $CK = BC$, and the segment $KL \parallel CB$ such that $KL = CB$. The line AH cuts CK at N and BL at M. We have

$$\text{area } (BMNC) = \text{area } (ABCH) \text{ and area } (BCKL) > \text{area } (MBCN);$$

thus

$$\text{area } (BCKL) > \text{area } (ABCH) > \text{area } (DEGI).$$

Now $LBCK$ is a square that has the same perimeter as the rhombus $ABCH$, and if $DG = AC$, the square also has the same perimeter as $DEGI$.

The square, as a regular polygon, has a greater area than that of every parallelogram of the same perimeter.[32]

Al-Khāzin thus returns to the general statement: *Of two convex polygons, one regular, the other arbitrary, having the same number of sides and the same perimeter, the regular polygon has the greatest area.*

[32] We can demonstrate that of all the convex quadrilaterals of the same perimeter, the square has the greatest area.

Let $ABCD$ be an arbitrary quadrilateral. Let us construct the points C' and A' on the perpendicular bisector of BD such that $C'D + C'B = CD + CB$ and $A'D + A'B = AD + AB$; the quadrilaterals $ABCD$ and $A'BC'D$ have the same perimeter. The triangles CDB and $C'DB$ on the one hand, and ADB and $A'DB$ on the other, have the same perimeter and, by the note from Lemma 5, we have area (CDB) < area ($C'DB$) and area (ADB) < area ($A'DB$); hence area ($ABCD$) < area ($A'BC'D$).

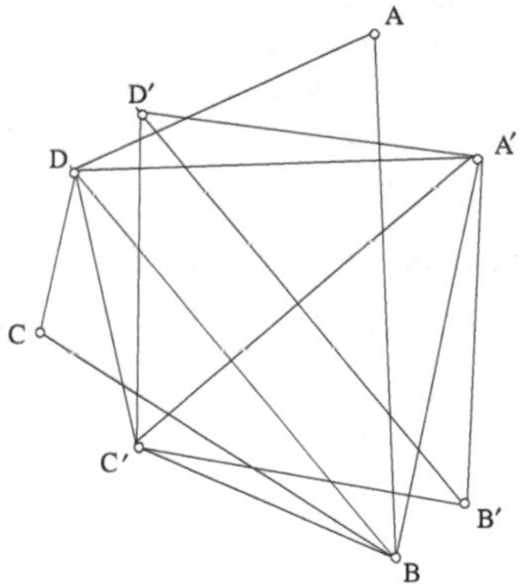

Fig. 4.2.9

By the same procedure, construct the points B' and D' on the perpendicular bisector of $A'C'$ such that $B'A' + B'C' = BA' + BC'$ and $D'A' + D'C' = DA' + DC'$. The quadrilateral $A'B'C'D'$ is then a rhombus and area ($A'BC'D$) < area ($A'B'C'D'$). Yet we know that the area of a rhombus is less than that of the square of the same perimeter; thus area ($ABCD$) < area of the square of the same perimeter.

Lemma 7.

Example: Let $ABCDE$ be a regular pentagon and G a point such that

$$GB + GE = AB + AE.$$

The pentagon $GBCDE$ has the same perimeter as $ABCDE$ and

$$\text{area}\ (ABCDE) > \text{area}\ (GBCDE).$$

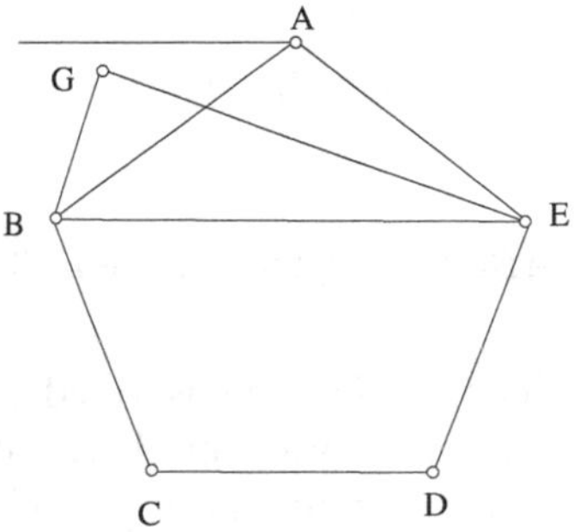

Fig. 4.2.10

Indeed, by the hypothesis for G, the point is between the line BE and the parallel to BE through A, as, by Lemma 3, if G is on this parallel, then

$$GB + GE > AB + AE.$$

Therefore

$$\text{area}\ (BAE) > \text{area}\ (BGE),$$

and as a result

$$\text{area}\ (ABCDE) > \text{area}\ (GBCDE).$$[33]

Lemma 8. — *The area of a polygon of perimeter* p *circumscribed about a circle of radius* r *is equal to the product* $\frac{1}{2}\mathrm{p} \cdot \mathrm{r}$.

Let ABC be a triangle circumscribed about a circle with centre I, with D, E, and G the points of contact. We have

$$\text{area}\ (AIC) = ID \cdot \tfrac{1}{2}AC,$$
$$\text{area}\ (AIB) = IE \cdot \tfrac{1}{2}AB,$$

[33] Starting with a regular pentagon, al-Khāzin produces an irregular pentagon of the same perimeter, and for which the area is smaller. But he does not demonstrate that an arbitrary pentagon has a smaller area than that of a regular pentagon of the same perimeter.

$$\text{area}\,(BIC) = IG \cdot \tfrac{1}{2}BC.$$

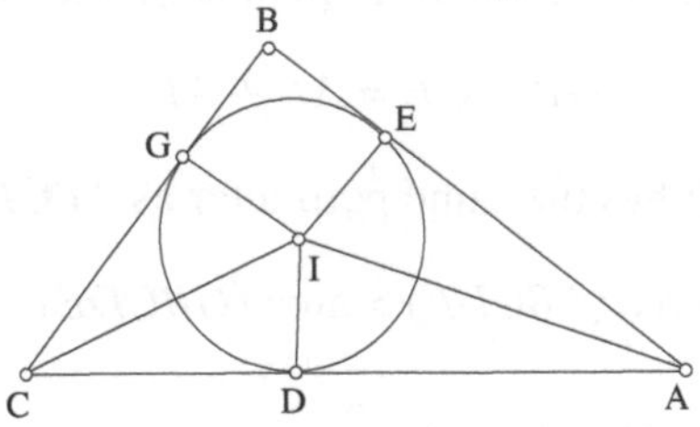

Fig. 4.2.11

From this, we deduce

$$\text{area}\,(ABC) = \frac{1}{2}\,(AB + AC + BC) \cdot r.$$

If the polygon has n sides and is circumscribed about a circle, we divide it into n triangles having the center of the circle as a common vertex and the radius r of circle for their common height. If p_n and S_n are respectively the perimeter and the surface area of the polygon,

$$S_n = \tfrac{1}{2}p_n \cdot r.$$

The area S_n of the polygon is greater than that of the inscribed circle, as p_n is greater than the circle's perimeter.

This demonstration is the same as that of the Banū Mūsā of the first part of Proposition 1. They, however, complete Proposition 1 with an extension into space, as an expression of the volume of a polyhedron circumscribed about a sphere of radius r.

If a polygon admits a circumscribed circle of radius R, then $R > r$ and $S_n < \tfrac{1}{2}p_n \cdot R$, and it follows that the area S_n is less than that of the circumscribed circle. These two inequalities are moreover evident from the inclusions of the figures.

Note that in this paragraph, al-Khāzin considers polygons admitting both an inscribed circle and a circumscribed circle, a condition that is true in the case of the triangle and of regular polygons, but is not true in general.

Proposition 9. — *Of two regular polygons having the same perimeter, that which has the most vertices has the greatest area.*

Example: Let ABC be an equilateral triangle and $DEGH$ be a square of the same perimeter p. Then

$$\text{area } (DEGH) > \text{area } (ABC).$$

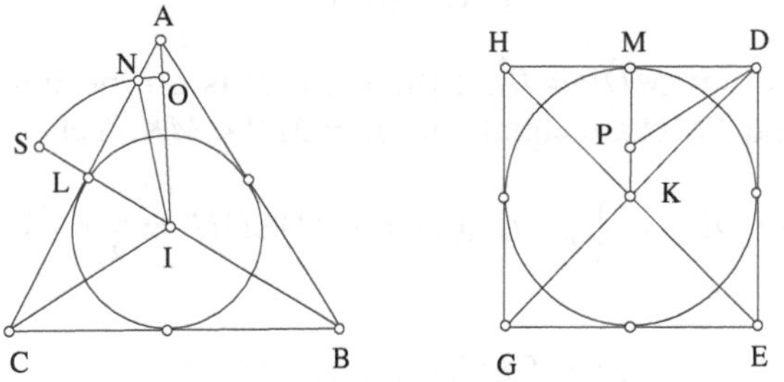

Fig. 4.2.12

If I and K are the centres of the inscribed circles, L the midpoint of AC and M the midpoint of HD, then

$$A\hat{I}C = 4\frac{\pi}{2}\cdot\frac{1}{3} \text{ and } AC = \tfrac{1}{3}p,$$

$$D\hat{K}H = 4\frac{\pi}{2}\cdot\frac{1}{4} \text{ and } DH = \tfrac{1}{4}p,$$

from which we deduce

$$\frac{A\hat{I}C}{D\hat{K}H} = \frac{AC}{DH} \text{ and } \frac{A\hat{I}L}{D\hat{K}M} = \frac{AL}{DM}.$$

Let us take N on AL such that $LN = DM$, and the circle (I, IN) meets IL at S and IA at O. Then

$$\frac{A\hat{I}N}{N\hat{I}L} = \frac{\text{area sector } (INO)}{\text{area sector } (INS)} < \frac{\text{area triangle } (AIN)}{\text{area triangle } (NIL)} = \frac{AN}{NL},$$

from which we deduce

$$\frac{A\hat{I}L}{N\hat{I}L} < \frac{AL}{DM};$$

hence

$$\frac{A\hat{I}L}{N\hat{I}L} < \frac{A\hat{I}L}{D\hat{K}M}$$

and

$$N\hat{I}L > D\hat{K}M,$$

and as a result

$$I\hat{N}L < K\hat{D}M.$$

Let us construct $M\hat{D}P = I\hat{N}L$; the point P is on the interval MK, the triangles ILN and PMD are equal, and $IL = MP < MK$. Yet

$$\text{area } (ABC) = \frac{1}{2}\, p \cdot IL \text{ and area } (DEGH) = \frac{1}{2}\, p \cdot MK;$$

thus

$$\text{area } (DEGH) > \text{area } (ABC).$$

We can extend this demonstration to regular polygons of the same perimeter for whatever numbers n and n' of vertices.

Note that this proposition will be taken up by Ibn al-Haytham (cf. proposition 2 of his treatise, Vol. 2). We find it next in Ibn Hūd's book (see Chapter VII).

Theorem 10. — *Of all the planar figures, regular convex polygons and the circle, having the same perimeter, it is the circle that has the greatest area.*

Let ABC be an equilateral triangle and DEG a circle having the same perimeter. Let MNS be an equilateral triangle circumscribed about the circle DEG. The perimeter of MNS is greater than that of the circle, which is equal to that of ABC; thus $MS > AC$.

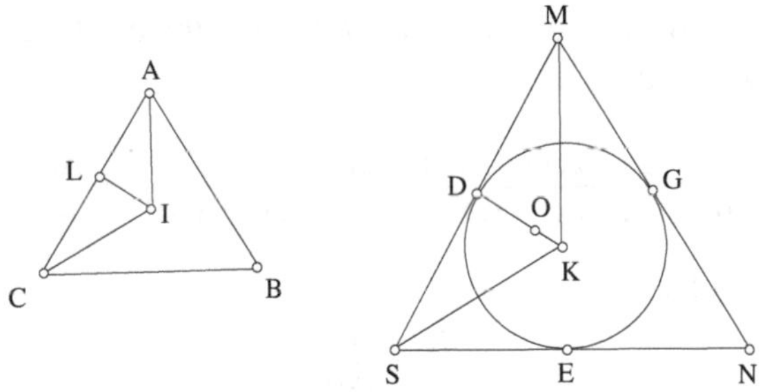

Fig. 4.2.13

Let I be the centre of the triangle ABC and K the centre of the circle, L the midpoint of AC and D the midpoint of MS; we have $DM > AL$. The triangles AIL and MKD are right-angled and similar, and as a result $DK > LI$. But the circle and the triangle ABC have the same perimeter p, and we have

$$\text{area of the circle} = \tfrac{1}{2} p \cdot DK,$$

$$\text{area }(ABC) = \tfrac{1}{2}p \cdot LI,$$

and as a result, the area of the circle is greater than that of the equilateral triangle *ABC*.

Al-Khāzin then indicates that the same reasoning applies to the square, to the pentagon and to every regular polygon and deduces from this the statement of his proposition.

This proposition is taken up by Ibn al-Haytham (cf. Proposition 1 of his treatise on isoperimetrics, Vol. 2).

Al-Khāzin even remarks that one can proceed in the same manner in the case of arbitrary polygons. This is true for the triangle, as, a triangle *ABC* and a circle being given, it is possible to construct a similar triangle to *ABC* circumscribed about the circle. But, in general, there does not exist a polygon similar to an arbitrary polygon that would be circumscribed by a given circle.

However, the reasoning applied to a regular polygon permits one to conclude in the case of an arbitrary convex polygon:

Let P be an arbitrary polygon, P' a regular polygon and C a circle having equal perimeters, P and P' having the same number of sides. We have

$$\text{area }(P) < \text{area }(P') \text{ and area }(P') < \text{area }(C),$$

whence

$$\text{area }(P) < \text{area }(C).$$

If a circle and a convex polygon have the same perimeter, the area of the circle is greater than that of the polygon.

Thus, for the isoperimetrics, al-Khāzin proceeds a) by comparing regular polygons of the same perimeter and of a different number of sides, and b) by comparing a regular polygon with a circle of the same perimeter by means of a similar polygon circumscribed about a circle. Compared with the approach of Ibn al-Haytham – cf. Vol. 2 – that of al-Khāzin might be qualified as static. One will see that Ibn al-Haytham uses a) to establish b), considering the circle as the limit of a sequence of regular polygons. In other words, even if al-Khāzin's method is different from that of Zenodorus or Pappus, it nonetheless falls in the same family, whilst that of Ibn al-Haytham is different from the rest.

4.2.3. *Isepiphanics*

The second part of al-Khāzin's treatise pertains to the same extremal problem, but in space: spatial isoperimetrics. It also consists of nine lemmas and a theorem. The first lemma pertains to the lateral area of a regular pyramid and the second to the volume of a pyramid admitting an inscribed sphere; in the third, al-Khāzin treats the lateral area of a cone of revolution and its volume. In the fourth lemma (Proposition 14), he considers the following problem: given a circle C, construct two similar polygons of area S_1 and S_2, one circumscribed about C, the other inscribed in C, and such that $S_1/S_2 < k$ (the given ratio). In the fifth lemma, al-Khāzin gives another expression of the lateral area of the cone, in order to pass, in the sixth lemma, to that of the frustum of the cone. From Lemma 6 (Proposition 16) is thus deduced Lemma 7:

If a regular polygonal line is inscribed in a circle of area S_1, *and circumscribed about a circle of area* S_2, *the area* S *of the surface generated by the rotation of that line about one of its axes satisfies* $4\,\mathrm{S}_2 < \mathrm{S} < 4\mathrm{S}_1$.

Al-Khāzin passes in Lemma 8 to the calculation of the area of the sphere, then in Lemma 9 to the volume of the sphere. It is in this lemma that al-Khāzin defines a polyhedron inscribed in a sphere, and admits the existence of a sphere tangent to all the faces of the solid, which is incorrect – *vide infra*. All the preliminaries are thus posed to establish the theorem:

Of all the solids having the same area, the sphere is that which has the greatest volume. The demonstration is made only for a solid that admits an inscribed sphere.

We now take in detail this path set by al-Khāzin.

Lemma 11. — *Regular triangular pyramid.* The base is an equilateral triangle ABC and the three lateral faces are equal isosceles triangles with vertex D. The height is DE, perpendicular to the plane ABC. If the triangles with vertex D are themselves equilateral, one obtains a regular tetrahedron.

Lateral area: The isosceles triangles with vertex D have equal heights, so let $DI = a$ be one of these:

$$\text{lateral area} = \frac{1}{2}\ \text{per. (ABC)} \cdot \text{a}.$$

Total area of the pyramid: The segment EI is the radius r of the circle inscribed in ABC:

$$\text{area (ABC)} = \frac{1}{2}\ \text{per. (ABC)} \cdot \text{r},$$

$$\text{total area} = (a + r) \cdot \tfrac{1}{2}\ \text{per.}\ (ABC).$$

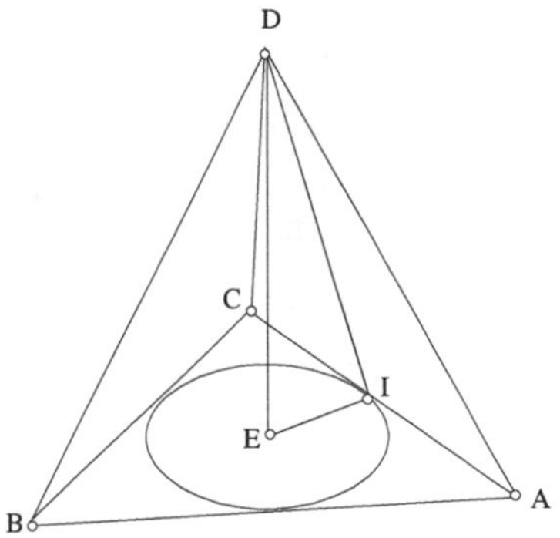

Fig. 4.2.14

The ratio of the lateral area to the area of the base is equal to $\frac{a}{r}$.

These results are valid for every regular pyramid, whatever the nature of the polygon at its base.

If a is the height of a lateral face, r the radius of the inscribed circle and p the perimeter of the polygon at the base, we have

$$\text{lateral area} = \tfrac{1}{2}\, p \cdot a,$$

$$\text{total area} = \tfrac{1}{2} p \cdot (a + r),$$

$$\frac{\text{lateral area}}{\text{area of the base}} = \frac{a}{r}.$$

Lemma 12. — *Volume of the pyramid ABCD.*

By *Elements* XII.6, this volume is one third the volume of the prism of base ABC and height DE; thus

$$V = \tfrac{1}{3}\ \text{area}\ (ABC) \cdot DE.$$

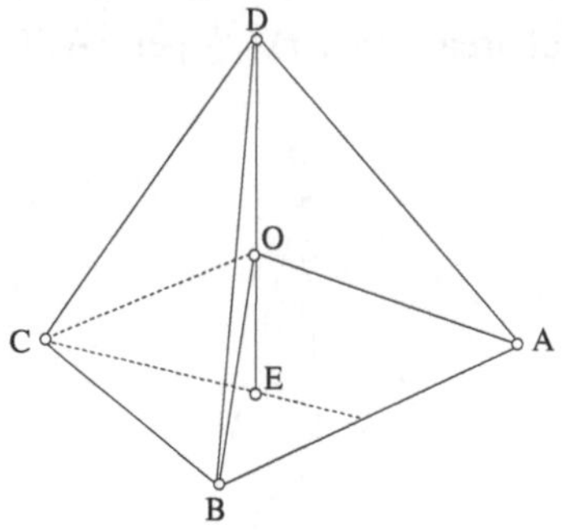

Fig. 4.2.15

Pyramid and inscribed sphere.

Let $DABC$ be a regular pyramid; there exists a sphere with center O inscribed in this pyramid. We can then decompose it into four pyramids having the sphere's centre O as a common vertex and heights equal to the sphere's radius r. The pyramid $OABC$ has volume

$$\frac{1}{3}\text{ area }(ABC)\cdot OE = \frac{1}{3}\text{ area }(ABC)\cdot r.$$

The pyramid $DABC$ has volume

$$V = \frac{1}{3}\text{ (sum of bases)}\cdot r,$$

$$(1) \qquad V = \frac{1}{3}\text{ total area}\cdot r.$$

Whatever the regular pyramid considered, there exists a sphere of centre O inscribed in this pyramid. We decompose this into $(n + 1)$ pyramids with vertex O having the radius r for heights, n being the number of sides of the polygon at the base. The result (1) remains true.

This pertains to the particular case of the extension of space made by the Banū Mūsā in the second part of their first proposition.

Generalisation.

We have shown that, for every polygon, regular or not, circumscribed about a circle of radius r, we have

$$\text{area polygon} = \frac{1}{2}\text{ perimeter polygon}\cdot r.$$

The same as for every pyramid, regular or not, if it is circumscribed about a sphere of radius r, then

$$\text{volume pyramid} = \frac{1}{3}\text{ total area} \cdot r.$$

Al-Khāzin then recalls some results relating to the cone of revolution.

Oblique or right circular cylinder.

Figure defined starting with two equal circles situated in parallel planes.

Height and axis of the right cylinder.

Generation of the right cylinder starting with a rectangle turning around one of its sides.

Lemma 13. — *Right circular cone – Lateral area.*

To a right cylinder there is associated a cone having as its base one of the bases of the cylinder and for its vertex the centre of the other base.

Let there be a cone whose base is the circle ($ABCD$) of diameter AC, with centre E, and whose vertex is the point G, with GE perpendicular to the base plane. The lateral area S of the cone is

$$S = \frac{1}{2}\text{ perimeter of the circle} \cdot AG$$

or

$$S = \text{length } \widehat{ABC} \cdot AG.$$

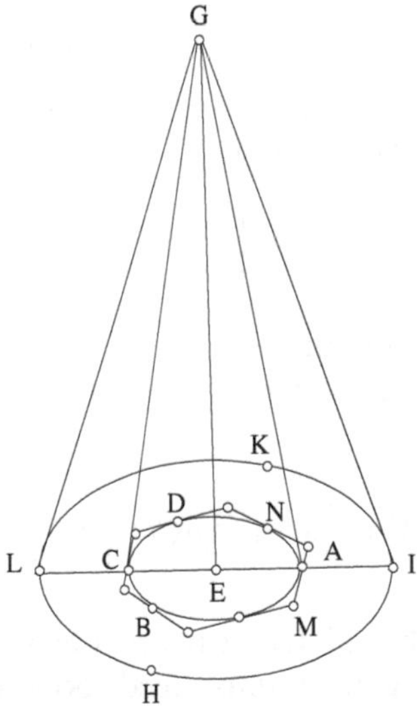

Fig. 4.2.16

Proof: by *reductio ad absurdum*

Suppose $S >$ length $\widehat{ABC} \cdot AG$ and let IL be the diameter of a circle ($IKLH$) such that $S =$ length $\widehat{IKL} \cdot AG$; one thus has $IL > AC$.

We then consider a regular polygon, circumscribed about the first circle and for which all the vertices are on the inside of the second circle; to this polygon is associated a pyramid with vertex G whose faces are tangent to the cone. The lateral area of this pyramid is greater than that of the cone.

Let us designate by p the perimeter of the circle ($ABCD$), by p_1 that of the circle ($IKLH$) and by p_2 that of the polygon; we have $p < p_2 < p_1$.

On the other hand, the lateral area of the pyramid is

$$S' = \frac{1}{2} p_2 \cdot AG$$

and we have by hypothesis

$$S = \frac{1}{2} p_1 \cdot AG;$$

$p_1 > p_2$ implies $S > S'$, which is absurd.

Suppose $S < \frac{1}{2} p \cdot AG$; then $\frac{1}{2} p \cdot AG$ is the lateral area of a cone with vertex G and whose base is a circle greater than ($ABCD$); let this circle be ($IKLH$).

We then consider as before a regular polygon circumscribed about ($ABCD$) and within ($IKLH$), and the associated pyramid whose lateral area is $\frac{1}{2}p_2 \cdot AG$. This area is greater than $\frac{1}{2}p \cdot AG$, which is that of the cone with base $IHLK$, which is absurd, for the pyramid is inside the cone.

The latcral area of the cone is thus

$$S = \frac{1}{2}p \cdot AG.$$

Volume of the right circular cone

By Euclid, *Elements* XII.9, the volume of the cone is one third that of the associated cylinder; thus

$$V = \frac{1}{3}\text{area } (ABCD) \cdot EG.$$

We have just seen that al-Khāzin accepts without justification the existence of a regular polygon circumscribed about the first circle and inside the second circle, a problem posed by the Banū Mūsā in the second part of their Proposition 3. Furthermore, the Banū Mūsā in the first part of Proposition 9 of their treatise use a regular polygon inscribed in the second circle and exterior to the first – that is, Proposition XII.16 of the *Elements*. Yet in the second part of the same proposition, they consider a regular polygon circumscribed about the smaller of the two circles and inside the larger; that is what al-Khāzin does here.

Lemma 14. — *Given a circle, construct two similar regular polygons, one circumscribed about the circle, the other inscribed in the circle such that the ratio of their areas is less than the ratio of two given magnitudes.*

Let EG and H be two magnitudes, $EG > H$. Let EI be their difference and let n be the smallest number of the form 2^p such that $EK = n \cdot EI > H$:

$$\frac{EI}{EK} < \frac{EI}{H}.$$

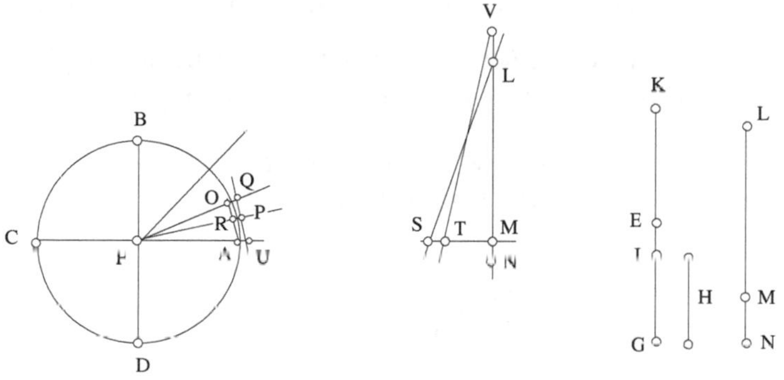

Fig. 4.2.17

Given a segment LM, we divide it into n parts and we produce it by $MN = \frac{1}{n} \cdot LM$:

$$\frac{MN}{LM} = \frac{EI}{EK}.$$

Thus

$$\frac{MN}{LM} < \frac{EI}{GI},$$

from which we deduce

$$\frac{LM + MN}{LM} < \frac{EI + GI}{GI},$$

that is,

$$\frac{LN}{LM} < \frac{EG}{H}.$$

We then construct the right angle *LMS*, with *LS* = *LN*.

Let *F* be the centre of the circle *ABCD*; we suppose $A\hat{F}B = 1$ right angle and we consider $\frac{1}{2}\ A\hat{F}B$, $\frac{1}{4}\ A\hat{F}B$, up to $A\hat{F}O = \frac{1}{2^k}\ A\hat{F}B$ such that

$$A\hat{F}O < 2\ M\hat{L}S\ .$$

The bisector of $A\hat{F}O$ meets the circle at *P* and the tangent at *P* cuts the lines *FA* and *FO* respectively at the points *U* and *Q*. The segments *AO* and *UQ* are the sides of two similar polygons having 2^{k+2} sides, one inscribed in the circle, the other circumscribed about that circle.

Let *R* be the midpoint of *AO*, $A\hat{F}R = \frac{1}{2}A\hat{F}O$; thus $A\hat{F}R < M\hat{L}S$, and as a result $\hat{S} < \hat{A}$. We construct on the right angle *LMS* a triangle *VMT* such that $\hat{T} = \hat{A}$ and *TV* = *LS*, so *MV* > *ML* and *MT* < *MS*. We have

$$\frac{PF}{RF} = \frac{AF}{RF} = \frac{VT}{VM} < \frac{LS}{LM} = \frac{LN}{LM} < \frac{EG}{H}.$$

But

$$\frac{PF}{RF} = \frac{UF}{AF} = \frac{UQ}{AO};$$

thus

$$\frac{UQ}{AO} < \frac{EG}{H}.$$

The ratio of the perimeters of the two polygons is equal to $\frac{UQ}{AO}$; it is thus less than $\frac{EG}{H}$.[34]

To find polygons whose ratio of areas is less than $\frac{EG}{H}$, we consider the length *X* such that $\frac{LN}{X} = \frac{X}{LM}$, and we do the same construction starting with

[34] Cf. Archimedes, *The Sphere and the Cylinder*, I.3 and 4.

the lengths LN and X. We then find two polygons with respective sides C_1 and C_2 such that

$$\frac{C_1}{C_2} < \frac{LN}{X};$$

hence

$$\frac{C_1^2}{C_2^2} < \frac{LN^2}{X^2}.$$

But

$$\frac{LN^2}{X^2} = \frac{LN^2}{LN \cdot LM} = \frac{LN}{LM};$$

thus

$$\frac{C_1^2}{C_2^2} < \frac{LN}{LM} < \frac{EG}{H}.$$

The ratio of the areas of the polygons is thus less than $\frac{EG}{H}$.[35]

Lateral area of a right circular cone (continued)

Lemma 15. — *The lateral area of the cone of revolution is equal to that of a circle whose radius is the mean proportional of the cone's generator and the radius of its base.*[36]

Let AG and AE be the generator and the radius of the base of the cone and IM such that $\frac{IM}{AG} = \frac{AE}{IM}$. We shall show that the area S' of the circle (M, IM) is equal to the lateral area S of the cone.

Al-Khāzin reasons by *reductio ad absurdum* and supposes at first that $S > S'$. By the previous proposition, we can construct a polygon circumscribed about the circle (M, IM) and a polygon inscribed in the circle. Let $INLS$ and $OKPH$ be hexagons such that

$$\frac{\text{area }(INLS)}{\text{area }(OKPH)} < \frac{S}{S'}.$$

Let $AUCQ$ and $XBJD$ be the hexagon circumscribed about the circle (E, AE) and the hexagon inscribed in that circle. We have

$$\frac{\text{area }(AUCQ)}{\text{area }(INLS)} = \frac{AE^2}{IM^2} = \frac{AE^2}{AE \cdot AG} = \frac{AE}{AG}.$$

[35] Archimedes, *The Sphere and the Cylinder*, I.5.
[36] Archimedes, *The Sphere and the Cylinder*, I.14.

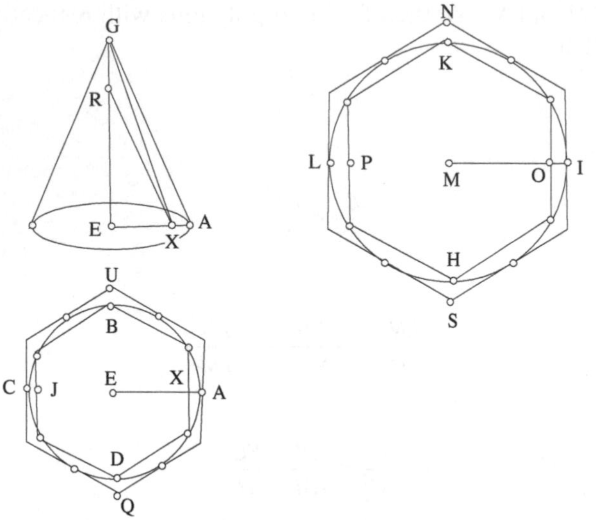

Fig. 4.2.18

But by Proposition 11

$$\frac{AE}{AG} = \frac{\text{area } (AUCQ)}{\text{lateral area of pyramid } (G,\ AUCQ)}.$$

Then

$$\text{area } (INLS) = \text{lateral area } (G, AUCQ);$$

hence

$$\frac{\text{area } (G,\ AUCQ)}{\text{area } (OKPH)} < \frac{S}{S'}$$

or

$$\frac{\text{area } (G,\ AUCQ)}{S} < \frac{\text{area } (OKPH)}{S'},$$

which is absurd, as area $(G, AUCQ) > S$ and area $(OKPH) < S'$.

If $S' > S$, then we construct the hexagons *INLS* and *OKPH* such that

$$\frac{\text{area } (INLS)}{\text{area } (OKPH)} < \frac{S'}{S},$$

and let the hexagon *XBJD* be inscribed in the circle (E, EA). We have

$$\frac{\text{area }(XBJD)}{\text{area }(OKPH)} = \frac{XE^2}{OM^2} = \frac{AE^2}{IM^2} = \frac{AE}{AG}.$$

In the plane AEG, let us produce XR parallel to AG. We have

$$\frac{AE}{AG} = \frac{XE}{XR} > \frac{XE}{XG}.$$

But

$$\frac{XE}{XG} = \frac{\text{area }(XBJD)}{\text{lateral area }(G,\ XBJD)};$$

hence

$$\frac{AE}{AG} > \frac{\text{area }(XBJD)}{\text{lateral area }(G,\ XBJD)},$$

and as a result

$$\text{lateral area }(G,\ XBJD) > \text{area }(OKPH)$$

and

$$\frac{\text{area }(INLS)}{\text{lateral area }(G,\ XBJD)} < \frac{S'}{S},$$

which is absurd as area $(INLS) > S'$ and lateral area $(G, XBJD) < S$. Consequently, $S = S'$.

Comment. — If we designate by p the perimeter of the circle at the base, by r its radius, and by l the length of the generator, we have established in Lemma 13

$$S = \tfrac{1}{2}p \cdot l.$$

We thus have

$$S = \pi r \cdot l.$$

Setting

$$\rho^2 = r \cdot l,$$

we have

$$S = \pi \cdot \rho^2, \qquad \text{area of the circle of radius } \rho.$$

Al-Khāzin does not use the expression for the perimeter of the circle as a function of its radius, whence the necessity of a new proof by *reductio ad absurdum*.

Lateral area of the frustum of a cone and application

Lemma 16. — Let *ABC* be an isosceles triangle with axis *BD*. A parallel to *CA* cuts *BA* at *E*, *BD* at *M* and *BC* at *G*. Let *I* be a point on the extension of *DB*; the triangle *IGE* is isosceles. We construct by the same procedure the isosceles *NKL*.

Turning around the line *BD*, the right-angled triangles *ABD*, *EIM*, *KLS* produce cones of revolution.

The lateral area of the frustum of the cone delimited by the circles (D, DA) *and* (M, ME) *is equal to the area of a circle of radius* O *such that*

$$O^2 = AE \cdot (AD + EM).$$[37]

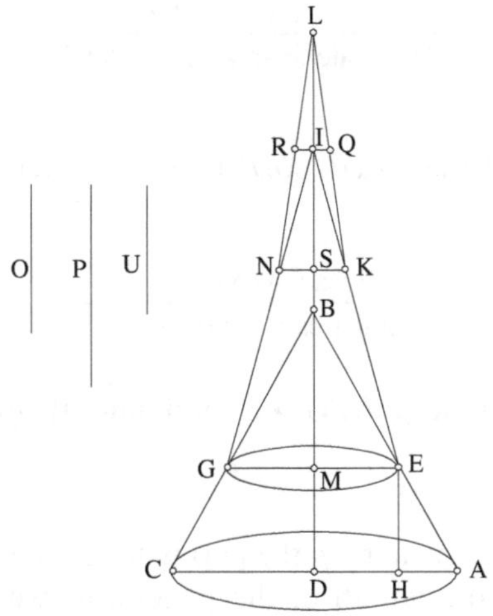

Fig. 4.2.19

Proof: Let *EH* be parallel to *BD*, *EH* = *MD*. Let us consider the segments *O*, *P* and *U* such that

$$O^2 = AE\,(AD + EM)$$

$$P^2 = AB \cdot AD$$

$$U^2 = EB \cdot EM.$$

By the previous proposition, the lateral areas of the cones *ABC* and *EBG* are respectively equal to those of the circles of radii *P* and *U*. Their difference is the desired area. But

[37] Archimedes, *The Sphere and the Cylinder*, I.16.

$$BA \cdot AD = BE \cdot AD + EA \cdot AD = BE \cdot EM + BE \cdot AH + EA \cdot AD.$$

The triangles BDA and EHA are similar; thus $BE \cdot AH = EA \cdot EM$ and

$$BA \cdot AD = BE \cdot EM + EA \cdot EM + EA \cdot AD,$$

$$BA \cdot AD = U^2 + O^2,$$

that is,

$$O^2 = P^2 - U^2.$$

The lateral area of the *frustum* of the cone is thus equal to that of a circle of radius O.

We see that al-Khāzin begins here with the expression for the lateral area of the cone found in the previous proposition, that is, $S = \pi \rho^2$, with $\rho^2 = r\, l$, r the radius of the base and l the generator. This expression is none other than the one established by the Banū Mūsā in Proposition 9, *i.e.* $S = \pi\, r\, l$, and which they will use in Proposition 11 for the lateral area of the *frustum* of the cone.

Likewise, the lateral area of the *frustum* of the cone defined by the trapezoid $EKNG$ is equal to a circle of radius O_1 such that $O_1^2 = KE \cdot (KS + EM)$, and so forth.

If we suppose $AE = EK = KQ$, the lateral area of the solid between the circle (I, IQ) and the circle (D, DA) is equal to the area of a circle of radius R_1 such that

(a) $$R_1^2 = AE \cdot (DA + 2ME + 2SK + IQ).$$

If we consider the solid with vertex L described by $LQKEAD$, with $LQ = KQ$, its area is equal to that of a circle of radius R_2 such that

(b) $$R_2^2 = AE \cdot (DA + 2ME + 2SK + 2IQ).$$

The approach of al-Khāzin is analogous to that of the Banū Mūsā in the second part of their Proposition 11.

Sphere

Let there be a sphere with centre H, $ABCD$ one of its great circles, AC and BD being two perpendicular diameters, and let $AEGBIKC$ be a regular polygonal line inscribed in the semi-circle ABC, and LMN the semi-circle inscribed in this line.

Lemma 17. — *The lateral area of the solid generated by the rotation of the line* AEGB *about the line* BH *is less than twice the area of the circumscribed circle* (H, HA) *and greater than twice the area of the inscribed circle* (H, HL).

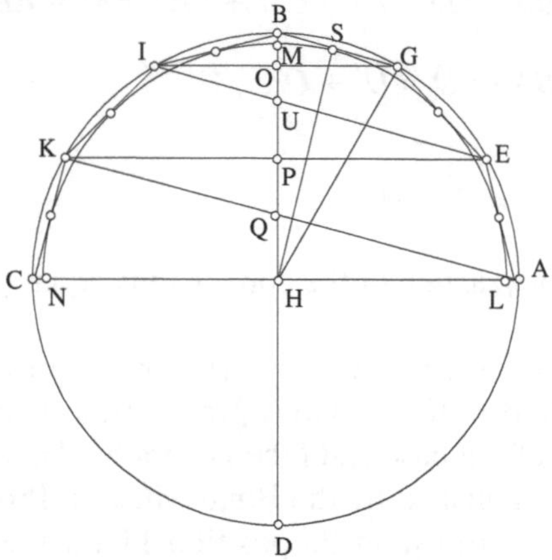

Fig. 4.2.20

Let P, O and S be the respective midpoints of EK, GI and GB, and let Q and U be the intersections of the line HB with the lines AK and EI. The lines GI, KE and AC are parallel, the lines EI and AK are also parallel, and so it follows that the triangles GBO, IUO, EPU, KPQ and AHQ are similar, and the triangle BSH is similar to them since HS is a bisector of the angle BHG and perpendicular bisector of BG. From this, we deduce

$$\frac{OG}{OB}=\frac{OI}{OU}=\frac{PE}{PU}=\frac{PK}{PQ}=\frac{AH}{HQ}=\frac{OG+OI+PE+PK+AH}{BO+OU+UP+PQ+HQ}=\frac{GI+EK+AH}{BH}.$$

But

$$\frac{OG}{OB}=\frac{SH}{SB};$$

thus

$$SB\cdot(GI+EK+AH)=BH\cdot SH.$$

And by the previous proposition, the lateral area of the solid generated by $AEGB$ is equal to the area of a circle of radius R such that

$$R^2=AE\cdot(GI+EK+AH).$$

We thus have

$$\frac{1}{2}R^2 = BH \cdot SH, \qquad \text{as } AE = 2SB.$$

As a result

$$2SH^2 < R^2 < 2BH^2;$$

thus the lateral area considered lies between twice the area of the great circle *ABCD* and twice the area of the inscribed circle (*H*, *HL*).

The reasoning is done on a polygonal figure *AEGBIKC* containing an even number of sides; we apply result (b) of the previous proposition.

Let us note that the figure there was concave, whereas we apply it here in a convex case; at any rate, this proposition is sufficiently general and applies irrespective of the two cases.[38]

If the polygonal line inscribed in the semi-circle of diameter *AC* has an odd number of sides, let these be *AEGIKC*, it doesn't have a vertex at *B*, the triangle *BSH* is replaced by the triangle *OGH* and we have

$$\frac{OI}{OU} = \frac{PE}{PU} = \frac{PK}{PQ} = \frac{AH}{HQ} = \frac{OI + EK + AH}{OH} = \frac{OH}{OG};$$

hence

$$OG\,(OI + EK + AH) = OH^2.$$

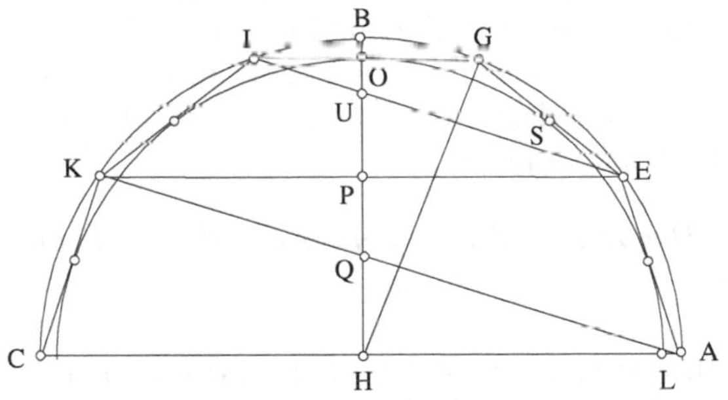

Fig. 4.2.21

By result (a), the lateral area described by the line *AEG* is that of a circle of radius R_1 such that

[38] Does this presentation manifest an intention of al-Khāzin? This demonstration is based on the same principles as that of Archimedes (*The Sphere and the Cylinder*, I.21 and subsequent), known in Arabic after the middle of the ninth century at least and already used by the Banū Mūsā.

$$R_1^2 = AE\,(OI + EK + AH);$$

thus

$$R_1^2 = 2\,OH^2.$$

The lateral area produced by *AEGO* is thus that of a circle of radius R_2 such that

$$R_2^2 = R_1^2 + OG^2 = 2OH^2 + OG^2;$$

thus

$$R_2^2 = OH^2 + HG^2$$

and

$$2\,OH^2 < R_2^2 < 2\,HG^2.$$

The lateral area considered thus lies between twice the area of the great circle *ABCD* and twice the area of the inscribed circle (*H*, *HO*).

Archimedes obtained the same results for a solid defined from a regular polygon whose number of sides is a multiple of 4, in Propositions 27–30 of *The Sphere and the Cylinder*. The Banū Mūsā next treated the same problem for a solid defined from a polygonal line, inscribed in a semi-circle and whose number of sides is even (Propositions 12 and 13). This is precisely the case treated by al-Khāzin. John of Tinemue studied in Proposition 5,[39] on the other hand, the same proposition starting with a regular polygon inscribed in a circle, the number of sides being either a multiple of 4 or just a multiple of 2.

Lemma 18. — *The area* S *of the sphere is equal to four times the area* s *of its great circle.*[40]

Let *ABCD* be a great circle of the sphere and *s* its area.

Suppose $4s < S$; then $4s = S'$, the area of a smaller sphere whose great circle is *LMN*. We then consider a regular polygon circumscribed about the circle *LMN* as in the previous study, a polygon whose vertices are inside the circle *ABCD* or on the circle. Let S'' be the area of the solid described by this polygon. We have

[39] See M. Clagett, *Archimedes in the Middle Ages*, vol. I: *The Arabo-Latin Tradition*, Madison, 1964, pp. 469 sqq.

[40] Archimedes, *The Sphere and the Cylinder*, I.33.

$$S' < S'' < S.$$

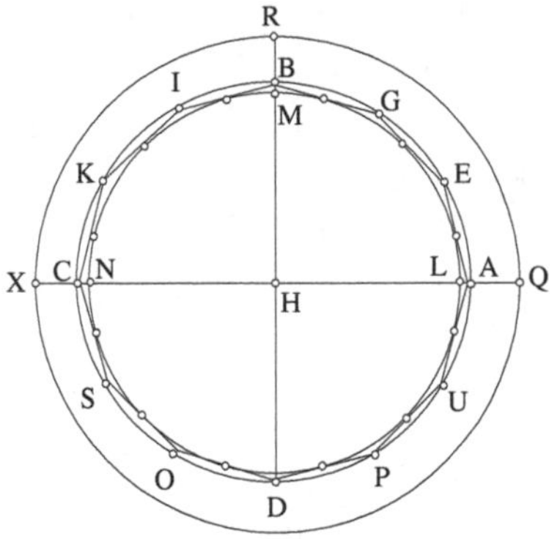

Fig. 4.2.22

By Lemma 17, $S'' < 4s$; thus $S' < 4s$, which is absurd because we have supposed that $S' = 4s$.

Suppose $4s > S$; then $4s = S_1'$, the area of a sphere greater than that of the sphere $ABCD$; let QRX be its great circle. We then consider a polygon circumscribed about the circle $ABCD$ and whose vertices are inside the circle QRX or on the circle. Let S_1'' be the area of the solid generated by this polygon; we have

$$S < S_1'' < S_1'$$

By Lemma 17, $S_1'' > 4s$, and as a result $S_1' > 4s$, which is absurd because by hypothesis $S_1' = 4s$.

Thus $S = 4s$: the area of the sphere is four times the area of its great circle, or again the product of the diameter of the great circle with its circumference.

Note that the Banū Mūsā for this same proposition use (cf. Proposition 14 of their treatise) in the two parts of the reasoning a solid inscribed in the larger of the two spheres, and not having any common points with the smaller, a solid obtained from the *Elements* XII.6.

Lemma 19. — *The volume* V *of the sphere is the product of the radius* R *of a great circle with a third of the surface* S *of the sphere.*

Let $ABCD$ be a great circle of the sphere. Suppose $V > \frac{1}{3} R \cdot S$; then there exists a smaller sphere whose volume is $V' = \frac{1}{3} R \cdot S$; let LMN be a great circle of that sphere.

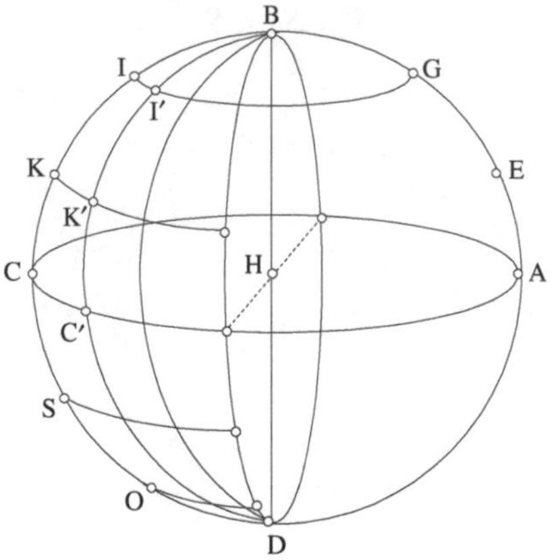

Fig. 4.2.23

We consider two perpendicular planes to the plane $ABCD$, one along AC, the other along BD; they intersect the sphere along the great circles. We consider the circles of diameter BD that divide each quarter of the circle of diameter AC in three equal parts. We thus have in all six circles of diameter BD. On each of these, we consider a polygon such as $AEGBIKC$. The vertices of all these polygons define a polyhedron whose faces are trapezoids or triangles. It is the polyhedron defined in the *Elements* XII.17 – see comment c). We associate with each of the faces a pyramid whose vertex is the centre H of the sphere.

Al-Khāzin supposes that the sphere LMN is tangent to each of these faces (see the comment at the end of this proposition); each pyramid of vertex H has in this case the radius R' of the sphere LMN for its height. The volume V_1 of the solid is thus the product of R' with a third of the total surface S_1 of the solid:

$$V_1 = R' \cdot \tfrac{1}{3} S_1 \text{ and } V_1 > V'.$$

But

$$R > R' \text{ and } S > S_1;$$

thus

$$R \cdot \tfrac{1}{3} S > V_1 > V',$$

which is absurd as we have supposed that $V' = R \cdot \frac{1}{3} S$.

If $V < \frac{1}{3} R \cdot S$, there exists a sphere greater than the sphere $ABCD$ whose volume is $V' = \frac{1}{3} R \cdot S$; let the sphere be QRX. In this sphere, we inscribe a polyhedron of the preceding type such that the sphere $ABCD$ is tangent to the faces of the polyhedron.[41] Let V', V_1 and V be respectively the volumes of the sphere QRX, of the polyhedron and of the sphere $ABCD$; we have

$$V_1 = \frac{1}{3} R \cdot S_1.$$

But

$$S_1 > S;$$

thus

$$V_1 > \frac{1}{3} R \cdot S,$$

which is absurd as $V_1 < V'$ and we have supposed that $V' = \frac{1}{3}R \cdot S$.

The volume of the sphere $ABCD$ is thus $V = \frac{1}{3} R \cdot S$. But if s designates the area of the great circle, one has

$$S = 4s,$$

whence

$$V = (1 + \frac{1}{3}) R \cdot s$$

and

$$2 s R = \frac{3}{2} V.$$

41 By the comment that concludes this proposition, we can suppose that the polyhedron inscribed in the sphere QRX is such that the sphere $ABCD$ does not intersect its faces. We have

$$\frac{1}{3} h_3 S_1 < V_1 < \frac{1}{3} h_1 S_1, \qquad \text{with } R \le h_3.$$

We know that $V_1 < V'$, whence

$$\frac{1}{3} h_2 S_1 < V'.$$

But $h_2 \ge R$ and $S_1 > S$; then

$$\frac{1}{3} h_2 S_1 > \frac{1}{3} R \cdot S;$$

but by hypothesis $V' = \frac{1}{3} R \cdot S$, which is absurd.

The cylinder associated with the sphere has as its volume

$$v = 2\,R \cdot s;$$

thus

$$v = \frac{3}{2}V.$$

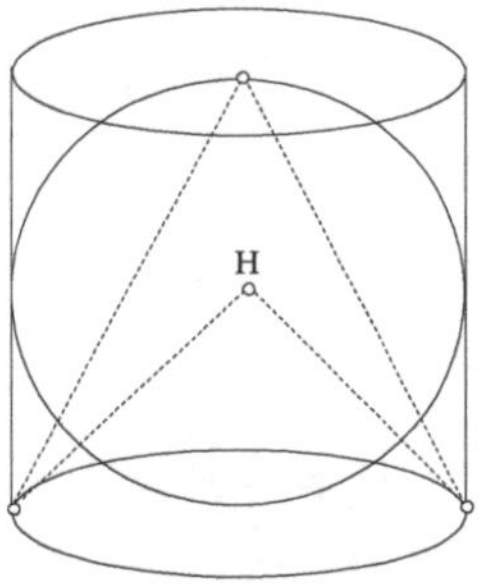

Fig. 4.2.24

The cone associated with this cylinder, a cone of height 2*R*, has as its volume

$$v_1 = \frac{1}{3}\,v = \frac{1}{2}V.$$

The cone of vertex *H* and of height *R* has as its volume

$$v_1' = \frac{1}{2}\,v_1;$$

thus

$$V = 4\;v_1'.$$

The cone whose base is a circle of area 4*s* (circle of radius 2 *R*) and whose height is *R* has the same volume as the sphere.

Comment. — Al-Khāzin's reasoning relies on the existence of a polyhedron inscribed in the sphere *ABCD* and circumscribed about the sphere *LMN*. Three remarks are called for:

a) The sphere *LMN* intersects the faces of the polyhedron. The circle *LMN* of Proposition 18 is tangent to the chords *BI, IK, KC* … and by construction in Proposition 19, the arcs defined on the equatorial circle of diameter *CA* and on the meridianal circles of diameter *BD* are all equal to the arc *BI*; their chords are thus equal to the chord *BI*, and the sphere *LMN* is tangent to all these chords. From this we deduce that the sphere *LMN* is intersected by all the faces of the polyhedron. Indeed, the midpoints of these equal chords are the points of contact of the sphere *LMN* with these chords;

each face of the polyhedron thus has at least two points on the sphere, and thus it intersects that sphere *LMN*.

b) The polyhedron does not admit an inscribed sphere. The distance from the point H to the faces of the polyhedron is variable. Let T be the midpoint of CC'; the plane HBT passes through the midpoints V and W of KK' and II'. $\widehat{CC'} > \widehat{KK'} > \widehat{II'}$; hence $HT < HV < HW$. If we designate by h_1, h_2, h_3 the respective distances from H to the plans of the faces BII', $II'K'K$, $KK'C'C$, we have $R > h_1 > HW > h_2 > HV > h_3 > HT$. Indeed, the triangle BII' and the trapezoids $II'K'K$ and $KK'C'C$ are isosceles with $IB > II'$, $IK > KK'$, $KC = CC'$; thus the angles IBI', KIK', CKC' are acute and the centres of the circumscribed circles are respectively the segments WB, VW and IV.

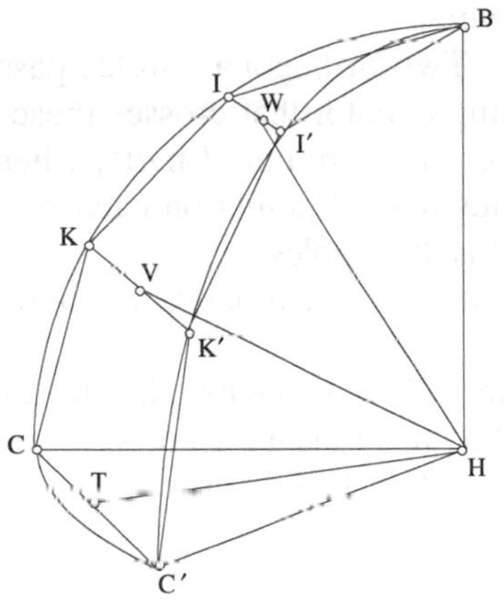

Fig. 4.2.25

More generally, whatever the number of sides, *i.e.* whatever the number n of subdivisions on each quarter of the circle, we have

$$R > h_1 > h_2 > \ldots h_{n-1} > h_n.$$

The polyhedron thus does not admit an inscribed sphere. Its volume V_1 satisfies

$$\frac{1}{3} h_n \cdot S_1 < V_1 < \frac{1}{3} h_1 \cdot S_1 < \frac{1}{3} R \cdot S_1.$$

To apply al-Khāzin's reasoning, we can consider that we have chosen n sufficiently large that the sphere *LMN* whose volume is $V' = \frac{1}{3} R \cdot S$ and whose radius is $R' < R$ does not intersect the faces of the polyhedron, *i.e.* n such that $h_n \geq R'$. We then have

$$V' < V_1 < \tfrac{1}{3}R \cdot S_1;$$

thus

$$\tfrac{1}{3}R \cdot S < \tfrac{1}{3}R \cdot S_1,$$

which is absurd as $S > S_1$.

c) Note as well that al-Khāzin imagines here a solid of the following type:

Polyhedron inscribed in the sphere

Let B and D be the poles. Two orthogonal circles pass through B and D; one traces the corresponding equator that crosses these circles, whence four points. Al-Khāzin divides each arc into 3 parts, whence there are 12 points on the equator; one thus has 12 points on each of the 6 associated meridians, that is, 10 points, plus the 2 poles.

If we divide each of the four arcs into n parts, we have $4n$ points on the equator, hence $2n$ meridians.

On each meridian there are $2(2n - 1)$ points, plus the 2 poles.

In all there are $4n(2n - 1)$ points, plus the 2 poles.

The polyhedron thus has $4n(2n - 1) + 2$ vertices:

$n = 1$	6 vertices
$n = 2$	26 vertices
$n = 3$	62 vertices, this is al-Khāzin's example
$n = 4$	114 vertices.

If A_n and V_n designate the area and volume of the solid Σ_n, and if A and V designate the area and volume of the sphere, then

A_n increases with n	$A_n < A$
V_n increases with n	$V_n < V$.

Theorem 20. — *Of all the convex solids having the same area, the sphere is that which has the greatest volume.*

Let there be a sphere with centre O, with R its radius, S its area and V its volume and let there be a polyhedron with the same area as S, with volume V_1; we suppose it circumscribed about a sphere *LMN* with centre *H*, of radius R', with area S'. We then have

$$V_1 = \frac{1}{3}S \cdot R'.$$

The area S' is less than that of the polyhedron; thus $S' < S$ and, moreover, $R' < R$. Therefore

$$\frac{1}{3}S \cdot R' < \frac{1}{3}S \cdot R,$$

i.e.

$$V_1 < V.$$

Note that the nature of the polyhedron is not specified, but the demonstration supposes that this polyhedron is circumscribed about a sphere, which is the case for a regular polyhedron, but the demonstration made here does not apply to an arbitrary polyhedron or solid.

Comment. — Examples of solids having the same area *S* as a sphere of radius *R*.

If a cylinder of radius *R* has height *R*, its lateral area is

$$2\pi R \cdot R;$$

its total area is thus

$$S = 4\pi R^2$$

and its volume is

$$V = \pi R^3 < \frac{4}{3}\pi R^3.$$

If a cone has a base of radius *R* and a generator with $l = 3R$, its total area is

$$S = \pi R\,(R + l) = 4\pi R^2;$$

its height is then *h* such that

$$h^2 = l^2 - R^2 = 8R^2,\ h = 2\sqrt{2}R$$

and its volume is

$$V = \frac{1}{3}\pi R^2 \cdot 2\sqrt{2}R = \frac{2\sqrt{2}}{3}R^3 < \frac{4}{3}\pi R^3.$$

As we have seen, al-Khāzin does not proceed by comparison of polyhedra, but he achieves the result using the formula that relates the volume of the sphere to its area, a formula that he obtains by approaching the sphere by non-regular polyhedra. Ibn al-Haytham's approach will be completely different: he tries to proceed by comparing regular polyhedra of the same area, and with a different number of faces, in order to be able to give a dynamic demonstration. This fails because of the finite number of regular polyhedra; consequently, instead of solving the initial problem, he develops an original theory of the solid angle. He is thus well within the family of Zenodorus and of Pappus that al-Khāzin belongs to once again, a family that is not Ibn al-Haytham's.

4.2.4. *The opuscule of al-Sumaysāṭī*

This text of al-Sumaysāṭī, widely circulated, contains a single result which had been demonstrated by al-Khāzin – this is the final step of the latter's reasoning in Theorem 10. All the results concerning irregular polygons are, in all evidence, absent from this text.

The area of the circle is greater than that of every regular polygon of the same perimeter.

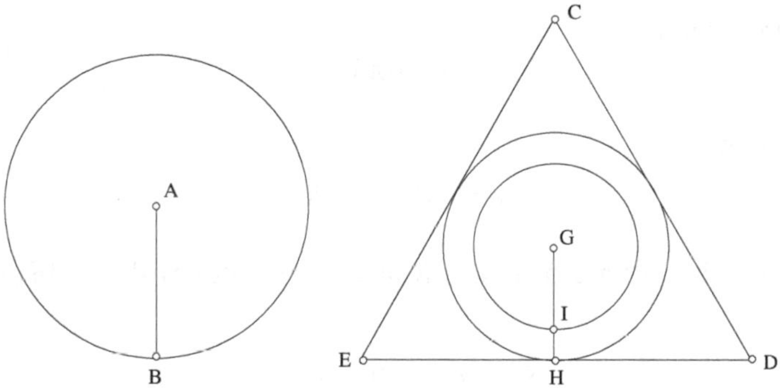

Fig. 4.2.26

Let there be a circle (A, AB) and a regular polygon CDE of the same perimeter p as the circle. Let G be the centre of the circle inscribed in CDE and GH a radius; H is for example the midpoint of DE. The product of the semi-perimeter p with GH is the area of the polygon.

If $GH = AB$, then the circle (G, GH) also has p as perimeter, and it has the same area as the polygon, which is absurd.

If $GH > AB$, then the perimeter of the circle G would be greater than that of the circle A, and the perimeter of CDE, which is greater than that of the circle G, would be even larger than that of the circle A, which is absurd. We thus have

$$GH < AB,$$

$$\text{area of the circle } (A) = \frac{1}{2} p \cdot AB,$$

$$\text{area of the polygon} = \frac{1}{2} p \cdot GH;$$

thus

$$\text{area of the circle } (A) > \text{area of the polygon } (CDE).$$

4.3. *Translated texts*

4.3.1. *Commentary on the First Book of the* Almagest

4.3.2. *The Surface of any Circle is Greater than the Surface of any Regular Polygon with the Same Perimeter* (al-Sumaysāṭī)

In the Name of God, the Merciful, the Compassionate

TRANSCRIPT[1] OF THE COMMENTARY BY ABŪ JAʿFAR MUḤAMMAD AL-ḤASAN AL-KHĀZIN

On the First Book of the *Almagest*

Ptolemy said that, of different figures with equal perimeters, that with the most angles is the largest. It is for this reason that it necessarily follows that the circle is the largest surface and the sphere is the largest solid.

He means that, for different polygons,[2] such as the triangle, the square, and the pentagon, and so on until infinity, if the sum of the sides of each of them is equal to that of the sides of the others, then that with the most angles has the largest area. Hence, for the triangle, the square and the pentagon, if the sum of the sides of each of them is ten, then the square has a greater area than the triangle, and the pentagon has a greater area than the square, and so on to infinity for the polygons.[3] Finally, the circle whose circumference is ten is greater than all of them. It is easy to verify this using arithmetic. To prove it geometrically, we first require a number of lemmas. We say that:

<Lemmas>

Of two polygons having the same number of sides and the same perimeter, that with equal sides and equal angles is greater than any of the others.

Some preliminary propositions are given below.

<1> The triangle *ABC* is equilateral and the triangle *ADE* is isosceles.

I say that the amount by which AB *exceeds* AD *is less than the amount by which* AD *exceeds* BE *and that <the sum of> these two amounts is equal to the amount by which* AB *exceeds* BE, *that is* BD.

[1] Lit.: We have transcribed.
[2] Lit.: figures having *latera recta.*
[3] Lit.: figures having many sides.

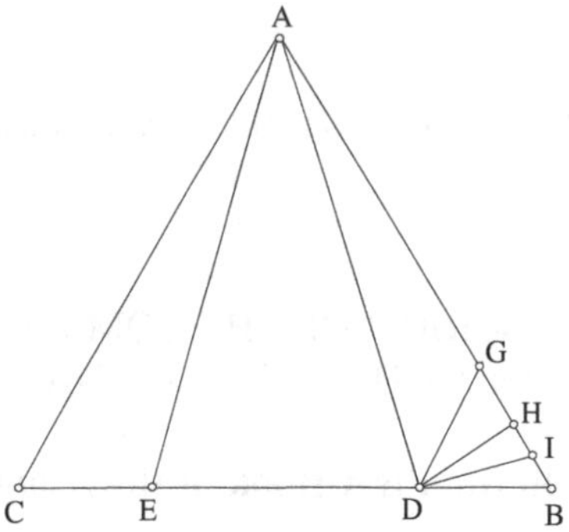

Fig. 1

Proof: We draw *DG* parallel to *AC*, and drop a perpendicular *DH* onto *AB*. The triangle *BGD* is then equilateral, and therefore *BH* is equal to *GH*, but *AH* is less than *AD*. We take *AI* as being equal to *AD*. Then *BI* is less than *IG*. But *BI* is the amount by which *AB* exceeds *AD*, and *IG* is the amount by which *AD* exceeds *BE*, as *GA* is equal to *DC* and *DC* is equal to *BE*, and the sum of *BI* and *IG* is equal to *BD*.

Consequently, the amount by which twice *AB* exceeds <the sum of> *BE* and *AD*, which is <the sum of> *BD* and *BI*, is less than the amount by which <the sum of> *AB* and *AD* exceeds twice *BE*, which is <the sum of> *BD* and *IG*. If we take *AB* as being common to both twice *AB* and <the sum of> *BE* and *AD*, and *BE* as being common to both <the sum of> *AB* and *AD* and twice *BE*, then the amount by which three times *AB* exceeds the sum of *AB*, *BE* and *AD*, that is *EA*, is less than the amount by which the sum of *AB*, *BE* and *EA* exceeds three times *BE*. The product of three times *BE* and three times *AB* – that is the product of *BE* and nine times *AB* – will be less than the square of the sum of *AB*, *BE* and *EA*, as the ratio of the first to the second is less than the ratio of the second to the third, since the excess subtracted from the first, which is <the sum of> *BD* and *BI*, in order for the remainder to be the second is less than the excess subtracted from the second, which is <the sum of> *BD* and *IG*, in order for the remainder to be the third.

– **2** – The triangle *ABC* is equilateral and the triangle *ADE* is isosceles.

I say that the ratio of the square of the perimeter ADE *to the square of the perimeter* ABC *is greater than the ratio of the triangle* ADE *to the triangle* ABC.

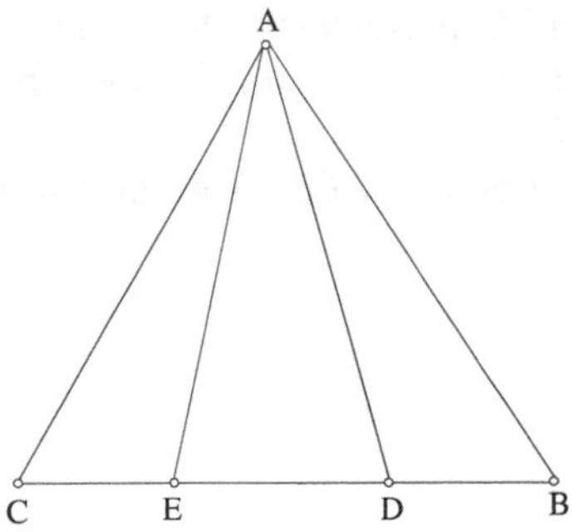

Fig. 2

Proof: The square of the perimeter ABE is greater than the product of BE and nine times BC, and also the square of the perimeter of ADC is greater than the product of DC and nine times BC. This is equal to the product of BC and nine times BC plus the product of DE and nine times BC. However, the product of BC and nine times itself is equal to the square of the perimeter ABC. Therefore, <the sum of> the squares of the perimeters ABE and ADC is greater than the square of the perimeter ABC plus the product of DE and nine times BC. But the line equal to <the sum of> the perimeters ABC and ADE has been divided into to equal parts which are ABE and ADC, and into two different parts which are ABC and ADE. Therefore, <the sum of> the squares of the perimeters ABC and ADE is greater than <the sum of> the squares of the perimeters ABE and ADC. But we have shown that <the sum of> the squares of ABE and ADC is greater than the square of the perimeter ABC plus the product of DE and nine times BC. <The sum of> the squares of the perimeters ABC and ADE is therefore much greater than the square of the perimeter ABC plus the product of DE and nine times BC. Subtracting the square of the perimeter ABC, which is common, the square of the perimeter ADE is greater than the product of DE and nine times BC. Therefore, the ratio of the square of the perimeter ADE to the square of the perimeter ABC is greater than the ratio of the product of DE and nine times BC to the square of the perimeter ABC. But the product of DE and nine times BC is equal to the product of three times DE and three times BC, and the ratio of the product of three times DE and three times BC to the square of the perimeter ABC is equal to the ratio of three times DE to three times BC, and is equal to the ratio of DE to BC, and is equal to the ratio of the triangle ADE to the triangle ABC. Therefore, the ratio of the square of the perimeter ADE to the square of the perimeter ABC is greater than the ratio of the triangle ADE to the triangle ABC

– **3** – The triangle ABC is isosceles. Draw the straight line DE through the point A parallel to BC. Draw two straight lines from the points B and C, such that they meet at G.

I say that the sum of BG *and* GC *is greater than the sum of* AB *and* AC.

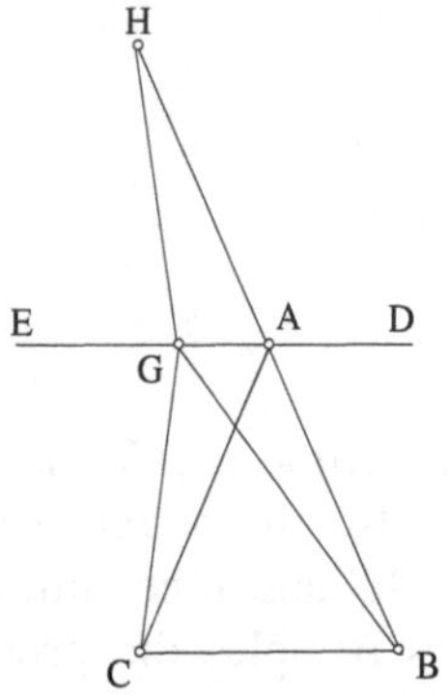

Fig. 3

Proof: We add to AB itself, let it be AH. We join GH. The angle DAB is then equal to the angle GAH, and the angle DAB is equal to the angle ABC which is equal to the angle ACB. But the angle ACB is equal to the angle GAC. Therefore, the angle GAC is equal to the angle GAH. AH is equal to AC, and AG is common to the two triangles AGC and AGH. Therefore, the side GH is equal to GC. The sum of GB and GH is greater than BH. Therefore, the sum of BG and GC is greater than the sum of AB and AC.

– **4** – The triangle ABC is equilateral, the triangle DEG has two equal sides, which are DE and DG, and their perimeters are equal.

I say that the triangle ABC *is greater than the triangle* DEG.

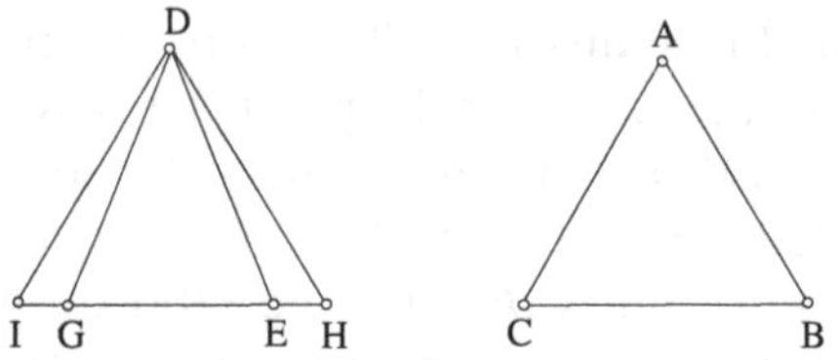

Fig. 4

Proof: We construct the equilateral triangle DHI. The ratio of the square of the perimeter DEG to the square of the perimeter DHI is therefore greater than the ratio of the triangle DEG to the triangle DHI. But the square of the perimeter *DEG* is equal to the square of the perimeter *ABC*. Therefore, the ratio of the square of the perimeter *ABC* to the square of the perimeter *DHI* is greater than the ratio of the triangle *DEG* to the triangle *DHI*. But the ratio of the square of the perimeter *ABC* to the square of the perimeter *DHI* is equal to the ratio of the triangle *ABC* to the triangle *DHI*. Therefore, the ratio of the triangle *ABC* to the triangle *DHI* is greater than the ratio of the triangle *DEG* to the triangle *DHI*. The triangle *ABC* is therefore greater than the triangle *DEG*

– **5** – The triangle *ABC* is equilateral, the triangle *DEG* is scalene, and their perimeters are equal.

I say that the triangle ABC is greater than the triangle DEG.

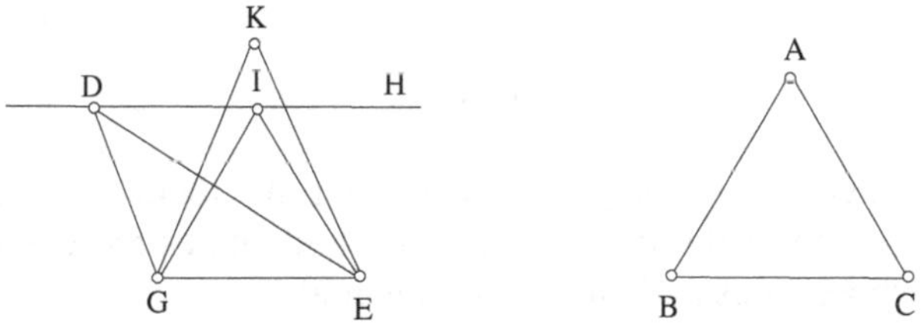

Fig. 5

Proof: Through the point *D*, we draw a straight line *DH* without limits and parallel to *EG*. We draw straight lines onto this straight line from the points *E* and *G* with the two straight lines *EI* and *GI* being equal. The sum of *IE* and *IG* is therefore less than the sum of *DE* and *DG*. From the two points *E* and *G* we draw two equal straight lines *EK* and *GK*, <the sum of which is> equal to <the sum of> the two straight lines *DE* and *DG*. The triangle *EKG* is therefore greater than the triangle *EIG*, and the triangle *EIG* is equal to the triangle *EDG* as they are <constructed> on the same base between two parallel straight lines. The triangle *EKG* is greater than the triangle *EDG*, and the triangle *EKG*, as we have shown, is not greater than the triangle *ABC* as their perimeters are equal. Therefore, the triangle *ABC* is greater than the triangle *DEG*.

From this, we have shown that an isosceles triangle is greater than a scalene triangle when their perimeters are equal.

– **6** – Let us take up the two triangles *ABC* and *DEG*, and double them by means of the straight lines *AH*, *CH* and *GI*, *DI*. The lozenge *BH* is greater than the parallelogram[4] *EI*, one of which has equal sides and one of

[4] Lit.: rectangle lozenge.

which has unequal sides, even though their perimeters are equal. We draw *CK* at a right angle and equal to *AC* from *BC*, and draw *KL* equal and parallel to *BC*. We join *LB* and draw *AH* to *M*. The triangle *AMB* is then equal to the triangle *CNH*, and the rectangle *BMNC* is equal to the lozenge *BH*. Therefore, the square *BK* is greater than the lozenge *BH*, and it is therefore very much greater than the parallelogram *EI*. One has equal sides and equal angles, and the other has unequal sides and unequal angles.

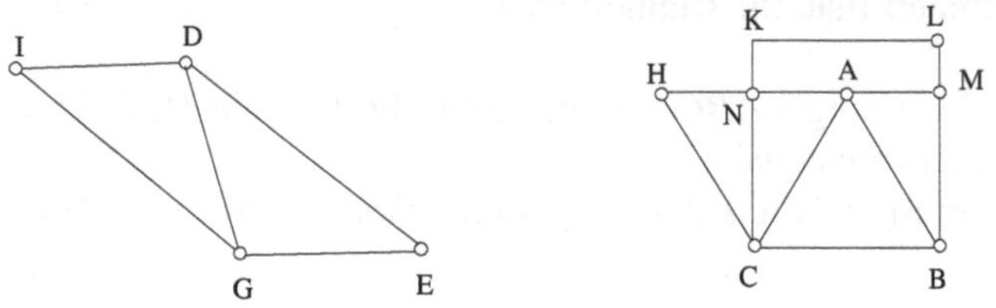

Fig. 6

This is the case of two polygons with equal numbers of sides and equal perimeters. The one of the two whose sides and angles are equal is greater than the one whose sides and angles are unequal.

– **7** – *Example*: The pentagon *ABCDE* has equal sides and equal angles. We join the straight line *EB*, and from the two points *E* and *B* we draw two straight lines which meet at *G* such that their sum is equal to the sum of *AE* and *AB*. The triangle *AEB* is therefore greater than the triangle *EGB*. We set the surface *EBCD* to be common, then the pentagon *ABCDE* is therefore greater than the pentagon *GBCDE*. If the same construction is made on the other sides, then the pentagon *ABCDE* will be much greater than the pentagon whose sides are not equal. Similarly, the angle *EAB* is equal to each of the angles *ABC*, *BCD*, *CDE* and *DEA*, and the angle *EGB* is different from the angle *EAB*. Therefore, the pentagon *ABCDE*, whose angles are equal, is greater than the pentagon *GBCDE* whose angles are not equal.

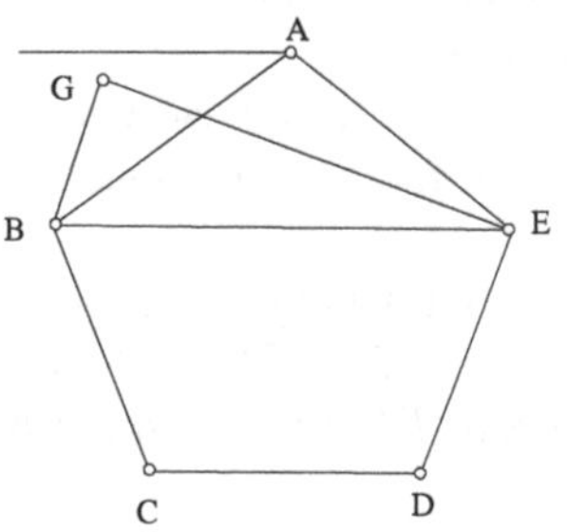

Fig. 7

– **8** – For any polygon circumscribed around a circle, the product of the half-diameter of the circle and half the sum of the sides is equal to the area of the polygon.

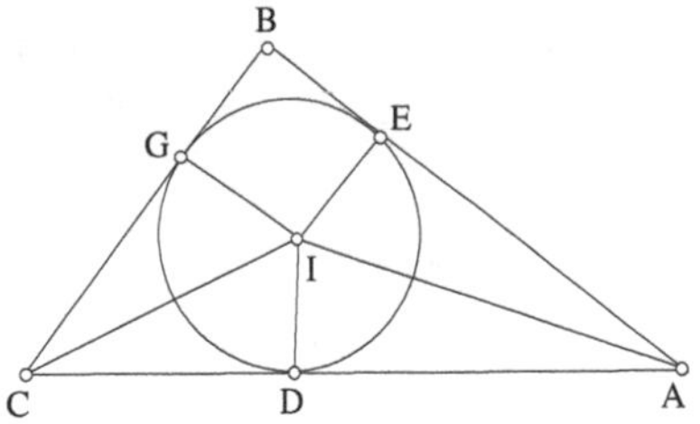

Fig. 8

Let *ABC* be the figure, and let the inscribed circle be *DEG* with its centre at *I*. We draw *ID*, *IE* and *IG*. These will be the perpendiculars to the sides. We join the straight lines *AI*, *BI* and *CI*. The product of *ID* and half of *AC* is the triangle *AIC*. The product of *IE* and half of *AB* is therefore the triangle *ABI* and the product of *IG* and half of *BC* is the triangle *BIC*. The product of the half-diameter of the circle and half the sum of the sides is therefore equal to the area of the triangle *ABC*.

If the polygon has four sides, it can be divided into four triangles. If it has many sides, it can be divided into as many triangles as the number of sides. The product of the half-diameter of the circle inscribed within the polygon and half of each of the sides is the area of each of the triangles. The sum of these triangles is the area of the polygon, and the area of the polygon is greater than the area of the circle – as the product of its half-diameter and half of its circumference is its area – and half of its circumference is less than half of the sum of the sides of the polygon as the polygon surrounds it. It is for this reason that the product of the half-diameter of the circle circumscribed around the polygon and half of the sum of its sides is greater than the area of the polygon, and the product of the half-diameter and half the circumference of the circle is the area of the circle, then the area of the circle is greater than the area of the polygon enclosed by the circle.

– **9** – Given two polygons with the same perimeter, with equal sides and with equal angles, but of two different species, then that with the largest number of angles is the greatest.

Example: Let *ABC* be a triangle; let there be a square *DEGH*, having equal sides and equal angles, and let their perimeters be equal. The square is then greater than the triangle.

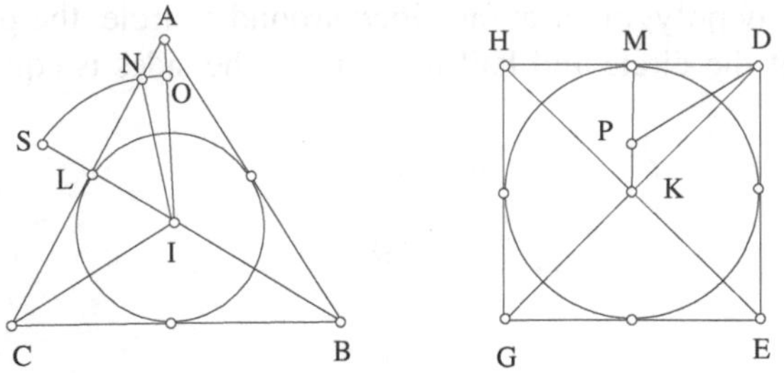

Fig. 9

Proof: We assume the points *I* and *K* to be the centres of the two circles inscribed in the two figures. We join *AI*, *BI*, *CI*, *HK*, *DK*, *EK* and *GK*. The sum of the three angles at the point *I* is equal to the sum of the four angles at the point *K*, as each of the two sums is equal to four right angles. The angle *AIC* is therefore the third of four right angles, and the angle *DKH* is a quarter of four right angles. *AC* is one third of the perimeter of *ABC*, and *DH* is a quarter of the perimeter of the square *DEGH*. The two perimeters are equal, and therefore the ratio of the angle *AIC* to the angle *DKH* is equal to the ratio of *AC* to *DH*. But the angle *AIC* is greater than the angle *DKH* and *AC* is greater than *DH*. We drop the two perpendiculars *IL* and *KM*. Each of the two angles *AIC* and *DKH*, and each of the two sides *AC* and *DH* are divided into two halves. The ratio of the angle *AIL* to the angle *DKM* is equal to the ratio of *AL* to *DM*. But the angle *AIL* is greater than the angle *DKL*, and *AL* is greater than *DM*. We take *NL* equal to *DM*, we join *NI*, and we draw the arc *SNO* at a distance of *IN* around the point *I*. We extend *IL* to <the point> *S*. Then the ratio of the angle *AIN* to the angle *NIL* is equal to the ratio of the sector *INO* to the sector *INS*. But the ratio of the sector *INO* to the sector *INS* is less than the ratio of the triangle *AIN* to the triangle *NIL*. But the ratio of the triangle *AIN* to the triangle *NIL* is equal to the ratio of *AN* to *NL*. The ratio of the angle *AIN* to the angle *NIL* is therefore less than the ratio of *AN* to *NL*. Composing, the ratio of the angle *AIL* to the angle *NIL* is less than the ratio of *AL* to *NL*. But *NL* is equal to *DM*. The ratio of the angle *AIL* to the angle *NIL* is therefore less than the ratio of *AL* to *DM*. But the ratio of the angle *AIL* to the angle *DKM* is equal to the ratio of *AL* to *DM*. The ratio of the angle *AIL* to the angle *NIL* is therefore less than the ratio of the angle *AIL* to the angle *DKM*. But the two angles *ALI* and *DMK* are right angles. Therefore the remaining angle *INL* in the triangle *ILN* must be less than the angle *KDM* in the triangle *KMD*. We construct the angle *MDP* equal to the angle *INL*. The triangle *MDP* is similar to the triangle *LNI*. But *DM* is equal to *NL*. Therefore, *MP* is equal to *LI*, and the product of half the perimeter of the square *DEGH* and *KM* is therefore greater than its product with *PM*. Now, its product with *KM* is the area of the square *DEGH*, and its product with *PM* is the area of the triangle *ABC*.

Using a similar procedure, it can be shown that, given two polygons having equal sides and equal angles, among polygons with the same perimeter, that with the largest number of angles has the greatest area.

– **10** – We take the triangle *ABC* given above, without the sector, and at the same time we draw a circle *DEG* with centre *K*; let their perimeters be equal.

I say that the circle is greater than the triangle.

Proof: We draw the equilateral triangle *MNS* circumscribed around the circle, and we join *KM* and *KS*. The ratio of the perimeter of the triangle *MNS* to the perimeter of the triangle *ABC* is therefore equal to the ratio of the side *MS* to the side *AC*. But the perimeter of the triangle *MNS* is greater than the perimeter of the triangle *ABC* as it is greater than the circumference of the circle *DEG*. Therefore, the side *MS* is greater than the side *AC*. We drop the perpendicular *KD*. *MD* is then greater than *AL*. The angle *MKS* is equal to the angle *AIC* as each of them is one third of four right angles, the angle *MKD* is half of the angle *MKS*, and the angle *AIL* is half of the angle *AIC*. Therefore, the angle *MKD* is equal to the angle *AIL*. But the angle *MDK* is a right angle, and equal to the angle *ALI*. Therefore, the triangle *MKD* is similar to the triangle *AIL*. But *MD* is greater than *AL*. Therefore *DK* is greater than *LI*. On *DK*, we mark off *DO* equal to *LI*. But the product of half of the circumference of the circle *DEG* and *DK* is the area of the circle, and its product with *DO* is the area of the triangle *ABC*. Therefore, the circle is greater than the triangle.

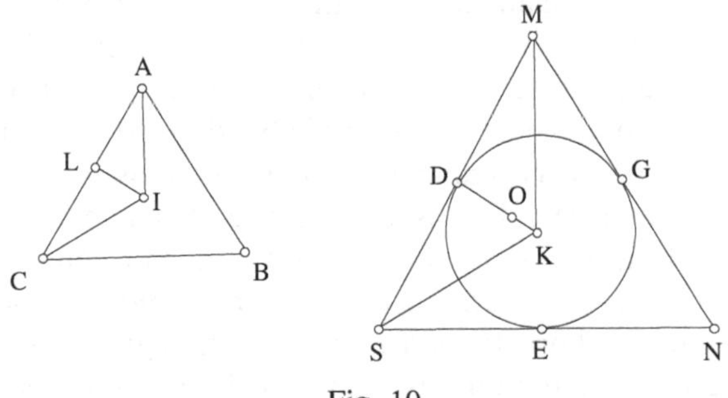

Fig. 10

We also compare this circle with the square *DEGH* from the previous proposition, taking the square in place of the triangle *ABC*, and we imagine that the sum of its sides is equal to the circumference of the circle *DEG*. We construct a square on this circle in place of the triangle *MNS* and we then show, using a similar proof to the first, that its area is greater than the area of the square *DEGH*.

Similarly, we compare it with a regular pentagon, the sum of whose sides is equal to its circumference, and with any figure taken from among the regular polygons beyond the pentagon regardless of the number of sides and angles, and show that the circle is the greatest of these polygonal figures having the same perimeter.

It is possible that we could have proved that which we have proved using two figures with unequal sides, providing that they are similar, and using a procedure similar to that which we have used, replacing the two triangles *ABC* and *MNS* with the two four-sided or many-sided figures with unequal sides but similar. We have, however, preferred to show this using two regular polygons, as each of them is greater than their homologue with different sides and with equal perimeters, as we have shown earlier.

Following this, we can show that the sphere is the greatest of the solid figures with equal surfaces,[5] regardless of whether these surfaces are plane as in the cube, the prism and the pyramid,[6] or whether they are curved as in the sphere, the cylinder and the cone.[7]

<11> We begin with the regular triangular pyramid.[8] This pyramid is the basic element of all these figures, in the same way as the triangle is the basic element in all plane figures with sides. We draw it according to this shape, and we imagine its base – the equilateral triangle *ABC* – located on a plane parallel to the horizon. The point *D*, at its vertex, is in the air, as are the triangles *ABD*, *ADC* and *BDC*. Each of these is isosceles, and the straight line *DE* is perpendicular to the plane of the base. If the sides of each of these triangles are equal and equal to the sides of the base *ABC*, then the pyramid is the first of the five figures mentioned at the end of the *Elements*, being that called the fire figure from its resemblance to the shape of a flame, such as the light from a candle or similar lights derived from fire, providing that the conical shape of the flame leans more towards the circular, even if its base has straight sides. This is so because this name is used for all pyramids whose base has sides that are straight and equal, regardless of whether the number of these sides is three, four, or more, up to as many as you wish, and all of whose faces are isosceles triangles. The rule for this species of pyramid is the same as that which we now explain for this

[5] Lit.: with equal limits, *i.e.* isepiphanic.

[6] Lit.: the cone whose base has straight sides.

[7] Lit.: the cone of the cylinder.

[8] Lit.: the cone whose base is triangular with equal straight sides.

pyramid: <To find> the area of the surface, excluding the area of the base, we multiply the perpendicular dropped from the point *D* on one of the sides *AB*, *BC* or *AC*, dividing it into two halves, by half of the sum of the sides, as the product of this perpendicular and half of one side is the area of a single triangle, and the product with three halves of the sides is the area of all three triangles making up the outside surface of the volume of the pyramid. But, as the <product of the> half-diameter of the circle inscribed within the triangle *ABC* and half the sum of its sides is the area of the triangle, the product of the sum of the perpendicular and the half-diameter of the circle multiplied by half the sum of the sides *AB*, *BC* and *AC* is the area of the surface of the entire pyramid.

<12> As the prism whose base is the triangle *ABC* and whose perpendicular is *ED* can be divided into three equal pyramids, as shown in Proposition Six of Book Twelve of the *Elements*, the pyramid *ABCD* is one third of the prism. But the product of the perpendicular *DE* and the surface *ABC* is the volume of the prism; therefore its product with one third of the surface *ABC* is the volume of the pyramid.

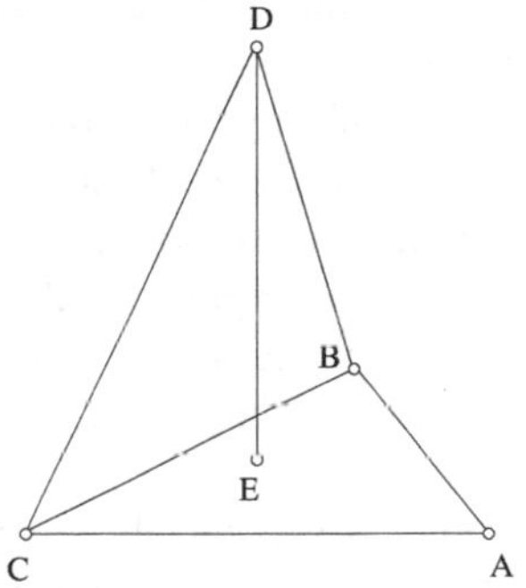

Fig. 11

From this, it can be shown that the ratio of the surface of the pyramid,[9] whose base is a figure with straight sides, to the area of the base is equal to the ratio of the perpendicular dropped along one of the sides to the half-diameter of the base, as the product of half of the sum of the sides of the base and this perpendicular is the surface of the pyramid, and its product with the half-diameter of the base is the area of the base.[10]

[9] This refers to the lateral area.

[10] The sequence of the text appears to be incorrect. The paragraph beginning with 'From this, …' should logically be placed before the previous paragraph relating to volume. And the implication 'It is for this reason …' suggests that there should be a

It is for this reason that the product of the half-diameter of the sphere inscribed within the pyramid having plane bases and one third of <the sum of> its bases is its volume, as it may be divided into pyramids whose vertices all meet at the centre of the sphere, and whose bases are the bases of the pyramid. The sphere is tangent to each of these bases and its half-diameter is perpendicular to the bases at the point[11] of contact, and as such is multiplied by one third of the base of each of these pyramids, as it is one third of the prism whose base is its base, and whose height is its height. The product of the height and the base is the volume of the prism, regardless of whether the sides of the base are equal or not. As we have already shown that the product of the half-diameter of a circle inscribed within a polygon and half the sum of its sides, whether equal or not, is the area of the polygon then, similarly, the product of the half-diameter of the sphere inscribed within this pyramid and one third of the sum of its bases, whether equal or not, is the volume of this pyramid.

<13> The circular cylinder is a solid figure bounded by two parallel circles and a curved surface joining them. Each of the circles is called the base of the cylinder. Any straight line joining the circumferences of the two bases and perpendicular to them[12] is called a side of the cylinder. The straight line joining the centres of the two bases is called the axis of the cylinder. If the axis stands on the surfaces of the two bases at angles that are not right angles, then the cylinder is said to be oblique. If it stands on them at right angles, then it is said to be a right cylinder, and it can be generated by a surface made up of parallel sides, of which one of the two sides enclosing the right angle is fixed and the surface rotated until it returns to its original position.

The cone of a right cylinder is a conically shaped solid figure extending from the circumference of one of the two bases of the cylinder until it disappears at the centre of the other base. This centre is the vertex of the cone. This shape is also called a pinecone from its resemblance to the fruit of the pine tree. The axis of the cylinder is a perpendicular, also called the height. Any straight line drawn perpendicularly from its vertex[13] to the circumference of its base is called a side of the cone.

---

paragraph relating to the sphere inscribed within a triangular pyramid before the text moves on to deal with the case of any pyramid (see Lemma 8 relating to the triangle and the inscribed circle, and the polygon circumscribed around a circle).

[11] Lit.: position.

[12] This assumes that the cylinder is a right cylinder.

[13] Any straight line joining the vertex of a cone to a point on the base circle is perpendicular to the tangent to the circle at that point.

We represent this figure as follows: We imagine its base as the circle *ABCD*, whose centre is at the point *E*, lying on a plane parallel to the horizon, and the point *G* in the air such that, when joined to <the point> *E* by a straight line, it rises perpendicularly to the plane of the circle. We draw the diameter *AC*, and join the two straight lines *GA* and *GC*.

I say that the product of AG *and the arc* ABC, *which is half of the circle* ABCD, *is the surface of the cone* ABCDG, *excluding the surface of its base.*

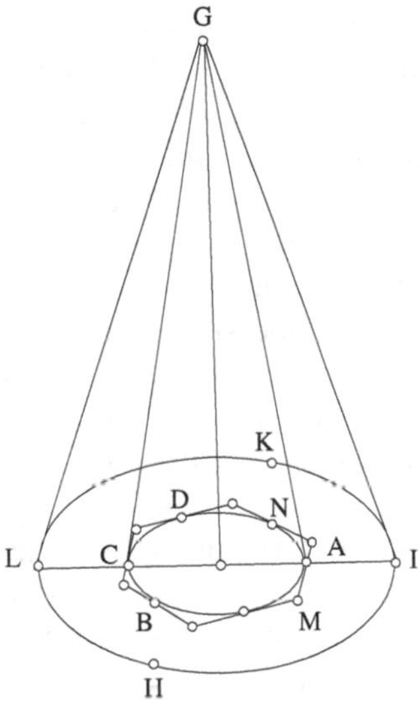

Fig. 12

Proof: It could not be otherwise. If it were possible, then let the product of *AG* and an arc greater than the arc *ABC* be the surface of the cone *ABCDG*. Let this arc be the arc *IKL*, which is half of the circumference of the circle *IKLH*. We construct a regular polygon on the circumference *ABCD* circumscribing the circle, which is the hexagon *AMCN*. We imagine straight lines dropping from the point *G* onto the extremities of the hexagon, generating a pyramid with a base having equal straight sides. This is greater than the cone *ABCDG*, as it surrounds it. We join the two straight lines *GI* and *GL*, forming the cone *IKLHG*. We multiply *AG* by the arc *IKL* to obtain the surface of the cone *ABCDG*. We multiply it by half the sum of the sides of the hexagon to obtain the surface of the pyramid *AMCNG*. The ratio of the arc *IKL* to half the sum of the sides of the hexagon is therefore equal to the ratio of the surface of the cone *ABCDG* to the surface of the pyramid *AMCNG*. But the arc *IKL* is greater than half the sum of the sides of the hexagon.[14] Therefore, the surface of the cone *ABCDG* is greater than the surface of the pyramid *AMCNG*. But we know that it is smaller; this is contradictory.

If the product of *AG* and <an arc> that is less than the arc *ABC* is the surface of the cone *ABCDG*, then its product with the arc *ABC* is the surface of a cone that is greater than the surface of the cone *ABCDG*. Let this be the surface of the cone *IKLHG*. This gives us the product of *AG* and the arc *ABC*, which is the surface of the cone *IKLHG*, and its product with half the sum of the sides of the hexagon, which is the surface of the pyramid *AMCNG*. The ratio of the arc *ABC* to half the sum of the sides of the hexagon is therefore equal to the ratio of the surface of the cone *IKLHG* to the surface of the pyramid *AMCNG*. But the arc *ABC* is less than half the sum of the sides of the hexagonal figure. Therefore, the surface of the cone *IKLHG* is less than the surface of the pyramid *AMCNG*. But it was greater than it, so this is contradictory and not possible.

Therefore we do not obtain the surface of the cone *ABCDG* by taking the product of *AG* and an arc that is greater than the arc *ABC*, and by taking its product with an arc that is less than that arc. Its product with the arc *ABC* is consequently the surface of the cone *ABCDG*.

We then draw the perpendicular *EG* and multiply it by one third of the surface of the base *ABCD*. This gives us the volume of the cone *ABCDG* as the product of the perpendicular *EG* and the surface of the base *ABCD* is the volume of the right cylinder. But the cone of the cylinder is one third of this, as shown by Euclid in Proposition 9 of Book 12 of the *Elements*. Similarly, <the product of> the perpendicular *EG* and one third of the surface of the hexagonal figure is the volume of the pyramid *AMCNG*, as it is one third of the volume of the cylinder[15] whose base is the surface of the hexagonal figure and whose height is the perpendicular *EG*, according to what has been mentioned in this proposition of the *Elements*.

<14> Consider the circle *ABCD* and the two given magnitudes *EG* and *H*, such that *EG* is greater than *H*; we wish to construct two similar polygons within and on the circle such that the ratio of that which is

[14] This assumes that the hexagon circumscribed around *ABCD* is inside the circle *IKL*. If the hexagon does not fulfill this condition, we know how to find a polygon which satisfies it. In fact, making use of Euclid's *Elements* XII.16, we obtain a polygon $\mathbf{P}_n$ inscribed inside *IKL* and that has no common point with the circle *ABCD*. Let a_n be its apothem and r the radius of the circle *ABCD*. The image of $\mathbf{P}_n$ in the homothety $(E, r/a_n)$ is a polygon $\mathbf{P}'_n$, which gives a solution to the problem.

[15] This is actually a prism.

constructed on the circle to that constructed within the circle is less than the ratio of *EG* to *H*.

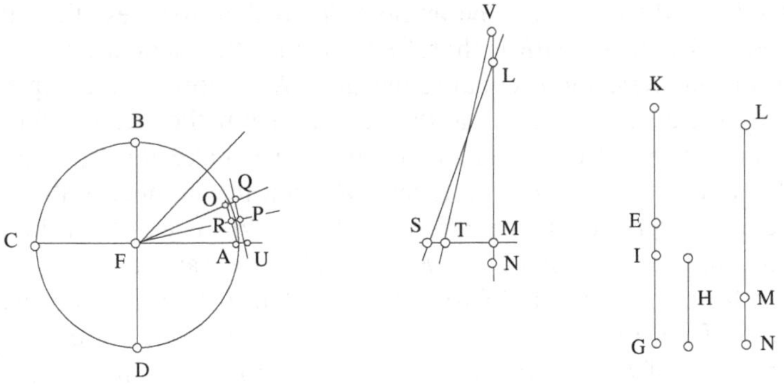

Fig. 13

We assume two different straight lines such that the ratio of the largest to the smallest is less than the ratio of *EG* to *H*. In order to find these, we take *GI* equal to *H* and we double *EI* until the multiple exceeds *H*. Let this multiple be *EK*. We assume any *LM* and we divide it by the number of times that *EK* includes *EI*. Let *MN* be equal to one of the parts of *LM*. Then the ratio of *MN* to *LM* is equal to the ratio of *EI* to *EK*. But *EK* is greater than *H*, *i.e.* greater than *GI*. The ratio of *EI* to *EK* is therefore less than the ratio of *EI* to *GI*. But the ratio of *EI* to *EK* is equal to the ratio of *MN* to *LM*. Therefore the ratio of *MN* to *LM* is less than the ratio of *EI* to *GI*. Composing, the ratio of *LN* to *LM* is less than the ratio of *EG* to *H*. If the two magnitudes *EG* and *H* were two surfaces or two solids, then it would be possible to define two straight lines *LN* and *LM* such that the ratio of *EG* to *H* was less than the ratio of *LN* to *LM*, as the multiplication[16] and division process is carried out in isolation according to which belongs to each genus. Having found *LN* and *LM*, they are placed in isolation according to this position, and a straight line *MS* is drawn perpendicular to the straight line *LM* such that if we join the straight line *LS*, it will be equal to *LN*. This is possible as *LN* is greater than *LM*. In the circle, we draw the two diameters *AC* and *BD* that cross each other at right angles. We divide the angle *AFB* into two halves, then we divide one half into two halves, and we continue this process until the remaining angle is less than twice the angle *MLS*, that is the angle *AFO*. We join the straight line *AO* to give one side of the polygon constructed within the circle. We divide the angle *AFO* into two halves by the straight line *FP*. Through <the point> *P*, we draw the straight line *UQ* tangent to the circle, and we draw *FA* and *FO* as far as the points

[16] Plural in the Arabic text.

U and *Q*. Then *UQ* will be one of the sides of the polygon constructed on the circle, which is similar to the <polygon> constructed within the circle. The angle *AFO*, which is twice the angle *AFR*, is therefore less than twice the angle *MLS*. The angle *AFR* is therefore less than the angle *MLS*. But the angle *M* is a right angle and is equal to the angle *R*. Therefore, the angle *S* is less than the angle *A*. If we draw a straight line from the straight line *MS* along <an angle> equal to the angle *A*, and if we make this straight line equal to *LS*, and if we extend it to meet *LM*, then it will meet it above the point *L*. Let it be equal to *TV*. Therefore its ratio to *VM* is less than the ratio of *LS* to *LM*, and the ratio of *TV* to *VM* is equal to the ratio of *AF*, *i.e.* *PF*, to *RF*. Therefore, the ratio of *PF* to *RF* is less than the ratio of *LS* to *LM*. But the ratio of *PF* to *RF* is equal to the ratio of *UF* to *AF*, and it is also equal to the ratio of *UQ* to *AO*. Therefore, the ratio of *UQ* to *AO* is very much less than the ratio of *EG* to *H*.

Now, we complete the two figures by adding the remaining sides. We can show from this that, if we wish the ratio of one polygon to the other to be less than the ratio of *LN* to *LM*, we define a straight line which is their mean in proportion. We then proceed using the straight line *LN* to determine two sides of a polygon as we have done previously with the two straight lines *LN* and *LM*. The ratio of one side to the other is then less than the ratio of *LN* to the mean straight line. But the ratio of the square of one side to the square of the other side[17] is less than the ratio of the square of *LN* to the square of the mean straight line, and the ratio of the square of one side to the square of the other side is equal to the ratio of one polygon to the other polygon, as shown in Proposition 19 of Book 6 of the *Elements*, and the ratio of the square of *LN* to the square of the mean straight line is equal to the ratio of *LN* to *LM*. Therefore, the ratio of one polygon to the other polygon is less than the ratio of *LN* to *LM*.

<15> The figure *ABCDG* is a cone of a right cylinder, and the half-diameter of the circle *IKLH*, which is *IM*, is the mean in proportion between the side of the cone, which is *AG*, and the half-diameter of its base, which is *AE*.

I say that the circle IKLH *– I mean its surface – is equal to the surface of the cone excluding its base.*

It this were not the case, let it be less than it. Then the surface of the cone and that of the circle *IKLH* would be two different magnitudes, the greatest of which is the surface of the cone. We construct two regular similar polygons, one within the circle and one lying on it, such that the ratio of that which was constructed on the circle to that constructed within <the

[17] Lit.: the ratio of the side to the side doubled by repetition.

circle> is less than the ratio of the surface of the cone to the circle *IKLH* – this is easy, given that which we have already introduced[18] – and let these be the two hexagons *INLS* and *OKPH*. The ratio of the hexagon *INLS* to the hexagon *OPKH* is therefore less than the ratio of the surface of the cone to the circle.

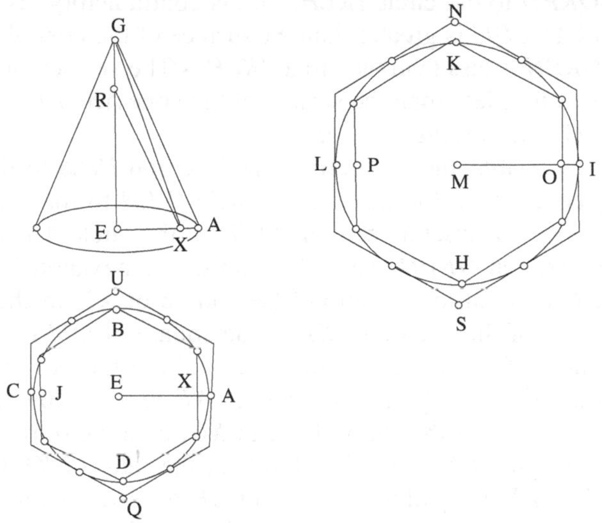

Fig. 14

Know that, if we speak of the ratio of one figure to another figure, be they circular or polygonal, we mean by that the areas of the two figures.

We construct a hexagon *AUCQ* on the circle *ABCD*. Its ratio to the hexagon *INLS* is then equal to the ratio of the square of *AE* to the square of *IM*, as shown in Book 12 of the *Elements*.[19] But the ratio of the square of *AE* to the square of *IM* is equal to the ratio of *AE* to *AG* and the ratio of *AE* to *AG* is equal to the ratio of the hexagon *AUCQ* to the surface of the pyramid *AUCQG*, as we have shown in Proposition 11 of these propositions. The ratio of the hexagon *AUCQ* to the hexagon *INLS* is therefore equal to the ratio of the hexagon *AUCQ* to the surface of the pyramid *AUCQG*. The surface of the pyramid is then equal to that of the hexagon *INLS*. But the ratio of the hexagon *INLS* to the hexagon *OKPH* is

[18] See the previous proposition.

[19] The first proposition of Book XII of the *Elements* stated that 'Similar polygons inscribed in circles are to one another as the squares on the diameters' (ed. Heath, vol. 3, p. 369). This same property will be proved for the circumscribed polygons with the help of a method that is similar to Euclid's one.

less than the ratio of the surface of the cone *ABCDG* to the circle *IKLH*. Applying a permutation, the ratio of the surface of the pyramid *AUCQG* to the hexagon *OKPH* is less than the ratio of the surface of the cone *ABCDG* to the circle *IKLH*. Applying a permutation, the ratio of the surface of the pyramid *AUCQG* to the surface of the cone *ABCDG* is less than the ratio of the hexagon *OKPH* to the circle *IKLH*. This is contradictory, as the surface of the pyramid *AUCQG* is greater than the surface of the cone *ABCDG* and the hexagon *OKPH* is less than the circle *IKLH*. <The surface of> the circle *IKLH* is not therefore less than the surface of the cone *ABCDG*.

I say that it is not greater than it.

If this were possible, then the ratio of the hexagon *INLS* to the hexagon *OKPH* would be less than the ratio of the circle *IKLH* to the surface of the cone *ABCDG*. We construct a hexagon *XBJD* in the circle *ABCD*, which is similar to the hexagon *OKPH*. Then the ratio of the hexagon *XBJD* to the hexagon *OKPH* is equal to the ratio of the square of *XE* to the square of *OM*. But the ratio of the square of *XE* to the square of *OM* is equal to the ratio of the square of *AE* to the square of *IM*, and the ratio of the square of *AE* to the square of *IM* is equal to the ratio of *AE* to *AG*. Now, the ratio of *AE* to *AG* is greater than the ratio of *XE* to *XG*, as if we draw *XR* parallel to *AG*, then the ratio of *XE* to *XG* is less than the ratio of *XE* to *XR*. But the ratio of *XE* to *XR* is equal to the ratio of *AE* to *AG*. Therefore the ratio of *XE* to *XG* is less than the ratio of *AE* to *AG*. Inverting, the ratio of *AE* to *AG* is greater than the ratio of *XE* to *XG*. But the ratio of *XE* to *XG* is equal to the ratio of the hexagon *XBJD* to the surface of the pyramid *XBJDG*. The ratio of *AE* to *AG* is therefore greater than the ratio of the hexagon *XBJD* to the surface of the pyramid *XBJDG*. Therefore, the ratio of the hexagon *XBJD* to the surface of the pyramid *XBJDG* is less than the ratio of the hexagon *XBJD* to the hexagon *OKPH*. Therefore, the surface of the pyramid *XBJDG* is greater than that of the hexagon *OKPH*. But we have assumed that the ratio of the hexagon *INLS* to the hexagon *OKPH* is less than the ratio of the circle *IKLH* to the surface of the cone *ABCDG*. Therefore, the ratio of the hexagon *INLS* to the surface of the pyramid *XBJDG* is very much less than the ratio of the circle *IKLH* to the surface of the cone *ABCDG*. This is contradictory, as the hexagon *INLS* is greater than the circle *IKLH* and the surface of the pyramid *XBJDG* is less than the surface of the cone *ABCDG*. We have already shown that <the surface of the circle> is not less than it. They must therefore be equal.

<16> The triangle *ABC* is a surface cutting a cone of a right cylinder along its axis *BD*, and the two triangles *EIG* and *KLN* are two surfaces cutting two cylindrical right cones along their axes *IM* and *LS*. The three axes are continuous along the same straight line, and the diameters of the bases of the cones, which are the straight lines *AC*, *EG* and *KN*, are parallel. It is because of this parallelism that the two bases of the upper cones are circles like the base of the lower cone as, if the straight line *DL* is held fixed and the triangles *ABC*, *EIG* and *KLN* rotated until they return to their original positions, then the straight line *AC* remains during this rotation within the circumference of the base, and to do this, the centres of the two circles are marked on the surfaces of the lower and middle cones. A straight line *QR* is also drawn parallel to *KN*, giving the base of a cone on which lies the triangle *QLR*.

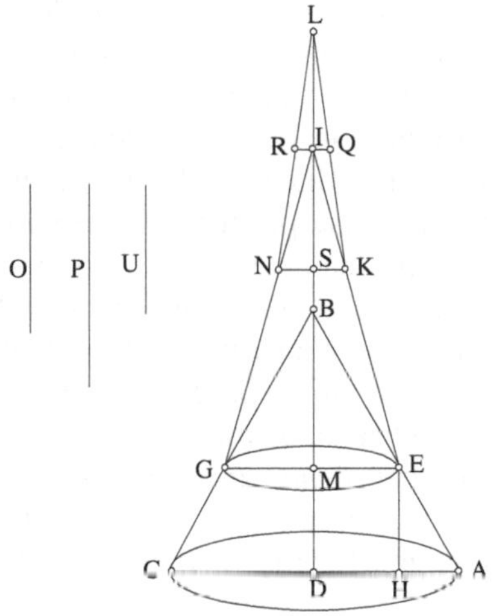

Fig. 15

I say that the straight line which is mean in proportion between AE *and the sum of* AD *and* EM *is the half-diameter of the circle equal to the surface of the portion* AEGC *of the lower cone.*

Proof: We draw *EH* parallel to *BD*, and suppose that the straight line *O* is equivalent in power to the product of *AE* and the sum of *AD* and *EM*, and that the straight line *P* is equivalent in power to the product of *BA* and *AD*, and that the straight line *U* is equivalent in power to the product of *BE* and *EM*. Then the straight line *P* is the half-diameter of the circle equal to the surface of the lower cone, and the straight line *U* is equal to the half-diameter of the circle equal to the surface of the cone on which lies the triangle *EBG*, as we have shown previously. But the product of *BA* and *AD* is equal to the product of *BE* and *AD* and that of *EA* and *AD*, and the product of *BE* and *AD* is equal to its product with *EM* plus <its product> with *AH*, and its product with *AH* is equal to the product of *EA* and *EM*, as the two triangles *BEM* and *AEH* are similar. The product of *BA* and *AD* is equal to the product of *BE* and *EM* plus <the product of> *EA* and *EM* and that with *AD*, from which is subtracted the product of *BE* and *EM*, which is equivalent in power to the straight line *U*. There remains, therefore, the product of *EA* and *EM* and that with *AD*, which is the straight line *O*.

Similarly, we can show that the straight line that is equivalent in power to the product of *KE* and the sum of *KS* and *EM* is the half-diameter of the circle equal to the surface of the portion of a cone on which lies the trapezium *EKNG*, and that the straight line that is equivalent in power to the product of *QK* and the sum of *QI* and *KS* is the half-diameter of the circle equal to the surface of the portion on which lies *KQRN*.

It is clear from this that, for any solid composed of portions of cones of right cylinders whose bases are parallel and such that two <contiguous> portions are joined by a common base and such that the straight lines that pass through their surfaces and which join the extremities of their bases such as *AE*, *EK* and *KQ* are equal, the product of one with the half-diameter of the lower base and with the diameter of any common base and with the half-diameter of the upper base is the square of the half-diameter of a circle equal to the surface of the solid excluding its base. If the vertex of the solid is a cone, as in this example, then the product of one of the straight lines and the half-diameter of the lower base, and with the diameters of the other bases, is the square of the half-diameter of the circle equal to the surface of the solid excluding its base. The same rule applies to a single cone cut in the same way as this section, and for the solid composed of portions as in this example.

<17> The circle *ABCD* is the great circle of a sphere. The great circle of a sphere is that which cuts it into two halves. The two diameters *AC* and *BD* cut each other at right angles and within is a polygon with an even number of sides. Let the half-polygon be *AEGBIKC*. It doesn't matter whether the number of sides is odd or even; the only reason for specifying an even number of sides is that it makes the process easier. If it is required to prove this, it is possible to follow the process using other polygons with odd numbers of sides, such as the pentagon, heptagon, and others up to infinity. We join the two straight lines *EK* and *GI*. These are parallel, and parallel to *AC*. We draw the circle *LMN* inscribed within the polygon and we imagine the two points *B* and *D* as being the two poles of a sphere. Its axis is then the diameter *BD*. If the sphere is rotated until it returns to its original position, then the two sides *AE* and *EG* will delineate two portions of two right cylindrical cones, the diameters of whose bases are *AC* and *EK*, and the side *GB* will delineate a right cylindrical cone, the diameter of whose base is *GI*. But the bases are parallel as their diameters are parallel.

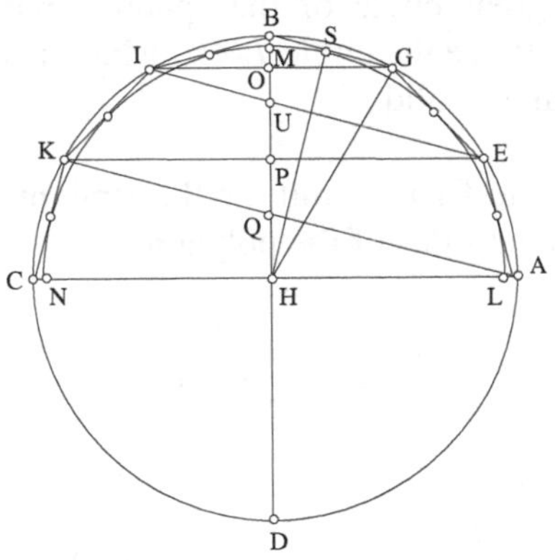

Fig. 16

I say that the surface of the solid composed of portions of cylindrical cones, excluding its base, is less than twice the surface of the great circle defining the hemisphere circumscribed around the solid, and greater than twice the surface of the great circle defining the hemisphere generated by rotating the semicircle LMN *inscribed within the solid.*

Proof: We place the point *S* on the point of contact of the side *GB* and the circle *LMN*; it also divides the side *GB* into two halves. We join the straight lines *SH*, *IE* and *KA*. Then, *IE* and *KA* are parallel and parallel to *GB*, the triangles *GBO*, *IUO*, *EPU*, *KQP* and *AQH* are similar, and the ratio of *GO* to *OB* is equal to the ratio of *IO* to *OU*, and equal to the ratio of *EP* to *PU*, and equal to the ratio of *KP* to *PQ*, and equal to the ratio of *AH* to *QH*. But the ratio of each of the antecedents to each of the successors is equal to the ratio of the sum to the sum. Therefore the ratio of *GO* to *OB* is equal to the ratio of the sum of *GI*, *EK* and *AH* to *BH*. But the ratio of *GO* to *OB* is equal to the ratio of *SH* to *SB* as the two triangles are similar.[20] Therefore, the ratio of *SH* to *SB* is equal to the ratio of the sum of *GI*, *EK* and *AH* to *BH*. Therefore, the product of *SB* and the sum of *GI*, *EK* and *AH* is equal to the product of *SH* and *BH*. But the product of *SH* and *BH* is less than the product of *BH* and itself, and greater than the product of *SH* and itself. But the product of *GB*, which is twice *SB*, and the sum of *GI*, *EK* and *AH*, as we have shown, is the square of the half-diameter of the circle equal to the surface of the composed solid <consisting of portions of cones>. Now, the ratio of the square of the half-diameter of any circle to the square

[20] Triangles *GOB* and *SHB*.

of the half-diameter of any other circle is equal to the ratio of one circle to the other circle. The surface of the solid is therefore less than twice the circle *ABCD*, which is the great circle of the sphere circumscribed around the solid, and greater than twice the circle *LMN*, which is the great circle of the sphere inscribed within the solid.

<18> We redraw the figure, omitting the straight lines *GI*, *EI*, *EK*, *AK* and *SH*, and complete the sides of the polygon.

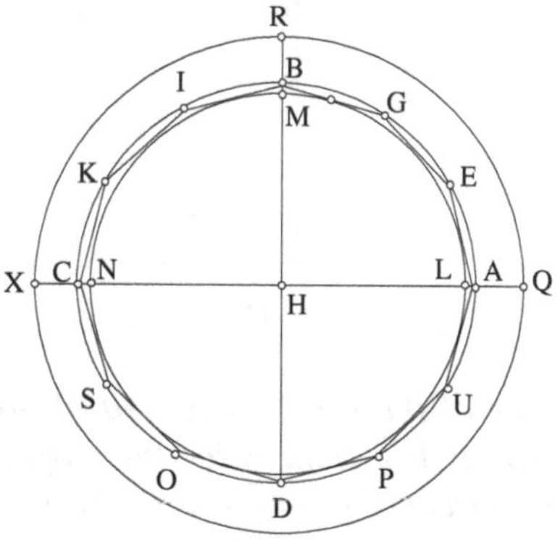

Fig. 17

We say that four times the circle ABCD – *by which I mean the surface of the circle – is equal to the surface of the sphere of which the circle is the great circle that lies upon it.*

If this were not the case, then let it[21] be less than the surface of the sphere, and let it be equal to the surface of a sphere that is smaller than that which the circle *ABCD* lies upon. This sphere is such that the circle *LMN* lies upon it, and the circle *LMN* is the great circle that lies upon this sphere. The surface of this sphere is then less than the surface of the solid composed of portions of cones similar to <the portions> of the first solid, which is tangent to the sphere on which lies the circle *LMN*, as the solid surrounds the circle.[22] We have shown that the surface of the solid is less than four

[21] 'it' here refers to four times the surface of the circle *ABCD*.

[22] In the first sentence of the statement and in the figure, it appears that the author is considering here a solid generated from a regular polygon circumscribed around the circle *LMN* and inscribed within the circle *ABCD* similar to the solid used in the previous study. This raises the question of the existence of such a polygon. If r and R are the respective radii of *LMN* and *ANCD*, the number n of the sides of the polygon must satisfy $r = R \cos \pi/n$. The data of r and R does not generally lead to an integer value of n. ($r = R/2$ gives $n = 3$, $(R\sqrt{3})/2$ gives $n = 6$). However, it is sufficient for the polygon

times the circle *ABCD* <and greater than four times the circle *LMN*>. Therefore, the surface of the sphere on which lies the circle *LMN* is very much less than four times the circle *ABCD*. But we have assumed it to be equal, which is contradictory.

Now, let four times the circle *ABCD* be greater than the surface of the sphere on which lies the circle *ABCD*, and let it[23] be equal to the surface of the sphere on which lies the circle *QRX*; this circle is the great circle which lies on this sphere. We imagine that this sphere surrounds a solid composed of portions of cylindrical cones.[24] Then, the surface of this solid will be greater than four times the circle *ABCD*. But the surface of the sphere on which lies the circle *QRX* is greater than the surface of this solid as the sphere surrounds it. Consequently, the surface of this sphere is greater than four times the circle *ABCD*. But we have assumed it to be equal, which is contradictory.

It follows that the surface of any sphere is four times the great circle that lies upon it. But, as the circle is the product of <one quarter of> its diameter and its circumference, the surface of the sphere is given by the product of the diameter of the great circle that lies upon it and its circumference.

<19> Take the same figure as it is.

We say that the product of BH, *which is the half-diameter of the circle* ABCD, *and one third of the surface of the sphere on which lies the circle* ABCD *is the volume of the sphere.*

Proof: Otherwise, let this product be the volume of a sphere that is smaller than this sphere, namely the sphere on which lies the circle *LMN*. We imagine a circle which passes through the two points *B* and *D* and which cuts the circle *ABCD* at right angles, and a circle which passes through it at right angles and which passes through the two points *A* and *C* so that the circle *ABCD* is divided into quarters, and two circles which fall[25] into each pair of quarters of the second circle and which pass through the two points *B* and *D*. We have thus divided each quarter, among the quarters of a circle, into three thirds. We imagine that each of the five circles[26]

---

circumscribed around the circle *LMN* to be inside the circle *ABCD*. The solid is then tangent to the sphere *LMN*, as the author states, and internal to the sphere *ABCD*.

[23] 'it' here refers again to four times the surface of the circle *ABCD*.

[24] This solid is either tangent to the sphere *ABCD* and internal to the sphere *QRX*, or inscribed within the sphere *QRX*, having no common point with the sphere *ABCD*.

[25] Lit.: between.

[26] These are the circles passing through *B* and *D* with the exception of the circle *ABCD*.

surrounds a polygon similar to that inscribed within the circle *ABCD*. We join the extremities of each pair of similar sides in each pair of polygons between two successive circles, forming a solid with plane bases. Those bases which are adjacent[27] to the two poles *B* and *D* are triangles. The others are all trapeziums and form the bases of the pyramids into which the solid is divided. The vertices of these pyramids all meet at the centre of the sphere, at the point *H*. The sphere on which lies the circle *LMN* is tangent to each of these bases[28] and its half-diameter is perpendicular at the point of contact. Therefore, the product of its half-diameter and one third of the sum of the bases is the sum of the pyramids forming the entire solid, and the sum of the bases is the surface of the solid. The product of *MH*, which is the half-diameter of the sphere on which lies the circle *LMN*, and one third of the surface of the solid is therefore the volume of the solid. But its product with one third of the surface of the solid is greater than its product with one third of the surface of the sphere on which lies the circle *LMN*, as the solid surrounds the sphere. The product of *BH* and one third of the surface of the sphere on which lies the circle *ABCD* is therefore very much greater than the volume of the sphere on which lies the circle *LMN*. But we have assumed them to be equal; this is contradictory.

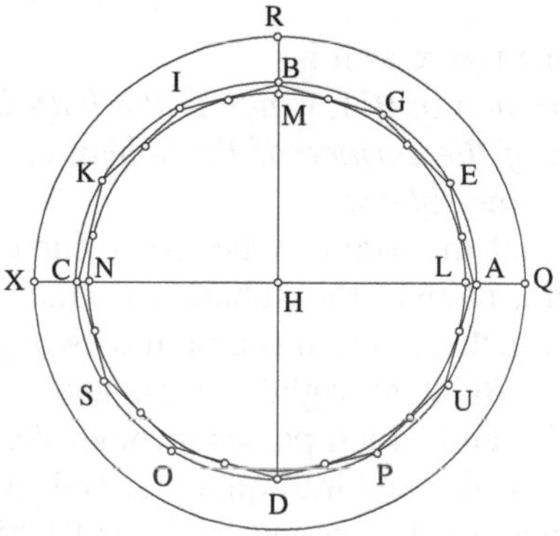

Fig. 18

Let us suppose, now, that the product of *BH* and one third of the surface of a sphere greater than that on which lies the circle *ABCD* is the volume of the sphere, and let this be the sphere on which lies the circle *QRX*, which is its great circle. We imagine that it surrounds a solid, with bases, similar to the first solid and tangent to a sphere on which lies the circle *ABCD*. Then the product of *BH* and one third of the surface of this

[27] These are the bases having a vertex at either *B* or *D*.
[28] See the mathematical commentary.

solid is greater than the sphere on which lies the circle *ABCD*. But, as we have supposed, let the product of *BH* and one third of the surface of this sphere be the volume of the sphere on which lies the circle *QRX*. This sphere is greater than the solid as it surrounds it. Therefore, one third of its surface is greater than one third of the surface of the solid, and therefore the product of *BH* and one third of the surface of the sphere is very much greater than the volume of the sphere on which lies the circle *ABCD*. But we have assumed them to be equal, this is contradictory.

The product of the half-diameter of the sphere on which lies the circle *ABCD* and one third of the surface of a sphere that is either less than or greater than it is not its volume. Consequently, its product with one third of its surface is its volume.

But we have shown that the surface of the greatest circle found on the sphere is one quarter of the surface of the sphere, and that the surface of the circle plus one third of it is one third of the surface of the sphere. Therefore, the product of the half-diameter of the sphere and one and one third <times> the surface of the greatest circle found upon this sphere is the volume of the sphere. The product of the half-diameter of a sphere and twice <the surface of> the greatest circle found upon it is therefore equal to one and a half times the <volume of> the sphere.

But for the cylinder that surrounds the sphere, the product of its axis, which is the diameter of the sphere, and its base, which is the greatest circle found upon the sphere, is the volume of the cylinder. Similarly, it is equal to the product of the half-diameter of the sphere and twice <the surface of> the greatest circle found on it. The cylinder surrounding the sphere is therefore equal to one and a half times the sphere.

But as the cone of a cylinder is one third of it, the cone whose base is equal to the greatest circle found on the sphere and whose axis is equal to the half-diameter of the sphere is one quarter of the sphere. Therefore, the sphere is four times this cone. But the ratio of the cone of a cylinder to the cone of another cylinder is equal to the ratio of one base to the other base, if they both have the same height, as shown in Proposition 11 of Book 12 of the *Elements*. The cone whose base is four times the greatest circle found upon the sphere and whose axis is equal to the half-diameter of the sphere is therefore four times the cone whose base is equal to the greatest circle found upon the sphere and whose axis is equal to the half-diameter of the sphere. This cone is therefore equal to the sphere.

<20> Now, let us draw the circle *TV* with its centre at *O* and let it be the greatest circle that can be found on a sphere. We draw the diameter *TV*. Let the surface of this sphere be equal to the surface of the solid circumscribed around the sphere on which lies the circle *LMN*. Then, <the volume of> the <first> sphere is greater than the solid, as if *MH*, which is the half-diameter of the sphere on which lies the circle *LMN*, were equal to *TO*, then the <first> sphere would be equal to the <second> sphere. But the surface of the sphere on which lies the circle *LMN* is less than the surface of the solid, and the surface of the solid is equal to the surface of the sphere on which lies the circle *TV*. Therefore, the surface of the sphere on which lies the circle *LMN* is less than the surface of the sphere on which lies the circle *TV*. Therefore, the sphere on which lies the circle *LMN* is less than the other sphere, and its half-diameter, which is *MH*, is shorter than *TO*. But the product of *MH* and one third of the surface of the solid is the volume of the solid, and the product of *TO* and one third of the surface of the sphere is the volume of the sphere on which lies the circle *TV*. This is therefore greater than the <volume of the> solid. Consequently, the sphere is greater than the solids with the same perimeter.[29]

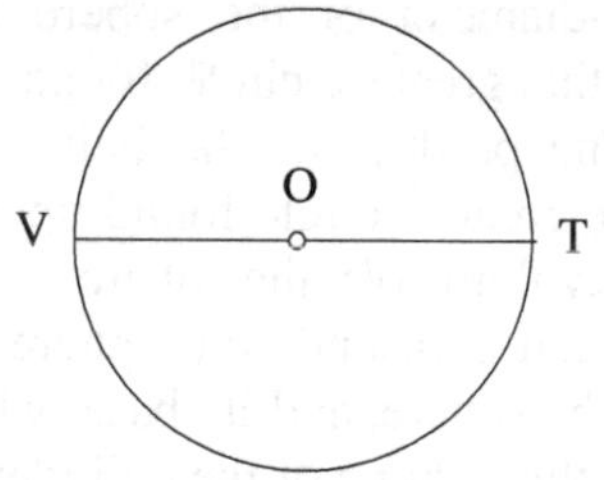

Fig. 19

[29] It is clear that it refers to the surface.

In the Name of God, the Merciful, the Compassionate

OPUSCULE

The surface of any circle is greater than the surface of any regular polygon with the same perimeter

We wish to show that the surface of any circle is greater than the surface of any regular polygon with the same perimeter.

Let a circle have its centre at *A*, its half-diameter *AB*, and its perimeter equal to the perimeter of the regular polygon *CDE*.

I say that the surface of the circle AB *is greater than the surface* CDE.

Proof: We draw an inscribed circle within the surface *CDE*, let its centre be at *G*. We extend its half-diameter to *H*, which is the point[30] of contact. If *GH* is equal to *AB*, then the circle *AB* is equal to the circle *GH*, and the product of *GH* and the half-circumference of the circle *GH* is the surface of the circle *GH*. But the product of *GH* and the half-perimeter of the figure *CDE* is the area of the surface *CDE*. Consequently, the circle *GH* is equal to the surface *CDE* and the smaller would be equal to the greater; this is contradictory.

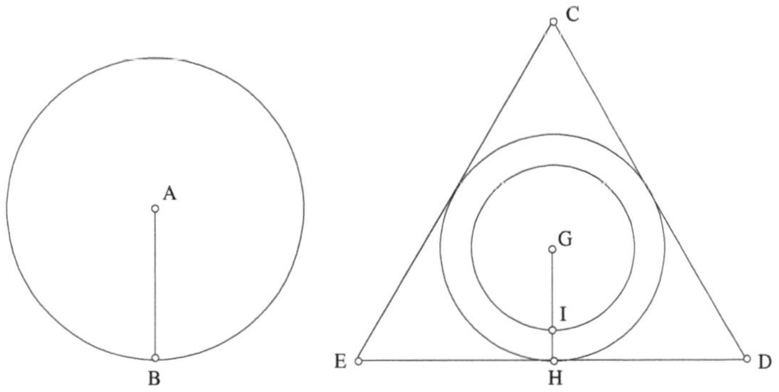

Similarly, *AB* is not less than *GH*, as if it were so we could remove from *GH* that which is equal to *AB*, which is *GI*. The circle *GI* would then be

[30] Lit.: position.

equal to the circle *AB*, and therefore the circumference of the circle *GH* would be longer than the circumference of the circle *GI* and the perimeter of the figure *CDE* would be greater than the circumference of the circle *GH*. The perimeter of the figure *CDE* would then be very much greater than the circumference of the circle *AB*. But have assumed them to be equal, so this is impossible. Therefore, *GH* is neither equal to *AB* nor longer than it. It is therefore shorter than it. But <the product of> *GH* and the half-perimeter of the figure *CDE* is the area of the surface *CDE*, and <the product of > *AB*, the longer, and half of the circumference of the circle *AB*, which circumference is equal to the perimeter of the figure *CDE*, is greater than <the product of> *GH*, the shorter, and the half-perimeter of the figure *CDE*, which perimeter is equal to the circumference of the circle *AB*. Therefore, the circle *AB* is greater than the figure *CDE*. That is what we wanted to prove.

End of the opuscule
May thanks be unto God

CHAPTER V

AL-QŪHĪ, CRITIQUE OF THĀBIT: VOLUME OF THE PARABOLOID OF REVOLUTION

5.1. INTRODUCTION

5.1.1. *The mathematician and the artisan*

Abū Sahl Wayjan (Bijān) ibn Rustam al-Qūhī (al-Kūhī) was one of the principal astronomers and mathematicians of the school of Baghdad, and in particular of the Buyid court. We can measure the importance of his works by the references made to them by his contemporaries, like al-Sijzī or Ibn Sahl, and by his successors, like Ibn al-Haytham and al-Bīrūnī. In his time, according to the report transmitted by the man of letters Abū Ḥayyān al-Tawḥīdī, al-Qūhī was presented as an eminent scholar who was concerned neither with theology nor with metaphysics.[1] This mathematician developed to their farthest point the epistemic characteristics that had distinguished this tradition since its foundation, a century earlier, by the Banū Mūsā, as well as throughout its successive transformations since Thābit ibn Qurra and his grandson Ibrāhīm ibn Sinān. Al-Qūhī was interested in the application of mathematics to astronomy and to statics, and in the study of mathematical instruments such as the perfect compass.[2] In addition, he took an active part in the broadening of research on geometrical transformations: in this regard, it suffices to mention his *Treatise on the Art of the Astrolabe by Demonstration.*[3] Al-Qūhī and the mathematicians of his tendency, among whom was Ibn Sahl, combined the two traditions of Greek geometry – that of Archimedes and that of Apollonius – in order to advance onto a terrain that was not truly Hellenistic: that of transformations. Al-Qūhī also had the

[1] In his *Kitāb al-Imtāʿ wa-al-muʾānasa,* ed. A. Amīn and A. al-Zayn, al-Tawḥīdī, after mentioning the philosopher Yaḥyā ibn ʿAdī, mentions an entire group, in which are included al-Qūhī, al-Ṣāghānī, al-Ṣūfī, and al-Sāmarrī, among others, to affirm that 'none of them pronounces a single word about the soul, the Intellect, or God, as if this was forbidden to them, or detestable' (First part, p. 38).

[2] See R. Rashed, *Geometry and Dioptrics in Classical Islam*, London, 2005, Chapter V.

[3] *Ibid.*, pp. 11–12.

advantage of his chronological situation, and he gathered the fruits of the already considerable accumulation of work carried out since the Banū Mūsā and Thābit ibn Qurra.

Who was al-Qūhī? Who were his teachers? On these questions, as on all the others we can raise, the historical and bio-bibliographical sources are quite silent. His name is that of a Persian, or at least of a family of Persian origin. His contemporary, the bio-bibliographer al-Nadīm, recalls that he was originally from Ṭabaristān, a mountainous region south of the Caspian Sea.[4] To this brief information, the other bio-bibliographers – except, as we shall soon see, for al-Qifṭī – add nothing substantial.[5] The only certainty comes to us later, from al-Bīrūnī.[6] Here, we encounter al-Qūhī in 359/969, already in the company of the luminaries of his time, for he was in the company of al-Sijzī, Naẓīf ibn Yumn, and Ghulām Zuḥal (*alias* Abū al-Qāsim ʿUbayd Allāh ibn al-Ḥasan) when he attended the astronomical observations ordered by the master of the province of Fārs – ʿAḍud al-Dawla in person – and carried out by the well-known ʿAbd al-Raḥmān al-Ṣūfī, from Wednesday the 2nd of Ṣafar to Friday the 4th of Ṣafar of the year 359/969.

At this date, therefore, al-Qūhī was a renowned and widely-cited mathematician. As an additional proof, let us recall that in that same year, as we learn from manuscript 2457/2 of the Bibliothèque Nationale, al-Sijzī had

[4] Al-Nadīm, *Kitāb al-Fihrist*, ed. R. Tajaddud, pp. 341–2. On the biography and bibliography of al-Qūhī, see also the article 'al-Qūhī', by Y. Dold-Samplonius, in *Dictionary of Scientific Biography*, 1975, vol. 11, pp. 239–41; C. Brockelmann, *Geschichte der arabischen Literatur*, B. I, Leiden, 1937, pp. 339–40; F. Sezgin, *Geschichte des arabischen Schrifttums*, B. V, Leiden, 1974, pp. 315–21, and B. VI, pp. 218–19.

[5] Among the ancient bio-bibliographers, al-Bayhaqī (462/1070–499/1105) sketches a highly colorful portrait of al-Qūhī (*Tārīkh ḥukamā' al-Islām*, ed. M. Kurd ʿAlī, Damascus, 1946, p. 88). If we can believe him, al-Qūhī was a kind of acrobat: 'he belonged to those who played in the markets by glass bottles (*qawārīr*); divine grace then touched him, and he distinguished himself in the science of ingenious procedures, like mechanics and mobile spheres; he was without rivals in these arts, and quite famous'. As always, al-Shahrazūrī took up this portrait in his book, and later diffused it (*Tārīkh al-ḥukamā', Nuzhat al-arwāḥ wa-rawaḍat al-afrāḥ*, ed. ʿAbd al-Karīm Abū Shuwayrib, Tripoli, Libya, 1988, p. 313). See also R. Rashed, 'Al-Qūhī *vs.* Aristotle: On motion', *Arabic Sciences and Philosophy*, 9.1, 1999, pp. 7–24.

[6] Al-Bīrūnī, *Kitāb Taḥdīd nihāyāt al-amākin li-taṣḥīḥ masāfāt al-masākin*, text established by P. Bulgakov and revised by Imām Ibrāhīm Aḥmad, *Majallat Maʿhad al-Makhṭūṭāt*, 8, fasc. 1–2, November 1962, pp. 99–100. Cf. the English translation of this work by Jamil Ali, *The Determination of the Coordinates of Positions for the Correction of Distances between Cities*, Beirut, 1967, pp. 68–9.

already copied his work on *The Centres of Tangent Circles*,[7] which was therefore composed considerably earlier. It was also around this date that he had written his treatise on *The Construction of the Regular Heptagon*.[8]

Our second encounter with al-Qūhī takes place 19 years later, at Baghdad, in the reign of Sharaf al-Dawla, son of ʿAḍud al-Dawla. The latter had ordered al-Qūhī to observe the motion of the seven planets, along with their displacement in their signs. With this in mind, al-Qūhī had built an observatory, fashioned an astronomical instrument, and set about his observations, in front of witnesses. The event was confirmed by the most diverse sources, not all of which were independent. We shall glance only at the testimony of an astronomer, al-Bīrūnī, a bio-bibliographer, al-Qifṭī, and of a historian, Ibn Taghrī Bardī. Al-Bīrūnī writes as follows:

> Sharaf al-Dawla ordered Abū Sahl al-Kūhī to make a new observation. So he constructed in Baghdad a house whose lowest part (*qarāruhu*) is a segment of a sphere, of diameter twenty-five cubits (13 and one-half meters), and whose center is in the ceiling of the house, at an aperture which admits the rays of the sun to trace the diurnal parallels.[9]

To confirm this testimony, al-Qifṭī bases his account on two crucial documents in the history of science: two notarized acts, intended to record scientific–technical results. Drawn up by two judges, these two acts were undersigned by them, as well as by all the witnesses present, that is, the astronomers Abū Isḥāq Ibrāhīm ibn Hilāl al-Ṣābi', Abū Saʿd ibn Būlis al-Naṣrānī, al-Qūhī himself, Abū al-Wafā' al-Būzjānī, Abū Ḥāmid al-Ṣāghānī, Abū al-Ḥasan al-Sāmarrī and Abū al-Ḥasan al-Maghribī. Al-Qūhī built this observatory in the gardens of the Royal Palace, and he proceeded to two series of observations in the month of Ṣafar, 378/988,[10] by means of his

[7] *Marākiz al-dawā'ir al-mutamāssa*, fols. 19–21.

[8] Cf. J. Dold-Samplonius, 'Die Konstruktion des regelmässigen Siebenecks', *Janus*, 50, 4, 1963, pp. 227–49. See R. Rashed, *Les Mathématiques infinitésimales du IXe au XIe siècle,* vol. 3, Appendix I, text no. 7.

[9] Al-Bīrūnī, *Taḥdīd nihāyāt al-amākin*, ed. Bulgakov, pp. 100–1; trans. Jamil Ali, *The Determination of the Coordinates of Positions*, p. 69.

[10] Al-Qifṭī, *Ta'rīkh al-ḥukamā'*, ed. J. Lippert, Leipzig, 1903, pp. 351–4. As usual, but this time very briefly, Ibn al-ʿIbrī takes up some of al-Qifṭī's information; see *Tārīkh mukhtaṣar al-duwal,* ed. O. P. A. Ṣāliḥānī, 1st ed., Beirut, 1890; repr. 1958, p. 176. On al-Qifṭī's description of the observatory built by al-Qūhī, see A. Sayili, *The Observatory in Islam and its Place in the General History of the Observatory*, 2nd edition, Ankara, 1988, pp. 112–17. Note that we find this same date given earlier: the 28th of Ṣafar 378/988. Al-Bīrūnī repeats this same testimony, with the same numbers and the same names, in his *al-Qānūn al-Masʿūdī*, Osmania Oriental Publications Bureau, Hyderabad, 1955, vol. 2, sixth book, pp. 642–3.

instrument, the exactitude and perfection of which were unanimously attested by those present.

The third source is the historian Ibn Taghrī Bardī,[11] who, relating the events of the same year 378/988, writes as follows:

> In the month of Muḥarram of that year, Sharaf al-Dawla, following the example of al-Ma'mūn, ordered the observance of the seven planets in their motion and their displacement in their signs. Ibn Rustam al-Kūhī took charge of this, for he knew astronomy and geometry, and with this in mind he built a house in the back of the gardens of the Royal Palace; he proceeded to observation for the two remaining nights of Ṣafar.

Although the chronicler Ibn Taghrī Bardī seems to depend on al-Qifṭī here, nevertheless the discoveries he reports, as well as the testimony of al-Bīrūnī, depend directly on the sources, and the historian also had available a letter from Naẓīf ibn Yumn,[12] relating one of the results of these observations.

Everything thus indicates that around this period, in the eighties of the tenth century, al-Qūhī was among the most prestigious mathematicians of Baghdad, and moved in the circle in which he met the scholars already cited, as well as others like Ibn Sahl. We can even add, without risk of error, that he had spent at least two decades in the limelight. Finally, let us note that from this date on, his former companions began to disappear: al-Ṣāghānī died one year later, in 379/989, and al-Ṣābi' six years after that, in 384/994, at the age of about 70. Yet who was al-Qūhī himself?

Without knowing anything about his fate after 378/988, we have seen that he was already a recognized author at least 20 years previously. Better yet, a piece of information which has so far gone unnoticed shows that he was scientifically active in the fifties of the tenth century: if this were truly the case, he would then be of the generation of his colleagues like al-Ṣābi', and of an equivalent age, finishing his scientific career, if not his life, with the end of the century. Al-Qifṭī, followed by Ibn Abī Uṣaybiʿa, reports in the article devoted to Sinān ibn Thābit ibn Qurra – father of Ibrāhīm ibn Sinān and of Thābit ibn Sinān – that he had 'corrected the expression of Abū Sahl al-Qūhī in all his books (*fī jamīʿi kutubihi*), since Abū Sahl has asked

[11] Ibn Taghrī Bardī, *al-Nujūm al-zāhira fī mulūk Miṣr wa-al-Qāhira*, introduction and notes by Muḥammad Ḥusayn Shams al-Dīn, vol. 4, Beirut, 1992, p. 156. Ibn Taghrī Bardī clearly takes up al-Qifṭī's text here, in the same words.

[12] Al-Bīrūnī writes in *Taḥdīd nihāyāt al-amākin*, ed. Bulgakov, p. 101: 'Naẓīf ibn Yumn informed me, in writing, that the summer solstice was found at the end of the first hour of the night whose morning was on Saturday, the twenty-eighth of Ṣafar, year three hundred seventy-eight of the Hijra [...]' (trans. Jamil Ali, *The Determination of the Coordinates of Positions*, pp. 69–70).

him to'.[13] Yet, still according to al-Qifṭī and Ibn Abī Uṣaybiʿa, Sinān ibn Thābit died in 331/943. Moreover, Ibn Abī Uṣaybiʿa specifies the month and the day of his death: the Friday of the beginning of Dhū al-Qaʿda. Now, the plural used does indeed designate several books, at least three, that were written by al-Qūhī before 943, and this takes his date of birth back to the end of the first or the second decade of the tenth century. We are also given a more important piece of information, infinitely more precious: we thereby learn that al-Qūhī was in direct relation at the time with the son of Thābit ibn Qurra and the father of Ibrāhīm ibn Sinān, whom he therefore cannot have failed to know. This indication by the two ancient bio-bibliographers needs to be matched up against that of other sources before it can be confirmed. For the moment, however, it contains nothing unbelievable; and it merely situates al-Qūhī in the place that mathematical analysis will show was truly his, within the tradition of the 'dynasty' of the Banū Qurra.

5.1.2. *The versions of the volume of a paraboloid*

The Volume of the Paraboloid is not the only contribution of al-Qūhī to the field of infinitesimal mathematics. He introduces this treatise as being a necessary part of a much larger project to investigate centres of gravity. This project also included a brief memorandum on *The Ratio of the Diameter to the Circumference*, which we shall discuss elsewhere as part of a section on Arabic commentaries on *The Measurement of the Circle* by Archimedes. Al-Quhi discusses the history behind his own research, and especially this project, in his introduction to the treatise. While engaged in writing a book on centres of gravity, which by all indications must have been a major work, he found that he needed first to be able to determine the volume of a paraboloid. At this point, he turned to the book by Thābit ibn Qurra, the only work that he knew on this topic. However, as we saw in Chapter II, Ibn Qurra began his work by proving 35 preliminary propositions before arriving at his goal. Al-Qūhī found this road a long and difficult one, so he sought a path that was both easier (*qarīb*) and shorter, requiring far fewer lemmas. This new method only involved two lemmas.

Al-Qūhī states in his own words that he began his study of the volume of the paraboloid as a requirement of his research into centres of gravity. While logically a precursor, it was in fact published later. In addition,

[13] Al-Qifṭī, *Ta'rīkh al-ḥukamā'*, p. 195; Ibn Abī Uṣaybiʿa, *'Uyūn al-anbā' fī ṭabaqāt al-aṭibbā'*, ed. A. Müller, 3 vols., Cairo/Königsberg, 1882–84, vol. 1, p. 324.

although he was studying the book by Thābit ibn Qurra, he wished to carry out his research more in the style of that author's grandson, Ibrāhīm ibn Sinān; that is, with economy and elegance. Taking a lead from Ibn Sinān, he abandoned the arithmetical lemmas in favour of a geometric approach, albeit in combination with the method of integral sums that had just been rediscovered. Al-Qūhī must therefore have written his treatise on the paraboloid either at the same time as he was writing his other book on *The Centres of Gravity* – which would have taken some time if the number of chapters mentioned by the author is anything to go by[14] – or very shortly thereafter. The larger book has sadly been lost, and the only indications we have as to the date when it was written appear in correspondence between al-Qūhī and Abū Isḥāq Ibrāhīm ibn Hilāl al-Ṣābi'. However, the dates of this correspondence are themselves by no means certain. We are therefore reduced to suggesting that this book was written between the early eighties and the early nineties in the tenth century.[15]

There are a number of surviving manuscripts of the treatise by al-Qūhī on the measurement of the paraboloid. We shall now examine these in order. They fall into three distinct groups. The first group, containing a single manuscript, consists of a 'rewriting' of the original text by al-Qūhī. The second group contains accurate copies, which allow us to get very close to the original. The third group also contains just one manuscript, similar to those in the second group, but showing signs of some rewriting of the introduction only. The first version appears in the manuscript Riyāḍa 41/2, fols 135ᵛ–137ᵛ, in the Dār al-Kutub in Cairo. This manuscript is in the

[14] From the correspondence between al-Qūhī and al-Ṣabi', we learn that this work consisted of six books, to which the author intended to add another four or five. See the following note.

[15] This correspondence has been published by J. L. Berggren, 'The correspondence of Abū Sahl al-Kūhī and Abū Isḥāq al-Ṣābī: A translation with commentaries', *Journal for the History of Arabic Science*, vol. 7, nos. 1 and 2, 1983, pp. 39–124. Note that a similar echo of his research into centres of gravity can not legitimately be found in the introduction by al-Qūhī to his treatise on *The Construction of the Regular Heptagon.*

This takes the form of a general list, including astronomy, numbers, weight, centres of gravity, and other topics. If these relatively vague expressions are to be read as a reference to his own research, then somewhere there should be similar works on number theory. Any search for these is bound to be a fruitless one.

See A. Anbouba, 'Construction of the Regular Heptagon by Middle Eastern Geometers of the Fourth (Hijra) Century', *Journal for the History of Arabic Science*, 1.2, 1977, pp. 352–84, especially pp. 368-369; 'Construction of the Regular Heptagon by Middle Eastern Geometers of the Fourth (Hijra) Century', *Journal for the History of Arabic Science*, 2.2, 1978, pp. 264–9.

Risāla fī istikhrāj ḍil' al-musabba' al-mutasāwī al-aḍlā', mss Paris 4821, fols 1–8; Istanbul, Aya Sofya 4832, fols 145ᵛ–147ᵛ; London, India Office 461, fols 182–189.

handwriting of the famous copyist Muṣṭafā Ṣidqī, whom we have had cause to mention several times previously. This copy was made in 1153/1740–41. An examination of the text reveals that it is a 'rewriting', an 'edition' (*taḥrīr*) and not the original by al-Qūhī.[16] It is written in the same style that we have seen in Chapter I in the case of the rewriting by al-Ṭūsī of the Banū Mūsā treatise. This is also the style used by Ibn Abī Jarrāda in rewriting the treatise by Thābit ibn Qurra *On the Sections of the Cylinder*, a style of writing that persisted until the late thirteenth century. This edition also omitted the historical and theoretical introduction, together with the final section in which al-Qūhī returns to a discussion of Proposition X.1 in the *Elements* of Euclid and its modification. In effect, he has removed anything that he did not consider to be strictly mathematical. Here, once again, the editor has removed everything that he considered to be redundant from the text, together with some of the intermediate steps in the proofs, relying on the perspicacity of the reader to supply them. In brief, he has deleted or summarised sections in line with certain rules of economy, didactically more efficient in his view, without in any way detracting from the spirit of the text. As to the letter of the text, he occasionally rewrites some of the expressions used by al-Qūhī in his own words. This linguistic difference is sufficient to indicate that al-Qūhī himself could not have been the author of this edition.

As to the identity of the editor, we know nothing for certain. We therefore feel constrained to offer this text and its translation so that the reader may compare it with al-Qūhī's treatise for themselves. We may, however, put forward a conjecture that one likely candidate could be Ibn Abī Jarrāda. This seems probable for two main reasons. Firstly, we feel we recognize aspects of his style as it appears in his edition of the treatise by Thābit ibn Qurra *On the Sections of the Cylinder*[17] – a treatise that itself appears in the collection copied by Muṣṭafā Ṣidqī. Secondly, Ibn Abī

[16] H. Suter has qualified this version as 'short' in order to distinguish it from that copied by the same Muṣṭafā Ṣidqī in 1159 [see manuscript Q], and he asks himself 'ob beide von Abû Sahl verfasst worden seien (es kam bei arabischen Gelehrten öfters vor, dass sie eine weiter ausgeführte und eine gekürzte Abhandlung über denselben Gegenstand veröffentlichten), oder ob die kürzere später von einem andern Gelehrten als Auszug aus der ersten verfasst worden sei, ist nicht zu entscheiden, doch ist das erstere wahrscheinlicher.' ['Die Abhandlungen Thâbit b. Ḳurras und Abû Sahl al-Kûhîs über die Ausmessung der Paraboloide', *Sitzungsberichte der phys.– med. Soz. in Erlangen*, 49, 1917, pp. 186–227, on p. 213]. The eminent historian has constructed an *ad hoc* argument by claiming that Arab academics often wrote a shorter treatise based on an earlier longer version. This is not the case, and any semblance of an attribution to al-Qūhī has no solid foundation in this case.

[17] Ms. Cairo, Riyāḍa 41, fols 36^v–64^v.

Jarrāda was interested in writings on infinitesimal mathematics, as is made evident by his notes on *The Measurement of the Circle* and *The Sphere and the Cylinder*[18] by Archimedes. This pure conjecture cannot be confirmed until such time as another manuscript tradition of the same text is discovered.

Let us now move on to the second version, the actual treatise written by al-Qūhī. This survives in the form of four manuscripts, all part of a single family, as we shall see. The first is found in collection 4832 in Aya Sofya, fols 125^v–129^r, a collection that we have mentioned more than once.[19] This copy appears to date from the fifth century of the Hegira (eleventh century), and is referred to here as manuscript A.

The second manuscript belongs to another famous collection, 4830 in Aya Sofya. It occupies fols 161^v–165^r, and the copy dates from 626/1228–1229. We shall refer to it as manuscript U. This manuscript also contains marginal annotations.[20]

The third manuscript belongs to the collection Riyāḍa 40 in Dār al-Kutub, Cairo, fols 187^v–190^v. It is in the handwriting of Muṣṭafā Ṣidqī, and was copied in 1159/1746. It is referred to here as manuscript Q.

Finally, the fourth manuscript belongs to collection 5648 in al-Ẓāhiriyya in Damascus, fols 166^r–171^r. We refer to it as manuscript D. This is a copy

[18] Ms. Istanbul, Fātiḥ 3414, *Kitāb fī Misāhat al-dā'ira*, fols 2^v–6^v; *Kitāb al-Kura wa-al-usṭūwāna*, fols 9^v–49^r.

[19] See the description of the manuscript, Chapter II, Section 2.1.3.

[20] These marginal annotations are in the handwriting of a certain Muḥammad Sartāq al-Marāghī, an unknown mathematician working in the eighth century of the Hegira. His writing appears in the margin of the treatise by al-Qūhī, fol. 165^r: 'Muḥammad Sartāq al-Marāghī – with the help of God the Highest – has read this illustrious treatise, has learned from it, and has written his comments, 1st of Ṣafar seven hundred and twenty-eight (17th December 1327) in the Maliki school al-Niẓāmiyya in Baghdad – may he be protected'.

Throughout this treatise, as throughout the other treatises by al-Qūhī in this Aya Sofya 4830 collection, al-Marāghī notes in the margin all the intermediate, and often elementary, steps in the proof, with references to his own book, *al-Ikmāl*, an edition of *al-Istikmāl* by Ibn Hūd. For example, in the margin of fol. 162^v, he writes: 'This has been shown in the proof of Proposition 6 in Chapter 1 of the third kind of species 4 of genus 1 of the two genera of mathematics in my book, *al-Ikmāl*: An edition of the *Istikmāl* in mathematics'. This particular book is cited throughout these marginal notes, *e.g.* fols 169^r, 171^r, 178^v, etc. Also see J.P. Hogendijk, 'The geometrical parts of the *Istikmāl* of Yūsuf al-Mu'taman ibn Hūd (11th century)', *Archives internationales d'histoire des sciences*, vol. 41, no. 127, 1991, pp. 207–81; on p. 219 he reads Baghdad instead of Nakīsār. On this town, see D. Krawulsky, *Īrān – Das Reich der Īlḫāne. Eine topographisch-historische Studie*, Wiesbaden, 1978, p. 407.

of the previous manuscript alone. We shall not, therefore, take it into account when establishing the text.

The third version appears in a manuscript copied in Mosul in 632/1234–5. It is held in collection 2519 of the Khuda-Bakhsh Library in Patna (Bankipore 2468, fols 191ᵛ–193ᵛ). This version differs from the previous version in that, while retaining the meaning, the introduction by al-Qūhī is expressed differently. The same expressions are there, as are the same ideas, but they are formulated differently. The only real difference is that the introduction to this version states that al-Qūhī also determined the centre of gravity of a portion of a hyperboloid. The remainder of the text is identical, except for the fact that B contains a surprisingly large number of omissions and errors considering the shortness of the text.

An examination of the final versions[21] results in the following stemma:

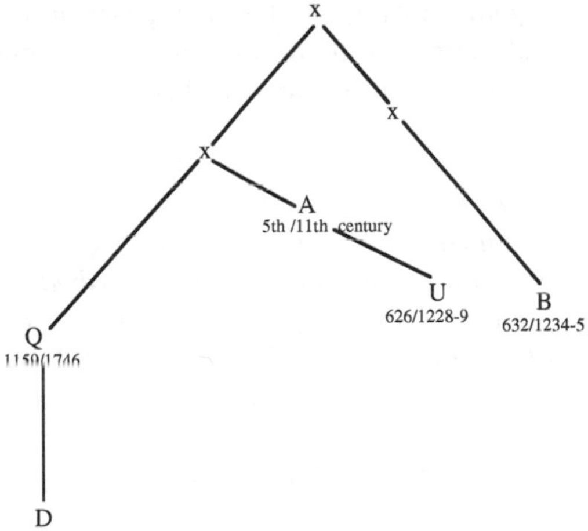

As far as I know, this treatise by al-Qūhī has never been published in a critical edition. Only manuscript B has ever been published, on three occasions. The first was in 1947,[22] the second in 1966[23] and the third, by

[21] For a detailed comparison of the manuscripts, see *Les Mathématiques infinitésimales*, vol. I, p. 841–2.

[22] Published in *al-Rasā'il al-mutafarriqa fī al-hay'a li-al-mutaqaddimīn wa-mu'āṣiray al-Bīrūnī*, ed. Osmania Oriental Publications Bureau, Hyderabad, 1947, sixth treatise.

[23] A.S. al-Dimirdāsh, 'Wayjan Rustam al-Qūhī wa-ḥajm al-mujassam al-mukāfi'', *Risālat al-'ilm*, 4, 1966, pp. 182–95.

ʿAbd al-Majīd Nuṣayr, in 1985.[24] H. Suter[25] produced a fairly free German translation of the introduction and final section of manuscript Q in order to supply the missing sections of a translation of the first version, which, as we have shown, is not the work of al-Qūhī.

5.2. MATHEMATICAL COMMENTARY

We shall now consider the mathematical content of al-Qūhī's memorandum. As we have already shown, it consists of three propositions. From the beginning, al-Qūhī distinguished between three different cases. In the first case, the inscribed and circumscribed cylindrical bodies are cylinders of revolution. In the second and third cases, these cylindrical bodies are generated by parallelograms. They are equivalent to the cylinders of revolution, as al-Qūhī explains in the first proposition. For the sake of simplicity, we shall refer to them as cylinders throughout. Taking an inscribed cylinder away from a circumscribed cylinder leaves a cylindrical ring.

Proposition 1. – *Consider a paraboloid with axis* XF *and let this axis be subdivided by a number of abscissal points* $(b_i)_{0 \leq i \leq n}$, *where* $b_0 = 0$ *and* $b_n = XF$. *Let* $(I_i)_{2 \leq i \leq n}$ *be the volumes of the inscribed cylinders associated with this subdivision, and let* $(C_i)_{1 \leq i \leq n}$ *be the volumes of the circumscribed cylinders associated with this subdivision. Let* V *be the volume of the cylinder associated with the paraboloid. Then*

$$\sum_{i=2}^{n} I_i < \frac{1}{2} V < \sum_{i=1}^{n} C_i \qquad \text{for all } n \in \mathbb{N}^*.$$

Proof:

(a) Let Z be a point on the axis and let EZ be the associated ordinate. From the fundamental properties of the parabola, we have $\frac{XF}{XZ} = \frac{AF^2}{EZ^2}$; hence, for all three cases, we have

$$\frac{XF}{XZ} = \frac{AD^2}{EI^2}.$$

[24] 'Risāla fī misāḥat al-mujassam al-mukāfī'', *Majallat Maʿhad al-Makhṭūṭāt*, 29, 1, 1985, pp. 187–208.

[25] 'Die Abhandlungen Thâbit b. Ḳurras und Abû Sahl al-Kûhîs über die Ausmessung der Paraboloide', *Sitzungsberichte der phys.– med. Soz. in Erlangen*, 49, 1917, pp. 186–227.

If S_1 is the area of the circle of diameter AD and σ_1 is that of the circle of diameter EI, then

$$XF \cdot \sigma_1 = XZ \cdot S_1.$$

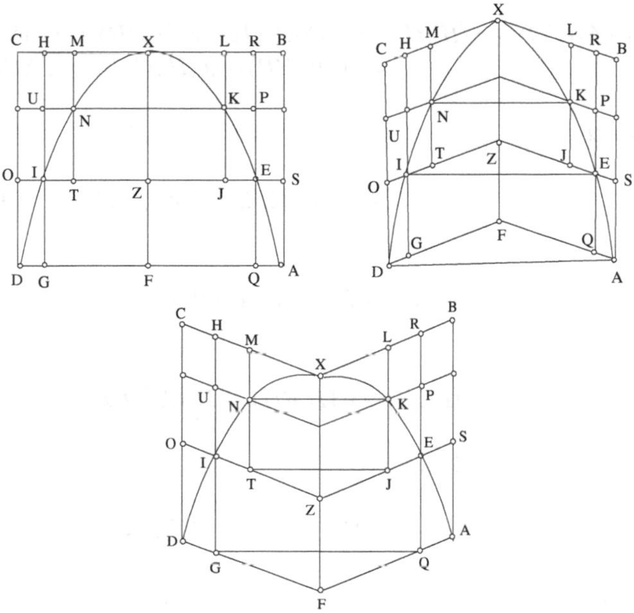

Fig. 5.1

Hence, for all three cases:

(1) $$v\,(QGHR) = v\,(SBCO).$$

The same argument applies for any pair of cylinders generated in the same way, for example

$$v\,(JLMT) = v\,(PRHU).$$

(b) Let $y = f(x)$ be the equation of the half-parabola used to generate the paraboloid. Each abscissal point b_i, $i \geq 1$, is associated with an ordinate at right angles, together with a parallelogram of dimensions b_i and $f(b_i)$. Let u_i be the volume of the cylinder generated by this parallelogram. If $u_0 = 0$, then

(2) $$u_i - u_{i-1} < 2\,C_i \qquad (1 \leq i \leq n).$$

Al-Qūhī first considers the case where $i = n$:

$$u_n = v\,(ABCD) \quad \text{and} \quad u_{n-1} = v\,(IERH).$$

We then have

$$\begin{aligned} v\,(ABCD) - v\,(IERH) &= v\,(SBCO) - v\,(IERH) + v\,(ASOD) \\ &= v\,(QGHR) - v\,(IERH) + v\,(ASOD), \end{aligned}$$

from (1); hence

(3) $$v\,(ABCD) - v\,(IERH) = v\,(QEIG) + v\,(ASOD).$$

However,

$$v\,(QEIG) < v\,(ASOD);$$

hence

$$v\,(ABCD) - v\,(IERH) < 2v\,(ASOD).$$

Similarly, for $i = n - 1$, we have

$$u_i = v\,(IERH),\ \ u_{i-1} = v\,(KLMN),$$

and we then have

$$v\,(IERH) - v\,(KLMN) < 2v\,(EPUI).$$

(c) The reasoning is identical for all $1 \leq i \leq n$, and we can deduce from (2) that

$$\sum_{i=1}^{n} C_i > \frac{V}{2}.$$

From (2)

$$\sum_{i=1}^{n} u_i - \sum_{i=1}^{n} u_{i-1} < 2\sum_{i=1}^{n} C_i.$$

However, $u_n = V$; hence

(4) $$\frac{V}{2} < \sum_{i=1}^{n} C_i.$$

(d) Similarly, we can show that

$$\sum_{i=1}^{n} I_i < \frac{V}{2}, \qquad \text{where } I_1 = 0.$$

Firstly,

(2′) $$u_i - u_{i-1} > 2I_i, \qquad 2 \le i \le n.$$

If we take $u_n = v\ (ABCD)$ and $u_{n-1} = v\ (IERH)$ and $I_n = v\ (QEIG)$, then, from (3),

$$v\ (ABCD) - v\ (IERH) = v\ (QEIG) + v\ (ASOD).$$

However,

$$v\ (QEIG) < v\ (ASOD);$$

hence

$$v\ (ABCD) - v\ (IERH) > 2v\ (QEIG).$$

Similarly, it can be shown that

$$v\ (IERH) - v\ (KLMN) > 2v\ (JKNT).$$

The reasoning is identical for all $1 \le i \le n$; hence

(2′) $$u_i - u_{i-1} > 2I_i, \qquad 2 \le i \le n.$$

From this, we can deduce that

$$\sum_{i=1}^{n} u_i - \sum_{i=1}^{n} u_{i-1} > 2\sum_{i=1}^{n} I_i$$

and

$$u_n > 2\sum_{i=2}^{n} I_i .$$

But $V = u_n$, so

(5) $$\frac{V}{2} > \sum_{i=1}^{n} I_i .$$

Combining (4) and (5) completes the proof.

Proposition 2. – *Let a portion of a paraboloid lie between any two ordinate surfaces, and let* I *and* C *be the volumes of the corresponding inscribed and circumscribed cylinders respectively. If this portion is cut by a third ordinate surface equidistant from the other two, and we construct two inscribed cylinders with volumes* I_1 *and* I_2 *respectively, and two homologous circumscribed cylinders with volumes* C_1 *and* C_2 *respectively, then*

$$(\mathrm{C}_1 - \mathrm{I}_1) + (\mathrm{C}_2 - \mathrm{I}_2) = \frac{1}{2}\,(\mathrm{C} - \mathrm{I}).$$

$C - I = v$ (ring $HGEC$), and $C_1 - I_1 = \mathrm{v}$ (ring $NLMC$)
$C_2 - I_2 = v$ (ring $LKGS$)

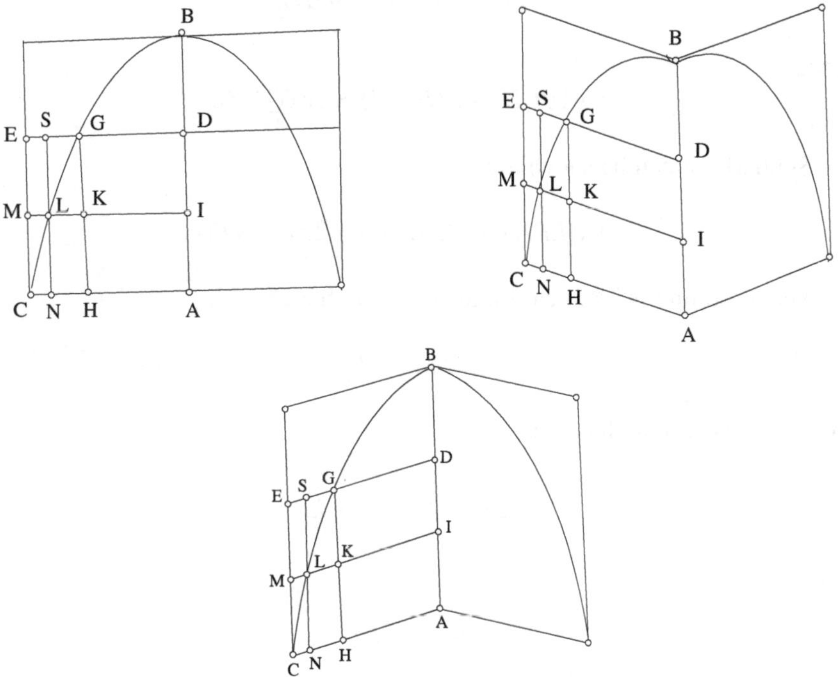

Fig. 5.2

However, *HGEC* is a parallelogram, and *KM* passes through the midpoint *L* of *NS*; therefore

$$v\,(\text{ring } NLMC) + v\,(\text{ring } LKGS) = \frac{1}{2}\,v\,(NSEC) + \frac{1}{2}\,v\,(\text{ring } NHSG)$$
$$= \frac{1}{2}\,v\,(\text{ring } HGEC);$$

and thence the result.

Comment. — In the mind of al-Qūhī, the meaning of this proof is as follows: If one begins with the subdivision of the axis XF by the abscissal points $(b_i)_{0\leq i\leq n}$, where $(I_i)_{1\leq i\leq n}$, $(C_i)_{1\leq i\leq n}$ and $I_1 = 0$ are the volumes of the corresponding cylinders, and if one then considers the series $(c_j)_{0\leq j\leq 2n}$, where $b_0 = c_0$, $b_n = c_{2n}$, $c_{2i+1} = \frac{b_i + b_{i+1}}{2}$, and $(I'_j)_{1\leq j\leq 2n}$ and $(C'_j)_{1\leq j\leq 2n}$ are the volumes of the corresponding cylinders associated with this subdivision, then

$$\sum_{j=1}^{2n}(C'_j - I'_j) = \frac{1}{2}\sum_{i=1}^{n}C_i - I_i).$$

Proposition 3. – *If* P *is the volume of a portion of a paraboloid and* V *the volume of the associated cylinder, then*

$$\mathrm{P} = \frac{\mathrm{V}}{2}.$$

Proof: If we suppose that $P \neq \frac{V}{2}$, then

$$P = \frac{V}{2} + \varepsilon \quad \text{or} \quad P = \frac{V}{2} - \varepsilon \qquad (\varepsilon > 0).$$

We can show that each of these cases results in a contradiction, regardless of the initial subdivision $(b_i)_{0\leq i\leq n}$ of the axis XF. Using the process described in the previous proposition, we can construct the subdivisions defined as follows:

$$(b_i^1)_{0\leq i\leq 2n},\ (b_i^2)_{0\leq i\leq n.2^2},\ \dots,\ (b_i^q)_{0\leq i\leq n.2^q}\ \dots.$$

If $(I_i^q)_{1\leq i\leq n.2^q}$ and $(C_i^q)_{1\leq i\leq n.2^q}$ are the volumes of the cylinders associated with subdivision $(b_i^q)_{1\leq i\leq n.2^q}$, we know from the previous proposition that

$$\sum_{i=1}^{n.2^q}(C_i^q - I_i^q) = \frac{1}{2}\sum_{i=1}^{n.2^{q-1}}(C_i^{q-1} - I_i^{q-1})$$

for a constant n and any q in $\mathbf{N}^*$. From this, al-Qūhī used an extension of Proposition X.1 of Euclid in order to show that, after a certain number of operations,

(6) $$\sum_{i=1}^{n.2^q}(C_i^q - I_i^q) < \varepsilon.$$

In other words, he showed that for all $\varepsilon > 0$, there exists N such that for all $q > N$ equation (6) is satisfied. However,

$$P - \sum_{i=1}^{n.2^q} I_i^q < \sum_{i=1}^{n.2^q}(C_i^q - I_i^q);$$

hence

$$P - \sum_{i=1}^{n.2^q} I_i^q < \varepsilon.$$

If $P = \frac{V}{2} + \varepsilon$, then $\frac{V}{2} < \sum_{i=1}^{n.2^q} I_i^q$, which is impossible by Proposition 1. Similarly, if $P = \frac{V}{2} - \varepsilon$, the same reasoning applies, as

$$\sum_{i=1}^{n.2^q} C_i^q - P < \sum_{i=1}^{n.2^q}(C_i^q - I_i^q) < \varepsilon;$$

hence

$$\sum_{i=1}^{n.2^q} C_i^q - \left(\frac{V}{2} - \varepsilon\right) < \varepsilon,$$

and hence

$$\sum_{i=1}^{n.2^q} C_i^q < \frac{V}{2},$$

which is also impossible by Proposition 1. Therefore

$$P = \frac{V}{2}.$$

Al-Qūhī's proof is effectively established here due to Proposition 1, in which he compares the sums of the inscribed and circumscribed cylinders with the volume of the large cylinder without needing to evaluate these sums; that is, as Archimedes did by summing an arithmetical progression. The proof of this proposition is based on the inequalities (2) and (2′), which derive from a consideration of equal cylinders such as *QGHR* and *SBCO* that are neither inscribed nor circumscribed, and which do not, therefore, constitute an *a priori* requirement.

Proposition 2 shows that, if the subdivision is made finer by dividing each interval by a factor of two, then the difference between the circumscribed and inscribed cylinders is also divided by two. This proposition serves the same purpose as Proposition 19 in Archimedes' book *The Conoids and Spheroids*.

In its use of integral sums, the method used by al-Qūhī appears to be similar to that of Archimedes. However, the way in which the proof proceeds is different. It appears to be more the case that al-Qūhī rediscovered the use of integral sums.

5.3. *Translated texts*

Abū Sahl al-Qūhī

5.3.1. *On the Determination of the Volume of a Paraboloid*

5.3.2. *On the Volume of a Paraboloid*

In the Name of God, the Merciful, the Compassionate

TREATISE BY ABŪ SAHL WAYJAN IBN RUSTAM AL-QŪHĪ

On the Determination of the Volume of a Paraboloid

As an understanding of the measurement of solids, figures, magnitudes and the ratios between them is a prerequisite to an understanding of their centres of gravity – the former being as an introduction to the latter, given that it is not possible to find the centres of gravity until after one has gained an understanding of measurement – we have needed to gain such a prior understanding of measurement from the book of Archimedes *On the Sphere and the Cylinder*, and from other books written on this subject. It was on completing this study that we began to write our book *On the Centres of Gravity*. In it, we undertook a minute analysis, in so far as is possible and within the limits of our capabilities, to the point where we have found the centres of gravity of many things having weight that have not been found previously by any of the ancients distinguished in geometry, let alone their modern inferiors, and the earlier discovery of which we are unaware of before our time, such as the centre of gravity of any given portion of a sphere or an ellipsoid.[1] Once we had discovered it, we were seized by an intense desire to find the centres of gravity of other solids which had not been previously found, such as the centre of gravity of a paraboloid. As we have explained earlier, its volume must be known before the centre of gravity can be found.

We state that: There is no other book in existence on this subject, other than the book by Abū al-Ḥasan Thābit ibn Qurra, which is a book that is well known and famed among geometers. But it is voluminous and long, containing around forty propositions, some numerical, some geometrical and some others, all of which are lemmas for a single proposition which is: How to know the volume of a paraboloid.

When we studied that book, we found it very difficult to understand it, while we found the book of Archimedes *On the Sphere and the Cylinder*

[1] The manuscript [B] mentions also the centre of gravity of a section of a hyperboloid.

much easier despite its difficulty and the multiplicity of its aims, and even though the aims of the two books are the same. We therefore thought that all those who have studied the book, from the moment it was written by Thābit ibn Qurra to the present day, would have found as we did, that it is difficult to understand. It was this belief that persuaded us to continue with our studies into the determination of the volume of this solid, that is, the paraboloid, beginning once more from the beginning. We have been able to make this determination by means of an accessible method, using none of these lemmas, and with no need for any of them. Anyone who examines this book and our book will see that it is so, as we have said.

But while, in the composition of our book *On the Centres of Gravity*, we did not find it necessary to know the volume of a paraboloid, and while we did know it and we did understand it from the book of Thābit, we did not take the time to pursue the determination of that which others before us have determined – regardless of the method they employed – and we did not speak of the methods of determination used by those who preceded us, whether they be long or short, difficult or easy, or requiring or not requiring lemmas, because that is not how we normally work, and especially because the byways of this science are many and wide ranging.

We say that: If a portion of a parabola rotates simultaneously with the parallelogram defined by the diameter of this portion and half of its base, together with the ordinates to this diameter and the straight lines passing through the extremities of these ordinates to this diameter, around and parallel to this same diameter, and if this rotation is continued until the portion of the parabola returns to its original position, then the solid generated by the rotation of the plane of this portion is a paraboloid. The solid generated by the rotation of the parallelogram defined by the diameter of the section and half of its base is the cylinder of the paraboloid. This diameter is also the diameter of the paraboloid. We call the surfaces generated by the rotation of the ordinate lines the ordinate surfaces of the paraboloid. We call the solids generated between the ordinate surfaces the cylindrical bodies of the paraboloid. Of all these cylindrical bodies, that which is generated by a parallelogram that can be wholly enclosed within the portion such that one of its angles lies on its boundary we call the cylindrical body inscribed within the paraboloid. And of all these cylindrical bodies, that which is generated by a parallelogram part of which lies outside the portion such that one of its angles lies on its boundary we call the cylindrical body circumscribed around the paraboloid. We use the term homologous to describe the pair of cylindrical bodies consisting of one inscribed within the paraboloid and one circumscribed around it, provided the inscribed body is separate from[2] the circumscribed one, by which we mean that they have the same height. We shall call any solid generated by the rotation of one of the surfaces on this portion around the diameter of this portion, whatever the surface, the solid of this surface or the solid formed from the surface, regardless of whether it resembles a ring, a cylinder or any other form.

– **1** – Any half-cylinder of a paraboloid is less than the sum of cylindrical bodies, whatever their number, circumscribed around the paraboloid, and is greater than the sum of the cylindrical bodies, whatever their number, inscribed within it.

Example: Let the cylinder of the paraboloid be *ABCD*, the paraboloid *AXD*, the circumscribed cylindrical bodies *ASOD, EPUI* and *KLMN,* and the cylindrical bodies inscribed within it *QEIG* and *JKNT.*

I say that half of the cylinder ABCD *is less than the sum of the cylindrical bodies* ASOD, EPUI *and* KLMN *circumscribed around the paraboloid, and that the sum of their analogues, however many their number, and is greater than the sum of the cylindrical bodies* QEIG *and* JKNT *inscribed within it and the sum of their analogues, however many their number.*

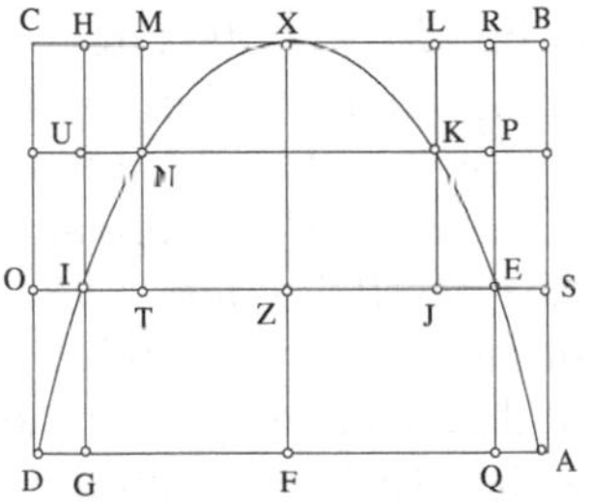

Fig. V.1.1a

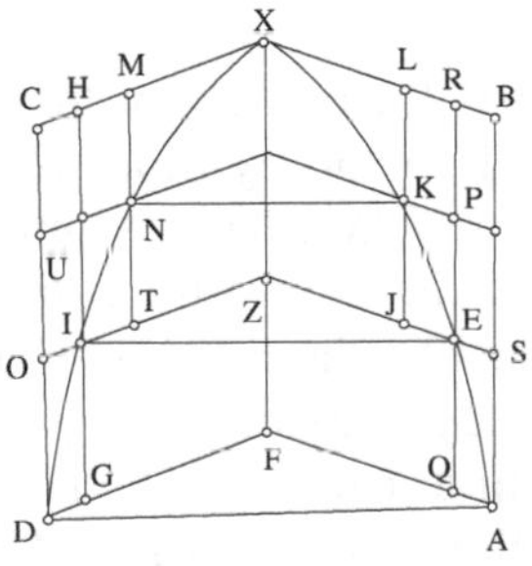

Fig. V.1.1b

Proof: Each of the straight lines *AF* and *EZ* is an ordinate to the diameter *XZF*. The ratio of the straight line *FX* to the straight line *XZ* is therefore equal to the ratio of the square of *AF* to the square of *EZ*, as the portion *AXD* is a portion of a parabola. But the ratio of the square of *AF* to the square of *EZ* is equal to the ratio of the square of *AD* to the square of *EI*, and the ratio of the square of *AD* to the square of *EI* is equal to the ratio of the circle of diameter *AD* to the circle of diameter *EI*. Therefore, the ratio of the circle of diameter *AD* to the circle of diameter *EI* is equal to the

[2] *i.e.* is a part of.

ratio of the straight line *FX* to the straight line *XZ*. The product of the straight line *FX* and the circle of diameter *EI* is therefore equal to the product of the straight line *XZ* by the circle of diameter *AD*. But the product of the straight line *FX* and the circle of diameter *EI* is equal to the cylinder *QRHG* generated by the rotation of the parallelogram *RQFX* about the diameter *XF*, regardless of whether or not the ordinate to the diameter is at right angle to it. If it were not at a right angle, the effect is identical to having taken a given cone away from one vertex of the cylinder and added it to the other vertex.

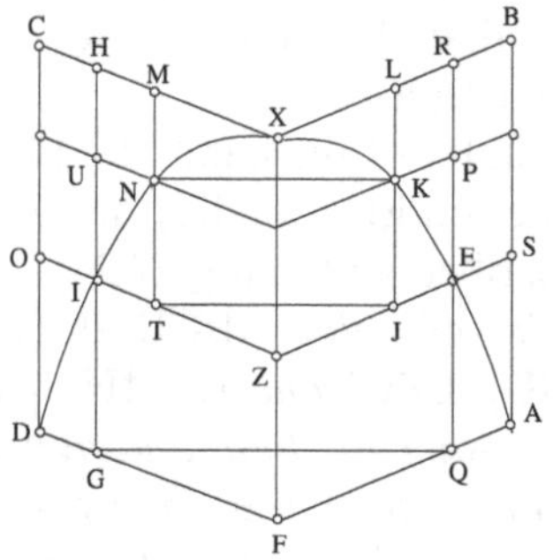

Fig. V.1.1c

Similarly, the product of the straight line *XZ* and the circle of diameter *AD* is equal to the cylinder *SBCO* generated by the rotation of the parallelogram *SBXZ*. The cylinder *QRHG* is equal to the cylinder *SBCO*. Therefore, if we remove the common cylinder *ERHI*, the remaining solid is that generated by the rotation of one of the parallelograms *SBRE* or *IHCO*, equal to the cylindrical body *QEIG*. But the cylindrical body *QEIG* is less than the cylindrical body *ASOD*. The solid generated by the rotation of one of the parallelograms *SBRE* or *IHCO* is therefore less than the cylindrical body *ASOD*.

Composing, the sum of this solid and this cylindrical body is less than twice the cylindrical body *ASOD*. But the solid and the cylindrical body together are the excess of the cylinder *ABCD* over the cylinder *ERHI*. The excess of the cylinder *ABCD* over the cylinder *ERHI* is therefore less that twice the cylindrical body *ASOD* circumscribed around the paraboloid. Similarly, the excess of the cylinder *ERHI* over the cylinder *KLMN* is less than twice the cylindrical body *EPUI*, which is circumscribed around the paraboloid. The same may be said of all the cylinders and cylindrical bodies circumscribed around it until one arrives at the remainder of the last part of the given cylinder *ABCD*; let this remainder be the solid *KLMN*. The excess of the cylinder *ABCD* over the solid *KLMN* is less than twice the sum of the cylindrical bodies circumscribed around the paraboloid, with the exception of the solid *KLMN*. If we take the solid *KLMN* to be common, the cylinder *ABCD* will be less than twice the sum of the cylindrical bodies circumscribed around the paraboloid, whatever their number. Half of it is therefore less than the sum of the cylindrical bodies circumscribed around the paraboloid, whatever their number.

Moreover, as the solid generated by the rotation[3] of the parallelograms *ABRQ* and *GHCD* is greater than the solid generated by[4] the parallelograms *SBRE* and *IHCO*, and as it is equal to the cylindrical body *QEIG*, as we have already shown, then the solid generated by[5] the two parallelograms *ABRQ* and *GHCD* is greater than the cylindrical body *QEIG*. Composing, the sum of the two <solids> is greater than twice the cylindrical body *QEIG*. But the sum is the excess of the cylinder *ABCD* over the cylinder *ERHI*. The excess of the cylinder *ABCD* over the cylinder *ERHI* is therefore greater than twice the cylindrical body *QEIG*. Similarly, the excess of the cylinder *ERHI* over the cylinder *KLMN* is greater than twice the cylindrical body *JKNT*, as we have shown. The same may be said of all the cylinders and cylindrical bodies inscribed within the paraboloid until one arrives at the final remainder of the given cylinder, which remainder is the solid *KLMN*. The excess of the cylinder *ABCD* over the solid *KLMN* is greater than twice <the sum of> the cylindrical bodies inscribed within the paraboloid, whatever their number. If we add the solid *KLMN* to the excess of the cylinder *ABCD* over it, we have the entire cylinder *ABCD* being much greater than twice <the sum of> the cylindrical bodies inscribed within the paraboloid, whatever their number. Half of the cylinder *ABCD* is therefore greater than the sum of the cylindrical bodies, whatever their number, inscribed within the paraboloid, and less than the sum of cylindrical bodies, whatever their number, circumscribed around the paraboloid. This is what we wanted to prove.

– **2** – If one of the cylindrical bodies between two of the ordinate surfaces of a paraboloid is divided into two halves by another of the ordinate surfaces such that the division results in two cylindrical bodies circumscribed around the paraboloid and two inscribed cylindrical bodies that are homologous to them, then the excess of <the sum of> the two circumscribed cylindrical bodies over their inscribed homologues is half the excess of the first circumscribed cylindrical body over its inscribed homologue prior to the division.

[3] Lit.: the solid which turns about.
[4] *Ibid.*
[5] *Ibid.*

Example: Let one of the cylindrical bodies circumscribed around the paraboloid *ABC* be that generated by the rotation of the parallelogram *ADEC*, and its homologue inscribed cylindrical body be that generated by the rotation of the parallelogram *ADGH*. The straight line *IKLM* is drawn so as to divide the straight lines *AD* and *EC*, and the lines between them and parallel to them, into two halves. This is why the straight line *IKLM* is parallel to the two straight lines *AC* and *DE*. Draw the straight line *NLS* parallel to the diameter *AB*.

I say that the excess of the two cylindrical bodies IDSL *and* AIMC *over the two homologous cylindrical bodies* IDGK *and* AILN, *that is, the two solids formed from the two parallelograms* KGSL *and* NLMC, *is half the excess of the cylindrical body* ADEC *over its homologous cylindrical body* ADGH, *that is, the solid formed from the parallelogram* HGEC.

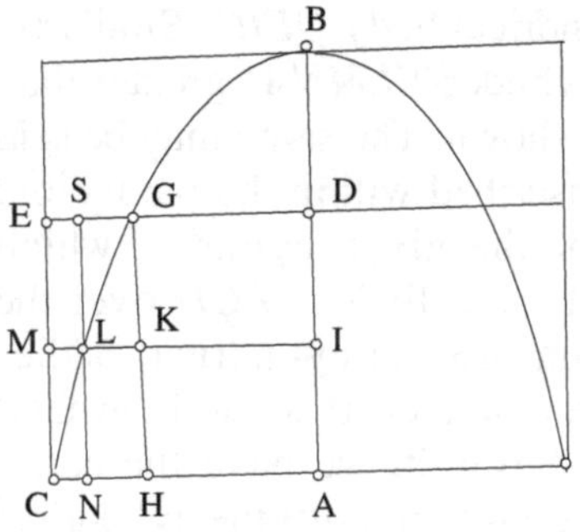

Fig. V.1.2a

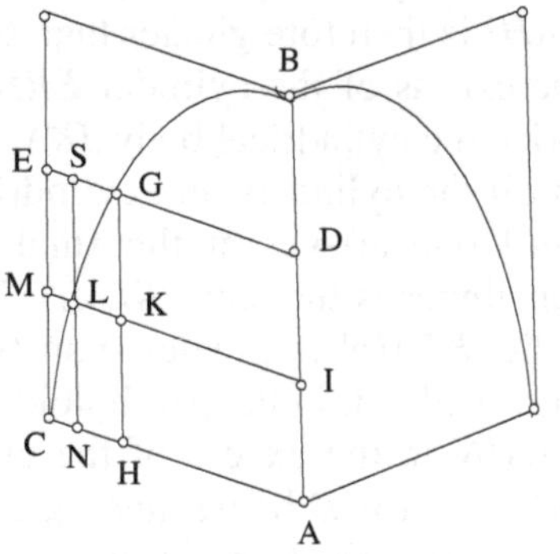

Fig. V.1.2b

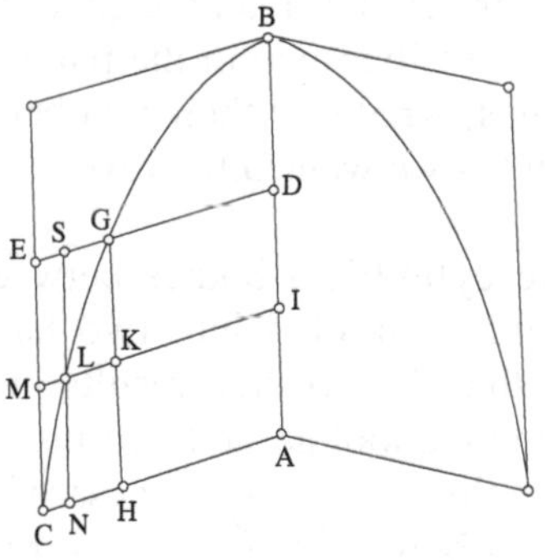

Fig. V.1.2c

Proof: As *HGSN* is a parallelogram and *GH* has been divided into two halves by the straight line *KL* parallel to the two straight lines *GS* and *HN*, then the parallelogram *KGSL* is equal to the parallelogram *HKLN*, and therefore the parallelogram *KGSL* is half the parallelogram *HGSN*. In the same way, we can show that the parallelogram *NLMC* is half the parallelogram *NSEC*. The two cylindrical bodies <generated by> the two surfaces *KGSL* and *NLMC* together – which are the excess of the two cylindrical bodies *IDSL* and *AIMC* over the two cylindrical bodies *IDGK* and *AILN* – are therefore equal to half the cylindrical body <generated by> the surface *HGEC*, which is the excess of the cylindrical body *ADEC* over the cylindrical body *ADGH*. This is what we wanted to prove.

– **3** – Any paraboloid is equal to half of its cylinder.

Example: Let *ABC* be the paraboloid and let *D* be a body equal to half the cylinder of the paraboloid *ABC*.

I say that the solid ABC *is equal to the body* D.

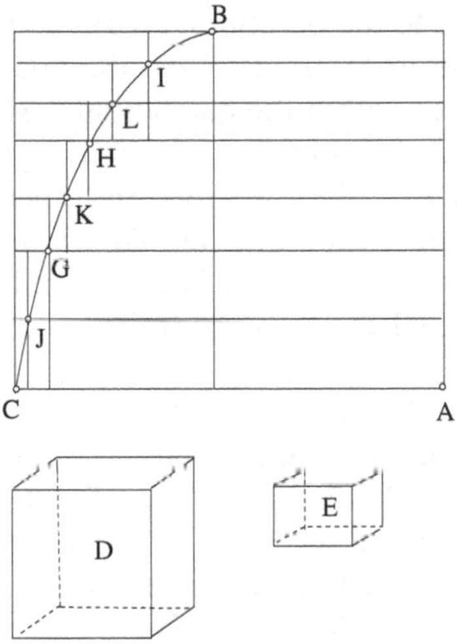

Fig. V.1.3

Proof: If the paraboloid *ABC* is not equal to the body *D*, then it must be either greater or less than it.

Let it be first of all greater than the body *D*, if that is possible. Now let the excess of the solid *ABC* over the body *D* be the body *E*. We construct any given number of cylindrical bodies circumscribed around the paraboloid *ABC*. Let us separate from each circumscribed cylindrical body a <corresponding> inscribed cylindrical body, *i.e.* its homologue. Let the excesses of the circumscribed cylindrical bodies over their inscribed homologues be the solids formed by the rotation of the parallelograms *CG*, *GH* and *HI*. Let us divide each of these cylindrical bodies into two halves by the ordinate surfaces such that the excesses of the cylindrical bodies circumscribed around the paraboloid over their inscribed homologues are equal to half the excesses that existed before the division, as we proved in the second proposition. Similarly, let us continue to divide these generated cylindrical bodies into two halves until the excesses of the cylindrical bodies circumscribed around the paraboloid over their inscribed homologues become less than the body *E*. The body *E* is therefore greater than the sum of these excesses. Let these excesses be the solids generated by the parallelograms *CJ, JG, GK, KH, HL* and *LI*. The body *E* is therefore greater than the sum of these solids, and it is therefore much greater than the solids formed from the triangles[6] contained in the paraboloid, as these constitute only part of these excesses. If we set the body *D* to be common, then the sum of the two bodies *E* and *D* is greater than the sum of the solids formed from all these triangles and the body *D*. But <the sum of> the two bodies *E* and *D* is equal to the paraboloid *ABC*, as we have supposed. The paraboloid *ABC* is therefore greater than the body *D* plus all the solids formed from the triangles within the paraboloid *ABC*. If we further remove the common solids formed from the common triangles, then the sum of the cylindrical bodies, whatever their number, inscribed within the paraboloid *ABC* remains greater than the solid *D*. This is impossible, as we have proved in the first proposition that it is less than half the cylinder of the paraboloid, which is equal to the body *D*. The paraboloid is therefore not greater than the body *D*.

If it is possible that the paraboloid *ABC* is less than the body *D*, with the difference between them being the body *E*, such that the solid *ABC* plus the body *E* is equal to the body *D*. We then divide each of the cylindrical bodies circumscribed around the solid *ABC* into two halves, as we have already said, so that the excesses arrive at <a sum that is> less than the body *E*, as we have shown. The <sum of the> solids of the triangles which are external to the paraboloid are very much less than the body *E*, as they form part of these excesses. If we take the paraboloid *ABC* to be common, then the solids of the triangles circumscribed around the paraboloid, that is, those which are external to it, plus the paraboloid *ABC* is less than the body *E* plus the paraboloid *ABC*. But the body *E* plus the paraboloid *ABC* is equal to the body *D*, which is as we had supposed. The solids of the triangles circumscribed around the paraboloid plus the paraboloid itself are the cylindrical bodies circumscribed around the paraboloid. The cylindrical bodies circumscribed around the paraboloid are therefore less than the body *D*, which is impossible, as we have proved in the first proposition that they are greater than half the cylinder of the paraboloid *ABC*, which is equal to the body *D*. The paraboloid *ABC* is therefore not less than the

[6] Implying: curvilinear triangles.

body *D*. As we have already shown that it is not greater, the paraboloid *ABC* is therefore equal to the body *D*, which is equal to half its cylinder. Any paraboloid is therefore equal to half of its cylinder. This is what we wanted to prove.

We have used the following in this proposition: If we have two different magnitudes and we separate out from the larger of these, half of it, half of the remainder, and half of that, and if we continue to proceed in the same way, we shall arrive at a magnitude that is less than the smaller of the <original> magnitudes. The larger magnitude in this case is the sum of the excesses of the cylindrical bodies circumscribed around the paraboloid over their inscribed homologues. Each of them is divided into two halves and the smaller magnitude is the body *E*. Euclid has shown that, if we separate out from the larger <magnitude>, more than half of it, more than half of the remainder, and if we continue to proceed in the same way, we shall arrive at a magnitude that is less than the smaller <of the original magnitudes>. The proof for both is the same. If it is as we have described, it would be better to say: If we have two different magnitudes and we separate out from the larger of these that[7] which is not less than half of it, and from the remainder, that which is not less than half of it, and if we continue to proceed in the same way, we shall arrive at a magnitude that is less than the smaller of the <original> magnitudes, so that the proof is general. All success derives from God.

The treatise of Abū Sahl al-Qūhī
on the volume of the paraboloid is completed.

[7] *i.e.* a part which.

In the Name of God, the Merciful, the Compassionate

THE BOOK OF ABŪ SAHL WAYJAN IBN RUSTAM AL-QŪHĪ

On the Volume of a Paraboloid

In a single treatise and in three propositions

Introduction

If a portion of a parabola limited by the arc of that portion, the diameter and half of the base is rotated about the diameter simultaneously with the parallelogram defined by the diameter of this portion and half of its base, together with the ordinates to this diameter and the straight lines passing through the extremities of these ordinates parallel to this same diameter, until it returns to its original position, then the solid generated by the rotation of this portion is a paraboloid and this diameter is its diameter. The solid generated by the rotation of this parallelogram is the cylinder of the paraboloid, the surfaces generated by the rotation of the straight line ordinates are the ordinate surfaces of the paraboloid, and the solids generated between them are the cylindrical bodies of the paraboloid. Of all these cylindrical bodies, that which is generated by a parallelogram that can be wholly enclosed within the portion such that one of its angles lies on its boundary is a cylindrical body inscribed within the paraboloid, and that which is generated by a parallelogram part of which is external to the portion such that one of its angles lies on its boundary is a cylindrical body circumscribed around the paraboloid. If one of these is separated from the other,[1] then they are two homologues. The solid generated by the rotation of one of the surfaces around this diameter is the solid of that surface and it is obtained from that surface, regardless of whether it <resembles> a cylinder, a ring, or any other form.

[1] *i.e.* is a part of the other.

The propositions

– **1** – Any half-cylinder of a paraboloid is less than the sum of cylindrical bodies circumscribed around the paraboloid, and is greater than the sum of the cylindrical bodies inscribed within it.

Let the paraboloid be *AXD*, its cylinder *ABCD*, the circumscribed cylindrical bodies *ASOD*, *EPUI* and *KLMN*, and the cylindrical bodies inscribed within it *QEIR* and *JKNT*.

I say that half of the cylinder ABCD *is less than the sum of the cylindrical bodies* ASOD, EPUI *and* KLMN, *whatever their number, circumscribed around the paraboloid, and greater than the sum of the cylindrical bodies* QEIR *and* JKNT, *whatever their number, inscribed within the paraboloid.*

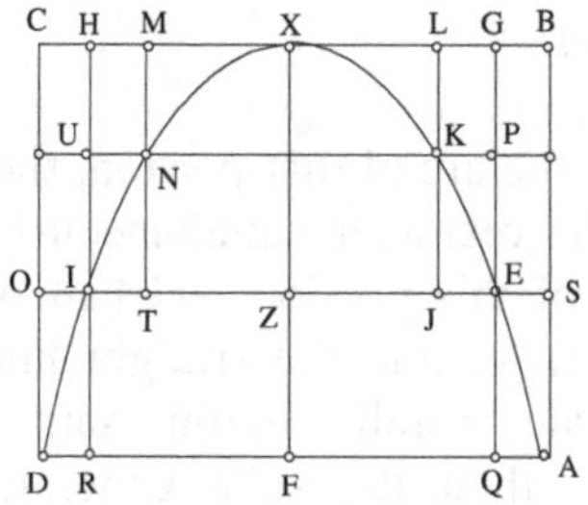

Fig. V.2.1a

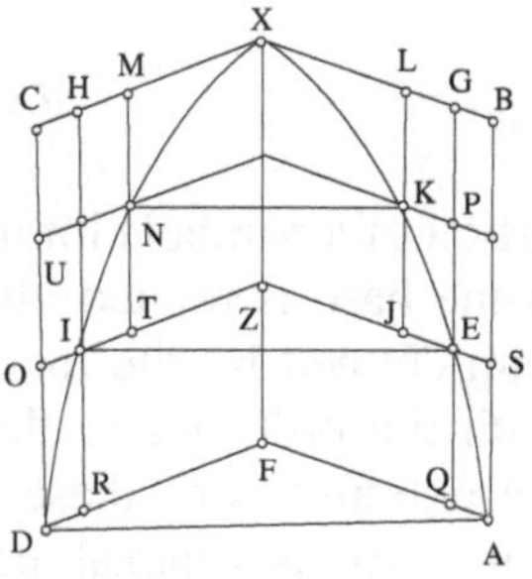

Fig. V.2.1b

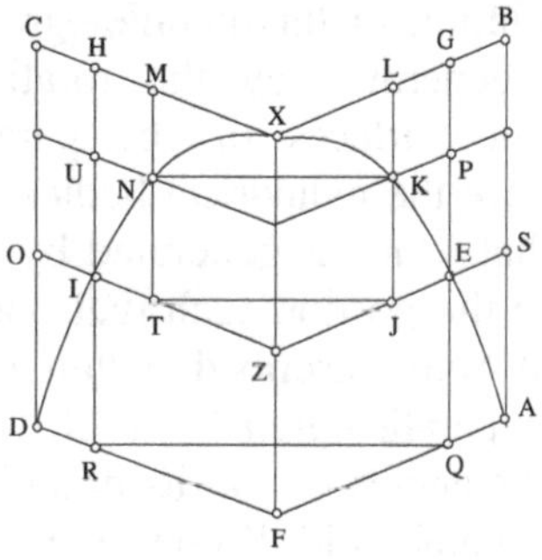

Fig. V.2.1c

Proof: We construct the diameter *XZF*. The ratio of *FX* to *XZ* is then equal to the ratio of the square of *AF* to the square of *EZ*, two straight line ordinates, that is the ratio of the square of *AD* to the square of *EI*, that is, the ratio of the circle of diameter *AD* to the circle of diameter *EI*. The cylinder *QGHR* defined by the rotation of the surface *QGXF* around the diameter *XF* is equal to the cylinder *SBCO* defined by the rotation of the surface *SBXZ* around the diameter *XF*, regardless of whether or not *XF* is an axis, as the excess generated at one extremity of the cylinder is equal to the missing section at the other extremity. If the common cylinder *EGHI* is removed, the remaining solid is that generated by one of the two surfaces *SBGE* and *IHCO* equal to the cylindrical body *QEIR*. It is therefore less than the cylindrical body *ASOD*. The solid mentioned and the cylindrical body *ASOD*, that is the excess of the cylinder *ABCD* over the cylinder *EGHI*, are <in sum> less than twice the cylindrical body *ASOD* circumscribed around the paraboloid. Similarly, we can show that the excess of the cylinder *EGHI* over the cylinder *KLMN* is less than twice the cylindrical body *EPUI*, and similarly for all the homologous cylinders and cylindrical bodies – for the reasons that we have just described – until one arrives at the remainder at the end of the cylinder *ABCD*; let this remainder be the solid *KLMN*. The excess of the cylinder *ABCD* over the solid *KLMN* is therefore less than twice the sum of the cylindrical bodies circumscribed around the paraboloid, with the exception of the solid *KLMN*. The entire cylinder *ABCD* is therefore less than the solid *KLMN* plus twice the sum of the cylindrical bodies circumscribed around the paraboloid, and half the cylinder *ABCD* is therefore less than the sum of the cylindrical bodies mentioned plus half the solid *KLMN*. It is therefore very much less than the sum of the cylindrical bodies mentioned plus the solid *KLMN*.

Similarly, the solid defined by the rotation of one of the two surfaces *ABGQ* and *RHCD* is greater than the solid defined by the rotation of one of the two surfaces *SBGE* and *IHCO*, that is, the cylindrical body *QEIR*. The solid defined by the rotation of one of the two surfaces *ABGQ* and *RHCD* plus the cylindrical body *QEIR*, that is, the excess of the cylinder *ABCD* over the cylinder *EGHI*, is therefore greater than twice the cylindrical body *QEIR*. Similarly, we can show that the excess of the cylinder *EGHI* over the solid *KLMN* is greater than twice the cylindrical body *JKNT*, and similarly for all the homologous cylinders and cylindrical bodies – for the reasons that we have just described – until one arrives at the remainder at the end of the cylinder *ABCD*; let the remainder be the solid *KLMN*. The excess of the cylinder *ABCD* over the solid *KLMN* is therefore greater than twice the sum of the cylindrical bodies inscribed within the paraboloid. The whole cylinder *ABCD* is therefore very much greater than twice the sum of the cylindrical bodies inscribed within the paraboloid. Half of this cylinder is therefore greater than the sum of the cylindrical bodies inscribed within the paraboloid; yet, it was less than the sum of cylindrical bodies circumscribed around the paraboloid. This is what we wanted to prove.

– **2** – If an ordinate surface is produced in any cylindrical body such that it is parallel to the two ordinate surfaces bounding the cylindrical body and divides the cylindrical body into two halves forming two cylindrical bodies circumscribed around a paraboloid and two inscribed homologues, then the excess of the two circumscribed cylindrical bodies over their inscribed homologues is equal to half the excess of the divided cylindrical body circumscribed around the paraboloid over its homologue inscribed within the paraboloid.

Let the paraboloid be that generated by the rotation of a portion *BC* of a parabola and a straight line ordinate *AC* about the diameter *AB*, and let the same rotation about the diameter generate the cylindrical body *ADEC* by the rotation of the parallelogram *ADEC,* and its homologue the cylindrical body *ADGH* by the rotation of the surface *ADGH*. We produce a straight line ordinate *IKLM* parallel to the two straight lines *DE* and *AC*, dividing the two straight lines *AD* and *EC* into two halves. Now, we produce an ordinate surface along the straight line *IM* parallel to the two surfaces *DE* and *AC*, which are also ordinates, such that this surface divides the cylindrical body *ADEC* into two halves and generates two cylindrical bodies *AIMC* and *IDSL* circumscribed around the solid, and two homologues *AILN* and *IDGK* inscribed within it.

I say that the excess of the two cylindrical bodies AIMC *and* IDSL *over the two homologous cylindrical bodies* AILN *and* IDGK *is equal to half the excess of the cylindrical body* ADEC *over the cylindrical body* ADGH.

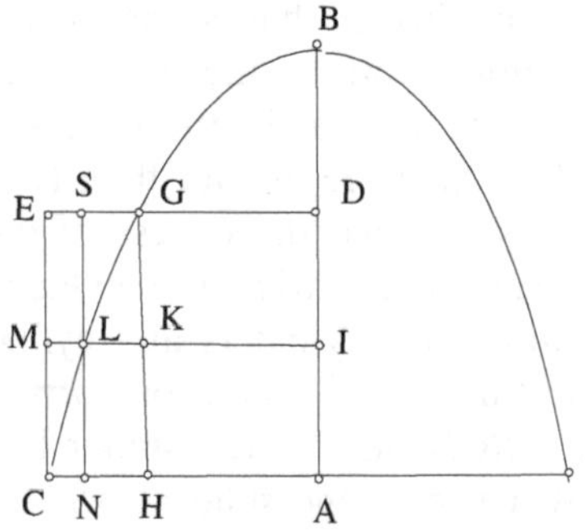

Fig. V.2.2

Proof: We produce a straight *SLN* from the point *L* parallel to the two straight lines *AD* and *EC*. As the straight line *IKLM* divides *AD* and its parallels into two halves, the surface *KLSG* is half the surface *GSNH* and the surface *NLMC* is half the surface *NSEC*. The same applies to the solids generated by their rotation. The solid generated by the rotation of the surface *KLSG* is therefore half that generated by the rotation of the surface *GSNH,* and that generated by the rotation of *NLMC* is half that generated by the rotation of *NSEC*. The <sum of the> two <solids> generated by the rotation of *KLSG* and the rotation of *NLMC*, that is, the excess of the two cylindrical bodies *IDSL* and *AIMC* over the two cylindrical bodies *IDGK* and *AILN*, is half of that generated by the rotation of *HGEC*, that is, the excess of the cylindrical body *ADEC* over the cylindrical body *ADGH*. This is what we required.

– **3** – Any paraboloid is equal to half of its cylinder.

Let *ABC* be a paraboloid.

I say that it is equal to half of its cylinder.

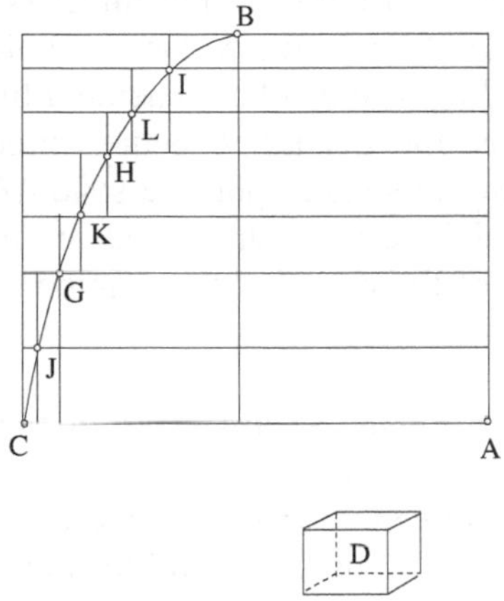

Fig. V.2.3

Proof: If this were not the case, let the solid *ABC* be greater than half of its cylinder by a magnitude equal to the solid *D*. Let us now circumscribe any number of cylindrical bodies around the solid *ABC* and separate them from their homologues inscribed with the solid. Let the excesses of the circumscribed cylindrical bodies over their inscribed homologues be the solids generated by the rotation of the surfaces *CG*, *GH* and *HI*. Let us divide each of the cylindrical bodies into two halves by ordinate surfaces. The excesses of the cylindrical bodies over their homologues are then equal to half the excesses that existed before the division, as was shown in the second proposition. We continue to proceed in this way until the excesses become less than the solid *D*. Let these excesses be the solids generated by the rotation of the surfaces *CJ*, *JG*, *GK*, *KH*, *HL*, and *LI*. The solid *D* is therefore greater than these solids, and it is therefore very much greater than the solids within the paraboloid and generated by the rotation of the triangles within the parabola whose lines consist of sections of the straight line ordinates, line parallel to the diameter and sections of the perimeter of the section. But half of the cylinder is greater than the cylindrical bodies inscribed within the paraboloid, and therefore half of the cylinder plus the solid *D*, that is, the paraboloid, is greater than the cylindrical bodies inscribed within the paraboloid plus the solids generated by the rotation of the triangles, that is the paraboloid. The paraboloid is therefore greater than itself; this is contradictory.

Now let the paraboloid *ABC* be less than half of its cylinder by the magnitude of the solid *D*. Then the paraboloid plus the solid *D* is equal to half of the cylinder. We now continue dividing the cylindrical bodies circumscribed around the paraboloid until the remainder is less than the solid *D*. The solids of the triangles found outside the paraboloid are therefore very much less than the solid *D*. These solids generated by the triangles plus the paraboloid *ABC*, that is, the cylindrical bodies circumscribed around the paraboloid, are less than the solid *D* plus the solid *ABC*, that is, half of the cylinder. The cylindrical bodies circumscribed around the paraboloid are therefore less than half of the cylinder; this is impossible. The paraboloid is equal to half of its cylinder.

Completed on the Blessed Saturday, the first night of the month of Rabīʿ al-awwal
in the year one thousand, one hundred and fifty three
by the humble al-Ḥājj Muṣṭafā Ṣidqī.
May God grant him pardon.

CHAPTER VI

IBN AL-SAMḤ

THE PLANE SECTIONS OF A CYLINDER AND THE DETERMINATION OF THEIR AREAS

6.1. INTRODUCTION

6.1.1. *Ibn al-Samḥ and Ibn Qurra, successors to al-Ḥasan ibn Mūsā*

Abū al-Qāsim Aṣbagh ibn Muḥammad ibn al-Samḥ died in Grenada on 'Tuesday, on the twelfth remaining night of Rajab, in the year four hundred and twenty-six, at the age of fifty-six solar years';[1] *i.e.* Tuesday, 27th May 1035,[2] implying that he was born in 979. While it appears that he was born in Cordoba, he came to Grenada to work with the Emir Ḥabbūs ibn Māksan (1019–1038 *ca*). We also know that he was a follower of the famous astronomer and mathematician Maslama al-Majrīṭī, who died in 398/1007–1008. A contemporary of mathematicians such as Ibn al-Haytham, Ibn al-Samḥ produced a substantial and important body of work in his own right in the fields of mathematics and astronomy. From the titles of his works as listed by Ṣāʿid,[3] it is clear that his interests included number theory, geometry, the geometry of the astrolabe, etc. His works encompass

[1] This date was given by the historian Ibn Jamaʿa, according to Lisān al-Dīn ibn al-Khaṭīb who quotes it in his *al-Iḥāṭa fī akhbār Gharnāṭa*, ed. Muḥammad ʿAbdallāh ʿInān, Cairo, 1955, p. 436. See also Ṣāʿid al-Andalusī, *Ṭabaqāt al-umam*, ed. H. Būʿalwān, Beirut, 1985, p. 170. See also the French translation by R. Blachère, *Livre des Catégories des Nations*, Paris, 1935, pp. 130–1. Finally, see Ibn al-Abbār, *al-Takmila li-Kitāb al-Ṣila*, ed. al-Sayyid ʿIzzat al-ʿAṭṭār al-Ḥusaynī, Cairo, 1955, vol. 1, pp. 206-207; and Ibn Abī Uṣaybiʿa, *'Uyūn al-anbā' fī ṭabaqāt al-aṭibbā'*, ed. A. Müller, 3 vols, Cairo / Königsberg, 1882-84, vol. II, p. 40, 4–6; ed. N. Riḍā, Beirut, 1965, p. 483, 23–5.

[2] The date is given as 'Tuesday, on the twelfth remaining night of Rajab'. Twelve complete nights remain before the end of the month of Rajab, Year 426 of the Hegira. Depending on the method of counting used, this corresponds to either the 27th or 28th May 1035 in Grenada. We have opted for the 27th May 1035 as this was a Tuesday.

[3] See Note 1.

also a commentary on Euclid's *Elements*, and 'a great book on geometry with an exhaustive discussion of the parts concerning the line: straight, arched and curved'.[4]

From this description we can deduce that this book by Ibn al-Samḥ was a voluminous work, including chapters on rectilinear figures, circles and arcs, conic sections, and possibly other topics as well. Of all the titles that have been listed by the early biobibliographers and historians, or at least those of which we are aware, this is the only work that could be expected to include a study of the cylinder and its sections. Although Ibn al-Samḥ could have written another book on the same field in geometry, it is likely that this 'great book' is the source of the text translated into Hebrew. This conjecture is supported by a further argument, taken from the Hebrew version itself.

In this version, Ibn al-Samḥ takes up a number of themes one after another, only to dispense with them equally rapidly. The text opens with a definition of a sphere, the same as that given by Euclid in the *Elements*. It would be logical to expect this to be followed by a study on the sphere, and Ibn al-Samḥ does promise one later (in Section 9), in which he intends to discuss 'the surfaces of spheres' and 'the volumes of these spheres'. However, one searches in vain for any trace of these questions in the Hebrew text that has survived. Another example of a 'forgotten' topic is that of the cone. Ibn al-Samḥ begins be restating Euclid's definition of a cone and, then, he refers later (in Section 4) to the 'first definitions' of the right and oblique cone in the *Conics* of Apollonius. Again, this is the only mention of Apollonius in the text. All these definitions lead on to nothing in the version of the text that has come down to us in the Hebrew tradition. These 'omissions' give us a clue to the topics that would have appeared in the 'great book of geometry' alongside to the studies of the cylinder. By this reasoning, there must have been a chapter on the circle, another on the sphere, and a further chapter on the cone, like that devoted by Ibn al-Samḥ to the cylinder. If this conjecture is true, it also throws light on another aspect of the work of Ibn al-Samḥ, namely a tendency to produce mathematical works that, while in essence are summaries, do not in any way exclude original research. This feature of the work of Ibn al-Samḥ is shared with that of other mathematicians in Muslim Spain, as was the case with Ibn Hūd (died 478/1085) in Saragossa. It also makes it possible to identify a *corpus* of work from which the Hebrew text was taken: the 'great book of geometry'.

[4] Taken from the *Ṭabaqāt* of Ṣā'id, ed. Bū'alwān, p. 170:

كتابه الكبير في الهندسة قصى منها أجزاءها من الخط المستقيم والمقوس والمنحني.

The translated text deals with the cylinder and its elliptical sections, a topic already addressed by one of the three Banū Mūsā brothers, al-Ḥasan, and later by his collaborator and pupil Thābit ibn Qurra in the book *On the Sections of the Cylinder*. As we pointed out in our earlier discussion of this book, Ibn Qurra based his work on the book by al-Ḥasan. This leads us to ask the question: did Ibn al-Samḥ belong to this tradition? And where exactly does he fit into the story?

If one examines his book and compares it with that of Ibn Qurra (as that of al-Ḥasan is not known to have survived), one is led to the inescapable conclusion that Ibn al-Samḥ was not familiar with Ibn Qurra's treatise, and that any points that appear to be common to the two works all derive from the treatise by al-Ḥasan. We shall show later that all the indications are that Ibn al-Samḥ based his book on the work of al-Ḥasan, and that he remained truer to the original than did Thābit ibn Qurra.

It is useful to consider first of all the differences that separate Ibn al-Samḥ and Ibn Qurra. Their aims were not the same, and their nomenclature and methods were different. Ibn al-Samḥ begins by showing that the figure obtained using the bifocal definition has the same properties as that obtained by taking a plane section of a cylinder. In contrast, Thābit ibn Qurra develops a theory of the cylinder and its plane sections inspired by Apollonius and his work on cones and conic sections. The terminology used by Ibn al-Samḥ includes terms that were never used by Thābit, such as the 'elongated circular figure' used to describe the figure obtained from the bifocal definition. Inversely, the terminology employed by Thābit includes many terms that do not appear in the treatise of Ibn al-Samḥ. The terminology used by Thābit is generally that of the *Conics* of Apollonius. The same can certainly not be said for the chapter written by Ibn al-Samḥ. The lexical divergence from the *Conics* is matched by a similar conceptual difference. To give but one example, consider the way in which Thābit approaches the case of a plane section of an oblique cylinder with a circular base by making use of a plane antiparallel to that of the base. This is identical to the approach taken by Apollonius for the cone. This concept, together with the associated terminology, does not appear in Ibn al-Samḥ's book. These differences also provide other clues. They distinguish Thābit's text from that of his older master, al-Ḥasan ibn Mūsā, at least according to the description given by his brothers that we have already translated.[5] All now becomes clear. As Thābit ibn Qurra had done before him, Ibn al-Samḥ based his text on the book by al-Ḥasan ibn Mūsā with, however, one crucial difference. While Thābit developed the work of his elder colleague in the light of the *Conics* of Apollonius, Ibn al-Samḥ continued in a straight line

[5] Chapter I, *supra*, p. 8.

from the original work without deviation. Remember that we have it from his own brothers that al-Ḥasan ibn Mūsā was engaged on research into the cylinder and its sections.[6]

As we have shown, Thābit wanted to develop a theory of the cylinder and its sections to stand in its own right in the same way as that of the cone and its sections developed by Apollonius. Ibn al-Samḥ, on the other hand, described his research in the chapter that has survived as an initial body of work leading on to a study of elliptical sections. Working at a later date, and geographically far from Baghdad, this Andalusian, who lived into the first decades of the eleventh century, is closer to al-Ḥasan ibn Mūsā than his collaborator and neighbour, Thābit ibn Qurra. However, there remains a narrow path between the treatise by Thābit and the surviving chapter by Ibn al-Samḥ that enables us to recognize, even from this distance, a number of topics covered in this important lost work by al-Ḥasan ibn Mūsā, and to wonder how Ibn al-Samḥ would have interpreted it.

6.1.2. *Serenus of Antinoupolis, al-Ḥasan ibn Mūsā, Thābit ibn Qurra and Ibn al-Samḥ*

Ibn al-Samḥ begins his text with a number of definitions, including that of an oblique cylinder with a circular base. As can easily be verified, this definition is similar to that given by Thābit. However, this definition also appears in the book by Serenus of Antinoupolis, *On the Section of the Cylinder*.[7] How are these texts related? And what role, if any, was played by al-Ḥasan ibn Mūsā in bringing them all together? We can answer some aspects of these questions with a fair degree of certainty. Others are more doubtful. To begin with, there can be no doubt whatsoever that Ibn al-Samḥ was unaware of the book by Serenus. Equally, we can be certain that Thābit ibn Qurra knew it well. In the absence of a definitive text by al-Ḥasan ibn Mūsā, we can only conjecture that he had a more or less direct awareness of the Serenus text, but that he did not make use of it as a basis

[6] Banū Mūsā, *Lemmas in the Book of Conics*, *vide supra*, Chapter I, p. 8.

[7] *Sereni Antinoensis Opuscula*. Edidit et latine interpretatus est I. L. Heiberg, Leipzig, 1896. See also the French translation by P. Ver Eecke, *Serenus d'Antinoë: Le livre de la section du cylindre et le livre de la section du cône*, Paris, 1969. This is the definition of the cylinder as given in French by Ver Eecke: 'Si, deux cercles égaux et parallèles restant immobiles, des diamètres constamment parallèles, qui tournent dans le plan des cercles autour du centre resté fixe, et font circuler avec eux la droite reliant leurs extrémités situées d'un même côté, reprennent de nouveau la même position, la surface décrite par la droite qu'ils ont fait circuler est appelée une surface cylindrique', pp. 2–3.

for his own work. So many relevant questions have never been asked, that it seems acceptable to risk a few digressions in attempting to answer them.

The Book on the Section of the Cylinder by Serenus opens with a collection of definitions of cylindrical surfaces and the cylinder on a circular base. The first three definitions[8] also appear in the introduction to the book by Thābit ibn Qurra, albeit with a few minor differences. Serenus defines the generator as a line 'which, being straight and located on the surface of the cylinder, touches each of the bases'. He adds that this is also the moving straight line that, as he describes it 'is also the straight line moving in a circle that we have spoken of as describing the cylindrical surface.[9] It is this latter phrase that Thābit gives as his definition. However, he goes on to show that the generator is parallel to the axis, and that the only straight lines on the surface of a cylinder are the generators. In Proposition 7, Serenus addresses the problem of how to move the generator of a cylinder passing through a given point, and in Proposition 8 he shows that any straight line joining two points on a cylinder that are not lying on a same generator falls *within* the cylinder, and is not therefore on the surface. These two propositions are similar to the first two propositions in the book by Thābit.

Serenus also gives four definitions taken from Apollonius which do not appear in Thābit's introduction; the diameters, conjugate diameters, centre, and similar ellipses,

In Propositions 2 and 3 Serenus discusses plane sections of a right or oblique cylinder with the plane lying either along the axis or parallel to the axis. These sections are parallelograms. At the end of Proposition 4, Thābit states that, if the cylinder is a right cylinder, the section is a rectangle. In Propositions 5 and 6, he goes on to establish the necessary and sufficient condition for the parallelogram to become a rectangle in the case of an oblique cylinder. These concepts relating to rectangles do not appear in the Serenus book.

Thābit, as we have seen, defines in Proposition 7 the cylindrical projection (translation) of a figure on a plane P onto another plane P' parallel to P, and then uses this in Proposition 8 to deduce the plane section by a plane parallel to the base of the cylinder. This section is discussed by Serenus in his Proposition 5, making use of his Proposition 2 and a lemma proved in Proposition 4, in which he establishes the 'equation of the circle'.

Serenus considers the section by a plane antiparallel to that of the base in Proposition 6, whereas Thābit uses the same method in his Proposition 9. Both make use of the 'equation of the circle'.

[8] *Ibid.*

[9] *Ibid.*, p. 3.

Thābit considers a section by a plane that cuts the axis and is neither parallel nor antiparallel to the plane of the base, and applies a cylindrical projection to show in Proposition 10 that this section is either a circle or an ellipse. In Proposition 11, he goes on to show that it must be an ellipse. Serenus addresses the same question in Propositions 9–17. He begins by showing that this section is not a circle and it is not composed of straight lines. He then introduces the principal diameter Δ (which becomes the major axis in two cases), the second diameter Δ', which is the conjugate diameter of Δ, and finally the properties of points on the ellipse relative to Δ and Δ'. In Propositions 17 and 18, he defines the *latus rectum* associated with the transverse diameter to arrive at Proposition 15 of Apollonius. The section is therefore an ellipse.

We can see that the paths begin to diverge as soon as Thābit introduces the explicit application of geometric projections. This is the point at which Thābit and Serenus go their separate ways. The geometry of Serenus is far from one in which projections and transformations are important instruments, despite the fact that there is just a hint of the concept of translation in his first proposition. From this point on, the divergence becomes a total break. Serenus and Thābit each move on to a completely different set of problems.

Ibn al-Samḥ deliberately chooses to take his direction from Proposition 7, where, in the case of a right cylinder with a circular base, the principal diameter Δ becomes the major axis, and the second diameter Δ' becomes the minor axis. From the seventh proposition onward, all the propositions of Ibn al-Samḥ make use of these two axes. From the properties of these two diameters, it is clear that we can place an ellipse with axes $2a$ and $2b$ ($a > b$) on a right cylinder of radius b; that is what Ibn al-Samḥ uses in Propositions 7, 10 and 19. Serenus, on the other hand, shows in Propositions 27 and 28 that there exists two families of ellipses with a major axis of $2a$ ($a > b$) on a cylinder of radius b.

These analogies make it possible to show that Thābit knew the book by Serenus. It can also be said that his knowledge of the *Conics* had a dual paradoxical effect. He used this knowledge to profit from the book of Serenus, bringing to it his own theoretical and non-essential technical contribution. Thābit had direct access to the definitions and results that Serenus borrowed from Apollonius and, as we have seen, he followed the path laid down by al-Ḥasan ibn Mūsā. The relationship between Ibn al-Samḥ and Serenus is a minor one, consisting of no more than the definition of the cylinder and one similar result obtained using a different method. The only link appears to be through the work of al-Ḥasan ibn Mūsā. And judging by the common basis of the work of Thābit and Ibn al-Samḥ, if al-

Ḥasan ibn Mūsā was aware of the work of Serenus, he certainly drew little profit from it.

In order to gain a better understanding of the role played by al-Ḥasan ibn Mūsā's book, we must make a brief comparison of the Ibn al-Samḥ chapter with the treatise by Thābit in the light of this hypothesis. Any methodological elements common to the two works may be considered to derive from the work by al-Ḥasan ibn Mūsā that they both used as a reference. We may begin by considering the aspects of their treatment of the ellipse in which they differ, in effect the topic as addressed by Ibn al-Samḥ.

Ibn al-Samḥ assumes that the results relating to the ellipse obtained from a section of a right cylinder are well known. He defines this ellipse in terms of the two axes, the smaller of which is equal to the diameter of the cylinder (see Property *A* below). Thābit ibn Qurra, on the other hand, considers plane sections of an oblique cylinder in Propositions 3–11 of his treatise. In Propositions 10 and 11, he examines the elliptical section using the cylindrical projection and Proposition I.21 of the *Conics*.

The two mathematicians have the following elements in common:

1.*The two orthogonal affinities*

Thābit first examines the affinity relating to the major axis, making use of Proposition I.21 of the *Conics*, and indicates that the same method may be used for the affinity relating to the minor axis. In Proposition 7, Ibn al-Samḥ discusses the affinity relating to the minor axis by making use of Property *A* and similar triangles. He then goes on to discuss the affinity relating to the major axis in Proposition 8.

2. *The area of an ellipse*

Thābit establishes the result in Proposition 14 using an apagogic method derived from XII.2 of the *Elements* and the orthogonal affinity relating to the major axis. Ibn al-Samḥ proceeds in a number of steps (see Propositions 12–17). In the most important of these, he shows that the ratio of the area of an ellipse to that of a circle of diameter $2b$ is equal to a/b, using an apagogic method, *Elements* XII.2, and the orthogonal affinity relating to the minor axis. We have it from Thābit himself (see the Introduction to his treatise) that al-Ḥasan ibn Mūsā had determined this area.

We can therefore conjecture as follows: Both Thābit and Ibn al-Samḥ drew on the work of al-Ḥasan ibn Mūsā for their ideas of projection and orthogonal affinity, combined with the application of the *Elements* XII.2 and an apagogic method. The formulation proposed by Thābit was influenced by his use of *Conics* I.21, while that of Ibn al-Samḥ remained truer to that of al-Ḥasan ibn Mūsā.

This close relationship between the works of Ibn al-Samḥ and al-Ḥasan ibn Mūsā provides further confirmation. We know from the brothers of al-

Ḥasan that he was interested in the diameters, chords and axes of cylindrical sections: 'he found out its science and the science of the fundamental proprieties relative to the diameters, the axes, and the chords, and he has found out the science of its area'.[10] Ibn al-Samḥ devotes his Propositions 19–21 to just these chords and axes.

6.1.3. *The structure of the study by Ibn al-Samḥ*

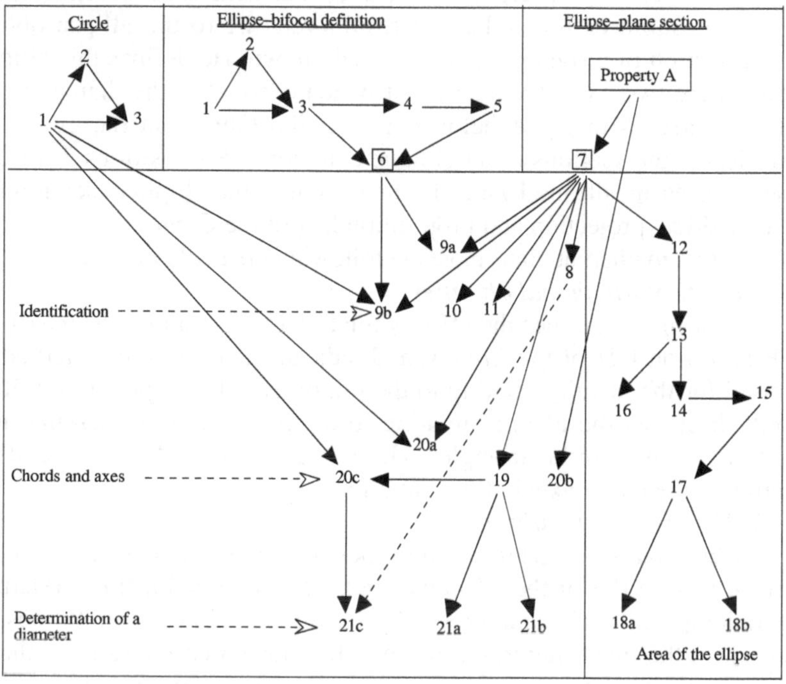

* Propositions 2 and 3 relating to the circle are not used.
* 20c is another demonstration of 8.

We now come to the chapter written by Ibn al-Samḥ as it appears in the surviving Hebrew version. The network of deductions demonstrates the consistency of the body of propositions, and raises no doubts as to the authenticity of the great majority of them. The only difficulties occur at the beginning and, to a greater extent, at the end of the text. It can be seen that Propositions 2 and 3 relating to the circle are not used in the remainder of the chapter, although they do follow naturally from the first proposition.

[10] See Apollonius, *Les Coniques*, tome 1.1: *Livre I*, ed. R. Rashed, pp. 504–5.

However, we do not believe that these propositions have been added to the original text of Ibn al-Samḥ. It does, however, appear likely that Propositions 20 and 21 have been added in the place of a number of propositions in the original. Parts of the text of these propositions have been lost, and the remainder have been collected together in a somewhat random manner to form the two propositions in the surviving text. We have no way of knowing whether this loss occurred in the original Arabic manuscript, or whether it was the fault of the translator or even a later copyist of the Hebrew text. However, it is clear that, at some point in its history, the text has been subject to the attentions of a glossator who added the obviously apocryphal Lemma 4.

6.2. MATHEMATICAL COMMENTARY

6.2.1. *Definitions and accepted results*

The first part of Ibn al-Samḥ's text consists of an introduction to the treatise, in which the author lists the definitions and prior results to be used without further proof. We do not know whether these prior results had been obtained by Ibn al-Samḥ himself in an earlier section of the treatise that was originally much longer than the surviving Hebrew translation, or whether they formed part of a body of accepted mathematical knowledge shared by his contemporaries. We have divided this introduction into separate sections for the purpose of this commentary. We shall now deal with each of these in turn, in the order in which they appear.

In Section 1, Ibn al-Samḥ begins by defining a sphere as a solid of revolution generated by the rotation of a semicircle about its diameter. He also defines the elements of a sphere: the surface area, diameter, centre, poles, and the great circle. Later, in Section 9, he states that he intends to discuss a number of problems relating to the sphere, including plane sections, the surface area and the volume. However, there is no further discussion of the sphere in the remainder of the text. This absence alone is sufficient to show that the surviving text is far from complete.

Ibn al-Samḥ then goes on to define a cylinder of revolution, a solid generated by rotating a rectangle about one of its sides, and its associated elements, the lateral surface and bases. This definition is the same as that given by Euclid in the *Elements* (Book XI, Definition 14), and different from that given by Serenus (pp. 2–3) and Ibn Qurra, who saw the cylinder of revolution as a special case of an oblique cylinder with circular bases. Ibn al-Samḥ does not mention the oblique cylinder until the end of this section. It should be noted that an oblique cylinder cannot be generated by revolution, which explains why Ibn al-Samḥ does not give a more general definition of the cylinder until later in the work.

The next definition given by Ibn al-Samḥ is that of the cone of revolution. This definition is deduced from that of the cylinder, with the lateral surface of the cone being generated by the diagonal of a cylinder, and the conic solid being generated by the rotation of a triangle about a fixed side. This definition is also taken from Euclid.

In other words, this confirms that the definition of a cylinder given by Ibn al-Samḥ in the first section is definitely that of Euclid. Unlike Ibn al-Samḥ, Euclid gives the definition of the cone of revolution before that of the cylinder. In addition, Euclid only mentions the cylinder and cone of revolution, while Ibn al-Samḥ continues with further definition in the following section.

In the second section, Ibn al-Samḥ gives a more general definition of the cylinder based on two curves, each with a centre, located on two parallel planes. It is clear that we are to assume that one of these curves is derived from the other by a process of translation. A movable straight line in contact with each of these two curves and parallel to the straight line joining the two centres generates the lateral surface of the cylinder. This cylinder may be right or oblique. It should be noted that, if these two curves are considered to be circles, then the definition corresponds to that given by Thābit ibn Qurra in his treatise *On the Sections of a Cylinder and its Lateral Surface*, and that given by Serenus. This case is discussed by Ibn al-Samḥ at the end of the section.

In the third section, it becomes clear that he assumes the curves in question to be either circles or ellipses. He then goes on to consider the solids obtained if these cylinders are intersected by two parallel planes, without actually specifying the shape of the base of the cylinder. However, it appears from the final sentence in the section that the base is assumed to be a circle. If the cylinder is a *right* cylinder on a circular base, then the sections formed by the two parallel planes will be two ellipses that Ibn al-Samḥ assumes to be equal (see the following comment below) and they define a *right* cylinder with elliptical bases.

Comment 1. — The results given are correct, but they are not proved in the surviving text. It should be noted that Thābit ibn Qurra, in Proposition 8 of his treatise *On the Sections of a Cylinder and its Lateral Surface*, gives a general proof that the sections of two parallel planes intersecting the axis of a cylinder with circular bases will always be equal. In Propositions 8–11, he shows that these sections are either circles or ellipses, using the characteristic properties of the circle and the ellipse given in Proposition I.21 of the

Conics. In Proposition 9, Thābit considers antiparallel circles. These are not mentioned by Ibn al-Samḥ.

Comment 2. – The final sentence in Section 3 by Ibn al-Samḥ reads as follows: 'Beginning with two species whose bases are ellipses, it is possible to generate the two species with circular bases by proceeding in the reverse manner.' It would appear from this sentence that the properties of plane sections of cylinders with elliptical bases were well known, and that some of these plane sections were known to be circular.

Ibn Abī Jarrāda, a thirteenth century commentator on Thābit's text, showed that the circular base specified in Thābit's Proposition 10 could be replaced by an elliptical base (see Supplementary note [3]).

In the fourth section of the introduction, Ibn al-Samḥ repeats the definition of a cone found in the *Conics* of Apollonius. As in the case of the sphere, no use is made of this definition elsewhere in the text and the cone is not considered further. This hints at another missing section of the text, the extent of which remains unknown.

In the fifth section, Ibn al-Samḥ provides a classification of the species of cylinder that he has considered. This classification may be summarized as shown in the diagram below:

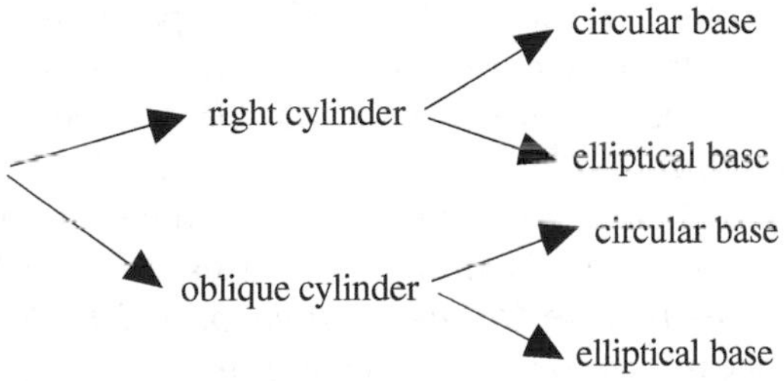

Ibn al-Samḥ notes that the right cylinder on a circular base was familiar to the Ancients. This comment indicates that he was only aware of the definition of a cylinder as given in Euclid's *Elements*, and that he did not know of the book by Serenus.

As regards to the cone, Ibn al-Samḥ takes up his general definition based on a circle and a point lying outside the plane of the circle before going on to distinguish between a right cone and an oblique cone. These are the 'First Definitions', 1–3, defined by Apollonius. It is worth repeating that these two references to the cone in the introduction are the only mentions of the cone in the surviving manuscript of this treatise. It appears that Ibn al-Samḥ was familiar with the *Conics* of Apollonius but, unlike Thābit ibn Qurra, he did not make use of it.

6.2.2. *The cylinder*

Ibn al-Samḥ continues by discussing the cylinder in more general terms. He begins with the concept of a closed curve, to remind the reader that the number of closed curves other than the circle is infinite and that they cannot all be listed (Section 6). Each of these closed curves can be associated with a corresponding cylinder, once we have a definition for *curves in similar positions* (Section 7).

Let there be two equal portions of planes P_1 and P_2, both having the same shape bounded by the closed curves C_1 and C_2. Let there be two points $M_1 \in P_1$ and $M_2 \in P_2$. Consider the straight lines joining M_1 to all the points on C_1 and the straight lines joining M_2 to all the points on C_2. Then, if each straight line from M_1 is equal to one of the straight lines from M_2 and if the angle between two straight lines from M_1 is equal to the angle between the two equal straight lines from M_2, then C_1 and C_2 are said to be at *similar positions*. We would say today that C_1 and C_2 have the same equation in polar coordinates relative to the two poles M_1 and M_2. Ibn al-Samḥ does not take the polar angle from a given axis; he compares the angles between two radial vectors.

In other words, Ibn al-Samḥ characterizes homologous points in the *displacement* from P_1 to P_2. If P_1 and P_2 are in two parallel planes and if a plane passing through M_1 and M_2 from 'similar positions' cuts them along equal straight lines, then the closed curves C_1 and C_2 are said to be in similar positions. In this case, either C_1 or C_2 may be derived from the other by means of a *translation*. This idea of two curves C_1 and C_2 'in similar positions' in parallel planes, and which may each be derived from the other by a process of translation, is similar to that used by Thābit ibn Qurra in Proposition 7 of his treatise *On the Sections of a Cylinder and its Lateral Surface*. The process is in some ways reciprocal.

Using these concepts, Ibn al-Samḥ then gives a general definition of a cylinder on any base (Section 8):

Let there be two plane figures bounded by the closed curves C_1 and C_2 in 'similar positions', and let M_1 and M_2 be two points 'at similar positions' lying within these figures. Then a straight line moving along and touching both C_1 and C_2 while remaining parallel to the line M_1M_2 will describe a cylindrical surface.

If M_1 and M_2 are the centres of symmetry of C_1 and C_2, then M_1M_2 is the axis of the cylinder. The movable straight line is called the side of the cylinder. If M_1M_2 is perpendicular to the planes of the two figures, then the cylinder is a right cylinder. If this is not the case, then the cylinder is an oblique cylinder.

It will be noted that the definition of a cylinder is a rigorous one, and that the general concept of closed curves explicitly goes beyond the class of conic sections characterised by the existence of conjugate diameters.

Comment. – This general definition does not appear again in the treatise until Proposition 20. In this proposition, Ibn al-Samḥ states that, in order to discuss it using the method described in Proposition 19, it is necessary to consider a cylinder on an elliptical base. The method would then involve placing a circular plane section on this cylinder. Ibn al-Samḥ simply proposes the method without developing it further.

However, this problem, of an elliptical cylinder and its plane sections, is not discussed by either Serenus or Thābit ibn Qurra. It is mentioned by Ibn Abī Jarrāda in his commentary on Thābit ibn Qurra's treatise.

In the final section of this chapter, Ibn al-Samḥ announces that he intends to discuss the plane sections of cylinders, the areas of these plane sections, spherical surfaces, sections, and the volumes of spheres (Section 9). However, none of these have survived in the version of the text available today.

6.2.3. *The plane sections of a cylinder*

In Section 10, Ibn al-Samḥ continues by summarizing the types of plane sections of a *cylinder of revolution* obtained by varying the position of the secant plane. If this plane passes through the axis or is parallel to the axis, then the plane section is a rectangle. Ibn al-Samḥ does not consider this case. If the secant plane is perpendicular to the axis, then the section is a circle. If the plane is not parallel to the bases and it intersects the axis, then the section is an ellipse.

Ibn al-Samḥ shows that the plane section generated by the rotation of a segment around one of its fixed extremities is 'necessarily' a circle. All the points on the boundary of this section are equidistant from the fixed point, satisfying the definition of a circle in terms of its centre and radius. In this way, Ibn al-Samḥ identifies the curve obtained as a plane section or circle defined by the locus of a set of points.

It should be noted, however, that Ibn al-Samḥ does not specify that the circular plane section is equal to the base circle. We should also note that in Proposition 8 of his treatise referred to above, Thābit ibn Qurra shows that the plane section of a right or oblique cylinder on a circular base is a circle equal to the base circle, a result derived from the translation considered in

Proposition 7. This constitutes a further argument that Ibn al-Samḥ did not base his work on that of Thābit.

6.2.4. *The properties of a circle*

Ibn al-Samḥ then considers certain properties of a circle in order to derive two lemmas needed in the subsequent propositions. The first properties to be considered are the following: Let C, C_1, and C_2 be circles with diameters d, d_1, and d_2 and circumferences p, p_1, and p_2 respectively. Let P_1 and P_2 be two regular similar polygons with sides l_1 and l_2 inscribed within C_1 and C_2. Ibn al-Samḥ then states the following:

a) $$\frac{\text{area } C_1}{\text{area } C_2} = \frac{d_1^2}{d_2^2} = \left(\frac{d_1}{d_2}\right)^2$$

and

b) $$\frac{\text{area } C_1}{\text{area } C_2} = \frac{\text{area } P_1}{\text{area } P_2} = \left(\frac{l_1}{l_2}\right)^2,$$

with the help of the *Elements* XII, Propositions 1 and 2.

c) $$\text{area } C = \frac{1}{2}\left(\frac{1}{2}\, d.p\right),$$

where $\frac{1}{2}d$ and p can be considered as the sides of a right angle of a right-angled triangle. This refers to Proposition 1 of Archimedes' *On the Measurement of the Circle.*

d) $$\frac{d_1}{p_1} = \frac{d_2}{p_2};$$

this proposition is the fifth of the treatise of the Banū Mūsā.[11]

e) $$3 + \frac{10}{71} < \frac{p}{d} < 3 + \frac{1}{7};$$

this is the third proposition of Archimedes' *On the Measurement of the Circle.*

[11] See Chapter I, *supra.*

f) $$\frac{\text{area } C}{d^2} \approx \frac{11}{14} = \frac{5}{7} + \frac{1}{14};$$

this is the second proposition within the same treatise by Archimedes.

Ibn al-Samḥ then establishes a number of properties that he claims are not mentioned by Euclid, or by Archimedes, or by anyone else.

Lemma 1. — *Let there be two circles of diameters* AB *and* EZ *and two points* G *and* H *on* AB *and* EZ *respectively, such that* $\frac{\mathrm{GA}}{\mathrm{GB}} = \frac{\mathrm{HE}}{\mathrm{HZ}}$; *their chords* DGT *and* KHL *being respectively perpendicular to* AB *and* EZ *such that* $\frac{\mathrm{DT}}{\mathrm{KL}} = \frac{\mathrm{AB}}{\mathrm{EZ}}$.

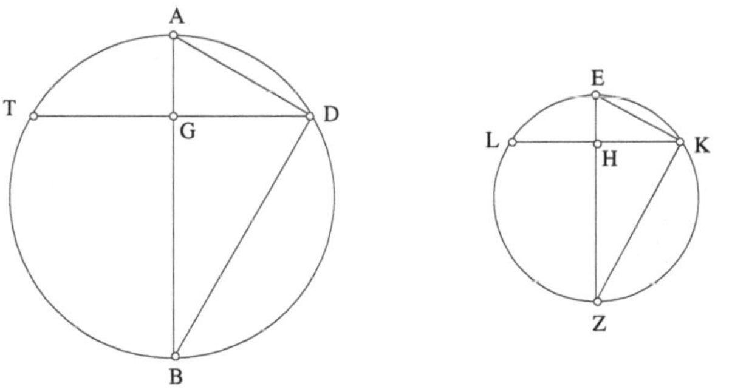

Fig. 6.1

We then have in triangles *ADB* and *EKZ*

$$GD^2 = GA \cdot GB \text{ and } HK^2 = HE \cdot HZ;$$

hence

$$\frac{GD^2}{HK^2} = \frac{GA}{HE} \cdot \frac{GB}{HZ}.$$

From the hypothesis we deduce

$$\frac{GA}{HE} = \frac{GB}{HZ} = \frac{AB}{EZ};$$

hence

$$\frac{GA}{HE} \cdot \frac{GB}{HZ} = \frac{AB^2}{EZ^2},$$

and consequently

$$\frac{AB}{EZ} = \frac{GD}{HK} = \frac{DT}{KL}.$$

The hypothesis led to the construction of two similar figures, hence the conclusion.

Lemma 2. — *Let there be two circles of diameters* AB *and* GD, *and two points* E *and* H *on* AB, *and two points* K *and* M *on* GD *such that* $\frac{\mathrm{AE}}{\mathrm{AB}} = \frac{\mathrm{GK}}{\mathrm{GD}}$ *and* $\frac{\mathrm{BH}}{\mathrm{AB}} = \frac{\mathrm{DM}}{\mathrm{GD}}$. *Let there be two semi-chords* $\mathrm{EZ} \perp \mathrm{AB}$, $\mathrm{HT} \perp \mathrm{AB}$, $\mathrm{KL} \perp \mathrm{GD}$, $\mathrm{MN} \perp \mathrm{GD}$. *Then, the triangles* ZHE *and* LMK *are similar, and the same obtains with regard to triangles* EHT *and* KMN.

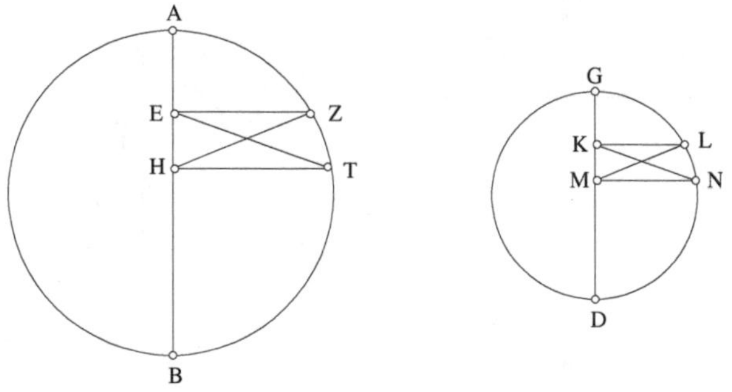

Fig. 6.2

From the hypotheses, we deduce

$$\frac{AB}{GD} = \frac{AE}{GK} = \frac{AH}{GM} = \frac{EH}{KM}.$$

Following Lemma 1, we have

$$\frac{AB}{GD} = \frac{HT}{MN};$$

hence

$$\frac{HT}{MN} = \frac{EH}{KM}.$$

The right-angled triangles *EHT* and *KMN* are accordingly similar. The same is the case with triangles *ZHE* and *LMK*.

Let us indicate that, as was the case with the preceding lemma, following the hypotheses, the two figures are similar; thus two homologous triangles – for example *EHT* and *KMN* – are similar.

Lemma 3. — *Let there be two circles with diameters* AB *and* KL, *and two points* G *and* N *dividing respectively these diameters according to the same ratio. Let there be* GH *and* NO *such that* $\mathrm{B\hat{G}H} = \mathrm{L\hat{N}O}$; *then* $\frac{\mathrm{HG}}{\mathrm{ON}} = \frac{\mathrm{AB}}{\mathrm{KL}}$.

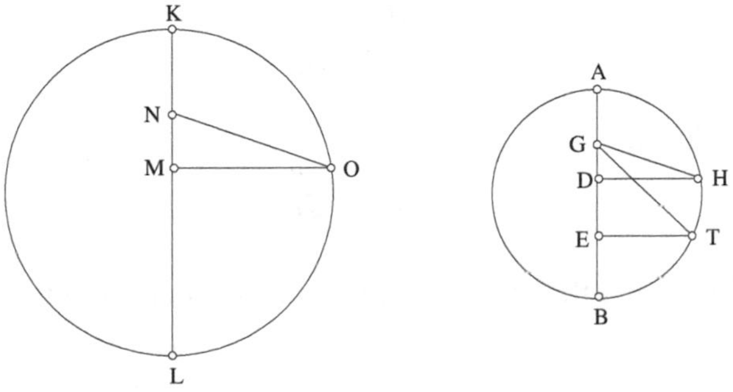

Fig. 6.3

This lemma constitutes a generalization of the first, in which we had

$$B\hat{G}H = L\hat{N}O = \frac{\pi}{2}.$$

Wc assume that

$$L\hat{N}O \neq \frac{\pi}{2};$$

therefore

$$B\hat{G}H \neq \frac{\pi}{2}.$$

If $OM \perp KL$ and $HD \perp AB$, we then have

$$\frac{AD}{DB} = \frac{KM}{ML}.$$

In fact, if this were not the case, then there would be a point E on AB, $E \neq D$ such that

$$\frac{AE}{EB} = \frac{KM}{ML}.$$

If we produced $ET \perp AB$, the triangles TGE and ONM would then be similar following Lemma 2, and hence $T\hat{G}E = L\hat{N}O$; yet $L\hat{N}O = H\hat{G}B$, which is absurd.

Following Lemma 1,

$$\frac{HD}{OM} = \frac{AB}{KL};$$

yet, on the other hand, based on similarity

$$\frac{HD}{OM} = \frac{HG}{ON};$$

hence the conclusion

$$\frac{HG}{ON} = \frac{AB}{KL}.$$

It should be noted that no use of this lemma is made in the remainder of the text.

6.2.5. *Elliptical sections of a right cylinder*

In the first paragraphs of this chapter, Ibn al-Samḥ states that he is going to establish that the plane section of a cylinder of revolution by a plane that is not parallel to the bases – *i.e.* an ellipse – is the same as the 'elongated circular figure' obtained from a triangle with a fixed base and two other sides whose sum is given. The locus of the moving vertex of the triangle is, in this case, the curves obtained using the bifocal definition. Ibn al-Samḥ intends to show that the curve obtained using these two procedures have common properties. He begins by using the same procedure that he used for the circular plane section. He defines each of the elements of the 'elongated circular figure': the vertices, centre, diameters, chord, axes, and the inscribed circle with a diameter equal to the minor axis, and the circumscribed circle with a diameter equal to the major axis. The first six propositions all relate to the 'elongated circular figure', that is the curve obtained from the bifocal definition $MF + MF' = 2a$. In classical notation,

$$AC = 2a,\ BD = 2b,\ FF' = 2c \ \ (\text{where } a^2 = b^2 + c^2).$$

Ibn al-Samḥ defines the invariant straight line FL and the separate straight line FK from the perpendicular to the major axis AC passing through F that cuts the great circle of diameter AC at L and the ellipse at K. He then proves the following propositions:

Proposition 1.

$$4FL^2 + FF'^2 = AC^2.$$

This equality is immediately deduced from the fundamental property of the circle:

$$FL^2 = FC \cdot FA = OA^2 - OF^2 \Rightarrow FL^2 = a^2 - c^2 = b^2,$$

which Ibn al-Samḥ shows in Proposition 2.

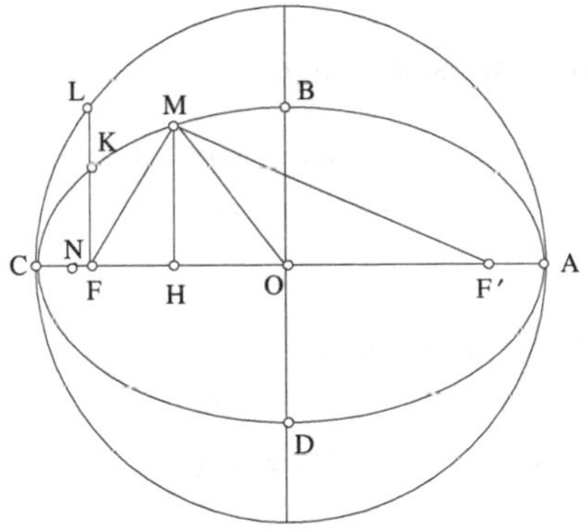

Fig. 6.4

Proposition 2.

i) $FL = b$;

ii) $OB^2 = OA \cdot FK \Rightarrow FK = \frac{b^2}{a} \Rightarrow \frac{FK}{FL} = \frac{b}{a}$.

These results are obtained through the preceding proposition and from the bifocal definition.

Proposition 3. — *The calculation of the radius vector* MF′ (MF′ > MF).

Ibn al-Samḥ presupposes point M on the arc BC, such that $M \neq C$, and distinguishes several cases of figures:

i) M between B and K, for which we have

$$\text{a) } F\hat{M}F' = \frac{\pi}{2}, \text{ b) } F\hat{M}F' > \frac{\pi}{2}, \text{ c) } F\hat{M}F' < \frac{\pi}{2};$$

ii) M at K;

iii) M between K and C.

In these last two cases, the angle $F\hat{M}F'$ is acute.

Let H be the projection of M on AB; Ibn al-Samḥ introduces point N of the semi-straight line HC, as defined by $HN = \frac{b^2}{c}$. The demonstration uses the bifocal definition of Propositions 1 and 2, Pythagoras' theorem for $F\hat{M}F'$ right, and Propositions II.12 and II.13 of the *Elements* respectively for $F\hat{M}F'$ obtuse, and $F\hat{M}F'$ acute.

For all the cases of figures, we have

$$\frac{MF'}{NF'} = \frac{OF'}{OA}.$$

Comments.

1) By positing $OH = x$, we can note

$$NF' = F'O + OH + HN = c + x + \frac{b^2}{c}.$$

Hence

$$MF' = \left(c + x + \frac{b^2}{c}\right) \cdot \frac{c}{a} = \frac{a^2 + cx}{a},$$

and we have

$$MF' = a + \frac{cx}{a},$$

and hence

$$MF = a - \frac{cx}{a}.$$

This relation is valid if M is in C; we then have

$$x = a,\ MF = a - c,\ MF' = a + c.$$

2) We obtain this result without distinguishing the cases of figures by using the bifocal definition and a metric relation in triangle MFM'. This relation is deduced from II.12 and 13 of the *Elements*. In fact we have

$$MF'^2 = MO^2 + OF'^2 + 2F'O \cdot OH \qquad (\textit{Elements}\ \text{II}.12)$$
$$MF^2 = MO^2 + OF^2 - 2FO \cdot OH \qquad (\textit{Elements}\ \text{II}.13);$$

hence

$$MF'^2 - MF^2 = 2OH \cdot (OF' + OF) = 2OH \cdot FF'.$$

We equally have

$$MF'^2 + MF^2 = 2OM^2 + 2OF^2,$$

which we will use in Proposition 4. We thus have

$$MF' + MF = 2a,$$
$$MF'^2 - MF^2 = 4cx;$$

therefore

$$MF' - MF = 2\frac{cx}{a},$$

and hencc

$$MF' = a + \frac{cx}{a} \text{ and } MF = a - \frac{cx}{a}.$$

Proposition 4. — *Product of radius vectors* MF *and* MF′.

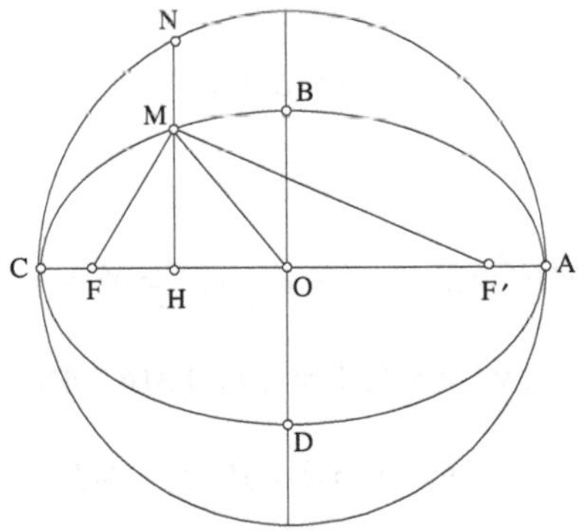

Fig. 6.5

With the preceding notations, if we designate by N the intersection of HM with the circle of diameter AC, we have

$$MF \cdot MF' = NH^2 - MH^2 + BO^2.$$

As in Proposition 3, Ibn al-Samḥ distinguishes between five cases of figures. The demonstration is done by way of using the power of a point with respect to a circle, and in each case of the figures a result is established in the course of Proposition 3. Like in this, he posits $M \neq C$. But the result is valid if M falls in C.

Comments.

1) As in the preceding proposition, we give only one proof, which will be valid for all the cases of figures by means of the bifocal definition and a metric relation in triangle MFF'. We have

$$MF + MF' = 2a,\ MF^2 + MF'^2 + 2MF \cdot MF' = 4a^2.$$

However, in triangle MFF', we have, following the *Elements* II.12 and 13,

$$MF^2 + MF'^2 = 2MO^2 + 2OF^2;$$

hence

$$MF \cdot MF' = 2a^2 - OM^2 - OF^2.$$

By designating x and y as coordinates of M, and Y as ordinate of N, we have

$$MF \cdot MF' = 2a^2 - (x^2 + y^2) - c^2.$$

However,

$$Y^2 = (a - x)\,(a + x) \qquad \text{(power of } H\text{);}$$

hence

$$MF \cdot MF' = Y^2 - y^2 + a^2 - c^2 = Y^2 - y^2 + b^2.$$

If M falls in B, we have $y = b$, $Y = a$ and $MF \cdot MF' = a^2$.

If M falls in C, we have $y = Y = 0$ and $MF \cdot MF' = b^2 = (a - c)\,(a + c)$.

2) If we take into account the results of the two preceding propositions, we have

$$\left(a - \frac{cx}{a}\right)\left(a + \frac{cx}{a}\right) = Y^2 - y^2 + b^2 \Leftrightarrow Y^2 - y^2 + b^2 = a^2 - \frac{c^2x^2}{a^2}.$$

However,

$$Y^2 = a^2 - x^2.$$

we then have

$$a^2 - x^2 - y^2 + b^2 = a^2 - \frac{c^2x^2}{a^2};$$

hence

$$b^2 = x^2\left(1 - \frac{c^2}{a^2}\right) + y^2.$$

Hence, on dividing the two members by b^2,

$$1 = \frac{x^2}{a^2} + \frac{y^2}{b^2}$$

the equation of the ellipse in relation to its axes.

Proposition 5. — *If to a point* M *of the elongated circular figure we associate, on the circle having the minor axis as diameter, a point* T *of the same ordinate* (MT ⊥ BD *to point* K), *we have*

$$MK^2 = KT^2 + (OA - MF)^2.$$

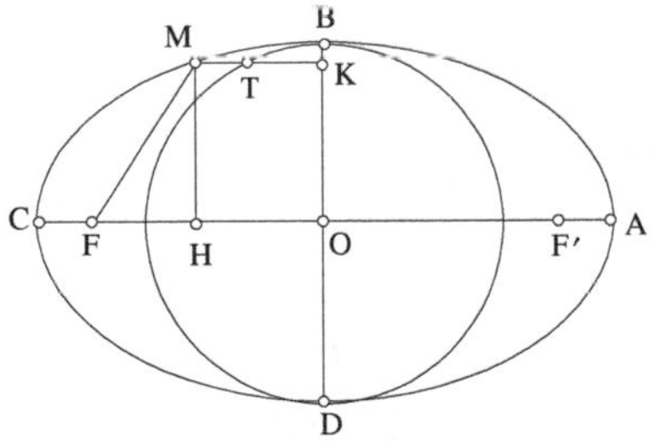

Fig. 6.6

Ibn al-Samḥ demonstrates this proposition with the help of that which precedes it, of the power of a point with respect to a circle and of the bifocal definition (at least implicitly).

Comment. — We established in the third proposition

$$MF = a - \frac{cx}{a}, \qquad \text{with } x = MK,$$

whence if we posit $KT = X$, the abscissa of T on the circle,

$$x^2 = X^2 + \frac{c^2}{a^2}x^2.$$

However, M and T have the same ordinate $y = MH$; and whence $X^2 = b^2 - y^2$, and then

$$x^2 - \frac{c^2}{a^2}x^2 + y^2 = b^2,$$

and, on dividing the two members by b^2, we obtain again the equation of the ellipse.

Proposition 6. — *Orthogonal affinity with regard to the minor axis.*

With the preceding notations, we have:

$$\frac{\mathrm{MK}}{\mathrm{TK}} = \frac{\mathrm{OA}}{\mathrm{OB}} \qquad [\textit{i.e.}\ \frac{\mathrm{x}}{\mathrm{X}} = \frac{\mathrm{a}}{\mathrm{b}}].$$

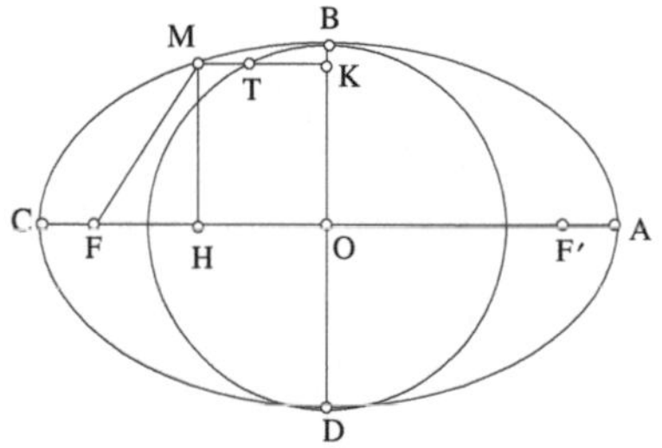

Fig. 6.7

Ibn al-Samḥ's demonstration is based on Propositions 3 and 5.

Comment. — We have seen that by accounting for Proposition 3, the obtained result in Proposition 5 is noted as

$$x^2 = X^2 + \frac{c^2}{a^2}x^2 \Leftrightarrow x^2\left(1 - \frac{c^2}{a^2}\right) = X^2 \Leftrightarrow b^2x^2 = a^2X^2;$$

hence

$$\frac{x}{X} = \frac{a}{b}.$$

Ibn al-Samḥ has thus defined an orthogonal affinity for axis *BD* of a ratio $\frac{a}{b} > 1$ in which the figure *ABCD* is the image of the circle of diameter *BD*; this affinity is a dilatation.

6.2.6. *The ellipse as a plane section of a right cylinder*

Ibn al-Samḥ continues by summarizing the results relating to plane sections of a right cylinder with circular bases. He presents these results as accepted fact, leading one to suppose that he arrived at them in some other part of the book that has now been lost. The most important is that given below:

<A> The section of a right cylinder with circular bases by a plane P_1 intersecting the axis and not parallel to the base is an ellipse, the centre of which lies on the axis of the cylinder. The diameter of the cylinder is equal to the minor axis of the ellipse.

The section of this same cylinder by a plane P_2 parallel to the base and intersecting the centre of the ellipse is a circle equal to the base circle and the inscribed circle of the ellipse, and having a diameter equal to the minor axis of the ellipse.

If the plane P_1 is rotated about this minor axis until it coincides with P_2, the circle inscribed within the ellipse becomes superimposed on the circle formed by the section through the cylinder of the plane P_2. It would appear that Ibn al-Samḥ shows that the circle of the plane P_2 is at the same time the rabattement of the small circle of the ellipse and the orthogonal projection of the ellipse.

Proposition 7. — *Orthogonal affinity with regard to the minor axis.*

Let there be an ellipse AGBD *with axes* AB *and* GD, AB > GD, *with centre* N *and an inscribed circle of diameter* GD. *If a parallel to* AB *cuts* GD *in* H, *the circle in* K *and the ellipse in* T, *we have*

$$\frac{\mathrm{HT}}{\mathrm{HK}} = \frac{\mathrm{AB}}{\mathrm{GD}} = \frac{\mathrm{a}}{\mathrm{b}}.$$

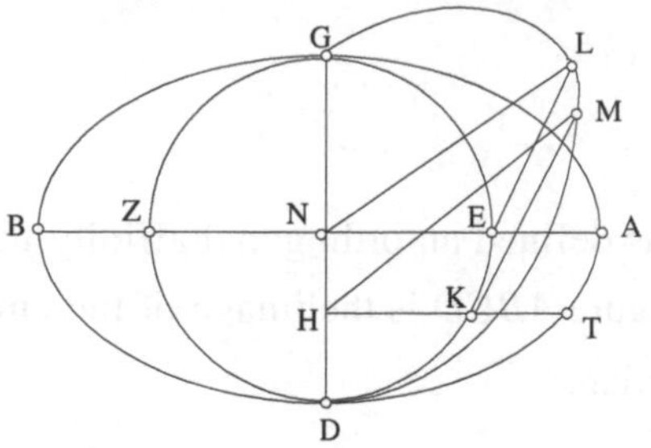

Fig. 6.8

If we revolve the ellipse around *DG*, point *A* describes a circle in the plane that is perpendicular in *N* to *DG*. This circle cuts the perpendicular in *E* at the plane of the ellipse in the point *L*. The ellipse is placed in the position *DLG*, the plane section of the right cylinder whose base is a circle.

Point *T* describes an arc of a circle with centre *H*, and falls into *M* on the generating line *MK*. *LNE* and *MHK* are similar right-angled triangles (since $\hat{N} = \hat{H}$, angles with parallel sides), and we have

$$\frac{LN}{NE} = \frac{MH}{KH} \Rightarrow \frac{AN}{NE} = \frac{HT}{HK} \Rightarrow \frac{HT}{HK} = \frac{AB}{GD}.$$

Porism. — *In the right-angled triangle* LEN, *we have* $LE^2 + NE^2 = LN^2$, *whence* $LE^2 = a^2 - b^2 = c^2$ *and* LE *is thus the distance from the centre to a focus.*

This porism will be used in Propositions 10 and 11.

Comment. — The ellipse *AGBD* constitutes the rabattement of the ellipse *DLG* onto the plane that is perpendicular in *N* to the axis of the cylinder, the circle *DEG* being the cylindrical projection of the ellipse *DLG* on this same plane.

Proposition 8. — *Orthogonal affinity with regard to the major axis.*

Let there be an ellipse AGBD *with axes* AB *and* GD, AB > GD, *and the centre* L *of a circumscribed circle with diameter* AB. *If a parallel to* GD *cuts* AB *in* H, *the ellipse in* E *and the circle in* Z, *we have*

$$\frac{ZH}{EH} = \frac{AL}{LG} = \frac{a}{b}.$$

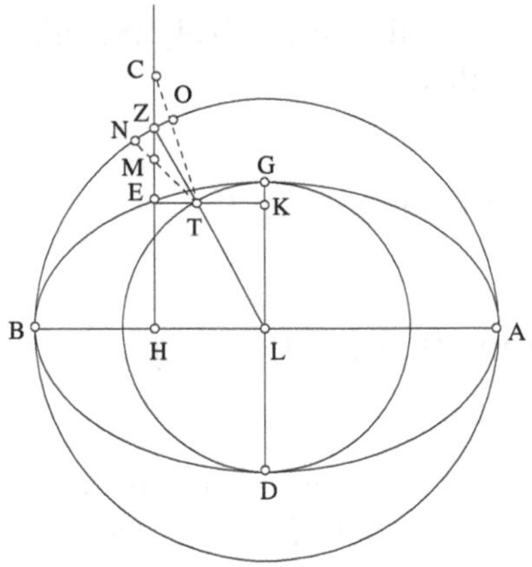

Fig. 6.9

The parallel to *AB* produced from *E* cuts *GD* in *K* and the inscribed circle in *T*. Ibn al-Samḥ shows by way of *reductio ad absurdum* and Proposition 7 that *L*, *T* and *Z* are aligned, hence deducing the result.

Comments.

1) Proposition 8 is treated by Ibn al-Samḥ like a corollary of 7. Let us note that in a same way he could have deduced, as a corollary of 6, a second affinity in the case of the bifocal definition.

2) Following Proposition 7, the ellipse *ABGD* is the image of the circle with diameter *GD* in an orthogonal affinity with a ratio $\frac{a}{b}$, which is a dilatation; and, following Proposition 8, the ellipse is the image of the circle with diameter *AB* in an orthogonal affinity with a ratio $\frac{b}{a}$, which is a contraction.

In other words, in an analytic language that was unknown to Ibn al-Samḥ, if within an orthogonal reference we consider ellipse **E** and circles $\mathbf{C}_1$ and $\mathbf{C}_2$ such as:

$$\mathbf{E} = \left\{(x, y),\ \frac{x^2}{a^2} + \frac{y^2}{b^2} = 1\right\}, \qquad \text{with } a > b,$$

$$\mathbf{C}_1 = \left\{(X, Y),\ X^2 + Y^2 = b^2\right\},$$

$$\mathbf{C}_2 = \left\{(X, Y),\ X^2 + Y^2 = a^2\right\},$$

and, if we designate by ψ and φ respectively the dilatation and contraction studied by Ibn al-Samḥ, then

$$\mathbf{E} = \psi\,(\mathbf{C}_1) \text{ with } \psi\text{: } (X, Y) \rightarrow (x, y)\text{: } \begin{cases} x = \dfrac{a}{b}X, \\ y = Y, \end{cases}$$

$$\mathbf{E} = \varphi\,(\mathbf{C}_2) \text{ with } \varphi\text{: } (X, Y) \rightarrow (x, y)\text{: } \begin{cases} x = X, \\ y = \dfrac{b}{a}Y. \end{cases}$$

3) It should be remembered that, in Proposition 3 of his treatise *On the Sections of a Cylinder*, Thābit ibn Qurra begins by considering the affinity relative to the major axis (in this case a contraction), taking as his point of departure the fundamental property (equation) of a circle of diameter equal to this major axis, $Y^2 = x\,(2a - x)$, and the equation of the ellipse defined in terms of its major axis, with d the *latus rectum* relative to it:

$$y^2 = \frac{d}{2a}\,x\,(2a - x).$$

He also shows that

$$\frac{y^2}{Y^2} = \frac{d}{2a} = \frac{b^2}{a^2}.$$

Thābit then indicates that the same technique may be used to show that the orthogonal affinity relative to the minor axis is a dilatation.

After having shown, in Propositions 6 and 7, that the elongated circular figure with axes $2a$ and $2b$ obtained by means of the bifocal definition and the ellipse with identical axes obtained by taking a plane section of a cylinder are both derived from a circle of radius b by a dilatation in the ratio of a/b, Ibn al-Samḥ states and proves their identity in the following proposition.

Proposition 9. — *Let there be 'an elongated circular figure'* AGBD *with axes* AB *and* DG, *and an ellipse obtained by a plane section* ZTHK *such that* AB = ZH *and* GD = TK. *The two figures will be superimposed on each other point by point.*

First method: The result is immediately obtained by superimposing the axes that are equal two to two, and in applying Propositions 6 and 7.

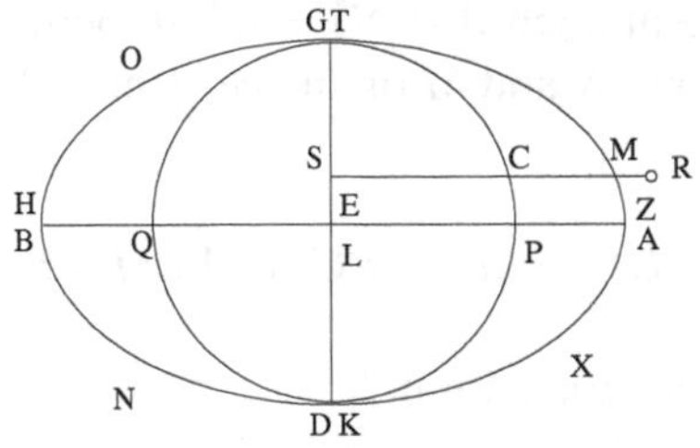

Fig. 6.10

Second method: This does not differ from the first and makes use of Propositions 6 and 7; yet, the axes are not superimposed. Ibn al-Samḥ takes on the minor axis of each figure a point that is equidistant to the centre and applies Lemma 1.

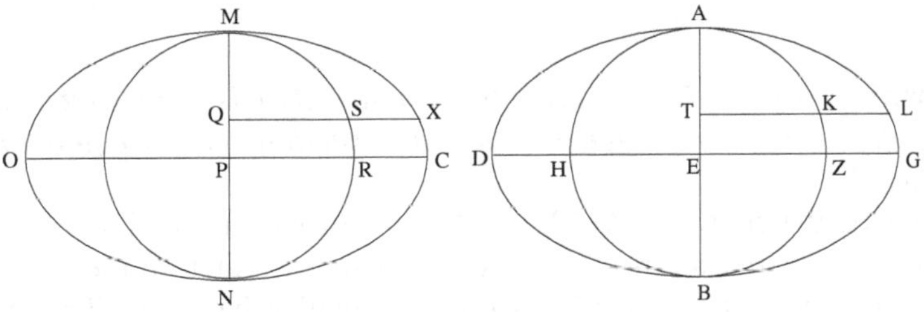

Fig. 6.11

Proposition 10. — *Let there be* AGBD, *a plane section with* E, *and axes* AB *and* GD, AB < GD. *How then can an equal curve be constructed by means of the bifocal method?*

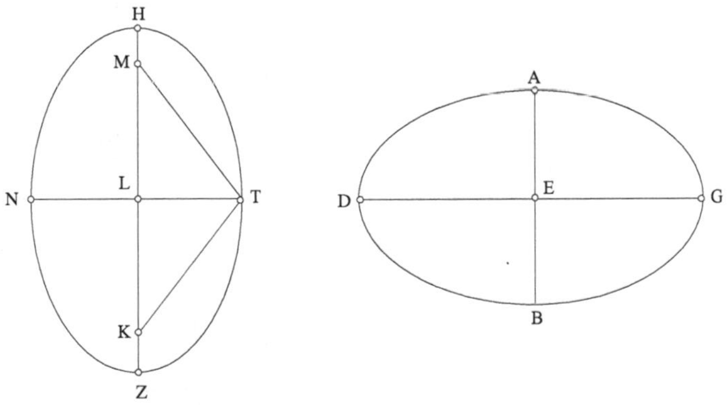

Fig. 6.12

Let NT be a segment such that $NT = AB$, L being the midpoint of NT. The foci of the ellipse are K and M on the perpendicular bisector of NT such that

$$LT^2 + LK^2 = EG^2 \text{ and } LM = LK.$$

Following the porism of Proposition 7,

$$b^2 + LT^2 = a^2 \Rightarrow LT^2 = a^2 - b^2 = c^2.$$

We thus have

$$TK = TM = EG.$$

Ibn al-Samḥ then notes the construction of the two other vertices Z and H.

Proposition 11. — *Let* AGBD *be a figure that is constructed by the bifocal definition. How then can a plane section that is equal to it be constructed?*

Let Z be the centre of $AGBD$ and E one of its foci, with $AB > GD$. In plane π we consider a circle with centre M that is equal to the circle with diameter GD. Let CN and HT be two perpendicular diameters. The required ellipse will be a plane section of a cylinder of revolution constructed on circle CHN, its minor axis being CN and its major axis PO, with P being constructed by applying the porism of Proposition 7:

$$PT \perp \pi \text{ and } TP = EZ.$$

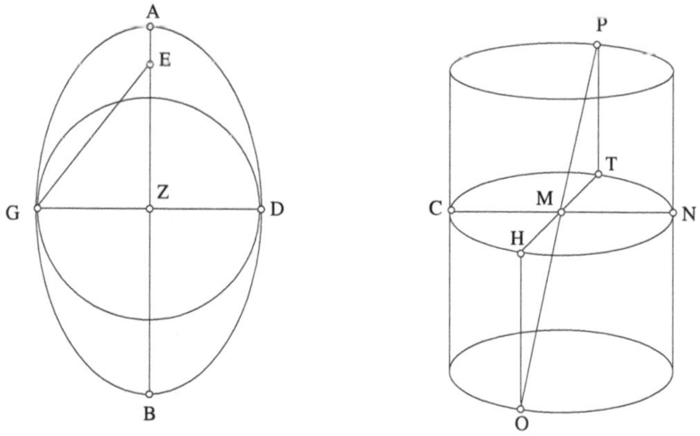

Fig. 6.13

We thus have

$$MP = EG = ZA.$$

Plane CPN cuts the cylinder along section $NPCO$, which is the required section.

6.2.7. *The area of an ellipse*

In the seven propositions in this chapter, Ibn al-Samḥ seeks to determine the area of an ellipse. The first of these, Proposition 12, is a lemma used in Proposition 13. Propositions 17 and 18 are effectively reformulations of the result established in Proposition 16 (Corollary 1). If S_1, S_2, E and Σ are the areas of the circles with diameters $2b$, $2a$, $2r$ and $2\sqrt{ab}$, if S is the area of the ellipse, and if P_1 and P_2 are the perimeters of S_1 and S_2 respectively, then the following results may be obtained:

13. $\frac{S}{S_1} = \frac{a}{b}$; 14. $\frac{S}{E} = \frac{ab}{r^2}$; 15. $\frac{S_2}{S} = \frac{S}{S_1}$;

16. $S = \frac{1}{2} P_1 . a$ and $S = \frac{1}{2} P_2 . b$, with corollary $S \approx \left(\frac{5}{7} + \frac{1}{14}\right) 2a . 2b$;

17. $S - \Sigma$; 18. is none other than the corollary of 16.

Let us successively take up these propositions.

Proposition 12. — *Let* ATK *be a quarter of an ellipse with centre* T, AT ⊥ TK *and* AT < TK, *and* ADBT *the quarter of an inscribed circle that is associated with it. Let* KZ *be a chord,* ZG ⊥ AT, *and* ZG *cuts the quarter of the circle in* D. *We have*

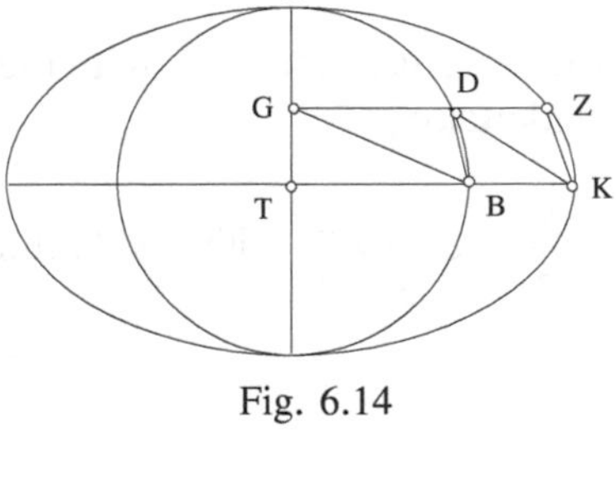

Fig. 6.14

$$\frac{\textit{area trapezoid}(\mathrm{KZGT})}{\textit{area trapezoid}(\mathrm{BDGT})} = \frac{\mathrm{TK}}{\mathrm{TA}} = \frac{\mathrm{a}}{\mathrm{b}}$$

The demonstration uses orthogonal affinity that is relative to the minor axis of the ellipse. Ibn al-Samḥ decomposes the trapezoids into triangles, which is not indispensable. In fact, the trapezoids with same height yield

$$\frac{\text{area }(KZGT)}{\text{area }(BDGT)} = \frac{TK+GZ}{TB+GD};$$

However, based on Proposition 6 (or 7), we have

$$\frac{GZ}{GD} = \frac{TK}{TB} = \frac{a}{b} = \frac{TK+GZ}{TB+GD},$$

whence

$$\frac{\text{area }(KZGT)}{\text{area }(BDGT)} = \frac{a}{b}.$$

We proceed in the same manner from another chord in the quarter of an ellipse that is being considered.

By reiterating the same with all the other quarters of the ellipse, we could show that the ratio of the area of an inscribed polygon within the ellipse to the area of a polygon inscribed within the circle, and associated with the former, is equal to the ratio of the major axis to the minor one.

Proposition 13. — *The ratio of the area* S *of an ellipse with axes* 2a *and* 2b *to the area* $\mathrm{S_1}$ *of the inscribed circle with diameter* 2b *is*

$$\frac{\mathrm{S}}{\mathrm{S_1}} = \frac{\mathrm{a}}{\mathrm{b}}.$$

Ibn al-Samḥ demonstrates this proposition with the help of the apagogic method. Following this undertaking:

a) Let us assume $\frac{b}{a} > \frac{S_1}{S}$. Let $\frac{b}{a} = \frac{S_1}{L}$ with $L < S$; thus $S = L + \varepsilon$.

Let P_1 be the area of a lozenge with its summits being the extremities of the axes of the ellipse; we have $P_1 > \frac{1}{2}S$.

We double the number of the sides of the inscribed polygon, and we reiterate this operation in such a way that we successively obtain the polygons with areas $P_2, \ldots, P_n$, P_n having 2^{n+1} sides. We have

$$P_1 > \frac{1}{2}S \Rightarrow S - P_1 < \frac{1}{2}S,$$

$$P_2 - P_1 > \frac{1}{2}(S - P_1) \Rightarrow S - P_2 < \frac{1}{2^2}S,$$

$$\ldots$$

$$P_n - P_{n-1} > \frac{1}{2}(S - P_{n-1}) \Rightarrow S - P_n < \frac{1}{2^n}S.$$

Hence, for a given $\varepsilon > 0$, there exists $N \in \mathbf{N}^*$ such that for $n > N$, we have $\frac{1}{2^n}S < \varepsilon$; thus

$$S - P_n < \varepsilon \text{ and } P_n > L.$$

Let P'_n then be the area of a polygon inscribed in the circle with area S_1 and deduce the polygon with area P_n by orthogonal affinity with ratio $\frac{b}{a}$. We have, following Proposition 12,

$$\frac{b}{a} = \frac{P'_n}{P_n};$$

hence

$$\frac{P'_n}{P_n} = \frac{S_1}{L}.$$

However

$$P_n > L \text{ and } P'_n < S_1;$$

hence

$$\frac{P'_n}{P_n} < \frac{S_1}{L},$$

which is absurd.

b) Let us assume $\frac{b}{a} < \frac{S_1}{S}$, namely $\frac{a}{b} > \frac{S}{S_1}$. Let $\frac{a}{b} = \frac{S}{L'}$ with $L' < S_1$; thus $S_1 - L' = \varepsilon$.

We then divide the circumference into 2^2, 2^3, …, 2^{n+1} parts, which brings us back to polygons P'_1, P'_2, …, P'_n. We successively have

$$S_1 - P'_1 < \frac{1}{2} S_1,$$

$$S_1 - P'_2 < \frac{1}{2^2} S_1,$$

$$\ldots$$

$$S_1 - P'_n < \frac{1}{2^n} S_1.$$

So, there exists $N \in \mathbf{N}^*$ such that for $n > N$, we have $\frac{1}{2^n} S_1 < \varepsilon$; thus

$$S_1 - P'_n < \varepsilon \text{ and } P'_n > L'.$$

However, if P_n is the area of the polygon inscribed in the ellipse that is associated with the polygon of area P'_n that is inscribed in the circle, we have by Proposition 12

$$\frac{a}{b} = \frac{P_n}{P'_n};$$

hence

$$\frac{P_n}{P'_n} = \frac{S}{L'}.$$

But

$$P'_n > L' \text{ and } P_n < S;$$

hence

$$\frac{P_n}{P'_n} < \frac{S}{L'},$$

which is absurd.

From a) and b) we deduce

$$\frac{S}{S_1} = \frac{a}{b}.$$

Comments.

1) The apagogic method that is applied here is not the usual method. To show that $\frac{S}{S_1}=\frac{a}{b}$, we assume that

a) $\frac{b}{a}=\frac{S_1}{L}$ with $L < S$, and hence $\frac{L}{S_1}<\frac{S}{S_1}$;

b) $\frac{b}{a}=\frac{L'}{S}$ with $L' < S_1$, and hence $\frac{S}{L'}>\frac{S}{S_1}$.

The two cases a) and b) lead to impossibility. However, we usually treat part b) by positing

$$\frac{b}{a}=\frac{S_1}{L'}, \qquad \text{with } L' > S.$$

After all, Ibn al-Samḥ notes in the Hebrew version, which we possess, that he has established that 'the ratio of the small diameter to the large diameter is not equal to the ratio of the circle to a surface that is either smaller than the ellipse, or greater than it', which does not exactly describe this undertaking.

2) From the property of orthogonal affinity, Ibn al-Samḥ shows that for all $n > N$, the ratio $\frac{P_n}{P'_n}$ of the areas of two homologous inscribed polygons, one that of the ellipse with area S and the other of the circle with area S_1, is equal to the ratio $\frac{b}{a}$ of the affinity. Proceeding from the equality $\frac{P_n}{P'_n}=\frac{a}{b}$, he shows that we also have $\frac{S}{S_1}=\frac{a}{b}$.

The ratio of the areas is preserved when reaching the limit (see the commentary on Proposition 14 of Thābit ibn Qurra's treatise *On the Sections of the Cylinder*).

Proposition 14. — *The ratio of the area* S *of the ellipse with axes* 2a *and* 2b *to the area* E *of the circle with diameter* 2r *is*

$$\frac{\mathrm{S}}{\mathrm{E}}=\frac{2\mathrm{a}}{\mathrm{Z}}, \qquad \textit{with Z such that } \frac{\mathrm{Z}}{2\mathrm{r}}=\frac{2\mathrm{r}}{2\mathrm{b}} \ \left(\textit{whence } \frac{\mathrm{S}}{\mathrm{E}}=\frac{\mathrm{ab}}{\mathrm{r}^2}\right).$$

Let S_1 be the area of the circle with diameter $2b$ that is inscribed within the ellipse. We have (*Elements* XII.2)

$$\frac{S_1}{E} = \frac{4b^2}{4r^2}.$$

However, by hypothesis $4r^2 = 2bZ$; hence

$$\frac{S_1}{E} = \frac{2b}{Z}.$$

By Proposition 13, we have

$$\frac{S}{S_1} = \frac{a}{b},$$

hence

$$\frac{S}{E} = \frac{2a}{Z}.$$

Comment. — We can immediately deduce $\frac{S}{E} = \frac{ab}{r^2}$, a result that corresponds with Proposition 5 of Archimedes' *The Sphere and the Cylinder*.

Proposition 15. — *The ratio of area* S_1 *of the circle with diameter* 2b *that is inscribed within an ellipse, to the area* S *of that ellipse is equal to the ratio of this area* S *to area* S_2 *of a circle with diameter* 2a *that is circumscribed within that ellipse*:

$$\frac{S_1}{S} = \frac{S}{S_2}.$$

The demonstration is immediately established and has recourse to Propositions 13 and 14. By Proposition 13 we have

$$\frac{S_1}{S} = \frac{b}{a},$$

and by Proposition 14

$$\frac{S}{S_2} = \frac{ab}{a^2}.$$

We thus have

$$\frac{S_1}{S} = \frac{S}{S_2}.$$

Ibn al-Samḥ deduces the corollaries

1) $\frac{S_1}{S_2}=\left(\frac{S}{S_2}\right)^2$ and $\frac{S_2}{S_1}=\left(\frac{S}{S_1}\right)^2=\left(\frac{S_2}{S}\right)^2$;

2) $\frac{S}{S_2}=\frac{b}{a}$

(which are immediately deduced from Proposition 14).

Proposition 16. — *The area* S *of the ellipse is equal to that of the right-angled triangle with one of the sides of its right angle equal to the perimeter* p_1 *of the inscribed circle with diameter* 2b *and the other side to the half-axis* a*:*

$$S = \frac{1}{2} p_1 \cdot a.$$

Based on Proposition 1 of Archimedes' *On the Measurement of the Circle*, we have

$$S_1 = \frac{1}{2} p_1 \cdot b,$$

and by Proposition 13

$$\frac{S}{S_1}=\frac{a}{b},$$

hence the result follows.

Similarly, if p_2 is the perimeter of the circumscribed circle with diameter $2a$, we have

$$S = \frac{1}{2} p_2 \cdot b.$$

Corollary 1. — $\frac{1}{2}p_2 \approx \frac{22}{7}a$, whence $S \approx \frac{22}{7}ab$, Ibn al-Samḥ's result given in the form

$$S \approx \left(\frac{5}{7}+\frac{1}{14}\right) 2a \cdot 2b.$$

Corollary 2. — If we know S and $2a$ (respectively $2b$), we find $2b$ (respectively $2a$).

Proposition 17. — *Every ellipse has an area equal to that of a circle with a diameter as the mean proportional between the two axes* 2a *and* 2b *of the ellipse.*

Let S_1 be the area of a circle with diameter $2b$, and S_2 that of a circle with diameter $2a$, and let Σ be the area of a circle with diameter $2r$ satisfying $\frac{2a}{2r}=\frac{2r}{2b}$ and hence $r=\sqrt{ab}$. Then it follows that

$$\frac{S_2}{\Sigma}=\frac{\Sigma}{S_1} \qquad (\textit{Elements}\ \text{XII.2 and VI.22}),$$

whence

$$\frac{S_2}{S_1}=\left(\frac{S_2}{\Sigma}\right)^2.$$

But in Proposition 15, we saw that

$$\frac{S_2}{S_1}=\left(\frac{S_2}{S}\right)^2.$$

Hence we have

$$S=\Sigma.$$

Comment. — In Proposition 14 of the treatise on the *Sections of the Cylinder*, Thābit establishes directly that the area S of the ellipse is equal to the area Σ of the circle with radius $r=\sqrt{ab}$. He proceeds with the aid of the apagogic method by successively considering the following:

a) $S>\Sigma$ and b) $S<\Sigma$.

He introduces the circle with area S_2 and diameter $2a$, the major axis of the ellipse, and the orthogonal affinity relative to that major axis. He associates the polygon with area P_n, which is inscribed in the ellipse, with a polygon having an area P'_n, which is inscribed in the circle with diameter $2a$. Thābit shows that

$$\frac{P_n}{P'_n}=\frac{b}{a}.$$

However, according to *Elements* XII.2, we have

$$\frac{\Sigma}{S_2}=\frac{ab}{a^2}=\frac{b}{a};$$

thus

$$\frac{P_n}{P'_n} = \frac{\Sigma}{S_2} = \frac{b}{a}.$$

It is this equality that allows him to show that a) and b) result in an absurdity; hence

$$S = \Sigma \text{ and } \frac{S}{S_2} = \frac{b}{a}.$$

Thābit does not introduce in this context anything but the circle with a diameter $2a$ and area S_2, and the circle Σ, while Ibn al-Samḥ introduces in addition a circle with a diameter $2b$ and area S_1.

However, we note that both mathematicians resort to Proposition XII.2 of the *Elements*, and evoke the ratio of the areas of the two polygons that are correlative with respect to an orthogonal affinity, which is a contraction for one and a dilatation for the other; this ratio is introduced by Thābit in the course of the demonstration, while it is given in the conclusion of Proposition 12 by Ibn al-Samḥ.

Proposition 18. — *Every ellipse consists of the* $\left(\frac{5}{7}+\frac{1}{14}\right)$ *of the rectangle that is circumscribed in it:*

$$S \approx \left(\frac{5}{7}+\frac{1}{14}\right) 2a \cdot 2b.$$

This result is established in the first corollary of Proposition 16. However, Ibn al-Samḥ presents here two demonstrations. Based on the preceding proposition, the first consists of using the result f) of the first lemmas; namely the second proposition of Archimedes' *On the Measurement of the Circle*. Based equally on the preceding proposition, the second consists of using *Elements* XII.2.

We could offer another form of the statement of this proposition:

The ratio $\frac{S}{2a \cdot 2b}$ *is the same for every ellipse,* $\frac{S}{2a \cdot 2b} = \frac{5}{7}+\frac{1}{14}.$

6.2.8. *Chords and sagittas of the ellipse*

In Propositions 19 and 20<a>, Ibn al-Samḥ studies the chords that are parallel to one of the axes of an ellipse and the sagittas that correspond to it.

Proposition 19. — *Let there be an ellipse* AGBD *with axes* AG *and* BD, AG > BD, *and the circle* $\mathbf{C}_1$ *with diameter* BD. *If to a chord* OT *of the ellipse that is perpendicular to* AG *at* W *we associate in* $\mathbf{C}_1$ *a chord* RN *that is equal and parallel, and cuts the diameter* EM *of* $\mathbf{C}_1$ in Q, *then* W *and* Q *divide respectively* AG *and* EM *in the same ratio:*

$$\frac{\mathrm{WG}}{\mathrm{WA}} = \frac{\mathrm{QE}}{\mathrm{QM}}.$$

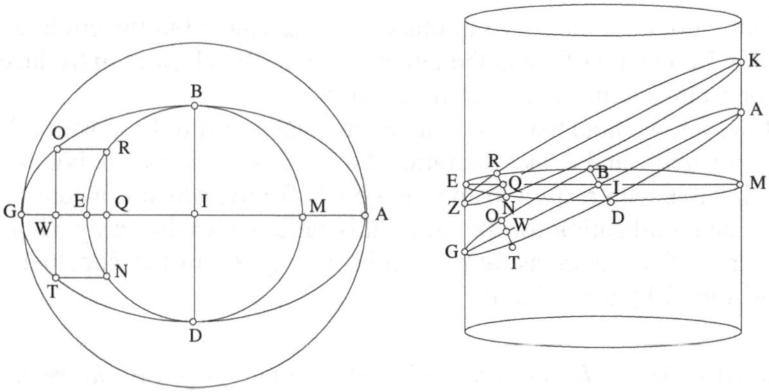

Fig. 6.15

Ibn al-Samḥ returns here to the method that has been already followed in Proposition 7, by placing the ellipse on the cylinder of revolution with a base $\mathbf{C}_1$; namely the method used in studying orthogonal affinity.

We then let the circle $\mathbf{C}_1$ rotate around its diameter *BD* to bring it to a plane parallel to that of the base. We pass by the chord *RN* a plane parallel to that of the ellipse. The section of the cylinder through this plane is an ellipse *KRZN* that is equal to ellipse *ABGD*. We have $RN = OT$, whence $GW = ZQ$ and $AW = KQ$. The right triangles *KQM* and *ZQE* are similar; we thus have

$$\frac{QZ}{QK} = \frac{EQ}{MQ},$$

and hence

$$\frac{GW}{AW} = \frac{EQ}{MQ}.$$

We see, however, that the result is an immediate consequence of orthogonal affinity ψ with a ratio $\frac{a}{b}$ relative to the minor axis. We have

$\psi(R) = O$ and $\psi(N) = T$; while W has the same abscissa T and O, Q has equally the same abscissa R and N, and we thus have

$$IW = \frac{a}{b}\, IQ.$$

We also have

$$IG = \frac{a}{b}\, IE \text{ and } IA = \frac{a}{b}\, IM;$$

thus

$$GW = \frac{a}{b}\, EQ \text{ and } WA = \frac{a}{b}\, QM,$$

and hence

$$\frac{WG}{WA} = \frac{QE}{QM}.$$

Comment. — The idea here is that the ratio of the two sagittas GW and EQ of the homologous chords OT and RN is equal to the ratio of the affinity:

$$\frac{GW}{EQ} = \frac{a}{b}.$$

Proposition 20. — Some sections of the text of Proposition 20 are not in the correct order, and it appears that some paragraphs have been omitted by either the copyist or the translator. At the start of this proposition, Ibn al-Samḥ notes that the previous problem may be addressed in the same way if one considers the circumscribed circle and two equal chords, one in the ellipse and the other in the circle, both of which are perpendicular to the minor axis.

It is possible to reconstruct the missing text in order to prove this statement.

Let there be an ellipse ABGD *and its circumscribed circle with diameter* AG *that cuts the straight line* BD *at points* M *and* E. *If the two chords* OT *in the ellipse, and* RN *in the circle, are in such a way that* OT = RN, OT ⊥ BD *at point* W, RN ⊥ EM *at point* Q, *we then have*

$$\frac{\mathrm{WB}}{\mathrm{WD}} = \frac{\mathrm{QM}}{\mathrm{QE}}.$$

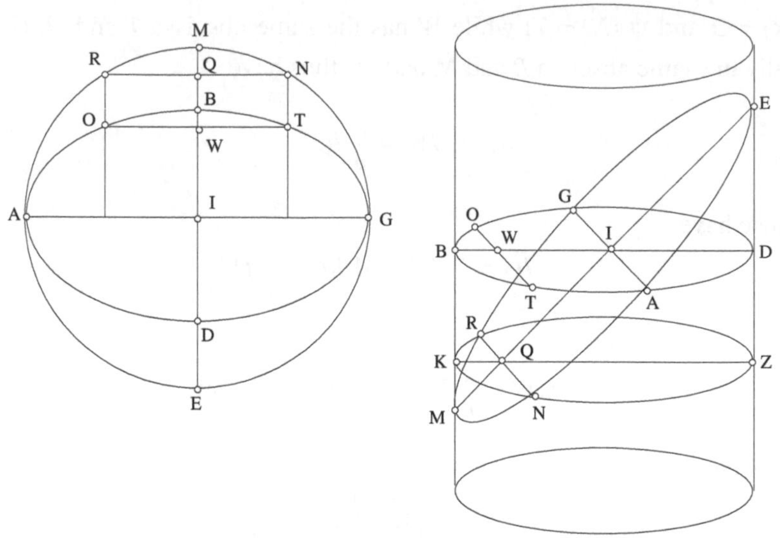

Fig. 6.16

We take the ellipse as the base of a right cylinder, and we rotate the circumscribed circle around AG until M reaches the generator passing by B; we obtain on oblique circular section of the cylinder, $AEGM$. From the chord RN, we pass a plane that is parallel to the plane $ABGD$, which cuts the cylinder in an ellipse $NZRK$, and we have $RN = OT$, $KQ = BW$, $QZ = WD$. The right triangles KMQ and QZE are similar, and we obtain the proof as in Proposition 19.

This result is a consequence of the orthogonal affinity φ of the ratio $\frac{b}{a}$ relative to the major axis, as already indicated by Ibn al-Samḥ; we thus have

$$\varphi(R) = O \text{ and } \varphi(N) = T.$$

However, W has the same ordinate as T and O, and Q has the same ordinate as R and N. We thus have

$$IW = \frac{b}{a} IQ;$$

we equally have

$$IB = \frac{b}{a} IM \text{ and } ID = \frac{b}{a} IE,$$

and hence

$$\frac{WD}{EQ}=\frac{WB}{QM}=\frac{b}{a},$$

and therefore

$$\frac{WB}{WD}=\frac{QM}{QE}.$$

The idea here is still that the ratio of two homologous sagittas is equal to the ratio $\frac{b}{a}$ of the affinity.

Ibn al-Samḥ returns later, in Proposition 20<a>, to this latter proposition, and he demonstrates it by *reductio ad absurdum*:

Let there be an ellipse AGBD *with major axis* AB, *with centre* I, *and a circle* ALBM *with diameter* AB. *Let there be in the ellipse and the circle the semi-chords* NE *and* HZ *such that* NE = HZ, NE ⊥ GD, HZ ⊥ LM. *Then*

$$\frac{\mathrm{ZL}}{\mathrm{ZM}}=\frac{\mathrm{EG}}{\mathrm{ED}}.$$

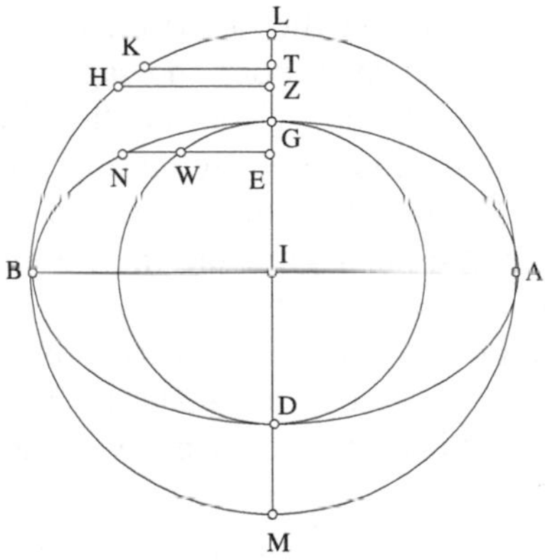

Fig. 6.17

If it were not such, then there exists $T \neq Z$ on LM such that

$$\frac{EG}{ED}=\frac{TL}{TM}.$$

If we take from the semi-chord TK, $TK \perp LM$, then, by Lemma 1, we have

$$\frac{EW}{TK}=\frac{IG}{IL}=\frac{b}{a}.$$

However, by Proposition 6 (or 7),

$$\frac{EW}{EN} = \frac{b}{a};$$

therefore

$$EN = TK,$$

which is absurd, since $EN = HZ$.

This demonstration is clearly more efficient.

It is worth noting that in Proposition 19, as well as in 20<a>, Ibn al-Samḥ established a ratio between the sagittas of two homologous chords, with one of the orthogonal affinities that associate an ellipse with one of the circles having its diameter as one of the axes of the ellipse.

To grasp paragraph <b>, let us take into account firstly that a circle is defined in a unique way by a chord and a sagitta. The equation $y^2 = x\,(d - x)$ shows that the given chord $2y$ and its associated sagitta x allow for the determination of d. However, the equation of an ellipse with axes $2a$ and $2b$ can be written as $y^2 = \frac{b^2}{a^2}\,x(2a - x)$, with the given chord $2y$ and its associated sagitta x not allowing the determination of a and b. Hence, it follows from this that such givens (chord and sagitta) do not characterize a unique ellipse *per se*. Supplementary givens are therefore needed, and this is precisely what Ibn al-Samḥ indicated by stating that the givenness of a chord, of its sagitta and of a diameter characterize an ellipse. Nonetheless, he added that it is 'possible that the sagitta and the chord are common to <this ellipse and to> another ellipse' (*infra*, p. 714).

The text of paragraph <b>, like that of <a>, is evidently incomplete. Was this lacuna due to a copyist or a translator? We do not really know. However, this is what we think might have been omitted:

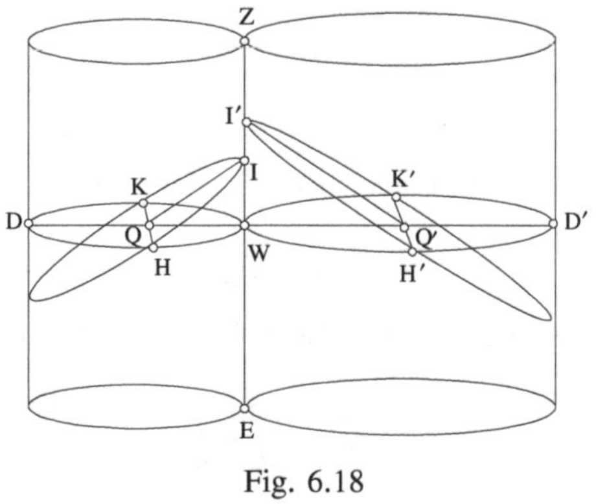

Fig. 6.18

Let l_1 and l_2 be the given lengths for the chord and the sagitta. Based on the figure, Ibn al-Samḥ seems to initially consider a cylinder having a base as a circle with diameter $WD > l_1$, in which the chord HK with a midpoint Q is placed, such that $HK = l_1$. We can thus place on this right cylinder, with a circular base, an ellipse HIK with a minor axis equal to WD, and having a sagitta QI of the chord HK with a length $QI = l_2$, whereby I is the generator of EZ that passes through point W (regarding the construction of I, refer to the 'comment' below). Let IJ be its major axis. Ibn al-Samḥ accounts for a second cylinder with diameter $WD' > WD$, that is tangent to the first cylinder following the generator EZ, along with a circle with a diameter WD' and a chord $H'K' = HK = l_1$. We thus have $WQ' < WQ$, whence $Q'I < QI$. The ellipse $H'IK'$ on the second cylinder does not answer to this problem.

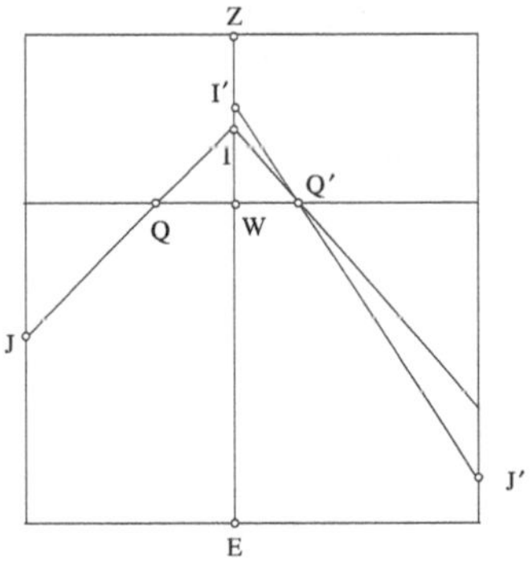

Fig. 6.19

Yet, there exists on WZ a point I' such that $Q'I' = QI = l_2$, and ellipse $H'I'K'$ solves the problem ($H'K' = l_1$, $Q'I' = l_2$). However, the two ellipses are not equal (their minor axes are different, since they are equal to the diameters of the cylinders).

We have thus shown that if a chord, its sagitta and the minor axis (as the diameter of the cylinder) are given, then the ellipse is determined; however, if only a chord and a sagitta are given, there are infinite ellipses that satisfy to this problem.

Comment. — The construction of I is not possible unless $QI = l_2 > QW$. This corresponds to the comment regarding the ratio of the sagittas in reference to Proposition 19. We thus need to have

$$\frac{l_2}{QW} = \frac{2a}{2b},$$

with $2b = WD$ and $2a$ equal to the major axis being sought.

The choice of having QW associated to the given HK, with $HK = l_1$ and $QW < l_2$, defines in a unique way the circle with diameter WD.

Paragraph <c>, which is inserted in the text of Proposition 20, pertains to another demonstration of Proposition 8 that has been already established by *reductio ad absurdum* from Proposition 7. This time we have a direct demonstration. Is this the reason that incited Ibn al-Samḥ to reconsider herein this proposition? Here is the demonstration:

Let $AGBD$ be an ellipse with a major axis AB, and AEB is its large circle. Let $TH \perp AB$; TH cuts the ellipse in K and the circle in H, and thus

$$\frac{HT}{TK} = \frac{AB}{PO}.$$

PO is the diameter of the small circle.

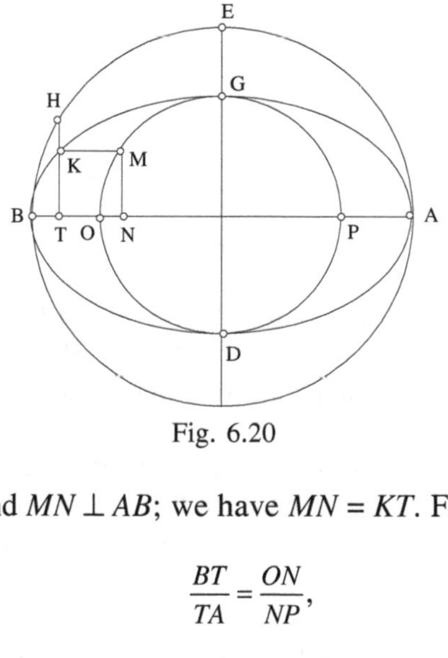

Fig. 6.20

Let $KM \parallel AB$ and $MN \perp AB$; we have $MN = KT$. From Proposition 19, we have

$$\frac{BT}{TA} = \frac{ON}{NP},$$

and, following Lemma 1,

$$\frac{HT}{MN} = \frac{AB}{PO};$$

hence

$$\frac{HT}{TK} = \frac{AB}{PO}.$$

The text of the proposition that follows has undoubtedly received some alterations. Certain indications suggest this, of which the first is that Ibn al-Samḥ informs us in the statement of this same proposition that he undertakes a line of calculation of the areas of the segments of the ellipse, which is not treated anywhere else.

Proposition 21. — *Given the chord of an ellipse, its sagitta and one of its axes, how can we derive its second axis, or the area of a segment of this ellipse or any other element associated with it?*

Consequently, Ibn al-Samḥ indicates that, to determine the second axis, three methods ought to be successively applied as per the case given later.

Problem: Let there be an ellipse *KATG*, with major axis *KT* and minor axis *AG*. Let there be a chord *OH* with midpoint *M*, and *KM* as its sagitta. Knowing that $OMH = 8$, $KM = 3$ and $KT = 15$, calculate *AG*.

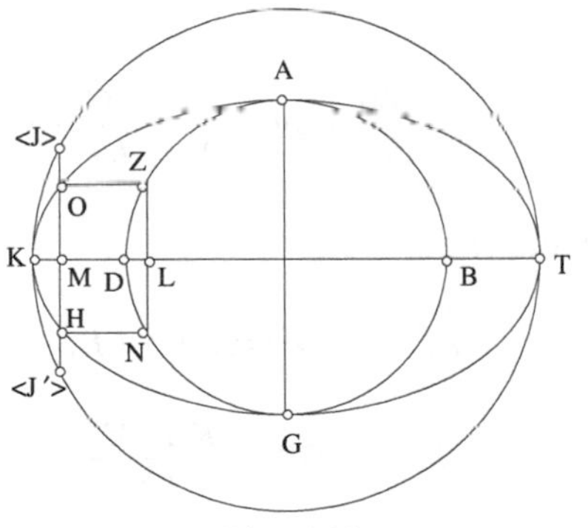

Fig. 6.21

First method:

$$OM^2 = 4^2, \quad \frac{KT}{KM} = 5, \quad \frac{KT}{MT} = \frac{5}{4}.$$

Thus, we obtain AG^2 by multiplying these three numbers; Ibn al-Samḥ applies here Proposition 19. In fact, since we have

$$OM = ZL = 4, \quad \frac{DB}{DL} = 5, \quad \frac{DB}{LB} = \frac{5}{4},$$

and since, within the circle, we have

$$ZL^2 = DL \cdot LB = \frac{DB}{5} \cdot \frac{4DB}{5} = 16,$$

then

$$DB^2 = 100, \quad AG = DB = 10.$$

Second method: The calculation that is proposed in the text is as follows:

$$\frac{KT}{4 \cdot KM} = \frac{15}{12} = \frac{5}{4}, \quad \frac{KT}{MT} = \frac{5}{4}, \quad \frac{KT^2}{4KM \cdot MT} \cdot 64 = AG^2,$$

whereby $OM = 4$, and this calculation yields the result $AG^2 = 100$, $AG = 10$.

Comment. — The expression of AG as a function of the given values is

$$AG^2 = \frac{OM^2 \cdot KT^2}{KM \cdot MT}.$$

In fact

$$\frac{DL}{DB} = \frac{KM}{KT} \quad \text{and} \quad \frac{LB}{DB} = \frac{MT}{KT} \text{ (by Proposition 19).}$$

Moreover, we have in the circle

$$ZL^2 = OM^2 = DL \cdot LB,$$

$$OM^2 = DB^2 \cdot \frac{KM . MT}{KT^2};$$

hence

$$DB^2 = AG^2 = \frac{OM^2 \cdot KT^2}{KM \cdot MT}.$$

Accordingly, we would not have $\frac{64}{4} = 16 = OM^2$ unless $OM = 4$.

Third method:

$$4KM \cdot MT = 4 \cdot 3 \cdot 12 = 144,\ KT^2 = 15^2,\ OH^2 = 8^2,$$

$$\frac{KT^2 \cdot OH^2}{4KM \cdot MT} = AG^2 = \frac{14 \cdot 400}{144} = 100.$$

And, if MO cuts the large circle in J and J', we have

$$KM \cdot MT = MJ^2,\ 4KM \cdot MT = JJ'^2,$$

$$\frac{TK^2 \cdot OH^2}{JJ'^2} = AG^2 \text{ and } \frac{TK}{AG} = \frac{JJ'}{OH} \qquad \text{(by Proposition 8).}$$

The text is eventually concluded with the following lemma:

Lemma 4. — *Let* A *be a number such that* A = B + G, B ≠ G. *We posit* $\frac{A}{B} = D$, $\frac{A}{G} = E$, $D \cdot E = H$, $B \cdot G = Z$ *and we want to show that* $Z \cdot H = A^2$.

This is immediately established, since

$$H = \frac{A^2}{BG} = \frac{A^2}{Z}.$$

Note that, in the course of the proof, and in two attempts, the conclusion is used. Neither the level of statement, nor that of demonstration, nor the place where the paragraph is located within the text, allow us to attribute this to course of inquiry to Ibn al-Samḥ, namely as the author of the remainder of the text. It is evident that from Proposition 20 onwards the text has been altered in several places.

6.3. *Translated text*

Ibn al-Samḥ

On the Cylinder and its Plane Sections

<FRAGMENT BY IBN AL-SAMḤ

On the Cylinder and its Plane Sections>*

Treatise on cylinders and cones

He[1] has said: In a text by the eminent[2] Ibn al-Samḥ,[3] I found these questions together,[4] with blank spaces left between them.[5] As far as I am aware, he included them in his work.[6] The intelligibles:[7]

Definition of spheres, cylinders and cones

<1> *Definition of the sphere*: A sphere is generated by a semi-circle, the diameter of which is fixed and unable to move and the arc of which is rotated until it returns to its original position. The solid described by the arc and the surface <bounded by it> is a sphere. The surface described by the arc is the surface of a sphere. The fixed straight line is its diameter. The extremities <of this straight line> are its poles. The midpoint <of the straight line> is its centre. The rotated arc is part of the greatest circle that can be carried on the sphere.

* This fragment has survived as a single manuscript in Hebrew; Neubauer Heb. 2008 [Hunt. 96] in the Bodleian Library, Oxford. This manuscript of 53 folios was written by Joseph b. Joel Bibas, who copied it in Constantinople in 1506 in a small cursive Spanish script. Ibn al-Samḥ's text occupies fols 46^{v}–53^{r}. The Hebrew translation is by Qalonymos b. Qalonymos, who completed it on 5th January 1312. He entitled it *Ma'amar ba-isṭewanot we-ha-meḥuddadim*, 'The Treatise on Cylinders and Cones'. The attribution of the text to Ibn al-Samḥ is beyond doubt, as is indicated in the *incipit*, and it is very likely that it is a fragment of his major work on geometry. Mr. Tony Lévy has transcribed the Hebrew text and translated it into French. I have since revised this translation. The notes accompanying the translation have been written by one or other of us.

[1] Almost certainly not the Andalusian sage himself, but a compiler. The other occurrences of the expression 'he has said' appear to refer to Ibn al-Samḥ himself.

[2] *ha-me'ulleh*. In Arabic, possibly: *al-fāḍil*.

[3] In the manuscript: A.L.S.M.A.Ḥ, which has therefore been read as al-Samāḥ.

[4] *ellu ha-she'elot mequbbaṣot*.

[5] *hinniaḥ beyneyhem ḥalaq*: One has left empty space between them.

[6] *be-ḥibburo*.

[7] *ha-muskalot*. In Arabic, almost certainly: *al-ma'qūlāt*.

Definition of the cylinder: A cylinder is obtained by fixing one side of a rectangle so that it cannot move, and then rotating the entire rectangle around the straight line until it returns to its original position. The rectangle describes the cylindrical solid, and the straight line parallel to the fixed straight line describes the surface of the cylinder. The two remaining straight lines, rotating about the extremities of the fixed side, describe the bases of the cylinder. If it is inclined, the cylinder is said to be oblique.[8]

The *definition of the cone* is similar to that of the cylinder. The axes are the same and the heights are equal. The upper end of the fixed side is the vertex of the cone and the surface of the cone is described by the diagonal <of the rectangle>. The conic solid is that described by (the triangle rotating about the) fixed side, and the base of the cylinder forms the base of the cone.

The cylinder and the cone were defined in this way by Euclid. However, Euclid only defined one species of each, that is, the cylinder with two circular bases and the axis perpendicular to the bases, and similarly for the cone, that is, the cone derived from this species of cylinder. Euclid had no need for anything else and this was the only species mentioned in his work.

<2> The general definition, which goes beyond that stated above, is as follows: Let two round[9] figures, with any contour, be located on two parallel planes. Let the centres of these figures be determined, and let them be joined by a straight line. Let a straight line move around the two figures, parallel to the axis joining their centres, until it returns to its original position. What this straight line parallel <to the axis> described is a cylinder. This definition includes all the species of cylinder studied in the books of the Ancients, together with all their properties. If the axis is inclined relative to the two bases, then the cylinder is oblique.

<3> Two further species may be derived from these two species by the use of <plane> sections arranged in a number of ways. If a right cylinder is sectioned by two parallel planes such that the sections are elliptical,[10] the two sections and the part of the cylinder lying between them form a cylinder whose bases are ellipses, with the cylinder inclined at an angle

[8] See the mathematical commentary: Section 6.2.1.

[9] *temunot ʿagolot*. It becomes clear from what follows that the author is using this expression to designate circles and ellipses.

[10] *ha-ḥatikhot kefufot*: the sections being curves. The term *kafuf* (adjective or noun) indicates an ellipse. We should highlight the absence of any reference to Apollonian terminology, which the translator Qalonymos knew and used in other texts.

relative to them. If an oblique cylinder is sectioned by two parallel planes at right angles to the axis, the two elliptical figures and the surface of the cylinder <lying between them> form a cylinder which is a right cylinder relative to them. These four species are in fact only two, with the other two being derived from them. Thus, beginning with two species whose bases are ellipses, it is possible to generate the two species with circular bases by proceeding in the reverse manner.

<4> The general definition of the cone is as follows: Take a circle and any point lying outside[11] the plane of the circle. Join this point to the centre of the circle with a straight line, and with an infinite number of straight lines from the point to the circumference of the circle. Hold the straight line joined to the centre of the circle fixed and rotate <one of> the others around the circle until it returns to its original position. The triangle describes a cone. The surface described by the side <touching> on the circumference is the surface of the cone. Its axis is the fixed straight line, its vertex is the point and its base is the circle. This is the definition given by Apollonius in the book of the *Conics*.[12] The cone whose axis is perpendicular <to the plane of the circle> is a right cone.[13] The cone whose axis is inclined is an oblique cone.

<5> The general definition of the cylinder is that given above. There are two genera of cylinder. The first one is the cylinder whose bases are two round figures that are equal and parallel, defining a regular surface between them.[14] This genus may be divided into two further species depending on whether the surface defined by these two bases stands at a right angle to the bases or does not stand at a right angle, but inclined on them. If the surface stands at a right angle to the bases, then the cylinder is a right cylinder; if it is inclined, then the cylinder is oblique. Each of these species may be subdivided further into two more species depending on whether the two bases are circles or ellipses. If the cylinder is of the first species and the bases are two circles, then the cylinder is the right circular cylinder mentioned by the Ancients. If it is oblique, then it is an oblique circular cylinder. If the two bases are ellipses, then the cylinder is either a right or oblique elliptical cylinder.

[11] *ba-awir, be-zulat sheṭaḥ ha-'agolah*: in the air, outside the plane of the circle.

[12] *Sefer ha-ḥaruṭim.*

[13] *yoṣe' min toshavto 'al zawiyyot niṣṣavot*: is built up from the base at right angles.

[14] For the second type, not defined here, see the mathematical commentary: Section 6.2.2.

From the point of view of their generation, all these species are covered by the definition that I have introduced above. If the two bases of the cylinder are circles and the surface is defined by them at a right angle, then the axis is perpendicular to the base of the cylinder, as we have said, and all its straight sides – those joining the bases – are equal, and any plane dividing the cylinder into two halves forms a rectangle whose diagonals are equal. These are the diameters of the cylinder and they are all equal.

The treatise on cylinders

<6> He has said: Cylinders, as we have shown, comprise a number of species, in one of which the bases are two circles and in another of which they are not two circles. There are many curves which are not circles and it is impossible to list all of them as they include the sections of right and oblique cylinders, the sections of cones, oval figures and others, and the figure bounded by a curved line which is not ordered.[15] For all these reasons, it is necessary to give a general definition that covers all cases. First, we must list the elements that must be specified prior to stating the definition itself.

<7> Given two round[16] figures that are equal and of the same shape, we consider a point within each of them from which we draw the same number of lines to the contour of the figure, such that each line is identical to its homologue, and such that each pair of lines in any one of the figures encloses an angle equal to that enclosed by the homologous pair of straight lines in the other figure. These two points are said to be similar[17] positions. If two round figures that are equal and of the same shape lie on two parallel planes, and if a common sectioning plane passing through the two points at similar positions cuts the figures along two straight lines that are equal, then the two figures are said to be at similar positions.

<8> Having established the above, the definition of the cylinder is as follows: Let us consider two equal round figures, having the same shape and lying on two parallel planes in similar positions. Let us determine two points at similar positions within these two figures, and let us join them with a straight line. Now let us rotate a straight line parallel to that joining

[15] *qaw 'aqum zulat seder*. See the mathematical commentary: Section 6.2.2.

[16] *shney me'uggalim*. The term is manifestly more general than 'round figures' in the sense of the circle or ellipse. As becomes clear from what follows, it can only refer to closed curves with a centre of symmetry.

[17] *mitdammot ha-maṣav*.

the two points at similar positions around the contour of the two figures until it returns to its original position. What is generated by this straight line is called a cylinder. The straight line joining the two points at similar positions is called the axis.[18] Any straight line from the contour of one of the two figures to the contour of the other and parallel to the axis is a side of the cylinder. If the axis is perpendicular to the planes of the two round figures, then the cylinder is a right cylinder. If this is not the case, then the cylinder is an oblique cylinder.

<**9**> He has said: We have previously stated that there are many different species of cylinder. As there is no single method encompassing all of them, we wish to mention a number of them that may act as guides in dealing with the others.

We begin by introducing a treatise concerning the figures obtained from the plane sections of a cylinder and the problems that are specific to these sections. This is followed by a treatise on their areas and the problems that are specific to these areas, including those relating to the ratios between them and their properties. We then offer a treatise on the surfaces of spheres obtained from semi-circles and their sections, and finally a treatise on the volume of these spheres.

We begin with a study of the right circular cylinder, as this is the simplest[19] of all cylinders: a right angle is the simplest of all angles, and a circle is the simplest of all round figures. From there, shall follow that which must follow. May the Creator, blessed be He, aid us in this endeavour.

<**10**> Treatise on the sections of right cylinders whose bases are two circles, and the definition applying specifically to this type of cylinder.

The definition applying to this type of cylinder, and to no other, is that given by Euclid. This is the definition based on fixing the side of a rectangle that we have mentioned previously. The sections of this type of cylinder may be divided into three species. If the plane of the section passes through the axis or is parallel to the axis, then the section is a rectangle. If the plane of the section is parallel to the bases, then the section is always a circle. If the plane of the section is not parallel to the bases, then the section is called an ellipse.[20]

[18] The straight line joining two points 'at similar positions' is only an axis if the two points are the centres of symmetry of the bases.

[19] *ha yoter yeshara*: the straightest. The same adjective is used to qualify the right angle and the circle.

[20] *kafuf*.

<11> The first species of sections of a right cylinder whose bases are two circles.

Cutting the cylinder by a plane parallel to the bases gives a section which can therefore be generated by the movement of a straight line, one extremity of which is fixed in the plane <of the section>, and which rotates in the plane until it returns to its original position. The portion of the plane swept by this straight line is called a circle, and the figure described by the other extremity is called the circumference. The moving straight line is called the half-diameter. The fixed point is called the centre of the circle. All the straight lines starting from this point and ending on the circumference are equal to each other. This section is necessarily a circle. Among its properties, we can find that it has an internal point such that all the straight lines drawn from this point to the periphery of the section are equal to each other. Also, in the figure generated by the movement of the straight line, which is the circle, we can also find a point such that all the straight lines drawn from this point to the periphery are equal to each other. Thus, if we apply this section on the circle whose half-diameter is equal to the half-diameter of the said section, then the two will coincide.

<12> We state the following lemmas in relation to the sections that are circles:

<a> The ratio of any circle to any other is equal to the ratio of the square of the diameter of the first to the square of the diameter of the second. This is the square[21] of the ratio of the diameter to the diameter.

<b> The ratio of any circle to any other is equal to the ratio of the polygon inscribed within the first to the polygon inscribed within the second. This ratio is equal to the square of the ratio of the side of the polygon to the side of the polygon.

All this has been shown by the proofs of Euclid in the 12th <book> of his work.

<c> Any circle is equal to the right-angled triangle having one of the sides enclosing the right angle equal to the circumference, and the second side enclosing the right angle equal to the half-diameter of the circle.

<d> The ratio of the diameter of any circle to its circumference is the same as the ratio of the diameter of any other circle to its circumference.[22]

<e> The ratio of the circumference of a circle to its diameter is less than three times the diameter plus one seventh of the said diameter added

[21] *shanuy be-kefel*: repeated twice. In Arabic: *muthannā bi-al-takrīr*.

[22] This property is stated by the Banū Mūsā: *On the Knowledge of the Measurement of Plane and Spherical Figures*, Proposition 5. See Chapter I.

to it, and greater than three times the diameter and ten seventy-firsts of the said diameter added to it.[23]

<f> The ratio of any circle to the square of its diameter is equal to the ratio of 11 to 14.

<13> All this has been proved by Archimedes.

The following questions relating to the circle have not been mentioned by Euclid, or by Archimedes, or by anyone else. They are included in the properties required for the study of the sections of a cylinder.

<Lemma 1> Given any two circles, the diameter of one is divided at any point other than the centre. A perpendicular is drawn from this point forming a chord of the circle. The diameter of the other circle is divided in a similar way. A perpendicular is drawn from the point <of division> forming a chord of the circle. The ratio of one chord to the other chord is equal to the ratio of one diameter to the other diameter.

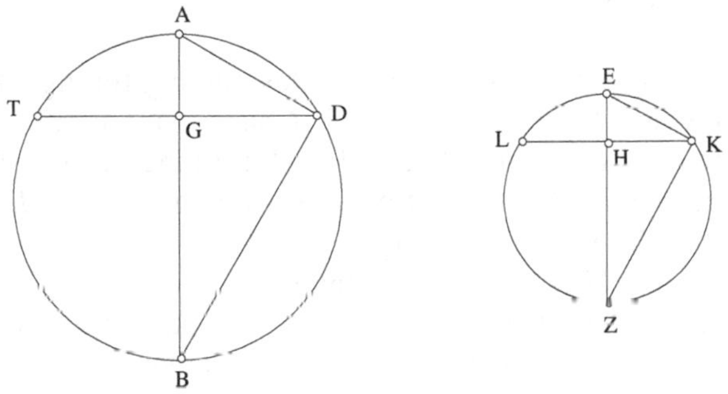

Fig. VI.1

Example: Consider two circles *AB* and *EZ*, having diameters *AB* and *EZ*. Divide *AB* at a point *G*, from which a perpendicular *GD* is drawn and extended to form the chord *DGT* of the circle. Now, divide *EZ* at a point *H*, such that the ratio of *EH* to *HZ* is equal to the ratio of *AG* to *GB*. Draw the chord *KHL* through the point *H* such that it is perpendicular to the diameter *EZ*.

I say that the ratio of DT *to* KL *is equal to the ratio of* AB *to* EZ.

[23] The formulation by Ibn al-Samḥ is similar to that using in the ninth century by the Banū Mūsā (*ibid.*, at the end of the proof of Proposition 6, see Chapter I) and al-Kindī (R. Rashed, 'Al-Kindī's Commentary on Archimedes' *The Measurement of the Circle*', *Arabic Sciences and Philosophy*, 3, 1993, pp. 3–53; Arabic: p. 50, 9–11; English: p. 41).

Proof: Let us join *AD*, *DB*, *EK* and *KZ*. The ratio of *AB* to *BG* is equal to the ratio of *EZ* to *ZH*, the triangle *ADB* is a right-angled triangle, and *DG* is a perpendicular. The ratio of *AG*[24] to *BG* is therefore equal to the ratio of the square of *AG* to the square of *GD*, as mentioned by Euclid in the sixth <book> of his work. Similarly, the ratio of *EH*[25] to *ZH* is equal to the ratio of the square of *EH* to the square of *HK*. The ratio of the square of *EH* to the square of *HK* is therefore equal to the ratio of the square of *AG* to the square of *GD*. The ratio of *AG* to *GD* is therefore equal to the ratio of *EH* to *HK*. <Therefore, the ratio of *AG* to *EH* is equal to the ratio of *GD* to *HK*, and the ratio of *AG* to *EH* is equal to the ratio of *GB* to *HZ*, and therefore to the ratio of *AB* to *EZ*.> Consequently, the ratio of *AB* to *EZ* is equal to the ratio of *DG* to *HK*, and *DT* is equal to twice *DG* and *KL* is equal to twice *HK*. We have therefore shown that a chord of the <first> circle, perpendicular to *AB*, relative to a chord of the <second> circle, perpendicular to *EZ*, has the same ratio as one diameter to the other, provided that these diameters are both divided in the same ratio. That is what we wanted to prove.

<Lemma 2> Let us consider two circles passing through *AB* and *GD*. Divide *GD* at *K* and *M*, and *AB* at *E* and *H*, such that the ratio of *AE* to *AB* is equal to the ratio of *GK* to *GD*, and such that the ratio of *BH* to *AB* is equal to the ratio of *DM* to *GD*. Draw the two straight lines *EZ* and *HT* at right angles, and draw the two straight lines *LK* and *NM* in the same way. The ratio of *TH* to *NM* will then be equal to the ratio of the diameter of one to the diameter of the other and, similarly, the ratio of *ZE* to *LK* will be equal to the ratio of the diameter of one to the diameter of the other. Let us join *TE*, *ZH*, *NK* and *LM*.

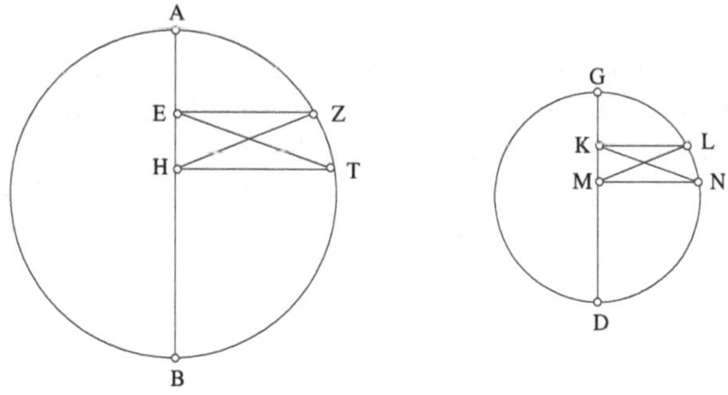

Fig. VI.2

[24] *AB* in the MS.
[25] *EZ* in the MS.

I say that the two triangles ZHE *and* LMK *are similar, and that* EHT *and* KMN *are also similar.*

Proof: The ratio of *AE* to *AB* is equal to the ratio of *KG* to *GD*, and the ratio of *AB* to *AH* is equal to the ratio of *GD* to *GM*. Then, considering the *ex-aequali* ratios,[26] the ratio of *AE* to *AH* will be equal to the ratio of *GK* to *GM*. If we separate,[27] the ratio of *AE* to *EH* will be equal to the ratio of *GK* to *KM*. If we permute,[28] the ratio of *AE* to *GK* will be equal to the ratio of *EH* to *KM*. But the ratio of *AE* to *GK* is equal to the ratio of one diameter to the other diameter, and therefore the ratio of *EH* to *KM* is equal to the ratio of one diameter to the other diameter, and the ratio of one diameter to the other diameter is equal to the ratio of *HT* to *MN*. The ratio of *EH* to *KM* is consequently equal to the ratio of *HT* to *MN* as the angles *H* and *M* are equal. The two triangles are therefore similar. The two triangles *EHZ* and *KML* may be shown to be similar in the same way. That is what we wanted to prove.

<Lemma 3> Let us consider two circles passing through *AB* and *KL*. Let us mark a point *G* at any position <on *AB*>. Let us draw a straight line from this point to *AB*, and let this be the straight line *HG*. Let us divide the diameter *KL* <in the same ratio> at the point *N*. Let us draw a line from this point to the circle, <namely the straight line *NO*>, enclosing with the straight line *NL* an angle equal to the angle *HGB*.

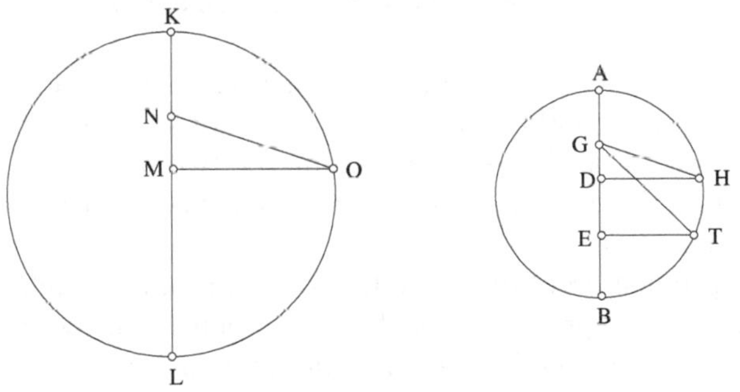

Fig. VI.3

[26] *ba-yaḥas ha-shiwwuy*: by the equality ratio; in Arabic: *fī nisbat al-musāwā*. This is a translation of the Greek expression *di'isou logos* (*Elements*, V, Definition 17), and indicates a consideration of the ratio of the extreme terms in each of the two sequences of magnitudes.

[27] *ka-asher hivdalnu*. The verb used refers to the Euclidian expression 'separation of the ratio' (V, Definition 15): *hevdel ha-yaḥas*; in Arabic: *tafṣīl al-nisba*.

[28] *ka-asher hamironu*. The verb refers to *temurat ha-yaḥas*: The permutation of the ratio (V, Definition 12); in Arabic: *tabdīl (*or *ibdāl) al-nisba*.

I say that the ratio of HG to ON is equal to the ratio of one diameter to the other diameter.

Proof: From the point *O* let us draw a perpendicular onto *KL*. If the angle *ONL* is acute, this will be the straight line *OM*. Similarly, from the point *H* let us draw a perpendicular *HD* onto *AB*. I say that the ratio of *AD* to *DB* is equal to the ratio of *KM* to *ML*. Proof: If this were not the case, <there would be a point *E*, other than *D* on *AB* such that> the ratio of *KM* to *ML* would be equal to the ratio of *A E* to *E B*. Let us draw the perpendicular *ET* and join *TG*. As we have shown earlier, the triangle *TGE* is similar to the triangle *ONM* and the angle *TGE* is therefore equal to the angle *N*. We have assumed that the angle *N* is equal to the angle *HGB*. Therefore, the angle *HGB* would be equal to the angle *TGB* – *i.e.* the smallest would be equal to the largest. This is impossible. It is then impossible that the ratio of *AD* to *DB* is not equal to the ratio of *KM* to *ML*.

From what has been proved in the first proposition, we can deduce that the ratio of *HD* to *OM* is equal to the ratio of one diameter to the other diameter. Also, the ratio of *HD* to *OM* is equal to the ratio of *HG* to *ON*, as the two triangles are similar. The ratio of *HG* to *ON* is therefore equal to the ratio of one diameter to the other diameter, and the same applies to the ratio of <any pair of> straight lines situated in a similar way.[29] That is what we wanted to prove.

The second species of sections of a right cylinder whose bases are two circles.

When a right cylinder with bases consisting of two circles is cut by a plane which is not parallel to its base, the resulting section is that generated by fixing one side of a triangle and rotating the two remaining sides in the plane of the triangle <such that their sum remains constant> until it returns to its original position.

We shall provide proof of this in the following <paragraphs>, when we indicate among those properties of the figure generated by the movement of the triangle the one that is characteristic, and among those properties of the oblique section of the cylinder the one that is characteristic, and <shall verify> that the latter accords well with the indications that we have given regarding the figure generated by the movement of the triangle. We proceed here in the same way as we did with the section of the cylinder parallel to its base, when we showed a property that accorded with the property of a circle, *i.e.* we found a point such that all straight lines drawn from this point to the circumference are equal.

[29] *ʿal zeh ha-ḥiqquy.*

Let us, then, introduce all the necessary lemmas relating to the figure obtained by the movement of the triangle.

We say that the figure obtained by the movement of the triangle is called the *elongated circular*[30] figure, a name derived from its shape. It has a circular contour which is elongated. Neither the circularity, nor the extension in length characterise it uniquely. This name is required by the act of generating the figure, as the process used to construct it combines both a circular movement and a rectilinear movement, this being the extension in length.

The movement of the <common> extremity of the two sides which turn generates that which is called *the contour of the elongated circular figure*. The fixed side of the triangle is called *the central side*,[31] and the other two sides, those which rotate, are called *the movable sides*.[32] The triangle itself is called *the triangle of movement*.

From our description of the construction, it follows that the two movable sides will coincide with the central side during their rotational movement, forming a single straight line. The rectilinear and circular extension is then at a maximum, as is the amount by which one exceeds the other, the difference between them being equal to the total <length> of the central side. It is also clear that, as they rotate, one becomes larger as the other becomes smaller. The one that rotates towards its starting end becomes smaller, while that which rotates away from its starting end becomes larger with its increased length being taken from the other side. As one becomes larger and the other smaller, it follows that they will be equal at certain positions. This equality only occurs at two positions, either side of the central side.[33] In this case, they are called *the equal movable sides*. The perpendicular <onto the central side> from their <common> extremity then cuts the central side into two halves. This point constitutes *the centre of the figure*. It is the centre of two circles. One of these passes through the <other> extremity of the said perpendicular, which forms its half-diameter, and this circle is tangent to the figure. In the case of the other circle, the end of its diameter is located at the point where the two sides of the triangle coincide making a single straight line. At this point, the distance of each end from the centre is at its greatest. It also appears that this diameter is equal to <the sum of> the two movable sides, as it can

[30] *temunat me'uggal 'arokh*. The expression is a perfect translation of that of the Banū Mūsā: *al-shakl al-mudawwar al-mustaṭīl*. See the mathematical commentary.

[31] *ṣela' ha-merkaz*: the side of the centre.

[32] *ṣal'ey ha-sibbuv*: the sides of the rotation.

[33] *mi-shtey ha-pe'ot*: in each of the two directions. In Arabic: *fī kiltā al-jihatayn*.

be built up by bringing them together, considering the excess at each end relative to the central side with each of these ends exchanged with the other. This larger circle described on the curve is also tangent to it; they have in common the greatest diameter. The large circle is said to be *circumscribed*, and the small circle *inscribed.*

All the straight lines passing through the centre and cutting the curved figure are divided into two halves at the centre. These are called *the diameters.*[34] The greatest of these is the diameter that is common to <both the figure and> the circumscribed circle. The smallest of these is the diameter that is common to <both the figure and> the inscribed circle.

Any straight line cutting the curve without passing through the centre is called *a chord.*[35] Those chords which are cut into two halves by one or other of the two diameters, the greatest or the smallest, do so at right angles. And if <one of these two diameters> cuts a chord at a right angle, then it cuts it into two halves.

The straight line drawn at a right angle from one extremity of the central straight line,[36] and crossing the large circle, is called *the invariant <straight line>.*[37] The portion of this straight line falling within the elongated circular figure is called *the separated <straight line>.*[38]

<Proposition 1> In any elongated circular figure, four times the square of the invariant straight line plus the square of the central straight line is equal to the square of the large diameter.

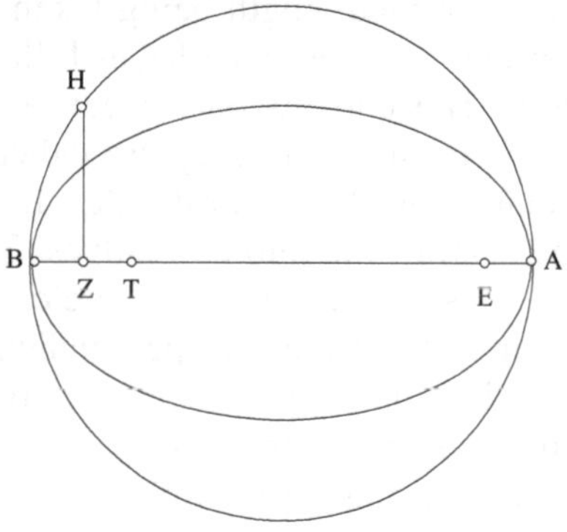

Fig. VI.4

[34] *qoṭer*; in Arabic: *quṭr*.

[35] *meytar*; in Arabic: *watar*.

[36] which we call a 'focus' of the ellipse.

[37] *<ha-qaw> ha-shaweh*: the equal straight line. In Arabic, almost certainly: *al-khaṭṭ al-musāwī*.

[38] *<ha-qaw> ha-nivdal*. In Arabic, almost certainly: *al-khaṭṭ al-munfaṣil*.

Example: Consider the elongated circular figure passing through *AB*. The central straight line is *EZ*, the circumscribed circle passes through *HB*, and the invariant straight line is *ZH*.

I say that four times the square of ZH *plus the square of* EZ *is equal to the square of* AB.

Proof: <On *ZE*>, mark *ZT* equal to *AE*. Then, *AT* is equal to *EZ* and *TZ* is equal to *ZB*. The product of *BZ* and *ZA*, taken four times, plus the square of *AT* is equal to the square of *AB*. The product of *ZB* and *ZA*, taken four times, is equal to four times the square of *ZH*. Four times the square of *ZH*, which is the invariant straight line, plus the square of *AT*, which is equal to the square of the central straight line, is therefore equal to the square of *AB*, which is the large diameter. That is what we wanted to prove.

<Proposition 2> In any elongated circular figure, the invariant straight line is equal to the half-small diameter, and the ratio of the separated straight line to the straight line forming proportion with the half-small diameter and the half-central straight line is equal to the ratio of the half-central straight line to the half-large diameter.

Example: Let the elongated circular figure be *ABGD*, of which the circumscribed circle is *ATD*, the central straight line is *EZ*, the invariant straight line is *ZT*, the separated straight line is *ZK*, the midpoint of the central straight line is the point *H*, the small diameter is *BHG*, and the ratio of *ZH* to *HB* is equal to the ratio of *BH* to *HL*,[39] such that the straight line *HL* forms proportion <with *BH* and *ZH*>.

I say that TZ *is equal to* HB, *and that the ratio of the separated straight line* ZK *to* HL *is equal to the ratio of* HZ *to* HD.

Proof: Let us join *TH* and *BZ*. Yet *TH* is half of the large diameter. These straight lines are therefore equal. It follows that <the sum of> the squares of *BH* and of *HZ* is equal to the <sum of the> squares of *TZ* and of *ZH*. Subtracting the square of *ZH*, which is common to both sides, it follows that the square of *TZ* is equal to the square of *BH*. In other words, *BH* is equal to *TZ*.

[39] The point *L* introduced in the statement only appears thereafter in the final paragraph of the proof.

It is stated that $HL = \frac{b^2}{c}$ and proved that $KZ = \frac{b^2}{a}$; hence $KZ = HL \cdot \frac{c}{a}$.

The position of *L* depends on the data:

If $b < c$, then *L* is between *H* and *B*.

If $b = c$, then *L* is at *B*.

If $b > c$, then *L* lies beyond *B*.

The straight line *HL* is in fact the third in the relationship *ZH*/*HB* = *HB*/*HL*.

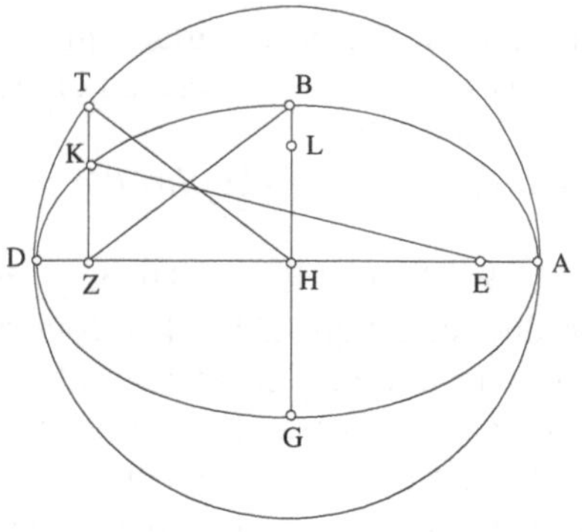

Fig. VI.5

Let us join *EK*. As *EK* plus *KZ* is equal to *AD*, the square of *EK* plus the square of *KZ* plus the double product of *EK* and *KZ* is equal to the product of *AD* by itself. Yet the product of *EK* by itself is equal to the product of *EZ* by itself plus the product of *ZK* by itself. Therefore the double product of *EK* and *KZ* plus twice the square of *KZ* plus the square of *EZ* is equal to the square of *AD*.

But the square of *AD* is equal to the quadruple square of *ZT*, which straight line is equal to *BH*, plus the square of *EZ*.[40] Therefore, the quadruple square of *ZT* plus the square of *EZ* is equal to the square of *EZ* plus the double square of *ZK* plus the double product of *EK* and *KZ*. Subtracting the square of *EZ*, which is common to both sides, it follows that the quadruple square of *ZT* is equal to the double product of *EK* and *KZ* plus the double square of *KZ*. The double square of *TZ* is therefore equal to one time the product of *EK* and *KZ* plus one time the square of *KZ*.

But the product of *EK* and *KZ* plus the square of *KZ* is equal to the product of *EK* and *KZ* together and *KZ*, and *EK* and *KZ* together is equal to *AD*. Consequently, the product of *AD* and *ZK*, the separated straight line, is equal to the double square of *TZ*, the invariant straight line.

As *TZ* is equal to *BH*, *BG* will be equal to twice *TZ*. It follows that the product of *TZ* and *BG* is equal to the double square of *TZ*. If this is so, then the product of *KZ* and *AD* is equal to the product of *TZ* and *BG*. In other words, the ratio of *KZ* to *ZT* is equal to the ratio of *BG* to *AD*. But *ZT* is equal to *HB*. Therefore, the ratio of *ZK* to *HB* is equal to the ratio of *BG* to *AD*, and also equal to the ratio of the half to the half, *i.e.* the ratio of *BH* to *HD*.

As the ratio of *ZK* to *HB* is equal to the ratio of *BH* to *HD*, and the ratio of *BH* to *HL* is equal to the ratio of *ZH* to *BH*, we have three magnitudes,

[40] See Proposition 1.

namely *KZ*, *BH* and *HL*, and an equal number of other magnitudes, namely *HZ*, *BH* and *HD*, where magnitudes taken in pairs from the first three are in the same ratio as magnitudes taken in pairs from the second three, in a perturbed order.[41] Therefore, considering the *ex-aequali* ratios, the ratio of *KZ*, which is the separated straight line, to *HL*, which is the proportion forming straight line, is equal to the ratio of *ZH*, which is half the central straight line, to *HD*, which is half the large diameter.

We have therefore shown that the invariant straight line is equal to half of the small diameter, and that the ratio of the separated straight line to the invariant straight line is equal to the ratio of the small diameter to the large diameter, and that the ratio of the separated straight line to the proportion forming straight line is equal to the ratio of the half-central straight line to the half-large diameter. That is what we wanted to prove.

<Proposition 3> In any elongated circular figure, if the two movable sides meet at any point other than the extremity of the small diameter, then the ratio of the largest movable side <to the straight line obtained by producing the largest of the straight lines cut from the central straight line by the foot of the perpendicular dropped from the point at which the two sides meet as far as> the proportion forming a straight line with the half-small diameter and the half-central straight line is equal to the ratio of the half-central straight line to the half-large diameter.

Example: Let *ABGD* be the elongated circular figure, *AB* the large diameter, *GD* the small diameter, *EZ* the central straight line, *EH* and *HZ* the movable sides, and *EL* and *ZK* the separated straight lines. A perpendicular *HT* is dropped from the point *H* <onto the large diameter>. Let the ratio of *TN* to *GM* be equal to the ratio of *GM* to *ME*. *TN*[42] is then the proportion forming the straight line.

[41] *yithallef ha-yahas <ba-shi'urim> ba-qedima we-'ihur*. In Arabic: *ikhtalafat al-nisba fī al-aqdār bi-al-taqdīm wa-al-ta'khīr*. This expression refers to the use of 'the perturbed proportion' (*Elements*, V, Definition 18): Two consecutive terms in the second sequence of magnitudes are in the same ratio as two consecutive terms in the first sequence, with the order of the terms in the second sequence being always offset relative to the order in the first sequence.

[42] The point *N* introduced in the statement is defined by $TN = \frac{GM^2}{ME} = \frac{b^2}{c}$. The length *TN* is equal to the length *HL* in the previous proposition. The length of *TN* does not depend on the point *H* chosen. The position of *N* is associated with the projection *T* of the point *H* on *AB*. The point *N* may be between *M* and *B*, at *B*, or beyond *B*.

The result obtained remains valid if the point *H* coincides with one of the vertices.

In this case as well, the straight line *TN* is obtained from the relationship *TN/GM* = *GM/ME*, where *M* is the centre of the ellipse.

I say that the ratio of EH *to* EN *is equal to the ratio of* EM *to* MA.

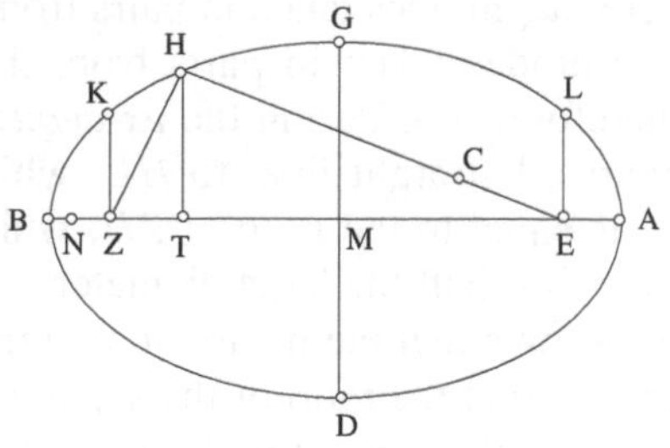

Fig. VI.6

Proof: The point H, at which the two movable sides meet, must lie either between the two points G and K, either on the point K, or between the two points K and B.

To begin, let this point lie between the points G and K. The angle H is then either a right angle, an obtuse angle, or an acute angle.

To begin, let this angle be a right angle.

<The product of> EH by itself plus the product of HZ by itself plus twice the product of EH and HZ is equal to the square of AB. Yet the product of EH by itself plus the product of HZ by itself is equal to the square of EZ, as the angle H is a right angle. Therefore, the double product of HE and HZ plus the square of EZ is equal to the square of EZ plus the quadruple square of the invariant straight line.[43] Subtracting the square of EZ, which is common to both sides, the double product of EH and HZ is equal to the quadruple square of the invariant straight line. One time the product of EH and HZ is therefore equal to the double square of the invariant straight line.

The product of LE, the separated straight line, and AB is the double square of the invariant straight line.[44] The product of EL and AB is therefore equal to the product of EH and HZ. In other words, the ratio of LE, which is equal to the straight line KZ, to EH is equal to the ratio of HZ to AB, which is equal to EH and HZ taken together.

Separating, inverting and composing,[45] the ratio of CH, <where C is a point on EH, such that EC is equal to EL>, to EH is equal to the ratio of EH to EH and HZ taken together. The product of CH and EH and HZ

[43] See Proposition 1.

[44] See Proposition 2.

[45] *ka-asher hivdalnu, ḥillafnu, hirkavnu*: when we have separated, inverted, composed. In Arabic, these operations on ratios are designated as *tafṣīl*, *'aks*, *tarkīb al-nisba*.

together is therefore equal to the product of *EH* by itself. But the product of *EH* by itself is equal to the product of *ZE* and *ET*, as the ratio of *ZE* to *EH* is equal to the ratio of *EH* to *ET*.

If this is the case, then the product of *CH* and *AB*, which is equal to *EH* and *HZ* together, is equal to the product of *ZE* and *ET*. In other words, the ratio of *CH* to *ET* is equal to the ratio of *EZ* to *AB*. But the ratio of *EZ* to *AB* is equal to the ratio of *EM* to *MA*. The ratio of *CH* to *ET* is therefore equal to the ratio of *EM* to *MA*. We have shown earlier that the ratio of *EM* to *MA* is equal to the ratio of *EL*, the separated straight line, to *TN* the proportion forming straight line. It follows that the ratio of *CH* to *ET* is equal to the ratio of *EL* to *TN*.

Composing, the ratio of *EH* to *EN* is then equal to the ratio of *CH* to *ET*,[46] which itself is equal to the ratio of *EM* to *MA*. Consequently, the ratio of *EH* to *EN* is equal to the ratio of *EM* to *MA*. That is what we wanted to prove.

Now let the angle *EHZ* be obtuse.

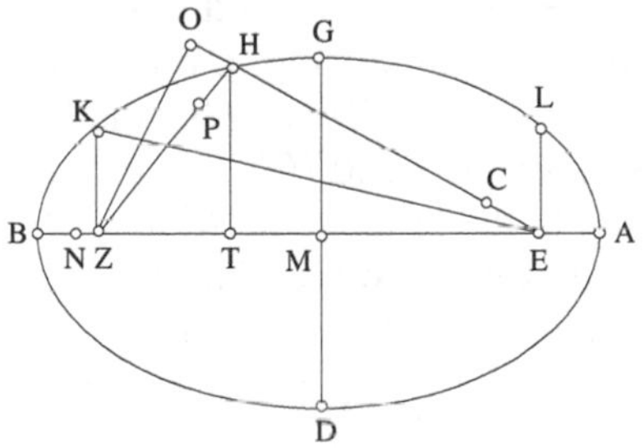

Fig. VI.7

Let us draw the perpendicular *ZO* <onto the extended straight line *EH*>. The quadruple square of the invariant straight line plus the square of *EZ* is equal to the square of *AB*.[47] The square of *AB* is equal to the square of *EH* plus the square of *HZ* plus twice the product of *EH* and *HZ*. Therefore, the quadruple square of the invariant straight line plus the square of *EH* plus the square of *HZ* plus twice the product of *EH* and *HO* is equal to the square of *EH* plus the square of *HZ* plus twice the product of *EH* and *HZ*.

[46] This is not, in fact, a composition of the ratios, but an application of Proposition V, 12: If $a/b = c/d$, then $ab = (a+c)/(b+d)$, antecedent with antecedent, consequent with consequent. We know that this particular operation on ratios does not have a specific designation in the *Elements*. Moreover, the author uses later a different term: he 'brings together' (= adds) the ratios.

[47] See Proposition 1.

Subtracting the squares of *EH* and *HZ*, which are common, the quadruple square of the invariant straight line plus twice the product of *EH* and *HO* is equal to twice the product of *EH* and *HZ*. In other words, the product of *EH* and *HZ* is equal to twice the square of the invariant straight line plus the product of *EH* and *HO*.

Now, let us make *HO* equal to *HP*, <where *P* is a point on *HZ*>. The product of *EH* and *HZ* is then equal to twice the square of the invariant straight line plus the product of *EH* and *HP*. Yet the product of *EH* and *HZ* is equal to the product of *EH*, *HP* and *PZ*. The product of *EH*, *HP* and *PZ* is accordingly equal to twice the square of the invariant straight line plus the product of *EH* and *HP*. Subtracting the product of *EH* and *HP*, which is common to both sides, the product of *EH* and *PZ* becomes equal to twice the square of the invariant straight line.

We have already shown that the product of *EL* and *AB* is equal to twice the square of the invariant straight line. The product of *EL* and *AB* is therefore equal to the product of *EH* and *PZ*; hence the ratio of *EL* to *EH* is equal to the ratio of *PZ* to <the sum of> *ZH* and *HE*.

Separating, inverting and composing, the ratio of *CH* to *EH* becomes equal to the ratio of <the sum of> *EH* and *HP*, the latter straight line being equal to *HO*, to <the sum of> *EH* and *HZ*. The product of *CH* and *EH* plus *HZ* is consequently equal to the product of *OE* and *EH*.

As the triangle *HET* is similar to the triangle *EOZ*, with both the angles *O* and *T* being right angles and the angle *E* being common to both triangles, the ratio of *OE* to *ET* is equal to the ratio of *ZE* to *EH*. For this reason, the product of *OE* and *EH* is equal to the product of *ZE* and *ET*. Under these conditions, the product of *CH* and *AB* is equal to the product of *ZE* and *ET*, and the ratio of *CH* to *ET* is equal to the ratio of *EZ* to *AB*, which is equal to the ratio of *EM* to *AM* <, the respective halves>.

We have already shown that the ratio of *LE* to *TN*, the proportion forming the straight line, is equal to the ratio of *EM* to *AM*. If we proceed by composition,[48] as before, the ratio of *EH* to *EN* becomes equal to the ratio of *EM* to *AM*. That is what we wanted to prove.

Now let the angle *EHZ* be acute.

Draw the perpendicular *ZO* <onto the straight line *EH*>. The quadruple square of the invariant straight line plus the square of *EZ* is equal to the square of *AB*, *i.e.* equal to the square of *EH* plus the square of *HZ* plus twice the product of *EH* and *HZ*. In other words, the product of each of the straight lines *EH* and *HZ* by itself plus twice the product of *EH* and *HZ* is

[48] See Note 46.

equal to the quadruple square of the invariant straight line plus the square of *EZ*.

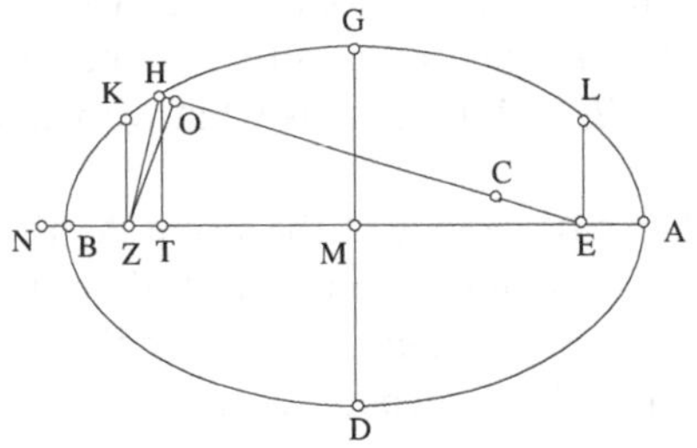

Fig. VI.8

Let us add twice the product of *EH* and *HO*. Then, twice the product of *EH* and *HO* plus twice the product of *EH* and *HZ* plus the square of *EH* plus the square of *HZ* is equal to the quadruple square of the invariant straight line plus the square of *EZ* plus twice the product of *EH* and *HO*. Now, the square of *EZ* plus twice the product of *EH* and *HO* is equal to <the sum of> the squares of *EH* and of *EZ*, as the angle *H* is acute. Therefore, twice the product of *EH* and *HZ* plus twice the product of *EH* and *HO* plus the square of *EH* plus the square of *HZ* is equal to the quadruple square of the invariant straight line plus the squares of *EH* and *HZ*.

Subtracting the squares of *EH* and *HZ*, which are common to both sides, twice the product of *EH* and *HZ* plus twice the product of *EH* and *HO* becomes equal to the quadruple square of the invariant straight line. In other words, twice the square of the invariant straight line is equal to the product of *EH* and *HZ* plus the product of *EH* and *HO*.

We have already shown that the product of *EL* and *AB* is equal to twice the square of the invariant straight line. Consequently, the product of *EH* and *HZ* plus the product of *EH* and *HO* is equal to the product of *EL* and <the sum of> *EH* and *HZ*. The ratio of *EL* to *EH* is therefore equal to the ratio of <the sum of> *HZ* and *HO* to <the sum of> *EH* and *HZ*.

Separating, inverting and composing, the ratio of *CH* to *EH* will be equal to the ratio of *OE* to <the sum of> *EH* and *HZ*. The product of *CH* and <the sum of> *EH* and *HZ* is therefore equal to the product of *EH* and *EO*.

Now, the product of *EH* and *EO* is equal to the product of *EZ* and *ET*, as the two triangles *EHT* and *EOZ* are similar. Consequently, the product of *CH* and <the sum of> *EH* and *HZ* is equal to the product of *EZ* and *ET*. Therefore, the ratio of *CH* to *ET* is equal to the ratio of *EZ* to *EH* and *HZ* together, which together are equal to *AB*, and also equal to the ratio of the half to the half. Therefore, the ratio of *CH* to *ET* is equal to the ratio of *EM* to *MA*.

We have already shown that the ratio of *LE* to *TN* is equal to the ratio of *ME* to *MA*. Bringing together <the ratios>,[49] the ratio of *EH* to *EN* will be equal to the ratio of *ME* to *MA*. That is what we wanted to prove.

If the two movable sides meet at the point *K*, then the perpendicular drawn <onto the large diameter> from *K* is *KZ*, as the angle *Z* is a right angle. The angle *K* is therefore acute. As the angle *Z* is a right angle, then the angle *E* is also acute. If a perpendicular <to the straight line *EK*> is drawn from the point *Z*, it will fall on the straight line *EK*. This perpendicular is *ZO*.

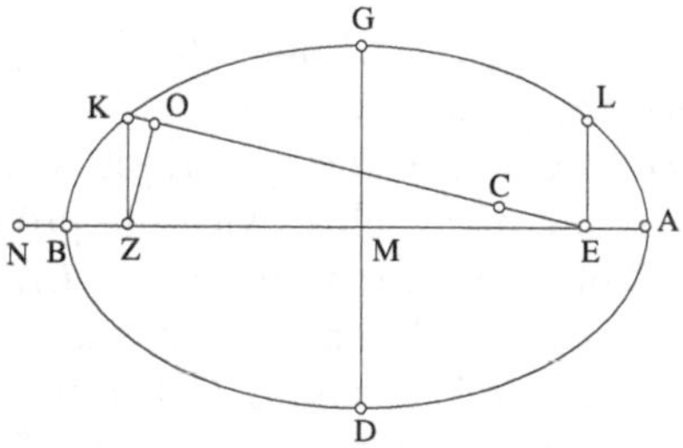

Fig. VI.9

We show, as was done in the previous case, that the product of *CK* and <the sum of> *EZ* and *KZ* is equal to the product of *KE* and *EO*. Yet the product of *KE* and *EO* is equal to the product of *EZ* by itself, as the two triangles *EOZ* and *EKZ* are similar. The angle *KZE* is a right angle, as is the angle *EOZ*, and the angle *E* is common to both. Consequently, the product of *CK* and *EK* and *KZ* together is equal to the product of *EZ* by itself. Therefore, the ratio of *CK* to *EZ* is equal to the ratio of *EZ* to *AB*, and also equal to the ratio of the half-central straight line to the half-large diameter.

But the ratio of *EL* to *ZN* is equal to the ratio of the half-central straight line to the half-large diameter.[50] Composing <the ratios>, the ratio of *KE* to *EN* will be equal to the ratio of *EZ* to *AB*, which is also the ratio of the half to the half. That is what we wanted to prove.

We have shown that if the two movable sides meet at the extremity of the separated straight line, then the central straight line relative to the large diameter is in the same ratio as the large diameter, less the separated

[49] *ka-asher qibbaṣnu*. In Arabic, almost certainly: *fa-idhā jamaʿnā*.

[50] See Proposition 2.

straight line relative to the central straight line, plus the proportion forming straight line.

If the two movable sides meet between the two points *K* and *B*, let us consider the two straight lines *EH* and *HZ*, the perpendicular *HT* drawn <onto the large diameter> from the point *H*, and the proportion forming the straight line *TN*. In this case too, the ratio of *EH* to *EN* will be equal to the ratio of the half-central straight line to the half-large diameter. The proof is the same as that given previously in the third case. As the angle *HZE* is obtuse, the angle *H* is acute and the angle *E* is also acute. Consequently, the perpendicular drawn <onto *EH*> from *Z*, *ZO*, falls on the straight line *HE*.

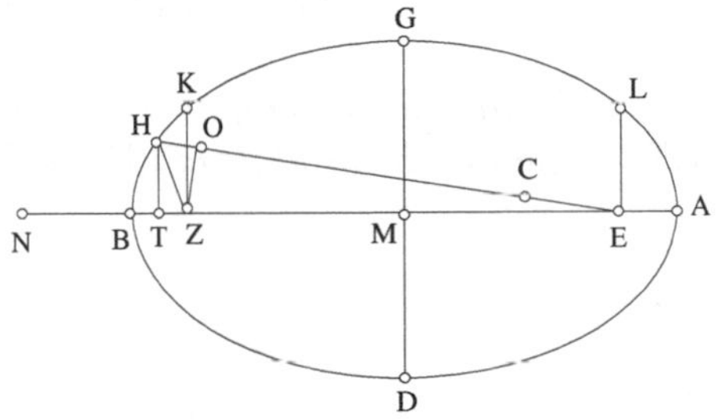

Fig. VI.10

In the same way, we can show that the product of *CH* and <the sum of> *EH* and *HZ* is equal to the product of *EH* and *EO*. The product of *EH* and *EO* is equal to the product of *EZ* and *ET*, as the triangle *EOZ* is similar to the triangle *EHT*. The product of *CH* and the sum of *EH* and *HZ* is therefore equal to the product of *EZ* and *ET*, and the ratio of *CH* to *ET* is equal to the ratio of *EZ* to the sum of *EH* and *HZ*, which sum is equal to the large diameter, this latter ratio being equal to the ratio of *EM* to *MA*, <the halves>. But the ratio of *EM* to *MA* is equal to the ratio of *EL* to *TN*.

Bringing <the ratios> together, the ratio of *EH* to *EN* is thus equal to the ratio of *EM* to *MA*. That is what we wanted to prove.

We have now completed our examination of this question in all its parts, as nothing remains beyond that which has been mentioned[*]. Praise be to God; may He be blessed, exalted and glorified.

* Gad Freudenthal, in his revision of the Hebrew text has proposed that this should read *ki lo hishlimuha bney shakir*: 'as the sons of Shākir did not complete it', in place of *ki [lo] ha-shlemut ke-fi she-zakhar*: 'as nothing remains beyond that which has been

<**Proposition 4**> In any elongated circular figure, if the two movable sides meet at any point other than the extremity of the small diameter, and if a chord of the circumscribed circle passing through the point at which the two movable sides meet falls at a right angle on the large diameter, then the square of half the chord plus the amount by which the half-small diameter exceeds the distance along the chord between the point at which the two movable sides meet and the foot of the chord on the large diameter is equal to the product of one of the two movable sides and the other.

Example: Let the elongated circular figure be *ABGD*, and let the circumscribed circle pass through *ANB*. Let the two movable sides be *EH* and *HZ*, the central straight line be *EZ*, the separated straight line be *ZK*, and the invariant straight line be *EL*. From the point *H*, draw a perpendicular onto *AB*, that is *HT*. Extend it in the circle until the two points *N* and *P*.

I say that the square of NT *plus the amount by which the square of* GM *exceeds the square of* HT *is equal to the product of* EH *and* HZ.

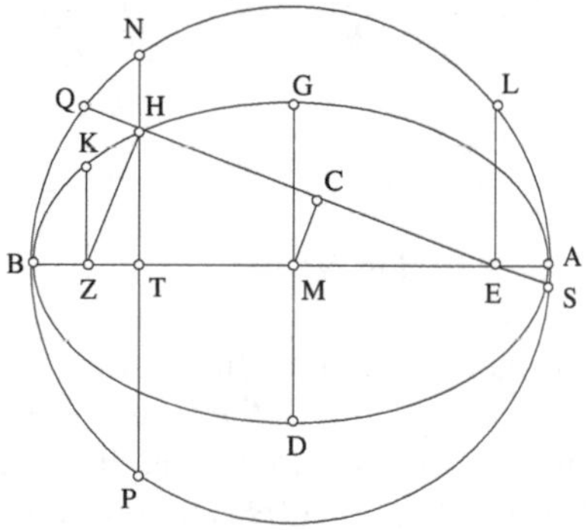

Fig. VI.11

Proof: The point *H*, at which the two movable sides meet, must lie either between the two points *G* and *K*, exactly on the point *K*, or between the two points *K* and *B*.

To begin, let this point lie between the points *G* and *K*. The angle *H* may be a right angle, an obtuse angle, or an acute angle.

To begin, let this angle be a right angle.

Extend *EH* to the points *S* and *Q*, so as to form a chord of the circumscribed circle. As shown in the first part of the preceding question, it can be proved that the product of *EH* and *HZ* is equal to twice the square of the invariant straight line, *EL*.

---

mentioned', translated in the text. This confirms the result obtained by the analysis of the text contents given in the mathematical commentary. [Note added later]

But the product of the invariant straight line *EL* by itself is equal to the product of *AE* and *EB*, which is equal to the product of *SE* and *EQ*.

Now, *ES* is equal to *HQ*. If the perpendicular *MC* <to *EH*> is drawn from the point *M*, it will be parallel to *ZH* because the angle *C* is a right angle as is the angle *H*. Therefore, the ratio of *EC* to *CH* is equal to the ratio of *EM* to *MZ*. Now, *EM* is equal to *MZ*, therefore *EC* is equal to *CH*, and also *CS* is equal to *CQ*. Subtracting *EC* and *CH*, it follows that *SE* is truly equal to *HQ*.

The product of *HQ* and *HS* is thus equal to the product of *SE* and *EQ*, and the product of *SE* and *EQ* is equal to the square of the invariant straight line.[51] The product of *HQ* and *HS* is therefore equal to the square of the invariant straight line. But the product of *HQ* and *HS* is equal to the product of *NH* and *HP*. Consequently, the product of *NH* and *HP* is equal to the square of the invariant straight line.

We have shown that the product of *EH* and *HZ* is equal to twice the square of the invariant straight line. In other words, the product of *EH* and *HZ* is equal to the square of the invariant straight line plus the product of *NH* and *HP*. But the square of the invariant straight line is the square of *HT* plus the amount by which the square of the invariant straight line exceeds the square of *HT*. Under these conditions, the product of *NH* and *HP* plus the square of *HT* plus the amount by which the square of the invariant straight line, which is equal to *GM*, exceeds <the square of *HT*> is equal to the product of *EH* and *HZ*. Yet, the product of *NH* and *HP* plus the square of *HT* is equal to the square of *NT*. The square of *NT* plus the amount by which the square of *GM* exceeds the square of *HT* is thus equal to the product of *EH* and *HZ*. That is what we wanted to prove.

Now let the angle *H* be obtuse.

The perpendicular onto the straight line *EH* from the point *Z* falls outside the point *H*. *ZO* is this perpendicular. As shown in the second part of the preceding proposition, it can be proved that the product of *EH* and *HZ* is equal to twice the square of the invariant straight line *EL* plus one time the product of *EH* and *HO*.

[51] *E* being the midpoint of the chord produced by extending *LE*.

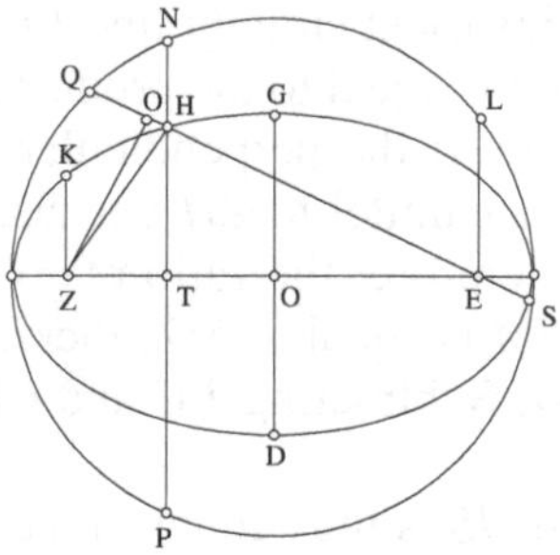

Fig. VI.12

As in the preceding section, we can show that the product of *EH* and *HZ* is equal to the square of the invariant straight line plus the product of *QO* and *OS* plus the product of *OH* and *HE*. But the product of *QO* and *OS* plus the product of *OH* and *HE* is equal to the product of *QH* and *HS*. The product of *QH* and *HS* plus the square of the invariant straight line is then equal to the product of *EH* and *HZ*.

Now, the product of *QH* and *HS* is equal to the product of *NH* and *HP*. The product of *NH* and *HP* plus the square of the invariant straight line is thus equal to the product of *EH* and *HZ*. But the product of *NH* and *HP* plus the square of *HT* is equal to the square of *NT*. Therefore, the square of *NT* plus the amount by which the square of *MG* exceeds the square of *HT* is truly equal to the product of *EH* and *HZ*. That is what we wanted to prove.

Now let the angle *H* be acute.

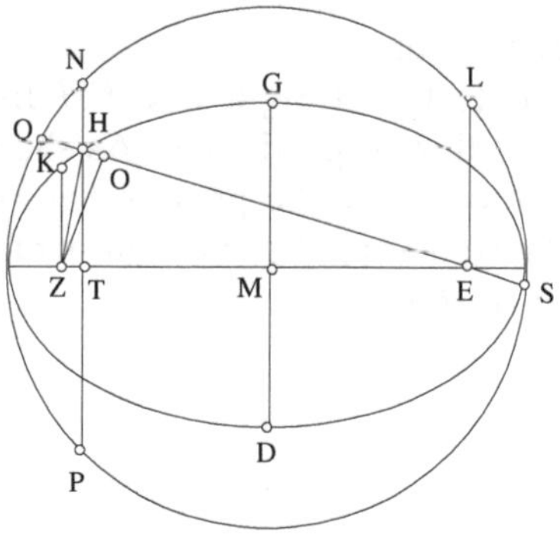

Fig. VI.13

As the angle *EZH* is acute, and less than the angle *KZE* <which is a right angle>, and the side *EH* is greater than the side *HZ*, then the angle *E* is acute. The perpendicular drawn onto the straight line *EH* from the point Z therefore falls on the straight line *EH* inside the triangle <*EHZ*>. *ZO* is this perpendicular.

As shown in the third part of the preceding question, it can be proved that one time the product of *HE* and *HZ* plus one time the product of *EH* and *HO* is equal to twice the square of the invariant straight line.

We can also show, as was done in the first part of this proposition, that the product of *QO* and *OS* is equal to the square of the invariant straight line, as *OQ* is equal to *ES*. The product of *QO* and *OS* plus the square of the invariant straight line is then equal to the product of *EH* and *HZ* plus the product of *EH* and *HO*. But the product of *QO* and *OS* is equal to the product of *QH* and *HS* plus the product of *OH* and *HE*. Therefore, the product of *QH* and *HS* plus the product of *OH* and *HE* plus the square of the invariant straight line is equal to the product of *EH* and *HZ* plus the product of *OH* and *HE*. Subtracting the product of *OH* and *EH*, which is common to both sides, it follows that the product of *QH* <and *HS* plus the square of the invariant straight line is equal to the product of *EH* and *HZ*>.

The proof is completed in the same way as in the two previous parts; the square of *NT* plus the amount by which the square of *MG* exceeds the square of *HT* is equal to the product of *EH* and *HZ*. That is what we wanted to prove.

If the movable sides meet at the point *K*, or at any point between the two points *K* and *B*, then the angle <*ZHE*> of the movable triangle is acute, and the perpendicular falls inside the triangle.[52] We then proceed as in the third part of this proposition. With the help of the Creator.

<Proposition 5> If a perpendicular is drawn from any point marked on the outline of any elongated circular figure onto the small diameter, then the square of this perpendicular is equal to the square of the part of this perpendicular that is contained within the inscribed circle plus the square of the amount by which the half-large diameter exceeds the smallest of the movable sides beginning at the point that was selected.

Example: Consider the elongated circular figure passing through *ABGD*. The inscribed circle is *BDT*. Draw a perpendicular from the point *K* on the small diameter and extend it until it reaches the elongated curve at the point *H*, cutting the circle at the point *T*. *EZ* is the central straight line. Join *EH* and *HZ*, which are the movable sides. Let the amount by which *AM* exceeds *HZ* be *HN*.

[52] See the mathematical commentary, Section 6.2.5, Proposition 4, for the case where *H* is at the vertex of the ellipse.

I say that the square of KH *is equal to the sum of the square of* KT *and the square of* HN.

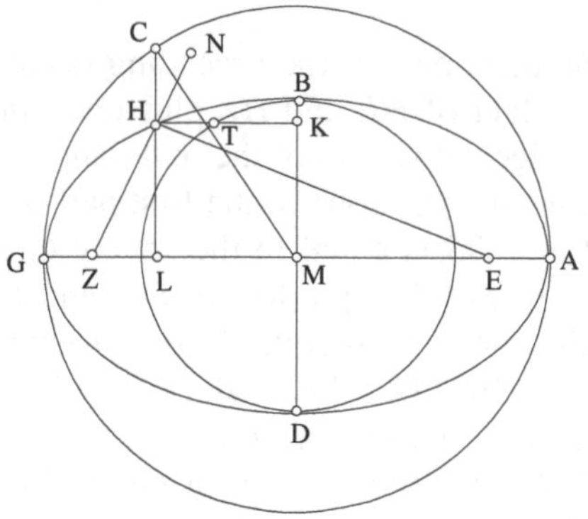

Fig. VI.14

Proof: Let us draw the circumscribed circle *ACG*. From the point *H*, we draw the perpendicular *HL* onto the diameter *AZ*. We extend this to *C*, and join *MC*.

By virtue of that which we have already proved,[53] the square of *CL* plus the amount by which the square of *BM* exceeds the square of *KM*, which is the square of *HL*, is equal to the product of *EH* and *HZ*. Now, the amount by which the square of *BM* exceeds the square of *KM* is equal to the square of *KT*; in fact, the product of *DK* and *KB* plus the square of *KM* is equal to the square of *BM*. Therefore, the amount by which the square of *BM* exceeds the square of *KM* is equal to the product of *DK* and *KB*. But the product of *DK* and *KB* is equal to the square of *KT*, and it follows that the square of *KT* is truly equal to the amount by which the square of *BM* exceeds the square of *KM*.

The square of *CL* plus the square of *KT* is equal to the product of *EH* and *HZ*. Adding the square of *HN*, the square of *CL* plus the square of *KT* plus the square of *HN* is equal to the product of *EH* and *HZ* plus the square of *HN*. But the product of *EH* and *HZ* plus the square of *HN* is equal to the square of *ZN*,[54] which is equal to half of the large diameter, as *ZN* is half of the large diameter. Consequently, the sum of the square of *CL*, the square of *KT*, and the square of *NH*, is equal to the square of *CM*, which is half the

[53] See Proposition 4.

[54] From the hypothesis $HN = AM - HZ$, we can derive $ZN = HZ + HN = AM$. We also have $HE + HZ = 2AM$; hence

$$HE \cdot HZ + HZ^2 = 2AM \cdot HZ \text{ and } HZ^2 - 2AM \cdot HZ + AM^2 + EH \cdot HZ = AM^2,$$

i.e. $(HZ - AM)^2 + EH \cdot HZ = AM^2$ and therefore $HN^2 + EH \cdot HZ = ZN^2$.

large diameter. But the square of *CM* is equal to <the sum of> the squares of *CL* and *LM*.

As this is so, <the sum of> the two squares of *CL* and *LM* is equal to the square of *CL* plus the square of *KT* plus the square of *NH*. Subtracting the square of *CL*, which is common to both sides, it follows that the square of *LM* is equal to the square of *KT* plus the square of *NH*. But the square of *LM* is equal to the square of *KH*. The square of *KH* is thus equal to <the sum of> two squares, that of *KT* and that of *HN*. That is what we wanted to prove.

<Proposition 6> If a perpendicular is drawn from any point marked on the outline of any elongated circular figure onto the small diameter, then the ratio of this perpendicular to its portion that is contained within the inscribed circle is equal to the ratio of the large diameter to the small diameter.

Example: Let *ABGD* be the figure, and let the inscribed circle be *GHD*. A point *E* is chosen at any point on the outline of the elongated circular figure, and a perpendicular *EHZ* is drawn from it <onto the small diameter>.

I say that the ratio of EZ *to* HZ *is equal to the ratio of the large diameter to the small diameter, that is equal to the ratio of the half-large diameter to the half-small diameter.*

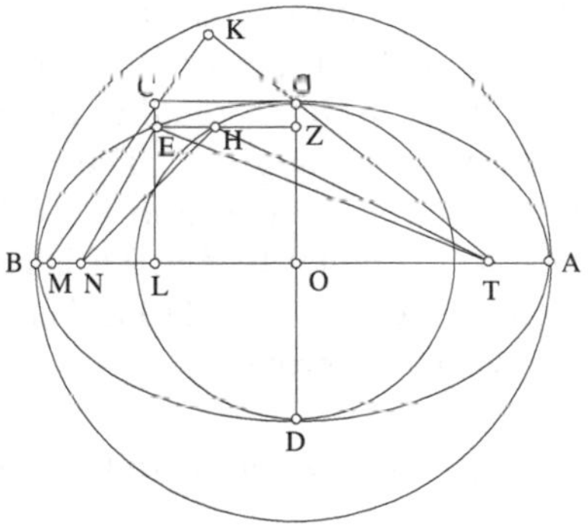

Fig. VI.15

Proof: *TN* is the central straight line and *O* is the centre of the circle. Let us join *TG*. It has been shown that *TG* is equal to half the large diameter. We extend it to *K*, such that *KT* is equal to *TE*. It is clear that *GK* is equal to the amount by which *AO* exceeds *EN*.[55] Let us draw the

[55] $ET + EN = 2AO$ and it is stated that $KT = TE$. Hence $KT = 2AO - EN$. But $KT = TG + GK$ and $TG = AO$; therefore $GK = AO - EN$.

perpendicular *EL* <onto *AB*>. Set <the point *M* on *AB* such that> the ratio of *ML* to *OG* is equal to the ratio of *OG* to *OT*. *ML* then forms proportion <with *OG* and *OT*>. Extend *LE* in a straight line as far as *C*, and join *CG*.

From that which has previously been proved, we can show that the ratio of *KT*, which is equal to *ET*, to *TM* is equal to the ratio of *TO*, which is half of the central straight line, to *TG*, which is equal to half of the large diameter.

These straight lines include the same angle. The triangle *GTO* is therefore similar to the triangle *TKM*. Consequently, the angle *K* is equal to the angle *O*. Now, the angle *O* is a right angle; the angle *K* is then a right angle. There remains the angle *TGO*, which is equal to the angle <of the vertex> *M*. The angle *L* is a right angle, the same as the angle *O*. The triangle *CML* is therefore similar to the triangle *GTO*. The ratio of *TO* to *OG* is thus equal to the ratio of *CL* to *LM*. But the ratio of *TO* to *OG* is equal to the ratio of *GO* to *LM*. Therefore, the ratio of *GO* to *LM* is equal to the ratio of *CL* to *LM*. Consequently, *OG* is equal to *CL*, and is parallel to it. The straight line *GC* is therefore equal to the straight line *OL* and is parallel to it. In addition, *OL* is equal to *ZE*, and therefore *GC* is equal to *ZE*. It has also been proven that the square of *ZE* is equal to <the sum of> the squares of *ZH* and *GK*.[56] But the square of *GC* is equal to <the sum of> the squares of *GK* and *KC* as the angle *K* is a right angle. Therefore, the <sum of the two> squares of *GK* and *KC* is equal to <the sum of> the two squares of *GK* and *ZH*. Let us remove the square of *GK*, which is common to both sides; the square of *KC* will be equal to the square of *ZH*. Yet, *GC* is parallel to *OL* and the angle *KGC* is equal to the angle *T*. Moreover, the angle *O* is a right angle, as is the angle *K*. The triangle *KGC* is therefore similar to the triangle *GOT*. Hence, the ratio of *GC* to *KC* is equal to the ratio of *TG* to *GO*; yet, *KC* is equal to *ZH*, *ZE* is equal to *GC*, and *GT* is equal to *AO*. Consequently, the ratio of *EZ* to *ZH* is equal to the ratio of *AO* to *OG*. That is what we wanted to prove.

<The ellipse as a plane section of a cylinder>

In this introduction, we have established all that is necessary in relation to the curve obtained by the movement of a triangle. We shall now proceed to all that is necessary in relation to the section of a cylinder.

If a right cylinder is cut <by a plane> not parallel to its base, the point at which this <sectioning> plane meets the axis of the cylinder is called *the*

[56] See Proposition 5.

centre of the figure, and the straight lines cutting the ellipse and passing through its centre are *the diameters*.

If the cylinder is cut by a plane parallel to its base and passing through the centre of the figure <obtained above>, the section will be a circle, and this circle will be inscribed within the ellipse. If the ellipse is rotated about the common straight line, that which is in the plane of the circle, then the circle will lie within the ellipse, and the straight line that is common to both sections is a diameter of the ellipse, the smallest of all the diameters. The diameter that crosses it a right angle is the largest of all the diameters. The diameters closest to the small diameter are smaller that those further away, closer to the large diameter.

The circle inscribed within the ellipse, and whose diameter is equal to the small diameter, is equal to the base circle of the cylinder of which the ellipse is a section. The base of the cylinder is the same as the section passing through the centre of the ellipse parallel to the said base. And the circle cutting the cylinder <and passing> through the centre of the ellipse will be inscribed within the ellipse as it has the smallest diameter in common with the latter.

<Proposition 7> Consider an ellipse and its inscribed circle. From a point on the small diameter that is not the centre, draw a straight line parallel to the large diameter until it reaches the outline of the ellipse. Then the ratio of this straight line that is contained within the ellipse to its portion that is contained within the circle is equal to the ratio of the large diameter to the small diameter.

Example: Let the ellipse be *ABGD* and let the inscribed circle be *GEZ*. The point *N* is the centre of the circle and the centre of the ellipse. *AENZB* is the large diameter and *DNG* is the small diameter, which is the same as the diameter of the circle. Mark anywhere on *DG* a point *H*. From this point, draw a straight line *HKT* parallel to the straight line *AB*. This line will then be perpendicular to the diameter *GD*.

I say that the ratio of HT *to* HK *is equal to the ratio of* AB *to* GD.

Proof: Let us imagine a cylinder on the circle *GEDZ* and imagine that the straight line *DG* is held fixed, so that it acts as a pivot.[57] Let us imagine the movement of the ellipse *DAG* around this pivot towards the surface of the cylinder so that it reaches the surface in the position of *DLMG*. The straight line *NL* is the same as the large diameter *NA*, and *HM* is the same as *HT*. Let us join *E* and *L*, *K* and *M*.

[57] *kush*. In Arabic, possibly: *miḥwar*.

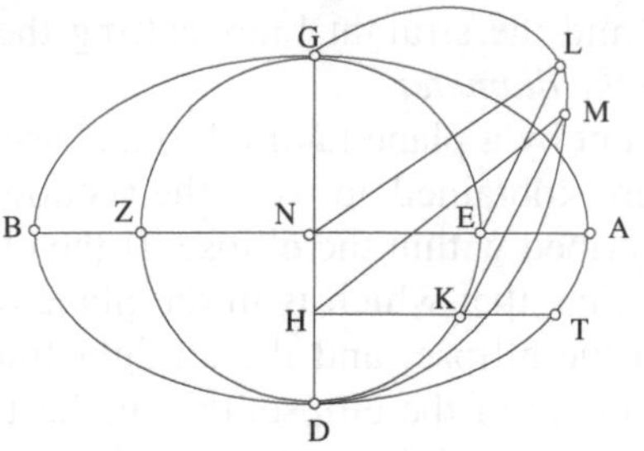

Fig. VI.16

As the angle *KHG* is a right angle, the angle *MHG* will be the same as the initial configuration does not change. The straight line *GN* is perpendicular to the plane *ENL*, and any plane passing through the straight line *NG* will be perpendicular to the plane *ENL*, as this has been mentioned by Euclid.[58] Similarly, any plane passing through the straight line *HG* will be perpendicular to the plane *KHM*.

Yet, the plane of the circle passes through the straight line *NG*, and therefore each of the planes *LEN* and *HKM* is perpendicular to the plane of the circle. The lateral surface of the cylinder is placed at a right angle on the plane of the circle, and that common sections such as *LE* and *MK* are perpendicular to the plane of the circle. Two perpendiculars to the same plane must be parallel to each other. *LE* is therefore parallel to *MK*. As the angle *LNG* is a right angle, as is the angle *MHG*, then the straight line *LN* is parallel to the straight line *MH*. Similarly, the straight line *EN* is parallel to the straight line *KH*, and both these lines are in the plane of the ellipse. Consequently, the sides of the triangle *LNE* are parallel to the sides of the triangle *MKH*, and the angles in these two triangles are equal. The two straight lines *LN* and *NE* contain the angle *ENL*, the two straight lines *MH* and *HK* contain the angle *KHM*, and the straight lines are not in the same plane. The two angles *ENL* and *KHM* are therefore equal. Hence, it can be shown that the angles in the triangle *LNE* are equal to the angles in the triangle *MKH*. The two triangles are therefore similar.

The ratio of *LN* to *NE* is therefore equal to the ratio of *MH* to *KH*. We know that *LN* is equal to *AN* and *MH* is equal to *HT*. But the ratio of *AN* to *NE* is equal to the ratio of *TH* to *HK*. That is what we wanted to prove.

<Porism> And we have shown that the square of the straight line from the extremity of the large diameter to the circle, which diameter cuts the

[58] *Elements*, XI, Definition 4.

circle into two halves and passes through the centre, plus the square of half the diameter of the circle is equal to the square of half the large diameter.

This may be deduced from the fact that the triangle *LEN* is a right-angled triangle and, as a result, the square of *NE*, which is half the small diameter, plus the square of *LE*, which is the straight line in question, is equal to the square of *LN*, which is half of the large diameter.

<Proposition 8> And I say:[59] If a <circumscribed> circle is constructed on the ellipse *ABGD*, and a point marked upon the outline of the ellipse from which a perpendicular is drawn onto the large diameter and extended as far as the circumference, as is the case of the perpendicular *HEZ*, then I say that the ratio of *ZH* to *EH* is equal to the ratio of *AL* to *LG*.

Proof: Let us fix an inscribed circle, that is circle *GTD*. From the point *E* <on the outline the ellipse>, draw a perpendicular *ETK* onto the small diameter. The point *L* is at the centre. Join *L* and *T*, and extend the straight line from that point to the point *N* on the circumference <of the circumscribed circle>. *LN* cuts *HZ* at the point *M*.

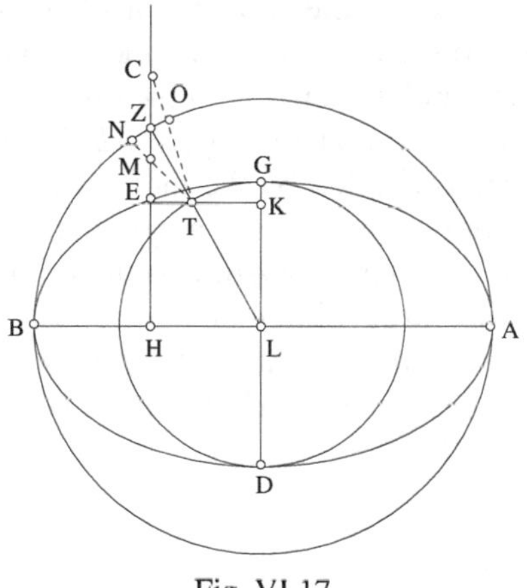

Fig. VI.17

The triangle *LKT* is similar to the triangle *TEM*, as each of them has a right angle and the two straight lines *LM* and *TE* cross each other. By composition, it follows that the ratio of *EK* to *KT* is equal to the ratio of *ML* to *LT*. We have already shown that the ratio of *EK* to *KT* is equal to the ratio of *AL*, which is half the large diameter, to *LT*, which is half the small

[59] The presence of this first person singular pronoun and the formulation that follows it, the rhetoric of which is closer to an exposition (ecthesis) rather than a statement (protasis) of the proposition, seems to indicate the absence of a statement as such.

diameter, in other words the ratio of *NL* to *LT*. As a result, *NL* is equal to *ML*, which is contradictory and impossible.[60]

Similarly, we can show that the straight line <*LT*> cannot pass above the point *Z*. If that were possible, it would pass through *O*. It could be extended and *HZ* could be extended such that they would meet at the point *C*. Proceeding as before, we can then shown that *CL* is equal to *OL*, which is contradictory and impossible.

It is therefore impossible that the extended *LT* could pass through any point other than the point *Z*.

TE is parallel to *LH*. Therefore, the ratio of *ZL*, which is half the large diameter, to *LT*, which is half the small diameter, is equal to the ratio of the perpendicular *ZH* to *EH*, the portion of the latter that lies within the ellipse. That is what we wanted to prove.

<Proposition 9> We wish to prove that the elongated circular figure generated by the movement of a triangle is equal to the oblique section of a cylinder when its large diameter is equal to the large diameter of the elongated figure generated by the movement of a triangle and when its small diameter is equal to the other small diameter, and that each coincides with the other at all parts and is identical to the other.

Example: Let us suppose that the figure *ABGD* is, in all its parts, an elongated circular figure generated by the movement of a triangle, and that the figure *ZHKT* is the oblique section of a cylinder. The straight line *ZH* is assumed to be equal to the diameter *AB*, and the diameter *TK* equal to the diameter *GD*. *AB* is the large diameter and, similarly, *ZH* is the large diameter. *GD* is the small diameter and, similarly, *TK* is the small diameter.

I say that the two round figures ABGD *and* ZHTK *are equal, and that each coincides with the other.*

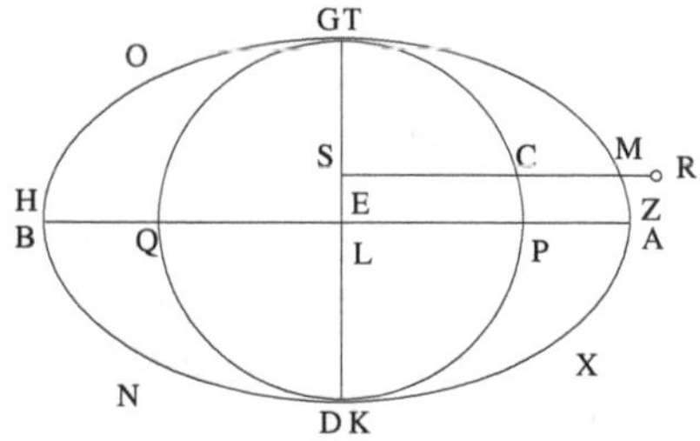

Fig. VI.18

[60] The point *M* cannot therefore lie between *T* and *N*. In other words, the straight line *LT* does not cut the straight line *HZ* below *Z*.

Proof: The diameters *AB* and *GD* cut each other at right angles and into two halves at <the point> *E*. Similarly, the diameters *ZH* and *TK* cut each other at right angles and into two halves <at the point *L*>. If we superimpose the figure *ABGD* on the figure *ZHTK*, with the straight line *AB* and the straight line *ZH* coinciding, as do the point *A* and the point *Z*, and the point *B* and the point *H*, then the point *E* will coincide with the point *L* as each figure is divided in half at *E* and at *L*. The straight line *GD* will coincide with the straight line *TK*, as each is perpendicular to the other diameter, and the point *G* coincides with the point *T* as *ET* is equal to *EG*. The point *D* will thus coincide with the point *K*.

The arc *AG* can be superimposed on the arc *ZT*, the arc *AD* on the arc *ZK*, the arc *DB* on the arc *KH*, and the arc *GB* on the arc *TH*. If they did not coincide when superimposed, the arcs *ZRT*, *TOH*, *ZXK*, and *KNH* would appear thus,[61] insofar as that were possible.

Let us therefore construct a circle on the diameter *TK*. This will be inscribed in both figures as they are both constructed on the small diameter. Let this be the circle *TPKQ*. Let us mark a point *S* at any point on the straight line *TL* and draw from it a straight line *SCM* that is parallel <to *PQ*>.

As the arc *TMZ* is part of an elongated circular figure generated by the movement of a triangle, the ratio of *MS* to *SC* is equal to the ratio of *ZL* to *LT*, that is to the ratio of the large diameter to the small diameter, as we have proved previously. Moreover, as the arc *TRZ* is part of the oblique section of a cylinder, then the ratio of *RS* to *RC* is also equal to the ratio of *ZL* to *LT*, as we have proved this previously. But the ratio of *ZL* to *LT* is equal to the ratio of *MS* to *SC*; therefore the ratio of *SC* to *RS* is the same as that to *MS*. It follows that *MS* is equal to *RS*. This is contradictory and impossible.

It is therefore impossible that the elongated circular figure *ABGD* does not coincide with the circular figure *ZHTK*. Each coincides perfectly with the other and is identical to it. That is what we wanted to prove.

Let us now prove this property in another way, different from the apagogic method.

Let *ABGD* be the <elongated> circular figure generated by the movement of a triangle, and let *MNCO* be that which is a section of a cylinder, the two <pairs of> diameters, small and large, are common to both. The large diameter *GD* is equal to the large diameter *CO*, and the small diameter *AB* is equal to the small diameter *MN*.

[61] See Fig. VI.18.

I say that the figure ABGD *coincides with the figure* MNCO.

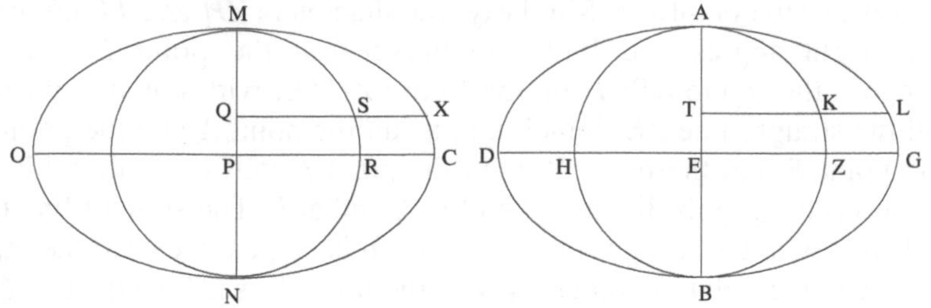

Fig. VI.19

Proof: Let us inscribe a circle in each of the two figures. These circles *ABH* and *NMR* are equal as the two <small> diameters are equal. Let us mark a point *L* anywhere on the arc *AG* and draw a straight line *LT* from it parallel to the straight line *GE*.

Let us cut the straight line *PQ* <on *PM*> in the same way as the straight line *ET*. From the point *Q*, let us draw a straight line *QSX* parallel to the straight line *CRP*. From that which we have proved previously in relation to the figure generated by the movement of a triangle, the ratio of *LT* to *TK* is equal to the ratio of *GE* to *EZ*, and equal to the ratio of *CP* to *PR*, as each <of the two first straight lines> is respectively equal to one of the <two> others. From that which we have proved previously in relation to a section of the cylinder, it is clear that the ratio of *CP* to *PR* is equal to the ratio of *XQ* to *QS*. Therefore, the ratio of *LT* to *TK* is equal to the ratio of *XQ* to *QS*. By permutation, the ratio of *XQ* to *LT* will be equal to the ratio of *QS* to *TK*. But *SQ* is equal to *TK* as the two circles are equal and *ET* is equal to *PQ*:[62] *XQ* is therefore equal to *LT*.

If we superimpose the circular figure *ABGD* on the circular figure *MNCO*, the points *ABGD* will coincide with the points *MNCO*, the point *T* will coincide with the point *Q*, and the straight line *LT* will coincide with the straight line *XQ* as each of them is perpendicular to the diameter of the circles. Therefore, the point *L* will coincide with the point *X* as the straight line *TL* is equal to the straight line *QX*.

We have thus proved that any point taken on the outline of the figure *ABGD* will coincide with a point on the outline of the figure *MNCO*. That is what we wanted to prove.

[62] See Lemma 1.

<Proposition 10> We wish to show how to construct an <elongated> circular figure from the movement of a triangle, such that it is equal to a given oblique section of a cylinder.

Consider the section *ABGD*, in which the small diameter is *AB* and the large diameter is *GD*. If we wish to construct an <elongated> circular figure from the movement of a triangle, such that it is equal to the section *ABGD*, we should draw any straight line *ZH* and divide it into two halves. A perpendicular *LT* is drawn from *L* that is equal to *EA*. Place the <point *K* on the straight line *ZH* such that the> square of *LT* plus the square of *LK* <are> equal to the square of *EG*.

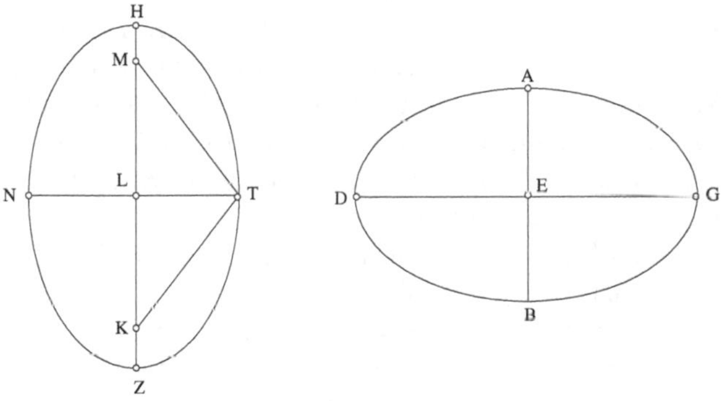

Fig. VI.20

From that which we have previously proved, it is clear that *LK* is equal to the perpendicular produced from the extremity of the large diameter <of the section> onto the <plane of the> circle which cuts the figure at its centre[64].

Let us join *T* and *K*. It is clear that *TK* is equal to *EG*. We make *LM* also equal to *LK*. We join *T* and *M*. It follows that *TM* is equal to *EG*. As a result, <the sum of> the straight lines *KT* and *TM* is equal to the straight line *GD*. Let us rotate the straight lines *KT* and *TM*, keeping the straight line *KM* fixed, until they return to their original position. This movement generates the figure *TZNH*.[63]

I say that the figure TZNH *is the same as the figure* ABGD.

Proof: At the end of the movement of the triangle *TKM* which brings the point *T* onto the point *Z*, <the sum of> the straight lines *MZ* and *ZK* is equal to <the sum of> the straight lines *KT* and *TM*. When the triangle

[63] The points *Z* and *H* introduced above thus lie on the ellipse. It was not rare, in writings of the time, to introduce certain magnitudes and define them later.

rotates <sufficiently> to bring the point *T* onto the point *H*, <the sum of> the straight lines *KH* and *HM* is equal to <the sum of> the straight lines *KT* and *TM*. The <sum of the> straight lines *KH* and *HM* is therefore equal to <the sum of> the two straight lines *MZ* and *ZK*. Subtracting *KM*, which is common to both sides, it follows that twice *MH* is equal to twice *ZK*, and therefore *ZK* is equal to *MH*. As <the sum of> *MZ* and *ZK* is equal to <the sum of> the straight lines *KT* and *TM* and *ZK* is equal to *MH*, then *ZH* is equal to <the sum of> *KT* and *TM*. But <the sum of> *KT* and *TM* is equal to *GD*. Therefore *GD* is equal to *ZH*. The large diameter is thus equal to the large diameter.

TL is produced in a straight line as far as *N*, such that <the sum of> *KN* and *NM* is equal to *KT* and *TM*. As *ML* is equal to *LK*, *LN* is common, and the angles at <the vertex> *L* are equal, it follows that *KN* is equal to *NM*. Therefore *KN* is equal to *KT*, and the angles to which they form the chords are right angles. Consequently, the squares of *KL* and *LT* are equal to the two squares of *KL* and *NL*. Subtracting the square of *KL*, which is common to both sides, there remains the square of *LN*, which will be equal to the square of *LT*. Therefore the straight line *LN* is equal to *LT*, and *LT* is equal to *AE*. It follows that *AB* is equal to *TN*.

The two circular figures *ABGD* and *TNZH*, the first being a section of a cylinder and the second being the figure generated by the movement of a triangle, have the same small and large diameters. They are therefore equal and they coincide.

We have shown that the circular figure *TZNH*, generated by the movement of a triangle, is truly equal to the section *ABGD*. That is what we wanted to prove.

<Proposition 11> We wish to show how to find a section of a cylinder such that it is equal to a <given elongated> circular figure generated from the movement of a triangle.

Let us fix *ABGD* the <elongated> circular figure generated by the movement of a triangle. Its large diameter is *AB*, the small diameter is *GD* and the centre is at the point *Z*. Let us imagine a circle, from all those circles that could, by superposition, be inscribed within the figure *ABGD*. Let us imagine a right cylinder on this circle. Let us imagine a plane *LK* which sections the cylinder and cuts its axis. Let us imagine a circular section *HTNC* parallel to the base with *HT* being its diameter and *M* its centre.

<On the perpendicular to the plane of the circle passing through *T*>, we cut off a straight line *TP* equal to *ZE*,[64] and let *NC* be <a diameter> perpendicular to *HT*. As any three points define a plane, there will be a plane passing through *NCP*. Extend this plane until it sections the cylinder. Let *PCON* define the outline <of this section>.

I say that the section PCON *is the same as the circular figure* ABGD.

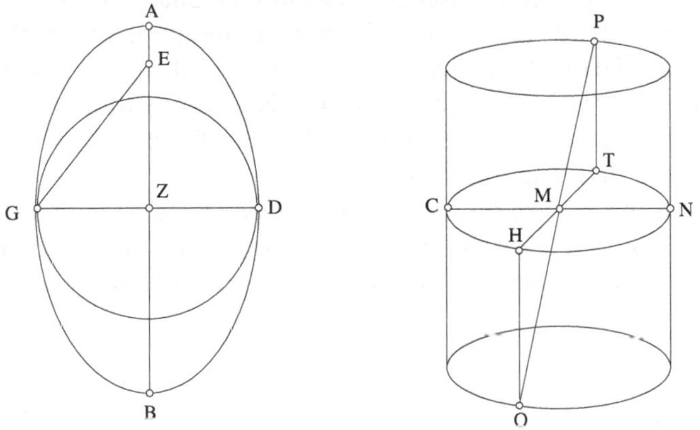

Fig. VI.21

Proof: The small diameter *GD* is equal to the small diameter *NC* <of the section>. The <sum of the> two squares of *TP* and *MT* is equal to the square of *PM*, and the two squares *MT* and *PT* are equal to the two squares *GZ* and *ZE*. We know that the squares of *GZ* and *ZE* are equal to the square of *GE*. The square of *GE* is therefore equal to the square of *PM*. It follows that *PM* is equal to *EG*. But *EG* is equal to *AZ*. Therefore *ZA* is equal to *PM*. Similarly, we can show that *ZB* is equal to *MO*.

The large diameter *AB* is thus equal to the large <diameter> *OP*. We have already obtained that the small diameters are equal. Consequently, the circular figure *ABGD* is the same as the section *PCON*. That is what we wanted to prove.

<Proposition 12> If any two consecutive chords are drawn in a quarter of an ellipse, beginning at the extremity of the large diameter and ending at <the extremity of> the small diameter, and if perpendiculars are drawn from <the extremities of these chords> onto the small diameter crossing the quarter circle inscribed within the ellipse, and if the chords associated with the arcs <of the circle> thus defined are also drawn, then two polygonal

[64] The point *E* is one of the foci of the ellipse.

surfaces will be generated, one inscribed within the ellipse and the other within the circle, such that the ratio of the area inscribed within the ellipse to the area inscribed within the circle is equal to the ratio of the large diameter to the small diameter.

Example: Let *ATK* be the quarter ellipse defined by two half-diameters. The half-large diameter is *KT* and the <half->small <diameter> is *AT*. Let *ADBT* be the quarter circle inscribed <within the quarter ellipse>. Let *T* be the centre. The chords are drawn within the ellipse, one of which is *KZ*. A perpendicular *ZDG* is drawn from *Z* onto *AT*, with the point *D* being on the circumference of the circle. The chord *DB* is drawn within the circle. Two <polygonal> surfaces are thus generated within the two figures, *KZGT* in the ellipse, and *BDGT* in the circle.

I say that the ratio of the polygonal surface inscribed within the ellipse, of which KZ *is one side, to the polygonal surface inscribed within the circle, of which* BD *is one side, is equal to the ratio of* TK *to* AT.

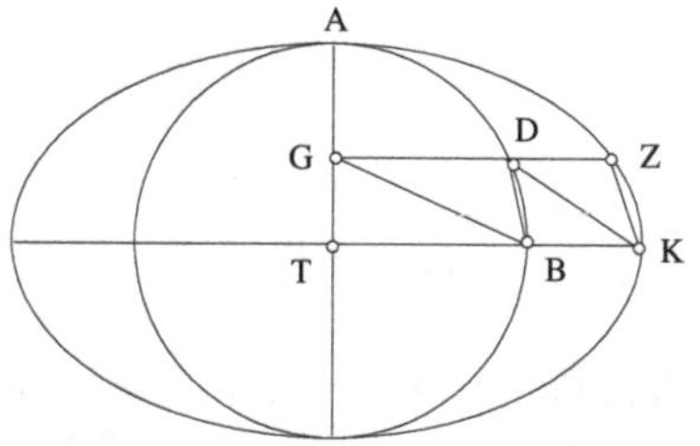

Fig. VI.22

Proof: Let us draw the straight lines *KD* and *BG*. These cut each of the two quadrilaterals[65] into two triangles. The two triangles *KZD* and *BDG* have equal heights. The ratio of the area of one to <the area of> the other is therefore equal to the ratio of the base *ZD* to the base *DG*. Similarly, the ratio of the triangle *KDB* to the triangle *BGT* is equal to the ratio of *KB* to *BT*.

Now, the ratio of *KB* to *BT* is equal to the ratio of *ZD* to *DG*, as this has been proved previously. It follows that the ratio of the <four> triangles, taken in pairs, is the same, and it remains the same when we added them. The ratio of the quadrilateral *KD* to the quadrilateral *BG* is therefore the same ratio, that is the ratio of *KB* to *BT*.

We proceed in the same way for all of the surfaces delimited by the chords and the perpendiculars. Their ratios, one to the other, will also be the same. Magnitudes that are in proportion remain in that proportion when

[65] *noṭeh*: trapezium, in the Euclidean sense (*Elements* I, Definition 22). In Arabic: *munḥarif*.

they are added together. The ratio of all the surfaces inscribed within the quarter ellipse *KTA* to all the surfaces inscribed within the quarter circle is thus equal to the ratio of *KT* to *BT*.

That which has been performed on the two quadrants may also be performed on the others which complete them. Hence, the ratio of the surface inscribed within the ellipse and contained within the chords defined by the arc of the half-ellipse and the large diameter to the surface inscribed within the inscribed half-circle and contained within the chords defined by the half-circumference and the diameter of the semi-circle is equal to the ratio of the large diameter to the small diameter.

The same holds true for the other half of the ellipse, which completes the whole figure, and for the remaining semi-circle inscribed in it and which completes the whole circle. The method is the same. The ratio of the entire figure inscribed within the ellipse to the entire figure inscribed within the circle is therefore equal to the ratio of the large diameter to the small diameter. That is what we wanted to prove.

<Proposition> 13. – We wish to show that the ratio of the area of the small circle, that which is inscribed within the ellipse, to the area of the ellipse itself is equal to the area of the small diameter to the large diameter.

Example: Let the ellipse be *ABGD*. Its large diameter is *AG*, and the small one is *BD*. The small circle, that which is inscribed within the ellipse, is *BWDE*, and its diameter is *WE*.

I say that the ratio of the surface ABGD *to the circle* EBWD *is equal to the ratio of* AG *to* BD.

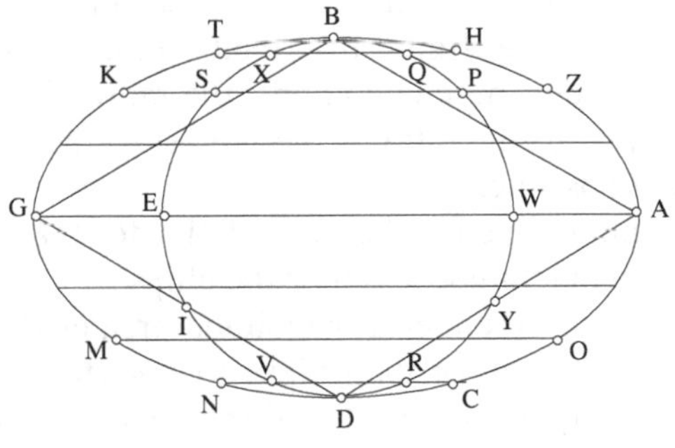

Fig. VI.23

Proof: The ratio of *WE*, the small diameter, to *AG*, the large diameter, is equal to the ratio of the circle *EBWD* to the ellipse *ABGD*, and it cannot be otherwise. If this were possible, this ratio would be equal to the ratio of the circle to a magnitude that is either smaller or larger than the ellipse.

To begin, let there be a magnitude that is smaller than the ellipse, and let this magnitude be the surface *L*.

The surface *L* is therefore less than the ellipse, the difference being the magnitude of the surface *U*. Let us join *AD*, *DG*, *GB* and *BA*. These straight lines on the elliptical surface enclose an area towards the centre that is greater than half of the latter, that is the lozenge *ABGD*. Now, let us divide each of the arcs so formed into two parts and draw the chords. These define surfaces from <the outline> of the surface towards the centre, that are greater than half <the elliptical segments defined by> the arcs. If we continue to proceed in the same way,[66] a surface will eventually be obtained that is smaller than the surface *U*. The <sum of the elliptical segments defined by the> arcs *AZ*, *ZH*, *HB*, *BT*, *TK*, *KG*, *GM*, *MN*, *ND*, *DC*, *CO* and *OA* is therefore smaller than the surface *U*. It follows that the polygonal surface thus generated and inscribed within the ellipse is greater than the surface *L*.

Draw a number of lines parallel to the large diameter from the extremities of the arcs that we have obtained by division. The circumference of the circle is then divided into an equal number of arcs at the points *P*, *Q*, *X*, *S*, *I*, *V*, *R* and *Y*. Now, draw in the chords as before. The ratio of the figure inscribed within the circle, and passing through *W*, *P*, *Q*, *B*, *X*, *S*, *E*, *I*, *V*, *D*, *R* and *Y*, to the <polygonal> figure inscribed within the ellipse, and passing through *Z*, *H*, *B*, *T*, *K*, *G*, *M*, *N*, *D*, *C*, *O* and *A*, is then equal to the ratio of *WE* to *AG*, which is in turn equal to the ratio of the circle to the surface *L*. The ratio of the figure inscribed within the circle to the figure inscribed within the ellipse is therefore equal to the ratio of the circle to the surface *L*. But the figure inscribed within the circle is smaller than the circle, and that the figure inscribed within the ellipse is greater than the surface *L*. Under these conditions, the ratio of the smallest to the largest would be equal to the ratio of the largest to the smallest, which is contradictory and impossible.

It is therefore impossible for the ratio of *WE* to *AG* to be equal to the ratio of the circle to a magnitude less than the ellipse.

I say that it can not be either a magnitude greater than the ellipse. If this were possible, the ratio of the ellipse to a surface smaller than the circle would then be equal to the ratio of the small diameter to the large

[66] The figure in the manuscript includes a number of errors, which have been corrected here. Moreover, it does not correspond to the text. There is a dividing point missing on each of the arcs *AZ*, *KG*, *GM* and *OA*. There are then 2^2 arcs on each quarter of the ellipse. The corresponding polygon has 2^4 sides. It should be noted that the figure is not essential to the argument made in the proof.

diameter. Let L be this magnitude, and let the circle exceed it by a magnitude U.

We proceed as before. The circumference of the circle is divided into parts and the chords are drawn. The <sum of the> surfaces delimited by these chords and the arcs will be less than the surface U. Under these conditions, the polygonal surface inscribed within the circle is greater than the surface L.

From the extremities of these arcs <on the circle>, we draw a number of straight lines parallel to the diameter, cutting the ellipse into a number of arcs. The chords defined by these elliptical arcs are then drawn. As before, we find that the ratio of the surface inscribed within the circle, which is greater than the surface L, to the polygonal surface inscribed within the ellipse, which is smaller than the ellipse, is equal to the surface L, which is smaller than the figure inscribed within the circle, to the ellipse, which is greater than the figure inscribed within it. Under these conditions, the ratio of the smallest to the largest would be equal to the ratio of the largest to the smallest. This is contradictory and impossible.

The ratio of the ellipse to a surface that is smaller than the circle cannot therefore be equal to the ratio of the small diameter to the large one.

We have thus proved that the ratio of the small diameter to the large diameter is not equal to the ratio of the circle to a surface that is either smaller than the ellipse, or greater than it.[67] Therefore, this ratio must be exactly equal to the ratio of the circle to the ellipse.

<Proposition> 14. – We wish to show that the ratio of an ellipse to any circle is equal to the ratio of the large diameter to a straight line whose ratio to the diameter of the circle is equal to the ratio of this diameter to the small diameter of the ellipse.

Let the ellipse be $ABGD$ and let the inscribed circle be AG. E is any other circle. <Let Z be a straight line such that> the ratio of Z to the diameter of E is equal to the ratio of this same diameter of E to AG.

I say that the ratio of the ellipse ABGD *to the circle* E *is equal to the ratio of the diameter* BD *to the straight line* Z.

Proof: The ratio of the circle AG to the circle E is equal to the ratio of the square of AG to the square of the diameter of E. Now, the ratio of the square of AG to the square of the diameter of E is equal to the ratio of the straight line AG to the straight line Z as all three straight lines are proportional.

[67] For more specific detail on the use of this apagogic method, see the mathematical commentary on Proposition 13 in Section 6.2.7, Comment 1.

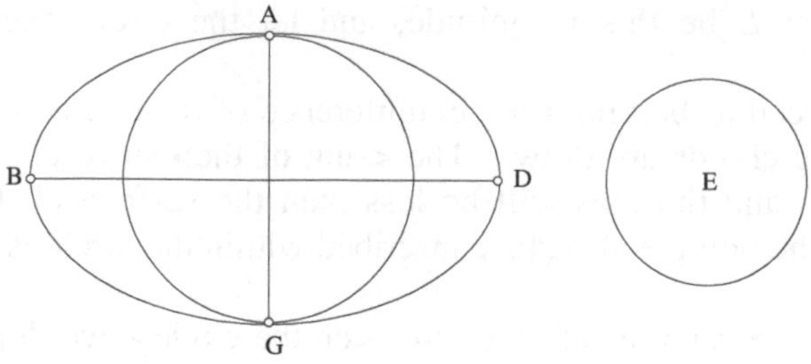

Fig. VI.24

The ratio of the circle *AG* to the circle *E* is therefore equal to the ratio of *AG* to *Z*, and the ratio of the ellipse *ABGD* to the circle *AG* is equal to the ratio of *BD* to *AG*. Under these conditions, and considering the equality ratio, the ratio of the ellipse *ABGD* to the circle *E* is equal to the ratio of *BD* to *Z*.

<Proposition> 15. – We wish to show that the ratio of the small circle to the ellipse is equal to the ratio of the ellipse to the large circle.

Example: Let *EGZD* be the large circle, *ABGD* the ellipse, and *AB* the small circle.

I say that the ratio of the circle AB *to the ellipse* ABGD *is equal to the ratio of the ellipse* ABGD *to the circle* EZGD.

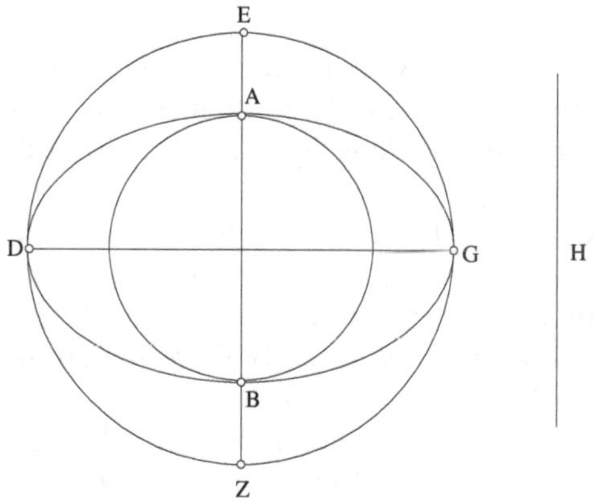

Fig. VI.25

Proof: We define <a straight line *H* such that> the ratio of the straight line *AB* to the straight line *GD* <is> equal to the ratio of *GD* to *H*. We have already proved that the ratio of *AB* to *H* is equal to the ratio of the square of *AB* to the square of *GD*. Now, the ratio of the square of *AB* to the square of

GD is equal to the ratio of the circle *AB* to the circle *EZ*. It follows that the ratio of the circle *AB* to the circle *EZ* is equal to the ratio of *AB* to *H*.

The ratio of the ellipse *ABGD* to a <certain> surface *T* is equal to the ratio of *GD* to *H*. Yet, we have proved that the ratio of the circle *AB* to the ellipse *ABGD* is equal to the ratio of *AB* to *GD*. Considering the equality ratios, it follows that the ratio of the circle *AB* to the surface *T* is equal to the ratio of *AB* to *H*. We have shown that the ratio of *AB* to *H* is equal to the ratio of the circle *AB* to the circle *EZ*. It follows that the ratio of the circle *AB* to the circle *EZ* and its ratio to the surface *T* are both the same. The surface *T* is therefore equal to the circle *EZ*. We have already stated that the ratio of the circle *AB* to the ellipse *ABGD* is equal to the ratio of the ellipse to the surface *T*. As the surface *T* is equal to the circle *EZ*, it follows that the ratio of the circle *AB* to the ellipse *ABGD* is equal to the ratio of the ellipse *ABGD* to the circle *EZ*. That is what we wanted to prove.

<Corollary 1> From this, it follows that the ratio of the small circle to the large circle is equal to the square of the ratio of the <small> circle to the ellipse and that the ratio of the large circle to the small circle is equal to the square of the ratio of the large circle to the ellipse.

<Corollary 2> It also follows that the ratio of the ellipse to the large circle is equal to the ratio of the small diameter to the large diameter. The ratio of the small circle to the ellipse is equal to the ratio of the small diameter to the large diameter, and it is also equal to the ratio of the ellipse to the large circle. Consequently, the ratio of the ellipse to the large circle is truly equal to the ratio of the small diameter to the large diameter.

<Proposition> 16. – Any ellipse is equal to the right-angled triangle having one of the sides enclosing the right angle equal to the circumference of the inscribed circle, and the second side equal to half the large diameter.

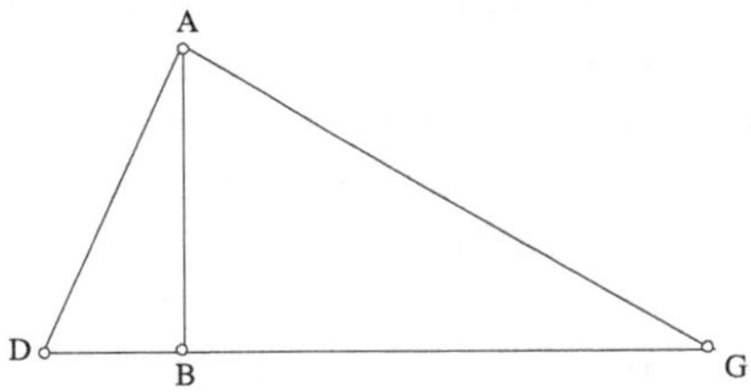

Fig. VI.26

Example: Let the circumference of the inscribed circle be *AB*, and let half the large diameter of the given ellipse be *BG*. The angle *ABG* is a right angle. Let us join *A* and *G*.

I say that the triangle ABC *is equal to the ellipse mentioned above.*

Proof: Let us extend *GB* in a straight line such that *BD* is equal to half the small diameter. As proved by Archimedes, the triangle *ABD* is equal to the small circle.[68] We have also proved that the ratio of the ellipse to the triangle *ABD* is equal to the ratio of the triangle *ABG* to the triangle *ABD*. The ellipse is therefore truly equal to the triangle *ABG*. That is what we wanted to prove.

We can use a similar proof to show that the ellipse is equal to the right-angled triangle having one of the sides enclosing the right angle equal to the circumference of the circle circumscribing the ellipse, and having the second side equal to half the small diameter. Understand this well.

<Corollary 1> From that which we have proved, it follows that if we take five and a half sevenths of the small diameter and multiply this by the large diameter, then we obtain the area of the ellipse.

The area of the triangle *ABG* is obtained by multiplying half of *AB* by *GB*. But half of *AB* is equal to three and one seventh times *BD*. Consequently, one quarter of half of *AB* is five and one half sevenths of *BD*. If this is so, then the product of five and one half sevenths of twice *BD* and twice *BG* is a measurement of the triangle *ABG*. That is what we wanted to prove.

<Corollary 2> If we know the area of an ellipse and one of the two diameters, then we know the other.

Let the known magnitudes be the larger of the two diameters and the area. Adding three elevenths of the area to itself and dividing the result by the known large diameter gives the unknown small diameter.

<Proposition> 17. – Any ellipse is equal to the circle whose diameter is the proportional mean of the two diameters of the ellipse.

Example: Let the small diameter be *A*, and the large diameter *G*. Take a straight line *B* which is the proportional mean between these two. The ratio of *A* to *B* is equal to the ratio of *B* to *G*. If three straight lines are proportional, then the circles to which these straight lines are diameters are also proportional. The ratio of the circle *A* to the circle *B* is therefore equal

[68] *On the Measurement of the Circle*, Proposition 1, and also Lemma c.

to the ratio of the circle *B* to the circle *G*, and the ratio of the circle *A* to the circle *G* is equal to the square of the ratio of the circle *A* to the circle *B*.[69]

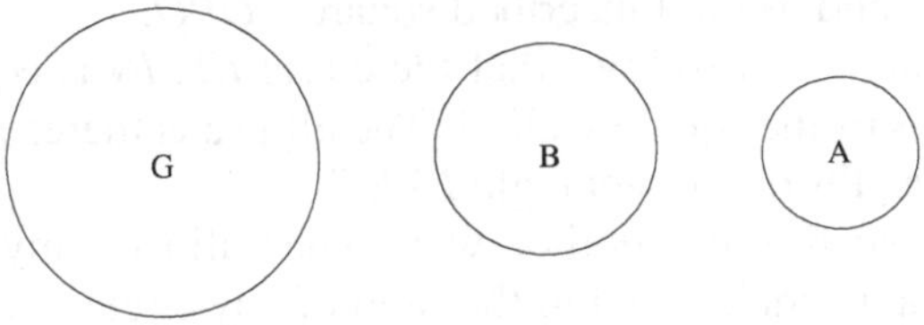

Fig. VI.27

But the circle *A* is that which is inscribed within the ellipse, and the circle *G* is that which circumscribes the ellipse. We have proved that the ratio of the circle inscribed within the ellipse to the circle circumscribed around the ellipse is equal to the square of the ratio of the circle inscribed within the ellipse to the ellipse itself.[70] It follows that the square of the ratio of the circle *A* to the circle *B* is equal to the square of the ratio of the circle *A* to the ellipse. Consequently, the ellipse is truly equal to the circle *B*. That is what we wanted to prove.

<Proposition> 18. – Any ellipse is equal to five and one half sevenths of the rectangle that is circumscribed around it.

Example: Let the ellipse be *ABGD*, the large diameter *BD*, and the small diameter *AG*. Let *EZHT* be the rectangle circumscribed around it.

I say that the ellipse ABGD *is equal to five and one half sevenths of the rectangle* EZHT.

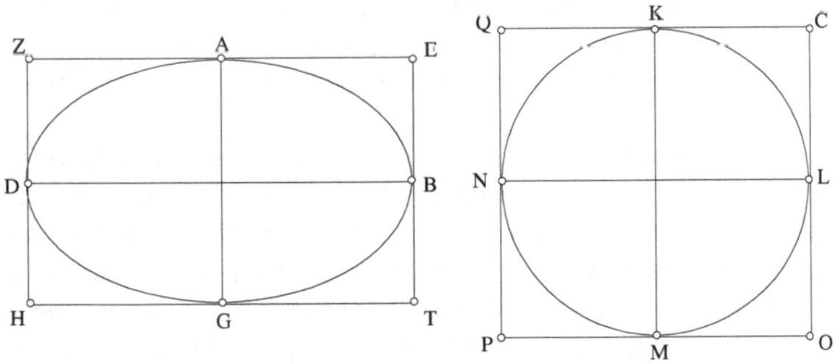

Fig. VI.28

[69] In the text, the letters *A*, *B* and *G* are used to designate both a segment and the circle having this segment as its diameter. The figure shows the three circles.

[70] See Proposition 15, Corollary 1.

Proof: Let us take a straight line *LN* that is the proportional mean between the straight lines *AG* and *BD*. Let us construct a circle *KLMN* on this straight line, and its circumscribed square *COPQ*.

From our lemmas, it follows that the circle *KLMN* is equal to five and one half sevenths of the square *COPQ*. The ellipse is therefore equal to five and one half sevenths of the rectangle *EZHT*.

We can prove this in another way. The ratio of any ellipse to the product of its diameters is equal to the ratio of any circle to the square of its diameter. The ratio of any ellipse to any circle is therefore equal to the ratio of the product of the diameters of the ellipse to the square of the diameter of the circle. The proof may be derived from these statements.

The ratio of any ellipse to the product of its diameters is equal to the ratio of any ellipse to the product of its diameters. That is what we wanted to prove.

He has formulated a premise in relation to the sections of an ellipse.

<Proposition> 19. – Given any ellipse and its inscribed circle, and equal chords in each <of these two figures> perpendicular to the large diameter, then the ratio of the segment cut by one of the two chords from the diameter that it crosses to the remainder of that diameter is equal to the ratio of the segment cut by the other <chord> from the other diameter that it crosses to the remainder of it.

Example: Let the ellipse be *ABGD*.[71] *MDEB* is the inscribed circle and *AMIEG* is the large diameter. *ME* is the diameter of the inscribed circle and *I* is its centre. Consider a chord *NR* in the circle and <a chord> *TO* in the ellipse, which are equal to each other and perpendicular to the <large> diameter. *NR* cuts the diameter at the point *Q*, and *TO* cuts the diameter at the point *W*.

I say that the ratio of WG *to* WA *is equal to the ratio of* EQ *to* QM.

Proof: <Let there be a cylinder whose base is a circle equal to the inscribed circle, and suppose that the ellipse *ABGD* is a plane section of this cylinder, as described above. Rotate the inscribed circle around the diameter *BD* to bring it into a plane parallel to the base>. Draw the straight line *KZ* parallel to the diameter *AG* from the point *Q* in the plane of the diameter *AG*, and divide the cylinder into two halves. The plane dividing the cylinder into two halves and passing through the diameter *AG* <therefore lies> between two parallel straight lines which themselves delimit two <other> parallel straight lines on the surface of the cylinder.

[71] The manuscript only shows one very confusing figure. The proof has been illustrated here with two figures, one in the space.

Consequently, the straight line *KZ* is equal to the straight line *AG* as the opposite sides of any parallelogram are equal.

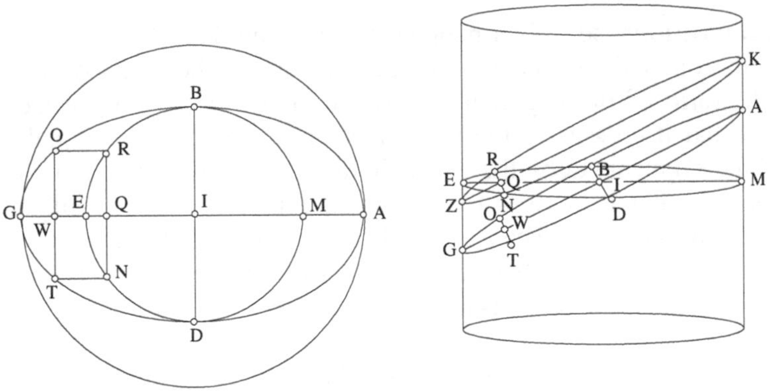

Fig. VI.29

Extend the plane in which the straight lines *KZ* and *QR* cross to form a plane sectioning the cylinder as the ellipse *KNZR*. We have therefore proved that the ellipse *KNZR* is equal to the ellipse *ABGD* and is parallel to it. In addition, the chord *NR* in one is equal to the chord *TO* in the other. The sagitta[72] *QZ* will therefore be equal to the sagitta *WG*, and the remainders of each of the two diameters will be equal. The section common to the plane mentioned above, which divides the cylinder into two halves and which contains the two parallel diameters, and to the surface of the cylinder consists of the straight line *KM* on one side and the straight line *EZ* on the other side.

The angle *KMQ* will be a right angle, as will the angle *ZEQ*. The surface of the cylinder stands at right angles to the plane of the circle and the common section, *i.e.* the straight line *KM*, is therefore perpendicular to the plane of the circle. Yet, any straight line drawn from *KM*, and which is in the plane of the circle, meets it at a right angle. It is for this reason that the angle *KMQ* is a right angle. Similarly, the angle *ZEQ* is also a right angle, for the same reason.

The two angles <at the vertex> *Q*, in the triangles are equal. Consequently, the triangles *KMQ* and *ZEQ* are similar. Their sides are therefore proportional. The ratio of *ZQ*, which is the chord of the right angle in one of the triangles, to *KQ* in the other is thus equal to the ratio of *EQ*, in the first triangle, to *QM* in the other. But *QK* is equal to *AW* and *QZ*

72 *i.e.* versed sine.

is equal to *WG*. The ratio of *GW* to *AW* is therefore truly equal to the ratio of *EQ* to *QM*. That is what we wanted to prove.

<Proposition> 20. – If it happens that the chords is drawn in the same way in the other direction, <*i.e.* perpendicular> to the <small> diameter, (then the ratio of the segment cut by the chord of the ellipse from the diameter that it crosses to the remainder of that diameter is equal to the ratio of the segment cut by the equal chord of the circumscribed circle from the diameter that it crosses to the remainder of it). The proof and the procedure are the same. There is no difference between the two cases. The ellipse would then be the base of a cylinder of which the <circumscribed> circle is a section. The proof is then completed <in the same way>.[73]

<b>[74] {There is no doubt that the knowledge of such arcs <of an ellipse> depends on knowing the sagitta, the chord, and one of the two diameters of the ellipse from which a segment has been taken. It is possible that the sagitta and the chord are common <to this ellipse and> to another ellipse.} **<b>**

<c> {From that which has already been proved, we shall now solve the following problem:

Let us assume that we have an elliptical circular figure passing through *ABGD*. Its large diameter is *AB*, and the circumscribed circle is *AEB*. A perpendicular *TH* is drawn from the large diameter, such that it cuts the ellipse at the point *K*.

I say that the ratio of HT *to* TK *is equal to the ratio of* AB *to* PO.

[73] See the mathematical commentary: Section 6.2.7, Proposition 20.

[74] From here until the statement of Proposition 21 (fol. 52^{v}, 9–22). The surviving manuscript text presents us with a problem. It appears to be a fairly incoherent collage of fragments. However, it is possible to distinguish three propositions, which we have designated <a>, <b> and <c>, and separated by braces without altering the layout of the text. <a> is an alternative proof of Proposition 20, the earlier proof being only a suggestion. <b> shows that there exist an infinite number of different ellipses having a given chord and a sagitta. <c> proposes an alternative proof of Proposition 8.

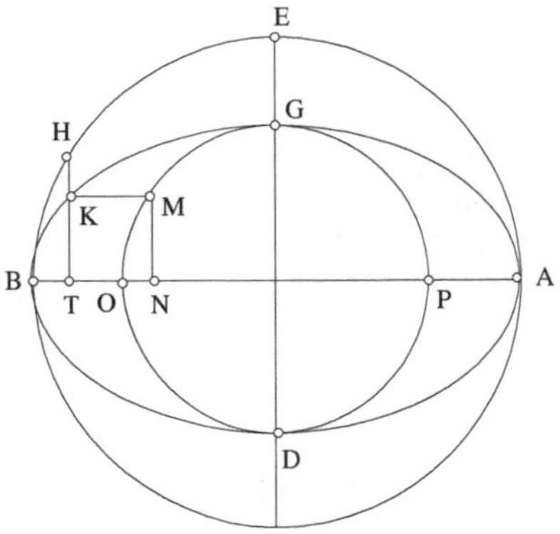

Fig. VI.30

Proof: Let us inscribe the circle *GODP* within the ellipse. From the point *K*, draw a straight line *KM* parallel to the straight line *AB*, and from *M*, draw *MN* perpendicular <to *AB*>. This is equal to *KT*. From that which we have already proved, the ratio of *BT* to *TA* is equal to the ratio of *ON* to *NP*. As the ratio of *BT* to *TA* is equal to the ratio of *ON* to *NP*, the ratio of *HT* to *MN* will be equal to the ratio of *AB* to *PO*, all this from that which we have established in relation to circles. But *MN* is equal to *KT*. It therefore follows that the ratio of *HT* to *TK* is equal to the ratio of *AB* to *PO*. That is what we wanted to prove.[75]}**<c>**

<b> {From what we have said, it follows thus. If the cylinder is another cylinder, greater than the first, then let it be a tangent to the first along the straight line *ZE*. Its circle is greater than <the circle> *DW* and tangent <to it> at the point *W*. The two <circles> are in the <same> plane.[76] In the larger circle, it is possible to draw a chord equal to the chord *HK* and it is possible to produce from the middle of this chord to the straight line *ZE*, the section common to the two surfaces, a straight line equal to the straight line *QI*. In other words, it is possible to determine a plane containing the two secant lines such that the oblique section of the <large> cylinder <by this plane> is <an ellipse> opposite[77] to the ellipse *KIH*, and having a chord and a sagitta equal to *KH* and *QI* respectively; and this can be performed in an infinite number of ways.} **<b>**

[75] This property of an ellipse has already been established by Proposition 8.
[76] The manuscript does not include a figure. We have included one as Fig. VI.31.
[77] *mitnaged*. In Arabic: *muqābil*.

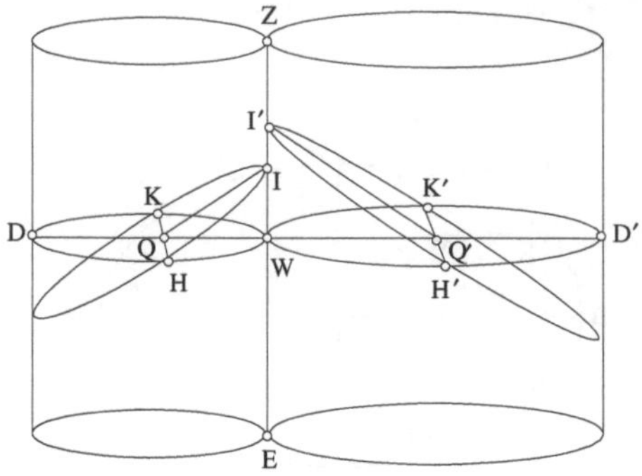

Fig. VI.31

<a> {<By another method>. Let *ABGD* be an ellipse, and let *ALB* be its circumscribed circle. Let there be two perpendiculars to the small diameter, *EN* and *ZH*, which are equal, one lying in the circle and the other in the ellipse.

I say then that the ratio of LZ *to* ZM *is equal to the ratio of* GE *to* ED.

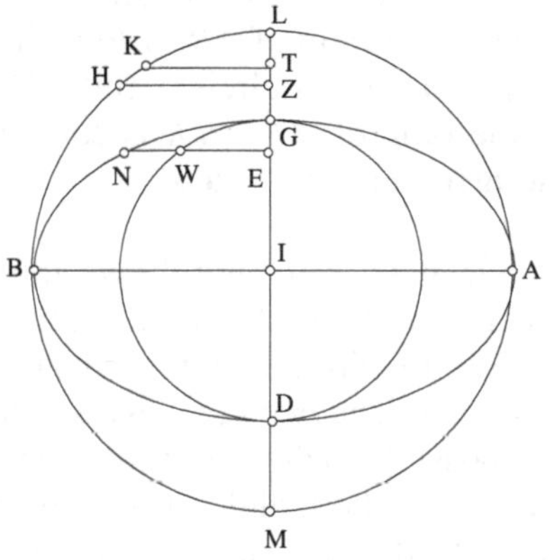

Fig. VI.32

Proof: If this were not the case, the ratio of *GE* to *ED* would be equal to the ratio of *LT* to *TM*. We therefore draw the chord *TK* <in the circle> perpendicular <to *LM*>.

From that which we have already proved, the ratio of *EW* to *TK* will be equal to the ratio of *GI* to *IL*. Consequently, *EN* will be equal to *TK*.[78] But *EN* is equal to *ZH*, so this is contradictory, and the ratio of *LZ* to *ZM* is

[78] From Proposition 6, *EW/EN* = *GI/IL*.

truly equal to the ratio of *GE* to *ED*. That is what we wanted to prove.}
<a>

<Proposition> 21. – After having established this premise, we wish to show how, given the chord and sagitta of an arc of an ellipse together with one on the two diameters, it is possible to determine the second diameter, so as to know the ellipse, the area of the elliptical segment and all the other elements.[79]

So, if someone says to you, 'We have an elliptical <segment> whose chord is eight, whose sagitta is three, and whose associated diameter in fifteen, how can we solve the problem?' In order to find the second diameter and know the area of the ellipse, we could proceed in a number of ways, the foundation of which is the premise mentioned above.

This is one of the procedures. Take half the chord, in this case four. Multiply it by itself, giving 16, and save the result. Then divide 15, *i.e.* the diameter, by each of its parts. One of these, the sagitta, measures three. We can then obtain the square of the second diameter (by multiplying the three numbers, 16, 153 and 1512).

If you wish, you could multiply either of the two <parts of the diameter by 4, then divide the diameter once by this product, and once by the remaining part>. Then multiply the two quotients by the square of the entire chord. This will give you the square of the sought diameter, and from which you extract the square root. Example: Multiply one part of the diameter, the sagitta, by four, giving 12. Divide the diameter once by this product, giving one and a quarter, and once by the remainder of the diameter, which also gives one and a quarter. Multiply the quotient by the quotient, which gives one and a half plus a half of an eighth. Multiply this by sixty-four, the square of the chord. This gives one hundred, which is the square of the second diameter.

If you wish, multiply one of the parts of the diameter by the other, then multiply this product by four, which gives 144, which you save as a total. Then multiply the diameter by itself, and the chord by itself, then multiply the two squares together, which gives fourteen thousand and 400. Divide this by the total, giving one hundred, which is the square of the required diameter.

Concerning the cause explaining these procedures, we shall illustrate it by an example.

[79] It can be seen that the area of an elliptical segment is not discussed, despite the statement.

Let there be an ellipse *KATG*. The circle *ABG* is tangent to it <internally>. The large diameter is *KDBT*, *DB* is the diameter of the circle, and *AG* is the small diameter, which is common to both the ellipse and the circle. <The straight line> *OMH* is a chord within the ellipse, <perpendicular> to the large diameter. This chord measures eight. *KM* is the sagitta. It measures three. *KT* is the entire diameter. It measures fifteen.

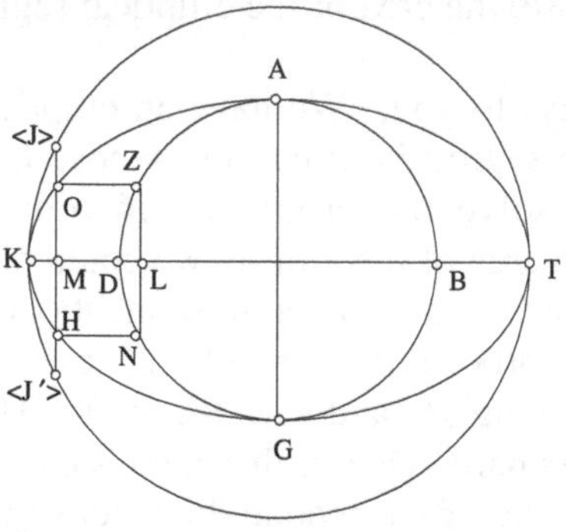

Fig. VI.33

We wish to know *AG*, the second diameter.

From *O*, draw a straight line *OZ*, parallel to the diameter *KT* and extending as far as the circumference. From *Z*, draw a perpendicular to this diameter. This is *ZL*. Let us produce it as far as the point *N* on the circumference on the other side. *ZN* is then the chord associated with the arc *ZDN*. As the straight line *OZ* is parallel to the diameter, and as the straight line *ZN* is perpendicular to the diameter and parallel to *OH*, then *OH* and *ZN* are equal. From the premise that we have already established, and the chords being equal, we know that the ratio of *KM* to *MT* is equal to the ratio of *DL* to *LB*. But the product of *DL* and *LB* is equal to the product of *LZ* by itself. The product of *LZ* by itself is known as *LZ* is known, and it measures four, the same as *OM*, as we have mentioned.

Under these conditions, the straight line *DB*, which is the unknown,[80] is divided into two parts whose product, one by the other, is known, <and whose quotient, one by the other, is also known>.

The result may be obtained in a number of ways. I have mentioned one of them, which leads to a determination of the square of the number *GA*, which can be known by approximation.[81]

[80] *ha-muskal*. In Arabic: *al-majhūl*.
[81] *be-qeruv*. In Arabic: *ʿalā al-taqrīb*.

We shall now establish a lemma, which is of value in the procedures that we have described.

<**Lemma 4**> Any number is separated into two different parts. The number is divided by each of these parts. The result of one division is multiplied by the result of the other division and the product saved. Then, each of the two parts is multiplied by the other. Then the product of this result and the number that was saved is the square of the number.

Example: The number A is separated into two numbers B and G. A is divided by B, which gives D. Then it is divided by G, which gives E. D is multiplied by E, giving the product H. B is multiplied by G, giving Z.

I say then that the product of Z *and* H, *i.e.* T, *is the square of the number* <A>.

Proof: Multiply D by A, which gives K. Hence, the number A has been divided by B, giving D, and D has been multiplied by A, giving K. This is therefore equal to the product of A by itself divided by B. But the product of A by itself is T. We have therefore divided T by B, giving K.

Similarly, we have divided A by G, giving E, and we have multiplied D by E, giving H, which is equal to the product of A and D divided by G, and <the product of> A and D is K. In other words, this product is the quotient of K by G, which is H.

T has been divided by B, which gives K. But the quotient of T by B, divided by G, is equal to the quotient of T by the product of B and G, and that the product of B and G is Z. The quotient of T by Z is thus H. Consequently, the product of H and Z is truly T. That is what we wanted to prove.

Having established that, the number A is the diameter of the circle in the previous proposition. It has been divided at L into two parts, DL and LB. The quotient of DB by each of these two parts DL and LB is known. It is equal to the ratio of KT to each of the two straight lines KM and MT, that is, the quotient of KT by each of these parts,[82] that is, five and one and a quarter. The product of each of these two <parts> by the other is also known. It is sixteen. It follows that the product of 5 by one and a quarter, which is six and a quarter, multiplied by sixteen is equal to the square of DB, as we have proved. That is what we wanted to prove.

[82] From $KM/MT = DL/LB$, we can derive by composition

$$(KM + MT)/MT = (DL + LB)/LB.$$

We have, by way of inversion, $MT/KM = LB/DL$; then, by composition,

$$(KM + MT)/KM = (DL + LB)/DL;$$

hence the indicated result.

This is all that I, Qalonymos, have found in Arabic, and I have translated all of it. I finished the translation on the 25 Ṭevet 72, according to the short reckoning <= 5 January 1312>. Praise be to God, the Highest.

I, Joseph ben Joel Bibas, completed <the copy>, here in Constantinople, at dawn on Friday 24 Ṭevet in the year 5267 of the Creation <= Friday 9 December 1506>. May the Holy Name be exalted and sanctified, and blessings be upon Him. Amen.

CHAPTER VII

IBN HŪD: THE MEASUREMENT OF THE PARABOLA AND THE ISOPERIMETRIC PROBLEM

7.1. INTRODUCTION

7.1.1. *Kitāb al-Istikmāl, a mathematical compendium*

Abū ʿĀmir Yūsuf ibn Hūd Aḥmad ibn Hūd, known as al-Mu'taman,[1] succeeded his father as King of Saragossa on the death of the former in 474/1081. His reign was not to be a long one, as al-Mu'taman died four years later in 478/1085.[2] The King is credited with the authorship of the

[1] Al-Mu'taman is not simply a nickname, meaning a man who can be relied upon. It carries much more significance as one of the titles held by a caliph, as in the case of al-Ma'mūn, al-Muqtadir etc. This custom of giving glorious titles to Andalusian kings and crown princes started to be disseminated towards the end of the Umayyad State. For more information, see ʿAbd al-Wāḥid al-Marākūshī, *al-Muʿjib fī talkhīṣ akhbār al-Maghrib*, ed. by M. S. al-ʿAryān and M. al-ʿArabī, 7th ed., Casablanca, 1978, p. 105. The author quotes the verses by the celebrated poet Ibn Rashiq that pour scorn on this usage:

مما يزهّدُني في أرض أندلس　　　سَمَاعُ مُقْتَدِرٍ فيها ومعتضِدِ
ألقابُ مملكة في غير موضعها　　　كالهرّ يحكي انتفاخًا صولة الأسدِ

[2] Ibn al-Abbār, *al-Ḥulla al-siyarā'*, ed. H. Mones, Cairo, n. d., vol. II, p. 248. H. Suter has translated a few brief extracts from an interesting correspondence between an Andalusian and an inhabitant of Tangier, reported by al-Maqqarī, in which each extols the advantages of their country. This correspondence confirms the high regard in which Ibn Hūd was held. H. Suter has also drawn attention to the work of Steinschneider on Yūsuf ibn Aknīn, the importance of which will be made clear later. See *Die Mathematiker und Astronomen der Araber und ihre Werke*, Leipzig, 1900, p. 108. See also Ibn al-Khaṭīb, *History of Islamic Spain (Kitāb aʿmāl al-aʿlām)*, Arabic text published with an introduction and index by E. Lévi-Provençal, Beirut, 1956, p. 172; Al-Maqqarī, *Nafḥ al-ṭīb min ghuṣn al-Andalus al-raṭīb*, ed. Iḥsān ʿAbbās, 8 vols, Beirut, 1968, vol. I, p. 441; Ṣāʿid al-Andalusī, *Ṭabaqāt al-umam*, ed. H. Būʿalwān, Beirut, 1985, p. 181. Ṣāʿid, it should be noted, places Ibn Hūd in context among his contemporaries together with the second mathematician discussed here, ʿAbd al-Raḥmān ibn Sayyid. This is essentially confirmed by the dates and sources. However, Ṣāʿid comments that Ibn Sayyid is a most distinguished mathematician, and that Ibn Hūd was also interested in logic, physics and metaphysics. He wrote: 'As for Abū ʿĀmir ibn al-

substantial mathematical work, the *Istikmāl*,[3] which he appears to have composed while still the Crown Prince. The wide range of topics covered by the book, together with its bulk, all suggest that it represents the sum total of a life devoted to mathematics. It could not possibly have been written during the few leisurely hours available to a king, regardless of the true extent of his kingdom. We must therefore consider him to have been a mathematical crown prince rather than a mathematical king, much as we would prefer to imagine the latter.

Although we do not yet possess the full text of this book, copies of it have been circulated in the past and several new sections have recently, and happily, come to light. The recently discovered geometrical sections[4] include a study of the measurement of the parabola and another addressing the isoperimetric problem. It is this work that we shall discuss here.[5] The attribution of the *Istikmāl* to Ibn Hūd is almost certain. However, in the absence of any direct proof, we are obliged to proceed with care. No known manuscript of the *Istikmāl*, or, more correctly, no known section of the work, mentions the name of Ibn Hūd.[6] We do, however, have a direct

---

Amīr ibn Hūd, while he collaborated with these (*i.e.* the mathematicians contemporary with Ṣā'id) in the science of mathematics, he was *distinguished from them* (our italics) by his interest in the science of logic, and by his work in the physical and metaphysical sciences', p. 181.

وأما أبو عامر بن الأمير بن هود فهو مع مشاركته لهؤلاء في العلم الرياضي منفرد دونهم بعلم المنطق والعناية بالعلم الطبيعي والعلم الإلهي.

This remark by Ṣā'id, a contemporary biobibliographer, has passed unnoticed, but is particularly important in understanding the project carried out by Ibn Hūd.

[3] See, *inter alia*, al-Akfānī, *Irshād al-qāṣid ilā asnā al-maqāṣid*, p. 54 of the Arabic text, in J. Witkam, *De egyptische Arts Ibn al-Akfānī*, Leiden, 1989, who quotes 'the *Istikmāl* of al-Mu'taman Ibn Hūd'.

[4] J.P. Hogendijk, 'The geometrical parts of the *Istikmāl* of Yūsuf al-Mu'taman ibn Hūd (11th century). An analytical table of contents', *Archives internationales d'histoire des sciences*, vol. 41, no 127, 1991, pp. 207–81. The author refers also to another article that he published in 1986 in *Historia Mathematica*, entitled 'Discovery of an 11th-century geometrical compilation: The *Istikmāl* of Yūsuf al-Mu'taman ibn Hūd, King of Saragossa', pp. 43–52.

[5] See our edition of the Arabic text in *Mathématiques infinitésimales*, vol. I, chapter VII.

[6] The following fragments of the *Istikmāl* are known to have survived at the present time: 1) The geometric sections, by far the most extensive, in manuscript Or. 82 in the Royal Library of Copenhagen, and manuscript Or. 123-a in Leiden. 2) The arithmetic fragment in the Cairo manuscript, Dār al-Kutub, Riyāḍa 40. A copy of this manuscript alone is also held, as we have shown, in Damascus, Ẓāhiriyya 5648. 3) Finally, the short fragment quoted by a commentator in a manuscript held in the Osmaniye Library in

citation in which the author attributes the work to Ibn Hūd's Andalusian predecessor, the famous mathematician 'Abd al-Raḥmān ibn Sayyid.[7] This important attribution is made by an anonymous author, namely, a commentator on the *Elements* of Euclid who was clearly familiar with the mathematical traditions that he was discussing. In relation to Proposition I.5 of the *Elements*: 'In isosceles triangle, the angles at the base are equal to one another, and, if the equal straight lines be produced further, the angles under the base will be equal to one another',[8] he writes: 'al-Nayrīzī proved this proposition through another course of demonstration in which he did not require this (argument by Euclid), and he was followed in this by Ibn Sayyid in a book known by the title *al-Istikmāl*'.[9] Yet, the commentary in the *Istikmāl* on the first book of the *Elements* and the early chapters of the second book, which must have been a substantial body of text, has not yet been discovered, depriving us of any direct verification. The fact remains that the anonymous author cites the *Istikmāl* around ten times and, in particular reproduces a long passage on amicable numbers to which we have already drawn attention.[10] Comparing this passage with the text from the *Istikmāl* leaves no room for doubt. They are both the same text, the one

---

Hyderabad that we have identified, see below. With the exception of this last fragment, in which the *Istikmāl* is quoted, none of these mentions either the title or the author.

[7] 'Abd al-Raḥmān ibn Sayyid was a contemporary of Ṣā'id (see Note 2). The latter was born in 420/1029. We also know from the philosopher Ibn Bājja that Ṣā'id was the disciple of Ibn Sayyid (see the letter from Ibn Bājja to the Vizir Abū al-Hasan ibn al-Imām, in *Rasā'il falsafiyya li-Abī Bakr ibn Bājja*, ed. Jamāl al-Dīn al-'Alawī, Beirut, 1983, p. 88). Ibn Bājja died in around 1139 and it is therefore possible to assume that Ibn Sayyid was a generation older and that he was active in the final decades of the eleventh century. Elsewhere, Ibn al-Abbār writes in his *Kitāb al-takmila li-Kitāb al-Ṣila*: 'Abd al-Raḥmān ibn 'Abd Allāh ibn Sayyid al-Kalbī of Valencia, whose surname is Abū Zayd, is an eminent scholar in numbers theory and arithmetic; and none of his contemporaries was his equal in geometry. Only Ṣā'id of Toledo mentioned him'. He then remarks that Ibn Sayyid composed in *farā'iḍ* and that he studied in 456/1064 (see *Complementum libri Assilah*, ed. F. Codera and Zaydin, 2 vols, Madrid, 1887–89, vol. II, p. 550), which confirms the dates given. He was therefore a contemporary of Ibn Hūd.

[8] T. Heath recounts the commentaries provoked by this proposition – Aristotle, Pappus and Proclus – see *The Thirteen Books of Euclid's Elements*, 3 vols, Cambridge, 1926; repr. Dover, 1956, vol. I, pp. 251–5.

[9] Ms. Hyderabad, Osmaniyye 992, fol. 46[r]:

وبرهن النيريزي على هذا الشكل برهانًا آخر لم يحتج فيه إلى ذلك، وتابعه على ذلك ابن سيد في كتابه المعروف بالاستكمال.

See also R. Rashed, 'Ibn al-Haytham et les nombres parfaits', *Historia Mathematica*, 16, 1989, pp. 343–52, in particular p. 351.

[10] See previous note.

that has survived.[11] The other references to the *Istikmāl* by the anonymous author either refer to sections of the book that have been lost or have been paraphrased.[12]

At the present time, this is the only source that attributes the work to Ibn Sayyid. It is important not to discount or ignore the fact that several independent sources agree in naming Ibn Hūd as the author of the *Istikmāl*. The oldest known attribution is that of al-Qifṭī,[13] who confirms the authorship of Ibn Hūd, citing an earlier attribution by Maimonides. Maimonides' pupil, Ibn Aknīn of Barcelona,[14] repeats this attribution in his *Ṭibb al-nufūs* (*The Medicine of the Souls*), and even includes a sort of diagrammatic list of the contents of the *Istikmāl*.[15] In fact, after he had written 'this is the book of the *Istikmāl* by al-Mu'taman Ibn Hūd, King of Saragossa', he enumerates the five topics that constitute the book.[16] The third source is a fourteenth century mathematician, Muḥammad Sartāq al-

[11] This is a fragment on amicable numbers, taken from Thābit ibn Qurra and included in the *Istikmāl*. This fragment is preserved in the Cairo manuscript, Dār al-Kutub, Riyāḍa 40, fols 36^r-37^v; it is cited in the Hyderabad manuscript, Osmaniyye 992, fols 295^r-297^r, which starts by noting: 'the author of the *Istikmāl* said… (*wa-qāla ṣāḥib al- Istikmāl…*)'. We shall account for this matter later.

[12] See, for example, fols 34^v, 36^r, 38^r, 46, 47, 50^r, 56^r, 57^r, 68^r, 151^r and 295^r.

[13] Al-Qifṭī, *Ta'rīkh al-ḥukamā'*, ed. J. Lippert, Leipzig, 1903, p. 319. Al-Qifṭī wrote in relation to Maimonides, that 'he rectified (*hadhdhaba*) the book of the *Istikmāl* in astronomy of the Andalusian Ibn Aflaḥ, and that he did it well; yet it showed a confusion in origination, since he (Maimonides) rectified the book of the *Istikmāl* of Ibn Hūd in the science of mathematics (*hadhdhaba Kitāb al-Istikmāl li-Ibn Aflaḥ al-Andalusī fī al-hay'a fa-aḥsana fīhi wa-qad kāna fī al-aṣl takhlīṭ wa-hadhdhaba Kitāb al-Istikmāl li-Ibn Hūd fī 'ilm al-riyāḍa*)'. It should be noted in this context that the title of the *Istikmāl* is not rare.

[14] The comments of Ibn Aknīn in relation to Ibn Hūd and the *Istikmāl* are particularly important and have been well known to historians since the nineteenth century. The major works in this area are M. Steinschneider, *Die hebraeischen Übersetzungen des Mittelalters und die Juden als Dolmetscher*, Berlin, 1893; repr. Graz, 1956, pp. 33–5; M. Steinschneider, *Die arabische Literatur der Juden*, Frankfurt, 1902; repr. Hildesheim/Zürich/New York, 1986, pp. 228–33.

[15] The importance of the work of Ibn Aknīn derives from the fact that he includes a schematic listing of the contents of the *Istikmāl* in his book in Arabic, but using Hebrew characters – *Ṭibb al-nufūs*, edited in the nineteenth century and translated into German by M. Güdemann, *Das jüdische Unterrichtswesen während der spanisch-arabischen Periode*, Vienna, 1873, see pp. 28–9 and 87–8. T. Langermann has also drawn attention to this text and has translated it into English; see 'The mathematical writings of Maïmonides', *The Jewish Quarterly Review*, LXXV, no 1, July 1984, pp. 57–65, in particular pp. 61–3. In 1986, J. Hogendijk included an English translation of the same text in *Archives internationales*, p. 210.

[16] Güdemann, *Das jüdische Unterrichtswesen während der spanisch-arabischen Periode*, p. 29.

Marāghī,[17] who wrote a commentary on the *Istikmāl* entitled the *Ikmāl*. No copy of this commentary has yet been found, but it is cited by the author in the glosses to manuscript 4830 in the Aya Sofya collection. Al-Marāghī also credits Ibn Hūd as the father of the *Istikmāl*. To these may be added a number of indirect references attributing to Ibn Hūd one or the other of the results found in the *Istikmāl*; one such example is found in a work by Ibn Haydūr.[18] Taken together, these clues enable us to state with a high degree of certainty that this treatise is definitely the work of Ibn Hūd. The mathematical content provides a further argument in favour of this hypothesis. All the references to the lost work of Ibn Sayyid indicate that he was at the forefront of mathematical research during his lifetime. We have already shown[19] that he may even have addressed questions relating to the use of generalised parabolas and skew curves. This is definitely not the level at which the *Istikmāl* is pitched. It results from a totally different project, as we shall see. It therefore appears to us that the attribution of the *Istikmāl* to Ibn Hūd is beyond reasonable doubt. However, the exact role of Ibn Sayyid remains a fundamentally important question. Could it be nothing more that a simple error? Or could it be a work with the same title, written by Ibn Sayyid and then included in a compilation and expanded by Ibn Hūd? Or is it simply an early confusion between two contemporary authors? The answers to these questions must await future research. For the moment, we can only reiterate our strong conviction that the attribution to Ibn Hūd is correct.

In order to arrive at an assessment of the *Istikmāl* project without either diminishing or amplifying its extent, consider the judgement found in the works of the thirteenth-century biobibliographer, al-Qifṭī, together with a single, incontestable historical fact. Al-Qifṭī wrote that the book was a 'compendium (*kitāb jāmiʿ*) that is elegant, yet, that necessitated verification'.[20] As to the historical fact, it is simply the wide distribution of the *Istikmāl*, especially among second-rate mathematicians and philosophers. The evidence points to a close relationship between this judgement and this historical fact. The opinion of al-Qifṭī – or quoted by him – corresponds perfectly to the surviving sections of the book and the project that is

[17] See Chapter V on al-Qūhī, Note 19; and also Hogendijk, 'The geometrical parts of the *Istikmāl* of Yūsuf al-Mu'taman ibn Hūd', p. 219.

[18] Ibn Haydūr (died in 816/1413), *al-Tamḥīṣ fī sharḥ al-talkhīṣ*, ms. Rabat, al-Ḥasaniyya 252, fol. 72; edited and analysed by R. Rashed in 'Matériaux pour l'histoire des nombres amiables et de l'analyse combinatoire', *Journal for History of Arabic Sciences*, 6, nos. 1 and 2, 1982, pp. 213 *sqq*.

[19] Sharaf al-Dīn al-Ṭūsī, *Œuvres mathématiques. Algèbre et géométrie au XII*[e] *siècle*, 2 vols, Paris, 1986, vol. I, pp. 128–9.

[20] Al-Qifṭī, *Ta'rīkh al-ḥukamā'*, p. 319.

revealed by their study. The *Istikmāl* provides a geometer's compendium, including arithmetic and Euclidean geometry (taken directly from the *Elements*, the *Data* and commentators such as al-Nayrīzī), the theory of amicable numbers (borrowed directly from the treatise by Ibn Qurra), the geometry of conics (from the *Conics* of Apollonius), spherical geometry and other topics, all derived in a similar manner. All this borrowing, often verbatim and at length, indicates that the *Istikmāl* must have been a kind of 'Encyclopaedia of Geometry', or, more accurately, an 'Encyclopaedia of Mathematics' in the sense of the ancient *quadrivium*, and that it was also designed to cover astronomy, optics and harmonics.[21] This 'Encyclopaedia of Mathematics', in this sense, would have been intended for a readership cultivated in mathematics but not necessarily research mathematicians in the pursuit of new knowledge. These would have included philosophers, as Ṣāʿid tells us, who had with Ibn Hūd many interests in common. This is how we, very briefly, see the *Istikmāl* project. It is important not to misinterpret the nature of this endeavour, or that of Ibn Hūd. The *Istikmāl* does not in any way aim to *unify* the mathematics of the period, as one may naively think.[22] It is simply a compilation of the mathematical works essential to a proper mathematical education. Ibn Hūd did not have the ability to conceive such a task, let alone to carry it out. To succeed, he would have needed an altogether different conception of algebra and its role, especially in terms of its relationship with geometry, something of which Ibn Hūd did not have the

[21] T. Langermann (pp. 63–5) has drawn attention to an enumeration and affirmation of al-Akfānī that implies that the *Istikmāl* was not completed in accordance with the plan laid down by Ibn Hūd, and that this plan called for several additional chapters that are not found in the *Istikmāl*. After having reviewed the ten sections on geometry (*Irshād al-Qāṣid*, p. 54), al-Akfānī wrote: 'I have not seen hitherto any book that contains these ten sections. Yet, if the composition of the *Istikmāl* by al-Mu'taman ibn Hūd – may God be merciful to him – were to be completed, then it would have been satisfying and sufficient…'

لكن لو كمل تصنيف الاستكمال للمؤتمن بن هود رحمه الله، لكان كافيًا مغنيًا

Let us furthermore note that al-Maqqarī cites the title as *Kitāb al-Istikmāl wa-al-manāẓir*; which indicates that *al-Istikmāl* contained also a part on optics (cited in Note 2, *supra*). Al-Akfānī is speaking here only of geometry, but he was aware that the book contained a major section on arithmetic that was not included in the ten sections in his list.

[22] The reader will come across similar affirmations throughout the text, some of which are even more excessive. Some claim Ibn Hūd to be the most brilliant of all the Andalusian geometers, while others, carried away by their enthusiasm, consider him to be a predecessor of Bourbaki … However, these claims all appear to be without foundation when one considers the work of other Andalusian mathematicians: One needs do little more that to read the pages of Ibn al-Samḥ, or the comments on Ibn Sayyid, or simply the comments of their contemporaries.

slightest idea. However, the question remains as to know when and how this style of encyclopaedic composition in mathematics, until then the preserve of philosophers such as Ibn Sīnā in his *al-Shifā'*, should have been taken up in the western Islamic world by mathematicians of the likes of Ibn Hūd. Following in the tradition of the great mathematicians Banū Mūsā, Ibn Qurra, Ibn Sinān, Ibn al-Haytham, Ibn al-Samḥ and others, he could have undertaken the task of preparing this encyclopaedia.

In any event, it is in the light of bearing the designation 'Compendium' that the *Istikmāl* includes two studies relating to infinitesimal mathematics. The encyclopaedic form of the work undoubtedly affects not only their presentation, but also their extent. The first of these deals with the measurement of the parabola and is based firmly on the treatise by Ibn Sinān on the same subject. The second addresses the isoperimetric problem and, as we shall see, is based on a proposition by Ibn al-Haytham. Presented in this form, these two studies are valued more for their historical interest than for the novelty of their mathematical results. In this work, Ibn Hūd gives the results in their logical order, rather than in the order in which they were discovered. The encyclopaedic style also limits the extent to which the results are developed, as we shall see. That statement may appear to be too restrictive, given the fact that Ibn Hūd retained some independence of spirit in his approach to the work. He is not afraid to change the formulation occasionally, often by making it more general. However, it is also evident, at least in the two cases discussed here, that this generalization did not always succeed and the proofs inspired by his predecessors are less rigorous than the originals. For example, while he succeeded in extending the result established by Ibn Sinān for the parabola (the comparison of sections of the parabola and triangles) to both the ellipse and the hyperbola, he was not able to use this comparison to extend Ibn Sinān's result for the area of a segment of a parabola, for which the first comparison is a precursor, to other conic sections.

7.1.2. *Manuscript transmission of the texts*

The text on the measurement of the parabola survives in a single manuscript, Or. 82 in the Royal Library of Copenhagen, while the second text on the isoperimetric problem survives both in that manuscript and another, Or. 123a in the Library of Leiden. Both these manuscripts have been used to make the first edition[23] of these two texts of the *Istikmāl* and to provide the first translation.

[23] See the edition of the Arabic text in *Mathématiques infinitésimales*, vol. I, pp. 1001–13 and 1023–27 respectively.

Arabic manuscripts are rarely described and catalogued as well as those in Copenhagen, as can be seen in the *Codices Orientales Bibliothecae Regiae Hafniensis Jussu et auspiciis regiis enumerati et descripti. Pars Altera: Codices Hebraicos et Arabicos Continens* (1851), vol. II, pp. 64–7. The author of this catalogue has carefully and accurately given all the information that he could possibly glean from the manuscript. He provides an extremely clear view of the plan followed in the *Istikmāl*, and gives citations in Arabic of the various subdivisions. He extracts all the significant data from the manuscripts themselves. In this way, we know that it came from the 'Coll. Paris. De la compagnie de Jésus', and one can see on fol. 1^{r} in the internal margin – as in codex 81 in the same library: 'Signed in accordance with the Decree dated 5th July 1763. Mesnil'. It should be noted that the same signature appears on the external margin of fol. 1^{r}. The manuscript therefore came from France after that date. In addition, a number of comments in Greek, noted by the author of the catalogue, and some writing that cannot be later than the Renaissance suggest that the manuscript spent some time in the Hellenist East before it arrived in Paris. All the comments (fols 12^{r}, 16^{r}, 21^{r}, 23^{v}, 32^{v} and 122^{r})[24] relate either to the titles of the chapters or in some way to their content. That is to say, they were written by a Hellenist who understood the content, at least in part. So, we can trace this early manuscript from its probable origin in Andalusia, through the Hellenistic East, to Paris and then on to Copenhagen.

The manuscript itself consists of 128 folios. In several places it has been damaged by insects or traces of damp. Several sections are missing, in particular the first section containing a commentary on Book I and part of Book II of the *Elements* of Euclid. We know from the anonymous author of the Osmaniyye manuscript that other important sections originally existed, including a commentary on the Postulate of Parallels. This early manuscript is written in a North African hand. Throughout, there appear here and there marginal notes in another more recent handwriting, which one must be

[24] Fol. 12^{r}: περὶ τῶν ἀριθμῶν ἀναλογίας καὶ πρὸς τὰ σώματα ἐπίπεδα καὶ γραμμὰς συγκρίσεως ('on the analogy of numbers and their comparison with bodies, surfaces and lines').

Fol. 16^{r}: περὶ ἀριθμῶν ἰδιότητος καὶ τοῦ πρὸς τὰ μέρη συγκρίσεως ('on the intrinsic character of numbers and the comparison of the same <thing> with parts').

Fol. 21^{r}: The catalogue (*Codices Orientales*, p. 65) notes the presence in the margin of the following words: ὧδε πολὺ λείπει ('he therefore leaves much'). However, it should be noted that πολὺ does not appear on the microfilm.

Fol. 23^{v}: περὶ τῶν κύκλων περιφερίας ('on the circumference of circles').

Fol. 32^{v}: διάπραξις τῶν σχημάτων καὶ ἡ ἐν αὐτοῖς ἄσκησις ('the drawing of figures and their study').

Fol. 122^{r}: περὶ στερεῶν ('on solids').

careful not to confuse with that of the copyist. He made his own notes in the margin, indicating that he had revised his copy by comparison with the source after he had completed the copy. Finally, the copyist wrote the letters in the mathematical propositions as they were pronounced (a: alif, b: bā', etc.).The text on the measurement of the parabola occupies fols 100^{v}–102^{v}.

The second manuscript is that held in the Library of Leiden, Or. 123-a. An accurate, albeit briefer, description is given in M.J. de Goeje, *Catalogus Codicum Orientalium Bibliothecae Academiae Lugduno-Batavae* (1873), vol. V, pp. 238–9.[25] This manuscript is a fragment of 80 folios of the *Istikmāl*. The writing is in eastern *naskhī* and the manuscript is undoubtedly more recent than the one discussed above. A comparison of the two manuscripts also reveals that each belongs to a different manuscript tradition. There is also no indication that the copyist of the Leiden manuscript had compared it with the source. The only marginal notes appear to have been added during the copying process – see fols 49^{v}, 55^{v} and 56^{r} – or with no relevance to the text – *e.g.* fol. 69^{v}, which is simply a verse from the Koran. The catalogue gives no information on the history of the manuscript, other than that it forms part of the collection of Golius.[26] The text on the isoperimetric problem occupies fols 7^{v}–11^{r} and folios 50^{r}–50^{v} of the Copenhagen manuscript.

7.2. THE MEASUREMENT OF THE PARABOLA

7.2.1. *Infinitesimal property or conic property*

Ibn Hūd's study of the measurement of the parabola forms part of one chapter of the *Istikmāl* relating to sections of the cylinder and cone of revolution. This chapter is itself divided into two parts. The first of these deals with 'sections and their properties, without these relating to one another', while the second covers 'the properties of lines, angles and surfaces of sections that relate to each other'.[27] These two titles provide a perfect indication of the background against which Ibn Hūd developed the work. His determination of the area of a section of a parabola did not constitute an end in itself; rather it was simply a step along the path to

[25] See also P. Voorhoeve, *Codices Manuscripti VII. Handlist of Arabic Manuscripts in the Library of the University of Leiden and Other Collections in the Netherlands*, 2nd ed., The Hague/Boston/London, 1980, p. 432.

[26] It would be interesting to know whether it was copied in the East or, like other collections in the time of Golius, in Holland.

[27] Ms. Copenhagen, Royal Library, Or. 82, fol. 90^{v}.

determining a property of the conic section. The infinitesimal aspects of the study interested him less than those of the conic sections. The importance of this point cannot be overemphasized, as it serves to distinguish the perspective of Ibn Hūd from that of his inspiration, Ibrāhīm ibn Sinān. We have already seen that the latter, in common with al-Māhānī and his grandfather, Thābit ibn Qurra, was interested in the measurement of the parabola as a metric question in its own right. This difference, together with the wholesale borrowing from the *Conics* of Apollonius, characterize the style of Ibn Hūd and the nature of his path. In order to understand this difference better, we must make the briefest possible examination of Ibn Hūd's work on the measurement of the parabola.

The structure of the text is as follows: After summarizing certain definitions of the three sections and their elements, he restates several propositions from the *Conics* of Apollonius, some paraphrased and some verbatim, before arriving eventually at the determination of the area of a section of a parabola in Propositions 18–21. Unlike the earlier propositions, these are based on the ideas of Ibrāhīm ibn Sinān. The work taken from the *Conics* is not only considerable, it also follows the order established by Apollonius.

In order to illustrate this context, we are obliged to consider both the preceding propositions leading up to Propositions 18–21 and those following them. It can clearly be seen that Ibn Hūd borrows his propositions from the sixth book of the *Conics*, in the original order, at least at the start, before returning to Ibrāhīm ibn Sinān.

The tenth proposition (as numbered in the manuscript) is nothing more than a paraphrased version of the first two propositions in the sixth book. In these propositions, Ibn Hūd shows that, 'If the *latera recta* of the parabolas are equal and the angles of their ordinates are also equal, then the sections are equal and similar. If these sections are equal and similar, then their *latera recta* are equal. If the sections are other than the parabola, and are such that the figures constructed on their transverse axes are equal and similar, then these sections will be equal and similar. Finally, if these sections are equal and similar, then the figures constructed on their transverse axes will be equal and similarly disposed'.[28]

However, it should be noted that, in the case of the parabola, Ibn Hūd considers the *latera recta* relative to any diameter, while Apollonius only considers those relative to the axes. This is also the reason why Ibn Hūd introduces the angles of the ordinates. It should also be noted that, when he discusses conics with centre, Ibn Hūd departs from the position taken with regard to the parabola and makes a distinction between a diameter lying on

[28] *Ibid.*, fol. 96^v.

the major axis and any other diameter. He deals with the latter in Proposition 15, corresponding to Proposition 13 of Apollonius.

The next proposition in the *Istikmāl* is a restatement of the sixth proposition in the same book of the *Conics*: 'If an arc of a conic section can be superimposed on an arc of another conic section, then the two sections are equal'.[29]

Proposition 13 in this chapter of the *Istikmāl* is the 11th in this book of the *Conics*. All the parabolas are similar. The following proposition in the *Istikmāl* is the same as Proposition 12 in the sixth book of the *Conics*. Here is the proposition as written by Ibn Hūd: 'the sections other than the parabola, of which the constructed figures on the axes are similar, would themselves be also similar; and, if the sections are similar, then the constructed figures on the axes are similar [and equal]'.[30]

In Proposition 18, Ibn Hūd then shows a consequence that he will need later for the measurement of the parabola:

Let there be two conic sections of the same kind, with respective diameters AB and GQ, and let points K and T be on BA, and points O and S on QG such that

$$\frac{BA}{AK} = \frac{QG}{GO} \quad \text{and} \quad \frac{BA}{AT} = \frac{QG}{GS}. \tag{1}$$

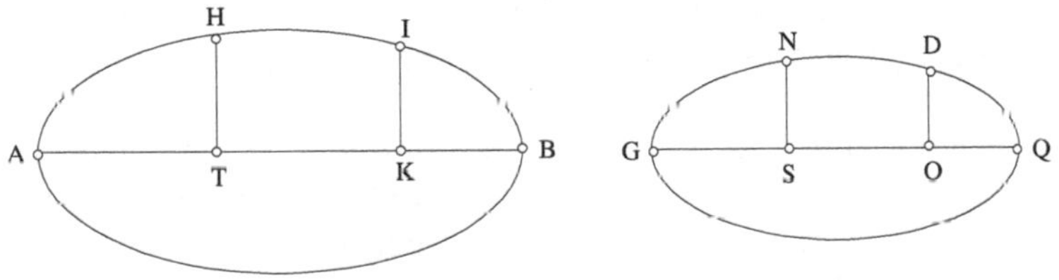

Fig. 7.1

Let KI, TH, OD and SN be the ordinates associated with these points. Then

$$\frac{KI}{TH} = \frac{OD}{SN}.$$

[29] *Ibid.*, fol. 97^{r}.

[30] *Ibid.*, fol. 97^{v}. Note that these figures are not equal, despite the assertion by Ibn Hūd that they are. It is for this reason that we have enclosed 'and equal' in square brackets.

We have

$$\frac{KI^2}{TH^2}=\frac{BK\cdot AK}{BT\cdot AT}=\frac{AK(AB\pm AK)}{AT(AB\pm AT)}=\frac{AK}{AB}\left(1\pm\frac{AK}{AB}\right)\Big/\frac{AT}{AB}\left(1\pm\frac{AT}{AB}\right);$$

similarly

$$\frac{OD^2}{SN^2}=\frac{GO}{GQ}\left(1\pm\frac{GO}{GQ}\right)\Big/\frac{GS}{GQ}\left(1\pm\frac{GS}{GQ}\right);$$

hence the result follows with the aid of (1).

The 15th proposition in this chapter of the *Istikmāl* is the same as Proposition VI.13 in the *Conics*. 'If the figures constructed on the diameters that are not axes, on sections that are not parabolas, are similar, and if the angles of their ordinates are equal, then the sections are similar'.[31] Ibn Hūd returns eventually to the reciprocal of this proposition. He then moves onto the 16th proposition, which is a restatement of Propositions 26 and 27 in the same book of the *Conics*: 'If parallel planes cut a cone, then the sections so generated are similar'.[32] Proposition 17 of the *Istikmāl* is inspired by Propositions 4, 7 and especially 8 of the same book by Apollonius, stated as follows: 'Let there be a conic section with an axis that separates its surface into two halves. If a segment is removed, it is possible to find another segment that is equal and similar to the one removed. Each of the diameters of an ellipse separates its surface into two halves and its contour also into two halves'.

These propositions are followed by Propositions 18–21, which we shall examine again in detail. These are then followed in turn by the following two propositions: 'to demonstrate how to construct a section that is equal to a known section and is also similar to another known section' – and this is Proposition 22 – in order to then establish that 'If there are two similar portions belonging to two sections of the same kind, then the ratio of the line surrounding one and forming part of the section to the line surrounding the other and forming part of the <other> section is equal to the ratio of the diameter of one to the diameter of the other'.[33]

This brief summary illustrates the background to the position of the determination of a portion of a parabola in the *Istikmāl*: the study of the properties of conic sections taken, for the most part, from Apollonius. The path followed by Ibn Hūd is that of an expository order rather than an order of discovery. This process is also found in his study of the measurement of

[31] *Ibid.*, fol. 98^{v}.
[32] *Ibid.*, fol. 99^{v}.
[33] *Ibid.*, fols 102^{v}–$103^{\mathrm{r-v}}$.

the parabola, *i.e.* in the four propositions discussed. The remainder of this chapter is devoted to an analysis of these propositions.

7.2.2. *Mathematical commentary on Propositions 18–21*

Ibn Hūd begins with a statement relating to the diameter and transverse diameter.

A segment of a parabola, ellipse or hyperbola is bounded by an arc and its chord. Let BAB' be that segment, and, through C, the midpoint of BB', passes AH, a diameter of the section that cuts the arc BB' at a point A, called the *summit* of the segment; the segment AC is then called the *diameter* of the segment having BB' as a *base*. In the case of a parabola, the diameter AH is parallel to the axis, and, in the case of an ellipse or a hyperbola, AH passes by K, the centre of the section, AH being a *transverse diameter*.

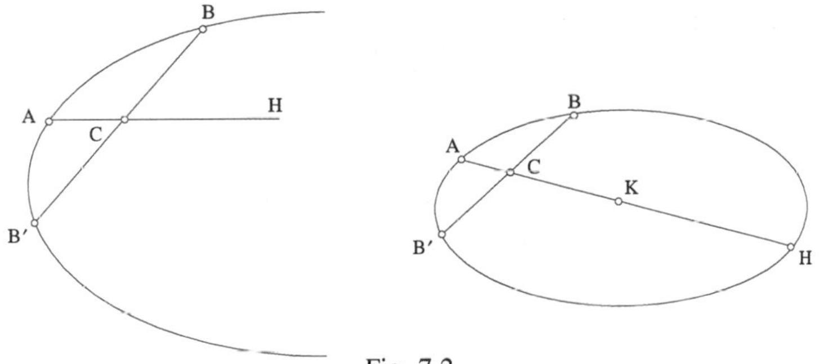

Fig. 7.2

In Propositions 18 and 19, the studied portions are not segments, rather portions such as ABC; the expressions *summit* A, *diameter* AC, *base* BC of the portion and *transverse diameter*, are nevertheless all preserved.

Proposition 18. — *We consider within two parabolas, two ellipses or two hyperbolas, the portions* ABC *and* DEG: *The first is delimited by a diameter passing through* A *and the ordinate* BC *with respect to that diameter, and the second is delimited by a diameter passing through* D *and the ordinate* EG *with respect to that diameter.*

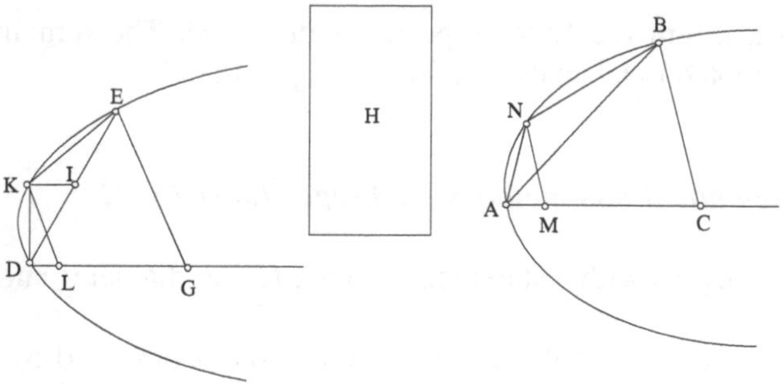

Fig. 7.3

We show that

a) *if* ABC *and* DEG *are portions of two parabolas*

or

b) *if* ABC *and* DEG *are the portions of two ellipses or of two hyperbolas, and that,* Δ *and* Δ′ *are the transverse diameters passing through* A *and* D, *assuming*

$$\frac{\Delta}{\mathrm{AC}} = \frac{\Delta'}{\mathrm{DG}},$$

then

$$\frac{\text{port. (ABC)}}{\text{port. (DGE)}} = \frac{\text{tr. (ABC)}}{\text{tr. (DGE)}}.$$

1) Let us assume that

$$(*) \qquad \frac{\text{tr.}(ABC)}{\text{tr.}(DGE)} = \frac{\text{port.}(ABC)}{H},$$

H being a surface such that $H <$ port. (DEG).[34] Let I be the midpoint of DE and IK the diameter of the section:

a) If the section is a parabola, then $IK \parallel DG$.

b) If the section is an ellipse or a hyperbola, then IK cuts DG at the centre of the section.

We know that

$$\text{tr. }(DEG) > \frac{1}{2}\text{ port. }(DEG)$$

and

[34] It must be assumed that tr. $(DEG) < H <$ port.(DEG), as if $H \leq$ tr. (DEG), the equality (*) is absurd.

$$\text{tr.}\,(DEK) > \frac{1}{2}\,\text{port.}\,(DEK).$$[35]

If we proceed in the same manner, by considering the midpoints of the chords *KD* and *KE*, we then have a polygonal surface larger than *H*. Let *DKEG* be the resulting polygon, let *KL* be the ordinate of *K*, and let *M* be on *AC* such that

$$(1) \qquad \frac{AM}{CM} = \frac{DL}{GL},$$

from which we deduce

$$(2) \qquad \frac{AC}{AM} = \frac{DG}{DL}.$$

[35] These two inequalities are given without justification.

a) $\text{tr.}(DEG) > \frac{1}{2}\,\text{port.}(DEG)$.

If we complete the parallelogram *DGEG′*, and no matter what type of section we consider, with *DG′* being tangent to that section, then we have

$$2 \cdot \text{area tr.}(DEG) = \text{area parall.}(DGEG') > \text{area port.}(DEG);$$

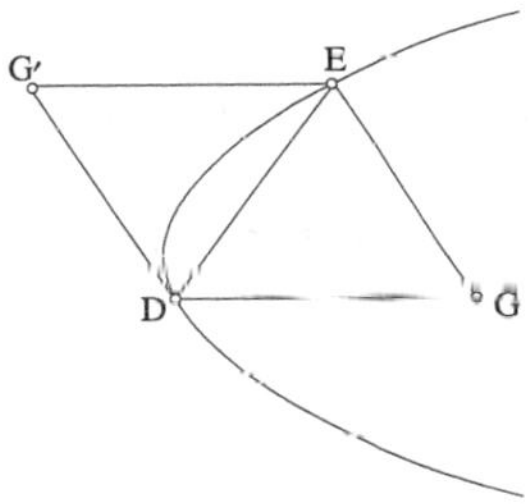

Fig. 7.4

hence the result.

b) $\text{tr.}\,(DEK) > \frac{1}{2}\,\text{port.}(DEK)$.

If the section is a parabola, *KI* is the diameter and the tangent at *K* is parallel to *DE* and cuts *DG* in *H*. If we complete the parallelogram *DHLE*, we have

$$2 \cdot \text{area}\,(DEK) = \text{area parall.}\,(DHLE) > \text{area port.}\,(DKE);$$

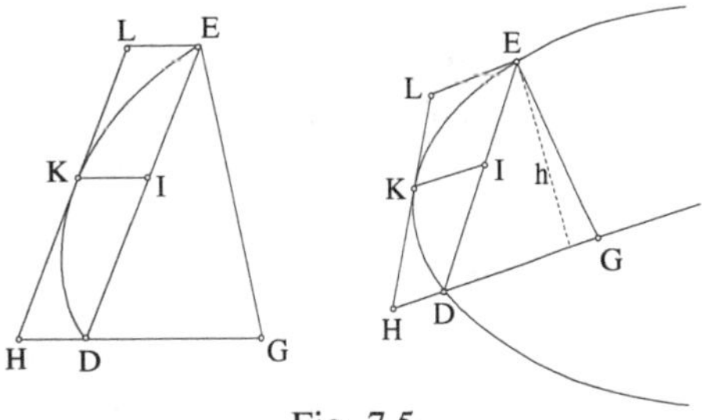

Fig. 7.5

hence the result.

Note that Ibn Sinān gives a demonstration of the inequality b) in the lemma of Proposition 2, in the case of the parabola. His reasoning applies to the ellipse and the hyperbola.

a) If the sections are parabolas, then, according to the *Conics* of Apollonius, I.20, we have

$$\frac{BC^2}{NM^2}=\frac{AC}{AM}$$

and

$$\frac{EG^2}{KL^2}=\frac{DG}{DL};$$

by (2) we have

$$\frac{BC}{NM}=\frac{EG}{KL}.$$

b) If the sections are ellipses or hyperbolas, then, according to the *Conics* of Apollonius, I.21, we have

$$\frac{BC^2}{NM^2}=\frac{AC\,(\Delta\pm AC)}{AM\,(\Delta\pm AM)}$$

and

$$\frac{EG^2}{KL^2}=\frac{DG\,(\Delta'\pm DG)}{DL\,(\Delta'\pm DL)}$$

(+ for a hyperbola; – for an ellipse).

However, by hypothesis, we have

$$(3)\qquad \frac{\Delta}{AC}=\frac{\Delta'}{DG}.$$

We can thus note

$$\frac{BC^2}{NM^2}=\frac{AC}{AM}\cdot\frac{\left(\frac{\Delta}{AC}\pm 1\right)}{\left(\frac{\Delta}{AC}\pm\frac{AM}{AC}\right)}\quad\text{and}\quad\frac{EG^2}{KL^2}=\frac{DG}{DL}\cdot\frac{\left(\frac{\Delta'}{DG}\pm 1\right)}{\left(\frac{\Delta'}{DG}\pm\frac{DL}{DG}\right)};$$

following (2) and (3), we have

$$(4)\qquad \frac{BC}{NM}=\frac{EG}{KL}.$$

Yet (1) implies

$$\frac{AC}{CM}=\frac{DG}{GL}$$

and (4) implies

$$\frac{BC}{BC+NM}=\frac{EG}{EG+KL};$$

hence

$$\frac{AC}{CM}\cdot\frac{BC}{BC+NM}=\frac{DG}{GL}\cdot\frac{EG}{EG+KL}.$$

We thus deduce

$$(5)\qquad \frac{\text{tr.}(ABC)}{\text{tp.}(BCMN)}=\frac{\text{tr.}(DEG)}{\text{tp.}(LKEG)}.^{*}$$

But

$$\frac{AC}{AM}\cdot\frac{BC}{NM}=\frac{DG}{DL}\cdot\frac{EG}{KL};$$

we thus deduce

$$(6)\qquad \frac{\text{tr.}(ABC)}{\text{tr.}(ANM)}=\frac{\text{tr.}(EDG)}{\text{tr.}(DLK)}.^{*}$$

We have

$$\frac{\text{tr.}(ABC)}{\text{polyg.}(ANBC)}=\frac{\text{tr.}(EDG)}{\text{polyg.}(DKEG)};$$

hence

$$\frac{\text{tr.}(ABC)}{\text{tr.}(EDG)}=\frac{\text{polyg.}(ANBC)}{\text{polyg.}(DKEG)}=\frac{\text{port.}(ABC)}{H},$$

and hence

* Justification of the equalities (5) and (6):
If we posit $A\hat{C}B=\alpha$ and $D\hat{G}E=\beta$, we have

$$\text{tr.}(ABC)=\frac{1}{2}BC\cdot AC\sin\alpha,\qquad \text{tp.}(BCMN)=\frac{1}{2}(BC+NM)\cdot CM\sin\alpha,$$
$$\text{tr.}(EGD)=\frac{1}{2}EG\cdot DG\sin\beta,\qquad \text{tp.}(LKEG)=\frac{1}{2}(EG+KL)\cdot GL\sin\beta;$$

hence the equality (5).

Likewise

$$\text{tr. }(ANM)=\frac{1}{2}MN\cdot AM\sin\alpha \text{ and tr. }(DLK)=\frac{1}{2}KL\cdot DL\sin\beta;$$

hence the equality (6).

$$\frac{\text{polyg.}(ANBC)}{\text{port.}(ABC)} = \frac{\text{polyg.}(DKEG)}{H},$$

which is impossible, since

$$\text{polyg. }(ANBC) < \text{port. }(ABC) \Rightarrow \text{polyg. }(DKEG) < H,$$

yet, we posited, by hypothesis, polyg. $(DKEG) > H$.

2) If we suppose

$$\frac{\text{tr.}(ABC)}{\text{tr.}(DGE)} = \frac{\text{port.}(ABC)}{H}, \quad \text{with } H > \text{port. }(DEG),$$

this entails that we assumed

$$\frac{\text{tr.}(DEG)}{\text{tr.}(ABC)} = \frac{\text{port.}(DEG)}{H_1}, \quad \text{with } H_1 < \text{port.}(ABC).$$

The preceding form of reasoning shows that this is absurd; hence we obtain the conclusion.

Comments:

1) Ibn Hūd's demonstration is undertaken by way of the quadrilaterals $DKEG$ and $ANBC$, which are obtained through first dividing the arcs DE and AB at points K and N respectively, while assuming that from this step we have

$$\text{area }(DKEG) > H.$$

He did not show that the same mode of reasoning can be applied, if necessary, to the division of arcs DE and AB in 2^n parts in order to obtain a polygon P_n such that

$$\text{area }(P_n) > H.$$

In the next step, to the midpoint I_1 of KD we associate K_1 as an intersection of the parallel to DG passing by I_1 and the arc KD, and L_1 on DG such that $K_1L_1 \parallel KL \parallel EG$; similarly, to point I'_1, as the midpoint of KE, we associate K'_1 as the intersection of the parallel to DG passing by I'_1 and the arc EK, and L'_1 on DG such that $K'_1L'_1 \parallel KL \parallel EG$. To the points L_1 and L'_1 we associate on AC the points M_1 and M'_1 such that

$$\frac{AC}{DG} = \frac{AM_1}{DL_1} = \frac{AM_1'}{DL_1'}.$$

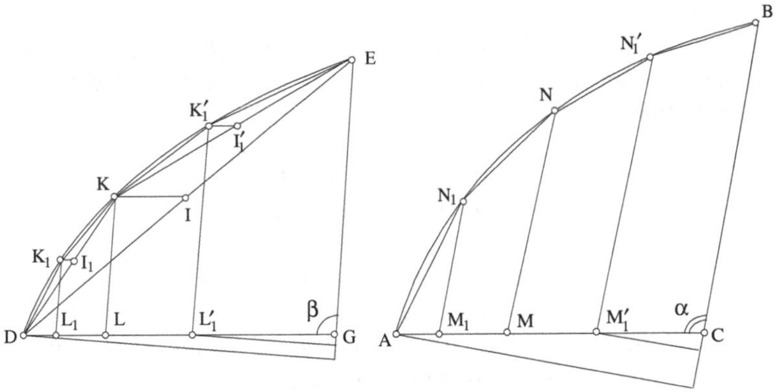

Fig. 7.6

We thus have

$$\frac{AC}{DG} = \frac{AM_1}{DL_1} = \frac{M_1M}{L_1L} = \frac{MM_1'}{LL_1'} = \frac{M_1'C}{I_1'G} = \lambda,$$

namely similar divisions into the segments AC and DG.

To points M_1 and M'_1 we associate on the arc AB the points N_1 and N'_1 such that $M_1N_1 \parallel MN \parallel M'_1N'_1 \parallel CB$.

Using the equations of the two conic sections, we show, as we did with points K and N, that

$$\frac{BC}{N_1M_1} = \frac{EG}{K_1L_1} \quad \text{and} \quad \frac{BC}{N_1'M_1'} = \frac{EG}{K_1'L_1'};$$

we thus have

$$\frac{BC}{EG} = \frac{N_1M_1}{K_1L_1} = \frac{NM}{KL} = \frac{N_1'M_1'}{K_1'L_1'} = \mu.$$

Each of the polygons P_2 and P'_2 obtained by dividing the arcs DE and AB into 2^2 equal parts is composed of a triangle and three trapezoids. If we designate by h_1, h_2, h_3 and h_4 the heights respective to the triangle DL_1K_1, and the trapezoids (K_1L), (KL'_1) and (K'_1G) with h'_1, h'_2, h'_3 and h'_4 the heights of their homologues in the second figure, we have

$$\frac{h_i'}{h_i} = \frac{AC \sin\alpha}{DG \sin\beta} = \lambda \frac{\sin\alpha}{\sin\beta}, \quad \text{for } i \in \{1, 2, 3, 4\}.$$

The properties of the polygons P_2 and Q_2 that are defined as such would thus be those of polygons A and B, which were studied by Ibn Sinān in his Proposition 1.[36]

It would be the same for the polygons P_n and Q_n obtained by way of dividing the arcs DE and AB into 2^n parts as per the indicated procedure; we thus show that

$$\frac{\text{area }(P_n)}{\text{area tr.}(DEG)} = \frac{\text{area }(Q_n)}{\text{area tr.}(ABC)}.$$ [37]

2) The portions ABC and DEG under consideration belong to segments BAB' and EDE', and are obtained by tracing the chords BB' and EE' with respective midpoints C and G. The triangles ABC and ACB' have equal areas, and the same applies to triangles DEG and DGE'.

It is clear that the result established in Proposition 18, for two portions belonging to distinct sections of the same kind, applies also to two portions belonging to a same section. We thus have

$$\frac{\text{tr.}(ABC)}{\text{tr.}(ACB')} = \frac{\text{port.}(ABC)}{\text{port.}(ACB')},$$

and, consequently,

$$\text{port. }(ABC) = \text{port. }(ACB').$$

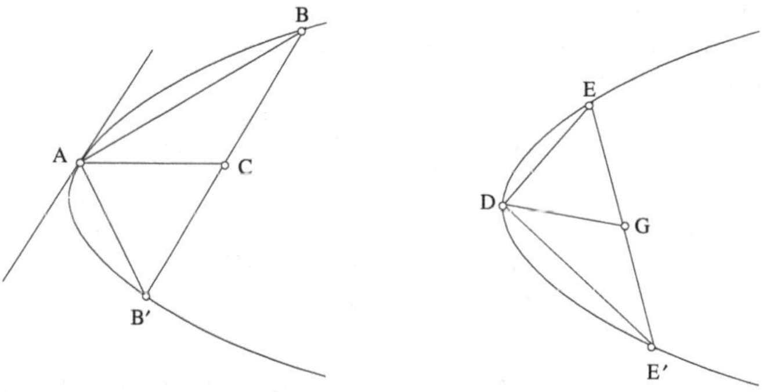

Fig. 7.7

Similarly

$$\text{port. }(EDG) = \text{port. }(DGE').$$

36 Refer to the mathematical commentary to Ibn Sinān, Chapter III.
37 *Ibid.*

Hence, the area of each of the portions ABC and DEG is equal to half the area of each of the segments BAB' and EDE'.

Proposition 19. — *The portions being studied are on the same section; namely,* AC *and* BE.

a) *If the section is a parabola, and if the diameter* AD, *which defines the first portion, is equal to the diameter* BG *that defines the second, then portions* ACD *and* BEG *are equal.*

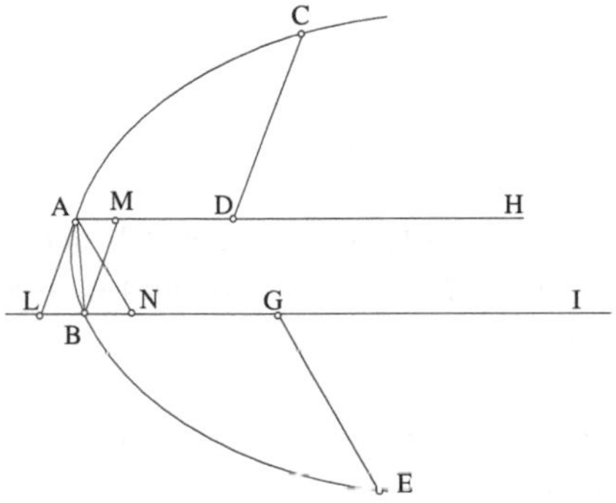

Fig. 7.8

b) *If the section is an ellipse or a hyperbola, the transverse diameters issued from A and B are respectively* Δ *and* Δ′; *hence, if*

$$\frac{\Delta}{AD} = \frac{\Delta'}{BG},$$

then the portions ACD *and* BEG *are equal.*

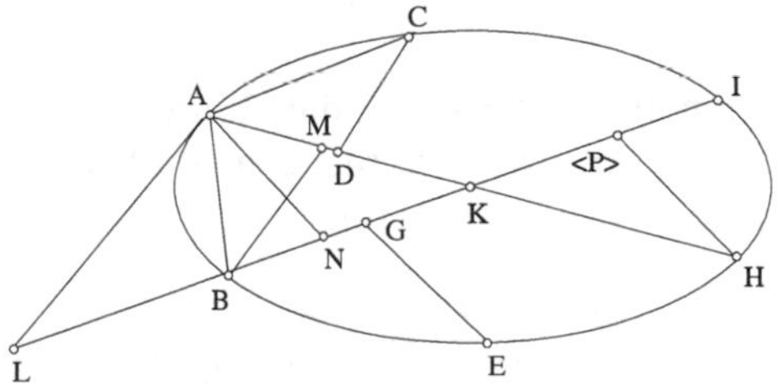

Fig. 7.9

Let AN be the ordinate of A relative to BG, and BM the ordinate of B relative to AD, and AL the tangent in A that cuts BG in L.

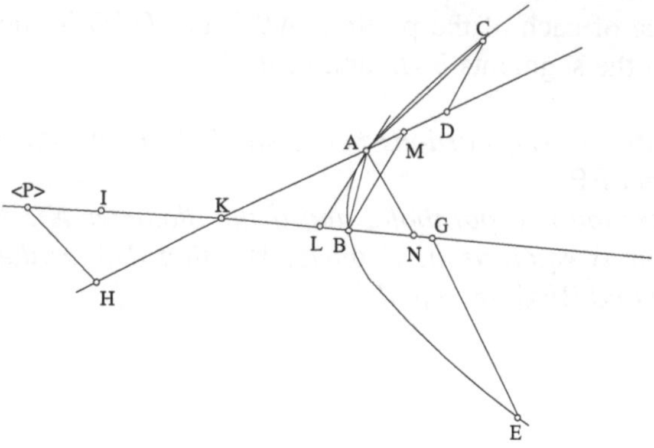

Fig. 7.10

a) If the section is a parabola, then $BL = BN$ according to Apollonius' *Conics* I.35. We thus have $BL = BN = AM$, and consequently the triangles ABM and ABN have equal areas.

b) If the section is an ellipse or a hyperbola, we have, following Apollonius' *Conics* I.36,

$$(1) \qquad \frac{IN}{NB} = \frac{LI}{LB}.$$

From (1) we deduce

$$\frac{BL}{BN} = \frac{LI}{IN};$$

hence

$$\frac{BL}{BN} = \frac{LI + \varepsilon BL}{IN + \varepsilon BN}$$

with $\varepsilon = +1$ for the hyperbola, $\varepsilon = -1$ for the ellipse.

Let there be $HP \parallel AN$, with K the centre of symmetry in the two sections; we then have $IP = BN$. Therefore

$$\frac{BL}{BN} = \frac{BI}{IN + \varepsilon IP} = \frac{BI}{PN} = \frac{BK}{KN};$$

thus

$$\frac{KN}{BN} = \frac{BK}{BL}.$$

Yet $BM \parallel AL$, so

$$\frac{KL}{LB} = \frac{KA}{AM};$$

we thus have

$$\frac{KB}{NB} = \frac{KA}{AM}.$$

Yet

$$\frac{AM}{AK} = \frac{\text{tr.}(ABM)}{\text{tr.}(ABK)}$$

and

$$\frac{BN}{BK} = \frac{\text{tr.}(ABN)}{\text{tr.}(ABK)};$$

the triangles ABM and ABN therefore have equal areas.

By hypothesis we have

$$\frac{AD}{AH} = \frac{BG}{BI}$$

or

$$\frac{AD}{AK} = \frac{BG}{BK};$$

we equally have

$$\frac{AK}{AM} = \frac{BK}{BN};$$

therefore

$$\frac{AD}{AM} = \frac{BG}{BN}.$$

This equality is verified for the three sections.

We deduce, as in Proposition 18 – namely, by using in each case the equation of the section being considered – that

$$\frac{CD}{BM} = \frac{EG}{AN}.$$

Consequently

$$\frac{CD}{BM} \cdot \frac{AD}{AM} = \frac{EG}{AN} \cdot \frac{GB}{BN};$$

hence

$$\frac{\text{tr.}(ACD)}{\text{tr.}(ABM)} = \frac{\text{tr.}(EGB)}{\text{tr.}(ANB)}.$$

However,

$$\text{tr.}(ABM) = \text{tr.}(ABN),$$

and accordingly

$$\text{tr.}(ACD) = \text{tr.}(EGB),$$

but, following Proposition 18,

$$\frac{\text{tr.}(ACD)}{\text{tr.}(EGB)} = \frac{\text{port.}(ACD)}{\text{port.}(BGE)};$$

thus, the portions ACD and BGE are equal.

Reciprocally – If two portions ACD and BEG of the same section have equal areas, the straight lines AD and BG, being the diameters Δ and Δ' that issue from A and from B, with CD and EG the ordinates of C and E relative to these diameters, then:

- if the section is a parabola, then $AD = BG$;
- if the section is an ellipse or a hyperbola, we have $\frac{AD}{\Delta} = \frac{BG}{\Delta'}$.

Comment. — Propositions 18 and 19 discussed here relate to two portions belonging to two parabolas, two ellipses or two hyperbolas, or to a single section. The area of each of the portions considered is half that of the segment with which it is associated (see Comment 2, Proposition 18).

It should be remembered that Ibn Sinān was only interested in the parabola, and that his second proposition considered the ratio of the areas of two segments of a parabola using Proposition 1 as a lemma.

Note that, in Proposition 18, Ibn Hūd makes use without justification of two inequalities relating to the areas of the portions and the triangles associated with them.[38] In Propositions 18 and 19 he uses, without acknowledgement, equalities that are either direct applications or consequences of propositions established by Apollonius.

However the study of the implication of

[38] See Note 35.

$$\frac{AC}{AM} = \frac{DG}{DL} \Rightarrow \frac{BC}{NM} = \frac{EG}{KL},$$

that is deployed in Proposition 18, has been examined – in the case of the ellipse and the hyperbola – in the last part of Proposition 14.

Proposition 20. — *Let* ABC *be a segment of a parabola, with vertex* B, *and base* AC; *we have*

$$\text{sg. (ABC)} = \frac{4}{3}\,\text{tr.(ABC)}.$$

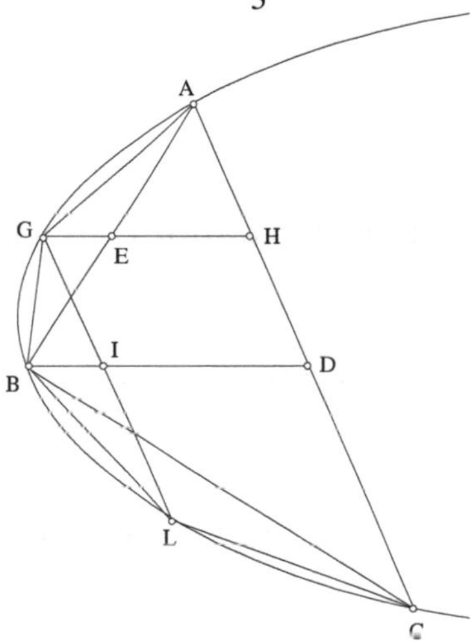

Fig. 7.11

Let BD be the conjugate diameter of AC. From the midpoint of AB we draw a parallel to BD, namely GEH, with ordinate GIL from G. Hence

$$AB = 2BE \Rightarrow AD = 2DH.$$

However,

$$HD = GI;$$

hence

$$AD = 2GI,$$

$$\frac{AD^2}{GI^2} = \frac{BD}{BI} = \frac{4GI^2}{GI^2} = 4,$$

and hence

$$BD = 4BI.$$

We thus have

$$\text{tr.}\ (BGI) = \frac{1}{8}\ \text{tr.}\ (ABD).$$

We equally have

$$GH = ID = 3BI$$

and

$$EH = \frac{1}{2}BD = 2BI,$$

so

$$GE = BI,$$

Hence

$$\text{tr.}\ (BGI) = \text{tr.}(BGE) = \frac{1}{2}\ \text{tr.}(AGB),$$

and therefore

$$\text{tr.}(AGB) = \frac{1}{4}\ \text{tr.}(ABD).$$

We also have

$$\text{tr.}(BLC) = \frac{1}{4}\ \text{tr.}(BDC),$$

so

$$\text{tr.}\ (AGB) + \text{tr.}(BLC) = \frac{1}{4}\ \text{tr.}(ABC),$$

and thence

$$\text{port.}(AGB) + \text{port.}(BLC) = \frac{1}{4}\ \text{port.}(ABC).$$

However,

$$\text{port.}(ABC) - [\text{port.}(AGB) + \text{port.}(BLC)] = \text{tr.}(ABC),$$

and then

$$\frac{3}{4}\ \text{port.}\ (ABC) = \text{tr.}(ABC)$$

or

$$\text{port.}(ABC) = \frac{4}{3}\ \text{tr.}(ABC).$$

Comments:

1) The demonstration barely differs from that of Ibn Sinān in Proposition 3. The comparison of the areas of two triangles of the same basis, which are BGA and BDA, is deduced here from the equalities $GE = BI = \frac{1}{4} BD$, without the use of heights, while Ibn Sinān shows that the ratio of the heights of the two triangles under consideration is $\frac{1}{4}$ and that the same applies to the ratio of the areas of the two triangles.

2) In Propositions 18 and 19 the studied properties are applied to the portions of parabolas, of ellipses or hyperbolas, while in Proposition 20 the property that is being studied concerns only the parabola.

In fact, it is clear that if we established for an ellipse of a diameter Δ the construction we indicated with regard to the parabola, we would have

$$\frac{AD^2}{GI^2} = 4 = \frac{BD(\Delta - BD)}{BI(\Delta - BI)}$$

and not

$$\frac{BD}{BI} = 4.$$

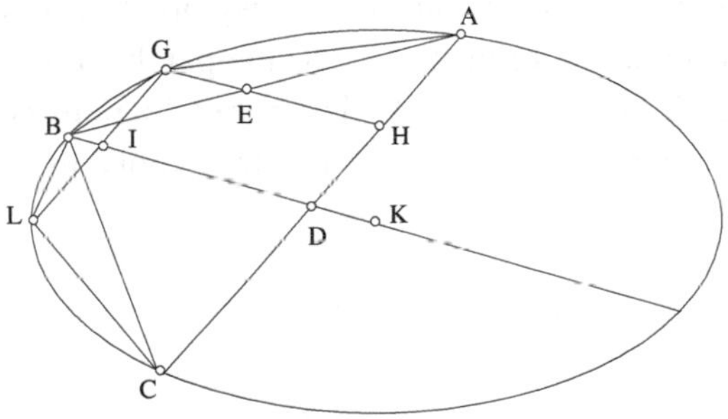

Fig. 7.12

Hence the remainder of this line of reasoning does not apply to the case of the ellipse.

Proposition 21. — *How to separate from a parabola* ABC *a portion with the vertex* B *that is equal to a given surface* D.

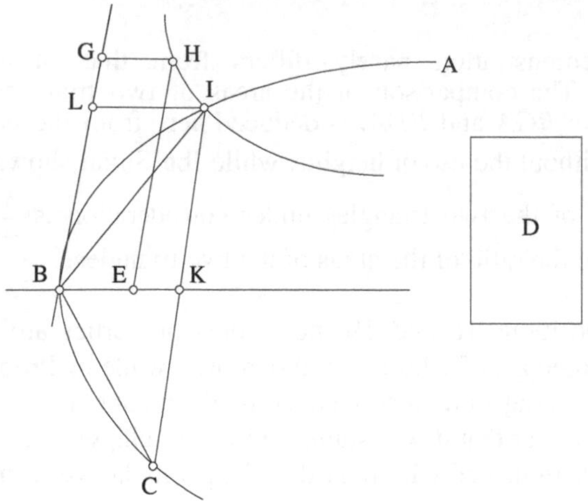

Fig. 7.13

Extend from B the diameter BE and the tangent BG, and construct upon BG a parallelogram $GBEH$ with an area equal to $\frac{3}{4}$ D. From point H, draw a hyperbola having BG and BE as asymptotes; it cuts the parabola in point I. The straight line ordinate IK cuts the parabola in C and the diameter in K, drawing IL as parallel to BK. The parallelograms (BH) and (BI) have equal areas (as per the property of the hyperbola in Apollonius' *Conics*, II.12). Then

$$(BH) = (BI) = \frac{3}{4} D,$$

$$\text{tr.}(IBC) = (BI) = \frac{3}{4} D.$$

But

$$\text{tr.}(IBC) = \frac{3}{4} \text{ port.}(IBC), \qquad \text{following Proposition 20;}$$

hence

$$\text{port. }(IBC) = D.$$

7.2.3. *Translation:* Kitāb al-Istikmāl

– **18** – Let there be two portions belonging to two parabolas, or two portions belonging to two hyperbolas or to two ellipses, such that the ratio of the transverse diameter of one to its diameter is equal to the ratio of the transverse diameter of the other to its diameter, then the ratio of the area of one of the two portions to the area of the other is equal to the ratio of the triangle whose base is the base of the portion and whose vertex is at its vertex to the triangle in the other portion whose base is the base of the section and whose vertex is at its vertex.[39]

Example: The two portions *AB* and *ED* belong to two homogeneous[40] sections. The diameter of the portion *AB* is the straight line *AC* and its ordinate is the straight line *BC*, and the diameter of the portion *DE* is the straight line *DG* and its ordinate is the straight line *EG*. If the two portions belonged to[41] two sections that were not parabolic, then the ratio of the transverse diameter of the portion *AB* to *AC* is equal to the ratio of the transverse diameter of the portion *DE* to the straight line *DG*. We join *AB* and *DE*.

I say that the ratio of the area of the portion ABC *to the area of the portion* DEG *is equal to the ratio of the triangle* ABC *to the triangle* DEG.

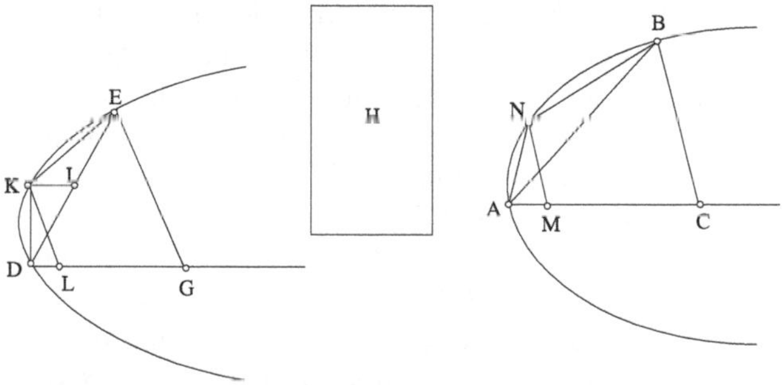

Fig. 7.14

Proof: It could not be otherwise. If this was possible, let the ratio of the triangle *ABC* to the triangle *DEG* be equal to the ratio of the portion *ABC* to an area less than or greater than the area of the portion *DEG*, and let this be the area *H*.

Let us assume first of all that it is less than the area of the portion *DEG*. Let us divide the straight line *DE* into two halves at the point *I*, and let us

[39] See the mathematical commentary.
[40] That is: of the same type.
[41] Lit.: if they were between.

produce the diameter *IK* through the point *I* until it meets the section at the point *K*, and let us join *EK* and *KD*. As the area of the triangle *DEG* is greater than half of the portion *DEG*, and as the triangle *DKE* is greater than half the portion *DEK*, if we continue to proceed in this way, we eventually arrive at a polygonal area that is greater than the area *H*. Let this area be *DKEG*. We draw the straight line ordinate *KL* from the point *K*. We divide the straight line *AC* at the point *M* such that the ratio of *AM* to *MC* is equal to the ratio of *DL* to *LG*. We draw the straight line ordinate *MN* from the point *M* and we join *AN* and *NB*. As the ratio of the transverse diameter to *AC* is equal to the ratio of the transverse diameter to *DG*, and as the ratio of *AC* to *CM* is equal to the ratio of *DG* to *GL*, then the ratio of *BC* to *NM* is equal to the ratio of *EG* to *KL*.[42] Therefore the ratio of *AC* to *CM* multiplied[43]* by the ratio of *BC* to the sum of *BC* and *NM* considered as a single straight line – which is equal to the ratio of the triangle *ABC* to the area of the quadrilateral *MNBC* – is equal to the ratio of *DG* to *GL* multiplied* by the ratio of *EG* to the sum of *EG* and *KL* considered as a single straight line – which is equal to the ratio of the triangle *DEG* to the area of the quadrilateral *LKEG*. But the ratio of *AC* to *AM* multiplied* by the ratio of *BC* to *NM* – which is equal to the ratio of the triangle *ABC* to the triangle *ANM* – is equal to the ratio of *DG* to *DL* multiplied* by the ratio of *EG* to *KL* – which is equal to the ratio of the triangle *DEG* to the triangle *DLK*. Therefore the ratio of the triangle *ABC* to the whole area of the polygon *ANBC* is equal to the ratio of the triangle *DEG* to the area of the polygon *DKEG*. If we apply a permutation, then the ratio of the triangle *ABC* to the triangle *DEG* – which is equal to the ratio of the area of the portion *ABC* to the area *H* – is equal to the ratio of the area of the polygon *ANBC* to that of the polygon *DKEG*. Therefore, the ratio of the area of the polygon *ANBC* to that of the polygon *DKEG* is equal to the ratio of the area of the portion *ABC* to the area *H*. If we apply a permutation, the ratio of the area of the polygon *ANBC* to the portion *ABC* is equal to the ratio of the area of the polygon *DKEG* to the area *H*. But the area of the polygon *ANBC* is less than the area of the portion *ABC*, and hence the area of the polygon *DKEG* is less than the area *H*. But we initially assumed it was greater, so this is contradictory and this is not possible.

Therefore, the ratio of the triangle *ABC* to the triangle *DEG* is not equal to the ratio of the portion *ABC* to an area less than the area of the portion *DEG*.

[42] See the mathematical commentary. This equality is obtained from the *Conics* of Apollonius, I.20 for the parabola and I.21 for the ellipse or hyperbola.

[43] Lit.: doubled (hereafter asterisked).

I also say: and neither to an area greater than it. If this were the case, the ratio of the triangle *DEG* to the triangle *ABC* would be equal to the ratio of the area of the portion *DEG* to an area less than the area of the portion *ABC*. We have already shown that this is contradictory, therefore the ratio of the triangle *ABC* to the triangle *DEG* is equal to the ratio of the area of the portion *ABC* to the area of the portion *DEG*. That is what we wanted to prove.

– **19** – Consider two portions of the same section. If the section is a parabola and the diameters of the two portions are equal, then the two portions are equal. If the section is not a parabola but the ratio of the transverse diameter of one to its diameter is equal to the ratio of the transverse diameter of the other to its diameter, then the two portions are equal.

Example: Let *AD* be the diameter of the portion *AC* of the parabola *AB*, equal to the diameter of the portion *BE*, which is *BG*. If it is not a parabola, then let the transverse diameter of the portion *AC* be the straight line *AH* and let the transverse diameter of the portion *BE* also be the straight line *BI* and the centre at the point *K*, and let the ratio of *HA* to *AD* be equal to the ratio of *IB* to *BG*.

I say that the portion AC *is equal to the portion* BE.

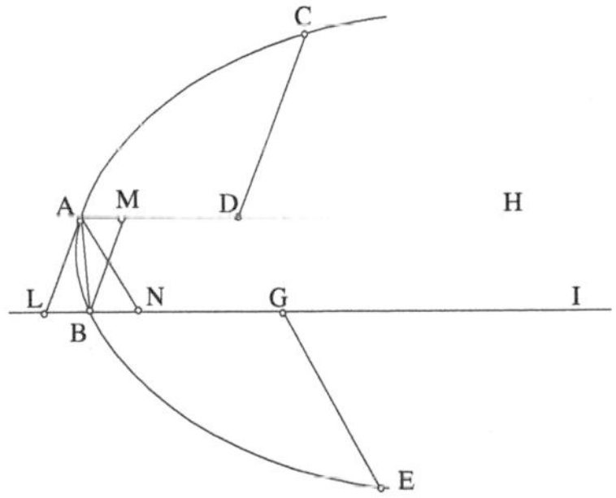

Fig. 7.15

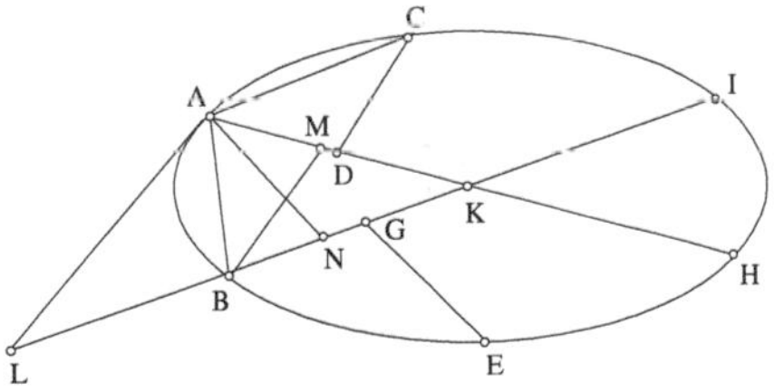

Fig. 7.16

Proof: From point *A* on the diameter *BG*, we draw the straight line ordinate *AN*, and from point *B* on the diameter *AD* we draw the straight line ordinate *BM*. We join *AB*, *AC* and *BE* and produce the straight line tangent *AL* from the point *A* meeting the diameter *BI* at the point *L*. If the section is a parabola, then the straight line *BL* will be equal to the straight line *BN*[44] and the straight line *BL* will be equal to the straight line *AM*. For this reason, the triangle *AMB* will be equal to the triangle *ABN*. If the section is not a

[44] See the *Conics* of Apollonius, I.35.

parabola, then the ratio of *IN* to *NB* is equal to the ratio of *IL* to *LB*.[45] If we compound (*componendo*), then the ratio of *IN* plus *NB* to *NB* is equal to the ratio of *IL* plus *BL* to *BL*. The halves of the antecedents are also proportional, and therefore the ratio of *KN* to *NB* is equal to the ratio of *KB* to *BL*.[46] If we separate (*separando*), then the ratio of *KB* to *BN* is equal to the ratio of *KL* to *LB*, which is equal to the ratio of *KA* to *AM*, which is equal to the ratio of the triangle *ABK* to each of the triangles *ABN* and *BAM*. Therefore, the two triangles *ABN* and *BAM* are equal. As the ratio of *DA* to *AH* is equal to the ratio of *GB* to *BI*, and the ratio of *HA* to *AK* is equal to the ratio of *IB* to *BK*, and the ratio of *KA* to *AM* is equal to the ratio of *KB* to *BN*, then by the equality (*ex aequali*) the ratio of *DA* to *AM* is equal to the ratio of *GB* to *BN*. That is why, for all sections, the ratio of *CD* to *BM* is equal to the ratio of *EG* to *AN*. <The ratio of > the straight line *CD* to the straight line *BM* multiplied* by the ratio of *DA* to *AM* – which is equal to the ratio of the triangle *ACD* to the triangle *BAM* – is equal to the ratio of *EG* to *AN* multiplied* by the ratio of *GB* to *BN*, which is equal to the ratio of the triangle *GEB* to the triangle *ANB*. But the two triangles *ABM* and *ANB* are equal,[47] and therefore the two triangles *ACD* and *BEG* are equal, and hence the two portions *AC* and *BE* are equal.

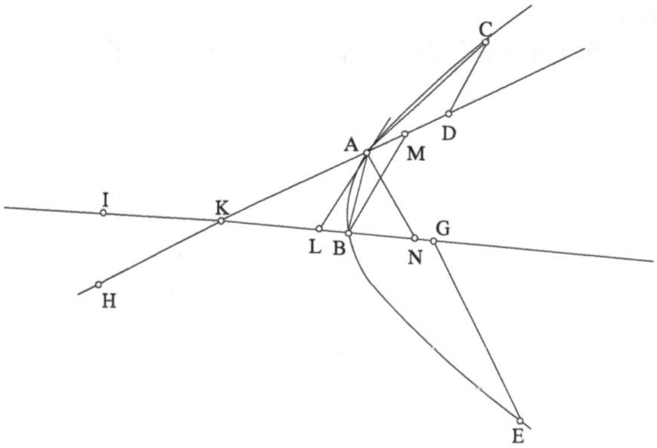

Fig. 7.17

If we now have two equal portions, we can go on to show that the ratio of the transverse diameter of one to its diameter is equal to the ratio of the transverse diameter of the other to its diameter, or furthermore, if we add a portion to one or remove a portion from one, how to add an equal portion to the other or to remove an equal portion from it.

[45] See the *Conics* of Apollonius, I.36.
[46] See the mathematical commentary.
[47] See the mathematical commentary.

– **20** – For any portion of a parabola, its area is one and one third times the area of the triangle whose base is the base of the parabola and whose vertex is its vertex.

Example: The vertex of the portion *ABC* is the point *B* and its base is the straight line *AC*. We join *AB* and *BC*.

I say that the area of the portion ABC *is equal to one and one third times the area of the triangle* ABC.

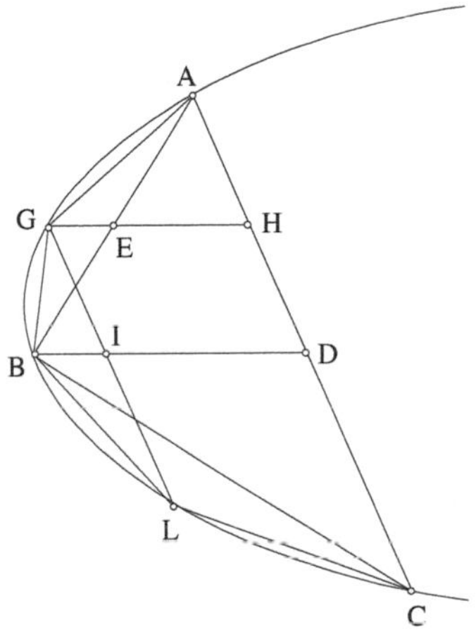

Fig. 7.18

Proof: Let the straight line *BD* be the diameter of the portion. We divide the straight line *AB* into two halves at the point *E*, and we produce the diameter *GEH* from it until it meets the section at the point *G* and the straight line *AC* at the point *H*. We produce the straight line ordinate *GIKL* from the point *G* until it meets the diameter *BD* at the point *I* and the section at the point *L*. We join *AG*, *GB*, *BL*, and *LC*. As *BA* is twice *BE*, then *AD* is twice *DH* which is equal to *GI*. But the ratio of the square of *AD* to the square of *GI* is equal to the ratio of *DB* to *BI*, and the square of *AD* is four times the square of *GI*. Therefore, *DB* is four times *BI*, and therefore the triangle *GBI* is one eighth of the triangle *ABD*. But the triangle *GBI* is equal to the triangle *GBE* and the triangle *AGB* is twice the triangle *GBE*. Therefore, the triangle *AGB* is one quarter of the triangle *ABD*. Similarly, the triangle *BLC* is one quarter of the triangle *BCD*, and therefore the two triangles *AGB* and *BLC* are one quarter of the triangle *ABC*, and the two portions *AGB* and *BLC* are one quarter of the portion *ABC*. Therefore the area of the entire portion *ABC* is equal to one and one third times the area of the triangle *ABC*. That is what we wanted to prove.

– **21** – We wish to show how to separate a portion from a parabola such that the area of this portion is equal to the area of a known rectangle and the vertex of the portion is at a known point.

Let the section be *ABC* and let the known area be area *D*. We wish to separate a portion from the section *ABC* such that the area of the portion is equal to area *D* and the vertex of the portion is at the point *B*.

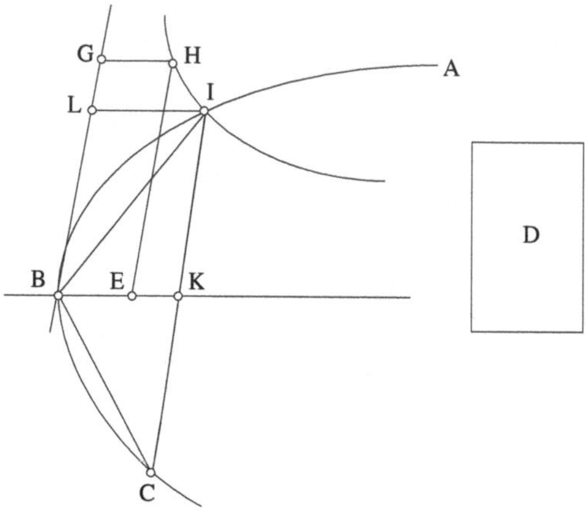

Fig. 7.19

From the point *B*, let us draw a diameter *BE* and a straight line tangent *BG*, and let us apply to the straight line *BG* an area equal to three-quarters of the area *D*. Let us draw the area *BGH* with parallel sides and with an angle *H* equal to the angle *B*. Let us make a hyperbola passing through the point *H* with asymptotes *GB* and *BE* such that it meets the section *AB* at the point *I*. From the point *I*, let us produce the straight lines *IK* and *IL* parallel to the two straight lines *GB* and *BE*. The area *BI* is then equal to the area *HB*, which is equal to three-quarters of the area *D*. Let us extend *IK* until it meets the section at the point *C*. We join *IB* and *BC*. The triangle *IBC* is then equal to three-quarters of the area *D* and it is equal to three-quarters of the portion *IBC*. Therefore, the portion *IBC* is equal to the surface *D*.

From this we can show how, given a portion of a parabola, another portion may be separated from the section such that its ratio to the given portion is any given ratio. That is what we wanted to prove.

7.3. THE ISOPERIMETRIC PROBLEM

7.3.1. *An extremal property or a geometric property*

Ibn Hūd's study of the isoperimetric problem forms part of one chapter of the *Istikmāl* on the properties of circles relating to 'the angles, surfaces, and lines that are inscribed in them'.[48] That is to say that, once again, it is not the extremal properties of circles that interest the author, but only those that arise from elementary geometry. Both these sections serve to illustrate the consistency of Ibn Hūd's approach. As in the case of the measurement of a parabola, the infinitesimal properties are not of interest in their own right, but rather as a means of gaining a better understanding of the geometrical figures. As before, he borrows considerably, but in a well-regulated way. In the case of the measurement of a parabola, his main sources are Apollonius and Ibn Sinān. For the isoperimetric problem, he turns instead to Archimedes, Ptolemy and Ibn al-Haytham.

However, to concentrate on exposing these borrowings would be to ignore the unifying aspects of the *Istikmāl* and to misunderstand the individual contribution of Ibn Hūd. This becomes clear with a better understanding of the aim of Ibn Hūd in producing this work. This was to make use of polygons inscribed or circumscribed on a circle to study the relationships between chords, or between chords and arcs, *i.e.* just those relationships that most often give rise to trigonometric relationships. In order to understand better the path taken by Ibn Hūd, we must make a brief examination of his work, especially from the point at which he first introduces polygons in the eleventh proposition. The isoperimetric problem itself is discussed in two propositions, the 16th and 19th,[49] which we shall translate later.

Proposition 11, taken from *The Sphere and the Cylinder* by Archimedes – I.3 – is stated as follows: 'Given two unequal magnitudes and a circle, show how to draw a polygon inscribed within the circle and a similar polygon circumscribed around it such that the ratio of the side of the circumscribed polygon to the side of the inscribed polygon is smaller that the ratio of the greater of the two magnitudes to the smaller of these magnitudes'.[50] This involves the construction of a regular polygon P with a side c that is inscribed within a circle and a similar polygon P' with a side c' that is circumscribed in this circle, such that

[48] Ms. Copenhagen, Royal Library, fol. 44^{v}.

[49] See the edition of the Arabic text in *Mathématiques infinitésimales*, vol. I.

[50] Mss Copenhagen, Royal Library, fols $48^{r\text{-}v}$ and Leiden, Or. 123, fols $3^{r\text{-}v}$.

$$1 < \frac{c'}{c} < k \qquad (k > 1 \text{ as given ratio}).$$

This demonstration refers to the consideration of an acute angle α such that $\cos\alpha > \frac{1}{k}$. We seek $\frac{c}{c'} \geq \cos\alpha$. Yet there exists N such that $n > N$, entailing $\frac{\pi}{2^n} \leq \alpha$;[51] hence $\cos\frac{\pi}{2^n} \geq \cos\alpha$. Therefore $\frac{\pi}{2^{n-1}}$ is the arc whose chord is the required side c. We thus have

$$c = 2R \sin \frac{\pi}{2^n}, \quad c' = 2R \tan \frac{\pi}{2^n};$$

hence

$$\frac{c}{c'} = \cos\frac{\pi}{2^n} \geq \cos\alpha \quad \text{or} \quad \frac{c'}{c} \leq \frac{1}{\cos\alpha} < k.$$

The polygons have 2^n sides. The second part of this same Proposition 11 deals with the ratio of the areas of two polygons that are obtained in the same manner as in the first part of the proposition, based on Archimedes' *The Sphere and the Cylinder*, I.5.

Proposition 12[52] is also taken from Archimedes (Propositions 21 and 22). However, it should be noted that, in his Proposition I.21, Archimedes considers a regular polygon with an even number of sides, rather than a multiple of four as is assumed by Ibn Hūd. This hypothesis is not used in the proof, which is identical to that of Archimedes. The final section of Ibn Hūd's proposition, corresponding to (2), is established by Archimedes in I.22 in the same book (see also Proposition 12 of the Banū Mūsā).

This is rewritten in the form: Let there be a regular polygon with $4n$ sides, $A_0A_1 \ldots A_{2n}A_{2n+1} \ldots A_{4n-1}$. The straight line A_0A_{2n} is an axis of symmetry, and the straight lines A_iA_{4n-i} $(1 \leq i \leq 2n-1)$ are perpendicular to A_0A_{2n} at the points $L_1, \ldots, L_{2n-1}$, with L_n as the midpoint of A_0A_{2n}; thus, we have

$$(1) \qquad \frac{\sum_{i=1}^{2n-1} A_iA_{4n-i}}{A_0A_{2n}} = \frac{A_1A_{2n}}{A_0A_1} \Leftrightarrow \left[\sum_{i=1}^{2n-1} \sin i\frac{\pi}{2n} = \operatorname{cotan}\frac{\pi}{4n}\right].$$

He shows that for $1 < \alpha \leq 2n - 1$, we have

[51] This is an application of the porism in Proposition X.1 of the *Elements* of Euclid discussed by Ibn al-Haytham (see *Les Mathématiques infinitésimales*, vol. II, pp. 499–500).

[52] Mss Copenhagen, Royal Library, fols 48^v–49^r and Leiden, Or. 123, fols 4^r–5^r.

$$(2)\qquad \frac{\sum_{i=1}^{\alpha-1} A_iA_{4n-i} + A_\alpha L_\alpha}{A_0L_\alpha} = \frac{A_1A_{2n}}{A_0A_1} \Leftrightarrow \left[\frac{2\left[\sum_{i=1}^{\alpha-1} \sin i\frac{\pi}{2n} + \frac{1}{2}\sin\alpha\frac{\pi}{2n}\right]}{1-\cos\alpha.\frac{\pi}{2n}} = \text{cotan}\frac{\pi}{4n} \right],$$

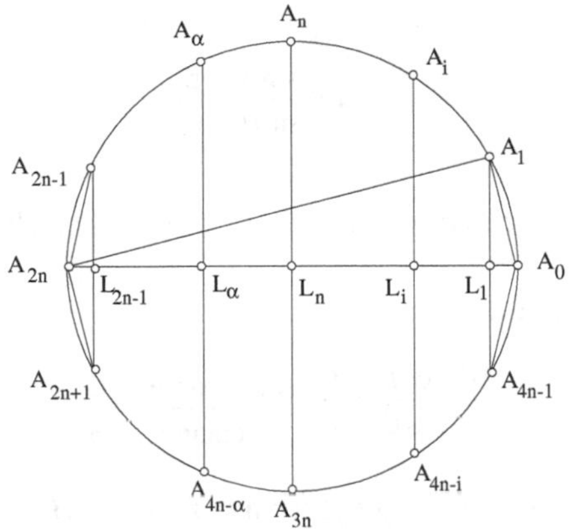

Fig. 7.20

Noting that for $\alpha = 1$ this equality yields

$$\frac{\sin\frac{\pi}{2n}}{1-\cos\frac{\pi}{2n}} = \text{cotan}\frac{\pi}{4n}.$$

The 13th proposition,[53] taken from the *Almagest*, states: 'For every quadrilateral that is inscribed in a circle, the sum of the products of each of its sides by that which is homologous and opposite to it is equal to the product of its diagonals with one another'.[54] The following proposition is a lemma to establish that, in a single circle or in two equal circles, the ratio of two angles at the centre (or two inscribed angles) is equal to the ratio of the arcs that they intercept. The 15th proposition,[55] taken from the *Almagest*,[56] shows that, if two arcs of a circle AB and BC are such that $\widehat{AB} + \widehat{BC} < 180°$ and $AB < BC$, then

[53] Mss Copenhagen, Royal Library, fol. 49^r and Leiden, Or. 123, fols 5^{r-v}.

[54] Heiberg, I.10, pp. 36–37; French trans. Halma I, p. 29.

[55] Mss Copenhagen, Royal Library, fols 49^v–50^r and Leiden, Or. 123, fols 6^v–7^v.

[56] Heiberg, I.10, pp. 43–5; French trans. Halma I, p. 34.

$$1<\frac{BC}{BA}<\frac{\widehat{BC}}{\widehat{BA}},$$

which is rewritten, by way of positing $\widehat{AB}=2\alpha$, $\widehat{BC}=2\beta$ with $\alpha+\beta<\frac{\pi}{2}$ and $\alpha<\beta$,

$$1<\frac{\sin\beta}{\sin\alpha}<\frac{\beta}{\alpha}.$$

Proposition 17[57] is rewritten: If, within a triangle ABC, we have $\hat{B}<\frac{\pi}{2}$, $\hat{C}<\frac{\pi}{2}$ and $AB>AC$, then

$$\frac{AB\ \cos\hat{B}}{AC\ \cos\hat{C}}>\frac{\hat{C}}{\hat{B}}\quad\text{or}\quad\frac{\operatorname{cotan}\hat{B}}{\operatorname{cotan}\hat{C}}>\frac{\hat{C}}{\hat{B}}.$$

In other words, the ratio of the projections of sides AB and AC onto the side BC is larger than the ratio of the angle C to angle B.

Finally, Proposition 18,[58] which is borrowed from the *Almagest*,[59] is rewritten in the form: Let there be a triangle ABC such that $AC<BC$, then, we have

$$\frac{AC}{CB-AC}>\frac{\hat{B}}{\hat{C}}\Leftrightarrow\left[\frac{\sin\hat{B}}{\sin\hat{A}-\sin\hat{B}}>\frac{\hat{B}}{\hat{C}}\right].$$

At the very least, this serves to illustrate the underlying path taken by Ibn Hūd and the reasons for including this series of borrowed propositions. We shall now consider Propositions 16 and 19.

7.3.2. *Mathematical commentary on Propositions 16 and 19*

Proposition 16. — *If within a triangle* ABC, *we have* AB > AC *and* AD ⊥ BC, *then* $\frac{BD}{DC}>\frac{B\hat{A}D}{D\hat{A}C}$.

Ibn Hūd takes D to be between B and C (see the comment).

[57] Mss Copenhagen, Royal Library, fol. 50^v and Leiden, Or. 123, fols 8^r–9^r.
[58] Mss Copenhagen, Royal Library, fol. 50^v and Leiden, Or. 123, fols 9^r–10^r.
[59] Heiberg vol. II, XII.1, pp. 456–8; French trans. Halma II, p. 317.

The circle (A, AC) cuts AD in H, AB in G, and BD in E. Thus, we have

$$\text{area tr. } (ABE) > \text{area sect. } (GAE),$$
$$\text{area tr. } (AED) < \text{area sect. } (EAH),$$

so

$$\frac{\text{area tr.}(ABE)}{\text{area tr.}(AED)} > \frac{\text{area sect.}(GAE)}{\text{area sect.}(EAH)}.$$

Hence

$$\frac{\text{area tr.}(ABD)}{\text{area tr.}(AED)} > \frac{\text{area sect.}(GAH)}{\text{area sect.}(EAH)},$$

and it follows that

$$\frac{BD}{DE} > \frac{B\hat{A}D}{E\hat{A}D}.$$

However,

$$DE = DC \text{ and } E\hat{A}D = D\hat{A}C;$$

hence

$$\frac{BD}{DC} > \frac{B\hat{A}D}{D\hat{A}C}.$$

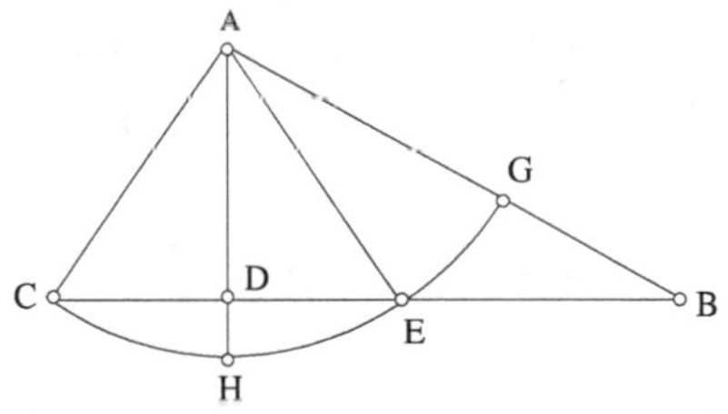

Fig. 7.21

Comment:

$$AB > AC \Rightarrow \hat{C} > \hat{B}.$$

D is between *B* and *C*, while taking *B* and *C* to be acute. Yet, we can have *C* as obtuse and the proposition would still be true, since, if we posit

$$D\hat{A}C = D\hat{A}E = \alpha,$$
$$D\hat{A}B = \beta,$$

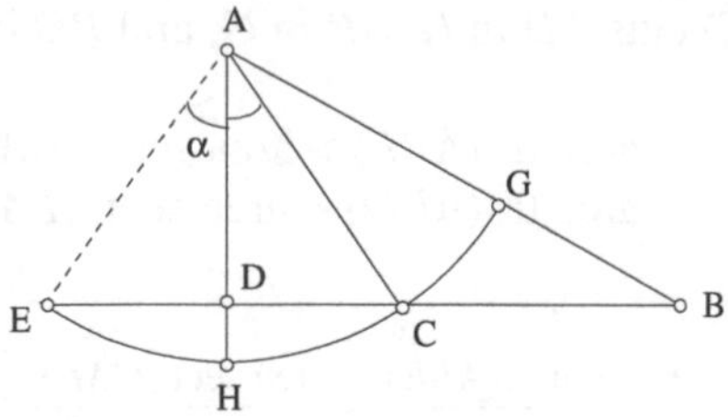

Fig. 7.22

then we have two cases,

$$DE = DC = AD \tan \alpha,$$
$$DB = AD \tan \beta,$$

and the result is noted as

$$\frac{\tan \beta}{\tan \alpha} > \frac{\beta}{\alpha}, \qquad \text{with } \alpha \text{ and } \beta \in \left]0, \frac{\pi}{2}\right[,$$

a lemma that is known and disseminated in Greek as well as in Arabic[60] (cf. al-Khāzin's treatise).

Proposition 19. — Let there be two regular polygons P_1 with n_1 sides at a length c_1, and P_2 with n_2 sides at a length c_2, such that $n_1c_1 = n_2c_2$ with $n_1 < n_2$; therefore $c_1 > c_2$. In order to compare their inscribed circles O_1 and O_2, Ibn Hūd uses a circle J equal to O_1. On the tangent at P to this circle, we take Q such that $PQ = \frac{1}{2} c_1$ and we extend the other tangent QZ. We thus have

$$P\hat{J}Z = \frac{2\pi}{n_1}, \qquad P\hat{J}Q = \frac{\pi}{n_1},$$

since the figure $PJZQ$ is equal to the figure associated with a summit A of polygon P_1. We then take on the tangent at P a point S and we extend the tangent SX, in order that figure $PSXJ$ becomes similar to the figure associated with a summit D of polygon P_2. It is sufficient to have $P\hat{J}S = \frac{\pi}{n_2}$, which determines point S.

[60] W.R. Knorr, 'The medieval tradition of a Greek mathematical lemma', *Zeitschrift für Geschichte der arabisch-islamischen Wissenschaften*, 3, 1986, pp. 230–64.

$$n_1 < n_2 \Rightarrow P\hat{J}Q > P\hat{J}S \Rightarrow PQ > PS.$$

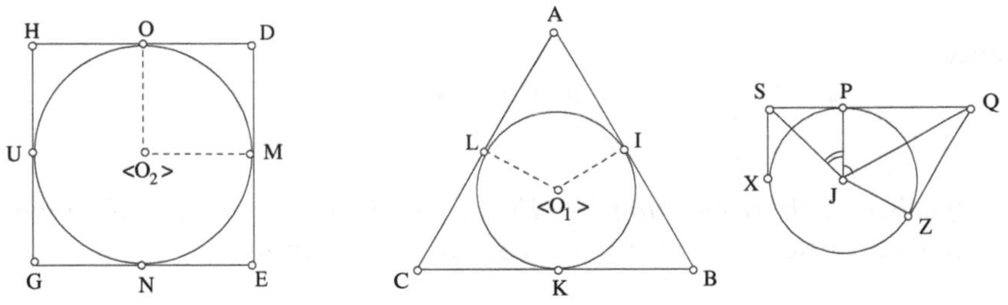

Fig. 7.23

The triangle *SJQ* satisfies the hypotheses of Proposition 16, therefore

$$\frac{PQ}{PS} > \frac{Q\hat{J}P}{S\hat{J}P},$$

and hence

$$\frac{2PQ}{2PS} > \frac{\widehat{PZ}}{\widehat{PX}}.$$

We thus have

$$\frac{PQ+QZ}{\widehat{PZ}} > \frac{PS+SX}{\widehat{PX}};$$

hence

$$\frac{c_1}{\widehat{PZ}} > \frac{c_2}{\widehat{PX}}.$$

If p_1 and p_2 are the respective perimeters of the inscribed circles, we have

$$\frac{n_1 c_1}{p_1} > \frac{n_2 c_2}{p_2}.$$

But

$$n_1 c_1 = n_2 c_2;$$

hence

$$p_1 < p_2.$$

If r_1 and r_2 are the respective radii of the inscribed circles, we then have $r_1 < r_2$. Yet

$$2 \text{ area } P_1 = n_1 c_1 r_1,$$
$$2 \text{ area } P_2 = n_2 c_2 r_2;$$

hence

$$\text{area } P_2 > \text{area } P_1.$$

Comparison with Proposition 2 of Ibn al-Haytham's treatise on the figures of equal perimeters and the solids of equal surface areas:

Despite the differences in the statement, both dealt with the same proposition (see also al-Khāzin, Proposition 9).

The two demonstrations introduce same properties without being identical.

In each of the polygons P_1 and P_2, to each side is associated an isosceles triangle, which is itself divided into two right triangles. The isosceles triangles have respective angles at the summit $\frac{2\pi}{n_1}$ and $\frac{2\pi}{n_2}$, and the right triangles associated to each of them have respective acute angles $\frac{\pi}{n_1}$ and $\frac{\pi}{n_2}$.

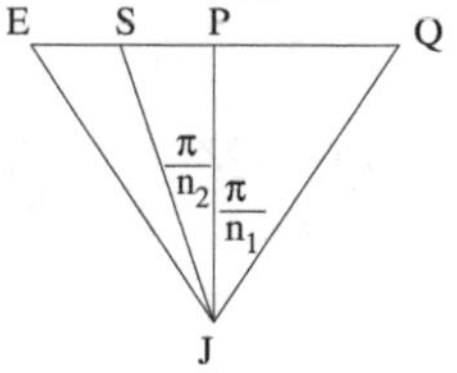

Fig. 7.24

The two authors consider a figure that comprises a right triangle equal to the triangle associated with P_1 and a right triangle similar to the one associated with P_2, both having in common a side of the right angle, which is the apothem of P_1.

We have $EP = PQ = \frac{1}{2}c_1$, $JP = a_1$ and $\frac{SP}{PJ} = \frac{\frac{1}{2}c_2}{a_2}$.

Ibn Hūd introduces the inscribed circles within each of the polygons; namely, circles with respective radii r_1 and r_2, as apothems a_1 and a_2. In applying Proposition 16, he shows that $a_1 < a_2$.

Ibn al-Haytham demonstrates[61] – like Ibn Hūd in Proposition 16 – through inequalities in the areas of the triangles and the areas of the sectors,

[61] See *Les mathématiques infinitésimales*, vol. II, p. 392.

that

$$\frac{EP}{PS} > \frac{c_1}{c_2};$$

thence $PS < \frac{1}{2}c_2$. Yet $JP = a_1$; hence apothem a_2 associated with c_2 satisfies $a_2 > a_1$, thus obtains the conclusion.

It is clear that here Ibn Hūd is following the proof of Ibn al-Haytham with a few slight variations. The original proof as written by Ibn al-Haytham remains the more elegant of the two. Ibn Hūd's introduction of the inscribed circles is not necessary.

One may wonder why Ibn Hūd stops here and fails to go on to consider isoperimetric polygons, in which the number of sides increases until they become a disc. This was the approach taken by Ibn al-Haytham and al-Khāzin, and both inspired his work as we have seen. Does this study appear in one of the missing sections of the book? Or did he believe that the problem, which in another language we would take to be going to the limit, was too complex for the level at which he pitched his compendium? Even if this was the case, he would certainly have taken this lemma from the isoperimetric theory developed in such a masterly fashion in the book by Ibn al-Haytham and treated it as an elementary geometric property following the style of composing a compendium.

7.3.3. *Translation*: Kitāb al-Istikmāl

– **16** – If a triangle has two unequal sides and a perpendicular is drawn from <the vertex of> the angle enclosed by the two unequal sides down to the base, then the ratio of the longest part of the base to the shortest part is greater than the ratio of the part of the angle, from which the perpendicular was drawn, that intercepts the longest part to the other part of the angle.

Example: Let the triangle be *ABC* and let the side *AB* be longer than the side *AC*. A perpendicular *AD* is drawn from <the vertex of> the angle *A* onto the side *BC*.

I say that the ratio of BD *to* DC *is greater than the ratio of the angle* BAD *to the angle* DAC.

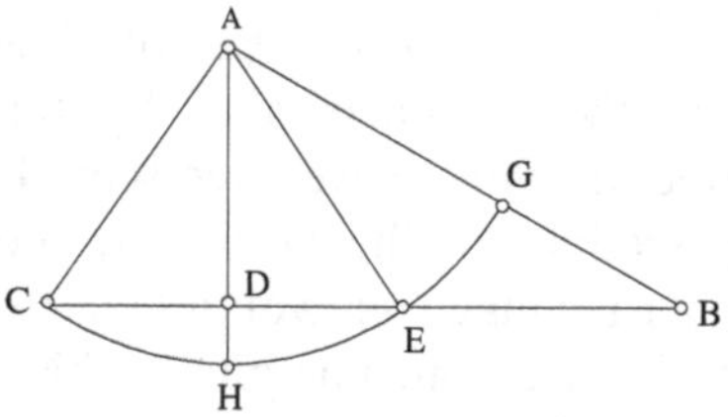

Fig. 7.25

Proof: We take the point *A* as a centre and with the shorter distance *AC* we describe a circle *CHEG* such that it cuts the straight line *BD* at the point *E* and *AB* at <the point> *G*, and let the perpendicular *AD* meet it at the point *H*. As the triangle *ABE* is greater than the sector *AGE*, its ratio to it will be greater than the ratio of the triangle *AED* to the sector *AEH*, as the triangle *AED* is less than the sector *AEH*. If we apply a permutation, then the ratio of the triangle *ABE* to the triangle *AED* is greater than the ratio of the sector *AGE* to the sector *AEH*. If we compound, then the ratio of the triangle *ABD* to the triangle *AED* is greater than the ratio of the sector *AGH* to the sector *AEH*. But the ratio of the triangle *ABD* to the triangle *AED* is equal to the ratio of the straight line *BD* to the straight line *DE* which is equal to the straight line *DC*, and the ratio of the sector *AGH* to the sector *AEH* is equal to the ratio of the angle *GAH* to the angle *EAD*, which is equal to the angle *DAC*. Therefore, the ratio of the straight line *BD* to the straight line *DC* is greater than the ratio of the angle *BAD* to the angle *DAC*. That is what we wanted to prove.

– **19** – If two polygons with equal perimeters are regular – within a circle – then the circle inscribed within the polygon with the greater number of sides is greater than the circle inscribed within the polygon with the lesser number of sides.

Example: Let there be two figures *ABC* and *DEGH*, and let the perimeter of the figure *ABC* be equal to the perimeter of the figure *DEGH*. Let each of the figures be regular within a circle, and let the figure *DEGH* have the greater number of sides.

I say that the circle inscribed within this figure is greater than the <circle> inscribed within the surface ABC.

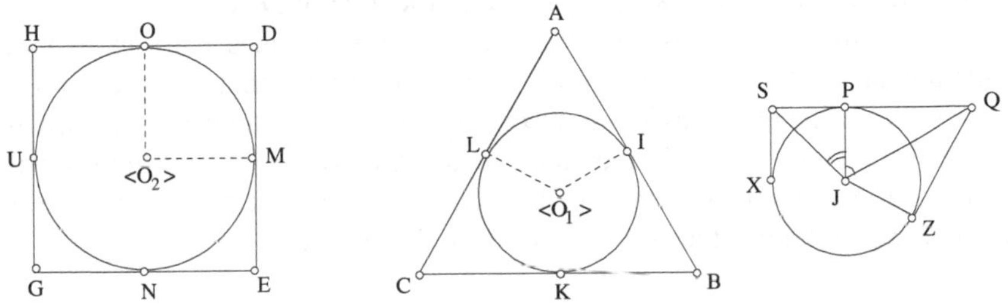

Fig. 7.26

Proof: We inscribe a circle within the figure *ABC*, that is, the circle *IKL*, and a circle within the figure *DEGH*, that is, the circle *MNUO*. We describe a circle *ZPX* equal to the circle *IKL* and draw from the point *P* a straight line *QPS* as a tangent to it. Make the straight line *PQ* equal to the straight line *AI*, and the straight line *PS* equal to half of one side of the figure circumscribed around the circle and similar to the figure *DEGH*. From the points *Q* and *S* draw two straight lines *QZ* and *SX* tangent to the circle, and let the centre of the circle be at the point *J*. We join *PJ*, *JQ* and *JS*. As the figure *DG* has a greater number of sides, the straight line *PS* is less than the straight line *PQ*,[62] and hence the ratio of the straight line *QP* to the straight line *PS* is greater than the ratio of the angle *QJP* to the angle *PJS*, which is equal to the ratio of half of the arc *ZP* to half of the arc *PX*. Therefore, the ratio <of the sum> of the two straight lines *ZQ* and *QP* to the arc *ZP* is greater than the ratio <of the sum> of the two straight lines *PS* and *SX* to the arc *PX*, and the ratio <of the sum> of the straight lines *ZQ* and *QP* to the arc *ZP* is equal to the ratio <of the sum> of the two straight lines *IA* and *AL* to the arc *IL*. But the ratio <of the sum> of the two straight lines *PS* and *SX* to the arc *PX* is equal to the ratio <of the sum> of the two straight lines *MD* and *DO* to the arc *MO*. The ratio <of the sum> of *IA* and *AL* to the arc *IL*, which is equal to the ratio of the perimeter of the figure *ABC* to the circumference of the circle *IKL*, is greater than the ratio <of the sum> of

62 See the commentary.

MD and *DO* to the arc *MO*, which is equal to the ratio of the perimeter of the figure *DEGH* to the circumference of the circle *MNUO*. But the perimeter of *ABC* is assumed to be equal to the perimeter of the figure *DEGH*, and therefore the circle *IKL* is less than the circle *MNUO*. Therefore, the half-diameter of the circle *IKL* is shorter than the half-diameter of the circle *MNUO*. So the product of the half-diameter of the circle *IKL*[63] and the half-perimeter of *ABC*, which is equal to the area of *ABC*, is less than the product of the half-diameter of the circle *MU* and the half-perimeter of *DEGH*, which is equal to the area of *DEGH*.

From this, it becomes clear that if two straight lines with an included angle are tangents to a circle, and if two straight lines that are shorter than them with an included angle are also tangents to the same circle, then the ratio <of the sum> of the longer lines to the arc lying between them on the circle is greater than the ratio <of the sum> of the shorter lines to the arc lying between them. That is what we wanted to prove.

[63] Lit.: its product.

SUPPLEMENTARY NOTES

The Formula of Hero of Alexandria according to Thābit ibn Qurra

[1] In his treatise on the measurement of plane and solid figures, Thābit ibn Qurra, who worked with the Banū Mūsā, mentions the formula and then discusses its origin. Reading the text, it appears that this formula was widely known, and that not all mathematicians attributed it to Hero. This is what Thābit wrote: 'The common rule to all types of triangles: Some attribute it to India, while others state that it comes from the Byzantines (*al-Rūm*). It is described as follows: Let the three sides of a triangle be added together, and then take half of the sum. Take the amount by which this half exceeds each of the sides, and multiply this half by the amount by which it exceeds one of the sides of the triangle. Then, multiply this product by the amount by which it exceeds one of the other sides of the triangle, and then multiply that product by the amount by which it exceeds the third side of the triangle. Take the square root of the product, which is the area of the triangle' (R. Rashed, 'Thābit et l'art de la mesure', in *Thābit ibn Qurra. Science and Philosophy in Ninth-Century Baghdad*, p. 182; Arabic p. 183, 16–21).

Commentary of Ibn Abī Jarrāda on *The Sections of the Cylinder* by Thābit ibn Qurra

[**2**, Proposition 6, p. 389] This conclusion is the same as that of Proposition 4 relating to the right cylinder. It is given here as a consequence of the conclusion relating to the oblique cylinder. In the case of a right cylinder, any plane containing the axis GH is a plane of symmetry of the cylinder and fulfils the same role as the plane GHI, the single plane of symmetry in an oblique cylinder.

[**3**, Proposition 10, p. 396] In re-writing this text, Ibn Abī Jarrāda considers the case where the circle ABC is replaced by an ellipse of which AB is a diameter and DC is an ordinate. In this case, he makes EH parallel to DC and continues with the proof in order to show that the two triangles FHE and GDC are similar, and to prove that

$$\frac{AH \cdot HB}{EH^2} = \frac{AD \cdot DB}{DC^2} \quad \text{and} \quad \frac{EH^2}{HF^2} = \frac{DC^2}{DG^2},$$

from which it follows that

$$\frac{AH \cdot HB}{HF^2} = \frac{AD \cdot DB}{DC^2}.$$

Ibn Abī Jarrāda writes: 'This proof also includes the circle and it is better. We give a general statement of the proposition and we prove it, even if the point D is not the centre'.

Ibn Abī Jarrāda has therefore generalized Proposition 10 by considering the cylindrical projection of an ellipse. This enables him to consider the plane section of a cylinder with an elliptical base in Proposition 11. See his text (ms. Cairo, Dār al-Kutub 41, fol. 40^{r})[1]:

> 'I say that: That which we seek may be shown by this proof if ABC is an ellipse. We draw DC as an ordinate and we draw EH parallel to it, we then continue the proof until it is shown that the two triangles FHE and GDC are similar. The ratio of the product of AH and HB to the square of EH will be equal to the ratio of the product of AD and DB to the square of DC, as has been shown in I.21 of the *Conics*. But the ratio of the square of EH to the square of HF is equal to the ratio of the square of DC to the square of DG. Using the equality ratio, the ratio of the product of AH and HB to the square of HF is equal to the ratio of the product of AD and DB to the square of DG and this gives us that which we were looking for.
>
> 'This proof also includes the circle and it is better. We give a general statement of the proposition and we prove it, even if the point D is not the centre; the proof is completed.'

[**4**, Proposition 11, p. 396] Ibn Abī Jarrāda introduces this proposition with a lemma (see fols 40^{r-v}) in order to justify his assertion that

$$GM \parallel SN,\ SG = MN \text{ and } G\hat{S}N = \frac{\pi}{2} \Rightarrow M\hat{N}S = \frac{\pi}{2}.$$

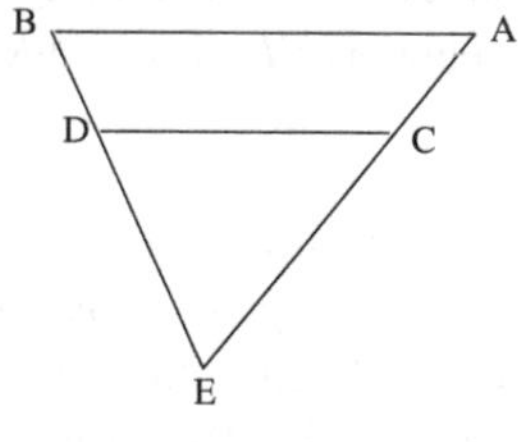

Fig. 1

[1] See the Arabic quotations in the French edition *Les mathématiques infinitésimales*.

'Lemma: Let there be two parallel straight lines *AB* and *CD* such that the two straight lines *AC* and *BD* are equal. I say that the two angles *A* and *B* are equal, or their sum is equal to two right angles.

'Proof: If *AC* and *BD* meet, let them meet at *E*. We have therefore drawn *CD* in the triangle *ABE* parallel to the base *AB*. The ratio of *EA* to *AC* is therefore equal to the ratio of *EB* to *BD*. But *AC* and *BD* are equal, and therefore the two straight lines *EA* and *EB* are equal. Therefore the two angles *A* and *B* are equal.

'If *AC* and *BD* are parallel, then <the sum of> the two angles *A* and *B* is equal to two right angles.'

[**5**, Proposition 11, p. 398] Ibn Abī Jarrāda notes that it is not necessary to draw *IR*. We know *MQ* = *ML*, hence $M\hat{L}Q = M\hat{Q}L$, and from that, $D\hat{E}L = I\hat{L}E$. An antiparallel plane therefore passes through *IL*. The remainder of the proof is unchanged (fol. 41^{r-v}):

'I say that you have no need to draw *IR*. Instead, you say that the part of the straight line intersecting the two circles that lies between <each of the points> of intersection and the point *M*, is a half diameter of each of them. It is therefore equal to *MQ*, a half diameter of the circle parallel to the two bases and equal to *ML*, a half diameter of the circle *IKL*. The two straight lines *ML* and *MQ* are therefore equal, the two angles *MLQ* and *MQL* are then equal, and the angle *DEL* is equal to the angle *MQL*. The two angles *DEL* and *ILE* are then equal and an antiparallel section passes through the straight line *LI*. The conclusion is as before'.

[**6**, Proposition 12, p. 399] Ibn Abī Jarrāda first proves a lemma using three methods (fol. 41^v–42^r).

1) If two ellipses have major axes *AP* and *CQ*, minor axes *BW* and *DH*, and centres *K* and *O* respectively, and if $\frac{PA}{BV} = \frac{CQ}{DH}$, then the ellipses are similar.

$$\frac{BV}{PA} = \frac{DH}{QC} \Leftrightarrow \frac{BK}{KA} = \frac{DO}{OC} \Rightarrow \frac{BK^2}{KA \cdot KP} = \frac{DO^2}{QO \cdot OC},$$

from I.21 of the *Conics*; we therefore have

$$\frac{\text{latus rectum of } PA}{PA} = \frac{\text{latus rectum of } QC}{QC}.$$

The proof would be the same if *PA*, *BV*, *QC* and *DH* were the conjugate diameters instead of the axes.

2) Let A, B and C, D be the axes of two ellipses where $\frac{A}{B} = \frac{C}{D}$, and let E and G be the *latera recta* relative to A and C respectively. From I.15 of the *Conics*, we have $A^2 = E \cdot B$ and $C^2 = G \cdot D$. Hence

$$\frac{E}{B} = \frac{E \cdot B}{B^2} = \frac{A^2}{B^2} \text{ and } \frac{G}{D} = \frac{G \cdot D}{D^2} = \frac{C^2}{D^2},$$

from which

$$\frac{E}{B} = \frac{C}{D},$$

and the ellipses are similar from VI.12 of the *Conics*.

3) From I.15 of the *Conics*, we have $\frac{E}{A} = \frac{A}{B}$ and $\frac{G}{C} = \frac{C}{D}$ and by hypothesis $\frac{A}{B} = \frac{C}{D}$, and therefore $\frac{E}{A} = \frac{G}{D}$. From VI.12 of the *Conics*, the ellipses are similar.

[**7**, Proposition 12, p. 399] If the intersecting plane under consideration is parallel or antiparallel to the planes of the base, the section of each cylinder is a circle equal to its base circle. Thābit mentions this in the course of the proof.

[**8**, Proposition 12, p. 401] In the expression 'the greatest diameter of any section is its largest axis, and its smallest diameter is its smallest axis', Ibn Abī Jarrāda is undoubtedly referring to the *Conics* V.11.

[**9**, Proposition 12, p. 401] In the *Conics* VI.12, Apollonius shows that if two ellipses have axes $2a$ and $2a'$ and associated straight sides c and c' such that $\frac{2a}{c} = \frac{2a'}{c'}$, then they are similar, and *vice versa*.

If $2b$ and $2b'$ are the second axes of the ellipses, then from Apollonius, second definitions III, we have

$$4b^2 = 2\,a \cdot c;$$

hence

$$\frac{a^2}{b^2} = \frac{2a}{c}.$$

Similarly,

$$\frac{a'^2}{b'^2} = \frac{2a'}{c'}.$$

Therefore

$$\frac{2a}{c} = \frac{2a'}{c'} \Leftrightarrow \frac{a}{b} = \frac{a'}{b'}.$$

It is this condition that Thābit uses. He makes no use here of the straight sides c and c', but in Proposition 24 he uses the property

$$\frac{2a}{c} = \frac{2a'}{c'}.$$

[**10**, Proposition 14, p. 402] In his edition, Ibn Abī Jarrāda includes two lemmas immediately prior to Proposition 14 (fol. 43^r).

Lemma 1. – Let *BAD* be a semi-ellipse with its centre at *L*, its major axis *BD*, and its vertex at *A*. *LO* and *LM* are half-diameters passing through the mid-points *K* and *G* of the chords *AD* and *AB*. Then *OM* is perpendicular to *AL*.

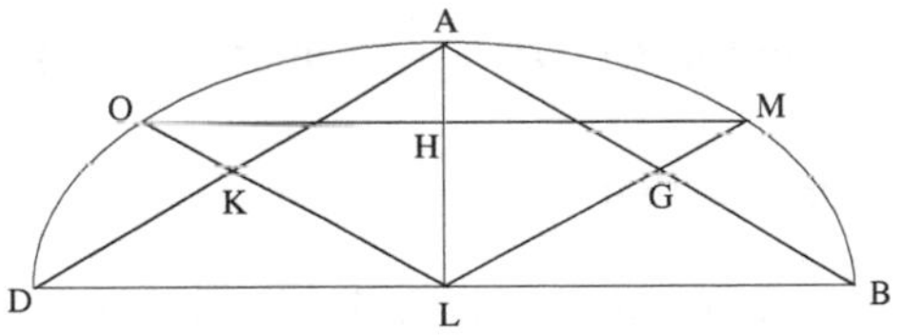

Fig. 2

The proof follows immediately from VI.8 of the *Conics*.

Lemma 2. – If two convex quadrilaterals *ABCD* and *EGHI* are located between the parallel lines *AD* and *BC*, with *EI* on *AD* and *HG* on *BC*, then

$$\frac{S\,(EGHI)}{S\,(ABCD)} = \frac{EI + GH}{AD + BC}.$$

The proof follows immediately from the fact that *ABCD* and *EGHI* are trapeziums or parallelograms with the same height.

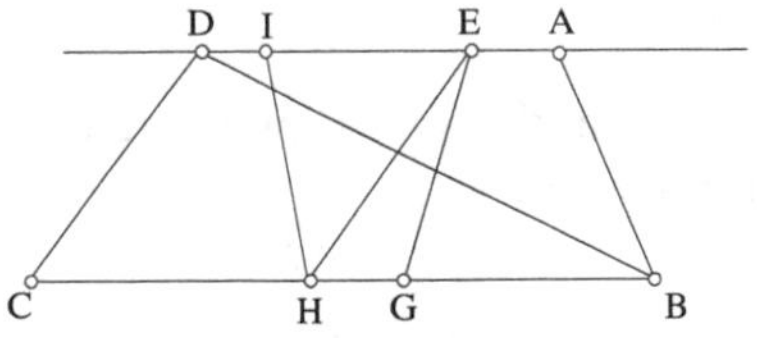

Fig. 3

[**11**, Proposition 14, p. 403] In order to double the number of sides, Thābit considers the diameters passing through the midpoints of the chords. Hence, if *G* is the midpoint of *AB*, then *LG* cuts the ellipse at *M*. The tangent at *M* is parallel to *AB*, and it cuts the tangents at *A* and *B* at the points *X* and *Y* such that *XY* < *AB*. *AXYB* is a trapezium and we have

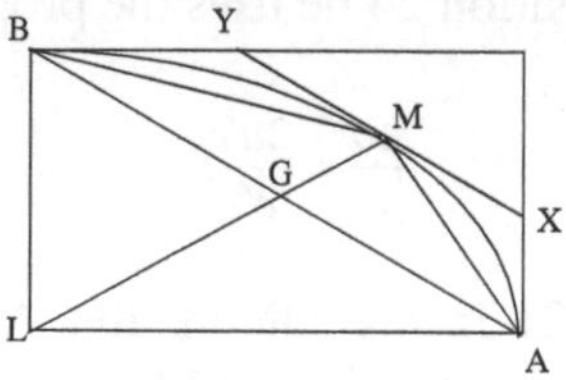

Fig. 4

$$\text{area of tr. } (AMB > \frac{1}{2} \text{ area of tp. } (ABYX).$$

Hence

$$\text{area of tr. } (AMB) > \frac{1}{2} \text{ area of sg. } (AB),$$

from which we can deduce that

$$\text{area of sg. } (AB) - \text{area of tr. } (AMB) < \text{area of sg. } (AB).$$

This is true regardless of whether $\widehat{AB}$ is an arc of an ellipse or an arc of a circle.

[**12**, Proposition 14, p. 405] If $2a$ and $2b$ are the axes of the ellipse, and $2r$ is the radius of the circle *E* equivalent to the ellipse, then $r^2 = ab$, from which

$$\frac{a}{r} = \frac{r}{b} \text{ and } \frac{a^2}{r^2} = \frac{r^2}{b^2}.$$

The area of an ellipse is the proportional mean of the areas of its major circle and its minor circle.

[**13**, Proposition 15, p. 405] Ibn Abī Jarrāda also includes two lemmas prior to Proposition 15 (fol. 44^{v}).

Lemma 1. – The arcs *AC* and *EH* of two circles of diameter d_1 and d_2 respectively are similar if and only if

$$\frac{AC}{d_1} = \frac{EH}{d_2}.$$

Lemma 2. – Let AC and EH be two chords whose midpoints are K and M respectively and that belong to two different circles. Let BL and GN be the diameters passing through K and M respectively. Then the arcs AC and EH are similar if and only if

$$\frac{KB}{KL} = \frac{MG}{MN}.$$

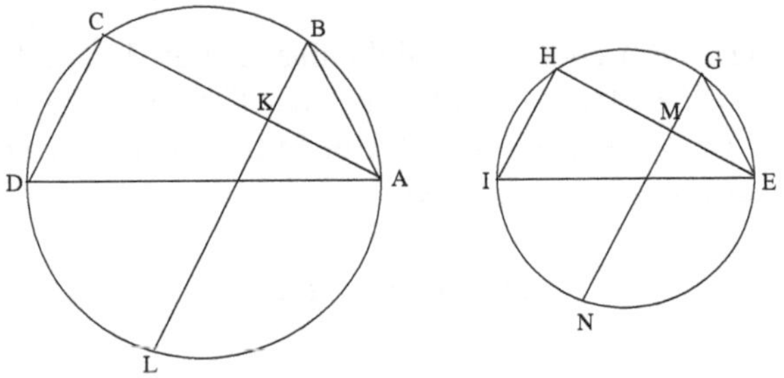

Fig. 5

The proofs follow immediately from the definitions of similar arcs. The included angles are equal and *vice-versa*.

[**14**, Proposition 18, p. 417] In I.15 of the *Conics*, Apollonius defines the conjugate diameter of a given diameter. These diameters are the axes if the angle between them is a right angle, according to Apollonius, first definitions VIII.

[**15**, Proposition 20, p. 420] Ibn Abī Jarrāda includes a lemma prior to Proposition 20, and follows it with a proof that is different from that of Thābit (fol. 49ʳ).

Lemma. – Let there be two parallel planes $\mathbf{P}_1$ and $\mathbf{P}_2$, a point G in $\mathbf{P}_1$, and a point I in $\mathbf{P}_2$ such that $GI \perp \mathbf{P}_2$. Let GL be a straight line in $\mathbf{P}_1$, and let HQ be a straight line in $\mathbf{P}_2$. Then $HQ \parallel GL$ and $I \notin HQ$. The orthogonal projection N from L onto $\mathbf{P}_2$ does not lie on the straight line HQ. The proof is by *reductio ad absurdum*.

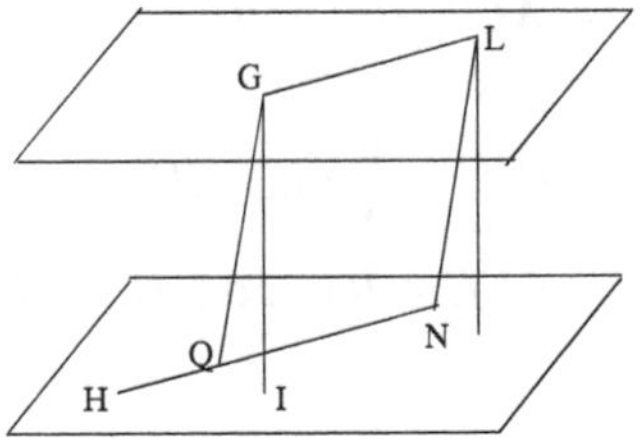

Fig. 6

Ibn Abī Jarrāda offers a simplified proof of Proposition 20 (fol. 50[v]):

1) Given a plane **P**, a point $C \notin \mathbf{P}$ and points $B, D, A \in \mathbf{P}$ such that $BC \perp \mathbf{P}$ and B, D, and A are aligned in that order, for any point $G \in \mathbf{P}$ not lying on BD, we have

$$B\hat{D}C < G\hat{D}C < A\hat{D}C.$$

The circle (D, DG) cuts the straight line BD at A and E. In all three cases, $BA > BG > BE$, from Euclid, *Elements* III.7 and III.8. From this, we can deduce that $CA > CG > CE \geq CB$. But in the triangles CDA, CDG, and CDE, we have $DA = DG = DE$; hence

$$A\hat{D}C > G\hat{D}C > E\hat{D}C.$$

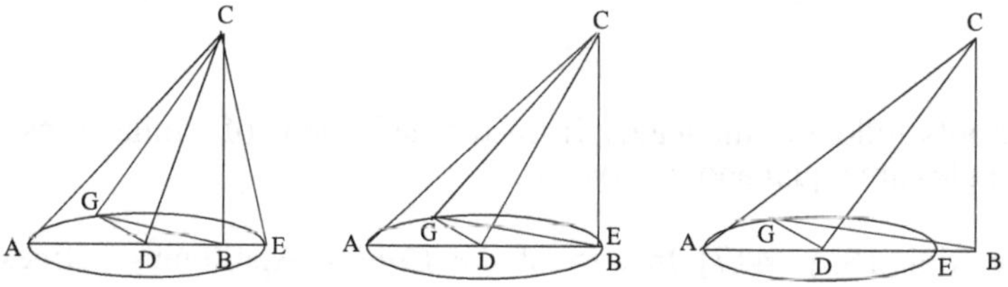

Fig. 7

2) If two parallelograms $ABCD$ and $IGEH$ satisfy $AB = CD = EG = HI$, $A\hat{D}C \geq E\hat{I}H \geq I\hat{E}G$, $A\hat{D}C > \frac{\pi}{2}$, and their areas are equal, then AC is the largest of all the segments whose ends lie one on AB and the other on CD, or one on EG and the other on IH.

If the areas and bases of the parallelograms are equal, then their heights must also be equal.

$$A\hat{D}C > \frac{\pi}{2} \Rightarrow A\hat{D}C > B\hat{A}D \text{ and } A\hat{D}C > A\hat{C}D$$

$$\Rightarrow AC > BD \text{ and } AC > AD,$$

$$E\hat{I}H \geq I\hat{E}G \Rightarrow E\hat{I}H \geq \frac{\pi}{2} \text{ and } E\hat{I}H > E\hat{H}I$$
$$\Rightarrow EH \geq IG \text{ and } EH > EI.$$

Let $L \in [CD]$ and $M \in [BA]$ such that $ML \parallel AD$, then $CM > BL$ and $CM < AC$ as $A\hat{M}C > \frac{\pi}{2}$.

With each segment having one end on AB and the other on CD, one can associate an equal segment having one end on B or C and the other on CD or AB. Therefore AC is the largest of these segments. Similarly, EH is the largest of all the segments having one end on EG and the other on IH.

On the other hand, $A\hat{D}C \geq E\hat{I}H \Rightarrow AD \geq EI$.

If $A\hat{D}C = E\hat{I}H$, then $AD = EI$, and one can deduce that $AC = EH$.

If $A\hat{D}C > E\hat{I}H \geq \frac{\pi}{2}$, then $AD > EI$. Let K be such that $IK = AD$; then $AC > HK$. But $HK > EH$ as $IK > IE$. Therefore $AC > EH$. Therefore AC is the largest of all the segments listed in the statement.

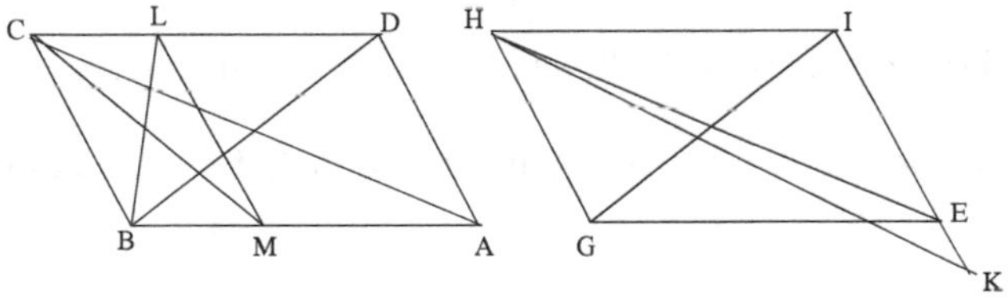

Fig. 8

3) Let us now return to the figure given by Thābit, Figure II.3.20. We have $G\hat{H}E > \frac{\pi}{2}$ by hypothesis, and $G\hat{H}E > G\hat{H}F > G\hat{H}D$, regardless of the position of the point F on the circle of diameter DE. The diagonal AE of the parallelogram $ABED$ is therefore greater than the diagonal LF in the parallelogram $LCFM$. Therefore, AE is the greatest segment joining a point on one generator to a point on the opposite generator.

The proof is then completed in the same way as that of Thābit.

[**16**, Proposition 23, p. 426] Ibn Abī Jarrāda (fol. 52[v]) proposes an alternative method using the equivalent circles to each of the two ellipses, from which

$$\frac{S_m}{S_M} = \frac{a_m b_m}{a_M b_M} = \frac{b_m}{a_M},$$

as $a_m = b_M = r$, the radius of a base.

[**17**, Proposition 25, p. 428] Ibn Abī Jarrāda includes the following lemma prior to Proposition 25 (fol. 53ʳ): Let there be three non-aligned points A, B, C. If D is the midpoint of BC, and $AB > AC$, then $D\hat{A}C > D\hat{A}B$.

It can be seen that $AB > AC$ and $DB = DC \Rightarrow \frac{AC}{AB} < \frac{DC}{DB}$.

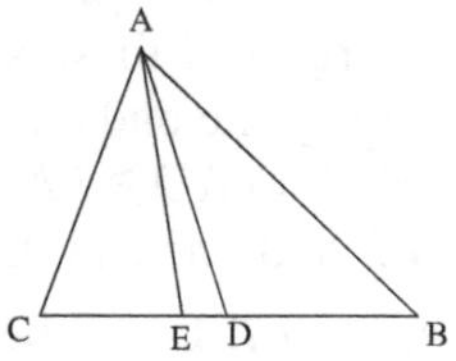

Fig. 9

Let $E \in [CB]$ such that $\frac{EC}{EB} = \frac{AC}{AB}$; then $EC < DC$, but AE is the bisector of $B\hat{A}C$. Therefore $E\hat{A}C = E\hat{A}B$, and hence the result follows.

[**18**, Proposition 31, p. 440] Ibn Abī Jarrāda includes two lemmas prior to Proposition 31 (fols 57ᵛ–58ʳ).

Lemma 1. – Given a segment AB, two surfaces c and d such that $c < d$, and two segments e and g such that $e < g$, there exists N on the segment AB such that

$$\frac{NB}{AB} > \frac{e}{g} \text{ and } \frac{NB^2}{AB^2} > \frac{c}{d}.$$

Lemma 2. – If $c > d$ and $e > g$, then there exists N on the extended BA such that

$$\frac{NB}{AB} < \frac{e}{g} \text{ and } \frac{NB^2}{AB^2} < \frac{c}{d}.$$

The proof given by Ibn Abī Jarrāda is based on the existence of a point L such that $\frac{LB}{AB} = \frac{h}{k}$ if h and k are two segments defined by $h^2 = c$ and $k^2 = d$.

However, his argument cannot be concluded successfully, as it requires the introduction of the point G such that $\frac{GA}{AB} = \frac{e}{g}$.

It should be noted that the points labelled A, B and N by Ibn Abī Jarrāda correspond to the points labelled A, K and M by Thābit, and the points L and G correspond to the points M_1 and M_2 that appear in the following note.

[**19**, Proposition 31, p. 441] Determination of the point M in this part of the proof.

We have $H < P$, hence $\frac{H}{P} < 1$, and we also have $\frac{S - \frac{1}{2}I}{S} < 1$.

If M_1 and M_2 satisfy $\frac{KM_1}{KA} = \frac{H}{P}$ and $\frac{KM_2^2}{KA^2} = \frac{S - \frac{1}{2}I}{S}$, then:

$KA > KM_1 \geq KM_2$: in this case, place M between M_1 and A.

$KA > KM_2 \geq KM_1$: in this case, place M between M_2 and A.

The method is the same if $\frac{H}{P} > 1$ and $\frac{S + \frac{1}{2}I}{S} > 1$.

[**20**, Proposition 32, p. 446] Ibn Abī Jarrāda includes a lemma prior to Proposition 32 in his edition (fol. 60^{r-v}) and follows it with three comments (fols 62^{r-v}).

Lemma. – Let a, b, c and d be positive numbers such that $\frac{a}{b} > \frac{c}{d}$ and $a < c$. Then $b < d$.

In other words, there exists e such that $\frac{a}{e} = \frac{c}{d}$. Therefore, $e < d$ and $\frac{a}{e} < \frac{a}{b}$, which implies that $e > b$, and consequently $b < d$.

Comments

1) We can show by *reductio ad absurdum* that it is impossible for $S > \frac{1}{2}p\ (IM + KN)$ to be true.

2) We can show by *reductio ad absurdum* that the opposite case is impossible; we can also show that if LS is the longest segment of the generator between the sections SMN and SL_cL_d, then these two sections cannot have any common point other than the point S.

3) Similarly, we can show by *reductio ad absurdum* that the polygon that is obtained in the plane MNS by the cylindrical projection of a polygon inscribed within the section IKL and that has no point in common with the section XYZ is itself inscribed within the section MNS and has no point in common with the section $O'L_aL_b$.

For the notation, see the text and the figures in Proposition 32.

[**21**, Proposition 35, p. 455] Ibn Abī Jarrāda comments that the same result can be obtained if the two sections are antiparallel circles (fol. 63^{v}).

This comment is unnecessary, as the result obtained by Thābit is equally valid for all sections regardless of their shape.

[**22**, Proposition 37, p. 457] Ibn Abī Jarrāda makes the same comment here as that above. It is also unnecessary for the same reason (fol. 64[r]).

[**23**, Proposition 37, p. 458] Ibn Abī Jarrāda proves (fol. 64[r]) that, in the case of a right cylinder on a circular base, the smallest of the sections is the base circle, and he states without explanation that the largest ellipse is that whose major axis is a diagonal of a rectangle whose plane passes through the axis.

One could prove, as in Proposition 20, that such a diagonal is the longest segment having its extremities on the two opposite generators. Therefore, if the two perpendicular planes passing through the diagonals are associated with each plane passing through the axis, then their intersections with the cylinder are the maximal ellipses.

To summarize, in both right and oblique cylinders, any plane perpendicular to the axis gives a minimal section, and while an oblique cylinder only has one maximal section, there are an infinite number in a right cylinder.

BIBLIOGRAPHY

1.1. MANUSCRIPTS

Banū Mūsā
Kitāb ma'rifat misāḥat al-ashkāl al-basīṭa wa-al-kuriyya
Hyderabad, Osmania University 992, fols 51^r–52^v (Propositions 12 and 18).

Banū Mūsā (edition by al-Ṭūsī)
Kitāb ma'rifat misāḥat al-ashkāl al-basīṭa wa-al-kuriyya
Berlin, Staatsbibliothek, or. quart. 1867/13, fols 156^v–164^v.
Cracovia, Biblioteka Jagiellonska, fols 183^v–194^v.
Istanbul, 'Atif 1712/14, fols 97^v–104^v.
Istanbul, Beşiraga 440/14, fols 162^v–171^v.
Istanbul, Carullah 1475/3, fols 1^v–14^v (without numbering).
Istanbul, Carullah 1502, fols 42^v–47^v.
Istanbul, Haci Selimaga 743, fols 71^v–81^v.
Istanbul, Köprülü 930/14, fols 214^v–227^r (other numbering 215^v–228^r).
Istanbul, Köprülü 931/14, fols 129^r–136^v.
Istanbul, Süleymaniye, Aya Sofya 2760, fols 177^r–183^v.
Istanbul, Süleymaniye, Esad Effendi 2034, fols 4^v–15^v.
Istanbul, Topkapi Sarayi, Ahmet III 3453/13, fols 148^r–152^v.
Istanbul, Topkapi Sarayi, Ahmet III 3456/15, fols 61^v–64^v.
Cairo, Dār al-Kutub, Riyāḍa 41, fols 26^v–33^v.
London, India Office 824/3 (no. 1043), fols 36^r–39^r, 50^r–52^v.
Manchester, John Rylands University Library 350.
Meshhed, Astān Quds 5598, fols 18–33.
New York, Columbia University, Plimpton Or 306/13, fols 116^r–122^v.
Oxford, Bodleian Library, Marsh 709/8, fols 78^r–89^v.
Paris, Bibliothèque nationale 2467, fols 58^v–68^r.
Tehran, Danishka 2432/13, fols 123–137 (other numbering 144^v–151^v)].
Tehran, Majlis Shūrā 209/3, fols 33–54.
Tehran, Majlis Shūrā 3919, fols 272–298.
Tehran, Milli Malik 3179, fols 256^v–261^v, 264^r–267^v.
Tehran, Sepahsalar 2913, fols 86^v–89^v.
Vienne, Nationalbibliothek, Mixt 1209/13, fols 163^v–173^r.

Ibn Abī Jarrāda
Taḥrīr Kitāb quṭū' al-usṭuwāna wa-basīṭihā li-Thābit ibn Qurra
Cairo, Dār al-kutub, Riyāḍa 41, fols 36^v–64^v.

Ibn Hūd
Kitāb al-Istikmāl
Copenhagen, Bibliothèque Royale Or. 82, fols 50^{r-v}, 100^v–102^v.
Leiden, Or. 123-a, fols 7^v–11^r.

Ibn al-Samḥ
Fragment on the cylinder and its plane sections
Ma'amar ba-iṣṭewanot we-ha-meḥuddadim
Oxford, Bodleian Library, Hunt. 96, fols 46^v–53^r.

Ibn Sinān, Ibrāhīm
Fī misāḥat qiṭ' al-makhrūṭ al-mukāfi'
London, India Office 461 (Loth 767), fols 191–197.
Paris, Bibliothèque nationale 2457, fols 134^v–136^r.
Patna, Khuda Bakhsh 2519; fols 132^r–134^v.

Fī misāḥat al-qiṭ' al-mukāfi'
Cairo, Dār al-kutub, Riyāḍa 40, fols 182^v–186^v.
Damascus, al-Ẓāhiriyya 5648, fols 159–165.
Istanbul, Aya Sofya 4832, fols 76v–79^r.

Al-Khāzin
Min al-sharḥ li-al-maqāla al-ūlā min al-Majisṭī
Paris, Bibliothèque nationale 4821/8, fols 47^v–68^v.

Al-Qūhī
Fī istikhrāj misāḥat al-mujassam al-mukāfi'
Cairo, Dār al-kutub, Riyāḍa 40, fols 187^v–190^v.
Damascus, al-Ẓāhiriyya 5648, fols 166^r–171^r.
Istanbul, Aya Sofya 4832, fols 125^v–129^r.
Istanbul, Aya Sofya 4830, fols 161^v–165^r.

Fī misāḥat al-mujassam al-mukāfi'
Cairo, Dār al-kutub, Riyāḍa 41/2, fols 135^v–137^r.

Thābit ibn Qurra
Fī misāḥat qiṭ' al-makhrūṭ alladhī yusammā al-mukāfi'
Istanbul, Aya Sofya 4832, fols 26^v–36^v.
Cairo, Dār al-kutub, riyāḍa 40, fols 165^v–181^r.
Meshhed, Astān Quds 5593 fols 26–42.
Paris, Bibliothèque nationale 2457, fols 122^v–134^v.

Fī misāḥat al-mujassamāt al-mukāfi'a
Paris, Bibliothèque nationale 2457, fols 110^v–113^r.

Fī quṭū' al-usṭuwāna wa-basīṭihā
Istanbul, Aya Sofya 4832, fols 4^r–26^r.

1.2 OTHER MANUSCRIPTS CONSULTED FOR THE ANALYSIS AND THE SUPPLEMENTARY NOTES

Anonymous
Commentary on Euclid's *Elements*
Hyderabad, Osmania University 992.

Apollonius

al-Makhrūṭāt, Tehran, Milli Malek 867.

Archimedes

Kitāb al-kura wa-al-usṭuwāna, Istanbul, Fātiḥ 3414, fols 9^v–49^r.

Kitāb fī misāḥat al-dā'ira, Istanbul, Fātiḥ 3414, fols 2^v–6^v.

Banū Mūsā

Muqaddamāt kitāb al-Makhrūṭāt

Istanbul, Aya Sofya 4832, fols 223^v–226^v (= II/32, fols 71^v–74^v).

Qawl fī tathlīth al-zāwiya al-mustaqīma al-khaṭṭayn (attributed to Aḥmad)

Oxford, Bodleian Library, Marsh 207, fol. 131^v and Marsh 720, fol. 260^v.

Ibn Hūd

Kitāb al-Istikmāl

Copenhagen, Royal Library Or. 82, 128 fols.

Leiden, Universiteitsbibliotheek, Or. 123-a, 80 fols.

Ibn Waḥshiyya

al-Filāḥa al-nabaṭiyya, Istanbul, Topkapi Saray, Ahmet III 1989.

Al-Khāzin

Mukhtaṣar mustakhraj min kitāb al-Makhrūṭāt

Oxford, Bodleian Library, Huntington 237, fols 82^r–104^v.

Algers, BN 1446, fols 126^v–153^r.

Tafsīr ṣadr al-maqāla al-'āshira min Kitāb Uqlīdis

Istanbul, Feyzullah 1359/6, fols 245^r–252^r.

Tunis, BN 16167, fols 65^v–72^r.

Al-Kindī

Fī al-ṣinā'at al-'uẓmā

Istanbul, Aya Sofya 4830/2, fols 53^r–80^v.

Ptolemy

al-Majisṭī, Leiden, Universiteitsbibliotheek, Or 680, 220 fols.

Al-Qūhī

Risāla fī istikhrāj ḍil' al-musabba' al-mutasāwī al-aḍlā'

Paris 4821, fols 1–8.

Istanbul, Aya Sofya 4832, fols 145^v–147^v.

London, India Office 461, fols 182–189.

Al-Samaw'al

Fī kashf 'uwār al-munajjimīn, Leiden, Universiteitsbibliotheek, Or 98.

Al-Sijzī

Fī rasm al-quṭū' al-makhrūṭiyya, Leiden, Universiteitsbibliotheek, Or 168(1), fols 1–22.

Thābit ibn Qurra

Kitāb fī annahu idhā waqaʿa khaṭṭ mustaqīm ʿalā khaṭṭayn mustaqīmayn …
Istanbul, Aya Sofya 4832, fols 51ʳ–52ʳ.
Istanbul, Carullah 1502, fols 13ʳ–14ᵛ.
Paris, Bibliothèque nationale 2457, fols 156ᵛ–159ᵛ.

Fī misāḥat al-ashkāl al-musaṭṭaḥa wa-al-mujassama
Istanbul, Aya Sofya 4832, fols 41ʳ–44ʳ.

Fragment on the amicable numbers (*incipit*: nurīd an najida ʿadadayn mutaḥābayn)
Cairo, Dār al-Kutub, Riyāḍa 40, fols 36ʳ–37ᵛ and Damascus, Ẓāhiriyya 5648 (quoted by Ibn Hūd in the *Istikmāl*).
Hyderabad, Osmania University 992, fols 295ʳ–297ʳ (quoted by the anonymous author from *Kitāb al-Istikmāl*)

2. BOOKS AND ARTICLES

W. Ahlwardt, *Handschriften der Königlichen Bibliothek zu Berlin XVII*, Arabische Handschriften 5, Berlin, 1893.

al-Akfānī, *Irshād al-qāṣid ilā asnā al-maqāṣid*, in J. Witkam, *De egyptische Arts Ibn al-Akfānī*, Leiden, 1989.

A. Anbouba

'Construction of the Regular Heptagon by Middle Eastern Geometers of the Fourth (Hijra) Century', *Journal for the History of Arabic Science* 1.2, 1977, pp. 352–84.

'Construction de l'heptagone régulier par les Arabes au 4ᵉ siècle de l'hégire', *Journal for the History of Arabic Science* 2.2, 1978, pp. 264–9.

'L'algèbre arabe aux IXᵉ et Xᵉ siècles: Aperçu général', *Journal for the History of Arabic Science* 2, 1978, pp. 66–100.

Apollonius

Les Coniques d'Apollonius de Perge, Œuvres traduites par Paul Ver Eecke, Paris, 1959.

Apollonius Pergaeus, ed. J. L. Heiberg, Stuttgart, 1974.

Les Coniques, commentaire historique et mathématique, édition et traduction du texte arabe par Roshdi Rashed, Berlin/New York, Walter de Gruyter; tome 1.1: *Livre I*, 2008; tome 3: *Livre V*, 2008; tome 2.2: *Livre IV*, 2009; tome 4: *Livres VI et VII*, 2009; tome 2.1: *Livres II et III*, 2010.

La section des droites selon des rapports, commentaire historique et mathématique, édition et traduction du texte arabe par Roshdi Rashed et Hélène Bellosta, Scientia Graeco-Arabica, vol. 2, Berlin/New York, Walter de Gruyter, 2009

Archimidis Opera Omnia, iterum edidit I.L. Heiberg, vol. 3 corrigenda adiecit E.S. Stamatis, Teubner, 1972.

Archimedes, *De la sphère et du cylindre/Sur les conoïdes et les sphéroïdes*, ed. and French trans. Charles Mugler, Collection des Universités de France, Paris, 1970, t. I.

Banū Mūsā, *Kitāb al-ḥiyal, The Book of Ingenious Devices*, ed. Ahmad Y. al-Hassan, Aleppo, 1981.

Al-Bayhaqī, *Tārīkh ḥukamā' al-Islām*, ed. M. Kurd 'Alī, Damascus, 1946.

E. Bessel-Hagen, O. Spies, 'Ṯābit b. Qurra's Abhandlung über einen halbregelmässigen Vierzehnflächner', *Quellen und Studien zur Geschichte der Math. und Phys.*, B. 1, Berlin, 1932, pp. 186–98.

J. L. Berggren, 'The correspondence of Abū Sahl al-Kūhī and Abū Isḥāq al-Ṣābī: A translation with commentaries', *Journal for the History of Arabic Science* 7.1-2, 1983, pp. 39–124.

Al-Bīrūnī

Al-āthār al-bāqiya 'an al-qurūn al-khāliya, Chronologie orientalischer Völker, ed. C.E. Sachau, Leipzig, 1923.

Al-Rasā'il al-mutafarriqa fī al-hay'a li-al-mutaqaddimīn wa-mu'āṣiray al-Bīrūnī, ed. Osmania Oriental Publications Bureau, Hyderabad, 1947.

'Kitāb taḥdīd nihāyāt al-amākin li-taṣḥīḥ masāfāt al-masākin', edited by P. Bulgakov and revised by Imām Ibrāhīm Aḥmad, in *Majallat Ma'had al-Makhṭūṭāt*, 8, fasc. 1–2, 1962. English translation by Jamil Ali, *The Determination of the Coordinates of Positions for the Correction of Distances between Cities*, Beirut, 1967.

al-Qānūn al-Mas'ūdī, ed. Osmania Oriental Publications Bureau, 3 vols, Hyderabad, 1954–1956.

A. Björnbo, 'Thābits Werk über den Transversalensatz', *Abhandlungen zur Geschichte der Naturwissenschaften und der Medizin,* 7, 1924.

C. Brockelmann, *Geschichte der arabischen Literatur*, I, 1st ed., Leiden, 1937; 2nd ed., Leiden, 1943.

F. Buchner, 'Die Schrift über den Qarastûn von Thabit b. Qurra', *Sitzungsberichte der physikalisch–medizinischen Sozietät in Erlangen,* Bd 52–53, 1920/21, pp. 141–88.

F. J. Carmody, *The Astronomical Works of Thābit b. Qurra*, Berkeley/Los Angeles, 1960.

D. Chwolson, *Die Ssabier und der Ssabismus*, vol. I, St. Petersburg, 1856; repr. Amsterdam, 1965.

M. Clagett, *Archimedes in the Middle Ages*, vol. I: *The Arabo-Latin Tradition*, Madison, 1964; vol. V: *Quasi-Archimedean Geometry in the Thirteenth Century*, Philadelphia, 1984.

M. Curtze, 'Verba Filiorum Moysi, Filii Sekir, id est Maumeti, Hameti et Hasen. Der Liber trium fratrum de Geometria, nach der Lesart des Codex Basileenis F. II. 33 mit Einleitung und Commentar', *Nova Acta der Ksl. Leop.-Carol. Deutschen Akademie der Naturförscher*, vol. 49, Halle, 1885, pp. 109–67.

J. al-Dabbagh, 'Banū Mūsā', *Dictionary of Scientific Biography*, vol. I, New York, 1970, pp. 443–6.

J. al-Dabbagh and B. Rosenfeld, *Matematitcheskie traktaty* (en russe), Coll. Nautchnoie Nasledstvo, t. 8, Moscow, 1984.

P. Dedron and Jean Itard, *Mathématiques et mathématiciens*, Paris, 1959.

Al-Dhahabī, *Tārīkh al-Islām* (years 281–290), ed. 'Umar 'Abd al-Salam Tadmūrī, Beirut, 1989–1993.

Y. Dold-Samplonius

'Die Konstruktion des regelmässigen Siebenecks', *Janus* 50, 4, 1963, pp. 227–49.

'Al-Khāzin', *Dictionary of Scientific Biography*, vol. VII, New York, 1973, pp. 334–5.

'Al-Qūhī', *Dictionary of Scientific Biography*, vol. XI, New York, 1975, pp. 239–41.

A.S. al-Dimirdāsh, 'Wayjan Rustam al-Qūhī wa-ḥajm al-mujassam al-mukāfi'', *Risālat al-'ilm* 4, 1966, pp. 182–95.

A. A. Duri, 'Baghdād', *Encyclopédie de l'Islam*, 2nd ed., Leiden, 1960, t. I, pp. 921–36.

K. Garbers, *Ein Werk Ṯābit b. Qurra's über ebene Sonnenuhren*, Dissertation, Hamburg/Göttingen, 1936.

T. M. Green, *The City of the Moon God*, Leiden, 1992.

Th. Heath, *The Thirteen Books of Euclid's Elements*, 3 vols, Cambridge, 1926; repr. Dover, 1956.

J.L. Heiberg, *Claudii Ptolemaei opera quae exstant omnia. I. Syntaxis mathematica*, Leipzig, 1898.

J.P. Hogendijk

'Discovery of an 11th-century geometrical compilation: The *Istikmāl* of Yūsuf al-Mu'taman ibn Hūd, King of Saragossa', *Historia Mathematica,* 1986, pp. 43–52.

'The geometrical parts of the *Istikmāl* of Yūsuf al-Mu'taman ibn Hūd (11th century). An analytical table of contents', *Archives internationales d'histoire des sciences*, 41.127, 1991, pp. 207–81.

Ibn al-Abbār

Al-Takmila li-Kitāb al-Ṣila, ed. al-Sayyid ʿIzzat al-ʿAṭṭār al-Ḥusaynī, Cairo, 1955, t. I; *Complementum libri Assilah*, ed. F. Codera and Zaydin, 2 vols, Madrid, 1887–1889.

Al-Ḥulla al-siyarā', ed. Hussain Monés, Cairo, n.d., vol. II.

Ibn Abī Uṣaybiʿa, *'Uyūn al-anbā' fī ṭabaqāt al-aṭibbā'*, ed. A. Müller, 3 vols, Cairo/Königsberg, 1882–84; ed. N. Riḍā, Beirut, 1965.

Ibn Aknīn, *Ṭibb al-nufūs al-alīma,* établi et traduit en allemand par M. Güdemann, *Das jüdische Unterrichtswesen während der spanisch-arabischen Periode*, Vienne, 1873; repr. Amsterdam, 1968.

Ibn al-Athīr, *al-Kāmil fī al-tārīkh*, ed. C.J. Tornberg under the title *Ibn-El-Athiri Chronicon quod perfectissimum inscribitur*, 12 vols, Leiden, 1851–71; repr. 13 vols, Beirut, 1965–67.

Ibn Bājja, *Rasā'il falsafiyya li-Abī Bakr ibn Bājja*, ed. Jamāl al-Dīn al-ʿAlawī, Beirut, 1983.

Ibn Haydūr, *al-Tamḥīṣ fī sharḥ al-talhkīṣ*, edited and analyzed by R. Rashed in 'Matériaux pour l'histoire des nombres amiables et de l'analyse combinatoire', *Journal for the History of Arabic Sciences*, 6. 1–2, 1982, pp. 209–78.

Ibn al-ʿIbrī, *Tārīkh mukhtaṣar al-duwal,* ed. O.P. A. Ṣāliḥānī, 1st ed., Beirut, 1890; repr. 1958.

Tārīkh al-zamān, trad. arabe Isḥāq Armala, Beirut, 1991.

Ibn al-ʿImād, *Shadharāt al-dhahab fī akhbār man dhahab*, ed. Būlāq, 8 vols, Cairo, 1350–51 H., year 288, vol. II.

Ibn ʿIrāq, 'Taṣḥīḥ Zīj al-Ṣafā'iḥ', in *Rasā'il Mutafarriqa fī al-hay'a*, Hyderabad, 1948.

Ibn al-Jawzī, *al-Muntaẓam fī tārīkh al-mulūk wa-al-umam,* 10 vols, Hyderabad, 1357–58/1938–40, vol. VI.

Ibn Juljul, *Ṭabaqāt al-aṭibbā' wa-al-ḥukamā'*, ed. F. Sayyid, Publications de l'Institut Français d'Archéologie Orientale du Caire. Textes et traductions d'auteurs Orientaux, 10, Cairo, 1955.

Ibn Kathīr, *al-Bidāya wa-al-nihāya*, ed. Būlāq, 14 vols, Beirut, 1966.

Ibn Khallikān, *Wafayāt al-a'yān*, ed. Iḥsān ʿAbbās, 8 vols, Beirut, 1978, vol. I.

Ibn al-Khaṭīb

Al-Iḥāṭa fī akhbār Gharnāṭa, ed. Muḥammad ʿAbdallāh ʿInān, Cairo, 1955.

Histoire de l'Espagne Musulmane (Kitāb a'māl al-a'lām), Arabic text with introduction and index by E. Lévi-Provençal, Beirut, 1956.

Ibn Khurdādhbih, *Al-Masālik wa-al-mawālik*, ed. M. J. de Goeje, Bibliotheca Geographorum Arabicorum VI, Leiden, 1889; repr. Bagdad, n.d.

Ibn Sinān, Ibrāhīm

Rasā'il Ibn Sinān, ed. Osmania Oriental Publications Bureau, Hyderabad, 1948. *See also* R. Rashed and H. Bellosta.

Ibn Taghrī Bardī, *al-Nujūm al-zāhira fī mulūk Misr wa-al-Qāhira*, introduced and annotated by Muḥammad Ḥusayn Shams al-Dīn, vol. IV, Beirut, 1992.

Ibn Waḥshiyya, *al-Filāḥa al-nabaṭiyya*, ed. Toufic Fahd, t. I, Damascus, 1993.

al-Khaṭīb al-Baghdādī, *Tārīkh Baghdād*, ed. Muḥammad Amīn al-Khānjī, 14 vols, Cairo, 1931; repr. Beirut with an additional index volume: *Fahāris Tārīkh Baghdād li-al-Khaṭīb al-Baghdādī*, Beirut, 1986.

Al-Khayyām: *see* R. Rashed.

W. R. Knorr

'Ancient sciences of the medieval tradition of mechanics', in *Supplemento agli Annali dell'Istituto e Museo di Storia della Scienza* Fasc. 2, Firenze, 1982.

'The medieval tradition of a Greek mathematical lemma', *Zeitschrift für Geschichte der arabisch-islamischen Wissenschaften* 3, 1986, pp. 230–64.

Textual Studies in Ancient and Medieval Geometry, Boston, Basel, Berlin, 1989, pp. 267–75.

D. Krawulsky, *Īrān – Das Reich der Īlḫāne. Eine topographisch-historische Studie,* Wiesbaden, 1978.

T. Langermann, 'The mathematical writings of Maïmonides', *The Jewish Quarterly Review* LXXV, no. 1, July 1984, pp. 57–65.

R. Lorch, 'Abū Ja'far al-Khāzin on isoperimetry and the Archimedian tradition', *Zeitschrift für Geschichte der arabisch-islamischen Wissenschaften* 3, 1986, pp. 150–229.

O. Loth, *A Catalogue of the Arabic Manuscripts in the Library of the India Office,* London, 1877.

A. G. Ma'ānī, *Fihrist kutub khaṭṭī Kitābkhāna Astān Quds*, Meshhed, 1350/1972, vol. VIII.

Al-Maqqarī, *Nafḥ al-ṭīb min ghuṣn al-Andalus al-raṭīb*, ed. Iḥsān 'Abbās, 8 vols, Beirut, 1968, vol. I.

Al-Marākūshī, *al-Mu'jib fī talkhīṣ akhbār al-Maghrib*, ed. M. S. al-'Aryān and M. al-'Arabī, 7th ed., Casablanca, 1978.

Al-Mas'ūdī

Al-Tanbīh wa-al-ishrāf, ed. M. J. de Goeje, Bibliotheca Geographorum Arabicorum VIII, Leiden, 1894.

Murūj al-dhahab (Les prairies d'or), ed. C. Barbier de Meynard and M. Pavet de Courteille, revue et corrigée par Charles Pellat, Publications de l'Université Libanaise, Section des études historiques XI, Beirut, 1966, vol. II.

Maulavi Abdul Hamid, *Catalogue of the Arabic and Persian Manuscripts in the Oriental Public Library at Bankipore*, volume XXII (Arabic MSS) Science, Patna, 1937.

Al-Nadīm, *Kitāb al-fihrist*, ed. R. Tajaddud, Tehran, 1971.

C. Nallino, *Arabian Astronomy, its History during the Medieval Times* [Conferences into Arabic at the Egyptian University], Roma, 1911.

ʿAbd al-Majīd Nuṣāyr, 'Risāla fī misāḥat al-mujassam al-mukāfi'', *Revue de l'Institut des manuscrits arabes* 29, 1, 1985, pp. 187–208.

Al-Nuwayrī, *Nihāyat al-arab fī funūn al-adab*, 31 vols, Cairo, 1923–93, vol. II.

Pappus d'Alexandrie

Pappi Alexandrini Collectionis quae supersunt e libris manu scriptis edidit latina interpretatione et commentariis instruxit F. Hultsch, 3 vols, Berlin, 1876-1878.

La collection mathématique, French trans. Paul Ver Eecke, Paris/Bruges, 1933.

Pappus of Alexandria, Book 7 of the Collection. Part 1. *Introduction, Text, and Translation*; Part 2. *Commentary, Index, and Figures*, Edited with Translation and Commentary by Alexander Jones, Sources in the History of Mathematics and Physical Sciences, 8, New York/Berlin/Heidelberg/Tokyo, Springer-Verlag, 1986.

Al-Qifṭī, *Ta'rīkh al-ḥukamā'*, ed. Julius Lippert, Leipzig, 1903.

R. Rashed

'L'induction mathématique: al-Karajī - as-Samaw'al', *Archive for History of Exact Sciences* 9, 1, 1972, pp. 1–21; repr. in *Entre arithmétique et algèbre: Recherches sur l'histoire des mathématiques arabes,* Paris, 1984, pp. 71–91; English translation: *The Development of Arabic Mathematics Between Arithmetic and Algebra*, Boston Studies in Philosophy of Science 156, Dordrecht/Boston/London, Kluwer Academic Publishers, 1994.

'La mathématisation des doctrines informes dans la science sociale', in *La mathématisation des doctrines informes*, under the direction of G. Canguilhem, Paris, 1972, pp. 73–105.

'L'analyse diophantienne au Xe siècle: l'exemple d'al-Khāzin', *Revue d'histoire des sciences* 32, 1979, pp. 193–222.

'Matériaux pour l'histoire des nombres amiables et de l'analyse combinatoire', *Journal for the History of Arabic Sciences* 6, nos 1 and 2, 1982.

Sharaf al-Dīn al-Ṭūsī, Œuvres mathématiques. Algèbre et géométrie au XIIe siècle, 2 vols, Paris, 1986.

'Ibn al-Haytham et les nombres parfaits', *Historia Mathematica*, 16, 1989, pp. 343–52.

'Problems of the transmission of Greek scientific thought into Arabic: examples from mathematics and optics', *History of Science*, 27, 1989, pp. 199–209; repr. dans *Optique et mathématiques: Recherches sur l'histoire de la pensée scientifique en arabe*, Variorum CS388, Aldershot, 1992, I.

'La philosophie mathématique d'Ibn al-Haytham. I: L'analyse et la synthèse', *M.I.D.E.O.* 20, 1991, pp. 31–231.

'Al-Kindī's Commentary on Archimedes' *The measurement of the Circle*', *Arabic Sciences and Philosophy* 3.1, 1993, pp. 7–53.

'Al-Qūhī *vs.* Aristotle: On motion', *Arabic Sciences and Philosophy* 9.1, 1999, pp. 7–24.

Géométrie et dioptrique au Xe siècle. Ibn Sahl, al-Qūhī et Ibn al-Haytham, Paris, 1993; English version: *Geometry and Dioptrics in Classical Islam*, London, al-Furqān, 2005.

Les Mathématiques infinitésimales du IXe au XIe siècle, vol. II: *Ibn al-Haytham*, London, 1993.

Œuvre mathématique d'al-Sijzī. Volume I: *Géométrie des coniques et théorie des nombres au Xe siècle*, Les Cahiers du Mideo, 3, Louvain-Paris, Éditions Peeters, 2004.

'Thābit et l'art de la mesure', in R. Rashed (ed.), *Thābit ibn Qurra. Science and Philosophy in Ninth-Century Baghdad*, Scientia Graeco-Arabica, vol. 4, Berlin/New York, Walter de Gruyter, 2009, pp. 173–209.

R. Rashed - H. Bellosta, *Ibrāhīm ibn Sinān. Logique et géométrie au Xe siècle*, Leiden, Brill, 2000.

R. Rashed - B. Vahabzadeh, *Al-Khayyām mathématicien*, Paris, Librairie Blanchard, 1999; English version (without the Arabic texts): *Omar Khayyam. The Mathematician*, Persian Heritage Series no. 40, New York, Bibliotheca Persica Press, 2000.

R. Rashed - Ch. Houzel, *Recherche et enseignement des mathématiques au IXe siècle. Le Recueil de propositions géométriques de Na'īm ibn Mūsā*, Les Cahiers du Mideo, 2, Louvain-Paris, Éditions Peeters, 2004.

Roberval, 'Observations sur la composition du mouvement et sur les moyens de trouver les touchantes des lignes courbes', in *Mémoires de l'Académie Royale des Sciences*, ed. 1730, vol. 6, pp. 1–79.

A. Rome, *Commentaires de Pappus et de Théon d'Alexandrie sur l'Almageste*, text edited and annoted, vol. II: *Théon d'Alexandrie, Commentaire sur les livres 1 et 2 de l'Almageste*, Vatican, 1936.

B. A. Rosenfeld and A. T. Grigorian, 'Thābit ibn Qurra', *Dictionary of Scientific Biography*, vol. XIII, 1976, pp. 288–95.
See also al-Dabbagh

Al-Ṣafadī, *al-Wāfī bi-al-Wafayāt*, 24 vols, 1931–1993; vol. X, ed. Ali Amara and Jacqueline Sublet, Wiesbaden, 1980.

Ṣā'id al-Andalusī, *Ṭabaqāt al-umam*, ed. H. Bū'alwān, Beirut, 1985. French trans. R. Blachère, *Livre des Catégories des Nations*, Paris, 1935.

A. Saidan

'Rasā'il of al-Bīrūnī and Ibn Sinān', *Islamic Culture* 34, 1960, pp. 173–5.

The Works of Ibrāhīm ibn Sinān, Kuwait, 1983.

G. Saliba

'Risālat Ibrāhīm ibn Sinān ibn Thābit ibn Qurra fī al-Ma'ānī allatī istakhrajahā fī al-handasa wa-al-nujūm', *Studia Arabica & Islamica,* Festschrift for Iḥsān 'Abbās, ed. Wadād al-Qāḍī, American University of Beirut, 1981, pp. 195–203.

'Early Arabic critique of ptolemaic cosmology: A ninth-century text on the motion of the celestial spheres', *Journal for the History of Astronomy* 25, 1994, pp. 115–41.

J. Samsó, 'Al-Khāzin', *Encyclopédie de l'Islam*, 2nd ed., Leiden, 1978, t. IV, pp. 1215–16.

J. Schacht and M. Meyerhof, *The Medico-Philosophical Controversy between Ibn Butlan of Baghdad and Ibn Ridwan of Cairo. A Contribution to the History of Greek Learning Among the Arabs*, Faculty of Arts no. 13, Cairo, 1937.

Serenus of Antinoupolis, *Sereni Antinoensis Opuscula*. Edidit et latine interpretatus est I. L. Heiberg, Leipzig, 1896. French transl. P. Ver Eecke, *Serenus d'Antinoë: Le livre de la section du cylindre et le livre de la section du cône*, Paris, 1969.

R. Şeşen, C. Izgi, Cemil Akpinar, *Catalogue of manuscripts in the Köprülü Library*, Research Centre for Islamic History, Art and Culture, 3 vols, Istanbul, 1986.

A. Sayili, *The Observatory in Islam and its Place in the General History of the Observatory*, 2nd ed., Ankara, 1988.

F. Sezgin, *Geschichte des arabischen Schrifttums*, B. V, Leiden, 1974 and B. VI, Leiden, 1978.

Al-Sijistānī, *The Muntakhab Ṣiwān al-ḥikmah*, Arabic Text, Introduction and Indices edited by D. M. Dunlop, The Hague/Paris/New York, 1979.

M. Le Baron de Slane, *Catalogue des manuscrits arabes de la Bibliothèque Nationale*, Paris, 1883–1895.

Al-Shahrazūrī, *Tārīkh al-ḥukamā', Nuzhat al-arwāḥ wa-rawaḍat al-afrāḥ*, ed. 'Abd al-Karīm Abū Shuwayrib, Tripoli-Libya, 1988.

M. Steinschneider

'Thabit ("Thebit") ben Korra. Bibliographische Notiz', *Zeitschrift für Mathematik u. Physik* XVIII, 4, 1873, pp. 331–8.

'Die Söhne des Musa ben Schakir', *Bibliotheca Mathematica* 1, 1887, pp. 44–8, 71–6.

Die hebraeischen Übersetzungen des Mittelalters und die Juden als Dolmetscher, Berlin, 1893; repr. Graz, 1956.

Die arabische Literatur der Juden, Frankfurt, 1902; repr. Hildesheim/Zürich/ New York, 1986.

H. Suter

Die Mathematiker und Astronomen der Araber und ihre Werke, Leipzig, 1900.

'Über die Geometrie der Söhne des Mûsâ ben Schâkir', *Bibliotheca Mathematica* 3, 1902, pp. 259–72.

'Über die Ausmessung der Parabel von Thâbit b. Ḳurra al-Ḥarrânî', *Sitzungsberichte der Physikalisch-medizinischen Societät in Erlangen* 48, 1916, pp. 65–86.

'Die Abhandlungen Thâbit b. Ḳurras und Abû Sahl al-Kûhîs über die Ausmessung der Paraboloide', *Sitzungsberichte der Physikalisch-medizinischen Societät in Erlangen* 49, 1917, pp. 186–227. Russian trans. J. al-Dabbagh and B. Rosenfeld, *Matematitcheskie traktaty,* pp. 157–96.

'Abhandlung über die Ausmessung der Parabel von Ibrāhīm b. Sinān b. Thābit', in *Vierteljahresschrift der Naturforschenden Gesellschaft in Zürich*, Herausgegeben von Hans Schinz 63, 1918, pp. 214–28.

Al-Ṭabarī, *Tārīkh al-rusul wa-al-mulūk*, ed. Muḥammad Abū al-Faḍl Ibrāhīm, Cairo, 1967, vol. 9.

Al-Tawḥīdī

Mathālib al-wazīrayn al-Ṣāḥib ibn 'Abbād wa-Ibn al-'Amīd, ed. Muḥammad al-Ṭanjī, Beirut, 1991.

Kitāb al-imtā' wa-al-mu'ānasa, ed. Aḥmad Amīn and Aḥmad al-Zayn, repr. Būlāq, n.d.

Y. E. Tessami, *Catalogue des manuscrits persans et arabes de la Bibliothèque du Madjless*, Publications de la Bibliothèque, Tehran, 1933, vol. II.

Thābit ibn Qurra

Thābit ibn Qurra. Œuvres d'astronomie, text edited and translated by Régis Morelon, Collection Sciences et philosophies arabes. Textes et études, Paris, Les Belles Lettres, 1987.

Thābit ibn Qurra. Science and Philosophy in Ninth-Century Baghdad, ed. R. Rashed, Scientia Graeco-Arabica, vol. 4, Berlin/New York, Walter de Gruyter, 2009.

J. Uri, *Bibliotecae Bodleianae Codicum Manuscriptorum Orientalium*, Oxonii, 1787.

Al-'Utbī

Sharḥ al-Yamīnī al-musamma bi-al-Fatḥ al-Wahbī 'alā Tārīkh Abī Naṣr al-'Utbī li-al-Shaykh al-Manīnī, Cairo 1286/1870, vol. 1.

G. Vajda

'Quelques notes sur le fonds de manuscrits arabes de la Bibliothèque nationale de Paris', *Rivista degli Studi Orientali*, 25, 1950.

Index général des manuscrits arabes musulmans de la Bibliothèque nationale de Paris, Publications de l'Institut de recherche et d'histoire des textes IV, Paris, 1953.

Luca Valerio, *De Centro Gravitatis Solidorum Libri Tres*, Bologna, 1661.

P. Voorhoeve, *Codices Manuscripti VII. Handlist of Arabic Manuscripts in the Library of the University of Leiden and Other Collections in the Netherlands*, 2nd ed., The Hague/Boston/London, 1980.

E. Wiedemann

'Die Schrift über den Qarasṭūn', *Bibliotheca Mathematica*, 12.3, 1911–12, pp. 21–39.

'Über Ṯābit ben Qurra, sein Leben und Wirken', in *Aufsätze zur arabischen Wissenschafts-Geschichte*, Hildesheim, 1970, vol. II.

F. Woepcke, 'Recherches sur plusieurs ouvrages de Léonard de Pise', *Atti Nuovi Lincei*, 14, 1861, pp. 301–24.

Yāqūt, *Kitāb Irshād al-arīb ilā ma'rifat al-adīb* (*Mu'jam al-udabā'*), ed. D. S. Margoliouth, vol. 7, London, 1926.

INDEX OF NAMES

ʿAbbās, I.: 7 n. 26, 114 n. 2, 721 n. 2
Abū Kāmil: 460
Abū Shuwayrib, ʿA: 580 n. 5
Abū al-Wafā' al-Buzjānī: 581
ʿAḍud al-Dawla: 460, 580, 581
Ahlwardt, W.: 37 n. 61
Aḥmad, I.: 7 n. 27, 505 n. 15, 580 n. 6
Aḥmad (Prince): 34
al-Akfānī: 722 n. 3, 726 n. 21
Akpinar, C.: 36 n. 51
al-ʿAlawī, J.: 723 n. 7
Ali, J.: 580 n. 6, 581 n. 9, 582 n. 12
Amara, A.: 114 n. 2
al-Amīn: 4
Amīn, A.: 579 n. 1
Anbouba, A.: 505, 505 n. 12, 584 n. 15
al-Anṭākī: 13
Apollonius of Perga: xiv, 3 n. 7, 9, 10, 119, 121, 128, 129, 131, 155, 161, 196, 201–203, 290, 333–336, 341, 349, 359, 360, 391 n. 11, 395, 401, 403, 413, 416, 417, 419, 422, 426–430, 460, 474, 477, 579, 616–621, 625, 669, 726, 730, 732, 736, 742, 748, 750 n. 42, 751 n. 44, 752 n. 45, 755
al-ʿArabī, M.: 721 n. 1
Archimedes: xiv, xxii, 10, 34, 38–40, 45–49, 59, 60, 68, 82, 120, 130, 131, 148, 209, 213, 235, 236, 333, 337, 347–350, 360, 361, 366–369, 371, 372, 377, 465, 480, 507, 530 n. 34, 531 n. 35–36, 534 n. 37, 537 n. 38, 538, 538 n. 40, 579, 583, 586, 594, 595, 599, 628, 629, 650, 651, 653, 673, 710, 755, 756
Archytas: 60
Aristotle: 505, 723 n. 8
Armala, I.: 119 n. 9
al-ʿAryān, M.S.: 721 n. 1
Asadabad: 507

Baghdad: xiv, 3, 5, 36 n. 47, 113, 113 n. 1, 115, 116, 118 n. 8, 119, 120, 459, 459 n. 1, 460, 508, 579–581, 586 n. 20
al-Balkhī, Abū Zayd: 505
Banū Mūsā: xiv, xx, 1–111, 115–117, 119, 120, 124, 333, 334, 460, 520, 526, 528, 529, 535, 537 n. 38, 538, 539, 579, 580, 585, 617, 621, 628, 672 n. 22, 673 n. 23, 677 n. 30, 727, 756, 767
Banū Qurra: 583
Barbier de Meynard, C.: 114 n. 2
Bar Hebraeus: *see* Ibn al-ʿIbrī
al-Bayhaqī: 580 n. 5
Bellosta, H.: 459 n. 1, 460 n. 3, 462 n. 8, 463 n. 9
Berggren, J.L.: 584 n. 15
Bernoulli, Jacob: 3
Bernoulli, Johann: 3
Bessel-Hagen, E.: 121 n. 13
al-Bīrūnī: 7 n. 27, 46, 46 n. 3, n. 5, 117, 117 n. 7, 503–506, 579, 580, 580 n. 6, 581, 581 n. 9–10, 582 n. 12
Björnbo, A.: 121 n. 12
Blachère, R.: 615 n. 1
Bourbaki, N.: 726 n. 22
Brockelmann, C.: 1 n. 2, 580 n. 4
Buchner, F.: 121 n. 12
Bukhārā: 505
Bulgakov, P.: 7 n. 27, 505 n. 15, 580 n. 6, 581 n. 9, 582 n. 12
Byzantines: 6, 113 n. 1, 116, 767
Būʿalwān, H.: 615 n. 1, 616 n. 4, 721 n. 2

Canguilhem, G.: 131 n. 1
Carmody, F.J.: 121 n. 12
Caussin de Perceval, A.-P.: 125
Chwolson, D.: 115 n. 2
Clagett, M.: 1 n. 3, 10 n. 38, 46 n. 4, 66 n. 10, 68 n. 12, 73 n. 1, n. 3, 74 n. 8, 538 n. 39
Codera, F.: 723 n. 7
Constantinople: 667 n. *, 720
Cordoba: 615
Curtze, M.: 10 n. 38

al-Dabbagh, J.: 1 n. 2, 45 n. 2, 128 n. 26–27, 129 n. 28
Damascus: 6
Dedron, P.: 66 n. 10
al-Dhahabī, Abu ʿUthmān: 114 n. 2
al-Dimashqī, Abū ʿUthmān: 13
al-Dimirdāsh, A.S.: 46 n. 3, 587 n. 23

Dold-Samplonius, Y.: 506 n. 19, 580 n. 4, 581 n. 8
Dunlop, D.M.: 114 n. 2
Duri, A.A.: 115 n. 3

Euclid: xxii, 2 n. 4, 11, 13, 74, 82 n. 37, 119, 121, 130, 131, 133, 160, 209, 237, 338, 349, 368, 372, 377, 383 n. 2, 504 n. 10, 564, 567 n. 19, 585, 593, 607, 616, 623–625, 629, 668, 671–674, 696, 723, 728, 756 n. 51, 774
Eudemus: 60
Eudoxus: 360, 377
Eutocius: 14, 60, 62, 427 n. 36

al-Fārābī, Abū Naṣr: 13
al-Fārisī, Kamāl al-Dīn: xx
Fārs: 580
Fahd, T.: 118 n. 8
Fibonacci: 46
Freudenthal, G.: 687 n.*

Galen: 9
Garbers, K.: 121 n. 13
Gerard of Cremona: 1, 10, 11, 13–33, 38, 62, 73 n. 1–107 n. 93
al-Ghāzī Maḥmūd Khān: 124
Ghulām Zuḥal: 580
De Goeje, M.J.: 6 n. 21–22, 729
Golius: 729, 729 n. 26
Green, Tamara M.: 118 n. 8
Grenada: 615, 615 n. 2
Grigorian, A.T.: 122 n. 17
Güdemann, M.: 724 n. 15–16

al-Ḥajjāj: 508 n. 24
Halma, N.: 757 n. 54, n. 56, 758 n. 59
Ḥamadān: 507
al-Harawī, Abū Bakr: 505
Ḥarrān: 113, 113 n. 1, 115–118 n. 8, 119 n. 9
Hārūn al-Rashīd: 4
al-Hasan, A.Y.: 1 n. 2
Heath, Th.: 42 n. 1, 567 n. 19, 723 n. 8
Heiberg, J.L.: 60 n. 8, 335 n. 2, 427 n. 36, 508 n. 24, 618 n. 7, 757 n. 54, n. 56, 758 n. 59
Hero of Alexandria: 38, 40, 46, 508, 767
Hogendijk, J.P.: 586 n. 20, 722 n. 4, 724 n. 15, 725 n. 17
House of Wisdom: 3, 5
Houzel, Ch.: 116 n. 6, 120 n. 10
Ḥubaysh: 6
Ḥunayn ibn Isḥāq: 3
Ḥusayn ibn Muḥammad ibn ʿAlī: 507
al-Ḥusaynī, ʿI.: 615 n. 1

Ibn al-Abbār: 615 n. 1, 721 n. 2, 723 n. 7
Ibn Abī Jarrāda: 129, 129 n. 28, 341, 585, 586, 625, 627, 767–778
Ibn Abī Manṣūr, Yaḥyā: 5
Ibn Abī Uṣaybiʿa: 1 n. 1, 2 n. 6, 6, 6 n. 23, 7, 7 n. 29, 113, 114, 114 n. 1, 119, 119 n. 9, 122, 123, 459 n. 1, 461, 508, 508 n. 27, 582, 583, 583 n. 13, 615 n. 1
Ibn ʿAdī, Yaḥyā: 579 n. 1
Ibn Aflaḥ, Jābir: 508, 724 n. 13
Ibn Aknīn, Yūsuf: 721 n. 2, 724, 724 n. 14–15
Ibn al-ʿAmīd: 506
Ibn al-Athīr: 114 n. 2, 505, 505 n. 18
Ibn Bājja: 723 n. 7
Ibn Būlis al-Naṣrānī, Abū Saʿd: 581
Ibn Farrūkhānshāh: 5
Ibn al-Ḥamāmī: 125
Ibn Haydūr: 725, 725 n. 18
Ibn al-Haytham, al-Ḥasan: xiv, xx, xxii, 13, 130, 210, 243, 333, 461, 508, 522, 523, 546, 579, 615, 727, 755, 756 n. 51, 762, 763
Ibn Hilāl al-Ḥimṣī, Hilāl: 3, 3 n. 7
Ibn Hilāl al-Ṣābiʾ, Abū Isḥāq Ibrāhīm: 581, 582, 584, 584 n. 14–15
Ibn Hūd: xiv, xxii, 13, 508, 522, 586 n. 20, 616, 721–766
Ibn Ibrāhīm al-Ḥalabī: 37 n. 58
Ibn al-ʿIbrī: 4, 4 n. 12, 5, 114, 114 n. 2, 118, 118 n. 8–9, 581 n. 10
Ibn al-ʿImād: 114 n. 2
Ibn al-Imān, Abū al-Ḥasan: 723 n. 7
Ibn ʿIrāq: 465, 503, 503 n. 3, 504
Ibn Jamaʿa: 615 n. 1
Ibn al-Jawzī: 114 n. 2
Ibn Juljul: 114 n. 2
Ibn Kathīr: 114 n. 2
Ibn Khallikān: 7, 7 n. 26, 114 n. 2, 116, 116 n. 5
Ibn al-Khaṭīb: 615 n. 1, 721 n. 2
Ibn Khurdādhbih: 4
Ibn Māksan, Ḥabbūs: 615
Ibn Maḥmūd al-Kunyānī: 36 n. 46
Ibn Muḥtāj, ʿAlī: 505
Ibn Mukhlad: 5
Ibn Mūsā, Aḥmad: 1–9, 117, 119, 120, 333, 334

Ibn Mūsā, al-Ḥasan: 1, 2, 2 n. 4–5, 4, 5 n. 13, 8, 34, 117, 119, 120, 333, 334, 381, 615–618, 621, 622
Ibn Mūsā, Muḥammad: 1–6, 8, 9, 34, 113 n. 1, 115–117, 119, 120, 333, 334
Ibn Qurra: *see* Thābit
Ibn Rashīq: 721 n. 1
Ibn Sahl: xx, xxii, 130, 333, 461, 579, 582
Ibn al-Samḥ: xiv, xxii, 8 n. 32, 9, 9 n. 34, 333, 334, 615–667, 726, 726 n. 22, 727
Ibn Sayyid, ʿAbd al-Raḥmān: 721 n. 2, 723–726 n. 22
Ibn Sīnā, Abū ʿAlī al-Ḥusayn ibn ʿAbd Allāh: 125, 727
Ibn Sinān, Ibrāhīm: xiv, xxii, 120, 126, 130, 333, 459–735, 579, 582, 584, 727, 730, 735 n. 35, 740, 744, 747, 755
Ibn Sinān, Thābit: 460, 582
Ibn Sulaymān, Aḥmad: 36 n. 49
Ibn Taghrī Bardī: 581, 582, 582 n. 11
Ibn Thābit ibn Qurra, Sinān: 459, 582, 583
Ibn Usayyid, Abū Mūsā ʿĪsā: 124
Ibn Waḥshiyya: 118 n. 8
Ibn Yumn, Naẓīf: 580, 582, 582 n. 12

Ibrāhīm, M. A.: 5 n. 15
ʿInān, M.A.: 615 n. 1
Isḥāq ibn Ḥunayn: 6
Itard, J.: 66 n. 10
Izgi, C.: 36 n. 51

al-Jaʿfariyya: 5
Jaouiche, Kh.: 123 n. 20
al-Jawharī: 460
John of Tinemue: 538
Joseph b. Joel Bibas: 667 n. *, 720

Kafr Tūtha: 116
al-Khāzin: xiv, xx, xxii, 36 n. 49, 40, 46, 59, 110, 503–551, 760, 762, 763
al-Khānjī, M.A.: 115 n. 3
al-Khatīb al-Baghdādī: 115 n. 3
al-Khayyām, ʿUmar: 503, 503 n. 1, 504
al-Khujandī: 503
Khurāsān: 4, 504
Kurd ʿAlī, M.: 580 n. 5
al-Khwārizmī: 460
al-Kindī: 2, 6, 117 n. 7, 505, 508, 508 n. 26, 673 n. 23
Knorr, W.R.: 10 n. 38, 123 n. 20
Krawulsky, D.: 586 n. 20
Kurio, H.: 37 n. 61

Langermann, Y.T.: 724 n. 15, 726 n. 21
Lévi-Provençal, E.: 721 n. 2
Lévy, T.: 667 n. *
Lippert, J.: 1 n. 1, 113 n. 1, 459 n. 1, 504 n. 6, 581 n. 10, 724 n. 13
Lorch, R.: 507, 507 n. 22
Loth, O.: 36 n. 49
Luca Pacioli: 46
Luca Valerio: 347

Maʿānī, A.G.: 36 n. 52
al-Maghribī, Abū al-Ḥasan: 581
al-Māhānī: 130, 462, 463, 495, 730
Maimonides: 724, 724 n. 13
al-Majrīṭī, Maslama: 615
al-Ma'mūn: 2 n. 4, 4, 5 n. 13, 7, 582, 721 n. 1
Manṣūr ibn Nūḥ: 505
al-Maqqarī: 721 n. 2, 726 n. 21
Marāgha School: xx
al-Marāghī, Muḥammad Sartāq: 586 n. 20, 724, 725
Margoliouth, D.S.: 505 n. 14
al-Marākūshī, ʿAbd al-Wāḥid: 721 n. 1
al-Masʿūdī: 4, 4 n. 21, 114 n. 2, 115 n. 4
Maulavi Abdul Hamid: 465 n. 15
Menelaus: 34, 39, 60, 60 n. 7, 62, 105
Mones, H: 721 n. 2
Morelon, R.: 115 n. 2, 121 n. 13, 123 n. 18
al-Muʿtaḍid: 113 n. 1
Mugler, Ch.: 68 n. 11, 347 n. 4, 361 n. 5
Müller, A.: 1 n. 1, 2 n. 6, 6 n. 23, 7 n. 29, 114 n. 1, 119 n. 9, 459 n. 1, 508 n. 27, 583 n. 13, 615 n. 1
al-Muntaṣir: 6
al-Muqtadir: 459, 721 n. 1
Mūsā ibn Shākir: 1, 4, 5, 5 n. 13
al-Muṣʿabī, Isḥāq ibn Ibrāhīm: 5
al-Mustaʿīn: 6
al-Mutanabbī: 506
al-Mutawakkil: 5, 6

al-Nadīm: 1 n. 1, 2 n. 5–6, 3 n. 7, 6, 6 n. 25, 7–9, 34, 60, 60 n. 7, 113–115, 118 n. 8, 122, 459 n. 1, 461, 461 n. 7, 462, 504, 504 n. 4–5, n. 8, 505 n. 13, 580, 580 n. 4
Nakīsār: 586 n. 20
Nallino, C.: 7, 7 n. 28
Naʿīm ibn Mūsā: 120
al-Nayrīzī: 13, 460, 465, 723, 726

Nicomachus of Gerasa: 121
Nūḥ ibn Naṣr: 505
Nuṣayr, ʿA.: 588
al-Nuwayrī: 114 n. 2

Pappus: 508, 508 n. 25, 509, 523, 546, 723 n. 8
Pascal, Etienne: 66 n. 10, 67, 68
Pavet de Courteille, M.: 114 n. 2
Pellat, Ch.: 114 n. 2, 115 n. 4
Plato: 14, 62, 65
Proclus: 723 n. 8
Ptolemy: 121, 462, 506–508, 755
Pythagoras (theorem): 634

al-Qāḍī, W.: 459 n. 1
al-Qāhir: 459
Qalonymos b. Qalonymos: 667 n. *, 668 n. 10, 720
al-Qifṭī: 1 n. 1, 2 n. 4, 4, 4 n. 11–12, 5, 5 n. 13, 7–9, 34, 113–115, 116 n. 6, 118, 118 n. 8–9, 119 n. 9, 121–123, 459, 459 n. 1, 460 n. 2, 461, 504, 504 n. 6, 580, 581–583, 724, 724 n. 13, 725, 725 n. 20
Qusṭā ibn Lūqā: 34, 47, 47 n. 6, 92 n. 49
Quṭb al-Dīn al-Shīrāzī: 36 n. 46
al-Qūhī: xiv, xxii, 126, 130, 210, 465, 579–599, 609, 725 n. 17

al-Rāḍī: 460
Rashed, M.: 124 n. 23
Rashed, R.: xiii–xviii, 3 n. 9, 4 n. 10, 8 n. 30–31, 34 n. 44, 116 n. 6, 120 n. 10, 126 n. 25, 131 n. 1, 215 n. 3, 335 n. 2, 459 n. 1, 460 n. 3, 461 n. 4–5, 462 n. 8, 463 n. 9, 465 n. 14, 475 n. 17, 477 n. 18, 503 n. 1–2, 579 n. 2, 580 n. 5, 581 n. 8, 673 n. 23, 723 n. 9, 725 n. 18–19
Reiche, Fritz: 125
Ritter, H.: 121 n. 13
Riḍā, N.: 1 n. 1, 2 n. 6, 6 n. 23, 7 n. 29, 114 n. 1, 119 n. 9, 459 n. 1, 508 n. 27, 615 n. 1
Roberval: 66, 66 n. 10, 68
Rome, A.: 507 n. 23
Rosenfeld, B.A.: 122 n. 17, 128 n. 26–27, 129 n. 28
Rukn al-Dawla: 505, 506

al-Ṣābi', Abū ʿAlī al-Muḥassin ibn Ibrāhīm ibn Hilāl: 116 n. 6, 121, 122, 460
Sabians: 113 n. 1, 118 n. 8, 119 n. 9, 459
Sachau, C.E.: 7 n. 27, 117 n. 7
al-Ṣafadī: 114 n. 2, 122, 122 n. 16, 122, 122 n. 16
al-Ṣāghānī, Abū Ḥāmid: 579 n. 1, 582
Ṣāʿid al-Andalusī: 615, 615 n. 1, 616 n. 4, 721-722 n. 2, 723 n. 7, 726
Saidan, A.S.: 459 n. 1, 460 n. 3, 466
Saliba, G.: 2 n. 4, 459 n. 1
Ṣāliḥānī, O.P. A.: 4 n. 12, 114 n. 2, 581 n. 10
Samanides: 505
al-Sāmarrī, Abū al-Ḥasan: 579 n. 1, 581
al-Samaw'al: xix, 504, 504 n. 9
Samsó, J.: 506 n. 19
Saragossa: 616, 721, 724
Sayili, A.: 581 n. 10
Sayyid, S.: 114 n. 2
Sebüktijīn: 505
Serenus of Antinoupolis: 8, 618–621, 623–625, 627
Şeşen, R.: 36 n. 51
Sezgin, F.: 1 n. 2, 580 n. 4
Shams al-Dīn, M. Ḥ: 582 n. 11
al-Shannī: 46
Sharaf al-Dawla: 581, 582
al-Shahrazūrī: 580 n. 5
Ṣidqī, Muṣṭafā: 126, 127, 129, 464, 585, 585 n. 16, 586, 614
al-Sijistānī: 114 n. 2
al-Sijzī: 8, 8 n. 30, 9, 125, 128, 332, 333, 465, 579, 580
de Slane, Baron: 36 n. 50
Spies, O.: 121 n. 13
Stamatis, E.S.: 60 n. 8
Steinschneider, M.: 1 n. 2, 114 n. 2, 721 n. 2, 724 n. 14
Sublet, J.: 114 n. 2
al-Ṣūfī, ʿAbd al-Raḥmān: 579 n. 1, 580
al-Sumaysāṭī: 546
Suter, H.: xx, 1 n. 2, 10 n. 38, 127, 128, 128 n. 26–27, 466, 585 n. 16, 588, 721 n. 2

al-Ṭabarī: 5–7
Ṭabaristān: 580
al-Tabrīzī, ʿAbd Allāh: 36 n. 47
al-Tabrīzī, Fatḥ Allāh: 36 n. 47
Tadmūrī, ʿU.: 114 n. 2
Tajaddud, R.: 1 n. 1, 113 n. 1, 459 n. 1, 504 n. 4, 580 n. 4
al-Ṭanjī, M.: 506 n. 19
Tannery, P.: 66 n. 10
Taqī al-Dīn al-Maʿrūf: 36 n. 51

al-Tawḥīdī, Abū Hayyān: 506, 506 n. 19, 579, 579 n. 1
Tessami, Y.E.: 37 n. 57
Thābit ibn Qurra: xiv, xxii, 2 n. 5, 3, 6, 7, 9, 13, 60, 60 n. 7, 113–460, 463, 495, 579, 580, 583, 584, 599, 600, 615, 617, 619–621, 623–628, 642, 649, 652, 653, 724 n. 11, 726, 727, 730, 767, 773, 776, 778
Theodosius: 39, 47, 47 n. 6, 92, 92 n. 49
Theon of Alexandria: 507, 508
Tornberg, C.J.: 114 n. 2, 505 n. 18
al-Ṭūsī, Naṣīr al-Dīn: 10–35, 47 n. 6, 73 n. 6–107 n. 93, 585
al-Ṭūsī, Sharaf al-Dīn: xix, 333, 461, 725 n. 19

Uri, J.: 36 n. 54

al-ʿUtbī, Abū Naṣr: 504, 504 n. 7, 505, 505 n. 16–17

Vahabzadeh, B.: 4 n. 10
Vajda, G.: 125 n. 24, 506 n. 21
du Verdus, François: 66 n. 10
Ver Eecke, P.: 508 n. 25, 618 n. 7
Voorhoeve, P.: 729 n. 25

al-Wāthiq: 6
Wiedemann, E.: 115 n. 2, 123 n. 20
Witkam, J.: 722 n. 3
Woepcke, F.: 504, 504 n. 11

Yāqūt: 116 n. 6, 505 n. 14

al-Zayn, A.: 579 n. 1
Zenodorus: 507, 509, 523, 546

SUBJECT INDEX

Abscissa: 156, 244, 247, 256, 473, 474, 477, 477 n. 19, 588, 589, 593, 638, 655
Affine: *see* application, transformation
Affine mapping: 470, 474
Affinity: 471, 642
– oblique: 471
– orthogonal: 336, 342–348, 351, 362, 471, 621, 638–641, 642, 646, 647, 649, 652–656, 658
see also axis, contraction and dilatation
Algebra: xv, xix, xx, 333, 503, 726
– geometric: 122
Analysis
– combinatorial: xxi
– Diophantine: xx, xxi, 503
– numerical: xxi
– and synthesis: xv, xxii, 460, 461, 463
Angle: 9, 14, 757
– acute: 66, 210, 211, 476, 543, 634, 756, 759, 762
– obtuse: 67, 210, 211, 510
– of the ordinates: 731
– polar: 626
– right: 45, 65, 68, 210, 471, 530, 628, 651, 671 n. 19, 762, 773
– solid: 546
Apagogic
– argument: 362
– method: 44, 54, 56, 60, 256, 621, 646, 648, 652, 707 n. 67 (*see also* Proof)
Apothem: 42, 43, 564 n. 14, 762
Approximation: 46, 69, 130, 131, 154
– of π: 38
– of the cubic root: 69
Arc: 50, 616, 633, 755, 757, 773
– of a circle: 45, 52, 368, 772
– of a conchoid: 67, 68
– of the ellipse: 359, 772
– of the parabola: 210, 228, 247
– of a conic section: 730, 733
Archimedes/Archimedean
– method: 46, 47, 60, 337
– tradition: xiv, xxii, 38, 40
Area:
– of circle: 38, 40–46, 56, 60, 344, 347–354, 369, 399, 522, 524, 529, 533–539, 541, 542, 546, 547, 589, 645, 649, 650, 652
– of curved surfaces: xxi, 10, 130, 366, 377
– of an ellipse: 333, 341–355, 362, 372, 621, 622, 645–652, 772
– of plane and spherical figures: 10, 34, 38; of plane, rectilinear and curvilinear: 124
– of a parabola: 130, 164, 212, 256, 459–480, 723, 727–755; of a portion:

130–132, 154, 164, 460, 471, 475, 678, 480, 729–730, 744
– of a parallelogram: 159, 161, 164, 256, 517 n. 31, 518, 735 n. 33, 774
– of a pentagon: 519
– of a polygon: 40, 45, 157, 165, 336, 347, 472, 480, 519, 524, 526, 529, 531, 546, 547, 646–649, 653, 756
– of a rectangle: 366, 368
– of a rhombus: 517 n. 31, 518 n. 32, 647
– of a right cone: 366
– of a section: elliptical 8, 341–355; maximal/minimal: 358, 359
– of a sector: 41, 45, 521, 762
– of a segment: of circle: 354; of the segments of the ellipse: 354, 661; of a parabola: 727, 744
– of a square: 517 n. 31, 518, 521, 522
– of a trapezoid: 157, 256, 353, 354, 364, 369, 646, 772
– of a triangle: 38, 40, 45–47, 159, 354, 471, 475, 480, 514–518, 521, 522, 742–744, 747, 759, 762, 767; equilateral 509, 523; isosceles 509, 516 n. 30; right-angled: 651
Area, lateral
– of a cylinder: 123, 337–338, 363–377, 434, 545, 624; of portions: 363–377, 434
– of a frustum of a cone: 38, 50, 524, 534, 535
– of a hemisphere: 56
– of a polyhedron: 57, 509, 545
– of a prism: 364, 364, 368–372, 375
– of a pyramid: 49, 524, 528, 529, 532
– of a right circular cone: 526, 531; of revolution: 38, 48–50, 524, 531, 624
– of a solid: 40, 55–57, 535–537, 544, 545; of revolution: 50
– of a sphere: 38, 47–54, 366, 524, 538, 539, 544–546, 623
Arithmetic: xv, 113 n. 1, 723 n. 7, 726
Arithmetisation: 130, 210
Astrolabe: 118 n. 8, 615
Astronomical
– calculations/observations: 3, 7, 581, 582
– tables: 117
Astronomy: xx, xxi, 2, 2 n. 4, 7, 10, 113 n. 1, 122, 503, 579, 582, 584 n. 15, 615, 724 n. 13, 726
Asymptotes: 748
Axis: 8, 335, 622, 632
– of affinity: 336, 351, 352, 639
– of the circle: 47
– collinear: 336, 359
– of the cone: 48, 49
– of a curve: 335
– of the cylinder: 337, 338, 341, 342, 356–358, 373, 377, 382, 526, 626
– of the dome: 251
– of the ellipse (major/minor): 336, 343, 344, 347, 350, 352, 353, 355–364, 621, 632, 637–639, 646, 647, 652, 653, 658–661, 769, 770, 771
– of the parabola: 210, 244, 247, 475, 477
– of the paraboloid: 251
– of rotation: 210
– of symmetry: 88 n. 46, 157, 158, 756
– transverse: 730
Axiom
– of Archimedes: 131, 148, 209, 235, 236
– of Eudoxus–Archimedes: 360, 377

Base: 335, 733
– elliptical: 371, 624–627, 768
– circular: 371, 618–620, 623–625, 627, 778
Bifocal:
– definition: 617, 622, 632–634, 636, 637, 641, 642
– method: 333, 334, 643
– property: 8, 8 n. 30
Bijection: 470
Bisector: 67, 83 n. 40, 85 n. 44, 88 n. 46, 530, 536, 776
Bound, upper/lower: 46, 130–132, 163, 165, 166, 210, 257, 362
Bounding: 142, 163, 209, 233, 362

Centre
– of gravity: 583, 584 n. 15, 587
– of symmetry: 670 n. 16, 671 n. 18, 742
Chord: 8, 55, 61, 156, 160, 210, 226, 228, 351, 352, 359, 471, 474, 479, 542, 622, 629, 630, 632, 645, 646, 653–659, 733, 735, 740, 755, 756, 771, 773
Circle: xv, xxi, 38, 40, 56, 60, 61, 68, 244, 336–355, 369, 509, 519–539, 541–547, 589, 616–660, 668 n. 9, 755, 757, 759–762, 767, 768, 772, 773, 775
– antiparallel: 339, 625, 777
– auxiliary: 223

– base: 49, 223–224, 252, 255, 336–341, 356–358, 368, 369, 371, 627, 639, 659, 770, 778
– elongated: 8, 8 n. 30 (*see* Figure)
– equatorial: 542
– homothetic: 373, 374
– meridianal: 542
– orthogonal: 544
Compass, perfect: 579
Concavity of the surfaces: 372
Concentric: 42, 351
Conchoid of a circle: 66–69
Cone: 38, 48, 55–60, 223, 225, 246, 247, 334, 339, 528, 531, 535, 541, 542, 545, 616, 618, 625, 732
– hollow: 209–211, 212, 224, 228, 229, 246, 247
– isosceles: 349
– oblique: 616, 625
– of revolution: 38, 48–50, 212, 223, 245, 524, 527, 531, 534, 624, 729
– right: 62, 366, 616, 625; circular: 527–528, 531
– scalene: 349
– truncated: *see* frustum
Configuration of orbs: 5 n. 13
Conics: xiv, xxii, 9, 10, 343, 503, 620, 726, 730 (*see also* section)
Conoids: 130
Construction
– geometric: 351, 352, 356
– mechanical: 4, 60
Continuity, principe: 372
Contraction: 342, 343, 471, 641, 642, 653
Convexity: 377, 509
Coordinates: 470, 473, 636
– polar: 67, 626
Coplanar: 38, 91 n. 48, 337
Cosine: 42, 357
Cube: 61, 124, 212
Curve: 10, 61, 66 n. 10, 335, 480, 624, 643
– closed: 626, 627, 670 n. 16
– convex: 49, 361, 372
– at similar positions: 626
– skew: 725
– trisecting: 66
Cylinder: xv, xvi, 8, 61, 123, 252, 253, 255, 256, 333, 335–342, 350, 356, 359, 363, 369–375, 381, 527, 528, 541, 545, 588–595, 615–621, 624–627, 639, 640, 644, 654, 656, 659, 729, 767, 778
– conic: 210, 229 n. 4
– hollow: 210, 229
– right: 62, 210, 229, 229, 335, 337, 349, 363, 366–369, 371, 382, 527, 619–621, 624–627, 632, 639, 640, 656, 659, 767, 778
– oblique: 335, 337–339, 341, 349, 356, 359, 363, 364, 366, 371, 382, 527, 618–621, 623–626, 767, 778
– of revolution: 61, 371, 588, 623, 624, 627, 632, 644, 654
Cylindrical bodies: 588

Descent, finite: 135, 136, 139, 154
Diagonal: 357, 509, 517 n. 31, 624, 757, 775, 778
Diameter(s)
– collinear: 341
– conjugate: 335–336, 474, 619, 620, 627, 745, 769, 773
– principal: 343, 620
– transverse: 620, 733, 741
Dilatation: 342, 343, 471, 639, 641, 642, 653
– orthogonal: 345
Disc: 507, 763
Displacement: 337, 341, 359, 362, 471, 626
Division
– of arcs: 738, 739
– of a diameter: 244
Dome, parabolic: 210, 211, 244, 247, 251, 252, 255
– with a pointed/regular/sunken vertex: 210, 211

Ellipse: xxi, 8, 123, 333–336, 340–363, 372, 377, 381, 619–622, 624, 625, 632, 633, 637–661, 668 n. 9–10, 727, 733, 734, 736, 741, 742, 744, 745, 747, 767, 769–772, 778
(*see also* Figure, elongated circular)
– homothetic: 337, 360–362, 372–374, 377
– maximal: 350, 358, 778
– minimal: 350, 357, 373, 375
Engineering: 7
Equalities
– numerical: 209
– between four magnitudes: 209
– of ratios: 131, 220, 221, 342, 359
– between sequences of integers: 131
Equation
– algebraic: 503
– of the circle: 619, 642
– cubic: 503

– of the ellipse: 637, 642, 658
– the half-parabola: 589
– of the parabola: 227, 244, 254
– of conic sections: 739, 743
Equator: 544
Exhaustion method: 45, 59, 130
Existence: 130
– of a circle: 42, 43
– of a cone of revolution: 49
– of a hemisphere: 56
– of *n*: 42, 235, 349, 360
– of a point: 776
– of a polygon: 42, 349, 361, 528, 572 n. 22
– of a polyhedron: 57, 58, 542
– of a sphere: 57, 524
Extremum: 509

Figure
– elongated circular: 2 n. 5, 8, 333, 618, 616, 632, 637, 642, 677, 677 n. 30
– convex: 509
– curved: 123
– curvilinear: 124
– equilateral and equiangular: 510
– isometric: 338
– isoperimetric: 510, 763
– plane: 54, 123, 124, 507, 626
– polygonal: 537
– rectilinear: 124, 616
– solid: 123, 508 n. 26
Focus: 640, 644, 678 n. 36
Frustum
– of a cone: 38, 50, 51, 54, 55, 209, 212, 524
– of a hollow cone: 209, 212, 246, 247
– of a cone of revolution: 50, 212
– of a solid rhombus: 209, 212, 246, 247
– of a prism: 373, 375
Function cosine: 42

Generator, generating line
– of a cone: 48, 50, 51, 531, 535, 545,
– of a cylinder: 61, 335, 337, 338, 358, 366–369, 375, 376, 619, 640, 656, 659, 775, 778
– opposite: 335, 358, 363, 364, 373, 376, 775
Geometry: xiv, 2, 2 n. 4, 5 n. 13, 7, 60, 113 n. 1, 122, 333, 503, 509, 579, 582, 616, 723 n. 7, 726, 755
– algebraic: xxi
– Archimedean: 460
– of the astrolabe: 615
– of conics: 9–10, 726
– Euclidean: 726
– infinitesimal: 123
– of measurement: 460
– plane: 38, 40, 52
– spherical: xv, 726
Groove, circular: 67

Harmonics: 726
Height
– of the cylinder: 253, 358, 527
– of the trapezium: 467, 469, 739
– of the triangle: 467, 469, 475, 479, 739, 747
Hero's formula: 38, 40, 46, 767
Hexagon: 531, 532, 564 n. 14
Homothety: 43, 44, 49, 223, 224, 235, 336, 341, 342, 346, 351, 352, 359–365, 373, 564 n. 14
Hydraulics: 3, 7
Hyperbola: 335, 727, 733–734, 741–744, 745, 747, 748
Hyperboloid: 587, 599 n. 1

Induction
– archaic: 134, 215, 243
– incomplete: 138
Inequalities
– numerical: 209
– sequence of segments: 132, 132
Infinitesimal:
– argument: 362
– calculus: xxi
– geometry: 123
– mathematics: xiv, xx–xxii, 9, 122, 130, 333, 463, 506, 583, 586, 727
– methods: 480
Infinity (concept): 124
Integers: 69, 131–137, 209, 211, 212, 215, 231, 236, 256
– consecutive: 132–134, 156, 212, 229, 233, 244, 256
– natural: 231
Integral: 54
– elliptic: 366
– sums: 333, 347, 377, 584, 595
Integration: 247
Intersection: 338
Isepiphanics: xxi, 503, 506, 507, 509, 523–546, 560 n. 5
Isometric: 338
Isometry: 471
Isoperimeters: xxi, 503, 506, 507, 509–523

Isoperimetric (problems): 503, 722, 727, 755–763
Iteration: 236

Latus rectum: 227, 342, 359, 473, 479, 620, 642, 730, 769, 770
Length
– of convex curves: 361, 372
– of the ellipse: 362
– of the generator: 50, 51, 373, 376, 531
Limaçon (of Pascal): 66, 67
Limit: 45, 60, 347, 649, 763
Line, polygonal: 54–56, 523, 535, 537, 538
Lozenge: *see* rhombus
Lunes: xxi

Magnitude: 12, 34, 60, 209, 222, 235–238, 241, 509, 755
Mathematics: xix, 2, 7, 60, 122, 579, 615, 727
– applied: 7
– Archimedean: 38
– astronomical: 122
– Hellenistic: 60
see also infinitesimal
Means
– the two means: 38, 60
– geometric/proportional: 351, 369, 531, 652, 772
Measurement: *see* area
Mechanical: apparatus: 65: device: 66; method: 62–65
Mechanics: 2, 7, 117, 580 n. 5
Meridians: 544
Meteorology: 7
Motion of the seven planets: 581, 582
Movement: 62–64
– of stars: 5 n. 13
– on the ecliptic: 123
Mutawassiṭāt: 10, 35

Numbers: *see also* sequence, theory
– amicables: 724 n. 11, 726
– even: 132, 136–142, 145, 148, 150, 156, 163, 212, 218
– odd: 132–142, 144, 148, 150, 151, 155, 161, 163, 212, 214, 218, 231, 233, 244, 247, 254
– real: 209

Optics: xx, 726
Ordinate: 154, 156, 158, 161–163, 244, 247, 253, 256, 340, 473, 474, 477, 477 n. 19, 588, 589, 636, 637, 656, 730–733, 735, 741, 745, 748, 767
Orthogonal: *see* affinity, dilatation, projection, symmetry, system of reference

Parabola: xv , xxi, 130–133, 154, 156–164, 209–211, 225, 227, 244, 247, 251, 256, 335, 466, 471, 473–480, 588, 589, 722, 727, 729–734, 741, 742, 744, 747, 748
Paraboloid: xv, xxi, 130, 209, 244, 251, 256, 583, 588–593
– of revolution: 210
Parallelism of the segments: 469
Parallelogramm: 159, 161, 164, 210, 228, 229, 256, 338, 475, 477, 509, 516–518, 588, 589, 592 619, 735 n. 33, 748, 771, 774, 775
Parallels, equidistant: 227
Partition: 150, 151, 154, 161, 163, 244, 251, 252, 255 (*see also* subdivision, division)
Pentagon: 510, 519, 523
Perimeter: 40–46, 361–364, 370–373, 376, 507–525, 527, 531, 533, 546, 645, 651, 761
Perpendicular bisector: 517 n. 31, 518 n. 32, 536, 644
Pin: 64, 67
Plane
– antiparallel: 335, 337, 339–341, 616, 619, 769, 770
– of the base: 335, 357, 616, 619, 654
– bisecting: 339
– meridian: 223, 224
– parallel: 49, 335, 335, 338, 341, 371, 376, 527, 624, 626, 654, 656, 773
– principal: 335, 338, 341, 356–358, 639
– secant/intersecting: 627, 770
– of symetry: 335, 339, 340, 357, 767
– of a right section: 339, 357, 372–376
Point
– of contact: 88 n. 46, 359, 519, 542
– equidistant: 46
– fixed: 627
– of intersection: 62
– non-coplanar: 38, 46, 47
Poles: 67, 544, 623, 626
Polygon: 40–45, 48, 132, 158, 160, 161, 164, 165, 244, 250, 251, 254, 256, 336, 344, 345, 347–351, 360–365, 369–373, 377, 468, 470, 472–474, 480, 509, 519, 523–526, 528–531, 539, 540, 546, 547, 564 n. 14,

646–649, 652, 653, 738–740, 755, 756, 760, 762, 777
– convex: 467, 509, 519, 523
– irregular: 42, 510
– isoperimetric: 763
– regular: 42, 49, 347, 348, 362, 510, 518–523, 526, 528, 538, 546, 572 n. 22, 628, 756, 760
Polyhedron: 58, 59, 524, 540–546
– convex: 509
– regular: 545, 546
Portion
– of an ellipse: 381, 399
– of a parabola: 130, 132, 154, 156, 158–161, 164, 471, 474, 475, 477–480, 727, 732, 733, 748
– of a paraboloid: 592, 593
Power of a point: 61, 105 n. 88, 636–637
Prism: 364, 364, 368–375, 525
Procedures
– infinitesimal: 122, 123, 480
– ingenious: 5 n. 13, 14, 580 n. 5
– mechanical: 14
Progression, arithmetical: 247, 594
Projection(s)
– cylindrical: 335, 338–342, 619, 621, 640, 768, 777
– geometric: 620
– orthogonal: 339, 340, 639, 773
Proof by *reductio ad absurdum*: 46, 137, 212, 255–257, 337, 338, 341, 372, 375, 463, 471, 528, 531, 533, 657, 660, 773, 777
Proportion
– continuous: 10, 236, 238, 241
– 'perturbed': 681 n. 41
Proportionality of the segments: 163
Property
– of angles: 729
– characteristic of the circle: xv, 335, 339, 342, 624, 628, 633, 642, 755; of the ellipse: 336, 624, 658
– of conic sections: xv, 119, 131, 729, 730, 732
– curved surfaces and solids: xv
– of diameters: 620
– of the ellipse: xv, 9, 333, 341, 342
– of equal ratios: 144
– extremal: 755
– of elementary geometry: 755, 763
– of the hyperbola: 748
– infinitesimal: 729, 755
– of the parabola: 131, 133, 588
– of perpendicular planes: 338, 356
– of points on the ellipse: 620
– of polygons: 510, 740
– of positions: 460
– of a segment: 363, 372; of four segments: 230
– of straight lines: 338, 356, 729
– of the subtangent: 249, 318 n. 51
– of the tangent: 210, 477
Pyramid: 40, 48, 49, 523–528, 532, 540
– regular: 49, 523–525

Quadrature: 377
Quadrilaterals: 229, 518 n. 32, 738, 757
– convex: 518 n. 32, 770

Rabattement: 640
Radial vector/radius vector: 626, 633–635
Ratio
– of the circumference to the diameter: 39
– of the diameter to the perimeter: 45
– of homothety: 223, 362
– of similarity: 361–363
– operations: 682 n. 45, 683 n. 46
Rectangle: 63, 64, 52, 335, 338, 366, 368, 527, 619, 627, 778
Relations
– metric: 359, 634, 636
– trigonometric: 755
Rhombus, solid: 209, 211, 212, 223, 225, 228, 246, 247, 344, 509, 509, 516–518, 647
Ring: 64, 67, 592
– cylindrical: 588
Rod: 63, 64, 67, 68
Roots, cubic: 12, 60, 69
Rotation: 359
– regular polygonal line: 524, 535
– of a semicircle: 623
– of a triangle: 211, 228

Sagittas: 653, 655–659
Section(s)
– antiparallel/of contrary position: 339, 391 n. 11, 393
– of a circle: 336, 354, 372; circular: 363, 627
– of a cone of revolution: 729
– conic: xv, 123, 460, 616, 627, 727, 729–732, 739
– cylindrical: 8, 334, 335, 339, 376, 621, 654, 729; maximal: 356–363, 415, 778; minimal: 339, 356–364,

415, 778; oblique circular: 656; plane section of a cylinder: 333–334, 337, 338, 341, 363, 382, 615–621, 625–627, 632, 639, 642–644, 767; of the lateral surface: 338
– elliptical: 8, 117, 333, 363, 617, 618, 621, 632
Segment
– of an ellipse: 336, 354, 373, 733
– of a hyperbola: 733
– a parabola: 733, 744
Sequences
– of consecutive even numbers: 136–142, 145, 147, 148, 150, 156, 218
– of consecutive odd numbers: 134, 135–142, 144, 147, 148, 150, 151, 155, 218
– of consecutive squares: 134–138, 155, 155
– increasing: 142, 144, 209, 235, 236
– decreasing: 139, 209, 236, 251
– of integers: 131, 134, 135, 137, 231
– of magnitudes: 675 n. 26, 681 n. 41
– numerical: 147
– of real numbers: 209
– of ordinates: 162, 163
– of segments: 10, 131, 132, 142, 144, 145, 150, 151, 154, 163, 218, 233–244
Set, convex: 130, 166
Side
– of a cylinder: 337, 382, 626
– fixed: 624
– opposite (of a cylinder): 337
– of a polygon: 755, 756, 760
– of a trapezium: 363
– of a triangle: 363
Similarity: 337, 340, 347, 356, 359–363, 372, 471, 480, 632
Slide: 64
Solid: 40, 41, 57–59, 123, 130, 210, 211, 223, 228, 229, 244–252, 256, 508, 524, 535–540, 544, 546
– conic: 247, 624
– convex: 544
– curved: xv, xxi
– homothetic: 223
– parabolic: 210
– polyhedral: 40, 41
– of revolution: 51, 54, 223, 255, 256, 623
Space (extension): 520, 524, 526
Sphere: xv, 38, 40, 41, 46, 47, 54–60, 130, 366, 507–509, 520, 524–526, 535, 538–546, 616, 623–625, 627
– concentric: 56–58
– hemisphere: 54–57
– mobile: 580 n. 5
– parabolic: 211; 'like an egg': 210; 'like a melon': 210
Sphericity: 507
Spheroids: 130
Square(s): 510, 517, 518, 520, 523
– consecutive: 132–134, 156, 212, 244
Statics: xxi, 122, 123, 579
Straight line(s)
– antiparallel: 339
– invariant: 632
– moveable: 619, 624, 626
– separate: 632
Structure, semantic/syntactic: 131, 132, 210
Subdivision
– of the axis: 252, 588, 593, 593
– of the diameter: 156–160, 247–249, 256, 472, 473
Subtangent: 249, 477 n. 19, 492 n. 7, n. 9
Summit: *see* vertex
Sundials: 460, 462
Surface (*see also* area)
– conic: 334
– convex: 49
– curved: xv, 10, 130, 366, 377
– of a cylinder: 61, 334, 375, 619, 626
– ordinate: 592
– polygonal: 735
– prismatic: 364
– spherical: 616, 627
Symmetry, orthogonal: 337
System
– of axes: 336, 339
– of reference: 355, 470; orthogonal: 342, 344, 641; orthonormal: 477

Taḥrīr (edition, rewriting): 11, 12, 584, 585
Tangent: 61, 83 n. 40, 210, 226, 227, 249–251, 336, 339, 360, 368, 369, 471, 477, 492 n. 8, 530, 735 n. 33, 741, 748, 760, 772
Tetrahedron, regular: 524
Theory
– of algebraic equations: 503
– of conics: 9
– of the cylinder and its plane sections: 334–337, 617, 618
– of proofs: 463
– of the ellipse and its elliptical sections: 334

– isoperimetric: 763
– of numbers: xiv, xx, 122, 503, 584 n. 15, 615, 723 n. 7, 726
Torus: 62, 210, 250
– triangular: 212
Transformation (geometrical): xv, 336, 341, 346, 362, 460, 470, 471, 579, 620
– affine: 235, 479, 480
– point-wise: 333–335, 377, 470
Translation: 359, 620, 624, 626, 627
– by the vector: 228, 229, 338
Trapezoid/trapezium: 157, 246, 256, 349, 354, 363, 369, 467, 469, 535, 540, 543, 646, 704 n. 65, 739, 771, 772
– rectangular: 245
Triangle: 38, 40, 46, 52, 158, 211, 228, 246, 250, 349, 354, 363, 467, 469, 471, 475, 480, 509, 514–518, 521–523, 530, 536, 537, 621, 629–631, 646, 727, 739, 740, 742–744, 747, 758, 762, 767, 774
– equilateral: 509, 510, 514–517, 520, 522, 524
– isosceles: 509, 510, 514–516 n. 30, 524, 534, 543, 723, 762
– right-angled: 45, 61, 63, 65, 88 n. 46, 245, 356, 476, 522, 534, 628, 630, 651, 654, 656, 762
– similar: 52, 61, 158, 621, 630, 632
Trigonometric: 52
Trisection of angle: 9, 14, 34, 38, 60, 66–69

Uniqueness:
– of the abscissa: 474
– of the upper bound: 165, 210, 257
– of a parallel: 338
– of the perpendicular: 341
– of a point: 254
– of a sphere: 38, 47
– of a sequence of segments: 163
Unit of length /measure: 147
Unit segment: 219, 221

Vertex
– moving: 632
– of a portion of parabola: 474, 475, 733, 745, 747
Visibility of crescents: 122
Volume:
– of a cone: 214, 223, 247, 248, 542, 545; right circular: 528; hollow: 224, 228; of revolution: 223, 245, 524
– of a parabolic dome: 244, 247, 251, 255
– of a cube: 61
– of curved surfaces and solids: xxi, 10, 130
– of a cylinder: 252, 255, 256, 528, 542, 545, 588–594; right: 253
– of the frustum of the cone: 209, 245; hollow: 209, 224, 225; of revolution: 223
– of the frustum of the solid rhombus: 209
– of paraboloids: 130, 210, 244, 256, 583, 588–594; of a portion: 592, 593
– of a polyhedron: 57, 58, 520, 540, 544
– of a polyhedral solid: 40
– of a pyramid: 40, 41, 524–527
– of a solid: 40, 41, 57–59, 124, 210, 223, 244, 245, 247, 248, 252, 544; conic: 256
– of solid rhombus: 223, 225, 228
– of a sphere: 38, 41, 47–60, 130, 524, 539–541, 544–546, 616, 623, 627
– of the torus: 250

INDEX OF WORKS

al-Akfānī
Irshād al-qāṣid: 722 n. 3, 726 n. 21

Anonymous [Pseudo-Jordanus]
Liber de triangulis: 68

Apollonius
Conics: 3 n. 7, 9, 119, 121, 128, 129, 131, 333–336, 488, 489, 492, 498, 499, 616–621, 669, 726, 730
I: 334–336, 477
I.5: 335, 339, 391 n. 11
I.13: 359, 426
I.15: 417, 426, 770, 773
I.17: 349, 403, 406, 428, 488 n. 2, 492 n. 8
I.20: 133, 155, 196, 474, 489, 492 n. 10, 498 n. 4, 499 n. 5, 736, 750 n. 42
I.21: 395, 401, 427, 621, 625, 736, 750 n. 42, 769
I.30: 430
I.33: 492 n. 9
I.35: 492 n. 9, 742, 751 n. 44
I.36: 742, 752 n. 45
I.46: 133, 201, 290
I.50: 359, 427
I.51: 133, 196
II.5: 133, 161, 202, 203, 290
II.12: 748
II.29: 360, 429
III.35: 105 n. 88
V.11: 360, 416, 419, 422, 429, 770,
VI: 730
VI.4: 430, 732
VI.7: 732
VI.8: 104 n. 87, 413, 732
VI.12: 341, 359, 401, 427 n. 35, 731, 770
VI.13: 731, 732
VI.26, 27: 732

Archimedes
On Conoids and Spheroids: 337, 347, 349, 595
The Measurement of the Circle: 34, 39, 349, 583, 586, 586 n. 18, 628, 629, 651, 653, 710
The Sphere and the Cylinder: 34, 38, 49, 59, 60, 121, 349, 350, 361, 366–369, 371, 372, 530 n. 34, 531 n. 35–36, 534 n. 37, 537 n. 38, 538, 538 n. 40, 586, 586 n. 18, 599, 650, 755, 756

Aristotle
De caelo: 505

Banū Mūsā
Kitāb ma'rifat misāḥat al-ashkāl al-basīṭa wa-al-kuriyya (*On the Knowledge of the Measurement of Plane and Spherical figures*): 1–111, 520, 526, 528, 529, 539, 628, 672 n. 22, 673 n. 23, 756
Kitāb al-shakl al-handasī alladhī bayyana Jālīnūs (*On a Geometric Proposition Proved by Galen*): 9
Muqaddamāt Kitāb al-Makhrūṭāt (*Lemmas of the Book of Conics*): 6 n. 24, 8, 8 n. 31, 9, 618 n. 6
Qawl fī tathlīth al-zāwiya al-mustaqīma al-khaṭṭayn (*On the Trisection of the Angle*) (attributed to Aḥmad): 9
See also Ibn Mūsā

al-Bīrūnī
al-Athār al-bāqiya 'an al-qurūn al-khaliya: 117 n. 7
al-Istī'āb: 117 n. 7
Istikhrāj al-awtār fī al-dā'ira: 46 n. 3, n. 5
al-Qanūn al-Mas'ūdī: 506 n. 20, 581 n. 10
Taḥdīd nihāyāt al-amākin: 504 n. 3, 505 n. 15

Euclid
Data: 11, 726
Elements: 2 n. 4, 11, 13, 39, 119 n. 9, 121, 560, 585, 616, 625, 723, 726, 728
I. def. 22: 704 n. 65
I.5: 723
I.33: 337, 383 n. 2
II.12, 13: 634, 635
III.7, 8: 774

III.35: 105 n. 88
V. def. 12: 675 n. 28
V. def. 15: 675 n. 27
V. def. 17: 675 n. 26
V. def. 18: 681 n. 41
V.12: 683 n. 46
VI: 674
VI.8: 104 n. 87
VI.19: 566
VI.22: 652
X: 504 n. 10
X.1: 130, 131, 133, 160, 210, 237, 349, 368, 372, 377, 480, 585, 593, 756 n. 51
XI. def. 4: 696 n. 58
XI. def. 14: 623
XI.16, 18, 19: 338
XI.21, 28: 349
XII: 41, 60, 567, 567 n. 19, 672
XII.1: 628
XII.2: 39, 46, 82 n. 37, 621, 628, 650, 652, 653
XII.6: 525, 539, 561
XII.9: 528, 564
XII.11: 575
XII.16: 39, 42–45, 56, 78 n. 27, 529, 564 n. 14
XII.17: 56, 58, 540

Eutocius
Commentary on the Sphere and the cylinder: 60

Ibn al-Abbār
Kitāb al-takmila li-Kitāb al-Ṣilā: 723 n. 7

Ibn Abī Jarrāda
Taḥrīr Kitāb quṭū' al-usṭuwāna wa-basīṭihā li-Thābit ibn Qurra (*Commentary on the Sections of the Cylinder by Thābit*): 767–778

Ibn Aknīn
Ṭibb al-nufūs (*The Medicine of the Souls*): 724, 724 n. 14–15

Ibn al-Haytham
Fī anna al-kura awsa' al-ashkāl al-mujassama (*On the Figures of Equal Perimeters*): 762
Fī khuṭūṭ al-sā'āt (*On the Lines of the Hours*): 461
Fī misāḥat al-kura (*On the Measurement of the Sphere*): 130
Fī misāḥat al-mujassam al-mukāfi' (*On the Measurement of the Paraboloid*): 130
Fī al-taḥlīl wa-al-tarkīb (*On Analysis and Synthesis*): 461

Ibn Hūd
al-Istikmāl: 586 n. 20, 721–748, 749–754, 755–763, 764–766

Ibn al-'Ibrī
Tārīkh mukhtaṣar al-duwal: 4, 4 n. 12, 5, 114, 114 n. 2, 118, 118 n. 8–9, 581 n. 10

Ibn Mūsa, Aḥmad
Kitāb al-ḥiyal (*Book of Ingenious Devices*): 1 n. 2, 2 n. 5

Ibn Mūsā, al-Ḥasan
al-Shakl al-mudawwar al-mustaṭīl (*The Elongated Circular Figure*): 2 n. 5, 8, 618

Ibn al-Samḥ
On the Cylinder and its Plane Sections: 616–663, *667–720*
Commentary on Euclid's *Elements*: 616

Ibn Sīnā
al-Shifā': 727

Ibn Sinān, Ibrāhīm
Aghrāḍ Kitāb al-Majisṭī (*Intentions of the Book of the* Almagest): 461
Fī ālāt al-aẓlāl (*On Shadow Instruments*): 462
Fī al-asṭurlāb (*On the Astrolabe*) (?): 462
Fī al-dawā'ir al-mutamāssa (*On the Tangent Circles*): 461, 462
Fī ḥarakāt al-shams (*On the Movements of the Sun*): 460 n. 3
Fī istikhrāj ikhtilāfāt Zuḥal wa-al-Mirrīkh wa-al-Mushtarī (*The Determination of the Anomalies of Saturn, Mars and Jupiter*): 462
Fī al-masā'il al-mukhtāra (*On Chosen Problems*): 462
Fī misāḥat al-qiṭ' al-makhrūṭ al-mukāfi' (*On the Measurement of the Para-*

bola) – 1st redaction: 462, 463, 483–493
Fī misāḥat al-qiṭ' al-mukāfi' (*On the Measurement of a Portion of a Parabola*) – 2nd redaction: 462, 463, 495–501
Fī rasm al-quṭū' al-thalātha (*On the Drawing of the Three Conic Sections*): 462
Fī al-taḥlīl wa-al-tarkīb (*On Analysis and Synthesis*): 462
Tafsīr li-al-maqāla al-ūlā min al-Makhrūṭāt (*Commentary on the First Book of the* Conics): 461
Fī waṣf al-ma'ānī (Autobiography): 460 n. 3, 461–464

Ibn Taghrī Bardī
al-Nujūm al-zāhira fī mulūk Miṣr wa-al-Qāhira: 582, 582 n. 11

Ibn Waḥshiyya
al-Filāḥa al-nabaṭiyya: 118 n. 8

al-Khāzin
Mukhtaṣar mustakhraj min Kitāb al-Makhrūṭāt: 504 n. 10
Min al-sharḥ li-al-maqāla al-ūlā min al-Majisṭī (*Commentary on the First Book of the Almagest*): 506–547, 551–576
Tafsīr ṣadr al-maqāla al-'āshira min Kitāb Uqlīdis (*Commentary on the Book X by Euclidis*): 504 n. 10
Zīj al-ṣafā'iḥ: 505

al-Kindī
Fī al-ṣinā'at al-'uẓmā: 508 n. 26
Fī al-ukar (*On the Spheres*): 508, 508 n. 26

al-Marāghī, Muḥammad Sartāq
al-Ikmāl: 586 n. 20, 725

Menelaus
On the Elements of Geometry: 60
Spherics: 34

Nicomachus of Gerasa
Arithmetical Introduction: 121

Pappus
Collection: 508, 508 n. 25

Ptolemy
Almagest: 121, 506, 507, 757, 758

al-Qūhī
Fī anna al-quṭr ilā al-muḥīṭ nisbat al-wāḥid ilā thalātha wa-sub' (*The Ratio of the Diameter to the Circumference*): 583
Fī istikhrāj ḍil' al-musabba' al-mutasāwī al-aḍlā' (*The Construction of the Regular Heptagon*): 581, 584 n. 15
Fī istikhrāj misāḥat al-mujassam al-mukāfi' (*On the Determination of the Volume of the Paraboloid*): 579, 583, 584, 599–607
Fī marākiz al-athqāl (*The Centres of Gravity*): 584, 599, 600
Marākiz al-dawā'ir al-mutamāssa (*The Centers of Tangent Circles*): 581
Kitāb misāḥat al-mujassam al-mukāfi' (*On the Volume of the Paraboloid*): 579, 583, 586, 609–614
Kitāb ṣan'at al-asṭurlāb bi-al-barāhīn (*Treatise on the Art of the Astrolabe by Demonstration*): 579

Ṣā'id al-Andalusī
Ṭabaqāt al-umam: 615, 615 n. 1, 616 n. 4, 721 n. 2

al-Samaw'al
Fī kasf 'uwār al-munajjimīn: 504 n. 9

Serenus of Antinoupolis
On the Section of a Cylinder: 8, 618–621, 623–625, 627

al-Shahrazūrī
Nuzhat al-arwāḥ wa-rawaḍat al-afrāḥ: 580 n. 5

al-Sijzī
Fī waṣf al-quṭū' al-makhrūṭiyya (*On the Description of Conic Sections*): 8 n. 30

al-Sumaysāṭī
Fī anna saṭḥ kull dā'ira awsa' min kull saṭḥ mustaqīm al-aḍlā'... (*The Surface of any Circle is Greater than the Surface of any Regular Polygon with the Same Perimeter*): 577–578

al-Tawḥīdī
Kitāb al-imtā' wa-al-mu'ānasa: 579 n. 1
Mathālib al-wazīrayn: 506 n. 19

Thābit ibn Qurra
Fī al-a'dād (*On the Numbers)*: 113 n. 1
Fī al-ḥujja al-mansūba ilā Suqrāṭ (*On the Proof Attributed to Socrates*): 113 n. 1
Fī istikhrāj al-masā'il al-handasiyya (*On Defining the Geometrical Problems*): 113 n. 1
Fī misāḥat al-ashkāl al-musaṭṭaḥa wa-al-mujassama (*On the Measurement of Plane and Solid Figures*): 123, 123 n. 21, 124, 767
Fī misāḥat al-mujassamāt al-mukāfi'a (*On the Measurement of the Paraboloids*): 123, 128, 130, 209–257, 261–332, 333
Fī misāḥat al-makhrūṭ alladhī yusammā al-mukāfi' (*On the Measurement of the Parabola*): 123–128, 130–166, 169–208, 266, 267, 302, 333
Fī misāḥat qiṭa' al-khuṭūṭ (*On the Measurement of Line Segments*): 124
al-Qarasṭūn: 123, 123 n. 20
Fī quṭū' al-usṭuwāna wa-basīṭihā (*On the Sections of the Cylinder and its Lateral Surface*): 2 n. 5, 9 n. 33, 123, 128, 129, 333–377, 381–458, 585, 617, 624–628, 642, 649, 652, 653, 767
Fī al-shakl al-qaṭṭā' (*On the Sector-Figure)*: 113 n. 1

Theodosius
Spherics: 39, 47, 47 n. 6, 92, 92 n. 49

Theon of Alexandria
Commentary on the First Book of the Almagest: 507, 508

Zenodorus
Isoperimetric Figures: 507

Milton Keynes UK
Ingram Content Group UK Ltd.
UKHW021937071024
449327UK00022B/1834

9 780367 865283